AF760498

EXTENDED FINITE ELEMENT AND MESHFREE METHODS

EXTENDED FINITE ELEMENT AND MESHFREE METHODS

TIMON RABCZUK
Bauhaus Universität Weimar
Weimar, Germany

JEONG-HOON SONG
University of Colorado at Boulder
Boulder, CO, United States of America

XIAOYING ZHUANG
Tongji University
Shanghai, China

Leibniz Universität Hannover
Hannover, Germany

COSMIN ANITESCU
Bauhaus Universität Weimar
Weimar, Germany

ELSEVIER

ACADEMIC PRESS
An imprint of Elsevier

Academic Press is an imprint of Elsevier
125 London Wall, London EC2Y 5AS, United Kingdom
525 B Street, Suite 1650, San Diego, CA 92101, United States
50 Hampshire Street, 5th Floor, Cambridge, MA 02139, United States
The Boulevard, Langford Lane, Kidlington, Oxford OX5 1GB, United Kingdom

Copyright © 2020 Elsevier Inc. All rights reserved.

MATLAB® is a trademark of The MathWorks, Inc. and is used with permission.
The MathWorks does not warrant the accuracy of the text or exercises in this book.
This book's use or discussion of MATLAB® software or related products does not constitute endorsement or sponsorship by The MathWorks of a particular pedagogical approach or particular use of the MATLAB® software.

No part of this publication may be reproduced or transmitted in any form or by any means, electronic or mechanical, including photocopying, recording, or any information storage and retrieval system, without permission in writing from the publisher. Details on how to seek permission, further information about the Publisher's permissions policies and our arrangements with organizations such as the Copyright Clearance Center and the Copyright Licensing Agency, can be found at our website: www.elsevier.com/permissions.

This book and the individual contributions contained in it are protected under copyright by the Publisher (other than as may be noted herein).

Notices

Knowledge and best practice in this field are constantly changing. As new research and experience broaden our understanding, changes in research methods, professional practices, or medical treatment may become necessary.

Practitioners and researchers must always rely on their own experience and knowledge in evaluating and using any information, methods, compounds, or experiments described herein. In using such information or methods they should be mindful of their own safety and the safety of others, including parties for whom they have a professional responsibility.

To the fullest extent of the law, neither the Publisher nor the authors, contributors, or editors, assume any liability for any injury and/or damage to persons or property as a matter of products liability, negligence or otherwise, or from any use or operation of any methods, products, instructions, or ideas contained in the material herein.

Library of Congress Cataloging-in-Publication Data
A catalog record for this book is available from the Library of Congress

British Library Cataloguing-in-Publication Data
A catalogue record for this book is available from the British Library

ISBN: 978-0-12-814106-9

For information on all Academic Press publications
visit our website at https://www.elsevier.com/books-and-journals

Publisher: Matthew Deans
Acquisition Editor: Brian Guerin
Editorial Project Manager: Isabella C. Silva
Production Project Manager: Surya Narayanan Jayachandran
Designer: Mark Rogers

Typeset by VTeX

ELSEVIER Book Aid International
Working together to grow libraries in developing countries
www.elsevier.com • www.bookaid.org

Contents

Preface

The objective of this book is to provide an overview and the theoretical/computational background of partition-of-unity based computational methods, their implementation and applications. The focus is on extended finite element and meshfree methods and their application with focus on modeling material failure. It is assumed that the readers are already familiar with finite element methods or similar computational approaches including their implementation. The content of this book is written from an engineering point of view. It explains concepts and formulations and provides details on the implementation through simple Matlab® codes. We provide classical benchmark problems for which state-of-the-art computational methods are tested at and present some interesting numerical examples to demonstrate the power and performance of the outlined methods. The book however does not contain mathematical proofs concerning for instance the convergence of the above methods. Convergence plots are just shown numerically for specific examples. Though some of the methods are implemented in commercial software such as ABAQUS, the book does not provide a description on the use of these functions within such commercial codes.

The book is aimed for students and researchers who are interested in learning and implementing partition-of-unity method, especially extended finite element and meshfree methods. It is well suited for students and post-doctoral fellows to start research in this direction and who are interested in method development or the application of described methods to challenging problems in engineering and materials science. It is also of interest to readers who are interested in state-of-the-art computational methods for linear and nonlinear fracture and choosing an adequate method for their problem of interest. The content of this book is too extensive to be covered in a single course though parts of it could be the basis for a 1-semester course on meshfree methods or extended finite elements.

Chapter 1 provides an introduction to computational challenges which occur in problems with moving boundaries such as fracture, fluid mechanics, fluid-structure interaction, inverse problems or optimization. It also presents the level set method which is commonly combined with partition-of-unity methods for those problems. Chapter 2 summarizes the governing equations for purely mechanical problems for applications in statics and dynamics as most of the methods are presented in such a setting. They are provided in strong and weak form including a Total Lagrangian and updated Lagrangian description of motion. Chapter 3 is focused on the extended finite element method (XFEM) and variations or improvements of it. Within this chapter, challenges related to so-called blending, numerical integration, enrichment and solution procedures are discussed and potential solutions are derived. The implementation of XFEM for static and dynamic fracture problems is described in detail and the representation of complex features such as crack nucleation, crack branching and crack coalescence is discussed. We also propose a variation of the classical XFEM for fracture, i.e. the smoothed extended finite element method (SXFEM). SXFEM avoids the integration of the singularity in case of asymptotic crack tip enrichments. It facilitates the subtriangulation commonly employed in cracked elements, and inherits certain superior properties of the smoothed finite element method including less sensitivity to mesh distortion and its high accuracy for triangular elements. We subsequently present the extended finite element formulations for coupled fracture problems including different enrichment strategies, implementation details and potential challenges related for instance to the ill-conditioning of the final system of equations to be solved. Thermo-mechanical, hydro-mechanical and electro-mechanical are described; the latter ones include piezo- as well as flexoelectric materials. Finally, two other important applications of XFEM are given: Inverse analysis and (topology) optimization. For those problems, XFEM allows an exact – implicit – representation of the topology through level set functions and hence employing always the same mesh during the iterations while maintaining optimal convergence rates. The performance of the method is demonstrated for several challenging problems in the associated section for selected problem. The phantom node method, another "variation" of XFEM, in Chapter 4 is not based on enrichment functions but overlapping elements. It has the advantage of being easily implementable into an existing finite element code but can be applied only to fracture problems. Concepts for how to incorporate multiple cracks and specific crack tip elements are devised in this chapter.

The topic of Chapter 5 is extended meshfree methods. Firstly, the basic concept of meshfree methods is explained. Subsequently, we present several popular meshfree approximations including the Smoothed Particle Hydrodynamics (SPH) and improvements of it such as the Reproducing Kernel Particle Method (RKPM). The mostly rational shape functions and lack of the so-called Kronecker-delta property impose additional challenges on meshfree methods compared to FEM. In this context, we provide different approaches of spatial numerical integration, imposition of essential boundary conditions as well as solutions to avoid instabilities of different sources which occur in many meshfree methods such as SPH. We subsequently present classical methods on how to incorporate strong discontinuities and finally model discrete fracture within meshfree methods including the visibility, diffraction and transparency method. Different extended meshfree methods are presented which are based either on an intrinsic or an extrinsic enrichment. In an intrinsic enrichment, the enrichment functions are included in the polynomial basis used to construct the meshfree shape functions while extrinsic enrichment schemes are mainly focused on partition-of-unity enrichments. Two classes of discrete fracture methods are presented. The first class ensures a continuous crack path and requires special approaches to represent the crack's topology and crack tracking algorithms while the second class, the so-called cracking particle methods, represent the crack as set of crack segments and avoid crack tracking algorithms and methods to represent the crack surface entirely. The performance of various extended meshfree methods are compared for several classical benchmark problems mostly in linear elastic fracture mechanics.

Chapter 6 presents formulations based on extended Isogeometric Analysis (XIGA). We first describe popular IGA basis functions including B-splines, Non Uniform Rational B-splines (NURBS) and PHT-splines, which are useful for h-adaptive refinement procedures within IGA. Implementation details of IGA for problems in linear elastostatics are provided before different XIGA approaches for weak discontinuities are discussed in detail. We dedicated an entire chapter to extended finite element and meshfree methods for modeling fracture in plates and shells. Formulations based on Mindlin-Reissner as well as Kirchhoff Love shells are devised. In the latter case, the higher-order continuity of the associated meshfree or IGA method has been exploited, so that no additional rotational degrees of freedom are needed, which drastically facilitates the enrichment strategy requiring fulfillment of a constraint condition. We also present a method that can efficiently deal with fluid-driven fracture due to fluid-structure interaction.

Each of the above mentioned methods are capable of dealing with discrete fracture efficiently. However, they all require a fracture criterion which determine the orientation and "length" of the crack. Therefore, Chapter 8 is related to state-of-the-art fracture criteria and crack tracking algorithms for methods requiring a continuous crack surface. Fracture criteria for problems in linear elastic fracture mechanics (LEFM) as well as nonlinear continua are described. While the fracture criterion provides the orientation of the crack surface, criteria for propagating the crack are needed as well. Different approaches for how to represent the crack surface based on triangular facets and level sets are explained and related to different computational methods. In this context, we describe in detail efficient crack tracking algorithms in three dimensions and discuss challenges and limitations.

Chapter 9 presents different multiscale methods for fracture which are useful for applications such as computational materials design. The focus will be on so-called concurrent multiscale methods for fracture where the geometry of a fine-scale model is directly integrated into the geometry of the coarse-scale model. All those methods are based on extended finite element methods to represent fracture either on one or two length scales. Two approaches are described: In the first approach, the fine-scale domain and the coarse-scale domain is coupled at a discrete interface. This approach seems promising for static applications while the second approach is better for dynamic fracture as artificial wave reflections are minimized through a handshake coupling which contains both the fine-scale and coarse-scale domain. Efficient strategies to coarse grain cracks are presented which are required in adaptive multiscale methods to guarantee computational efficiency. The multiscale methods for fracture are described for coupling two continuum models as well as coupling atomistic and continuum models.

Chapter 10 briefly gives a short introduction to competitive and popular alternative methods for fracture, i.e. the numerical manifold method (NMM), peridynamics (PD) as well as phase field models for fracture. The NMM shares some features of the phantom node method though it has been proposed much earlier, even before the extended finite element method. PD is a very efficient method for dynamic fracture as – similarly to the cracking particles method – it does not require crack tracking procedures. The crack path in PD is a natural outcome of the simulation. Phase field models are somehow related to gradient damage models and smear the crack over a certain width. The beauty of the phase field model lies in the thermodynamic consistent framework, which allows a straightforward implementation into a finite element framework. Though they belong to the class of continuous approaches for fracture, we included them in our book due to their growing popularity.

The last chapter of the book is dedicated to the implementation details of the presented approaches in this book, i.e. XFEM, extended meshfree methods and XIGA. The focus is on the development of CAD-compatible formulations in the framework of (X)IGA is briefly explained. Some numerical examples are included and a link to an open-source repository containing corresponding Matlab code with additional documentation and explanations is provided.

We would like to thank our collaborators and current and former students, whose research contributed to this book, among whom are: Fatemeh Amiri, Stéphane Bordas, P.R. Budarapu, C.L. Chan, Thanh Chau-Dinh, Lei Chen, G.R. Liu, Mohammed Msekh, S.S. Nanthakumar, Vinh Phu Nguyen, Nhon Nguyen-Thanh, Hung Nguyen-Xuan, Harold S. Park, Huilong Ren, Mohammad Silani, Hossein Talebi, Navid Valizadeh, Nam Vu-Bac and Goangseup Zi. We would also like to express our gratitude to the Elsevier editors and technical staff, in particular Brian Guerin, Sabrina Webber, Thomas van der Ploeg, and Isabella C. Silva for their support in the realization of this project.

Weimar, Germany — T.R. and C.A.
Boulder, Colorado — J.H.S.
Hannover, Germany — X.Z.
September, 2019

Nomenclature

The following symbols are standard in engineering literature and are also used in this book. However, the notation might vary somewhat among the different chapters and sections.

Greek symbols

Δ	increment
δ	variation or Kronecker delta or Dirac delta function
$\epsilon, \boldsymbol{\epsilon}$	strain
Γ	boundary
λ	Lagrange multiplier or eigenvalue or Lamé's first parameter
Ω	domain
ϕ	level set
Π	potential
ψ	enrichment function
ρ	density
$\sigma, \boldsymbol{\sigma}$	Cauchy stress

Latin symbols

$\mathbf{B}$	B-operator
$\mathbf{b}$	body force
$\mathbf{C}$	(tangent) material matrix
$\mathbf{d}$	displacement vector
$\mathbf{F}$	deformation gradient
$\mathbf{f}$	force
$\mathbf{I}$	identity tensor
$\mathbf{K}$	stiffness matrix
$\mathbf{M}$	mass matrix
$\mathbf{n}$	normal
A	area
$a, \mathbf{a}$	accelerations
E	Young's modulus
G	shear modulus
H	Heaviside function
h	mesh size parameter
J	Jacobian
K_I, K_{II}	stress intensity factor
N	shape function
p	polynomial degree of basis
$P, \mathbf{P}$	first Piola Kirchhoff stress
R	support size
$t, \mathbf{t}$	traction
$u, \mathbf{U}$	displacement
V	volume

$v, \mathbf{v}$	velocities
w	quadrature weight
$X, \mathbf{X}$	material coordinates
$x, \mathbf{x}$	spatial coordinates

CHAPTER ONE

Introduction

1.1. Partition of unity methods

The Finite Element Method (FEM) was developed in the 1950s and 1960s as a convenient way to solve partial differential equations arising from various scientific and engineering applications. Rigorous mathematical analysis of the method started in 1970s and since then thousands of papers have been published on this topic. The finite element method is a Galerkin method that approximates the solution of the partial differential equation (PDE), posed in a variational form. The method involves partitioning the domain into a finite number of "elements", and an approximate solution is sought in a finite dimensional space of piecewise polynomials, defined relative to the elements. Naturally, the quality of the approximation depends on the fineness (or coarseness) of the discretization, i.e. on the size of the elements, the degree of the underlying polynomials in the finite element space and on the regularity of the exact solution.

During the 1970s and 1980s, as computational resources became less expensive, the popularity of the finite element method grew rapidly and a large number of both mathematical results and computer programs were developed [25]. Nevertheless, even with increasing computational capabilities, certain problems (e.g. crack propagation problems, multi-scale problems, problems with complex boundaries, etc.) were, and still remain too expensive (due to size or complexity) to be solved satisfactorily by the classical finite elements. Therefore interest grew in taking a more general approach, with essentially two features. These features involve: (a) either not using a mesh at all or using a very simple mesh to discretize the domain, and (b) suitably choosing approximation spaces that are not based on polynomials. The associated methods are broadly denoted as *meshless* or *meshfree methods*.

These methods can also be grouped into two classes – the classical particle methods [52,53,55,56] and the methods based on the idea of data fitting techniques [11,18]. The classical particle methods were developed for time dependent problems or conservation laws. They involve a discrete set of points called particles to discretize the underlying domain, and the solutions of a system of time-dependent PDEs for all the particles are sought. On the other hand, the meshless methods based on data-fitting techniques, which

Extended Finite Element and Meshfree Methods
https://doi.org/10.1016/B978-0-12-814106-9.00007-3
Copyright © 2020 Elsevier Inc.
All rights reserved.

were initially developed for stationary problems, also discretize the domain by particles. Here, each particle is associated with a "patch" (an open set), such that the union of the patches covers the underlying domain. Suitable finite dimensional spaces (which may not be based on polynomials and could be obtained by data-fitting techniques) are defined on each of these patches. Finally, the associated "shape functions" are used in a Galerkin or collocation method to obtain a matrix equation. The solution of this linear system is then used to obtain the solution of the meshless method.

Several meshless methods, based on the idea of data fitting techniques described above, have been developed primarily by engineers and they differ essentially in the choice of the finite dimensional spaces on the patches. The first of these methods is known as Shepard's method [61]. This idea was further generalized into a method called Smoothed Particle Hydrodynamics (SPH), [27,28,43,49,50,71] and the Method of Clouds [16,17]. Method of Clouds used moving least-squares (MLS) ideas based on polynomials to construct the finite dimensional space. Many methods, with slight variations of these ideas, were also developed; for example the Diffuse Element Method [54] and Method of Spheres [15]. Later, a class of methods were introduced where the finite dimensional spaces were constructed using the ideas of reproducing kernels [38,40–42]. Also, Radial Basis Functions were used to construct meshless methods, and the approach was developed by both the mathematicians and engineers [23,24,35,36,73,74]. We mention that though the methods described above (other than those based on RBFs) were not mathematically investigated in detail, the generic ideas and the associated mathematical analysis were given in 70's in [3,4,64]. For a modern mathematical treatment, we refer to [8].

The *Partition of Unity Finite Element Method* (PUFEM) could also be viewed as a meshless method in a broad sense. On the other hand, it is not a specific method, but is a general flexible framework. In fact, the classical FEM and many meshless methods, discussed before, could be cast into the PUFEM framework in certain situations.

The idea of PUFEM was partially introduced in [6], where piecewise L-splines [72] instead of piecewise polynomials were used to approximate locally the solution of a boundary value problem with "rough" coefficients. However, this method could not be directly generalized to higher dimensions. In 1994, Babuška, Caloz and Osborn [7] introduced the so-called "Special Finite Element Method" where they used certain non-polynomial functions for local approximation on "finite element stars" (which served as patches). These local approximations were "pasted together" using the finite element hat functions (which form a partition of unity) to obtain a global approximation. This idea was used to approximate the solution of a PDE, modeling a particular unidimensional composite. It was further refined in [5,44] and a general abstract framework was developed. It was shown that the local approximations (in the finite dimensional spaces defined locally on a patch) could be pasted together using any partition of unity. A basic approximation result was obtained in [44], which indicated that "accurate" local approximations, pasted together using a partition of unity, yields an accurate global approximation of the unknown solution of the PDE. These ideas were used to approximate solutions of Laplace and Helmholtz equations. The basic approximation result for PUFEM was further elaborated in [9]. The framework presented in [5,44] was successfully implemented [65–67] by Strouboulis et al. on engineering applications involving composites made of fibers and domains involving multiple voids and cracks. The name of *Generalized Finite Element Method* (GFEM) was first used in these papers. Several issues related to the implementation of GFEM, e.g. hierarchical construction of the patches (open cover), numerical integration, and solving the linear system, were addressed in a series of papers [30–33] by Griebel and Schweitzer. We note however, that the local approximation in these papers were based on polynomials. Ideas similar to GFEM and PUFEM have also been developed, to some extent independently and in parallel, under the name of *Extended Finite Element Method* (XFEM), [62], [68]. This method is also an extension of the standard FEM, where certain "enrichment" functions (based on the available information on the solution) are used for approximation, locally in specially chosen finite element stars. These enrichment functions are then pasted together using standard finite element hat functions. XFEM has been successfully implemented for various crack propagation problems [1,22,46–48, 62,69,70]. We mention that XFEM could also be cast in the framework of PUFEM, where standard finite elements are used as the partition of unity.

The PUFEM framework can be described by the following discretization steps:

- We consider an overlapping finite open cover $\{\omega_i\}$ of the underlying domain Ω. The sets ω_i, called *patches*, could be the interiors of the finite element stars, with respect to a simple finite element mesh that "triangulates" a region containing Ω. We note that precisely these patches are used in XFEM and were also used by Strouboulis et al. in [65,66] and [67]. They could also be spherical, as used in the Method of Spheres and Method of Clouds.

- Relative to the open cover $\{\omega_i\}$, the PUFEM uses a *partition of unity* $\{\phi_i\}$. Standard finite element hat functions, with respect to a finite element mesh, could serve as the partition of unity, as are used in XFEM and [65,66] and [67]. On the other hand, the "reproducing kernel particle (RKP) shape functions" [38,40–42], can also be used as a partition of unity.
- Special, problem dependent, finite dimensional spaces V_i, defined on ω_i's, are used for *local approximation* in PUFEM. These finite dimensional spaces may contain constant functions together with singular functions, harmonic polynomials or other special functions based on the available information [9,44]. In fact, a one-dimensional V_i containing only constants, together with reproducing kernel particle functions as a partition of unity, define many meshless methods. The functions in V_i could also be constructed using data-fitting ideas, e.g. MLS functions, as it was done in the Method of Clouds.

  We finally mention that PUFEM is a Galerkin method, where the trial space is constructed as

$$S^{PUFEM} = \sum_j \phi_j V_j = \{v \in H^1(\Omega) : v = \sum_j \phi_j v_j,\ v_j \in V_j\}.$$

  If $\{v_{ji}\}_{i=1}^{n(j)}$ is a basis for V_j, then

$$S^{PUFEM} = \text{span}\{\phi_j v_{ji}\};$$

  the functions $\phi_j v_{ji}$ will be referred to as shape functions of PUFEM.

  With appropriate choice of ϕ_j and V_j, the space S^{PUFEM} yields precisely the trial spaces used in various meshless methods and XFEM. See [8,9].

It is apparent that PUFEM is essentially an extension of the standard finite element method, but some added flexibility is obtained by the selection of the partition of unity functions ϕ_j and local approximation spaces V_j. In fact it has been shown that the PUFEM/GFEM using the hat-function partition of unity and $V_j = \mathcal{P}^k(\omega_i)$ is equivalent to the finite element method using Lagrange shape functions in certain situations [9].

One of the main advantages of PUFEM is the ability to use a discretization scheme that is easy to construct compared to the "meshing" step of the standard FEM. While tremendous advances have been made in the area of mesh generation, generating a satisfactory mesh for problems with complicated boundaries or discontinuities, especially in three dimensions, remains a time-consuming process which often requires human interaction. This is particularly inconvenient for problems where the meshing needs to change at each time-step, such as crack propagation or certain fluid flow problems. The difficulties of mesh-generation have led to the development of meshless or "mesh-free" methods, which were briefly discussed above.

Another attractive feature of PUFEM is that it allows the use of local approximation spaces which can be adapted to the problem at hand. In general, the better the functions in V_j approximate locally the exact solution of the PDE, the better will be the global approximation quality of the PUFEM solution. Therefore it is possible (and desirable) to tailor the approximation spaces V_j to the information available about the exact solution. For example, if it is known that the exact solution is harmonic, one could choose V_j to be the space of harmonic polynomials rather than all polynomials. Also, if the exact solution has singularities due to the boundary geometry, one could select a V_j whose functions accurately approximate the singular behavior of the exact solution (as in XFEM, see [9,62,68]).

Conversely, if the exact solution is known to be smooth, one can select a smooth partition of unity and local approximation spaces that contain smooth functions (e.g. polynomials). Then the resulting global shape functions of PUFEM are also smooth and will approximate well the unknown exact solution. This is an improvement over the standard finite elements, where it is more difficult to obtain an approximation which is C^1-continuous or smoother.

The added flexibility of PUFEM is not without some costs. The main drawback, which is shared with most of the other mesh-free methods, is the added complexity of the numerical integration. In PUFEM, instead of integrating over shape-regular elements, one has to consider the support of each ϕ_j which can have a more complicated structure. For example, if the ϕ_j have radial supports, then their intersection is a lens-shaped region which is difficult to handle by numerical integration routines. Moreover, PUFEM and XFEM allow the use of approximating functions that may not be piecewise polynomials of low degree, therefore accurate numerical integration will require the use of a large number of quadrature points. In general, numerical integration required in the assembly of the linear system becomes the most time-consuming part of the solution process. On the other hand, this step is more amenable to parallelization and the recent advent of parallel CPU architectures is likely to make computational costs manageable in a majority of cases [33].

Another potential pitfall is that certain choices of partition of unity functions and local approximation spaces can give rise to (almost) linearly dependent PUFEM shape functions. If that is the case, the resulting linear system is consistent but the stiffness matrix will be singular and special algorithms will be needed to deal with these situations. Furthermore, multi-grid solvers and other efficient methods available for the linear systems arising from the standard finite elements may need to be modified for use with PUFEM [32].

We also mention that solving a Dirichlet boundary value problem by PUFEM will require that the functions in the trial space S^{PUFEM} satisfy the essential boundary conditions. This in turn requires that the functions in V_j, for $\bar{\omega}_j \cap \partial\Omega \neq \emptyset$, satisfy the essential boundary conditions, which may be very difficult to accomplish. Several different approaches have been proposed to overcome this problem, such as coupling to mesh-based methods close to the boundary [37], penalty or perturbation methods [2,9,39], the Lagrange multiplier method [10,39,51], and a method due to Nitsche [34, 59]. Most of these approaches either add an extra layer of complexity to the problem, or lose some of the advantages of PUFEM, or result in a loss of the optimal rate of convergence. The most promising idea in the literature seems to be Nitsche's method which works by modifying the variational formulation of the problem to account for the boundary data. This approach retains the optimal convergence rate and the mesh-free character of the method and was implemented in [59].

1.2. Moving boundary problems

Moving boundary problems are of major importance in engineering and particularly challenging for computational methods. They include multi-phase flow, fluid-structure interaction, biofilm/tumor growth, shape/topology optimization, inverse analysis or image registration, to name some of the applications. Fracture is another challenging moving boundary problem which will be the focus in the next section. There are two subtle differences to the aforementioned problems: (1) A crack is not a closed boundary but an open boundary and (2) boundary does not propagate orthogonal to its interface but from the crack front.

There are two ways to model the boundary: (1) as smeared interface and (2) as sharp interface. The first category commonly simplifies the implementation of computational methods on cost of accuracy and/or computational efficiency. A good example is the 88 line code for topology optimization based on Solid Isotropic Material with Penalization (SIMP). In SIMP, the material properties are interpolated and therefore continuous.

Hence, the SIMP method cannot capture jumps in the strain/stress field as they occur at material interfaces. Though SIMP and its improved versions have been successfully applied to numerous challenging problems, it has its limitations. For instance, lack of robustness and mesh dependency has been reported for multi-material and nonlinear optimization problems. Methods based on level sets for instance capture the sharp interface but the implementation effort is higher; the 129 line implementation of the level set method from [14] is probably among the shortest available 'level set' codes. Note that also the solution procedure of the forward problem might be more complex. In FEM for instance, the interface can be represented 'explicitly' or 'implicitly'. In other words: When the interface is aligned to the discretization, we call it an 'explicit representation'; otherwise an 'implicit' one. The first scenario requires the generation of new meshes when the topology changes during the iterations which in turn deteriorates computational efficiency when the entire stiffness matrix has to be reassembled. When the interface is to be captured within an element by the level set function, enrichment techniques as described in this book are needed. It is of course possible to avoid enrichment schemes by employing a simple sub-triangulation techniques for integration and the outcome of the topology optimization might not differ much from enrichment schemes in many cases. However, the jump in the stress/strain field cannot be captured and there might be applications where this is important, for instance when interface phenomena needs to be accounted for as in topology optimization of nano-structures. It is imperative to say that without enrichment, the convergence rates will be sub-optimal when material interfaces are inside an element.

Another issue is how to deal with *moving* interfaces. This apparently depends on the application and method employed. In level set based topology optimization where the interface is represented 'implicitly' by the level set, the new interface (i.e. the new level set) is obtained through the solution of the (stabilized) Hamilton-Jacobi equation where commonly shape derivatives are used as velocity normal to the interface. The procedure is quite similar in fluid mechanics problems when level sets are used to implicitly represent the interface between two fluids. However, in this case, the velocity normal to the interface is the velocity of the fluid which is interpolated from the nodes. In problems involving fluid-structure-interaction, one can distinguish between *interphase capturing* methods and *interface tracking* methods. The most classical interface capturing method is probably ALE (Arbitrary Lagrangian Eulerian) which adapts the Eulerian (fluid) mesh to the deforming Lagrangian mesh of the solid. Classical interface tracking methods are the marker and cell method, volume of fluid method and the level set method.

1.3. Fracture mechanics

Material failure, or more precisely the nucleation and propagation of cracks, is another challenging moving boundary problem. We devoted a separate section on this topic as it is (1) one of the key applications in this book and (2) different in several ways to above mentioned moving boundary/interface problems. As already mentioned in the previous section, cracks are open surfaces. Furthermore, the crack consists actually of two crack surface (interfaces) which propagate from their crack front. Criteria for crack propagation (and crack nucleation) is another topic itself and will be addressed in Chapter 8 of this book.

Computational methods for fracture can be classified into continuous approaches to fracture and discrete approaches. Popular continuous approaches are gradient models, nonlocal models, viscous models or phase field approaches. They all smear the crack over a certain width and avoid representing the crack topology and crack tracking algorithms. If the global response is of interest, such approaches often work as well as discrete crack approaches. However, the computational cost is commonly higher as a finer discretization is needed in order to resolve the crack width. In certain applications such as fracture in heterogeneous structures, the implementation effort and the computational cost might raise drastically. For instance, the interphase in polymer-matrix-composites is of the order of several nanometers and the length scale parameter of a phase field model should be less than approximately 10% of the characteristic length of the composite; in this case the interphase thickness. It requires *at least* one element to resolve the characteristic length. On the other hand, similar results can be achieved by discrete crack approaches where it might be sufficient to resolve the interphase with a single element. Also, the application to dynamic fracture is challenging for continuous approaches for fracture. The micro-branch instability problem from [19–21] which could be modeled by discrete fracture approaches [58] is a good example which has never been solved successfully by a continuous crack approach. Fluid-driven fracture problems such as fluid flow through fracturing thin structures is another example where continuous approaches for fracture are prone to fail. Though there are continuous approaches for fluid-driven fracture, they cannot model the fluid flow through the propagating cracks, i.e. through the solution of the Navier-Stokes equation which can be easily done by discrete crack approaches.

The discrete approach for fracture requires two key ingredients:

- a method to capture the crack kinematics (the discontinuous displacement field) and
- a fracture criterion for crack propagation/nucleation.

Classical representatives of computational methods for discrete fracture are element deletion, cohesive elements, remeshing techniques, boundary element methods, embedded elements, extended finite element and meshfree methods, cracking particles methods and peridynamics. Cohesive elements, boundary elements and remeshing techniques align the crack to the discretization while embedded elements and extended finite element and meshfree methods allow for arbitrary crack propagation through the discretization by exploiting partition of unity enrichment. Usually, they model the crack as continuous surface which requires (1) a crack surface representation (e.g. by triangles in 3D) and (2) crack tracking algorithms. Note that ensuring a continuous crack surface in 3D is nontrivial. This issue will be discussed in Chapter 8.

Fracture can be categorized as *brittle fracture*, *quasi-brittle fracture* and *ductile fracture*. Brittle fracture can be modeled by linear elastic fracture mechanics (LEFM). Brittle materials are characterized by linear elastic material behavior in the bulk and a small fracture process zone. However, LEFM cannot be used when the failure process zone is of the order of the size of the structure. The relative size of the fracture process zone l_{pz} with respect to the smallest critical dimension D of the structure is important for the choice of the fracture model [26]. For $D/l_{pz} > 100$, LEFM is valid while for $5 < D/l_{pz} < 100$, a quasi-brittle fracture approach is required which accounts for the energy dissipation at postlocalization. Gdoutos [26] recommends the use of nonlocal damage models for $D/l_{pz} < 5$. The length of the fracture process zone is of the order of the characteristic length [46]:

$$l_{ch} = \frac{E\,G_f}{f_t^2} \quad \text{or} \quad l_{ch} = \frac{E\,G_f}{(1-\mu^2)f_t^2} \tag{1.1}$$

where E is the Young's modulus, μ the Poisson's ratio, G_f the fracture energy and f_t denotes the tensile strength of the material. Carpinteri [13] introduced a non-dimensional brittleness number $s_E = \frac{G_f}{f_t b}$ where b is a geometric measure. He tested pre-notched beams under 3-point bending where b denotes the distance between the notch and the upper boundary of

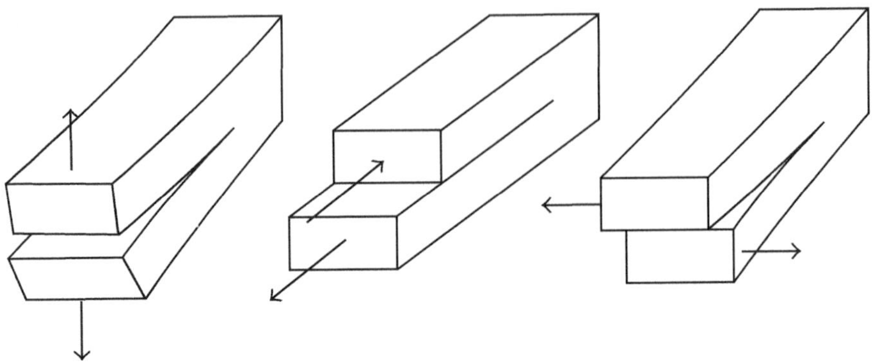

Figure 1.1 Different failure modes.

the beam. He concluded if the process zone is small compared to b, the failure is brittle and LEFM is applicable. Otherwise, the energy dissipation at postlocalization cannot be neglected and a quasi-brittle approach is needed. A common approach to account for the energy dissipation which is well suited for discrete crack methods are cohesive zone models. When large plastic deformations occur before the material looses stability, the fracture is ductile.

There are three failure modes, see Fig. 1.1. Mode-I failure is related to crack opening, mode-II failure (sliding) is a pure shear failure mode, and mode-III failure (tearing) can be considered as out-of-plane shearing. In many applications, materials will fail due to a mixed mode failure.

1.4. Level set methods

In the modeling of moving boundaries problems such as crack propagation, phase-changing, topology optimization and inverse analysis, a geometry description of interfaces is necessary. Therefore, a numerical scheme that can be incorporated with the XFEM to track and update the geometry should be devised. The level set method is perhaps one of the most widely used method apart from other method such as parametric description. The level set method (sometimes abbreviated as LSM) is a recently developed numerical method to capture the motion of interfaces and shapes in an implicit way. The name "level sets" refers to the sets which collect points having the same certain level, i.e. value of signed distance. It provides an implicit way of describing the geometry of a surface by measuring the

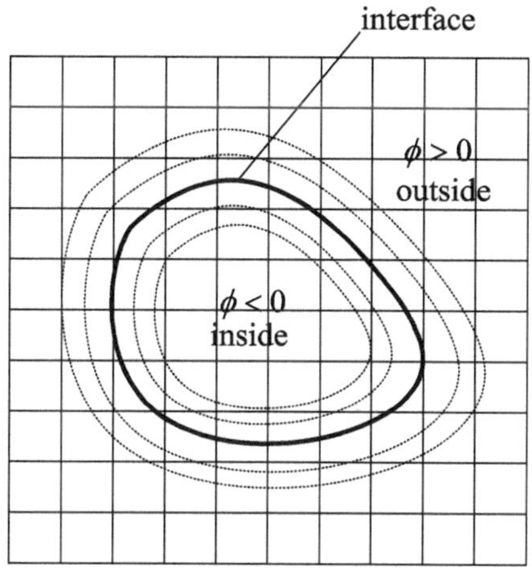

Figure 1.2 Signed distance function to describe an implicit surface.

shortest distance from any point inside the domain to the boundary of the surface. The LSM has proven efficient and powerful in simulating problems such as burning flames, image enhancement, film etching and crystal growth [60]. In the following we will present some basic concepts of the LSM, i.e. implicit interface and signed distance functions.

1.4.1 Implicit interface and signed distance functions

In an explicit interface representation method, a number of points belonging to the interface are sampled and recorded. Alternatively, an implicit interface representation defines the interface as isocontours of a function.

For example, consider an interface shown as a closed solid curve in Fig. 1.2. A domain enclosed by the interface is noted as Ω and its boundary (noted as $\partial\Omega$) is the interface itself. In an implicit method, a set of isocontours, shown as the dashed curve, are used to describe the interface. Contours inside Ω take negative values while those outside are positive. Each isocontour collects all the points having the same distance value to the interface by

$$\phi(\mathbf{X}) = \text{sign}\left(\vec{N}(\mathbf{X}_C) \cdot (\mathbf{X} - \mathbf{X}_C)\right) \|\mathbf{X} - \mathbf{X}_C\| \tag{1.2}$$

where $\mathbf{X}$ contains the coordinates of an arbitrary point, $\mathbf{X}_C$ is the closest point on the interface to $\mathbf{X}$, $\phi(\mathbf{X})$ is termed the implicit signed distance function and $\vec{N}(\mathbf{X}_C)$ is the unit outer normal at $\mathbf{X}_C$ calculated by [45]:

$$\vec{N}(\mathbf{X}_C) = \frac{\nabla\phi}{\|\nabla\phi\|} \tag{1.3}$$

where $\|\cdot\|$ is the L_2 norm of a vector which defined as follows. Suppose a vector $\mathbf{X} = \{x_1, x_2, ..., x_n\}$ where n is the length of the vector and the L_2 norm of the vector is calculated by

$$\|\mathbf{X}\| := \sqrt{x_1^2 + x_2^2 + \cdots + x_n^2}\ . \tag{1.4}$$

The zero contour $\phi(\mathbf{X}) = 0$ represents the interface. Normally, a set of data points are used to discretize the implicit functions ϕ. It is usually convenient to generate the points from the grid lines based on Cartesian coordinates, i.e. shown as the square mesh in Fig. 1.2.

1.4.2 Discretization of the level set

In the general case, the signed distance to the interface is not known analytically. The level set is usually discretized. The discretization is based on a finite element mesh where $N_I(\mathbf{X})$ is the shape function associated to the node I. The set of nodes which belong to the mesh is denoted by $\mathcal{S}$. Note that the same finite element shape functions as for the mechanical properties can be used that facilitates implementation though this is not mandatory [57]. The discretized level set is:

$$\phi(\mathbf{X}) = \sum_{I\in\mathcal{S}} N_I(\mathbf{X})\phi_I, \tag{1.5}$$

where ϕ_I is the value of the level set on node I. This discretization becomes usefull when the value of the level set is needed at the element level. It can be evaluated by interpolation. Moreover, the derivative of the level set involves only the well known derivatives of the shape functions:

$$\phi(\mathbf{X})_{,i} = \sum_{I\in\mathcal{S}} N_{I,i}(\mathbf{X})\phi_I \tag{1.6}$$

1.4.3 Capturing motion interface

The advantage of the LSM is that one can perform numerical computations involving curves and surfaces on a fixed Cartesian grid without having to parameterize these objects (this is called the *Lagrangian approach* [60]). Explicit methods such as the marker particle method, shown in Fig. 1.3 suffer

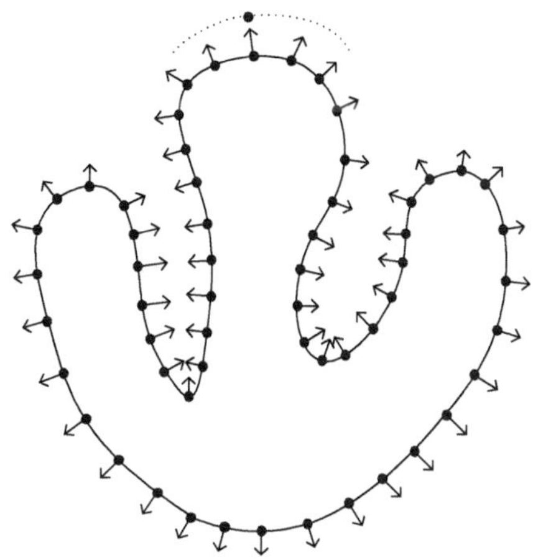

Figure 1.3 Marker particles used to discretize the front.

from problems of unstable results and the loss of information as the interface moves. For example, consider an interface which moves along its outer normal direction as shown in Fig. 1.3. The marker particle method arranges scattered points along the interface shown as the black dots in the figure. As the interface grows it can be proved by [60] that by using the LSM the motion of the interface can be tackled satisfying the entropy condition, i.e. the geometric information is not lost as the interface moves. By assuming that the discontinuity moves with a velocity field $\mathbf{v}$, the previous equation can be written as

$$\frac{\partial \phi(\mathbf{X}, t)}{\partial t} + \nabla \phi(\mathbf{X}, t) \cdot \mathbf{v}(\mathbf{X}, t) = 0 \tag{1.7}$$

or

$$\dot{\phi} + \phi_{,i} v_i = 0 \tag{1.8}$$

This equation is known as the Hamilton-Jacobi equation. The gradient of the interface function is defined as

$$\nabla \phi(\mathbf{X}) = \left(\frac{\partial \phi}{\partial x}, \frac{\partial \phi}{\partial y}, \frac{\partial \phi}{\partial z} \right). \tag{1.9}$$

The gradient $\nabla \phi$ is orthogonal to the isosurfaces of ϕ and points in the direction of an increasing ϕ.

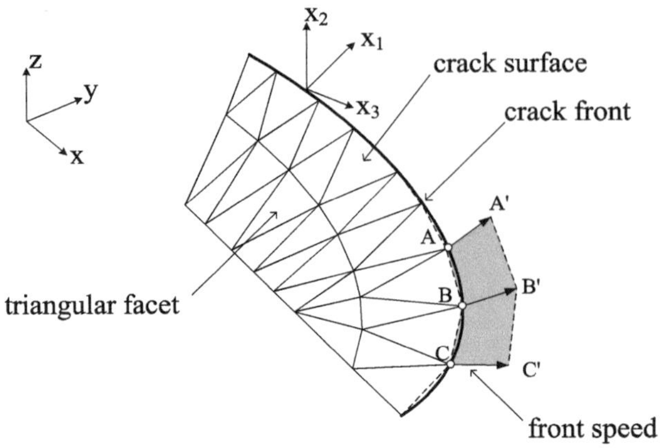

Figure 1.4 Piece-wise triangular facets used to describe a crack surface.

1.4.4 Level sets for 3D fracture modeling

The accurate description of a crack's surface and front is important since the stress field around a crack front is strongly dependent on the front curvature and surface curvature. For example, an elliptic crack, which has a varying curvature along its front (the rim of the crack surface), has significant variations in the singular stress along the crack front. However conventional explicit ways (meaning parametric description of geometry) for describing the motion interface are incapable of dealing with the curvature accurately and suffer from problems in retaining the existing interface and show decreasing accuracy as the crack evolves. For example, in [12] a crack surface in 3D is represented by piecewise triangular facets as shown in Fig. 1.4, an explicit way of describing geometry like the marker particle method in Fig. 1.3. Like many other explicit methods, new points will be added in this method as the crack front moves and thus the entropy condition cannot be guaranteed. Furthermore, to obtain the local coordinate systems using the coordinates of the points is cumbersome and lacks accuracy. The solutions to this problem is to use level sets to describe the crack geometry.

To use level sets in fracture modeling requires specific adaptation and implementation. The first use of the LSM for crack description is given by [63] in 2D with XFEM. Two orthogonal level sets were used to describe a crack line (or combinations of lines) in 2D. A number of examples were tested and showed the feasibility of using level sets in describing and tracking crack propagation. The idea was later extended and developed in 3D crack modeling in a number of papers using the XFEM [29,46,70]. In this book, a numerical framework will be presented for coupling a meshless method, the element-free Galerkin method (EFG), and the level set method (LSM) for 3D crack modeling. More details about crack propagation methods are given in Chapter 8.

References

[1] P.M.A. Areias, T. Belytschko, Analysis of three-dimensional crack initiation and propagation using the extended finite element method, International Journal for Numerical Methods in Engineering 63 (2005) 760–788.

[2] S.N. Atluri, S. Shen, The Meshless Local Petrov Galerkin Method, Tech. Sci. Press, 2002.

[3] I. Babuška, Approximation by Hill functions, Commentationes Mathematicae Universitatis Carolinae 11 (1970) 787–811.

[4] I. Babuška, Approximation by Hill functions. II, Commentationes Mathematicae Universitatis Carolinae 13 (1972) 1–22.

[5] I. Babuška, J.M. Melenk, The partition of unity finite element method, International Journal for Numerical Methods in Engineering 40 (1997) 727–758.

[6] I. Babuška, J. Osborn, Generalized finite element methods and their relation to mixed methods, SIAM Journal on Numerical Analysis 20 (1983) 510–536.

[7] I. Babuška, G. Caloz, J. Osborn, Special finite element methods for a class of second order elliptic problems with rough coefficients, SIAM Journal on Numerical Analysis 31 (1994) 945–981.

[8] I. Babuška, U. Banerjee, J. Osborn, Survey of meshless and generalized finite element methods, Acta Numerica 12 (2003) 1–125.

[9] I. Babuška, U. Banerjee, J. Osborn, Generalized finite element methods: main ideas, results, and perspective, International Journal of Computational Methods 1 (1) (2004) 1–37.

[10] T. Belytschko, Y.Y. Lu, L. Gu, Element-free Galerkin methods, International Journal for Numerical Methods in Engineering 37 (1994) 229–256.

[11] T. Belytschko, Y. Krongauz, D. Organ, M. Fleming, P. Krysl, Meshless methods: an overview and recent developments, Computer Methods in Applied Mechanics and Engineering 139 (1996) 3–47.

[12] S. Bordas, T. Rabczuk, G. Zi, Three-dimensional crack initiation, propagation, branching and junction in non-linear materials by an extended meshfree method without asymptotic enrichment, Engineering Fracture Mechanics 75 (5) (2008) 943–960.

[13] A. Carpinteri, Notch sensitivity in fracture testing of aggregative materials, Engineering Fracture Mechanics 16 (1982) 467–481.

[14] V. Challis, A discrete level-set topology optimization code written in Matlab, Structural and Multidisciplinary Optimization 41 (2010) 453–464.

[15] S. De, K. Bathe, The method of finite spheres, Computational Mechanics 25 (4) (2000) 329–345.

[16] C.A. Duarte, J.T. Oden, An h-p adaptive method using clouds, Computer Methods in Applied Mechanics and Engineering 139 (1996) 237–262.

[17] C.A. Duarte, J.T. Oden, *H-p* clouds—an *h-p* meshless method, Numerical Methods for Partial Differential Equations 12 (6) (1996) 673–705.

[18] C.A.M. Duarte, A Review of Some Meshless Methods to Solve Partial Differential Equations, Technical Report 95-06, TICAM, University of Texas at Austin, 1995.

[19] J. Fineberg, M. Marder, Instability in dynamic fracture, Physics Reports 313 (1999) 1–108.
[20] J. Fineberg, S.P. Gross, M. Marder, H.L. Swinney, Instability in dynamic fracture, Physical Review Letters 67 (1991) 457–460.
[21] J. Fineberg, S.P. Gross, M. Marder, H.L. Swinney, Instability in the propagation of fast cracks, Physical Review. B 45 (1992) 5146–5154.
[22] M. Fleming, Y.A. Chu, B. Moran, T. Belytschko, Enriched element-free Galerkin methods for crack tip fields, International Journal for Numerical Methods in Engineering 40 (1997) 1483–1504.
[23] C. Franke, R. Schaback, Convergence order estimates of meshless collocation methods using radial basis functions, Advances in Computational Mathematics 8 (1998) 381–399.
[24] C. Franke, R. Schaback, Solving partial differential equations by collocation using radial basis functions, Applied Mathematics and Computation 93 (1) (1998) 73–82.
[25] B. Fredriksson, J. Mackerle, Partial list of major finite element programs and description of some of their capabilities, in: State-of-the-Art Surveys on Finite Element Technology, Amer. Soc. Mech. Engrs. (ASME), New York, 1983 (Technical report).
[26] E. Gdoutos, Fracture Mechanics: An Introduction, vol. 123, Kluwer Academic, 2005.
[27] R.A. Gingold, J.J. Monaghan, Smoothed particle hydrodynamics — theory and application to non-spherical stars, Monthly Notices of the Royal Astronomical Society 181 (Nov. 1977) 375–389.
[28] R.A. Gingold, J.J. Monaghan, Kernel estimates as a basis for general particle methods in hydrodynamics, Journal of Computational Physics 46 (3) (1982) 429–453.
[29] A. Gravouil, N. Moes, T. Belytschko, Non-planar 3D crack growth by the extended finite element and level sets — part II: level set update, International Journal for Numerical Methods in Engineering 53 (2002) 2569–2586.
[30] M. Griebel, M.A. Schweitzer, A particle-partition of unity method for the solution of elliptic, parabolic, and hyperbolic PDEs, SIAM Journal on Scientific Computing 22 (3) (2000) 853–890 (electronic).
[31] M. Griebel, M.A. Schweitzer, A particle-partition of unity method. II. Efficient cover construction and reliable integration, SIAM Journal on Scientific Computing 23 (5) (2002) 1655–1682 (electronic).
[32] M. Griebel, M.A. Schweitzer, A particle-partition of unity method. III. A multilevel solver, SIAM Journal on Scientific Computing 24 (2) (2002) 377–409 (electronic).
[33] M. Griebel, M.A. Schweitzer, A particle-partition of unity method. IV. Parallelization, in: Meshfree Methods for Partial Differential Equations, Bonn, 2001, in: Lect. Notes Comput. Sci. Eng., vol. 26, Springer, Berlin, 2003, pp. 161–192.
[34] M. Griebel, M.A. Schweitzer, A particle-partition of unity method. V. Boundary conditions, in: Geometric Analysis and Nonlinear Partial Differential Equations, Springer, Berlin, 2003, pp. 519–542.
[35] E. Kansa, Application of Hardy's multiquadric interpolation to hydrodynamics, in: Proc. Simulation Conf., vol. 4, San Diego, California, 1986, pp. 111–116.
[36] E. Kansa, Multiquadrics—a scattered data approximation scheme with applications to computational fluid dynamics—I. Surface approximation and partial derivative estimates, Computers & Mathematics with Applications 19 (8/9) (1990) 127–145.
[37] Y. Krongauz, T. Belytschko, Enforcement of essential boundary conditions in meshless approximations using finite elements, Computer Methods in Applied Mechanics and Engineering 131 (1–2) (1996) 133–145.

[38] S. Li, H. Lu, W. Han, W.K. Liu, D.C. Simkins, Reproducing kernel element method. Part II: globally conforming I^m/C^n hierarchies, Computer Methods in Applied Mechanics and Engineering 193 (2004) 953–987.
[39] G. Liu, Mesh Free Methods: Moving Beyond the Finite Element Method, CRC Press, 2002.
[40] W. Liu, S. Jun, Y. Zhang, Reproducing kernel particle methods, International Journal for Numerical Methods in Engineering 20 (1995) 1081–1106.
[41] W.K. Liu, Y. Chen, S. Jun, J.S. Chen, T. Belytschko, C. Pan, R.A. Uras, C.T. Chang, Overview and applications of the reproducing kernel particle methods, Archives of Computational Methods in Engineering: State of the Art Reviews 3 (1996) 3–80.
[42] W.K. Liu, W. Han, H. Lu, S. Li, J. Cao, Reproducing kernel element method. Part I: theoretical formulation, Computer Methods in Applied Mechanics and Engineering 193 (2004) 933–951.
[43] L. Lucy, A numerical approach to the testing of the fission hypothesis, The Astronomical Journal 82 (1977) 1013–1024.
[44] J.M. Melenk, I. Babuška, The partition of unity finite element method: theory and application, Computer Methods in Applied Mechanics and Engineering 139 (1996) 289–314.
[45] I.M. Mitchell, The flexible, extensible and efficient toolbox of level set methods, Journal of Scientific Computing 35 (2) (Jun 2008) 300–329.
[46] N. Moës, T. Belytschko, Extended finite element method for cohesive crack growth, Engineering Fracture Mechanics 69 (7) (2002) 813–834.
[47] N. Moës, J. Dolbow, T. Belytschko, A finite element method for crack without remeshing, International Journal for Numerical Methods in Engineering 46 (1999) 131–150.
[48] N. Moës, A. Gravouil, T. Belytschko, Non-planar 3D crack growth by the extended finite element and level sets, part I: mechanical model, International Journal for Numerical Methods in Engineering 53 (2002) 2549–2568.
[49] J.J. Monaghan, Why particle methods work, SIAM Journal on Scientific and Statistical Computing 3 (4) (1982) 422–433.
[50] J.J. Monaghan, An introduction to SPH, Computer Physics Communications 48 (Jan. 1988) 89–96.
[51] Y.X. Mukherjee, S. Mukherjee, On boundary conditions in the element-free Galerkin method, Computational Mechanics 19 (4) (1997) 264–270.
[52] K. Nanbu, Direct simulation scheme derived from the Boltzmann equation, Journal of the Physical Society of Japan 49 (Nov. 1980) 2042–2058.
[53] K. Nanbu, Theoretical basis on the direct simulation Monte Carlo method, in: V. Boffi, C. Cercignani (Eds.), Rarefied Gas Dynamics, vol. 1, Teubner, 1986.
[54] B. Nayroles, G. Touzot, P. Villon, Generalizing the finite element method: diffuse approximation and diffuse elements, Computational Mechanics 10 (1992) 307–318.
[55] H. Neunzert, J. Struckmeier, Particle methods for the Boltzmann equation, Acta Numerica 4 (1995) 417–457.
[56] H. Neunzert, A. Klar, J. Struckmeier, Particle Methods: Theory and Applications, Technical report, Arbeitsgruppe Technomathematik, Kaiserslautern Univ., 1995.
[57] B. Prabel, A. Combescure, A. Gravouil, S. Marie, Level set XFEM non-matching meshes: application to dynamic crack propagation in elasto-plastic media, International Journal for Numerical Methods in Engineering 69 (8) (2007) 1553–1569.
[58] T. Rabczuk, J.H. Song, T. Belytschko, Simulations of instability in dynamic fracture by the cracking particles method, Engineering Fracture Mechanics 76 (2009) 730–741.

[59] M.A. Schweitzer, A Parallel Multilevel Partition of Unity Method for Elliptic Partial Differential Equations, Springer, 2000.
[60] J. Sethian, Fast marching methods, SIAM Review 41 (1999) 199–235.
[61] D. Shepard, A two-dimensional interpolation function for irregularly spaced data, in: Proc. ACM Nat. Conf., 1968, pp. 517–524.
[62] F.L. Stazi, E. Budyn, J. Chessa, T. Belytschko, An extended finite element with higher-order elements for curved cracks, Computational Mechanics 31 (2003) 38–48.
[63] M. Stolarska, D.L. Chopp, N. Moes, T. Belytschko, Modeling crack growth by level sets in the extended finite element method, International Journal for Numerical Methods in Engineering 51 (2001) 943–960.
[64] G. Strang, G. Fix, A Fourier analysis of the finite element variational method, in: Constructive Aspects of Functional Analysis, Edizioni Cremonese, 1973, pp. 795–840.
[65] T. Strouboulis, L. Zhang, I. Babuška, Generalized finite element method using mesh-based handbooks: application to problems in domains with many voids, Computer Methods in Applied Mechanics and Engineering 192 (2003) 3109–3161.
[66] T. Strouboulis, L. Zhang, I. Babuška, *p*-version of the generalized FEM using mesh-based handbooks with applications to multiscale problems, International Journal for Numerical Methods in Engineering 60 (10) (2004) 1639–1672.
[67] T. Strouboulis, L. Zhang, D. Wang, I. Babuška, A posteriori error estimation for generalized finite element methods, Computer Methods in Applied Mechanics and Engineering 195 (9–12) (2006) 852–879.
[68] N. Sukumar, N. Moës, B. Moran, T. Belytschko, Extended finite element method for three dimensional crack modelling, International Journal for Numerical Methods in Engineering 48 (11) (2000) 1549–1570.
[69] N. Sukumar, D.L. Chopp, N. Moës, T. Belytschko, Modeling holes and inclusions by level sets in the extended finite-element method, Computer Methods in Applied Mechanics and Engineering 190 (46–47) (2001) 6183–6200.
[70] N. Sukumar, D.L. Chopp, E. Béchet, N. Moës, Three-dimensional non-planar crack growth by a coupled extended finite element and fast marching method, International Journal for Numerical Methods in Engineering 76 (5) (2008) 727–748.
[71] J.W. Swegle, S.W. Attaway, M.W. Heinstein, F.J. Mello, D.L. Hicks, An Analysis of Smoothed Particle Hydrodynamics, NASA STI/Recon Technical Report N. 95, Mar. 1994.
[72] R.S. Varga, Functional analysis and approximation theory in numerical analysis, in: Conference Board of the Mathematical Sciences Regional Conference Series in Applied Mathematics, No. 3, Society for Industrial and Applied Mathematics, Philadelphia, Pa, 1971.
[73] H. Wendland, Numerical solution of variational problems by radial basis functions, in: Approximation Theory IX, vol. 2, Nashville, TN, 1998, in: Innov. Appl. Math., Vanderbilt Univ. Press, Nashville, TN, 1998, pp. 361–368.
[74] H. Wendland, Gaussian interpolation revisited, in: K. Koptun, T. Lyche, M. Neamtu (Eds.), Trends in Approximation Theory, Vanderbuilt University Press, 2001, pp. 1–10.

CHAPTER TWO

Weak forms and governing equations

2.1. Strong form for pure mechanical problems

2.1.1 One dimensional model problem

We consider an elastic bar of length L, with cross-section area $A(x)$, subject to a distributed load $f(x)$, where $x \in (0, L)$. The quantities of interest are the axial stress $\sigma(x)$, axial strain $\epsilon(x)$, and (axial) displacement $u(x)$. The stress is defined as

$$\sigma(x) = \frac{P(x)}{A(x)}, \tag{2.1}$$

where $P(x)$ is the axial force.

The governing equations are:

$$\frac{dP}{dx} + f = 0, \quad \textbf{Equilibrium equation} \tag{2.2}$$

$$\sigma(x) = E(x)\epsilon(x), \quad \textbf{Constitutive (stress-strain) equation}, \tag{2.3}$$

$$\epsilon = \frac{du}{dx}, \quad \textbf{Kinematic (strain-displacement) equation}, \tag{2.4}$$

where $E(x) > 0$ is Young's (elastic) modulus. By substituting (2.1), (2.3) and (2.4) into (2.2), we obtain the strong form of the model problem:

$$-\frac{d}{dx}\left(AE\frac{du}{dx}\right) = f, \text{ for } x \in (0, L). \tag{2.5}$$

In order to solve (2.5) for the unknown function $u(x)$, we also need to consider some boundary conditions. To ensure uniqueness of the solution, the bar should have prescribed displacement at one or more points on the boundary (also called a Dirichlet boundary condition). Commonly considered are "fixed end" conditions where the displacement is zero, e.g. $u(0) = 0$. Another possible boundary conditions are the "prescribed traction" (also called Neumann or "natural") type, where the stresses or the derivatives of the displacement are given at an endpoint. For example, in

Extended Finite Element and Meshfree Methods
https://doi.org/10.1016/B978-0-12-814106-9.00008-5
Copyright © 2020 Elsevier Inc.
All rights reserved.

the case of the boundary $x = L$,

$$\sigma(L) = \frac{P(L)}{A(L)} = G$$

or equivalently

$$\frac{du}{dx} = \frac{P(L)}{A(L)E(L)} = G,$$

where G is the value of the applied traction. In the case $G = 0$, we say the boundary is "free".

In the case of time-dependent problems, such as elasto-dynamics, we consider an additional term to the equilibrium equation (2.2), which becomes:

$$\frac{\partial}{\partial x}\left(E\frac{\partial u}{\partial x}\right) + f = \rho\frac{\partial^2 u}{\partial t^2}, \text{ for } x \in (0, L) \text{ and } t \in (0, T), \tag{2.6}$$

where $u(x, t)$ is the displacement at time t and $\rho(x)$ is the material density. This requires some additional boundary conditions with respect to time of the form:

$$u(x, 0) = u_0 \text{ or}$$
$$u_t(x, 0) = g_0,$$

where $u_0(x)$ and v_0 are the initial displacement and initial velocity at $t = 0$, and $u_t(x, t)$ denotes the partial derivative of u with respect to t.

2.1.2 Model problem in higher dimensions

The governing equations for linear elasticity involve the stress tensor $\boldsymbol{\sigma}(\mathbf{x})$, the strain tensor $\boldsymbol{\epsilon}(\mathbf{x})$, the displacement vector $\mathbf{u}(\mathbf{x})$, the traction vector $\mathbf{g}(\mathbf{x})$. For an isotropic solid, the material properties are usually given either in terms of Young's modulus E and Poisson's ration ν or in terms of Lamé constants μ and λ. The material properties are related by the equalities:

$$\lambda = \frac{\nu E}{(1+\nu)(1-2\nu)},$$
$$\mu = \frac{E}{2(1+\nu)}.$$

The governing equations consist of the equilibrium equation:

$$-\nabla \cdot \boldsymbol{\sigma} = \mathbf{f}, \tag{2.7}$$

the strain-displacement equation

$$\boldsymbol{\epsilon} = \frac{1}{2}(\nabla \mathbf{u} + \nabla \mathbf{u}^T), \tag{2.8}$$

and the constitutive equation:

$$\boldsymbol{\sigma} = 2\mu\boldsymbol{\epsilon} + \lambda(\nabla \cdot \mathbf{u})\mathbf{I}. \tag{2.9}$$

We furthermore consider the Dirichlet (displacement) or Neumann (traction) boundary conditions of the form:

$$\mathbf{u}(\mathbf{x}) = \mathbf{u}_0 \text{ for } \mathbf{x} \in \Gamma_D, \tag{2.10}$$
$$\mathbf{g}(\mathbf{x}) = \boldsymbol{\sigma}\mathbf{n} = \mathbf{g}_0 \text{ for } \mathbf{x} \in \Gamma_N. \tag{2.11}$$

For elastodynamics, the equations are similar as in the one-dimensional case with the addition of a time-dependent term, so that the equilibrium equation becomes:

$$\nabla \cdot \boldsymbol{\sigma} + \mathbf{f} = \rho \mathbf{u}_{tt}, \tag{2.12}$$

where as before ρ is the material density and $\mathbf{u}_{tt} = \ddot{\mathbf{u}}$ is the acceleration. Furthermore, the initial boundary conditions are:

$$\mathbf{u}(\mathbf{x}, 0) = \mathbf{u}_{00}(\mathbf{x}), \text{ or} \tag{2.13}$$
$$\dot{\mathbf{u}}(\mathbf{x}, 0) = \mathbf{u}_t(\mathbf{x}, 0) = \mathbf{g}_{00}(\mathbf{x}). \tag{2.14}$$

2.1.3 Total Lagrangian formulation

Let us consider a body Ω in $\mathbb{R}^3$ with boundary Γ; their images in the initial state are Ω_0 and Γ_0, respectively. The initial state will also serve as the reference state. The motion is described by

$$\mathbf{x} = \boldsymbol{\phi}(\mathbf{X}, t), \tag{2.15}$$

where $\mathbf{x}$ are the spatial (Eulerian) coordinates and $\mathbf{X}$ the material (Lagrangian) coordinates. The displacement is then given by

$$\mathbf{u}(\mathbf{X}, t) = \mathbf{x} - \mathbf{X} = \boldsymbol{\phi}(\mathbf{X}, t) - \mathbf{x}. \tag{2.16}$$

Neglecting thermomechanical and frictional forces, the conservation equations in the total Lagrangian formulation are given by:

$$\varrho J = \varrho_0 J_0 \tag{2.17}$$

$$\ddot{\mathbf{u}} = \frac{1}{\varrho_0}\nabla \cdot \mathbf{P} + \mathbf{b} \quad \text{on } \Omega_0 \setminus \Gamma_0^c \tag{2.18}$$

$$\dot{e} = \frac{1}{\varrho_0}\,\dot{\mathbf{F}} : \mathbf{P}^{\mathrm{T}} \tag{2.19}$$

where J and J_0 are the Jacobian determinant and initial Jacobian determinant (of the deformation gradient), respectively, ϱ_0 is the initial density, $\mathbf{P}$ are the nominal stresses (note $\mathbf{P}$ is the transpose of the first Piola Kirchhoff stress tensor), $\mathbf{b}$ are the body forces per mass unit, e is the internal energy and $\mathbf{F} = \nabla\mathbf{u} + \mathbf{I}$ denotes the deformation gradient where $\mathbf{I}$ is the second order identity tensor. Note that the mass conservation equation is written in algebraic form since it is integrable for a Lagrangian description.

The boundary conditions are:

$$\mathbf{u}(\mathbf{X}, t) = \bar{\mathbf{u}}(\mathbf{X}, t) \quad \text{on} \quad \Gamma_0^u \tag{2.20}$$

$$\mathbf{n}_0 \cdot \mathbf{P}(\mathbf{X}, t) = \bar{\mathbf{t}}_0(\mathbf{X}, t) \quad \text{on} \quad \Gamma_0^t \tag{2.21}$$

$$\mathbf{n}_0 \cdot \mathbf{P}^- = \mathbf{n}_0 \cdot \mathbf{P}^+ = \mathbf{t}_{c0} \quad \text{on} \quad \Gamma_0^c \tag{2.22}$$

$$\mathbf{t}_{c0} = \mathbf{t}_{c0}(\llbracket \mathbf{u} \rrbracket) \quad \text{on} \quad \Gamma_0^c \tag{2.23}$$

where $\bar{\mathbf{u}}$ and $\bar{\mathbf{t}}_0$ are the prescribed displacement and traction, respectively, $\mathbf{n}_0$ is the outward normal to the domain and $\Gamma_0^u \cup \Gamma_0^t \cup \Gamma_0^c = \Gamma_0$, $(\Gamma_0^u \cap \Gamma_0^t) \cup (\Gamma_0^t \cap \Gamma_0^c) \cup (\Gamma_0^c \cap \Gamma_0^u) = \emptyset$. Moreover, we assume that the stresses $\mathbf{P}$ at the crack surface Γ_0^c are bounded. Since the stresses are not well defined in the crack, the crack surface Γ_0^c is excluded from the domain Ω_0 which is considered as an open set. Note that Eq. (2.21) is only valid for cohesive cracks where traction continuity is assumed. Eq. (2.23) defines the cohesive law which depends on the jump in the displacement field.

In statics, the inertia term is omitted so that the equilibrium equation in the total Lagrangian description is given by

$$\nabla_0 \cdot \mathbf{P} - \mathbf{b} = \mathbf{0} \quad \forall \mathbf{X} \in \Omega_0 \setminus \Gamma_0^c \tag{2.24}$$

2.1.4 Updated Lagrangian formulation

In the updated Lagrangian formulation the equations are expressed in the current domain of the body. It should be mentioned that the governing equations in the total and updated Lagrangian formulation are identical, they are only different descriptions (or transformations) of the same equations. Neglecting thermomechanical and frictional forces, the conservation equations in the strong form are

$$\dot{\varrho} = -\varrho \nabla \cdot \mathbf{v} \tag{2.25}$$

$$\dot{\mathbf{v}} = \frac{1}{\varrho} \nabla \cdot \boldsymbol{\sigma} + \mathbf{b} \quad \text{on} \quad \Omega \setminus \Gamma^c \tag{2.26}$$

$$\dot{e} = \frac{\boldsymbol{\sigma}}{\varrho} : (\nabla \otimes \mathbf{v})^S \tag{2.27}$$

where ϱ is the current density, $\mathbf{v}$ the velocity vector, $\boldsymbol{\sigma}$ the Cauchy stress tensor, e the internal specific energy, ∇ is the gradient or divergence operator expressed in spatial derivatives and a superimposed S denotes the symmetric part of a tensor or a vector. The energy conservation equation is only necessary as a PDE if heat transfer is relevant. The continuity equation can also be written as

$$\varrho J = \varrho_0 \tag{2.28}$$

The boundary conditions are:

$$\mathbf{v}(\mathbf{X}, t) = \bar{\mathbf{v}}(\mathbf{X}, t) \quad \text{on} \quad \Gamma^u \tag{2.29}$$

$$\mathbf{n} \cdot \boldsymbol{\sigma}(\mathbf{X}, t) = \bar{\mathbf{t}}(\mathbf{X}, t) \quad \text{on} \quad \Gamma^t \tag{2.30}$$

$$\mathbf{n} \cdot \boldsymbol{\sigma}^- = \mathbf{n} \cdot \boldsymbol{\sigma}^+ = \mathbf{t}_c \quad \text{on} \quad \Gamma^c \tag{2.31}$$

$$\mathbf{t}_c = \mathbf{t}_c([\![\mathbf{v}]\!]) \quad \text{on} \quad \Gamma^c \tag{2.32}$$

where $\bar{\mathbf{v}}$ and $\bar{\mathbf{t}}$ are the prescribed velocity and traction, respectively, $\mathbf{n}$ is the outward normal to the domain and $\Gamma^u \cup \Gamma^t \cup \Gamma^c = \Gamma$, $(\Gamma^u \cap \Gamma^t) \cup (\Gamma^t \cap \Gamma^c) \cup (\Gamma^c \cap \Gamma^u) = \emptyset$.

In the updated Lagrangian formulation the constitutive equations are commonly formulated in rate form for large deformations:

$$\boldsymbol{\sigma}^\nabla = \dot{\boldsymbol{\sigma}}(\mathbf{D}, \mathfrak{Q}) \tag{2.33}$$

where $\mathfrak{Q}$ are internal state variables which depend on the material behavior, $\mathbf{D}$ is the rate of deformation or velocity strain and $\boldsymbol{\sigma}^\nabla$ is a frame invariant rate. The velocity strain is the symmetric part of the velocity gradient $\mathbf{L}$:

$$\mathbf{D} = \frac{1}{2}(\nabla \otimes \dot{\mathbf{x}} + \dot{\mathbf{x}} \otimes \nabla) = \frac{1}{2}\left(\mathbf{L} + \mathbf{L}^T\right) \tag{2.34}$$

The antisymmetric part

$$\mathbf{W} = \frac{1}{2}(\nabla \otimes \dot{\mathbf{x}} - \dot{\mathbf{x}} \otimes \nabla) = \frac{1}{2}\left(\mathbf{L} - \mathbf{L}^T\right) \tag{2.35}$$

is the spin-tensor and is needed to calculate certain frame-indifferent rates.

2.2. From the strong form to the weak form

In this section we will briefly discuss the derivation of the weak forms from the strong form of the governing equations introduced previously. This procedure is a key step in the development of finite element solvers, which include those related to XFEM and meshfree methods, as it allows the approximation of a 2nd order differential equation using functions of reduced continuity.

2.2.1 Weak form for the one-dimensional model problem

We first consider the strong form of the equilibrium equation (2.5). For simplicity, we assume the mixed boundary conditions:

$$u(0) = 0 \text{ and } u'(L) = G. \tag{2.36}$$

Instead of requiring that the equality is satisfied point-wise, we consider instead an averaged form of the equation. This is obtained by multiplying both sides by a *test function* v and integrating both sides over the computational domain:

$$\int_0^L -\frac{d}{dx}\left(AE\frac{du}{dx}\right) v\, dx = \int_0^L fv\, dx, \quad \text{for all } v \in V, \tag{2.37}$$

where V is a suitable space of test functions, satisfying:

$$V = \{v : v \in C^0(\Omega), v(0) = 0\},$$

where $C^0(\Omega)$ is the space of continuous functions on $\Omega = (0, L)$. We now integrate by parts on the left-hand side to obtain:

$$-AE\left(\frac{du}{dx}v\Big|_0^L - \int_0^L \frac{du}{dx}\frac{dv}{dx}\, dx\right) = \int_0^L fv\, dx. \tag{2.38}$$

Rearranging terms, the equality becomes:

$$\int_0^L AE\frac{du}{dx}\frac{dv}{dx}\, dx = \int_0^L fv\, dx + AEu'(L)v(L) - AEu'(0)v(0). \tag{2.39}$$

After substituting the boundary conditions from (2.36), we finally obtain the *weak form* of the governing equation as:

$$\int_0^L AE\frac{du}{dx}\frac{dv}{dx}\,dx = \int_0^L fv\,dx + AEGv(L), \quad \text{for all } v \in V. \tag{2.40}$$

In this derivation, we assume that the functions AE, u, and v are sufficiently smooth so that the integration by parts can be performed. In the case that these functions have jump discontinuities, the jump terms must be considered in the variational formulation (i.e. the integral is performed piecewise over the smooth subdomains).

Since the left-hand side of (2.40) is linear in u and v, we define the following bilinear and form corresponding to (2.5):

$$\mathcal{B}(u, v) = \int_0^L AE\frac{du}{dx}\frac{dv}{dx}\,dx. \tag{2.41}$$

Similarly for the right-hand side, which is linear v, we define the linear form:

$$\mathcal{L}(v) = \int_0^L fv\,dx + AEGv(L). \tag{2.42}$$

The functions u are called "trial functions", while the functions v are called "test functions". A practical procedure for solving the weak form $\mathcal{B}(u, v) = \mathcal{L}(v)$ is provided by the *Galerkin method*. The idea is to select a finite-dimensional subspace $V_n \subset V$:

$$V_n = \text{span}\{\phi_1, \phi_2, \ldots, \phi_n\}. \tag{2.43}$$

Then we seek $u_n \in V_n$ such that:

$$\mathcal{B}(u_n, \phi_i) = \mathcal{L}(\phi_i), \quad \text{for } i = 1, 2, \ldots, n. \tag{2.44}$$

By setting $u_n = \sum_j \phi_j$, it is easy to see this gives rise to a linear system of the form $\mathbf{Ku} = \mathbf{f}$, where

$$K_{ij} = \mathcal{B}(\phi_j, \phi_i), \quad \text{and } f_i = \mathcal{L}(\phi_i). \tag{2.45}$$

Here K_{ij} denotes the entry in the ith row and jth column of the matrix $\mathbf{K}$ and f_i denotes the ith entry of the vector $\mathbf{f}$. The functions ϕ_i are commonly taken to be piecewise polynomials in finite elements. However, better accuracy can be obtained by using more general basis functions which incorporate some information available about the exact solution u, such as the location of discontinuities or material interfaces. Weak forms can be developed in a similar way for higher-dimensions and other PDEs as will be discussed in more detail in later sections.

2.2.2 Weak form for the total Lagrangian formulation

Let us consider at least C^0 test and trial functions in the entire domain except across the crack surface. The test and trial functions belong to the following spaces:

$$\begin{aligned}\mathcal{V} &= \left\{\mathbf{u}(\cdot,t)|\mathbf{u}(\cdot,t) \in \mathcal{H}^1,\ \mathbf{u}(\cdot,t) \text{ discontinuous on } \Gamma_0^c,\ \mathbf{u}(\cdot,t) = \bar{\mathbf{u}}(t) \text{ on } \Gamma_0^u\right\} \\ \mathcal{V}_0 &= \left\{\delta\mathbf{u}|\delta\mathbf{u} \in \mathcal{H}^1,\ \delta\mathbf{u} \text{ discontinuous on } \Gamma_0^c,\ \delta\mathbf{u} = 0 \text{ on } \Gamma_0^u\right\}\end{aligned} \tag{2.46}$$

In the following, we will only consider the Total Lagrangian description. The derivation of the weak form for the updated Lagrangian description is straightforward. The weak form of the linear momentum equation is obtained by multiplying (2.18) with test functions:

$$\int_{\Omega_0\backslash\Gamma_0^c} \nabla_0\cdot\mathbf{P}\cdot\delta\mathbf{u}\ d\Omega_0 + \int_{\Omega_0\backslash\Gamma_0^c} \varrho_0\left(\mathbf{b}-\ddot{\mathbf{u}}\right)\cdot\delta\mathbf{u}\ d\Omega_0 = 0 \tag{2.47}$$

The first term on the RHS of the momentum equation can be transformed by integration by parts

$$\int_{\Omega_0\backslash\Gamma_0^c} \nabla_0\cdot\mathbf{P}\cdot\delta\mathbf{u}\ d\Omega_0 = \int_{\Omega_0\backslash\Gamma_0^c} \nabla_0\cdot(\mathbf{P}\cdot\delta\mathbf{u})\ d\Omega_0 - \int_{\Omega_0\backslash\Gamma_0^c} (\nabla_0\otimes\delta\mathbf{u})^T : \mathbf{P}\ d\Omega_0 \tag{2.48}$$

The Gauss theorem applied on the first term of the right-hand side of (2.48) gives:

$$\begin{aligned}\int_{\Omega_0\backslash\Gamma_0^c} \nabla_0\cdot(\mathbf{P}\cdot\delta\mathbf{u})\ d\Omega_0 &= \int_{\Gamma_0^t} \mathbf{n}\cdot\mathbf{P}\cdot\delta\mathbf{u}\ d\Gamma_0 + \int_{\Gamma_0^{+c}} \mathbf{n}^+\cdot\mathbf{P}^+\cdot\delta\mathbf{u}^+ d\Gamma_0 \\ &+ \int_{\Gamma_0^{-c}} \mathbf{n}^-\cdot\mathbf{P}^-\cdot\delta\mathbf{u}^- d\Gamma_0\end{aligned} \tag{2.49}$$

With the relation $\mathbf{t} = \mathbf{n}\cdot\mathbf{P}$ and assuming that $\mathbf{n}^+ = \mathbf{n}^-$, we obtain

$$\int_{\Omega_0\backslash\Gamma_0^c} \nabla_0\cdot(\mathbf{P}\cdot\delta\mathbf{u})\ d\Omega_0 = \int_{\Gamma_0^t} \mathbf{t}\cdot\delta\mathbf{u}\ d\Gamma_0 + \int_{\Gamma_0^{+c}} \mathbf{t}_{c0}^+\cdot\delta\mathbf{u}^+ d\Gamma_0$$

$$+\int_{\Gamma_0^{-c}} \mathbf{t}_{c0}^- \cdot \delta\mathbf{u}^- \, d\Gamma_0 \tag{2.50}$$

Let as also assume that $\mathbf{t}_{c0}^+ = \mathbf{t}_{c0}^-$, so that we finally obtain

$$\int_{\Omega_0 \backslash \Gamma_0^c} \nabla_0 \cdot (\mathbf{P} \cdot \delta\mathbf{u}) \, d\Omega_0 = \int_{\Gamma_0^t} \mathbf{t} \cdot \delta\mathbf{u} \, d\Gamma_0 + \int_{\Gamma_0^c} \mathbf{t}_{c0} \cdot \delta[\![\mathbf{u}]\!] \, d\Gamma_0 \tag{2.51}$$

which yields to

$$\begin{aligned}\int_{\Omega_0 \backslash \Gamma_0^c} (\nabla_0 \otimes \delta\mathbf{u})^T : \mathbf{P} \, d\Omega_0 - \int_{\Omega_0 \backslash \Gamma_0^c} \varrho_0 \, \mathbf{b} \cdot \delta\mathbf{u} \, d\Omega_0 + \int_{\Gamma_0^t} \bar{\mathbf{t}}_0 \cdot \delta\mathbf{u} \, d\Gamma_0 \\ + \int_{\Gamma_0^c} \mathbf{t}_{c0} \cdot \delta[\![\mathbf{u}]\!] \, d\Gamma_0 + \int_{\Omega_0 \backslash \Gamma_0^c} \varrho_0 \, \delta\mathbf{u} \cdot \ddot{\mathbf{u}} \, d\Omega_0 = 0\end{aligned} \tag{2.52}$$

It is also possible to consider the crack surface as part of the domain, i.e. the domain as closed set:

$$\int_{\Omega_0} \nabla_0 \cdot \mathbf{P} \cdot \delta\mathbf{u} \, d\Omega_0 - \int_{\Omega_0} \varrho_0 \, \mathbf{b} \cdot \delta\mathbf{u} \, d\Omega_0 + \int_{\Omega_0} \varrho_0 \, \delta\mathbf{u} \cdot \ddot{\mathbf{u}} \, d\Omega_0 = 0 \tag{2.53}$$

Let us consider only the first term of Eq. (2.53). After integration by parts, we obtained

$$\int_{\Omega_0} \nabla_0 \cdot \mathbf{P} \cdot \delta\mathbf{u} \, d\Omega_0 = \int_{\Omega_0} \nabla_0 \cdot (\mathbf{P} \cdot \delta\mathbf{u}) \, d\Omega_0 + \int_{\Omega_0} (\nabla_0 \otimes \delta\mathbf{u})^T : \mathbf{P} \, d\Omega_0 \tag{2.54}$$

Applying the divergence theorem to the first term on the RHS of Eq. (2.54) we obtain

$$\begin{aligned}\int_{\Omega_0} \nabla_0 \cdot \mathbf{P} \cdot \delta\mathbf{u} \, d\Omega_0 = \int_{\Gamma_0^t} \bar{\mathbf{t}}_0 \cdot \delta\mathbf{u} \, d\Gamma_0 + \int_{\Gamma_0^c} [[\mathbf{t}_{c0} \cdot \delta\mathbf{u}]] \, d\Gamma_0 \\ + \int_{\Omega_0} (\nabla_0 \otimes \delta\mathbf{u})^T : \mathbf{P} \, d\Omega_0\end{aligned} \tag{2.55}$$

where the second term on the RHS of Eq. (2.55) appears due to the interior (crack) discontinuity. With the assumption of traction continuity $\mathbf{t}_{c0}^- = \mathbf{t}_{c0}^+$, we regain Eq. (2.52) except that the crack boundary is not excluded from the domain Ω_0.

2.3. Variational formulation

The governing equations can also be given in a (single-field) variational formulation that states that the variation of the energy has to be zero.

The problem is then: find $\mathbf{u} \in \mathcal{V}$ such that

$$\delta W = \delta W_{int} - \delta W_{ext} - \delta W_{coh} + \delta W_{kin} = 0 \quad \forall \delta\mathbf{u} \in \mathcal{V}_0 \tag{2.56}$$

where

$$\mathcal{V} = \left\{\mathbf{u}(\cdot, t) | \mathbf{u}(\cdot, t) \in \mathcal{H}^1,\ \mathbf{u}(\cdot, t) \text{ discontinuous on } \Gamma_0^c,\ \mathbf{u}(\cdot, t) = \bar{\mathbf{u}}(t) \text{ on } \Gamma_0^u\right\}$$
$$\mathcal{V}_0 = \left\{\delta\mathbf{u} | \delta\mathbf{u} \in \mathcal{H}^1,\ \delta\mathbf{u} \text{ discontinuous on } \Gamma_0^c,\ \delta\mathbf{u} = 0 \text{ on } \Gamma_0^u\right\} \tag{2.57}$$

and with

$$\delta W_{\text{int}} = \int_{\Omega_0 \setminus \Gamma_0^c} (\nabla_0 \otimes \delta\mathbf{u})^T : \mathbf{P}\ \mathrm{d}\Omega_0 \tag{2.58}$$

$$\delta W_{\text{kin}} = \int_{\Omega_0 \setminus \Gamma_0^c} \varrho_0\, \delta\mathbf{u} \cdot \ddot{\mathbf{u}}\ \mathrm{d}\Omega_0 \tag{2.59}$$

$$\delta W_{\text{ext}} = \int_{\Omega_0 \setminus \Gamma_0^c} \varrho_0\, \delta\mathbf{u} \cdot \mathbf{b}\ \mathrm{d}\Omega_0 + \int_{\Gamma_0^t} \delta\mathbf{u} \cdot \bar{\mathbf{t}}_0\ \mathrm{d}\Gamma_0 \tag{2.60}$$

$$\delta W_{\text{coh}} = \int_{\Gamma_0^c} [\![\delta\mathbf{u}]\!] \cdot \mathbf{t}_{c0}\ \mathrm{d}\Gamma_0 \tag{2.61}$$

Note again, that the equations are given only for a Total Lagrangian description. The equations in updated Lagrangian description are straightforward. We would like to mention that we have used a mixed formulation and employed terms in updated Lagrangian description where convenient. As can be seen, the equations in this section are equivalent to the equations derived in the previous section.

CHAPTER THREE

Extended finite element method

3.1. Formulation and concepts

"Standard" finite element approximations are approximations with piecewise differentiable polynomials that are obviously not well suited for problems with strong and weak discontinuities. The only opportunity to accurately model these kind of discontinuities is to conform the finite element mesh with the line of discontinuity. This becomes a major difficulty for moving discontinuities[1] as already mentioned at the very beginning of this manuscript.

The extended finite element method (XFEM) is a finite element approximation that is able to handle arbitrary strong and weak discontinuities. XFEM is based on a local[2] extrinsic PU enrichment that idea was used before for strong discontinuities, i.e. cracks, in meshfree methods, see Section 5.5. Hence, it is not surprising that the first XFEM approximations were developed for cracks. The XFEM concept was later extended to interface problems and weak discontinuities. The basic idea is the same as described in Section 5.5: To enrich the approximation space such that it is able to reproduce certain features of the problem of interest, e.g. cracks or interfaces.

One important aspect of problems with moving interfaces is the tracking of these interfaces. A powerful tool for tracking interfaces is the level set method that will be explained in the next section. Though it is not mandatory to use level sets in XFEM, many XFEM formulations take advantage of the level set method.

3.1.1 Standard XFEM

XFEM is based on a local partition of unity. It uses an extrinsic enrichment to model the weak or strong discontinuity within a finite element. The XFEM approximation can be decomposed into a usual part and into an

[1] That can occur for moving interfaces in two phase-flow problems or evolving cracks for example.

[2] Local means that the enrichment is only employed in a certain *local* subdomain.

Extended Finite Element and Meshfree Methods
https://doi.org/10.1016/B978-0-12-814106-9.00009-7
Copyright © 2020 Elsevier Inc.
All rights reserved.

enriched part

$$\mathbf{u}^h(\mathbf{X}) = \sum_{J \in \mathcal{S}} N_J(\mathbf{X})\ \mathbf{u}_J + \sum_{K \in \mathcal{E}} \sum_{J \in \mathcal{S}_c} \tilde{N}_J^K(\mathbf{X})\ \psi^K(\mathbf{X})\ \mathbf{a}_J^K \tag{3.1}$$

where the first term on the RHS of Eq. (3.1) is the usual approximation and the second term on the RHS of Eq. (3.1) is the enrichment; $\mathcal{S}$ is the set of nodes in the entire discretization and $\mathcal{S}_c$ is the set of enriched nodes that are influenced by the interface; N_J and $\tilde{N}_J$ are the shape functions[3]; $\psi(\mathbf{X})$ is an enrichment function that is chosen according to the problem of interest[4] and $\mathbf{a}_J$ are additional degrees of freedom introduced into the variational formulation. The set $\mathcal{E}$ is the number of interfaces.[5] If we have only a single interface, then the first sum of the second term of the right hand side of Eq. (3.1) can be omitted (as well as the superimposed K).

3.1.1.1 Application to strong discontinuities

Let us use the XFEM enrichment with $N_J(X) = \tilde{N}_J(X)$ to be able to model one single strong discontinuity in one dimension. Therefore, the enrichment function ψ is chosen to be the step function S:

$$S(\xi) = \begin{cases} 1 & \forall \xi > 0 \\ -1 & \forall \xi < 0 \end{cases} \tag{3.2}$$

The approximation (3.1) then reads:

$$u^h(X) = \sum_{J \in \mathcal{S}} N_J(X)\ u_J + \sum_{J \in \mathcal{S}_c} N_J(X)\ S(\phi(X))\ a_J \tag{3.3}$$

Moreover, we assume linear shape functions $N_1 = 0.5(1 - r)$, $N_2 = 0.5(1 + r)$, where r is a local scaled element coordinate. Let us consider three finite elements as shown in Fig. 3.1. The shape functions $N_2(X)$ of node 2 and $N_3(X)$ of node 3 are shown in Fig. 3.1 as well. The nodes are numbered in ascending order from 1 to 4. The element in the middle supposed to have a strong discontinuity, i.e. a crack, at an arbitrary location X_c between node 2 and node 3. Therefore, the nodes 2 and 3 will be enriched. The nodes 1 and 4 will not be enriched since they are not influenced by the crack.

[3] Note that generally different shape functions can be used for the standard part and for the enrichment.

[4] $\psi(\mathbf{X})$ will differ for a strong and weak discontinuity.

[5] E.g. cracks.

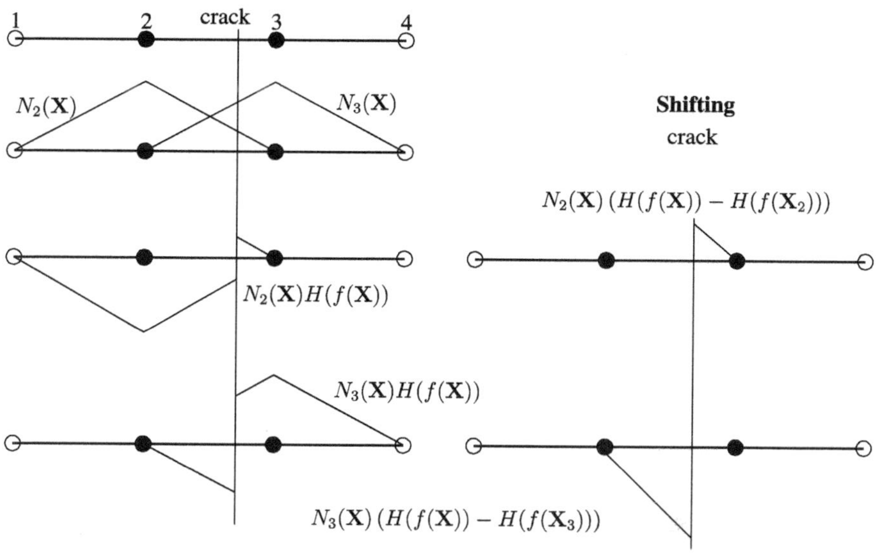

Figure 3.1 Principle of XFEM in 1D for a strong discontinuity.

The crack is defined with the level set $\phi(X_c) = 0$ where X_c is the global position of the crack and $\phi(X) < 0$ when $X < X_c$ and $\phi(X) > 0$ when $X > X_c$. Since for the coordinate of node 2, $X_2 < X_c$, it follows that $S(\phi(X_2)) = -1$. Accordingly, $S(\phi(X_3)) = 1$ since it lies on the opposite side of the crack with $X_3 > X_c$. The resulting enriched shape functions $N_J(X)\ S(X)$ for nodes 2 and 3 are shown on the LHS of Fig. 3.1 as well. From (3.3), it is obvious that the value of $u(X)$ on an enriched node $K \in \mathcal{S}_c$ is

$$u(X_K) = u_K + S(\phi(X_K))\ a_J \tag{3.4}$$

Thus, the nodal parameter u_K is not the real displacement value on the node. In order to satisfy this relation, the enriched shape functions are shifted around the node of interest:

$$u^h(X) = \sum_{J \in \mathcal{S}} N_J(X)\ u_J + \sum_{J \in \mathcal{S}_c} N_J(X)\ \big(S(\phi(X)) - S(\phi(X_J))\big)\ a_J \tag{3.5}$$

such that $u(X_K) = u_K$. The result of this shifting is illustrated on the RHS of Fig. 3.1. We note that the enriched region is getting narrower. This shifting is standard in XFEM. The jump in the displacement field is given by

$$[\![u^h(X)]\!] = u(X^+) - u(X^-)$$

$$
\begin{aligned}
&= \sum_{J\in\mathcal{S}} N_J(X^+)\ u_J + \sum_{J\in\mathcal{S}_c} N_J(X^+)\ \left(S(\phi(X^+))\right)\ a_J \\
&- \sum_{J\in\mathcal{S}} N_J(X^-)\ u_J + \sum_{J\in\mathcal{S}_c} N_J(X^-)\ \left(S(\phi(X^-))\right)\ a_J \\
&= \sum_{J\in\mathcal{S}_c} N_J(X) \left(\left[S(\phi(X^+)) - S(\phi(X^-))\right]\right)\ a_J \\
&= 2\sum_{J\in\mathcal{S}_c} N_J(X)\ a_J
\end{aligned} \tag{3.6}
$$

Here, we have used the continuity of the shape functions across the crack, i.e. $N_J(X^-) = N_J(X^+)$. Several authors prefer to use the Heaviside function instead of the step function to model the jump in the displacement field. In that case, Eq. (3.6) becomes:

$$
\begin{aligned}
[\![u^h(X)]\!] &= \sum_{J\in\mathcal{S}_c} N_J(X) \left(\left[H(\phi(X^+)) - H(\phi(X^-))\right]\right)\ a_J \\
&= \sum_{J\in\mathcal{S}_c} N_J(X)\ a_J
\end{aligned} \tag{3.7}
$$

and hence the jump is $\sum_{J\in\mathcal{S}_c} N_J(X)\ a_J$ instead of $2\sum_{J\in\mathcal{S}_c} N_J(X)\ a_J$. Note that any enrichment function which is discontinuous across the crack surface can be exploited. The outcome will be the same though the additional nodal parameters will differ.

3.1.1.2 Application to weak discontinuities

For a weak discontinuity, the enrichment function ψ is chosen to be the signed distance function ϕ:

$$
\psi_J(\mathbf{x}, t) = |\phi(\mathbf{x}, t)| - |\phi(\mathbf{x}_J, t)| \tag{3.8}
$$

The approximation (3.1) then reads:

$$
v^h(x) = \sum_{J\in\mathcal{S}} N_J(x)\ v_J(t) + \sum_{J\in\mathcal{S}_c} N_J(x)\ \psi_J(\phi(x), t)\ a_J(t) \tag{3.9}
$$

where $\mathcal{S}_c$ denotes the interface. Note that we have written the approximation in terms of spatial coordinates since weak discontinuities are often associated in fluid mechanics (two-phase flow). Hence, we have also used the letter v instead of u that should indicate velocity instead of displacement. We consider again 3 elements in one dimension that are numbered from 1

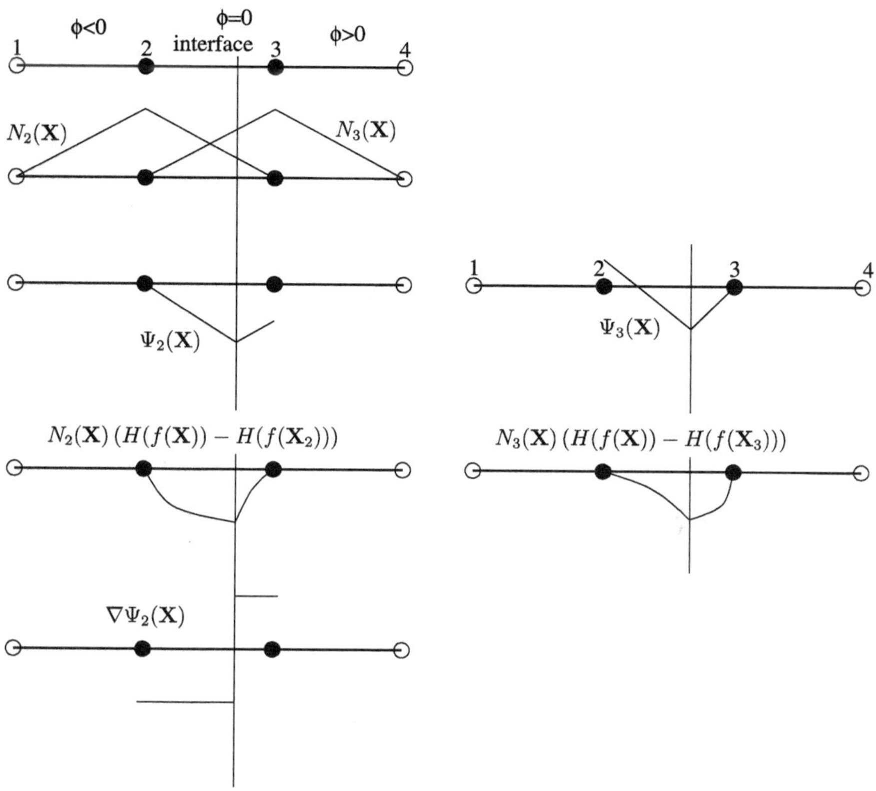

Figure 3.2 Principle of XFEM in 1D for a weak discontinuity.

to 4 in ascending order and with the interface between node 2 and node 3, see Fig. 3.2. The enrichment function ψ for node 2 and node 3 is illustrated as well. As can be seen, a kink is introduced that will cause the jump in the gradient of the function since the jump occurs in the derivatives of ψ, see the bottom on the LHS of Fig. 3.2. The resulting shape function $N_2(\mathbf{x}, t)\ \psi_2(\mathbf{x}, t)$ for node 2 and $N_3(\mathbf{x}, t)\ \psi_3(\mathbf{x}, t)$ for node 3 are shown in Fig. 3.2, too. The velocity gradient is obtained by formal differentiation of Eq. (3.9):

$$
\begin{aligned}
\nabla v^h(x) = & \sum_{J \in \mathcal{S}} \nabla N_J(x) v_J(t) \\
& + \sum_{J \in \mathcal{S}_c} \left(\nabla N_J(x)\ \psi_J(\phi(x), t) + N_J(x)\ \nabla \psi_J(\phi(x), t) \right)\ a_J(t)
\end{aligned} \tag{3.10}
$$

with

$$\nabla \psi_J(\mathbf{x}, t) = \text{sign}(\phi)\ \nabla\phi = \text{sign}(\phi)\mathbf{n}_{int} \tag{3.11}$$

where $\mathbf{n}_{int}$ denotes the normal to the interface. The only term that can cause the jump is $\nabla \psi_J(\mathbf{x}, t)$. Similar to the jump in the displacement field, we obtain the jump in the strain field

$$[\![\nabla v^h(X)]\!] = 2 \sum_{J \in \mathcal{S}_c} N_J(X)\ a_J\ \mathbf{n}_{int} \tag{3.12}$$

or the jump normal to the interface:

$$[\![\nabla v^h(X)\mathbf{n}_{int}]\!] = 2 \sum_{J \in \mathcal{S}_c} N_J(X)\ a_J \tag{3.13}$$

where the factor 2 results from the step size from -1 to 1, see Fig. 3.2.

3.1.2 Hansbo-Hansbo XFEM

An alternative to the standard XFEM for strong discontinuities was proposed by [23]. They do not model the crack kinematics with additional degrees of freedom but by overlapping elements, see Fig. 3.3. The [23] XFEM-version can be derived from standard XFEM. Therefore, let us consider the approximation of the discontinuous displacement field for a linear finite element in one dimension:

$$\begin{aligned} u^h(X) &= \sum_{I=1}^{2} N_I(X)\,[u_I + a_I\,(H(X - X_c) - H(X_I - X_c))] \\ &= u_1\ N_1 + u_2\ N_2 + a_1\ N_1\ H(X - X_c) \\ &\quad + a_2\ N_2\ [H(X - X_c) - 1] \end{aligned} \tag{3.14}$$

where H is the Heaviside function. With $N_I = N_I\ H(X - X_c) + N_I(1 - H(X - X_c))$, $I = 1, 2$, we can rewrite Eq. (3.14)

$$\begin{aligned} u^h(X) &= (u_1 + a_1)\,N_1\ H(X - X_c) + u_1\ N_1\,(1 - H(X - X_c)) \\ &\quad + (u_2 - a_2)\,N_2\,(1 - H(X - X_c)) + u_2\ N_2\ H(X - X_c) \end{aligned} \tag{3.15}$$

Let us define

$$\text{element 1} \left\{ \begin{aligned} u_1^1 &= u_1 \\ u_2^1 &= u_2 - a_2 \end{aligned} \right. \tag{3.16}$$

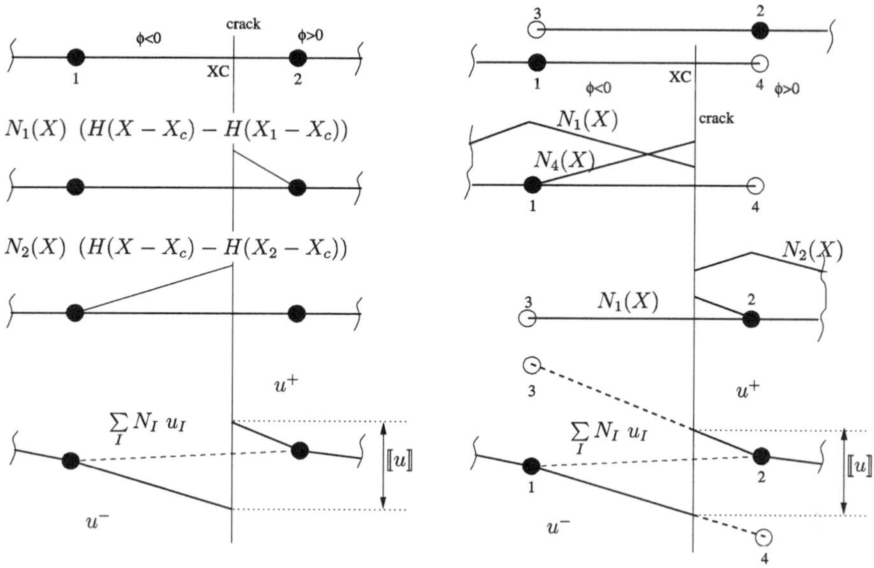

Figure 3.3 Standard XFEM vs. Hansbo XFEM.

$$
\text{element 2}\begin{cases} u_1^2 = u_1 + a_1 \\ u_2^2 = u_2 \end{cases} \tag{3.17}
$$

where superscripts and subscripts denote the element and node numbers, respectively. Eq. (3.15) can then be rewritten as

$$
\begin{aligned}
u^h(X) = {} & u_1^1\ N_1\,(1 - H(X - X_c)) + u_2^1 N_2\,(1 - H(X - X_c)) \\
& + u_1^2\ N_1\ H(X - X_c) + u_2^2\ N_2\ H(X - X_c)
\end{aligned} \tag{3.18}
$$

Thus, we can consider the displacement field to consist of the displacement fields of two elements: element 1, which is only active for $X < X_c$ due to $(1 - H(X - X_c))$ and element 2, which is only active for $X > X_c$ due to $H(X - X_c)$. The displacement jump across the crack is then:

$$
\begin{aligned}
[\![u^h(X)]\!]_{X=X_c} &= \underbrace{\lim}_{\epsilon \to 0}\,[u(X+\epsilon) - u(X-\epsilon)]_{X=X_c} \\
&= N_1(X_c)\ \left(u_1^2 - u_1^1\right) + N_2(X_c)\ \left(u_2^2 - u_2^1\right) \\
&= a_1\ N_1(X_c) + a_2\ N_2(X_c)
\end{aligned} \tag{3.19}
$$

From Eq. (3.18), we can see that the discontinuous field can be constructed by adding an extra element, element 2 in that case, as shown in Fig. 3.3.

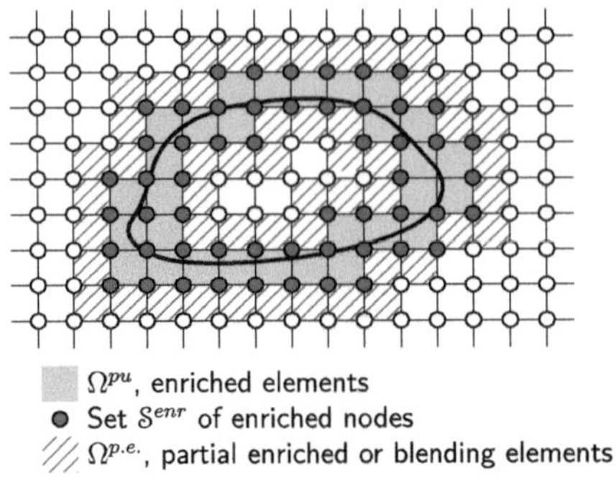

Figure 3.4 Partial enrichment around the enriched area.

Then, two additional nodes are added (u_2^1 and u_1^2). The two parts of the model are completely disjoint. The Hansbo-Hansbo version of XFEM has later been implemented in statics by [38] and in dynamics by [56]; the latter approach is also well known under the name phantom-node method, which will be described in higher dimensions and more detail in Chapter 4.

3.2. Blending, integration and solvers

3.2.1 Blending

Around the enriched domain Ω^{pu} which is the support of the partition of unity, there is a partial enriched area where the elements do not have all their nodes enriched. These elements are called blending (or partial enriched) elements (Fig. 3.4). The approximation in a 4-node element which has less than 4 enriched nodes, 3 for instance, is

$$u_i^h(\mathbf{x}) = \sum_{I=1}^{4} N_I(\mathbf{x}) u_{Ii} + \sum_{J=1}^{3} N_J(\mathbf{x}) \psi(\mathbf{x}) a_{Ji} \tag{3.20}$$

In this element the function $\psi(\mathbf{x})$ can not be recovered by taking $u_{Ii} = 0$ and $a_{Ji} = 1$ because (N_1, N_2, N_3) is not anymore a partition of unity, in other terms,

$$\sum_{J=1}^{3} N_J(\mathbf{x}) \neq 1. \tag{3.21}$$

The fact that the additional function can not be recovered in these elements is not important since these elements do not contain the discontinuity. The

Table 3.1 A few standard combinations for shape functions, partition of unity and enrichment.

Standard shape functions $N_I(\mathbf{x})$	**Partition of unity** $f_i(\mathbf{x})$	**Enrichment** $\psi(\mathbf{x})$	**Order of** $f_i(\mathbf{x}) \times \psi(\mathbf{x})$	**Spurious terms**
4-node element 1st order	4-node element 1st order	Heaviside zero order	1	No
4-node element 1st order	4-node element 1st order	Ramp 1st order	2	Yes
9-node element 2st order	4-node element 1st order	Ramp 1st order	2	No

main point is that it may introduce spurious terms in the approximation which produce an error in the solution. The spurious terms can be automatically corrected by the standard part of the approximation if the order of the standard part is more or equal than the order of the partition of unity times the enrichment. Table 3.1 shows a few possible combinations.

The spurious terms can be corrected by an assumed strain method [15]. The finite element shape functions form a partition of unity

$$\sum_{I \in N} N_I(\mathbf{x}) = 1 \tag{3.22}$$

It follows from the above that for an arbitrary function $\Psi(\mathbf{x})$, the following satisfies

$$\sum_{I \in N} N_I(\mathbf{x}) \Psi(\mathbf{x}) = \Psi(\mathbf{x}) \tag{3.23}$$

Therefore any function Ψ can be reproduced by a set of functions $N_I \Psi$. This is the key property of enriched finite element methods based on a partition of unity. Although one could enrich the entire domain, only a sub-domain is usually enriched since the features need to be modeled local – for instance, a crack compared to the plate containing it. Moreover, keeping enrichment local permits keeping the matrix banded. This is why X-FEM can be considered as *a local partition of unity enriched finite element method*. A partitioning of a typical domain into its non-enriched sub-domains and enriched sub-domains is shown in Fig. 3.4. In this local enrichment scheme, three types of elements are distinguished. The first types are the classical finite elements, those in which none of its nodes are enriched, these elements are grouped in Ω^{std}. The second type are fully

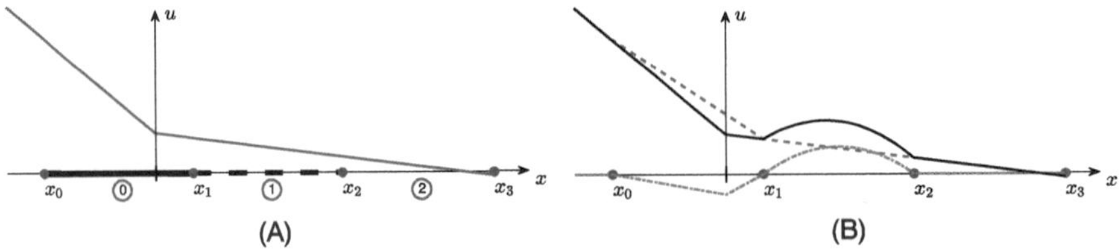

Figure 3.5 A 1D example of how a locally enriched finite element method fails to be able to reproduce a linear field. The desired piecewise linear field is shown in (A) and the enriched part (*dash-dot line*), standard part (*dash line*), and the total (*solid line*) approximation are shown in (B). The discretization is shown in (A), where the enriched nodes denoted by filled circles, the blending element is denoted by a thick dashed line and the fully enriched element by a thick solid line (adapted from [15]).

enriched elements, i.e. all of its nodes are enriched. These elements are denoted as Ω^{enr}. The third type of elements, called partially-enriched elements, are those for which only some – but not all – of the nodes are enriched. These elements form the blending sub-domain Ω^{blnd}.

Let $\mathbf{u}_I = 0$ and $\mathbf{a}_J = 1$ in the enriched finite element approximation, then we have

$$\mathbf{u}^h(\mathbf{x}) = \sum_{J \in \mathcal{N}^{enr}} \begin{cases} \widetilde{N}_J(\mathbf{x})\Psi(\mathbf{x}) = \Psi(\mathbf{x}) & \forall \mathbf{x} \in \Omega^{enr} \\ \widetilde{N}_J(\mathbf{x})\Psi(\mathbf{x}) \neq \Psi(\mathbf{x}) & \forall \mathbf{x} \in \Omega^{blnd} \\ \widetilde{N}_J(\mathbf{x})\Psi(\mathbf{x}) = 0 & \forall \mathbf{x} \in \Omega^{std} \end{cases} \tag{3.24}$$

Therefore the approximation can reproduce the enrichment in Ω^{enr} and it vanishes in Ω^{std}. However, in the blending domain, it consists of the product of a subset of the enriched shape functions $\widetilde{N}_J$ and the enrichment function Ψ so this enrichment function cannot be reproduced. The *blending elements* or *transition elements* lead to a lower convergence rate for enriched finite element methods compared to standard finite element methods. The following example, extracted from [15], shows the reason. Consider a one-dimensional mesh as illustrated in Fig. 3.5 with a discontinuity in the derivative in element 0. The enrichment function is the ramp function

$$\Psi(\mathrm{x}) = xH(x) \tag{3.25}$$

where H is the Heaviside step function. This enrichment adds a discontinuity in the gradient of the approximation at $x = 0$. Linear shape functions are used for both the standard approximation and the partition of unity. Let element 0 be the fully enriched element and element 1 be the blending element to the right. The approximation of element 1 is given by

$$\mathrm{u}^h(\mathrm{x}) = \sum_{I=1}^{2} N_I(\mathrm{x}) + N_1(\mathrm{x})(\mathrm{x}H(x) - \mathrm{x}_1 H(x_1))a_1 \tag{3.26}$$

$$\mathrm{u}^h(\xi) = u_1(1-\xi) + u_2\xi + a_1\xi h(1-\xi) \tag{3.27}$$

where

$$\xi = \frac{x - x_1}{h} \tag{3.28}$$

and h is the length of element 1. Let the finite element interpolation to the solution be given by u^h and denote the error in the interpolation by e, we have

$$e \equiv u - u^{int} \tag{3.29}$$

The maximum error occurs at the point $\overline{x}$ where

$$e_{,x|\overline{x}} \equiv \frac{d}{dx} e(\overline{x}) = 0 \tag{3.30}$$

Then a Taylor expansion about $\overline{x}$ gives

$$e(x) = e(\overline{x}) + e_{,x|\overline{x}}(x - \overline{x}) + \frac{1}{2} e_{,xx|\overline{x}}(x - \overline{x})^2 + O(h^3) \tag{3.31}$$

or

$$e(x) = e(\overline{x}) + \frac{1}{2} e_{,xx|\overline{x}}(x - \overline{x})^2 \tag{3.32}$$

If we let $x = x_1$, then $e(x_1) = 0$ since u^h is the finite element interpolation of u, i.e. $u^h(x_I) = u(x_I)$. Therefore, we obtain

$$e(x) = -\frac{1}{2} e_{,xx|\overline{x}}(x - \overline{x})^2 \tag{3.33}$$

Since

$$e(x) = u_{,xx} + \frac{2a_1}{h} \tag{3.34}$$

and since

$$\frac{1}{2}(x - x_1)^2 \leq \frac{1}{8} h^2 \tag{3.35}$$

it follows that

$$e(\overline{x}) \leq \frac{1}{8} h^2 \max(u_{,xx} + \frac{2a_1}{h}) \tag{3.36}$$

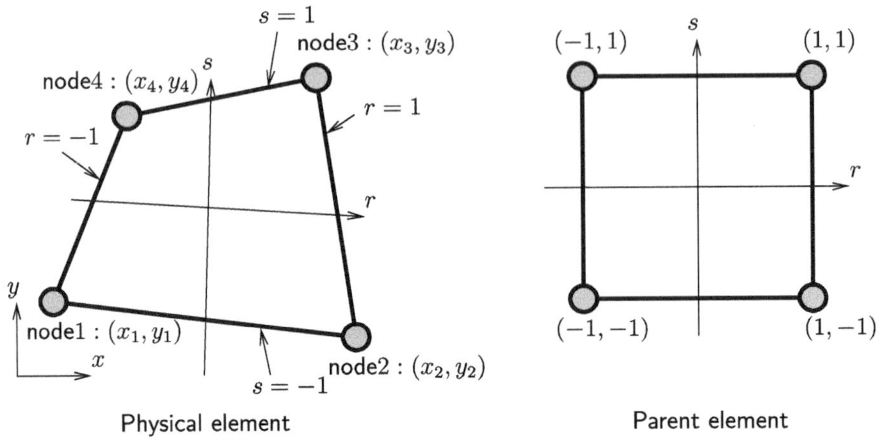

Figure 3.6 Physical and parent 4-node elements.

The last term, $2a_1/h$, does not appear for standard finite elements. It increases the interpolation error in the blending elements from order h^2 to h. Although this occurs only in few elements, it reduces the rate of convergence of the entire approximation. The reason for this is that the partition of unity property (completeness) of the approximation is not verified in the whole domain. Therefore, the theoretical rate of convergence cannot be attained. When the enrichment is a polynomial of order n, i.e. ξ^n, then for $n > 1$ the interpolation error in the blending elements is increased even further. If we go through the same steps as before, we find that

$$e(\overline{x}) \leq \frac{1}{8}h^2 \max(u_{,xx} + \frac{2a_1}{h^n}) \tag{3.37}$$

At this time, we can understand why in the quadratic XFEM formulation, the partition of unity shape functions have been chosen as linear. To get improved rate of convergence, Chessa proposed the enhanced strain formulation for the blending elements. By properly choosing an enhanced strain field, the undesired terms in the enriched approximation can be eliminated.

3.2.2 Isoparametric 2D quadrilateral XFEM element for linear elasticity

The implementation of XFEM will be explained for a 4-node quadrilateral element with linear shape functions, see Fig. 3.6.

3.2.3 Shape functions

The shape functions N_I, $I = 1...4$ are given by:

$$N_1(r,s) = \frac{1}{4}(1-r)(1-s)$$
$$N_2(r,s) = \frac{1}{4}(1+r)(1-s)$$
$$N_3(r,s) = \frac{1}{4}(1+r)(1+s)$$
$$N_4(r,s) = \frac{1}{4}(1-r)(1+s)$$

where r and s are scaled coordinates in the parent element, see Fig. 3.6. The standard finite element approximation of the displacement is

$$\mathbf{u}^e(M) = \begin{bmatrix} u_x \\ u_y \end{bmatrix} = \begin{bmatrix} N_1 & N_2 & N_3 & N_4 & 0 & 0 & 0 & 0 \\ 0 & 0 & 0 & 0 & N_1 & N_2 & N_3 & N_4 \end{bmatrix} \begin{bmatrix} u_{x1} \\ u_{x2} \\ u_{x3} \\ u_{x4} \\ u_{y1} \\ u_{y2} \\ u_{y3} \\ u_{y4} \end{bmatrix}$$

$$= \mathbf{N}^e_{std}(M)\ \mathbf{q}^e$$

The enriched finite element approximation of the displacement is

$$\mathbf{u}^e(M) = \begin{bmatrix} u_x \\ u_y \end{bmatrix} = \begin{bmatrix} N_1 & N_2 & N_3 & N_4 & 0 & 0 & 0 & 0 \\ 0 & 0 & 0 & 0 & N_1 & N_2 & N_3 & N_4 \end{bmatrix} \cdots$$

$$\cdots \begin{matrix} N_1\psi_1 & N_2\psi_2 & N_3\psi_3 & N_4\psi_4 & 0 & 0 & 0 & 0 \\ 0 & 0 & 0 & 0 & N_1\psi_1 & N_2\psi_2 & N_3\psi_3 & N_4\psi_4 \end{matrix} \Bigg] \begin{bmatrix} u_{x1} \\ u_{x2} \\ u_{x3} \\ u_{x4} \\ u_{y1} \\ u_{y2} \\ u_{y3} \\ u_{y4} \\ a_{x1} \\ a_{x2} \\ a_{x3} \\ a_{x4} \\ a_{y1} \\ a_{y2} \\ a_{y3} \\ a_{y4} \end{bmatrix}$$

$$\mathbf{u}^e(M) = [\ \mathbf{N}^e_{std}(M)\ \ \mathbf{N}^e_{enr}(M)\ ]\ \mathbf{q}^e$$

In other terms, this leads to

$$\mathbf{u}^e(M) = \mathbf{N}^e(M)\ \mathbf{q}^e$$

where $\mathbf{N}^e(M) = [\mathbf{N}^e_{std}(M)\ \mathbf{N}^e_{enr}(M)]$. In this last expression, it is assumed that the 4 nodes of the element are enriched by a single additional function $\psi(\mathbf{x})$ where ψ_I denotes the shifted function:

$$\psi_I(\mathbf{x}) = \psi(\mathbf{x}) - \psi(\mathbf{x}_I)$$

3.2.4 The B-operator

The strain tensor components in Voigt notation are

$$\boldsymbol{\epsilon} = \begin{bmatrix} \epsilon_{xx} \\ \epsilon_{yy} \\ 2\epsilon_{xy} \end{bmatrix} = \mathbf{D}\mathbf{u}^e(M)$$

with

$$\mathbf{D} = \begin{bmatrix} \frac{\partial}{\partial x} & 0 \\ 0 & \frac{\partial}{\partial y} \\ \frac{\partial}{\partial y} & \frac{\partial}{\partial x} \end{bmatrix}$$

By replacing $\mathbf{u}^e(M)$ by its approximated form, it gives

$$\boldsymbol{\epsilon} = \mathbf{D}\mathbf{N}^e(M)\ \mathbf{q}^e = \mathbf{B}^e(M)\ \mathbf{q}^e$$

where $\mathbf{B}^e(M)$ is the discretized gradient operator. It contains both the standard part and the enriched part and it can be written as

$$\mathbf{B}^e(M) = [\mathbf{B}^e_{std}(M)\ \mathbf{B}^e_{enr}(M)]$$

In the last expression, the discretized gradient operator $\mathbf{B}^e_{std}(M)$ is equal to:

$$\mathbf{B}^e_{std} = \begin{bmatrix} N_{1,x} & N_{2,x} & N_{3,x} & N_{4,x} & 0 & 0 & 0 & 0 \\ 0 & 0 & 0 & 0 & N_{1,y} & N_{2,y} & N_{3,y} & N_{4,y} \\ N_{1,y} & N_{2,y} & N_{3,y} & N_{4,y} & N_{1,x} & N_{2,x} & N_{3,x} & N_{4,x} \end{bmatrix}$$

and the enriched discretized gradient operator $\mathbf{B}^e_{enr}(M)$ is equal to

$$\mathbf{B}^e_{enr} = \begin{bmatrix} (N_1\psi_1)_{,x} & (N_2\psi_2)_{,x} & (N_3\psi_3)_{,x} & (N_4\psi_4)_{,x} & \dots \\ 0 & 0 & 0 & 0 & \dots \\ (N_1\psi_1)_{,y} & (N_2\psi_2)_{,y} & (N_3\psi_3)_{,y} & (N_4\psi_4)_{,y} & \dots \end{bmatrix}$$

$$\begin{bmatrix} \dots & 0 & 0 & 0 & 0 \\ \dots & (N_1\psi_1)_{,y} & (N_2\psi_2)_{,y} & (N_3\psi_3)_{,y} & (N_4\psi_4)_{,y} \\ \dots & (N_1\psi_1)_{,x} & (N_2\psi_2)_{,x} & (N_3\psi_3)_{,x} & (N_4\psi_4)_{,x} \end{bmatrix}$$

In the case of a **Heaviside enrichment**, the derivative of the approximation can be written as

$$\begin{aligned} \mathbf{u}^h_{i,j} &= \sum_{I\in\mathcal{S}} N_{J,i}(\mathbf{x})\ u_{iJ} + \sum_{I\in\mathcal{S}} \left(N_J(\mathbf{x})H(\phi(\mathbf{x}))\right)_{,i} a_{iJ} \\ &= \sum_{I\in\mathcal{S}} N_{J,i}(\mathbf{x})\ u_{iJ} + \sum_{I\in\mathcal{S}} \left(N_{J,i}(\mathbf{x})H(\phi(\mathbf{x})) + N_J(\mathbf{x})H_{,i}(\phi(\mathbf{x}))\right) a_{iJ} \end{aligned}$$

The derivatives of the Heaviside is the Dirac delta function $H_{,i}(\phi(\mathbf{x})) = \boldsymbol{\delta}$. That means that $H_{,i} = 1$ at the crack interface and $H_{,i} = 0$ otherwise. If we assume the crack to be traction-free, we can omit the derivatives of the Heaviside function and can give the enriched B-operator as

$$\mathbf{B}^e_{enr} = \begin{bmatrix} N_{1,x}\psi_1 & N_{2,x}\psi_2 & N_{3,x}\psi_3 & N_{4,x}\psi_4 & \dots \\ 0 & 0 & 0 & 0 & \dots \\ N_{1,x}\psi_1 & N_{2,x}\psi_2 & N_{3,x}\psi_3 & N_{4,x}\psi_4 & \dots \end{bmatrix}$$

$$\begin{bmatrix} \dots & 0 & 0 & 0 & 0 \\ \dots & N_{1,y}\psi_1 & N_{2,y}\psi_2 & N_{3,y}\psi_3 & N_{4,y}\psi_4 \\ \dots & N_{1,y}\psi_1 & N_{2,y}\psi_2 & N_{3,y}\psi_3 & N_{4,y}\psi_4 \end{bmatrix}$$

In the case of a **ramp enrichment**, $\psi(\mathbf{x}) = |\phi(\mathbf{x})|$, the computation of the derivative of $\psi(\mathbf{x})$ is needed:

$$\Big(\psi(\mathbf{x})\Big)_{,i} = \text{sign}(\phi(\mathbf{x}))\ \phi_{,i}(\mathbf{x})$$

The derivative of $\phi(\mathbf{x})$ is then needed. Since

$$\phi(\mathbf{x}) = [\ N_1 \quad N_2 \quad N_3 \quad N_4\ ] \begin{bmatrix} \phi_1 \\ \phi_2 \\ \phi_3 \\ \phi_4 \end{bmatrix}$$

the derivative according to x is

$$\phi(\mathbf{x})_{,x} = [\, N_{1,x} \;\; N_{2,x} \;\; N_{3,x} \;\; N_{4,x} \,] \begin{bmatrix} \phi_1 \\ \phi_2 \\ \phi_3 \\ \phi_4 \end{bmatrix}$$

and the derivative according to y is

$$\phi(\mathbf{x})_{,y} = [\, N_{1,y} \;\; N_{2,y} \;\; N_{3,y} \;\; N_{4,y} \,] \begin{bmatrix} \phi_1 \\ \phi_2 \\ \phi_3 \\ \phi_4 \end{bmatrix}$$

3.2.5 The element stiffness matrix

The previous computations involve the derivative of the shape functions in terms of the physical coordinates. The relations between the derivative in the parent and in the physical coordinates are

$$\frac{\partial N_I}{\partial x} = \frac{\partial N_I}{\partial r}\frac{\partial r}{\partial x} + \frac{\partial N_I}{\partial s}\frac{\partial s}{\partial x}$$
$$\frac{\partial N_I}{\partial y} = \frac{\partial N_I}{\partial r}\frac{\partial r}{\partial y} + \frac{\partial N_I}{\partial s}\frac{\partial s}{\partial y}$$

which can be written, for each function N_I:

$$\begin{bmatrix} N_{,x} & N_{,y} \end{bmatrix} = \begin{bmatrix} N_{,r} & N_{,s} \end{bmatrix} \underbrace{\begin{bmatrix} \frac{\partial r}{\partial x} & \frac{\partial r}{\partial y} \\ \frac{\partial s}{\partial x} & \frac{\partial s}{\partial y} \end{bmatrix}}_{=\mathbf{J}^{-1}}$$

where $\mathbf{J}$ is the Jacobian. The derivatives of the functions $N_I(r,s)$ in terms of the parent coordinates r and s are:

$$N_{1,r} = -\frac{1}{4}(1-s) \qquad N_{1,s} = -\frac{1}{4}(1-r)$$
$$N_{2,r} = \frac{1}{4}(1-s) \qquad N_{2,s} = -\frac{1}{4}(1+r)$$
$$N_{3,r} = \frac{1}{4}(1+s) \qquad N_{3,s} = \frac{1}{4}(1+r)$$
$$N_{4,r} = -\frac{1}{4}(1+s) \qquad N_{4,s} = \frac{1}{4}(1-r)$$

The expression of the Jacobian is

$$\mathbf{J} = \begin{bmatrix} \frac{\partial x}{\partial r} & \frac{\partial x}{\partial s} \\ \frac{\partial y}{\partial r} & \frac{\partial y}{\partial s} \end{bmatrix}$$

with

$$x = \sum_{I=1}^{4} N_I x_I \ , \quad \frac{\partial x}{\partial r} = \sum_{I=1}^{4} \frac{\partial N_I}{\partial r} x_I \ , \quad \frac{\partial x}{\partial s} = \sum_{I=1}^{4} \frac{\partial N_I}{\partial s} x_I$$

which is also

$$\frac{\partial x}{\partial r} = \begin{bmatrix} N_{1,r} & N_{2,r} & N_{3,r} & N_{4,r} \end{bmatrix} \begin{bmatrix} x_1 \\ x_2 \\ x_3 \\ x_4 \end{bmatrix}$$

$$\frac{\partial x}{\partial s} = \begin{bmatrix} N_{1,s} & N_{2,s} & N_{3,s} & N_{4,s} \end{bmatrix} \begin{bmatrix} x_1 \\ x_2 \\ x_3 \\ x_4 \end{bmatrix}$$

and

$$y = \sum_{I=1}^{4} N_I y_I \ , \quad \frac{\partial y}{\partial r} = \sum_{I=1}^{4} \frac{\partial N_I}{\partial r} y_I \ , \quad \frac{\partial y}{\partial s} = \sum_{I=1}^{4} \frac{\partial N_I}{\partial s} y_I$$

which is also

$$\frac{\partial y}{\partial r} = \begin{bmatrix} N_{1,r} & N_{2,r} & N_{3,r} & N_{4,r} \end{bmatrix} \begin{bmatrix} y_1 \\ y_2 \\ y_3 \\ y_4 \end{bmatrix}$$

$$\frac{\partial y}{\partial s} = \begin{bmatrix} N_{1,s} & N_{2,s} & N_{3,s} & N_{4,s} \end{bmatrix} \begin{bmatrix} y_1 \\ y_2 \\ y_3 \\ y_4 \end{bmatrix}$$

The expression of the stiffness matrix for an enriched element is

$$\mathbf{k}^e = \int_{\Omega^e} \mathbf{B}^{e^T}(M)\ \mathbf{C}^e\ \mathbf{B}^e(M)\ \mathrm{d}\Omega = \int_{-1}^{1}\int_{-1}^{1} \mathbf{B}^{e^T}(r,s)\ \mathbf{C}^e\ \mathbf{B}^e(r,s)\ \det\mathbf{J}\ \mathrm{d}r\ \mathrm{d}s$$

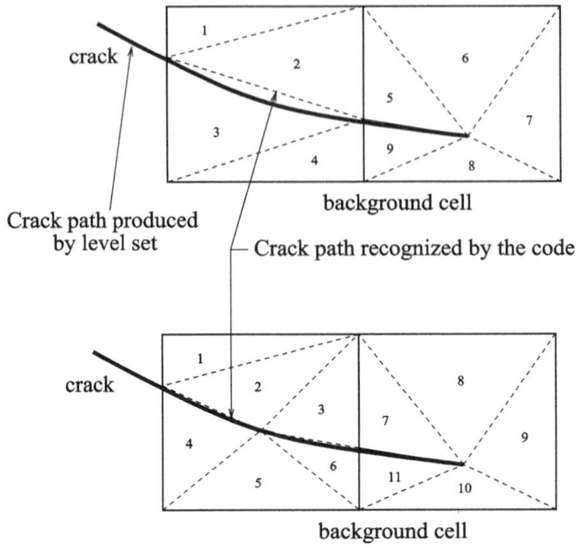

Figure 3.7 Sub-triangulation of finite elements.

where $\mathbf{C}^e$ is the constitutive tangent operator. The stiffness matrix can be decomposed into four 8×8 blocks:

$$\mathbf{k}^e = \begin{bmatrix} \int_{\Omega^e} \mathbf{B}_{std}^{e^T}(M)\mathbf{C}^e\mathbf{B}_{std}^e(M) & \int_{\Omega^e} \mathbf{B}_{std}^{e^T}(M)\mathbf{C}^e\mathbf{B}_{enr}^e(M) \\ \int_{\Omega^e} \mathbf{B}_{enr}^{e^T}(M)\mathbf{C}^e\mathbf{B}_{std}^e(M) & \int_{\Omega^e} \mathbf{B}_{enr}^{e^T}(M)\mathbf{C}^e\mathbf{B}_{enr}^e(M) \end{bmatrix}$$

The up-left block corresponds to the standard stiffness matrix. The elemental enriched stiffness matrix has a 16×16 size.

3.2.6 Integration

In FEM, Gauss quadrature is usually applied. However, standard Gauss quadrature cannot be used if the element is crossed by a discontinuity. One integration strategy is based on sub-dividing existing elements into several smaller triangular elements as shown in Fig. 3.7 for the 2D case. Difficulties occur for high curvatures of interface ϕ. For example: Consider linear finite elements. If an interface severely curves as shown for the two-dimensional case in Fig. 5.72 and the interface is discretized with level sets where for the discretization of the level set, the linear FE shape functions are used, then the level set cannot capture the curvature of the crack correctly. Therefore, an additional node is introduced to maintain the accuracy of the integration. Adding one more point, the error is reduced due to the higher order quadrature.

After the element with the interface is sub-triangulated, the integration of a function $\mathcal{F}$ can be done as follows:

$$\begin{aligned}\mathcal{F} &= \int_{\Omega^-} \mathcal{F}(\mathbf{X})d\Omega + \int_{\Omega^+} \mathcal{F}(\mathbf{X})d\Omega \\ &= \int_{\bar{\Omega}^-} \mathcal{F}(\mathbf{X}(\boldsymbol{\xi}))\ \det\mathbf{J}^-(\boldsymbol{\xi})d\Omega + \int_{\bar{\Omega}^+} \mathcal{F}(\mathbf{X}(\boldsymbol{\xi}))\ \det\mathbf{J}^+(\boldsymbol{\xi})\ d\Omega\end{aligned} \tag{3.38}$$

where the Jacobian maps the domains $\bar{\Omega}^{\pm}$ onto $\Omega^{\pm}$. With the mapping $\tilde{\Omega}^{\pm} \rightarrow \bar{\Omega}^{\pm}$, we finally obtain:

$$\begin{aligned}\mathcal{F} = &\int_{\tilde{\Omega}^-} \mathcal{F}(\mathbf{X}(\boldsymbol{\xi})(\boldsymbol{\eta}))\ \det\mathbf{J}^-(\boldsymbol{\xi}(\boldsymbol{\eta}))\ \det\bar{\mathbf{J}}^-(\boldsymbol{\eta})d\Omega \\ &+ \int_{\tilde{\Omega}^+} \mathcal{F}(\mathbf{X}(\boldsymbol{\xi})(\boldsymbol{\eta}))\ \det\mathbf{J}^+(\boldsymbol{\xi}(\boldsymbol{\eta}))\ \det\bar{\mathbf{J}}^+(\boldsymbol{\eta})d\Omega\end{aligned} \tag{3.39}$$

and in its discrete form using Gauss quadrature:

$$\mathcal{F} = \sum_{I=1}^{n_{GP}^-} \mathcal{F}(\boldsymbol{\eta}_I)\ \det\mathbf{J}^-(\boldsymbol{\eta})\ \det\bar{\mathbf{J}}^-(\boldsymbol{\eta})w_I + \sum_{I=1}^{n_{GP}^+} \mathcal{F}(\boldsymbol{\eta}_I)\ \det\mathbf{J}^+(\boldsymbol{\eta})\ \det\bar{\mathbf{J}}^+(\boldsymbol{\eta})w_I \tag{3.40}$$

where n_{GP}^- and n_{GP}^+ are the Gauss points in Ω^- and Ω^+, respectively, and $\boldsymbol{\eta}$ are the local coordinates of the Gauss points and w_I are their quadrature weights.

An alternative approach is the modification of the quadrature weights, that is illustrated in Fig. 3.8. In that method, the quadrature weights crossed by the crack are computed according to their areas A^+ and A^-:

$$\begin{aligned}w_I^+ &= w\frac{A_I^+}{A_I} \\ w_I^- &= w\frac{A_I^-}{A_I}\end{aligned} \tag{3.41}$$

The method requires subdivision of the element; we note that a certain number of Gauss points are needed in order to obtain accurate results.

In crack problems, it is preferable to have an accurate stress field around the crack tip in order to model the propagation of the crack as accurate as possible. In a sub-triangulation procedure, state variables have to be mapped from the original Gauss points to the new Gauss points generated by the sub-triangulation procedure *at the time* the crack enters the element. Therefore, it is preferable to use the second approach with higher numbers of

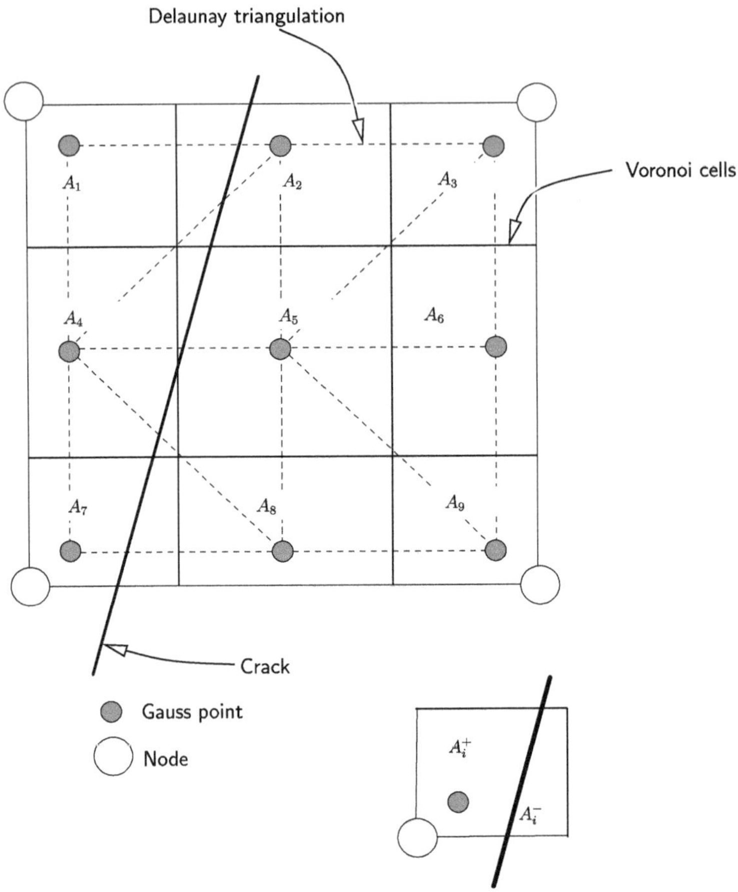

Figure 3.8 Integration with modified quadrature weights.

Gauss points adjacent to crack-tip elements. The mapping is then done *before* the crack approaches the element that will be cracked.

In LEFM, the elements that contain the crack tip do not only have a discontinuity but also a singularity at the crack tip. Hence, a sub-triangulation procedure might not be accurate enough if Gauss points of the sub-triangles are close to the stress singularity. This drawback can be circumvented by expressing the elementary integrals of the type

$$\int \nabla \left(F_j \psi_i\right) \cdot \nabla \left(F_l \psi_k\right) \; dx \tag{3.42}$$

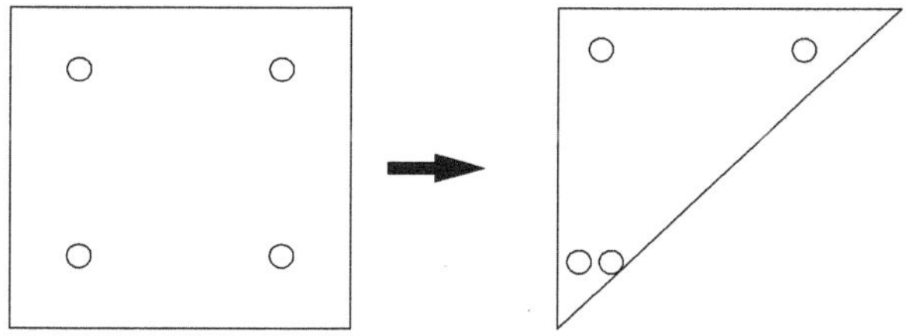

Figure 3.9 Transformation of an integration method on a square into an integration method on a triangle for crack tip functions.

in polar coordinates that will remove the $r^{-0.5}$ singularity of ∇F_i. The geometric transformation

$$\mathcal{G}: \left\{ \begin{array}{c} x \\ y \end{array} \right\} \leftarrow \left\{ \begin{array}{c} x\,y \\ y \end{array} \right\} \tag{3.43}$$

maps the unit square onto a triangle, see Fig. 3.9. With this transformation, it is possible to build a quadrature rule on the triangle from a quadrature rule on the unit square. The new integration points $\bar{\xi}$ and their weights $\bar{w}$ are obtained from those of the original quadrature rule with

$$\bar{\xi} = \mathcal{G}(\xi) \quad , \bar{w} = w \det(\nabla \mathcal{G}) \tag{3.44}$$

An effective integration strategy for curved discontinuity in the context of higher order elements was proposed by [14]. The key idea is to use special designed sub-elements which account for the curvature of the crack. In this context, isogeometric extended finite elements might have a severe impact on modeling curved discontinuities since isogeometric, i.e. NURBS, shape functions are capable of modeling a wider class of geometries compared to polynomial based finite elements.

For nonlinear problems, the crack tip enrichment can be easily omitted. This is achieved by 'proper' enrichment as shown in Fig. 3.10. Instead of using a tip enrichment and ensuring that the crack tip is in the interior of the element, only a step enrichment (or Heaviside enrichment) is used and the nodes at the edge of the crack tip are not enriched as illustrated in Fig. 3.10. Omitting the tip enrichment circumvents difficulties in blending and integration of non-polynomial functions. However, the crack tip will always be located at the side spanned by the two non-enriched nodes.

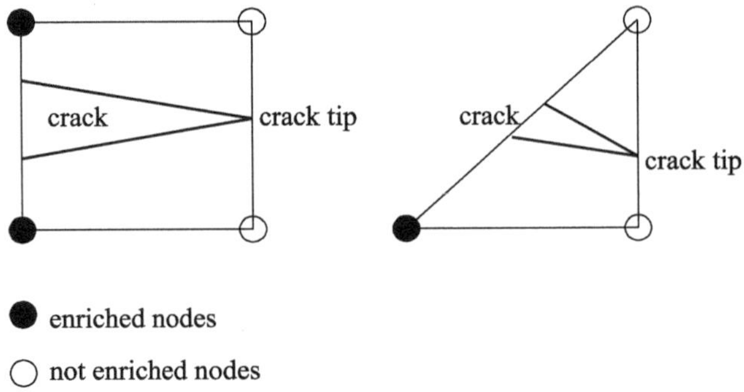

Figure 3.10 One opportunity to close the crack in XFEM.

3.3. XFEM for static/quasi-static fracture modeling in 2D and 3D

3.3.1 XFEM approximation for cracks

The application of XFEM to cracks poses additional challenges onto the method:

- It has to be ensured that the crack closes at its crack tip.
- During a loading cycle, a crack can open and close. It has to be ensured that the two crack surfaces do not overlap.
- While in fluid-fluid or fluid-structure interaction problems, the interface is given at the beginning of the computation, cracks can be initiated at any time during a loading cycle. Hence, we need also a criterion that tells us when a crack is initiated, how the crack will be oriented and how the shape of the crack looks like, e.g. how long is the crack?
- Crack path tracking algorithms are more complex than interface tracking algorithms in two-phase flow or fluid-structure interaction problems.

We will discuss these aspects only for a single crack. Moreover, we consider two-dimensional triangular and quadrilateral finite elements with linear shape functions. The crack line is approximated with level sets where the level set is discretized with the same shape functions as the mechanical properties. Hence, the curvature of the crack can be maximal bi-linear for the quadrilateral element and linear for the triangular element. We do not consider geometrical and material nonlinearities. For now, we assume that the crack does not propagate and study only the crack kinematics. The subsequent issues will be discussed in the following sections.

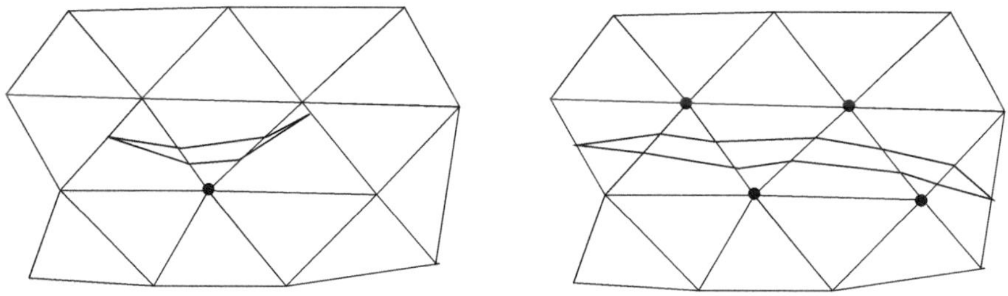

Figure 3.11 Example of enriched nodes when crack closure is realized by closing the crack tip at the element sides.

The Heaviside enrichment is able to capture the jump in the displacement field. However, it does not necessarily guarantee the closure of the crack at its crack tip. The simplest possibility to ensure crack closure at the crack tip is to locate the crack tip at one of the element edges and not to enrich the adjacent nodes, see Fig. 3.10. Two potential crack pathes including the enriched nodes are illustrated in Fig. 3.11. The black dots denote the enriched nodes. The standard nodes are not shown. This kind of crack closure has the drawbacks that the crack length is governed by the element size since it is required to grow the crack through the entire element. Another opportunity is to close the crack within an element. However, the enrichment procedure has to be modified. Therefore, let us consider a triangular element as shown on the RHS in Fig. 3.12. The crack is assumed to pass through the side $\overline{23}$ and intersect side $\overline{12}$. Other relationships can be obtained by permuting the node numbers. Due to compatibility, the enrichment has to vanish on the sides $\overline{12}$ and $\overline{13}$ and is continuous across $\overline{23}$ with the field in the adjacent element. To meet this condition, only node 3 is enriched and the discontinuous displacement field in the tip element is

$$\mathbf{u}_{disc} = \boldsymbol{\xi}_3^* \; \Psi_3(\boldsymbol{\xi}^*) \; \mathbf{a}_3 \tag{3.45}$$

where $\boldsymbol{\xi}^* = \left[\xi_1^* \; \xi_2^* \; \xi_3^*\right]$ are the parent coordinates of the sub-triangle $\overline{23P}$ in Fig. 3.12. The shaded parent area coordinates are related by $\xi_3^* = 1 - \xi_1^* - \xi_2^*$ and $\Psi_3(\boldsymbol{\xi}^*) = \text{sign}\left(\phi(\boldsymbol{\xi}^*)\right) - \text{sign}\left(\phi_3\right)$. The relation between $\boldsymbol{\xi}^*$ and $\boldsymbol{\xi}$ is given by

$$\xi_1^* = \frac{\xi_1}{\xi_{1P}}, \quad \xi_2^* = \xi_2 \tag{3.46}$$

where ξ_{1P} is the area coordinate of point P. When the direction of the crack intersects side $\overline{31}$, see the LHS of Fig. 3.12, then the discontinuous

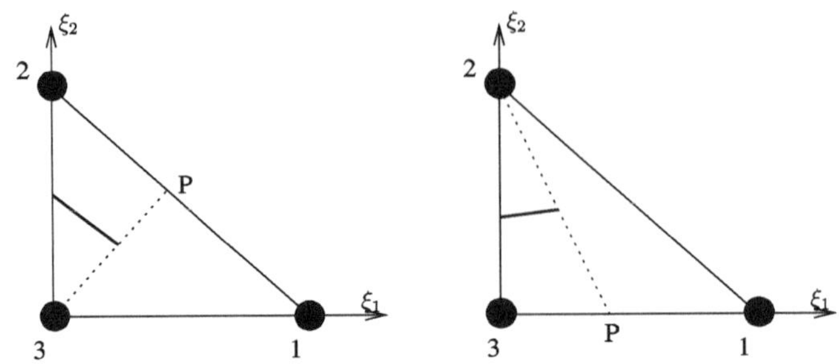

$$N_3(\boldsymbol{\xi}) = 1 - \xi_1 - \xi_2$$
$$N_1(\boldsymbol{\xi}) = \xi_1$$
$$N_2(\boldsymbol{\xi}) = \xi_2$$

Figure 3.12 XFEM enrichment for a crack tip that ends in an element.

part of the displacement field is

$$\mathbf{u}_{disc} = \boldsymbol{\xi}_2^* \; \Psi_2(\boldsymbol{\xi}^*) \; \mathbf{a}_2 \tag{3.47}$$

with

$$\xi_1^* = \xi_1 - \frac{\xi_{1P}}{\xi_{2P}} \xi_2, \quad \xi_2^* = \frac{\xi_2}{\xi_{2P}} \tag{3.48}$$

and $\Psi_2(\boldsymbol{\xi}^*) = \text{sign}\left(\phi(\boldsymbol{\xi}^*)\right) - \text{sign}\,(\phi_2)$ and $\mathbf{a}_3 = \mathbf{a}_P = 0$. The enrichment can be implemented easier by letting

$$\mathbf{u}_{disc} = \sum_I \boldsymbol{\xi}_I^* \; \Psi_I(\boldsymbol{\xi}^*) \; \mathbf{a}_I \tag{3.49}$$

and constructing $\mathbf{a}_I$ to vanish for the nodes on the edge towards which the crack is heading. These enrichment displacements $\mathbf{u}_{disc}$ vanish on the boundary of Ω_{enr}. Hence, only the elements in Ω_{enr} need a special treatment of the same type. This is a partition of unity in Ω_{enr} and there is no inner blending between different enrichments. Moreover, since the enrichment vanishes on the boundary of Ω_{enr}, blending outside the enriched subdomain does not occur. Thus, although it is a local partition of unity, it is indistinguishable from a global partition of unity.

Another opportunity to close the crack is by use of branch functions $\mathbf{B}$ which can be defined as

$$\mathbf{B} = [B_1 \; B_2 \; B_3 \; B_4]$$

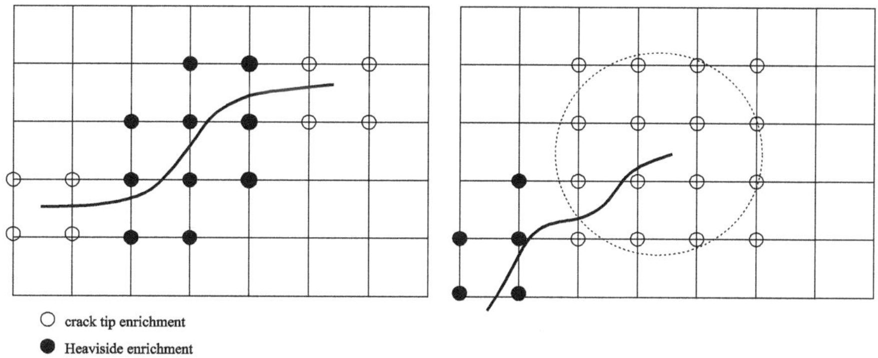

Figure 3.13 Crack with enriched elements.

$$= \left[\sqrt{r}\sin\frac{\theta}{2}, \sqrt{r}\cos\frac{\theta}{2}, \sqrt{r}\sin\frac{\theta}{2}\sin(\theta), \sqrt{r}\cos\frac{\theta}{2}\sin(\theta) \right] \tag{3.50}$$

Here, we will denote the branch functions with **B** according to the LEFM literature. It is obvious that for $r = 0$, the crack closes at the crack tip. Moreover, the solution will be more accurate since we put the information of the analytical solution into our approximation. The approximation of the displacement field is now given by:

$$\begin{aligned} \mathbf{u}^h(\mathbf{X}) = \sum_{I \in \mathcal{S}} N_I(\mathbf{X})\, \mathbf{u}_I + \sum_{I \in \mathcal{S}_c(\mathbf{X})} N_I(\mathbf{X})\, \bar{H}(f_I(\mathbf{X}))\, \mathbf{a}_I \\ + \sum_{I \in \mathcal{S}_t(\mathbf{X})} N_I(\mathbf{X}) \sum_K \bar{\mathbf{B}}_K(\mathbf{X})\, \mathbf{b}_{KI} \end{aligned} \tag{3.51}$$

where $\mathcal{S}_t$ are the set of nodes that are influenced by the crack tip and the superimposed bar denotes a shifting as described in Section 3.1.1.1. The first term on the RHS of Eq. (3.51) is the standard approximation, the second term and the third term is the enrichment, see also Fig. 3.13. When only the element containing the crack tip is enriched with the branch functions **B**, it is referred to as topological enrichment. However, this will not lead to optimal convergence rates which requires a geometrical enrichment as illustrated on the RHS of Fig. 3.13. In a geometrical enrichment, the enriched area around the crack tip is kept constant once the mesh is refined. However, geometrical enrichments increase the number of degrees of freedom and lead to an increased ill-conditioning of the system's stiffness matrix which might require pre-conditioning or special solution techniques.

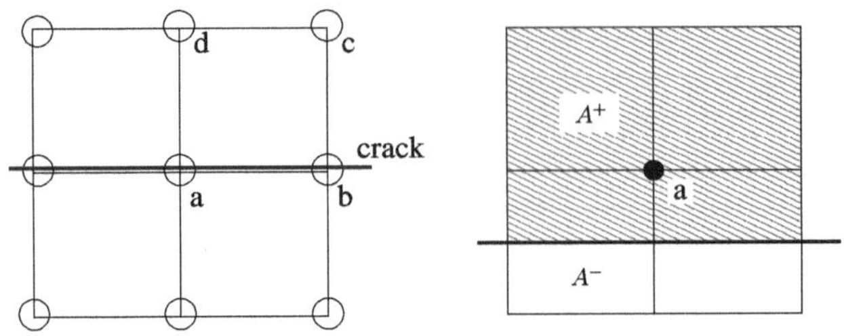

Figure 3.14 Enrichment criteria for cracks that are close to a node.

Finally, we would like to mention the special case, where a crack is located close to a node[6] as shown in Fig. 3.14. Therefore, let us consider the element spanned by the nodes a, b, c, d. The support of the node is defined by its adjacent neighbor elements. Node a for example has four neighboring elements. Hence, the support size is equal to the area of these four neighboring finite elements. Let us define the parameters r^+ and r^- between the area on both sides of the crack and the total area of the support:

$$r^+ = \frac{A^+}{A^+ + A^-}, \quad r^- = \frac{A^-}{A^+ + A^-} \tag{3.52}$$

If one of these ratios is smaller than a given threshold, the support node is no longer enriched. In the case of our element a, b, c, d, only nodes a and b will be enriched.

3.3.2 Discrete equations

To obtain the discrete equations, the test and trial functions have to be substituted into the weak form of the equilibrium equation. We will consider the approximation with crack tip enrichment as defined in Eq. (3.51). The approximation of the test functions looks identical. Then we obtain the final system of equations:

$$\begin{bmatrix} \mathbf{K}_{IJ}^{uu} & \mathbf{K}_{IJ}^{ua} & \mathbf{K}_{IJK}^{ub} \\ \mathbf{K}_{IJ}^{au} & \mathbf{K}_{IJ}^{aa} & \mathbf{K}_{IJK}^{ab} \\ \mathbf{K}_{IJK}^{bu} & \mathbf{K}_{IJK}^{ba} & \mathbf{K}_{IJK}^{bb} \end{bmatrix} \begin{Bmatrix} \mathbf{u}_J \\ \mathbf{a}_J \\ \mathbf{b}_{JK} \end{Bmatrix} = \begin{Bmatrix} \mathbf{f}_I^{ext} \\ \mathbf{f}_I^{ext} \\ \mathbf{f}_{IK}^{ext} \end{Bmatrix} \tag{3.53}$$

or

$$\mathbf{K}\,\mathbf{d} = \mathbf{f}^{ext} \tag{3.54}$$

[6] The node cannot be enriched with a crack tip enrichment.

where $\mathbf{K}$ is the stiffness matrix, $\mathbf{d} = \{\mathbf{u}\ \mathbf{a}\ \mathbf{b}\}^T$ is the vector with the nodal parameters, $\mathbf{f}^{ext} = \{\mathbf{f}^u\ \mathbf{f}^a\ \mathbf{f}^b\}^T$ is the external force vector with $\mathbf{f}^b = \{\mathbf{f}^{b1}\ \mathbf{f}^{b2}\ \mathbf{f}^{b3}\ \mathbf{f}^{b4}\}$ and

$$\mathbf{f}_I^u = \int_\Omega N_I\ \mathbf{b}\ d\Omega + \int_{\Gamma^t} N_I\ \bar{\mathbf{t}}\ d\Gamma \tag{3.55}$$

$$\begin{aligned}\mathbf{f}_I^a = &\int_\Omega N_I\ (H(\phi(\mathbf{X})) - H(\phi(\mathbf{X}_I)))\ \mathbf{b}\ d\Omega \\ &+ \int_{\Gamma^t} N_I\ (H(\phi(\mathbf{X})) - H(\phi(\mathbf{X}_I)))\ \bar{\mathbf{t}}\ d\Gamma\end{aligned} \tag{3.56}$$

$$\mathbf{f}_I^{bl} = \int_\Omega N_I\ \left(B_I^l(\mathbf{X}) - B_I^l(\mathbf{X}_I)\right)\ \mathbf{b}\ d\Omega + \int_{\Gamma^t} N_I\ \left(B_I^l(\mathbf{X}) - B_I^l(\mathbf{X}_I)\right)\ \bar{\mathbf{t}}\ d\Gamma \tag{3.57}$$

The stiffness matrix is

$$\mathbf{K} = \int_\Omega \bar{\mathbf{B}}^T\ \mathbf{C}\ \bar{\mathbf{B}}\ d\Omega \tag{3.58}$$

where $\bar{\mathbf{B}}$ is the B-operator defined by:

$$\mathbf{B}_I^u = \begin{bmatrix} N_{I,X} & 0 \\ 0 & N_{I,Y} \\ N_{I,Y} & N_{I,X} \end{bmatrix} \tag{3.59}$$

$$\mathbf{B}_I^a = \begin{bmatrix} N_{I,X}\ (H(\phi(\mathbf{X})) - H(\phi(\mathbf{X}_I))) & 0 \\ 0 & N_{I,Y}\ (H(\phi(\mathbf{X})) - H(\phi(\mathbf{X}_I))) \\ N_{I,Y}\ (H(\phi(\mathbf{X})) - H(\phi(\mathbf{X}_I))) & N_{I,X}\ (H(\phi(\mathbf{X})) - H(\phi(\mathbf{X}_I))) \end{bmatrix} \tag{3.60}$$

$$\mathbf{B}_I^{bl}|_{l=1,2,3,4} = \begin{bmatrix} \left[N_I\ (B_K^l(\mathbf{X}) - B_K^l(\mathbf{X}_I))\right]_{,X} & 0 \\ 0 & \left[N_I\ (B_K^l(\mathbf{X}) - B_K^l(\mathbf{X}_I))\right]_{,Y} \\ \left[N_I\ (B_K^l(\mathbf{X}) - B_K^l(\mathbf{X}_I))\right]_{,Y} & \left[N_I\ (B_K^l(\mathbf{X}) - B_K^l(\mathbf{X}_I))\right]_{,X} \end{bmatrix} \tag{3.61}$$

In Eq. (3.60), we have already omitted the derivatives of the Heaviside function, see Section 3.2.4. The partial derivatives for the branch function are

$$\left[N_I\ B_K^l(\mathbf{X})\right]_{,i} = N_{I,i}\ B_K^l(\mathbf{X}) + N_I\ B_K^l(\mathbf{X})_{,i} \tag{3.62}$$

To obtain the derivatives of the branch functions, let us define a local crack coordinate system, see Fig. 3.15. The angle α denotes the slope of the local

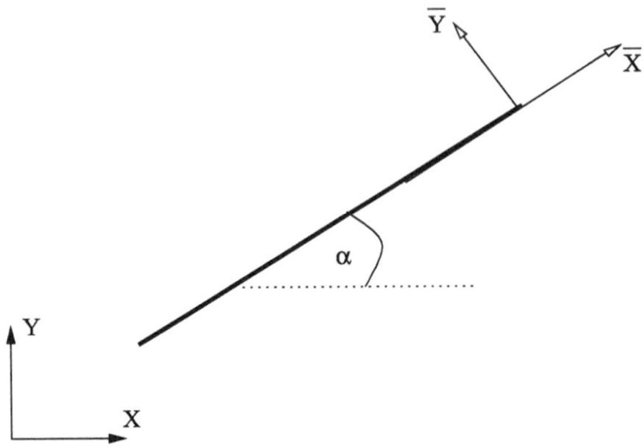

Figure 3.15 Definition of a local crack coordinate system.

crack coordinate system with respect to the global coordinate system. The derivatives of the branch functions in the local crack coordinate system are

$$B^l_{,\bar{i}} = B^l_{,r}\, r_{,\bar{i}} + B^l_{,\theta}\, \theta_{,\bar{i}} \tag{3.63}$$

where θ and r are defined in Fig. 5.33 and the subscript "$,\bar{i}$" denotes derivatives with respect to the local crack coordinate system. The terms $B^l_{,r}$ and $B^l_{,\theta}$ are obtained by formal differentiation:

$$\begin{aligned}
B^1_{,r} &= \frac{\sin(\theta/2)}{2\sqrt{2}} \quad B^1_{,\theta} = \frac{\sqrt{2}\cos(\theta/2)}{2} \\
B^2_{,r} &= \frac{\cos(\theta/2)}{2\sqrt{2}} \quad B^2_{,\theta} = -\frac{\sqrt{2}\sin(\theta/2)}{2} \\
B^3_{,r} &= \frac{\sin(\theta/2)\,\sin(\theta)}{2\sqrt{2}} \quad B^3_{,\theta} = \sqrt{r}\left(\frac{\cos(\theta/2)\,\sin(\theta)}{2} + \sin(\theta/2)\,\cos(\theta)\right) \\
B^4_{,r} &= \frac{\cos(\theta/2)\,\sin(\theta)}{2\sqrt{2}} \quad B^4_{,\theta} = \sqrt{r}\left(\frac{\sin(\theta/2)\,\sin(\theta)}{2} + \cos(\theta/2)\,\cos(\theta)\right)
\end{aligned} \tag{3.64}$$

The derivatives of r and θ with respect to the local crack coordinate system are

$$\begin{aligned}
r_{,\bar{X}} &= \cos(\theta) \quad \theta_{,\bar{X}} = -\sin/r \\
r_{,\bar{Y}} &= \sin(\theta) \quad \theta_{,\bar{Y}} = \cos/r
\end{aligned} \tag{3.65}$$

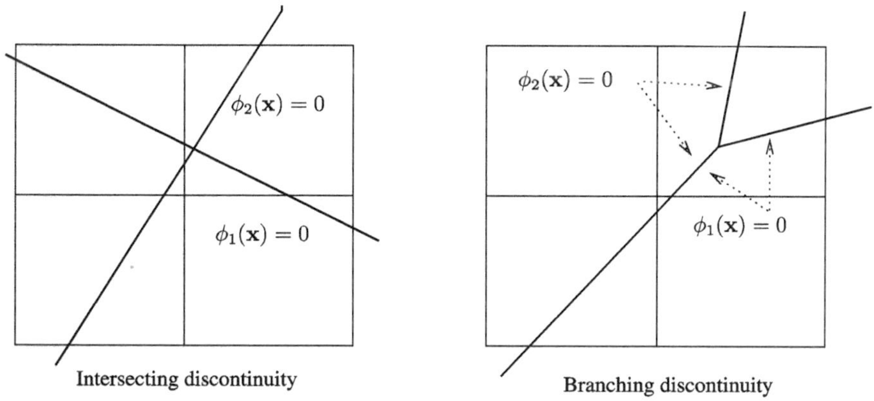

Figure 3.16 XFEM element with intersecting discontinuities and branching discontinuities.

With (3.64) and (3.65) in (3.63), we have the derivatives of the branch functions in the local crack coordinate system:

$$\begin{aligned}
B^1_{,\bar{X}} &= \frac{\sin(\theta/2)}{2\sqrt{2}} \quad B^1_{,\bar{Y}} = \frac{\cos(\theta/2)}{2\sqrt{2}} \\
B^2_{,\bar{X}} &= \frac{\cos(\theta/2)}{2\sqrt{2}} \quad B^2_{,\bar{Y}} = \frac{\sin(\theta/2)}{2\sqrt{2}} \\
B^3_{,\bar{X}} &= -\frac{\sin(3\theta/2)\,\sin(\theta)}{2\sqrt{2}} \quad B^3_{,\bar{Y}} = \frac{\sin(\theta/2) + \sin(3\theta/2)\,\cos(\theta)}{2\sqrt{2}} \\
B^4_{,\bar{X}} &= -\frac{\cos(3\theta/2)\,\sin(\theta)}{2\sqrt{2}} \quad B^4_{,\bar{Y}} = \frac{\cos(\theta/2) + \cos(3\theta/2)\,\cos(\theta)}{2\sqrt{2}}
\end{aligned} \tag{3.66}$$

Finally, the derivatives in the global coordinate system are obtained by:

$$\begin{aligned}
\mathbf{B}_{,X} &= \mathbf{B}_{,\bar{X}}\,\cos(\alpha) + \mathbf{B}_{,\bar{Y}}\,\sin(\alpha) \\
\mathbf{B}_{,Y} &= \mathbf{B}_{,\bar{X}}\,\sin(\alpha) + \mathbf{B}_{,\bar{Y}}\,\cos(\alpha)
\end{aligned} \tag{3.67}$$

where α is the inclination angle of the local crack coordinate system versus the global coordinate system.

3.3.3 Crack branching and crack junction

Let us consider the case of crack branching and crack intersection in a single finite element as shown in Fig. 3.16. Let $\mathcal{S}_c^1$ be the set of nodes whose domain of influence is completely cut by the discontinuity described with the signed distance function $\phi_1(\mathbf{X}) = 0$ and $\mathcal{S}_c^2$ the corresponding set for $\phi_2(\mathbf{X}) = 0$; $\mathcal{S}_c^3 = \mathcal{S}_c^1 \bigcap \mathcal{S}_c^2$. The same applies accordingly for nodes whose domain of influence is cut by the crack tip enrichment. We will denote this set of nodes with $\mathcal{S}_t^1$ and $\mathcal{S}_t^2$. Then the approximation of the displacement may be given by

$$\begin{aligned}\mathbf{u}^h(\mathbf{X}) = &\sum_{I \in \mathcal{S}(\mathbf{X})} N_I(\mathbf{X})\ \mathbf{u}_I + \sum_{I \in \mathcal{S}_c^1(\mathbf{X})} N_I(\mathbf{X})\ H(\phi_1(\mathbf{X}))\ \mathbf{a}_I^{(1)} \\ &+ \sum_{I \in \mathcal{S}_c^2(\mathbf{X})} N_I(\mathbf{X})\ H(\phi_2(\mathbf{X}))\ \mathbf{a}_I^{(2)} \\ &+ \sum_{I \in \mathcal{S}_c^3(\mathbf{X})} N_I(\mathbf{X})\ H(\phi_1(\mathbf{X}))\ H(\phi_2(\mathbf{X}))\ \mathbf{a}_I^{(3)} \\ &+ \sum_{I \in \mathcal{S}_t^1(\mathbf{X})} N_I(\mathbf{X}) \sum_K \mathbf{B}_K^{(1)}(\mathbf{X})\ \mathbf{b}_{KI}^{(1)} \\ &+ \sum_{I \in \mathcal{S}_t^2(\mathbf{X})} N_I(\mathbf{X}) \sum_K \mathbf{B}_K^{(2)}(\mathbf{X})\ \mathbf{b}_{KI}^{(2)}\end{aligned} \tag{3.68}$$

Note, that crack branching requires the introduction of another level set. Crack junction can be treated similarly. A computationally more efficient approach was proposed by [70] by modifying the signed distance functions so that no cross terms are needed for junction or branch problems.

When two cracks are joining, the crack tip enrichment is removed. By using the signed distance functions of the pre-existing and approaching crack, the signed distance function of the approaching crack is modified, see Fig. 3.17. Three different subdomains have to be considered: ($\phi_1 < 0$, $\phi_2 < 0$), ($\phi_1 > 0, \phi_2 > 0$), ($\phi_1 > 0, \phi_2 < 0$) as in Fig. 3.17B or ($\phi_1 > 0$, $\phi_2 < 0$), ($\phi_1 > 0, \phi_2 > 0$), ($\phi_1 < 0, \phi_2 < 0$) as in Fig. 3.17D. The signed distance function of crack 1 of a point $\mathbf{X}$ is then obtained by:

$$\phi_1(\mathbf{X}) = \begin{cases} \phi_1^0(\mathbf{X}), & \text{if } \phi_2^0(\mathbf{X}_1)\,\phi_2^0(\mathbf{X}) > 0 \\ \phi_2^0(\mathbf{X}), & \text{if } \phi_2^0(\mathbf{X}_1)\,\phi_2^0(\mathbf{X}) < 0 \end{cases} \tag{3.69}$$

where the superimposed 0 denotes the sign distance function before crack junction. Therefore, the final approximation without the cross term reads:

$$\begin{aligned}\mathbf{u}^h(\mathbf{X}) = &\sum_{I \in \mathcal{S}(\mathbf{X})} N_I(\mathbf{X})\ \mathbf{u}_I + \sum_{n=1}^{n_c} \sum_{I \in \mathcal{S}_c(\mathbf{X})} N_I(\mathbf{X})\ H(\phi_I^{(n)}(\mathbf{X}))\ \mathbf{a}_I^{(n)} \\ &+ \sum_{m=1}^{m_t} \sum_{I \in \mathcal{S}_t(\mathbf{X})} N_I(\mathbf{X}) \sum_K \mathbf{B}_K^{(m)}(\mathbf{X})\ \mathbf{b}_{KI}^{(m)}\end{aligned} \tag{3.70}$$

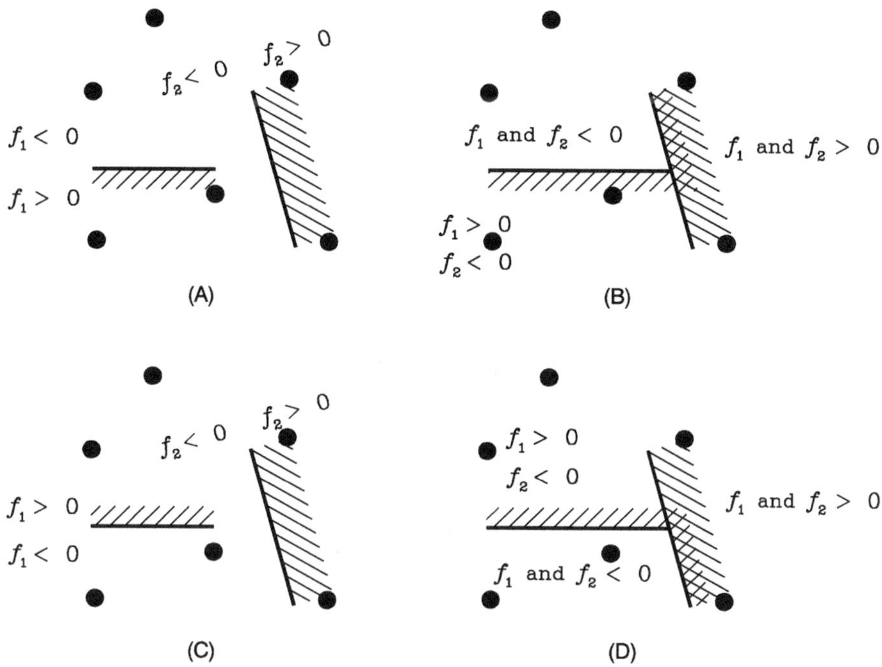

Figure 3.17 Sign functions for crack junction.

where n_c and m_t are the number of cracks that completely cross the element or contain the crack tip, respectively.

3.3.4 Crack opening and crack closure

Let us consider a pure mechanical problem with for traction-free cracks. In Chapter 2, we stated a similar problem only for the case of crack opening. The case of crack closure that can arise for example under cyclic loading conditions was not considered in that formulation. To generalize the crack problem in elastostatics, let us rephrase the problem:

$$\nabla_0 \cdot \mathbf{P} - \mathbf{b} = \emptyset \quad \forall \mathbf{X} \in \Omega_0 \setminus \Gamma_0^c \tag{3.71}$$

with boundary conditions

$$\mathbf{u}(\mathbf{X}, t) = \bar{\mathbf{u}}(\mathbf{X}, t) \quad \text{on} \quad \Gamma_0^u \tag{3.72}$$

$$\mathbf{n}_0 \cdot \mathbf{P}(\mathbf{X}, t) = \bar{\mathbf{t}}_0(\mathbf{X}, t) \quad \text{on} \quad \Gamma_0^t \tag{3.73}$$

$$\mathbf{n}_0 \cdot \mathbf{P}(\mathbf{X}, t) = 0 \quad \text{on} \quad \Gamma_0^c \quad \text{if not in contact} \tag{3.74}$$

$$t_{0t}^{+} = t_{0t}^{-} = 0, \quad t_{0N}^{+} = -t_{0N}^{-} \quad \text{on} \quad \Gamma_0^c \quad \text{if in contact} \tag{3.75}$$

$$[\![\mathbf{u}_N]\!] \leq 0 \quad \text{on} \quad \Gamma_0^c \tag{3.76}$$

$$[\![\mathbf{n} \cdot \mathbf{P}]\!] = 0 \quad \text{on} \quad \Gamma_0^c \tag{3.77}$$

where Eqs. (3.71) to (3.74) were already formulated in Chapter 2 and Eqs. (3.75) to (3.77) guarantee no interpenetration where $t_{0N} = \mathbf{n} \cdot \mathbf{P} \cdot \mathbf{n}$ is the normal traction and t_{0t} is the traction acting in tangential direction of the crack. The inequality (3.76) with $[\![\mathbf{u}_N]\!] = \mathbf{u}^+ \cdot \mathbf{n}^+ = \mathbf{u}^- \cdot \mathbf{n}^- \leq 0$ guarantees that the crack surfaces do not interpenetrate in the case of crack closure and Eq. (3.77) ensures traction continuity. The superimposed plus and minus sign indicates the different sides of the crack surfaces. Note that it is often assumed that $\mathbf{n}^+ = \mathbf{n}^-$ though this assumption is not mandatory.

If the Lagrange multiplier method is used to enforce no-interpenetration conditions of the adjacent crack surfaces, the weak form of the equilibrium equation reads:

$$\int_{\Omega_0} (\nabla \otimes \delta \mathbf{u})^T : \mathbf{P} \; d\Omega_0 - \int_{\Omega_0} \delta \mathbf{u} \cdot \mathbf{b} \; d\Omega_0 - \int_{\Gamma_0^t} \delta \mathbf{u} \cdot \bar{\mathbf{t}}_0 \; d\Gamma_0$$
$$+ \delta \int_{\Gamma_0^c} \boldsymbol{\lambda} \; [\![\mathbf{u}_N]\!] \; d\Gamma_0 \geq 0 \tag{3.78}$$

where it is sufficient to choose $\mathcal{C}^{-1}$ approximation functions for the Lagrange multiplier field if the approximating functions for the test and trial function are $\mathcal{C}^0$.

3.4. XFEM for dynamic fracture modeling in 2D and 3D

3.4.1 Diagonalized mass matrix

Methods that are based on an explicit time integration generally use a diagonalized mass matrix that is obtained by a simple row-sum technique. A diagonalized mass matrix facilitates the solution of the system of equations enormously. However, if additional degrees of freedom are introduced as in XFEM, it can be shown that a standard row-sum technique will lead to an erroneous lumped mass matrix.[7] A row-sum technique for standard XFEM can be obtained based on the assumption that for a rigid body motion, the discrete kinetic energy has to be exact. A lumped mass matrix can

[7] Note that for the Hansbo-Hansbo XFEM, a standard row sum technique is sufficient.

then be obtained by:

$$m_{diag} = \frac{m}{n_{nodes}} \frac{1}{mes(\Omega_{el})} \int_{\Omega_{el}} \psi^2 \, d\Omega_{el} \tag{3.79}$$

where Ω_{el} is the element being considered, m is its mass, $mes(\Omega)_{el}$ its length, n_{nodes} the number of nodes in Ω and ψ is the enrichment function. If the step function is used, the mass matrix can be diagonalized by one of these two procedures:

$$\mathbf{M}_{II}^{lumped} = \sum_J \mathbf{M}_{IJ}^{consistent}, \quad \text{or}$$

$$\mathbf{M}_{II}^{lumped} = m \frac{\mathbf{M}_{II}^{consistent}}{\sum_J \mathbf{M}_{IJ}^{consistent}} \tag{3.80}$$

One difficulty with a diagonalized mass matrix in an XFEM formulation is that the critical time step $\Delta t \le \Delta t_c = 2/\omega_{max}$ is reduced drastically when a crack is located close to a node.[8] Within the approach proposed above, the critical time step is not diminished so drastically when the discontinuity is close to a node; drastically means by a factor around 2 (compared to the CFL criterion of the element without discontinuity). Note that the lumping procedure is slightly different when the Heaviside function is used as enrichment.

Example. Let us consider a one-dimensional element with two nodes. The approximation is given by

$$\mathbf{u}^h(\mathbf{X}) = N_1 \, \mathbf{u}_1 + N_1 \, \phi_1 \, \mathbf{a}_1 + N_2 \, \mathbf{u}_2 + N_2 \, \phi_2 \mathbf{a}_2 \tag{3.81}$$

and the lumped mass matrix of the element by

$$\mathrm{M}_{lumped} = \begin{bmatrix} m_1 & 0 & 0 & 0 \\ 0 & m_2 & 0 & 0 \\ 0 & 0 & m_3 & 0 \\ 0 & 0 & 0 & m_4 \end{bmatrix} \tag{3.82}$$

The coefficients m_i have to be determined such that $E_{kin}^h = 0.5 \dot{\mathbf{u}}^T \, \mathbf{M}_{lumped} \, \dot{\mathbf{u}}$ equals $E_{kin} = 0.5 \int_{\Omega_{el}} \varrho \, \mathbf{v}^2 \, d\Omega$. Let us consider that the element moves with

[8] Note that ω_{max} is the largest solution of $\det(\mathbf{K} - \omega \, \mathbf{M})$ where $\mathbf{K}$ is the stiffness matrix and $\mathbf{M}$ the mass matrix.

a constant velocity $\dot{\bar{\mathbf{u}}}$ in the same direction. Hence, we set $\dot{\mathbf{a}}$ equal zero and obtain

$$E^h_{kin} = 0.5\left(m_1\ \dot{\mathbf{u}}_1^2 + m_2\ \dot{\mathbf{u}}_2^2\right) = 0.5\ \dot{\bar{\mathbf{u}}}^2\ (m_1 + m_2) \tag{3.83}$$

and obtain with $E_{kin} = 0.5\ m\ \dot{\bar{\mathbf{u}}}^2 = E^h_{kin}$ that $m_1 = m_2 = 0.5\ m$ where m is the mass of the element. Next let us consider the separation of the element into two parts, i.e. $\dot{\bar{\mathbf{u}}} = \bar{\mathbf{a}}\phi_1(\mathbf{x})$. Thus, we set $\dot{\mathbf{u}}$ equal zero and obtain

$$E^h_{kin} = 0.5\left(m_3\ \dot{\mathbf{a}}_1^2 + m_4\ \dot{\mathbf{a}}_2^2\right) = 0.5\ \bar{\mathbf{a}}^2\ (m_3 + m_4) \tag{3.84}$$

and

$$E_{kin} = 0.5\ \varrho\ \bar{\mathbf{a}}^2 \int_{\Omega_{el}} \psi_1^2\ d\Omega_{el} \tag{3.85}$$

so that finally the mass m_3 and m_4 are

$$m_3 = m_4 = \frac{m}{2\ mes(\Omega)_{el}} \int_{\Omega_{el}} \psi_1^2\ d\Omega_{el} \tag{3.86}$$

Let us now determine the minimal critical time step for the one-dimensional XFEM element and compare it to the critical time step of a standard element. If the length of the element is denoted by l, the linear shape function can be given by

$$\begin{aligned} N_1(x) &= 1 - \frac{x}{l} \\ N_2(x) &= \frac{x}{l} \end{aligned} \tag{3.87}$$

The consistent mass matrix and the stiffness matrix of the standard element is given by

$$\mathrm{M}_{FE} = \varrho\ A\ l \begin{bmatrix} 1/3 & 1/6 \\ 1/6 & 1/3 \end{bmatrix},\quad \mathrm{K}_{FE} = \frac{E\ A}{l} \begin{bmatrix} 1 & -1 \\ -1 & 1 \end{bmatrix} \tag{3.88}$$

where E is the Young's modulus and A the cross section. The critical time step is easily computed by

$$\Delta t_{c,FE} = \frac{2}{\omega_{max}} = l\sqrt{\frac{\varrho}{3E}} \tag{3.89}$$

With the lumped mass matrix

$$\mathrm{M}_{FE}^{lumped} = \varrho\, A\, l \begin{bmatrix} 1/2 & 0 \\ 0 & 1/2 \end{bmatrix} \tag{3.90}$$

the critical time step is

$$\Delta t_{c,FE}^{lumped} = l\sqrt{\frac{\varrho}{E}} = \sqrt{3}\Delta t_{c,FE} \tag{3.91}$$

Now, let us study the critical time step for the XFEM approximation. The discontinuity is located at position s and with the generalized step function centered in s, the approximation of the displacement field is

$$\begin{aligned} \mathbf{u}^h(x) &= N_1(x)\ \mathbf{u}_1 + N_1(x)\ S(x-s)\ \mathbf{a}_1 \\ &+ N_2(x)\ \mathbf{u}_2 + N_2(x)\ S(x-s)\mathbf{a}_2 \end{aligned} \tag{3.92}$$

and the consistent mass matrix and stiffness matrix

$$\mathrm{M}_{XFEM} = \varrho\, A\, l \begin{bmatrix} 1/3 & 1/6 \\ 1/6 & 1/3 \\ 2s^2-2s+1/3-2/3s^3 & 1/6-s^2+2/3s^3 \\ 1/6-s^2+2/3s^3 & 1/3-21 \end{bmatrix} \cdots$$

$$\cdots \begin{bmatrix} 2s^2-2s+1/3-2/3s^3 & 1/6-s^2+2/3s^3 \\ 1/6-s^2+2/3s^3 & 1/3-2/3s^3 \\ 1/3 & 1/6 \\ 1-2s & 2s-1 \\ 1/6 & 1/3 \end{bmatrix} \tag{3.93}$$

$$\mathrm{K}_{XFEM} = \frac{E\,A}{l} \begin{bmatrix} 1 & -1 & 1-2s & 2s-1 \\ -1 & 1 & 2s-1 & 1-2s \\ 1-2s & 2s-1 & 1 & -1 \\ 2s-1 & 1-2s & -1 & 1 \end{bmatrix} \tag{3.94}$$

The lumped mass matrix for the XFEM approximation is

$$\mathrm{M}_{XFEM}^{lumped} = 0.5\ \varrho\, A\, l \begin{bmatrix} 1 & 0 & 0 & 0 \\ 0 & 1 & 0 & 0 \\ 0 & 0 & 1 & 0 \\ 0 & 0 & 0 & 1 \end{bmatrix} \tag{3.95}$$

The critical time step for the XFEM-approximation depends on the location of the discontinuity s in the element. The smallest critical time step is obtained when the discontinuity is located at $x=0$ or $x=l$. For the consistent XFEM mass matrix, the critical time step goes to zero when the discontinuity approaches 0 or l while the critical time step at $x=0$ and $x=l$ for the lumped XFEM mass matrix is $\Delta t_{c,XFEM}^{lumped} = \frac{1}{\sqrt{2}} \Delta t_{c,FE}^{lumped}$. Hence, even when the discontinuity is located very close to a node, the critical time step is not destroyed. In contrast, for the Hansbo-Hansbo approach, a standard row sum technique can be employed. However, the critical time step will tend to zero when the discontinuity approaches a node. In practice, a certain minimum mass is assigned to a node such that the computation can be proceeded.

3.4.2 Limitations

For particular discretizations and crack configurations, the XFEM approximation cannot accurately represent the discontinuity in the near-tip displacement fields. Difficulties occur for the non-physical case of too close parallel cracks (within a single element) and when the extent of the crack approaches the support size of the nodal shape functions. Then, the asymptotic branch functions for each tip may extend beyond the length of the crack, resulting in a non-conforming approximation. Although the XFEM approximation is capable of representing crack geometries that are independent of element boundaries, it also relies on the interaction between the mesh and the crack geometry to determine the sets of enriched nodes. This leads to particular crack configurations that cannot be accurately represented by Eq. (3.51). Such cases are shown in Fig. 3.18. As the crack size approaches the local nodal spacing, the set $\mathcal{S}_c$ of nodes for the Heaviside or step enrichment is empty, Fig. 3.18B. Moreover, node 1 for the cracking case in Fig. 3.18A or nodes 1 to 4 for the cracking case in Fig. 3.18B, respectively, contain two branch enrichments. Thus, the standard approximation gets difficulties with this crack configuration since the discontinuous function $\sqrt{2}\sin(\theta/2)$ extends too far. This problem always arises when one or more nodal supports contain the entire crack geometry. Usually, this kind of problem arises whenever a crack nucleates. Similar difficulties occur for approximations without any crack tip enrichment, see Fig. 3.18D. In order to close the crack within a single element, the set $\mathcal{S}_c$ is empty as well. One solution is to refine the mesh locally such that the characteristic element size falls below that of the crack.

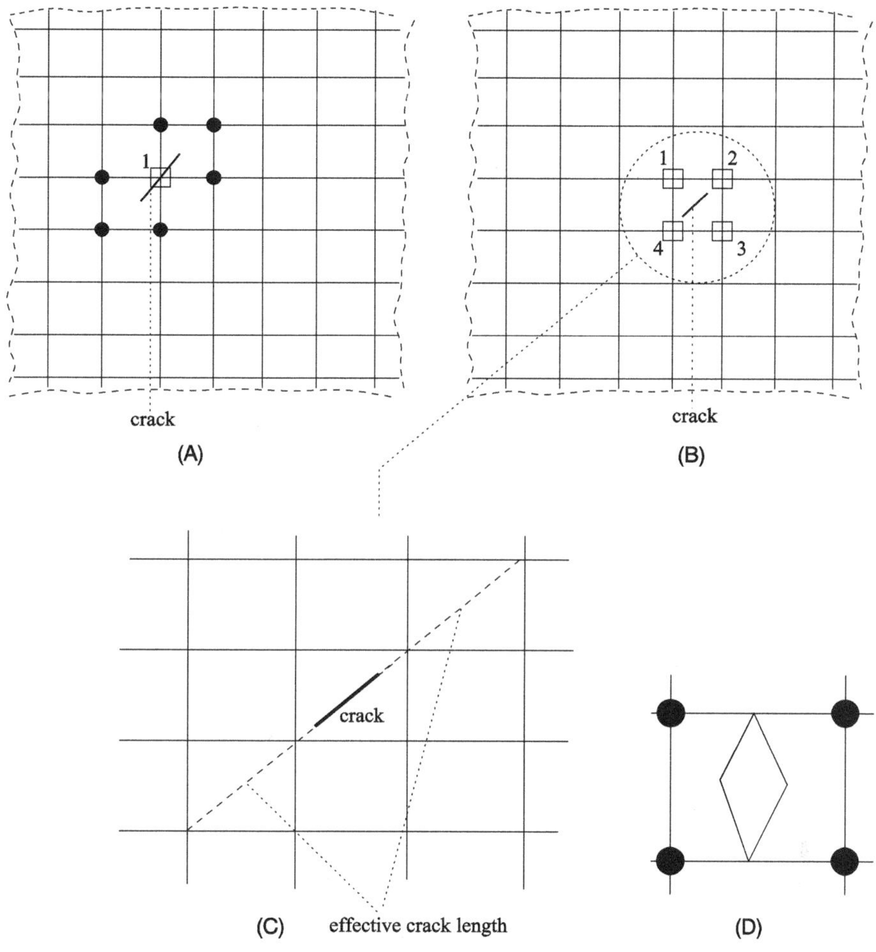

Figure 3.18 (A), (B) Crack length that approach the local element size cannot be accurately represented by the standard XFEM approximation. Dots denote single enriched nodes and squares denote double (in our case, the node will contain the enrichment of two crack tips) enriched nodes; (C) the dashed line shows the effective crack length; (D) even if no crack tip enrichment is used, in order to close the crack within a single element, no nodes have to be enriched with a step function.

3.5. Smoothed extended finite element method

The Smoothed Finite Element Method was proposed by the group around Prof. G.R. Liu [33]. It originates from the stabilized conforming nodal integration (SCNI) originally developed in the context of meshfree methods in order to stabilize nodal integration [12]. The idea behind SCNI is to use a strain measure calculated as the spatial average of the compatible strain field. It has been shown that smoothed finite elements are especially suitable for low-order and triangular/tetrahedra elements as they significantly improve the accuracy compared to associated pure displacement based formulations. The salient properties of SFEM can be summarized as follows:

- Insensitivity to mesh distortion due to the absence of isoparametric mapping.
- Derivatives of the shape functions are not required.
- Lower computational cost than FEM for the same accuracy level.
- Insensitivity to locking for low numbers of subcells.
- Flexibility, offering elements ranging from the standard FEM to quasi-equilibrium FEM, within a single framework.
- Higher convergence rate in the H_1 norm compared to displacement based finite elements.

SFEM has been discussed in many publications and details can be found in the excellent book by G.R. Liu and T. Nguyen-Thoi [32]. Therefore, only the basic concept will be summarized subsequently. Besides above mentioned features, SFEM has also advantages in the context of partition of unity enriched methods such as XFEM which suffers from the following difficulties – among others:

- The XFEM based on triangular elements is too stiff.
- The XFEM requires sub-triangulation for integration increasing complexity.
- The XFEM requires the derivatives of the shape function and requires many Gauss points for integrating the crack tip singularity.

The combination of SFEM and XFEM, the so-called smoothed extended finite element method (SXFEM), alleviates therefore some of the following difficulties:

1. It simplifies integration of discontinuous functions by transforming domain integration on Gauss points into boundary integration by using the divergence theorem. Consequently, there is no need to integrate the $1/\sqrt{r}$ term.
2. The functions to be integrated remain non-polynomial, and optimized one-dimensional integration techniques for these functions are promising routes to increase the accuracy of XFEM.
3. It remains insensitive with respect to mesh distortion.
4. No subtriangulation is needed for integration reducing complexity.

5. It inherits the robustness, accuracy and insensitivity with respect to mesh distortion of SFEM.

3.5.1 Introduction to SFEM

As mentioned before instead of compatible strains, so-called "smoothed" strains are employed. Therefore, the domain Ω is divided into a set of smoothing domains Ω^k as shown in Fig. 3.19. There are different ways in constructing the smoothing domains. The original version [33] divides the domain into n subcells as illustrated in Fig. 3.19 (top) yielding to the so-called cell based smoothed finite element method (CS-FEM). If the smoothing domains are constructed around the edges (or faces in 3D) as illustrated in Fig. 3.19 (middle), the method is referred to edge based smoothed finite element method (ES-FEM) while Fig. 3.19 (bottom) shows the concept of the nodal based smoothed finite element method (NS-FEM) where the smoothing cells are constructed around a node. It was shown that NS-FEM yields similar to equilibrium elements an upper bound solution while ES-FEM (and CS-FEM) provide a lower bound solution [32]. Note that this property does not hold any more once an enrichment in the context of NS-XFEM is introduced as shown in [64]. It was also reported that ES-FEM outperforms CS-FEM in terms of accuracy. There is also a so-called α-FEM which tries to combine the advantages of different smoothing techniques. However, the choice of the parameter α is tricky.

The smoothed strain is obtained from the compatible strain $\boldsymbol{\epsilon} = \nabla_s \mathbf{u}_k^h$ through a smoothing operation over the smoothing domain Ω^k:

$$\bar{\boldsymbol{\epsilon}}_k = \int_{\Omega_k^s} \boldsymbol{\epsilon}(\mathbf{x}) \Phi_k(\mathbf{x}) \, d\Omega = \int_{\Omega_k^s} \nabla_s \mathbf{u}^h(\mathbf{x}) \Phi_k(\mathbf{x}) \, d\Omega \tag{3.96}$$

where $\Phi_k(\mathbf{x})$ is a smoothing function which has to satisfy the following property:

$$\int_{\Omega_s^k} \Phi_k(\mathbf{x}) \, d\Omega = 1 \tag{3.97}$$

In most SFEM formulations, a constant smoothing function is used:

$$\Phi = \begin{cases} 1/A_k^s & \mathbf{x} \in \Omega_k^s \\ 0 & \mathbf{x} \notin \Omega_k^s \end{cases} \tag{3.98}$$

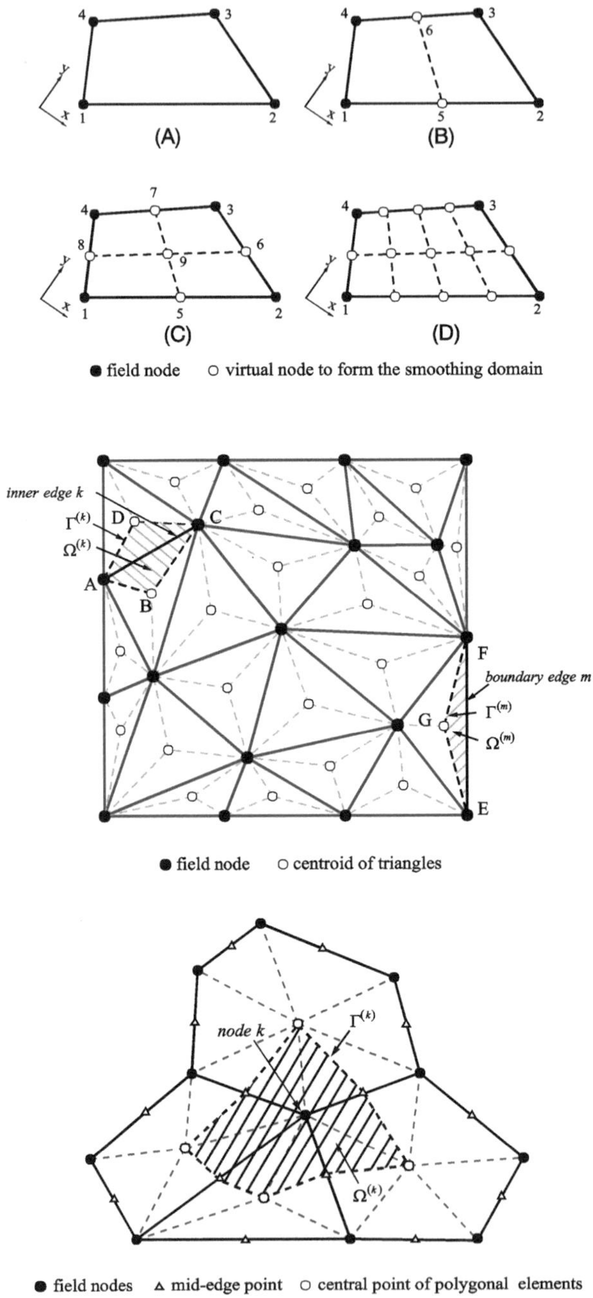

Figure 3.19 Smoothing domains for CS-FEM (top), ES-FEM (middle) and NS-FEM (bottom).

In the next step, the domain integral is transformed into a boundary integral using the Gauss divergence theorem

$$\bar{\boldsymbol{\epsilon}}_k = \frac{1}{A_k^s} \int_{\Omega_k^s} \nabla_s \mathbf{u}^h(\mathbf{x})\, \mathrm{d}\Omega = \frac{1}{A_k^s} \int_{\Gamma_k^s} \mathbf{L}_n \mathbf{u}^h(\mathbf{x})\, \mathrm{d}\Gamma \tag{3.99}$$

where $A_k^s = \int_{\Omega_k^s} \mathrm{d}\Omega$ is the area of the smoothing domain Ω_k^s, Γ_k^s is the boundary of the smoothing domain Ω_k^s, and $\mathbf{L}_n$ is a matrix comprising of normal components which are expressed in 2D as follows:

$$\mathbf{L}_n = \begin{bmatrix} n_x & 0 \\ 0 & n_y \\ n_y & n_x \end{bmatrix} \tag{3.100}$$

The discretized strain field $\bar{\boldsymbol{\epsilon}}_k$ is computed through the so-called smoothed discretized gradient operator or smoothed strain displacement operator, $\bar{\mathbf{B}}$.

$$\bar{\boldsymbol{\epsilon}}_k = \sum_{I \in n_k^s} \bar{\mathbf{B}}_I\,(\mathbf{x}_k)\,\bar{\mathbf{d}}_I \tag{3.101}$$

where $\bar{\mathbf{d}}_I$ are the unknown displacement coefficients defined at the nodes of the finite element, n_k^s is the set of nodes associated to the smoothing domain Ω_k^s. The smoothed element stiffness matrix for element e is computed by the sum of the contributions of the subcells

$$\bar{\mathbf{K}}_{IJ} = \sum_{k=1}^{N_s} \bar{\mathbf{K}}_{IJ,k}^s = \sum_{k=1}^{N_s} \int_{\Omega_k^s} \bar{\mathbf{B}}_I^{\mathrm{T}} \mathbf{D} \bar{\mathbf{B}}_J \mathrm{d}\Omega = \sum_{k=1}^{N_s} \bar{\mathbf{B}}_I^{\mathrm{T}} \mathbf{D} \bar{\mathbf{B}}_J A_k^s \tag{3.102}$$

where $\bar{\mathbf{B}}_I(\mathbf{x}_k)$ is the smoothed strain gradient matrix:

$$\bar{\mathbf{B}}_I\,(\mathbf{x}_k) = \begin{bmatrix} \bar{b}_{Ix}\,(\mathbf{x}_k) & 0 \\ 0 & \bar{b}_{Iy}\,(\mathbf{x}_k) \\ \bar{b}_{Iy}\,(\mathbf{x}_k) & \bar{b}_{Ix}\,(\mathbf{x}_k) \end{bmatrix} \tag{3.103}$$

with

$$\bar{b}_{Ih}\,(\mathbf{x}_k) = \frac{1}{A_k^s} \int_{\Gamma_k^s} n_h\,(\mathbf{x})\, N_I\,(\mathbf{x})\, \mathrm{d}\Gamma \;;\; h = x,\, y \tag{3.104}$$

Eq. (3.104) is now evaluated by line integration along the boundary Γ_k^s of the smoothing domain Ω_k^s. We note that only the shape function itself is needed to compute the strain displacement matrix leading to simple computations for integration of discontinuous functions in XFEM. We also note that no isoparametric mapping is needed which makes the formulation insensitive with respect to large deformations.

3.5.2 Enrichment in SXFEM and selection of enriched nodes

Let us focus on problems in linear elastic fracture mechanics (LEFM) for sake of simplicity where we employ the Heaviside function asymptotic branch (near-tip) enrichment functions:

$$\mathbf{u}^h(\mathbf{x}) = \underbrace{\sum_{I\in N^{is-fem}} N_I(\mathbf{x})\,\mathbf{d}_I}_{\mathbf{u}^{standard}} + \underbrace{\sum_{J\in N^{is-c}} N_J(\mathbf{x})H(\mathbf{x})\,\mathbf{a}_J + \sum_{K\in N^{is-f}} N_K(\mathbf{x})\sum_{\alpha=1}^{4}\Phi_\alpha(\mathbf{x})\,\mathbf{b}_K^\alpha}_{\mathbf{u}^{enr}} \tag{3.105}$$

where $N_I(\mathbf{x})$, $N_J(\mathbf{x})$ and $N_K(\mathbf{x})$ are finite element shape functions whose support domain is shown in Fig. 3.20 (for ES-FEM and NS-FEM, respectively), while $\mathbf{d}_I$ are nodal degrees of freedom associated with node I, $\mathbf{a}_J$ and $\mathbf{b}_K$ are additional nodal degrees of freedom corresponding to the Heaviside function $H(\mathbf{x})$ and the near-tip functions, $\{\Phi_\alpha\}_{1\leqslant\alpha\leqslant4}$, respectively. Due to the nonlocality introduced through the strain smoothing, the 'support domain' for a nodal shape function is larger than in the standard FEM. This influences, the selection of enriched nodes. As in standard XFEM, there are five types of element as illustrated in Fig. 3.20C, D:

- Tip elements, which either contain the tip, or are within a fixed distance of the tip in case a geometrical enrichment is used. All nodes belonging to a tip element are enriched with the near-tip fields.
- Split elements, which are elements completely cut by the crack. Their nodes are enriched with the discontinuous function H of Eq. (3.105).
- Tip-blending elements, which are elements neighboring tip elements. Note that some of their nodes are enriched with the near-tip fields while others are not enriched at all.
- Split-blending elements, which are elements neighboring the split elements. Some of their nodes are enriched with H while others are not enriched at all.
- Standard elements, which are elements that belong in neither of the above categories. Accordingly, none of their nodes are enriched.

Different from standard XFEM where the 'support' of a node I is defined as the collection of elements that include node I, the support in

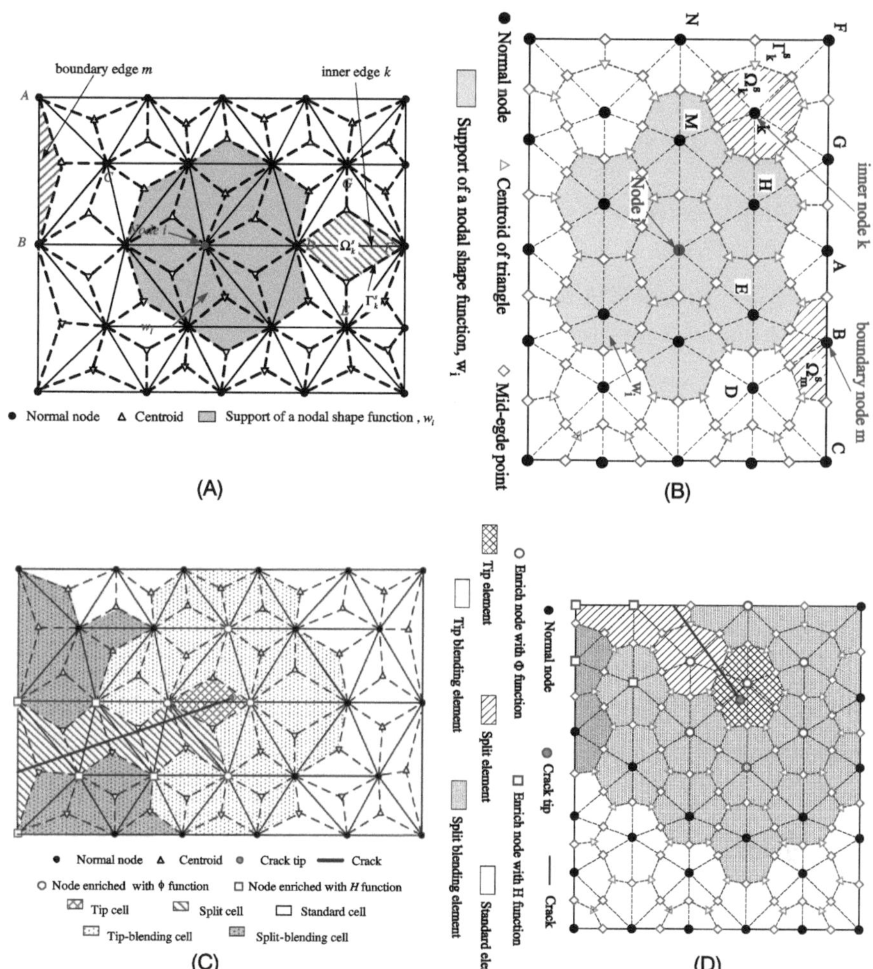

Figure 3.20 Nodal support for (A) ES-FEM, (B) NS-FEM and different element types for (C) ES-FEM and (D) NS-FEM.

SXFEM is defined as the collection of smoothing domains that are associated with this specific node (namely, the associated smoothing domain). The support of the nodes in the set N^{is-c} are split by the crack while the nodes in set N^{is-f} belong to the smoothing domains that contain a crack tip. These nodes are enriched with the Heaviside and asymptotic branch function fields depicted with squares and circles, respectively, in Fig. 3.20 for NS-XFEM and ES-XFEM; the superscript i in front of s in the node sets indicate the different SFEM versions (i.e. $i = n, es, c$ for node based, edge-based and cell-based, respectively). A set of nodes N^{is-fem} whose support domain associated with a node of ES-FEM and NS-FEM is illustrated in Fig. 3.20A/B, respectively.

3.5.3 Displacement-, strain field approximation and discrete equations

It is common in SXFEM to employ a simple shifting technique to approximate the displacement field as discussed earlier:

$$\begin{aligned}\mathbf{u}^h(\mathbf{x}) = &\sum_{I \in N^{ns-fem}} N_I(\mathbf{x})\mathbf{d}_I + \sum_{J \in N^{ns-c}} N_J(\mathbf{x})(H(\mathbf{x}) - H(\mathbf{x}_J))\mathbf{a}_J \\ &+ \sum_{K \in N^{ns-f}} N_K(\mathbf{x}) \sum_{\alpha=1}^{4} (\Phi_\alpha(\mathbf{x}) - \Phi_\alpha(\mathbf{x}_K))\mathbf{b}_K^\alpha\end{aligned} \tag{3.106}$$

The shifting circumvents problems due to blending for the Heaviside enrichment but not for the tip enrichment. Applying the node-based smoothing operation, the smoothed strain associated with node k can be written as:

$$\begin{aligned}\bar{\boldsymbol{\epsilon}}_k = &\sum_{I \in N_k^s} \bar{\mathbf{B}}_I^u(\mathbf{x}_k)\bar{\mathbf{d}}_I + \sum_{J \in N^{is-c}} \bar{\mathbf{B}}_J^a(\mathbf{x}_k)(H(\mathbf{x}) - H(\mathbf{x}_J))\mathbf{a}_J \\ &+ \sum_{k \in N^{is-f}} \bar{\mathbf{B}}_K^b(\mathbf{x}_k) \sum_{\alpha=1}^{4} (\Phi_\alpha(\mathbf{x}) - \Phi_\alpha(\mathbf{x}_K))\mathbf{b}_K^\alpha\end{aligned} \tag{3.107}$$

where N_k^s is the set of nodes associated with the smoothing domain Ω_k^s, $\bar{\mathbf{B}}_I^{\mathbf{u}}(\mathbf{x}_k)$ is the smoothed strain gradient matrix and $\bar{\mathbf{B}}_I^{\mathbf{a}}(\mathbf{x}_k)$, $\bar{\mathbf{B}}_I^{\mathbf{b}}(\mathbf{x}_k)$ correspond to the enriched parts of the smoothed strain gradient matrix associated with the Heaviside and branch functions, respectively. These matrixes operations can be written as follows:

$$\bar{\mathbf{B}}_I^r(\mathbf{x}_k) = \begin{bmatrix} \bar{b}_{Ix}^r(\mathbf{x}_k) & 0 \\ 0 & \bar{b}_{Iy}^r(\mathbf{x}_k) \\ \bar{b}_{Iy}^r(\mathbf{x}_k) & \bar{b}_{Ix}^r(\mathbf{x}_k) \end{bmatrix} \quad r = u,\ a,\ b \tag{3.108}$$

$$\bar{\mathbf{B}}_I^u = \int_{\Gamma_k^s} \frac{1}{A_k^s} \begin{bmatrix} n_x N_I & 0 \\ 0 & n_y N_I \\ n_y N_I & n_x N_I \end{bmatrix} \mathrm{d}\Gamma$$

$$\bar{\mathbf{B}}_I^a = \int_{\Gamma_k^s} \frac{1}{A_k^s} \begin{bmatrix} n_x\left[N_I\left(H(\mathbf{x}) - H(\mathbf{x}_I)\right)\right] & 0 \\ 0 & n_y\left[N_I\left(H(\mathbf{x}) - H(\mathbf{x}_I)\right)\right] \\ n_y\left[N_I\left(H(\mathbf{x}) - H(\mathbf{x}_I)\right)\right] & n_x\left[N_I\left(H(\mathbf{x}) - H(\mathbf{x}_I)\right)\right] \end{bmatrix} \mathrm{d}\Gamma$$

$$\bar{\mathbf{B}}_I^b = \int_{\Gamma_k^s} \frac{1}{A_k^s} \begin{bmatrix} n_x\left[N_I(\mathbf{x}_{m,n})\left(\Phi_\alpha(x) - \Phi_\alpha(\mathbf{x}_I)\right)\right] \\ 0 \\ n_y\left[N_I(\mathbf{x}_{m,n})\left(\Phi_\alpha(\mathbf{x}) - \Phi_\alpha(\mathbf{x}_I)\right)\right]_{,y} \end{bmatrix} \cdots$$

$$\cdots \begin{bmatrix} 0 \\ n_y\left[N_I(\mathbf{x}_{m,n})\left(\Phi_\alpha(\mathbf{x}) - \Phi_\alpha(\mathbf{x}_I)\right)\right]_{,y} \\ n_x\left[N_I(\mathbf{x}_{m,n})\left(\Phi_\alpha(\mathbf{x}) - \Phi_\alpha(\mathbf{x}_I)\right)\right]_{,x} \end{bmatrix} \mathrm{d}\Gamma$$

$$(\alpha = 1,\ 2,\ 3 \text{ and } 4) \tag{3.109}$$

Using Gauss-Legendre integration along Γ_k^s, we obtain:

$$\bar{\mathbf{B}}_I^u = \frac{1}{A_k^s} \begin{bmatrix} \sum_{m=1}^{N_{seg}} \left(\sum_{n=1}^{N_{gau}} n_x(\mathbf{x}_{m,n}) N_I(\mathbf{x}_{m,n}) w_{m,n} \right) \\ 0 \\ \sum_{m=1}^{N_{seg}} \left(\sum_{n=1}^{N_{gau}} n_y(\mathbf{x}_{m,n}) N_I(\mathbf{x}_{m,n}) w_{m,n} \right) \end{bmatrix} \cdots$$

$$\cdots \begin{bmatrix} 0 \\ \sum_{m=1}^{N_{seg}} \left(\sum_{n=1}^{N_{gau}} n_y(\mathbf{x}_{m,n}) N_I(\mathbf{x}_{m,n}) w_{m,n} \right) \\ \sum_{m=1}^{N_{seg}} \left(\sum_{n=1}^{N_{gau}} n_x(\mathbf{x}_{m,n}) N_I(\mathbf{x}_{m,n}) w_{m,n} \right) \end{bmatrix}$$

$$\bar{\mathbf{B}}_I^a = \frac{1}{A_k^s} \begin{bmatrix} \sum_{m=1}^{N_{seg}} \left(\sum_{n=1}^{N_{gau}} n_x(\mathbf{x}_{m,n}) N_I(\mathbf{x}_{m,n}) \left(H(\mathbf{x}_{m,n}) - H(\mathbf{x}_I) \right) w_{m,n} \right) \\ 0 \\ \sum_{m=1}^{N_{seg}} \left(\sum_{n=1}^{N_{gau}} n_y(\mathbf{x}_{m,n}) N_I(\mathbf{x}_{m,n}) \left(H(\mathbf{x}_{m,n}) - H(\mathbf{x}_I) \right) w_{m,n} \right) \end{bmatrix} \cdots$$

$$\cdots \begin{bmatrix} 0 \\ \sum_{m=1}^{N_{seg}} \left(\sum_{n=1}^{N_{gau}} n_y(\mathbf{x}_{m,n}) N_I(\mathbf{x}_{m,n}) \left(H(\mathbf{x}_{m,n}) - H(\mathbf{x}_I) \right) w_{m,n} \right) \\ \sum_{m=1}^{N_{seg}} \left(\sum_{n=1}^{N_{gau}} n_x(\mathbf{x}_{m,n}) N_I(\mathbf{x}_{m,n}) \left(H(\mathbf{x}_{m,n}) - H(\mathbf{x}_I) \right) w_{m,n} \right) \end{bmatrix}$$

$$\bar{\mathbf{B}}_I^b = \frac{1}{A_k^s} \begin{bmatrix} \sum_{m=1}^{N_{seg}} \left(\sum_{n=1}^{N_{gau}} n_x(\mathbf{x}_{m,n}) N_I(\mathbf{x}_{m,n}) \left(\Phi_\alpha(bfx_{m,n}) - \Phi_\alpha(\mathbf{x}_I) \right) w_{m,n} \right) \\ 0 \\ \sum_{m=1}^{N_{seg}} \left(\sum_{n=1}^{N_{gau}} n_y(\mathbf{x}_{m,n}) N_I(\mathbf{x}_{m,n}) \left(\Phi_\alpha(\mathbf{x}_{m,n}) - \Phi_\alpha(\mathbf{x}_I) \right) w_{m,n} \right) \end{bmatrix} \cdots$$

$$
\begin{matrix}
 & 0 \\
\dots & \sum_{m=1}^{N_{seg}}\left(\sum_{n=1}^{N_{gau}} n_y(\mathbf{x}_{m,n})N_I(\mathbf{x}_{m,n})\left(\Phi_\alpha(x)-\Phi_\alpha(\mathbf{x}_I)\right)w_{m,n}\right) \\
 & \sum_{m=1}^{N_{seg}}\left(\sum_{n=1}^{N_{gau}} n_x(\mathbf{x}_{m,n})N_I(\mathbf{x}_{m,n})\left(\Phi_\alpha(\mathbf{x}_{m,n})-\Phi_\alpha(\mathbf{x}_I)\right)w_{m,n}\right)
\end{matrix}\Bigg]
$$

$$(\alpha = 1,\ 2,\ 3 \text{ and } 4) \tag{3.110}$$

where N_{neg} is the number of segments of the boundary Γ_k^s, N_{gau} is the number of Gauss points used in each segment, $w_{m,n}$ is the corresponding Gauss weights, n_x, n_y are the outward unit normal components to each segment on the smoothing domain boundary and $\mathbf{x}_{m,n}$ is the n-th Gaussian point on the m-th segment of the boundary Γ_k^s. The final discrete system of equations is obtained by substituting the trial and test function into the weak form which yields the well known form:

$$\bar{\mathbf{K}}\bar{\mathbf{d}} = \mathbf{f} \tag{3.111}$$

where $\mathbf{f}$ is the nodal force vector that is identical to that in the standard XFEM. The smoothed enriched stiffness matrix $\bar{\mathbf{K}}$ for all sub-cells is computed by:

$$
\begin{aligned}
\bar{\mathbf{K}}_{IJ} &= \sum_{k=1}^{N_s}\bar{\mathbf{K}}_{IJ,k}^s \\
&= \sum_{k=1}^{N_s}\begin{bmatrix}
\int_{\Omega_k^s}(\bar{\mathbf{B}}_I^u)^{\mathrm{T}}\mathbf{D}\bar{\mathbf{B}}_J^u\mathrm{d}\Omega & \int_{\Omega_k^s}(\bar{\mathbf{B}}_I^u)^{\mathrm{T}}\mathbf{D}\bar{\mathbf{B}}_J^a\mathrm{d}\Omega & \int_{\Omega_k^s}(\bar{\mathbf{B}}_I^u)^{\mathrm{T}}\mathbf{D}\bar{\mathbf{B}}_J^b\mathrm{d}\Omega \\
\int_{\Omega_k^s}(\bar{\mathbf{B}}_I^a)^{\mathrm{T}}\mathbf{D}\bar{\mathbf{B}}_J^u\mathrm{d}\Omega & \int_{\Omega_k^s}(\bar{\mathbf{B}}_I^a)^{\mathrm{T}}\mathbf{D}\bar{\mathbf{B}}_J^a\mathrm{d}\Omega & \int_{\Omega_k^s}(\bar{\mathbf{B}}_I^a)^{\mathrm{T}}\mathbf{D}\bar{\mathbf{B}}_J^b\mathrm{d}\Omega \\
\int_{\Omega_k^s}(\bar{\mathbf{B}}_I^b)^{\mathrm{T}}\mathbf{D}\bar{\mathbf{B}}_J^u\mathrm{d}\Omega & \int_{\Omega_k^s}(\bar{\mathbf{B}}_I^b)^{\mathrm{T}}\mathbf{D}\bar{\mathbf{B}}_J^a\mathrm{d}\Omega & \int_{\Omega_k^s}(\bar{\mathbf{B}}_I^b)^{\mathrm{T}}\mathbf{D}\bar{\mathbf{B}}_J^b\mathrm{d}\Omega
\end{bmatrix}
\end{aligned} \tag{3.112}
$$

In SXFEM, the stiffness matrix in Eq. (3.112) can therefore be rewritten as:

$$
\bar{\mathbf{K}}_{IJ} = \sum_{k=1}^{N_s}\bar{\mathbf{K}}_{IJ,k}^s = \sum_{k=1}^{N_s}\begin{bmatrix}
(\bar{\mathbf{B}}_I^u)^{\mathrm{T}}\mathbf{D}\bar{\mathbf{B}}_J^u A_k^s & (\bar{\mathbf{B}}_I^u)^{\mathrm{T}}\mathbf{D}\bar{\mathbf{B}}_J^a A_k^s & (\bar{\mathbf{B}}_I^u)^{\mathrm{T}}\mathbf{D}\bar{\mathbf{B}}_J^b A_k^s \\
(\bar{\mathbf{B}}_I^a)^{\mathrm{T}}\mathbf{D}\bar{\mathbf{B}}_J^u A_k^s & (\bar{\mathbf{B}}_I^a)^{\mathrm{T}}\mathbf{D}\bar{\mathbf{B}}_J^a A_k^s & (\bar{\mathbf{B}}_I^a)^{\mathrm{T}}\mathbf{D}\bar{\mathbf{B}}_J^b A_k^s \\
(\bar{\mathbf{B}}_I^b)^{\mathrm{T}}\mathbf{D}\bar{\mathbf{B}}_J^u A_k^s & (\bar{\mathbf{B}}_I^b)^{\mathrm{T}}\mathbf{D}\bar{\mathbf{B}}_J^a A_k^s & (\bar{\mathbf{B}}_I^b)^{\mathrm{T}}\mathbf{D}\bar{\mathbf{B}}_J^b A_k^s
\end{bmatrix} \tag{3.113}
$$

The key difference between the different SXFEM versions (NS-XFEM, ES-XFEM and CS-XFEM) is in the construction of the smoothing domains.

3.5.4 Numerical integration

As mentioned before, SXFEM avoids the integration of the singularity. Due to the constant Jacobian, Gauss quadrature will be exact in the case of $T3$ elements. However, the elements containing the crack tip enrichment require integration of a non-polynomial function. It was suggested in [13, 64] to employ 7 Gauss points for tip elements. In summary, the following Gauss quadrature rules are suggested in [8]

1. *Tip elements*: 7 Gauss points for each sub-element.
2. *Split elements*: 1 Gauss points for each sub-element.
3. *Tip-blending elements*: 7 Gauss points.
4. *Split-blending elements*: 1 Gauss point.
5. *Standard elements*: 1 Gauss point.

In contrast to numerical integration in the standard XFEM, where the polygonal sub-domains are usually divided into triangles that permit the use of well known quadrature rules, strain smoothing in the SXFEM provides an elegant solution to this problem by transforming interior integration into boundary integration which was first proposed in [8]. Therefore, decomposition of polygonal sub-domains into triangles is not needed as shown in Fig. 3.21. Furthermore, no isoparametric mapping is necessary, which further simplifies numerical integration.

Since the product of linear shape functions and a discontinuous function, i.e., $N_I\ H$, is also linear along all boundary segments, one Gauss point is thus sufficient along each boundary segment for split smoothing domains. No partition is required for the split-blending smoothing domains and one Gauss point will be sufficient on each boundary segment.

Tip smoothing domains, in which the discontinuity is still present, contain also a singularity. In that case, a set of branch functions (non-polynomial functions) is used to model the asymptotic features of the displacement fields for capturing the high stress gradients in the vicinity of the crack tip. However, simply splitting such smoothing domains into polygonal sub-domains may lead to poor numerical results because of the presence of non-polynomial functions. In order to ensure accuracy, a higher integration point density should be used close to the crack tip. The steps involved are: (1) Split the smoothing domain into triangular sub-domains, e.g., from sub-sd1 to sub-sd6, as shown in Fig. 3.22; (2) partition sub-domain (triangle) into n_{sc} sub-cells (also triangles) according to the rules depicted in Fig. 3.23. Fig. 3.22 illustrates the schematic of subcells after partitioning sub-sd1 and sub-sd4 with $n_{sc} = 3$, e.g., sub-sd1 is split as sc1, sc2 and sc3; and sub-sd4 is split into sc4, sc5 and sc6; (3) the numerical

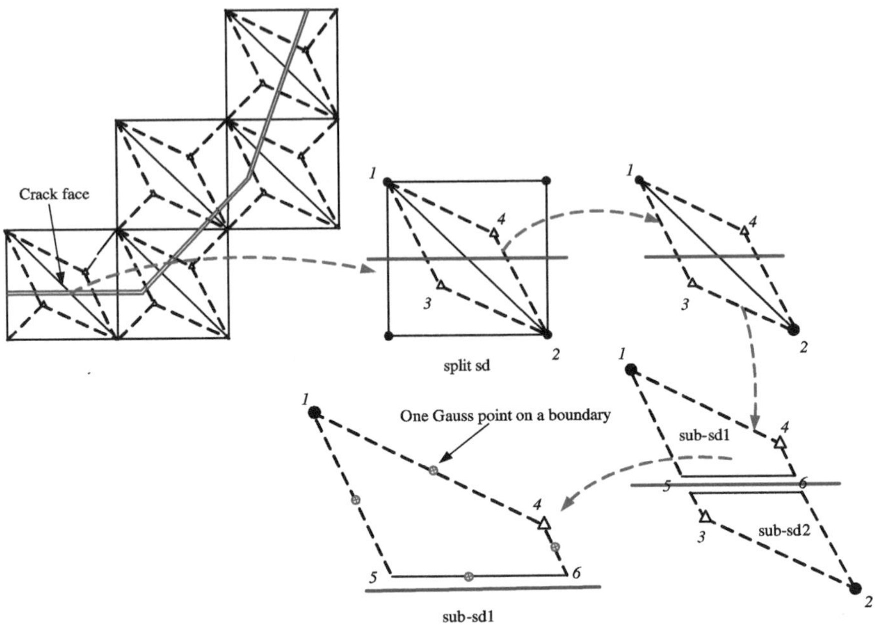

Figure 3.21 Partition of split smoothing domains (sd) in the ESm-XFEM. The decomposition of polygonal domains (sub-sd1 and sub-sd2) into triangles is not necessary. Integration is performed on the boundary of sub-sd1 and sub-sd2 instead. The bold line represents the elements, and the dashed line denotes the boundaries of smoothing domains.

integration is performed by integrating along the boundary of each subcell. Note that extra care is needed when using boundary integration along the crack face. A sub-cell with one or more boundary segments aligned with the crack surface, denoted by c-sub-cell, accounts for this issue, e.g., sc2 and sc3 from sub-sd1, as well as sc4 and sc6 from sub-sd4, as shown in Fig. 3.22. When we perform integration on two c-sub-cells that share a boundary segment with the crack surface (e.g., sc3 and sc6 share boundary segment 69) for the Gauss points falling on this boundary but belonging to two c-sub-cells, the calculated values of the discontinuous function or branch functions are the same for both c sub-cells, if the calculation is based on the coordinates of Gauss points. However, when we come back to the displacement field of the crack surface, it is clear that there is a jump across the crack faces. To solve this problem, we propose in this case that the values of the enrichment functions are computed not based on the coordinates of Gauss points, but the center of this sub-cell. Compared to standard smoothing domains, more Gauss points should be used for each boundary

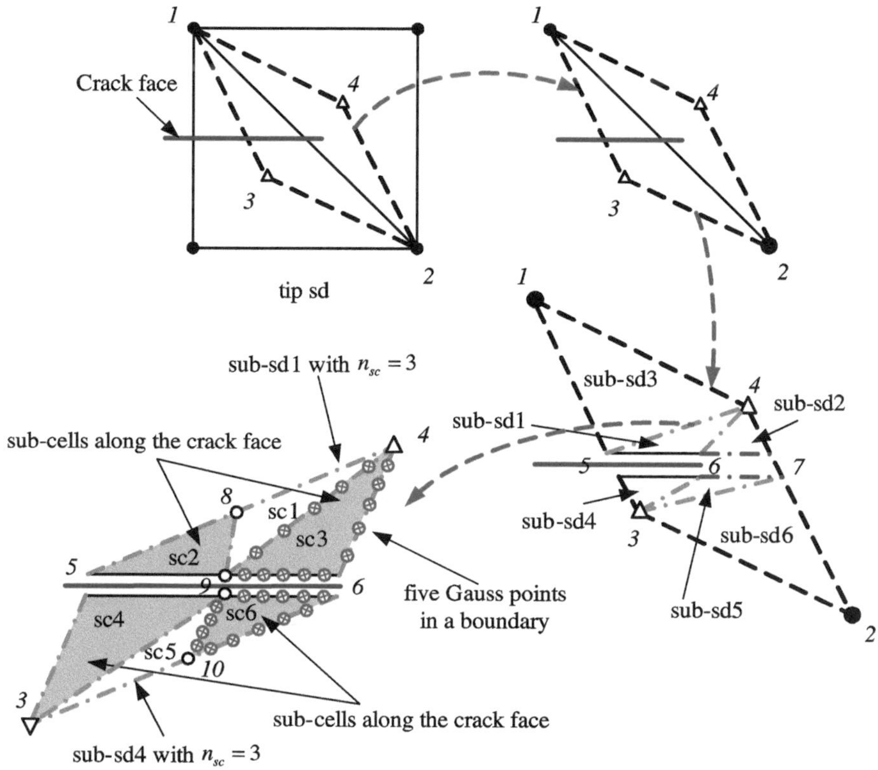

Figure 3.22 Partition of tip smoothing domains (sd) in the ESm-XFEM. Note that sub-sds and sub-cells do not carry any degrees of freedom and are solely used for smoothing and numerical integration.

segment as the gradients in the vicinity of the crack tip are large in the radial direction. Numerical experiments [13,64] demonstrated that eight smoothing cells in a smoothing domain ($n_{sc} = 8$), and five Gauss points on a segment of smoothing cells ($n_{gauss} = 5$) are sufficient.

No partition is required for tip-blending smoothing domains and as for tip smoothing domains, eight smoothing cells in a smoothing domain and five Gauss point are suggested on each boundary segment [13,64].

3.6. XFEM for coupled problems

3.6.1 Hydro-mechanical problems

Hydro-mechanical problems are relevant in many engineering disciplines, for instance in geotechnical problems where the material can often be con-

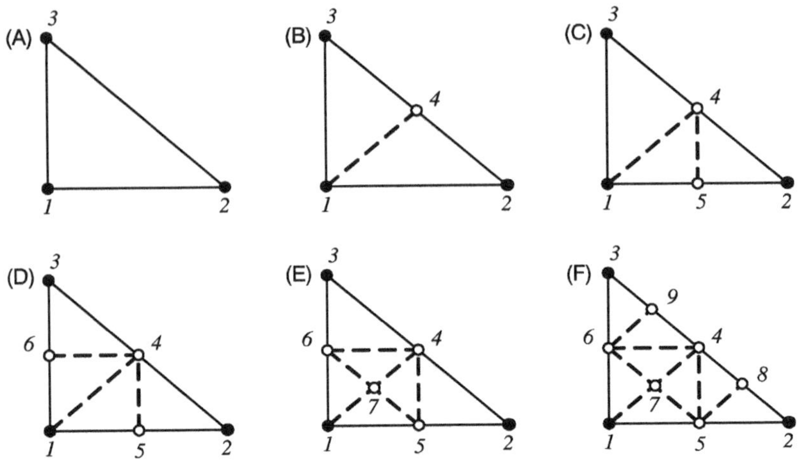

Figure 3.23 Division of a sub-smoothing domain into sub-smoothing cells: (A) $n_{SC} = 1$; (B) $n_{SC} = 2$; (C) $n_{SC} = 3$; (D) $n_{SC} = 4$; (E) $n_{SC} = 6$; (F) $n_{SC} = 8$.

sidered as a multi-phase material consisting of a solid and a fluid phase. Such problems are commonly treated with the theory of porous media. A very timely topic is hydraulic fracturing which consists commonly of 1 solid phase and 2 fluid phases. Subsequently, a coupled hydro-mechanical XFEM formulation for a two-phase material will be summarized.

3.6.1.1 *Strong and weak form of the coupled hydro-mechanical problem*

Balance of momentum

We assume small displacement theory, no mass transfer between the constituents and isothermal conditions [1]. Hence, the balance equation of momentum for a two phase material reads:

$$\nabla \cdot \boldsymbol{\sigma}_\pi + \hat{\boldsymbol{p}}_\pi + \rho_\pi \boldsymbol{g} = \frac{\partial(\rho_\pi \boldsymbol{v}_\pi)}{\partial t} + \nabla \cdot (\rho_\pi \boldsymbol{v}_\pi \otimes \boldsymbol{v}_\pi) \tag{3.114}$$

where the subscripts $\pi = s, f$, denote the solid and fluid phase, respectively, ρ represents the densities, $\boldsymbol{v}$ the absolute velocities and $\boldsymbol{\sigma}$ the stress of the constituents. Furthermore, $\hat{\boldsymbol{p}}_\pi$ represents the source of the momentum from the other constituent, which accounts for the load drag between the solid and fluid phase. For a closed system, the source of the momentum must fulfill:

$$\hat{\boldsymbol{p}}_s + \hat{\boldsymbol{p}}_f = 0 \tag{3.115}$$

Neglecting body forces and the inertia terms, Eq. (3.114) is reduced to

$$\nabla \cdot \boldsymbol{\sigma}_\pi + \hat{\boldsymbol{p}}_\pi = 0 \tag{3.116}$$

Balance of mass

For a two phase medium, the continuity equation is given by [30]

$$\frac{\partial \rho_\pi}{\partial t} + \nabla \cdot (\rho_\pi \boldsymbol{v}_\pi) - grad\ \rho_\pi \cdot \boldsymbol{v}_\pi = 0 \tag{3.117}$$

where ρ is the mass density and $\boldsymbol{v}$ the absolute velocity. The volume ratio of the solid and fluid phase n_s and n_f, respectively, have to fulfill the following equation:

$$n_s + n_f = 1 \tag{3.118}$$

The apparent mass density for each constituent is obtained by

$$\rho_\pi = n_\pi \times \bar{\rho}_\pi \tag{3.119}$$

where $\bar{\rho}$ is the absolute mass density. Substituting Eq. (3.119) into Eq. (3.117) gives:

$$n_s \nabla \cdot \boldsymbol{v}_s + \frac{1}{\bar{\rho}_s}\frac{\partial \rho_s}{\partial t} = 0 \tag{3.120}$$

$$n_f \nabla \cdot \boldsymbol{v}_f + \frac{1}{\bar{\rho}_f}\frac{\partial \rho_f}{\partial t} = 0 \tag{3.121}$$

Combining Eq. (3.118) with Eq. (3.120) and Eq. (3.121) yields:

$$\nabla \cdot \boldsymbol{v}_s + n_f \nabla \cdot (\boldsymbol{v}_f - \boldsymbol{v}_s) + \frac{1}{\rho_s}\frac{\partial \rho_s}{\partial t} + \frac{1}{\rho_f}\frac{\partial \rho_f}{\partial t} = 0 \tag{3.122}$$

For a compressible solid, the time derivative of the density of the solid phase is obtained from the mass conservative equation [30] by

$$\frac{\partial (\rho_s \boldsymbol{v}_s)}{\partial t} = 0 \tag{3.123}$$

From the entropy inequality [25] for unsaturated flow accounting for interfaces, the pressure in the solid phase is:

$$p_s = p_f \times n_f \tag{3.124}$$

Assuming the solid density is a function of pressure p_s and temperature, yields

$$\frac{1}{\rho_s}\frac{\partial \rho_s}{\partial t} = -\frac{1}{\boldsymbol{v}_s}\frac{\partial \boldsymbol{v}_s}{\partial t} = \frac{1}{K_s}\frac{\partial p_s}{\partial t} + \beta_s \frac{\partial T_s}{\partial t} \tag{3.125}$$

where K_s designates the bulk modulus of solid phase, β_s the thermal expansion coefficient and T_s the temperature. Since the whole process is under isothermal, the last item can be omitted:

$$\frac{1}{\rho_s}\frac{\partial \rho_s}{\partial t} = -\frac{1}{\boldsymbol{v}_s}\frac{\partial \boldsymbol{v}_s}{\partial t} = \frac{1}{K_s}\frac{\partial p_s}{\partial t} \tag{3.126}$$

Let us define the Biot' constant [30] as

$$1 - \alpha = \frac{K_T}{K_s} \tag{3.127}$$

with K_T being the overall bulk modulus of the two phase medium and α the Biot coefficient. The change of the solid mass density is related to its volume change by

$$\nabla \cdot \boldsymbol{v}_s = -\frac{K_s}{K_T}\frac{n_s}{\rho_s}\frac{\partial \rho_s}{\partial t} \tag{3.128}$$

Substituting now Eq. (3.127) into Eq. (3.128) leads to

$$(1-\alpha)\nabla \cdot \boldsymbol{v}_s = \frac{n_s}{\rho_s}\frac{\partial \rho_s}{\partial t} \tag{3.129}$$

For the fluid phase, the relationship between the incremental change of fluid density and of the fluid pressure reads:

$$\frac{1}{Q}\,dp = \frac{n_f}{\rho_f}\,dp_f \tag{3.130}$$

with the compressibility modulus:

$$\frac{1}{Q} = \frac{\alpha - n_f}{K_s} + \frac{n_f}{K_f} \tag{3.131}$$

where K_f is the bulk modulus of the fluid phase. Inserting Eq. (3.130) and Eq. (3.131) into Eq. (3.122) the balance equation of mass:

$$\alpha \nabla \cdot \boldsymbol{v}_s + n_f \nabla \cdot (\boldsymbol{v}_f - \boldsymbol{v}_s) + \frac{1}{Q}\frac{\partial p}{\partial t} = 0 \tag{3.132}$$

Kinematic relation

Assuming a linear elastic solid, the kinematic relation for small strain theory reads:

$$\boldsymbol{\epsilon}_s = \nabla^s \boldsymbol{u}_s \tag{3.133}$$

with $\boldsymbol{\epsilon}_s$ and $\boldsymbol{u}_s$ being the linear strain tensor and displacement field of the solid phase.

3.6.1.2 Constitutive relation

The effective stress increment $d\boldsymbol{\sigma}'_s$ in the solid reads:

$$d\boldsymbol{\sigma}'_s = \frac{d\boldsymbol{\sigma}_s}{n_s} \tag{3.134}$$

the incremental stress-strain relationship for the solid media:

$$d\boldsymbol{\sigma}_s = \boldsymbol{D} : d\boldsymbol{\epsilon}_s \tag{3.135}$$

with $\boldsymbol{D}$ being the fourth-order elasticity tensor.

Boundary conditions

The boundary conditions for the two phase media are given by

$$\boldsymbol{n}_\Gamma \cdot \boldsymbol{\sigma} = \boldsymbol{t}_p \text{ on } \partial\Omega_t; \quad \boldsymbol{u} = \boldsymbol{u}_p \text{ on } \partial\Omega_u \tag{3.136}$$

with the von Neumann and Dirichlet boundaries $\partial\Omega_t$ and $\partial\Omega_u$ and $\partial\Omega = \partial\Omega_t \cup \partial\Omega_u$, $\boldsymbol{t}_p$ and $\boldsymbol{u}_p$ describe the traction and displacement, respectively (see Fig. 3.24). According to Darcy's law [17] for an isotropic media, the fluid velocity ($\boldsymbol{v}$) is related to the Darcy flux ($\boldsymbol{q}$) by the porosity (ϕ).

$$\phi \boldsymbol{v} = \boldsymbol{q} \tag{3.137}$$

On complementary part of the boundary $\partial\Omega_q$ and $\partial\Omega_p$, with $\partial\Omega = \partial\Omega_q \cup \partial\Omega_p$ hold:

$$n_f(\boldsymbol{v}_f - \boldsymbol{v}_s) \cdot n_\Gamma = \boldsymbol{q}_p \cdot n_\Gamma \text{ on } \partial\Omega_q; \quad p = p_p \text{ on } \partial\Omega_p \tag{3.138}$$

with $\boldsymbol{q}_p \cdot n_\Gamma$ and p are describing the outflow of pore fluid and pressure, respectively.

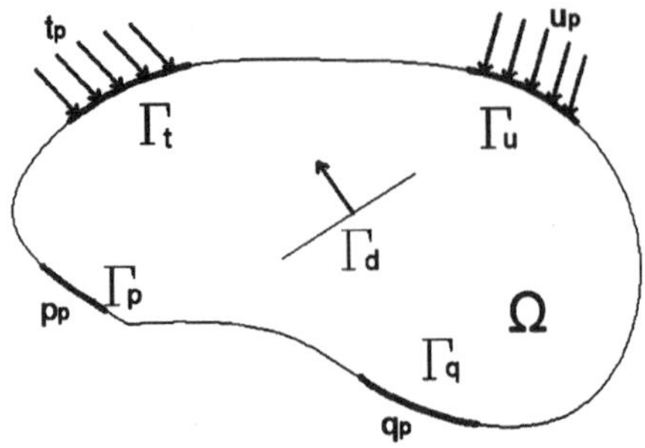

Figure 3.24 Boundary conditions.

Weak form and coupling

To derive the weak form from the strong form, we multiply the momentum balance equation and the mass balance equation with kinematically admissible test functions for the displacements $\boldsymbol{\eta}$ and pressure $\boldsymbol{\xi}$. Integrating by parts, Gauss divergence theorem and applying Darcy's law finally leads to the well known weak form:

$$\int_{\Omega} (\nabla \cdot \boldsymbol{\eta}) \cdot \boldsymbol{\sigma} d\Omega + \boxed{\int_{\Gamma_d} [\![\boldsymbol{\eta}]\!] \cdot \boldsymbol{\sigma} \cdot \boldsymbol{n}_{\Gamma_d} d\Omega} = \int_{\Gamma} \boldsymbol{\eta} \cdot \boldsymbol{t}_p d\Omega \tag{3.139}$$

and

$$\begin{aligned} &- \int_{\Omega} \alpha \boldsymbol{\xi} \nabla \cdot \boldsymbol{v}_s d\Omega + \int_{\Omega} k_f \nabla \boldsymbol{\xi} \cdot \nabla p d\Omega - \int_{\Omega} \boldsymbol{\xi} Q^{-1} \dot{p} d\Omega \\ &+ \boxed{\int_{\Gamma_d} \boldsymbol{n}_{\Gamma_d} \cdot [\![\boldsymbol{\xi} n_f (\boldsymbol{v}_f - \boldsymbol{v}_s)]\!] d\Gamma} = \int_{\Gamma} \boldsymbol{\xi} \boldsymbol{n}_{\Gamma} \cdot \boldsymbol{q}_p d\Omega \end{aligned} \tag{3.140}$$

where $\dot{p} = \frac{\partial p}{\partial t}$ denotes the time derivatives. The traction force on Γ_d is induced by the flow pressure inside cavity. Due to the presence of the discontinuity inside the domain, the traction force $\boldsymbol{\sigma} \cdot \boldsymbol{n}_{\Gamma_d}$ on Γ_d and fluid flux $[\![\boldsymbol{\xi} n_f (\boldsymbol{v}_f - \boldsymbol{v}_s)]\!]$ through the discontinuity face are essential parts of the weak form. Since the length to width ratio of the cavity is considerable large, one could assume that the traction force on each side of the cavity is equal. Because of the continuity from the bulk to cavity, this gives:

$$\boldsymbol{\sigma} \cdot \boldsymbol{n}_{\Gamma_d} = -p \boldsymbol{n}_{\Gamma_d} \quad \text{on } \Gamma_d \tag{3.141}$$

Substituting Eq. (3.141) into Eq. (3.139), the final weak form of the balance equation of momentum gives

$$\int_{\Omega} (\nabla \cdot \boldsymbol{\eta}) \cdot \boldsymbol{\sigma}\, d\Omega - \int_{\Gamma_d} [\![\eta]\!] \cdot p\boldsymbol{n}_{\Gamma_d} d\Omega = \int_{\Gamma} \boldsymbol{\eta} \cdot \boldsymbol{t}_p d\Omega \tag{3.142}$$

Darcy's law is expressed by

$$\boldsymbol{v} = \frac{\boldsymbol{q}}{\phi} \tag{3.143}$$

where ϕ is the porosity of the bulk and $\boldsymbol{v} = \boldsymbol{v}_f - \boldsymbol{v}_s$ the velocity. The pressure values for both faces of the cavity are identical leads to the following coupling term of the weak form for the mass balance equation:

$$\int_{\Gamma_d} \boldsymbol{n}_{\Gamma_d} \cdot [\![\xi n_f (\boldsymbol{v}_f - \boldsymbol{v}_s)]\!] d\Gamma = \int_{\Gamma_d} \xi n_f \boldsymbol{n}_{\Gamma_d} \cdot [\![\boldsymbol{v}_f - \boldsymbol{v}_s]\!] d\Gamma = \int_{\Gamma_d} \xi \boldsymbol{n}_{\Gamma_d} \cdot \boldsymbol{q}_d d\Gamma \tag{3.144}$$

with $\boldsymbol{q}_d$ being the flow flux through the discontinuity.

Fluid flow inside the cavity

Assuming the fluid flow inside the cavity is a Newtonian fluid, the general balance equation of momentum reads [4]:

$$\underbrace{\frac{\partial \boldsymbol{u}}{\partial t}}_{\text{Variation}} + \underbrace{(\boldsymbol{u} \cdot \nabla)\boldsymbol{u}}_{\text{Convection}} - \underbrace{\mu \nabla^2 \boldsymbol{u}}_{\text{Diffusion}} = \underbrace{-\nabla w}_{\text{Internal source}} + \underbrace{g}_{\text{External source}} \tag{3.145}$$

with μ being the viscosity of the fluid, $\boldsymbol{u}$ being the velocity of the fluid.

Assuming small displacements, no mass exchange and neglecting body forces and inertia forces, the balance equation of momentum is simplified to:

$$\underbrace{-\mu \nabla^2 \boldsymbol{u}}_{\text{Diffusion}} = \underbrace{-\nabla w}_{\text{Internal source}} \tag{3.146}$$

In the two phase medium, the interface pressure on the two sides of the cavity serves as internal source of momentum, which gives the final balance equation of momentum:

$$\mu \nabla^2 \boldsymbol{v}_{\bar{f}} = \nabla p \tag{3.147}$$

with $\boldsymbol{v}_{\bar{f}}$ denotes the fluid flow velocity, the subscript $\bar{f}$ denotes fluid inside cavity.

Due to the high length to cross-section ratio (Fig. 3.25), the fluid flow inside cavity can be considered as quasi one-dimensional flow given by

$$\frac{\partial p_{\bar{f}}}{\partial \bar{y}} = 0 \tag{3.148}$$

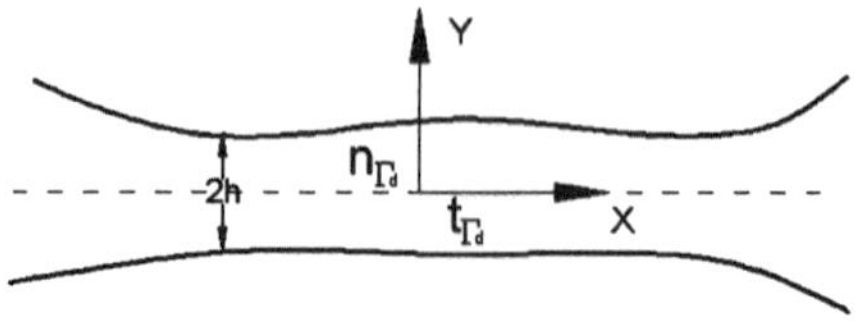

Figure 3.25 Cavity geometry.

in the normal direction $\boldsymbol{n}_{\Gamma_d}$ and in the tangential direction of $\boldsymbol{t}_{\Gamma_d}$:

$$\frac{\partial p_{\bar{f}}}{\partial \bar{x}} = \mu \frac{\partial^2 v_{\bar{f}}}{\partial \bar{y}^2} \tag{3.149}$$

with $\bar{x}$ and $\bar{y}$ beeing the coordinates with respect to the normal and tangent directions of the cavity, respectively. In the normal direction $\boldsymbol{n}_{\Gamma_d}$ of the cavity, the pressure $p_{\bar{f}}$ is constant through the cross section of the cavity, for continuity restriction, $p_{\bar{f}}$ must be equal to p.

$$p_{\bar{f}} = p \tag{3.150}$$

In order to derive the axial velocity, Eq. (3.149) is integrated twice with respect to the $\bar{y}$ coordinate:

$$\int_{-h}^{h} \int_{-h}^{h} \frac{\partial p}{\partial \bar{x}} = \int_{-h}^{h} \int_{-h}^{h} \mu \frac{\partial^2 v_{\bar{f}}}{\partial \bar{y}^2} \tag{3.151}$$

yielding

$$v(y) = \frac{1}{2\mu} \frac{\partial p}{\partial \bar{x}} (\bar{y}^2 - h^2) + v_f \tag{3.152}$$

where $v_f = \boldsymbol{v}_f \cdot \boldsymbol{t}_{\Gamma_d}$ serves as the essential boundary conditions on both sides of the cavity.

Under the assumption of smaller changes in concentrations, the balance of mass for fluid inside cavity reads:

$$\frac{\partial \rho}{\partial t} + \rho \nabla \cdot \boldsymbol{u}_{\bar{f}} = 0 \tag{3.153}$$

Assuming the fluid inside cavity as mono-phase (no mass transfer between cavity and bulk), the balance equation of mass simplifies to

$$\nabla \cdot \boldsymbol{u}_{\bar{f}} = 0 \tag{3.154}$$

Obviously, the flow velocity inside the cavity is much higher than in the bulk. The mass balance equation can be rewritten as:

$$\frac{\partial v}{\partial \bar{x}} + \frac{\partial w}{\partial \bar{y}} = 0 \tag{3.155}$$

where $w = \boldsymbol{v}_{\bar{f}} \cdot \boldsymbol{n}_{\Gamma_d}$ denotes the normal velocity of the fluid flow inside the cavity.

The mass balance equation is averaged over the width of the cavity:

$$\int_{-h}^{h} \frac{\partial v}{\partial \bar{x}} d\bar{y} + \int_{-h}^{h} \frac{\partial w}{\partial \bar{y}} d\bar{y} = 0 \tag{3.156}$$

The difference of the fluid flow between two sides of the cavity is given by

$$[\![w_{\bar{f}}]\!] = w(h) - w(-h) = -\int_{-h}^{h} \frac{\partial v}{\partial \bar{x}} d\bar{y} \tag{3.157}$$

Substituting Eq. (3.152) into Eq. (3.157) gives:

$$[\![w_{\bar{f}}]\!] = \frac{2}{3\mu} \frac{\partial}{\partial \bar{x}} \left(\frac{\partial p}{\partial \bar{x}} h^3\right) - 2h \frac{\partial v_{\bar{f}}}{\partial \bar{x}} \tag{3.158}$$

This equation describes the total amount of fluid attracted in the tangential flow. It can be included in the weak form for the mass coupling, which ensures the coupling between fluid inside the cavity and fluid in the bulk. Indeed, the coupling term $n_f \boldsymbol{n}_{\Gamma_d} \cdot [\![v_f - v_s]\!]$ can be written as:

$$n_f \boldsymbol{n}_{\Gamma_d} \cdot [\![v_f - v_s]\!] = n_f [\![w_{\bar{f}} - w_s]\!] = n_f \left(\frac{2}{3\mu} \frac{\partial}{\partial \bar{x}} \left(\frac{\partial p}{\partial \bar{x}} h^3\right) - 2h \frac{\partial v_f}{\partial \bar{x}} - 2\frac{\partial h}{\partial t}\right) \tag{3.159}$$

with w_s the normal velocity of the solid skeleton, and the difference between the two sides of the cavity gives:

$$[\![w_s]\!] = 2\frac{\partial h}{\partial t} \tag{3.160}$$

Following the Darcy's law, the tangential velocity reads:

$$v_{\bar{f}} = \left(\boldsymbol{v}_s - \frac{k_f}{n_f} \nabla p\right) \cdot \boldsymbol{t}_{\Gamma_d} \tag{3.161}$$

3.6.1.3 Discretization and discrete system of equations

Approximation of the primary fields

The crack (or cavity) leads to a discontinuous displacement field while the pressure field across the cavity is continuous. Moreover, the spatial derivatives of the pressure orthogonal to the cracks is also discontinuous. Hence, the discretization of the displacement field is given by

$$\boldsymbol{u}_s(x) = \sum_{i \in N} N_i(x)\bar{\boldsymbol{u}}_i + \sum_{i \in N_{cut}} N_i(x)\aleph_{\Gamma_d}(x)\hat{\boldsymbol{u}}_i + \sum_{i \in N_{tip}} \sum_{j \in [1,4]} N_i(x)\Psi_j(x)\tilde{u}_{ij} \quad (3.162)$$

where N_i are the standard finite element shape functions, $\bar{u}_i$, $\hat{u}_i$ and $\tilde{u}_i$ are nodal parameters and the Heaviside function is defined by

$$\aleph_{\Gamma_d}(x) = \begin{cases} 0 & \text{if } \phi(x) \leq 0 \\ 1 & \text{if } \phi(x) \geq 0 \end{cases} \quad (3.163)$$

where $\phi(x)$ being the level-set function.

For the node set N_{tip}, we include the well known crack tip enrichment functions in Ψ_j:

$$\psi_1(\boldsymbol{x}) = \sqrt{r}\frac{\sin\theta}{2} \quad (3.164)$$

$$\psi_2(\boldsymbol{x}) = \sqrt{r}\frac{\sin\theta}{2}\sin\theta \quad (3.165)$$

$$\psi_3(\boldsymbol{x}) = \sqrt{r}\frac{\cos\theta}{2} \quad (3.166)$$

$$\psi_4(\boldsymbol{x}) = \sqrt{r}\frac{\cos\theta}{2}\sin\theta \quad (3.167)$$

These functions depend on a local coordinate system (r, θ) shown in Fig. 3.26. Eq. (3.162) can be rewritten as

$$\boldsymbol{u}_s = \boldsymbol{NU} \quad (3.168)$$

where the matrix $\boldsymbol{N}$ contains the standard and enriched shape functions. Accordingly, the array $\boldsymbol{U}$ includes the displacement for the standard and enriched degrees of freedom $\bar{u}_i$, $\hat{u}_i$, $\tilde{u}_{ij}$.

For the discretization of the pressure, the node set N_{pres} is enriched with the signed distance function $\boldsymbol{D}_{\Gamma_d}$. The enriched distance function $\boldsymbol{D}_{\Gamma_d}$ is

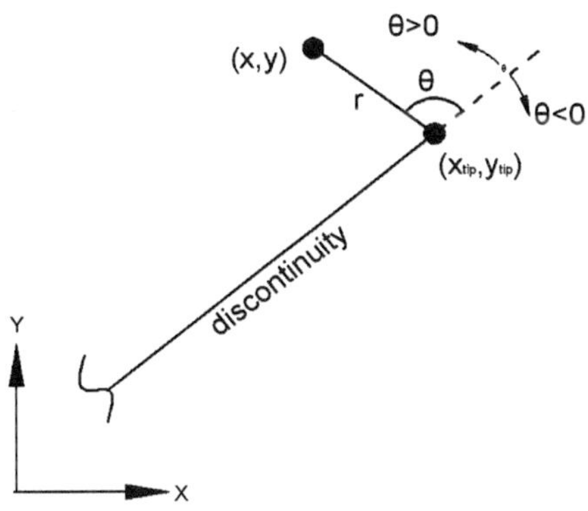

Figure 3.26 Local coordinates (r, θ) at tip element.

continues to the discontinues, but its normal derivative is discontinues.

$$\boldsymbol{D}_{\Gamma_d} = \begin{cases} -d & \text{if } \phi(x) \leq 0 \\ d & \text{if } \phi(x) \geq 0 \end{cases} \tag{3.169}$$

with d the absolute distance to the discontinuity.

The node set N_{pres} includes the nodes affected by the crack. Hence, the discretization of the pressure field can be expressed as

$$p(\boldsymbol{x}) = \sum_{i \in N} H_i(\boldsymbol{x})\bar{p}_i + \sum_{i \in N_{pres}} H_i(\boldsymbol{x})\boldsymbol{D}_{\Gamma_d}(\boldsymbol{x})\hat{p}_i \tag{3.170}$$

where H_i is the standard FE shape function for the pressure. We can also rewrite this expression in matrix-vector form by

$$p = \boldsymbol{HP} \tag{3.171}$$

where $\boldsymbol{H}$ contains the standard as well as the enriched shape functions and $\boldsymbol{P}$ contains the degrees of freedom for the pressure $\bar{p}_i$, $\hat{p}_i$. The order of the shape function N_i and H_i should be adequate to fulfill the modeling requirements. For the consideration of the consistency of the momentum balance equation, the order of the displacement shape function N_i should be greater than or equal to the order of pressure shape function H_i. This study used the quadrilateral elements equipped with linear shape functions.

Discrete system of equations

The vector of external force $\boldsymbol{F}_{ext}$ and external fluid flux $\boldsymbol{Q}_{ext}$ are given by:

$$\boldsymbol{F}_{ext} = \int_{\Gamma} \boldsymbol{N}^T \boldsymbol{t}_p d\Gamma \tag{3.172}$$

$$\boldsymbol{Q}_{ext} = \int_{\Gamma} \boldsymbol{H}^T \boldsymbol{n}_t \boldsymbol{q}_p d\Gamma \tag{3.173}$$

Using backward finite difference approximation.

$$(\frac{\partial \cdot}{\partial t})_i = \frac{\cdot_i - \cdot_{i-1}}{\Delta t} \tag{3.174}$$

with Δx the time increment, $\cdot_i$ and $\cdot_{i-1}$ denoting the unknown at time step i and $i-1$, respectively.

The coupling force vector $F_{coupling}$ (on the crack boundary) is derived from Eq. (3.142):

$$F_{coupling} = -\left(\int_{\Gamma_d} [\![\eta]\!]^T \boldsymbol{n}_{\Gamma_d} \boldsymbol{H}\right) \boldsymbol{P} \tag{3.175}$$

Integrating Eq. (3.144) along Γ_d gives the internal fluid flux $Q_{coupling}$:

$$Q_{coupling} = \int_{\Gamma_d} \boldsymbol{H}^T \boldsymbol{n}_{\Gamma_d}^T \boldsymbol{q}_d d\Gamma_d \tag{3.176}$$

The formation of the internal fluid flux $Q_{coupling}$ is nonlinear, hence, an iteration procedure must be conducted at each time step Δt in order to control the accuracy of the resolution. The iteration residual vector $\boldsymbol{R}^i$ at i^{th} iteration is defined:

$$\boldsymbol{R}^i = \begin{bmatrix} 0 & 0 \\ \boldsymbol{K}_{up}^T & \boldsymbol{K}_{pp}^{(1)} \end{bmatrix} \begin{pmatrix} \Delta \boldsymbol{U} \\ \Delta \boldsymbol{P} \end{pmatrix}^i + \begin{bmatrix} \boldsymbol{K}_{uu} & \boldsymbol{K}_{up} \\ 0 & \Delta t \boldsymbol{K}_{pp}^{(2)} \end{bmatrix} \begin{pmatrix} \boldsymbol{U} \\ \boldsymbol{P} \end{pmatrix}^i + \begin{pmatrix} \boldsymbol{F}_{coupling} \\ \boldsymbol{Q}_{coupling} \end{pmatrix}^i - \begin{pmatrix} \boldsymbol{F}_{ext} \\ \boldsymbol{Q}_{ext} \end{pmatrix} \tag{3.177}$$

with stiffness matrix:

$$\boldsymbol{K}_{uu} = \int_{\Omega} \boldsymbol{B}^T \boldsymbol{D}^{tan} \boldsymbol{B} d\Omega \tag{3.178}$$

$$\boldsymbol{K}_{up} = -\int_{\Omega} \alpha \boldsymbol{B}^T \boldsymbol{m} \boldsymbol{H} d\Omega \tag{3.179}$$

$$\boldsymbol{K}_{pp}^{(1)} = -\int_{\Omega} Q^{-1}\boldsymbol{H}^T\boldsymbol{H}d\Omega \tag{3.180}$$

$$\boldsymbol{K}_{pp}^{(2)} = -\int_{\Omega} k_f \nabla\boldsymbol{H}^T\nabla\boldsymbol{H}d\Omega \tag{3.181}$$

where $\boldsymbol{m} = [1, 1, 0]$ for two-dimensional problem.

In the Newton-Raphson algorithm, the iteration matrix K^i derived from:

$$K^i = \frac{\partial f}{\partial \bullet} \tag{3.182}$$

with f and $\bullet$ denote the residual function and unknowns, respectively. In this study, the iteration matrix K^i has the form:

$$K^i = \begin{bmatrix} \boldsymbol{K}_{uu} & \boldsymbol{K}_{up} + \frac{\partial \boldsymbol{F}_{coupling}}{\partial \boldsymbol{P}} \\ \boldsymbol{K}_{up}^T + \Delta t \frac{\partial \mathbf{Q}_{coupling}}{\partial \boldsymbol{U}} & \boldsymbol{K}_{pp}^{(1)} + \Delta t \boldsymbol{K}_{pp}^{(2)} + \frac{\partial \mathbf{Q}_{coupling}}{\partial \boldsymbol{P}} \end{bmatrix} \tag{3.183}$$

with all items evaluated at iteration i.

The coupling term $\boldsymbol{F}_{coupling}$ and $\mathbf{Q}_{coupling}$ cause the Jacobian matrix of the residual R^i to become asymmetric, to regain the symmetric the coupling terms are omitted in the Jacobian matrix. This may decrease the convergence rate of the Newton-Raphson algorithm. Nevertheless, the symmetric matrix allows flexible implementation as well as better condition of the matrix structure. The simplified Jacobian matrix reads:

$$K^i = \begin{bmatrix} \boldsymbol{K}_{uu} & \boldsymbol{K}_{up} \\ \boldsymbol{K}_{up}^T & \boldsymbol{K}_{pp}^{(1)} + \Delta t \boldsymbol{K}_{pp}^{(2)} \end{bmatrix} \tag{3.184}$$

3.6.2 Thermo-mechanical problems

The thermoelastic governing equations can be formulated coupled or uncoupled. In the theory of uncoupled thermoelasticity, the temperature and heat distribution are described by a parabolic partial differential equation which does not contain any strain or displacement terms. The uncoupled theory of thermoelasticity has two unrealistic assumptions: (1) The displacement field does not affect the temperature field and (2) it yields an infinite wave propagation speed. These two shortcomings have been overcome by the coupled theories of thermoelasticity that account for wave propagation effects. Nonetheless, for sake of simplicity, we will focus in this section on XFEM applied to fracture for uncoupled thermoelasticity. In uncoupled thermoelasticity, the governing equations are given by

$$\begin{aligned} \sigma_{ij,j} + b_i &= 0 \quad \forall \mathbf{X} \in \Omega \\ -q_{i,i} + Q &= 0 \quad \forall \mathbf{X} \in \Omega \end{aligned} \tag{3.185}$$

where σ_{ij} denotes the components of the Cauchy stress tensor, b_i the components of the body forces, q_i is the heat flux Q a heat source. The constitutive and kinematic equations for a linear elastic isotropic solid are given by

$$\begin{aligned} \sigma_{ij} &= C_{ijkl}\epsilon_{kl} \quad \forall \mathbf{X} \in \Omega \\ q_i &= -k\nabla T \quad \forall \mathbf{X} \in \Omega \\ \epsilon_{ij} &= \epsilon_{ij}^M - \epsilon_{ij}^T \quad \forall \mathbf{X} \in \Omega \\ \epsilon_{ij}^T &= \alpha_T \Delta T \delta_{ij} \quad \forall \mathbf{X} \in \Omega \end{aligned} \tag{3.186}$$

where C_{ijkl} indicates the components of the fourth order elasticity tensor, k is the thermal conductivity, ΔT the temperature difference and the superscripts T and M distinguish between the thermal and mechanical strains. Above equations are complemented with the following boundary conditions:

$$\begin{aligned} \sigma_{ij}n_j = t_i &= \bar{t}_i \quad \text{on } \Gamma_t \\ u_i &= \bar{u}_i \quad \text{on } \Gamma_u \\ q_i n_j &= \bar{q} \quad \text{on } \Gamma_q \\ T &= \bar{T} \quad \text{on } \Gamma_T \end{aligned} \tag{3.187}$$

where the superimposed bar indicates imposed boundary conditions on the boundary $\Gamma = \Gamma_T \bigcup \Gamma_q = \Gamma_u \bigcup \Gamma_t$ with $\Gamma_T \bigcap \Gamma_q = \Gamma_u \bigcap \Gamma_t = \emptyset$. The weak form can be derived by the method of weighted residuals and is given by: Find $u_i \in \mathcal{U}\ \forall \delta u_i \in \mathcal{U}_0$ and $T \in \mathcal{V}\ \forall \delta T \in \mathcal{V}_0$ such that

$$\begin{aligned} \int_\Omega \delta\epsilon_{ij}^M C_{ijkl}\left(\epsilon_{kl}^M - \delta T \alpha_T \delta_{ij}\right) d\Omega - \int_\Omega \delta u_i b_i \, d\Omega - \int_{\Gamma_t} \delta u_i \bar{t}_i = 0 \\ \int_\Omega \nabla \delta T \, k \, \nabla T \, d\Omega + \int_\Omega \delta T \, Q \, d\Omega - \int_{\Gamma_q} \delta T \bar{q} \, d\Gamma = 0 \end{aligned} \tag{3.188}$$

with the approximation spaces

$$\mathcal{U} = \left\{u_i(\mathbf{X}) | u_i(\mathbf{X}) \in \mathcal{H}^1(\Omega \setminus \Gamma_c), \;\; u_i(\mathbf{X}) = \bar{u}_i\right\}$$

$$\mathcal{U}_0 = \{\delta u_i(\mathbf{X}) | \delta u_i(\mathbf{X}) \in \mathcal{H}^1(\Omega \setminus \Gamma_c),\ \delta u_i(\mathbf{X}) = 0\}$$
$$\mathcal{V} = \{T(\mathbf{X}) | T(\mathbf{X}) \in \mathcal{H}^1(\Omega \setminus \Gamma_c),\ T(\mathbf{X}) = \bar{T}\}$$
$$\mathcal{V}_0 = \{\delta T(\mathbf{X}) | \delta T(\mathbf{X}) \in \mathcal{H}^1(\Omega \setminus \Gamma_c),\ \delta T(\mathbf{X}) = 0\} \tag{3.189}$$

For adiabatic as well as isothermal cracks, the displacement field will be discontinuous and the strain/stress field singular as the crack tip as discussed in the previous sections. The difference will be in the discretization of the temperature field which is discontinuous for adiabatic cracks but continuous for isothermal cracks. In isothermal cracks, the flux will be discontinuous around the crack tip. Duflot [18] suggested to use only the first term of the crack tip enrichment for the temperature field which yields the following approximation

$$T^h(\mathbf{X}) = \sum_{I \in \mathcal{S}} N_I(\mathbf{X})\mathbf{u}_I + \sum_{I \in \mathcal{S}_c} N_I(\mathbf{X}) H_I(f\mathbf{X})\mathbf{a}_I + \sum_{I \in \mathcal{S}_t} N_I(\mathbf{X})\sqrt{r} sin\ (\frac{\theta}{2})\mathbf{b}_I \tag{3.190}$$

where $\mathcal{S}$, $\mathcal{S}_c$ and $\mathcal{S}_t$ denote the set of all nodes, the set of nodes influenced by elements completely cut by the crack and the set of 'crack tip' nodes, respectively; $H_I(f\mathbf{X})$ is the shifted Heaviside function. For isothermal cracks, the modified abs enrichment can be used in order to guarantee a discontinuous flux. Furthermore, the crack tip enrichment needs to be modified accordingly since the angular variation of the singularity is different to adiabatic cracks. This can be accomplished by using the second term of the crack tip enrichment function which finally yields

$$T^h(\mathbf{X}) = \sum_{I \in \mathcal{S}} N_I(\mathbf{X})\mathbf{u}_I + \sum_{I \in \mathcal{S}_c} N_I(\mathbf{X}) F(f\mathbf{X})\mathbf{a}_I + \sum_{I \in \mathcal{S}_t} N_I(\mathbf{X})\sqrt{r} cos\ (\frac{\theta}{2})\mathbf{b}_I \tag{3.191}$$

where $F(f\mathbf{X})$ denotes the modified abs enrichment.

Let us exemplary focus on adiabatic cracks. Substituting the discretization of the test and trial functions into the weak form, Eq. (3.187), yields the final system of equations:

$$\begin{bmatrix} \mathbf{K}^{uu} & \mathbf{K}^{ua} & \mathbf{K}^{ub} & \mathbf{K}^{u\theta} & \mathbf{K}^{u\alpha} & \mathbf{K}^{u\beta} \\ \mathbf{K}^{au} & \mathbf{K}^{aa} & \mathbf{K}^{ab} & \mathbf{K}^{a\theta} & \mathbf{K}^{a\alpha} & \mathbf{K}^{a\beta} \\ \mathbf{K}^{bu} & \mathbf{K}^{ba} & \mathbf{K}^{bb} & \mathbf{K}^{b\theta} & \mathbf{K}^{b\alpha} & \mathbf{K}^{b\beta} \\ 0 & 0 & 0 & \mathbf{K}^{\theta\theta} & \mathbf{K}^{\theta\alpha} & \mathbf{K}^{\theta\beta} \\ 0 & 0 & 0 & \mathbf{K}^{\alpha\theta} & \mathbf{K}^{\alpha\alpha} & \mathbf{K}^{\alpha\beta} \\ 0 & 0 & 0 & \mathbf{K}^{\beta\theta} & \mathbf{K}^{\beta\alpha} & \mathbf{K}^{\beta\beta} \end{bmatrix} \begin{Bmatrix} \mathbf{u} \\ \mathbf{a} \\ \mathbf{b} \\ \boldsymbol{\theta} \\ \boldsymbol{\alpha} \\ \boldsymbol{\beta} \end{Bmatrix} = \begin{Bmatrix} \mathbf{f}^u \\ \mathbf{f}^a \\ \mathbf{f}^b \\ \mathbf{f}^\theta \\ \mathbf{f}^\alpha \\ \mathbf{f}^\beta \end{Bmatrix} \tag{3.192}$$

3.6.3 Piezoelectric materials

Piezoelectricity is the generation of electricity or electric polarity in dielectric crystals subjected to mechanical stress, or the generation of stress in such crystals subjected to an applied voltage. The piezoelectric effect was discovered in 1880 by Pierre and Jacques Curie. They found that certain materials had the ability to convert mechanical stress to electrical charge. The converse effect was predicted one year later and subsequently confirmed by the Curie brothers. In 30 years the discovery was put into practice with the development of a piezoelectric based sonar or hydrophone. The cause of this phenomenon was found to lie in the atomic structure.

The structure of certain crystalline materials has negative and positive polarization that neutralize along polar axis. When an external mechanical stress disturbs this charge balance, electric charge carriers create current in the crystal. This is called the direct effect. On the other hand, a mechanical stress is created when an external charge creates an imbalance in neutral charge state, which is termed as the converse effect. Piezoelectric materials under converse effect can be used as an actuator and under direct effect can be used as a sensor or energy transducer. The crystal structure of Zinc Oxide (ZnO), a common piezoelectric material is shown in Fig. 3.27. In the figure the all the ZnO tetrahedra has the same orientation with the Zinc having one Oxygen directly above it along the *c* axis and the three other oxygen neighbors below it. When ZnO is compressed along the *c* axis, the material deforms by change in angle of O-Zn-O bond. When the tetrahedron deforms, the positive and negative charge of the unit are displaced leading to polarization. On the other hand, if a tensile stress is applied parallel to *c* axis, the tetrahedra elongates, developing a surface charge of opposite polarity as shown in Fig. 3.27. In case of mono crystals, the polarization is set by the crystal orientation. In case of polycrystals each domain has it own orientation and so a collective piezoelectric effect is attained by subjecting such polycrystals to a process called poling.

In practice, poling usually involves heating the polycrystal above the Curie point, application of the electric field, cooling below the Curie point, and finally removal of the electric field. Upon heating the material above the Curie point, the crystal structure becomes centrosymmetric, and all electric dipoles vanish. When the material is cooled in the presence of a sufficiently large electric field, the dipoles tend to align with the applied field, all together giving rise to a non-zero net polarization. After cooling and removal of the electric field, not all dipoles can return to their original direction. The result is a remanent polarization throughout the ceramic as

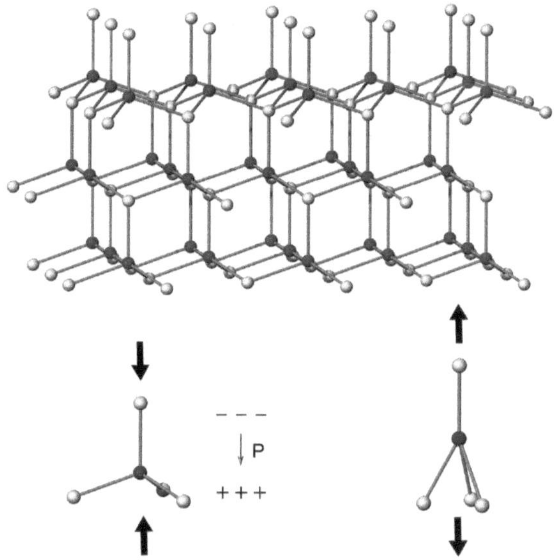

Figure 3.27 The crystal structure of Zinc Oxide and the deformation of the ZnO tetrahedra under compressive and tensile stresses (adapted from [54]).

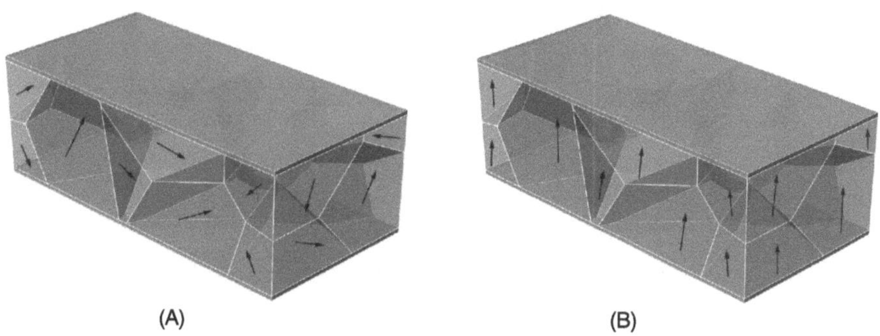

Figure 3.28 The orientation of crystal domains: (A) before poling and (B) after poling [10].

well as a permanent deformation. The polycrystalline ceramic now does exhibit an artificial anisotropy, enabling piezoelectric behavior. The piezoelectricity is maintained as long as the material is not depoled, due to for example a temperature above the Curie point, or extreme electric or mechanical conditions. The poling direction in a polycrystal before and after poling is shown in Fig. 3.28.

The material possessing piezoelectric effect when deformed along the poling direction produces high voltage of same or opposite polarity as that of the poling voltage, depending on load direction. When a voltage difference is applied across the electrodes placed above and below the piezoelectric material, deformation takes place. An input AC signal leads to vibration of the piezomaterial at the same frequency as the AC signal.

The intrinsic electromechanical coupling behavior of piezoelectric materials has important applications in sensors (e.g., sonars), actuators (e.g., ultrasonic cleaners, ultra-precision positioners, ink jetprint heads), signal transmitters (e.g., cellular phone, remote car opener), and surface acoustic wave devices to mention a few. Recently, piezoelectric materials are used in several applications in aircraft industry, for example, shaping the wing of aircraft to improve aerodynamic performance [31]. In diesel engines, solenoid injectors are replaced by piezo actuated injectors [47]. In such applications, piezoelectric materials may experience high stress and electric field concentrations as a result of which they may fail due to fracture or dielectric breakdown. Besides these materials are inhomogeneous, inherently brittle and have low fracture toughness. So assessment of defects like cracks and voids is needed to ensure the reliability of piezoelectric components. Numerical simulation of fracture in piezoelectric ceramics is primarily based on a linear elastic fracture mechanics model [45]. The fundamentals of piezoelectric fracture mechanics can be found in [48]. The analytical work to study the fracture mechanics of piezoelectric ceramics are based on Stroh and Lekhnitskii Formalism. Suo et al. [60] extended the Stroh formalism to piezoelectric problems, considering a semi infinite piezoelectric ceramic with a crack inside. Sosa [57] considered an elliptic hole with major axis perpendicular to the polarization direction inside piezoelectric ceramic, and obtained the field variables around the cavity. Xu and Rajapakse [68] extended this work considering an arbitrarily oriented elliptic hole and concluding that the highest concentrations occur when the elliptical hole is 33° with respect to the polarization direction. A short overview and a critical discussion about the present state in the field of piezoelectric fracture mechanics is given by [29]. FEM analysis of cracks in piezoelectric structures under dynamic electro-mechanical loading considering the influence of dielectric medium inside the crack is presented in [19,27]. A survey on numerical algorithms for crack analyses in piezoelectric structures to be used along with FEM for determining fracture parameters is presented in [28].

3.6.3.1 Strong form and weak form

The electro-elastic response of a piezoelectric body of volume Ω and regular boundary surface S, is governed by the mechanical and electrostatic equilibrium equations,

$$\sigma_{ij,j} + f_i = 0 \text{ in } \Omega \tag{3.193}$$

$$D_{i,i} - q = 0 \text{ in } \Omega \tag{3.194}$$

where f_i, q are mechanical body force components and electric body charges, respectively; σ_{ij} and D_i are the Cauchy stress tensor and electric displacement vector components. They are related to the linear strain tensor ε_{kl} and electric field vector E_k through the converse and direct linear piezoelectric constitutive equations,

$$\sigma_{ij} = C^E_{ijkl}\varepsilon_{kl} - e_{kij}E_k \tag{3.195}$$

$$D_i = e_{ikl}\varepsilon_{kl} + \kappa^{\varepsilon}{}_{ik}E_k \tag{3.196}$$

where C^E_{ijkl}, e_{kij} and $\kappa^{\varepsilon}{}_{ik}$ denote fourth-order elastic stiffness tensor at constant electric field, piezoelectric coupling tensor and dielectric permittivity tensor at constant strain respectively. The strain-charge form of the piezoelectric constitutive equation is given by

$$\varepsilon_{ij} = S^E{}_{ijkl}\sigma_{kl} - d_{kij}E_k \tag{3.197}$$

$$D_i = d_{ikl}\sigma_{kl} + \kappa^{\sigma}{}_{ik}E_k \tag{3.198}$$

These two commonly used forms are related to each other by

$$C^E{}_{ijkl} = S^E{}_{ijkl}{}^{-1} \tag{3.199}$$

$$e_{ikl} = d_{ikl} \cdot S^E{}_{ijkl}{}^{-1} \tag{3.200}$$

$$\kappa^{\varepsilon}{}_{ik} = \kappa^{\sigma}{}_{ik} - d_{ikl} \cdot S^E{}_{ijkl}{}^{-1} \cdot d_{ikl}{}^T \tag{3.201}$$

The strain tensor and electric field vector components are linked to mechanical displacement components u_i and electric field potential ϕ, respectively, by the following relations,

$$\varepsilon_{ij} = \frac{1}{2}\left(\left(u_{i,j}\right) + \left(u_{j,i}\right)\right) \tag{3.202}$$

$$E_i = -\phi_{,i} \tag{3.203}$$

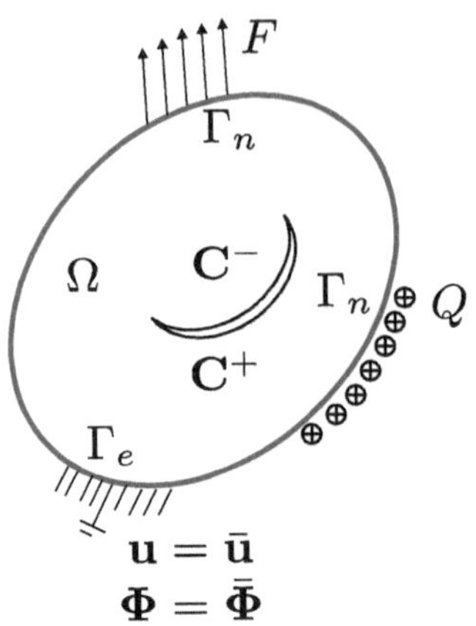

Figure 3.29 Piezoelectric domain with a crack.

The piezoelectric body Ω, is subjected to the following essential and natural boundary conditions:

Essential boundary conditions

$$u_i = \bar{u}_i \text{ (or) } \phi = \bar{\phi} \text{ on } \Gamma_e \tag{3.204}$$

Natural boundary conditions

$$\sigma_{ij} n_j = F_i \text{ (or) } D_i n_i = -Q \text{ on } \Gamma_n \tag{3.205}$$

where $\bar{u}_i$, $\bar{\phi}$, F_i, Q and n_i denote the mechanical displacement, electric potential, surface force components, surface charge and outward unit normal vector components, respectively (see Fig. 3.29). The crack faces C^+ and C^- are assumed to be both traction-free and electrically impermeable.

It can be shown that the weak form is given by: Find $u_i \in \mathcal{U}\ \forall \delta u_i \in \mathcal{U}_0$ and $\phi \in \mathcal{V}\ \forall \delta\phi \in \mathcal{V}_0$ such that

$$\begin{aligned}
\delta W &= \delta W_{int} - \delta W_{ext} = 0 \\
\delta W_{int} &= \int_\Omega \delta\epsilon_{ij} C_{ijkl} \epsilon_{kl}\, d\Omega - \int_\Omega \delta\epsilon_{ij} e_{ijk} E_k\, d\Omega \\
&\quad - \int_\Omega \delta E_i e_{ijk} \epsilon_{jk}\, d\Omega - \int_\Omega \delta E_i \kappa_{ij} E_j\, d\Omega \\
\delta W_{ext} &= \int_\Omega \delta u_i b_i\, d\Omega + \int_\Gamma \delta u_i \bar{t}_i\, d\Gamma
\end{aligned}$$

$$+\int_{\Omega}\delta\phi q\,d\Omega+\int_{\Gamma}\delta\phi Q\,d\Gamma \tag{3.206}$$

with the approximation spaces

$$\begin{aligned}
\mathcal{U}&=\left\{u_i(\mathbf{X})|u_i(\mathbf{X})\in\mathcal{H}^1(\Omega\setminus\Gamma_c),\ \ u_i(\mathbf{X})=\bar{u}_i\right\}\\
\mathcal{U}_0&=\left\{\delta u_i(\mathbf{X})|\delta u_i(\mathbf{X})\in\mathcal{H}^1(\Omega\setminus\Gamma_c),\ \ \delta u_i(\mathbf{X})=0\right\}\\
\mathcal{V}&=\left\{\phi(\mathbf{X})|\phi(\mathbf{X})\in\mathcal{H}^1(\Omega\setminus\Gamma_c),\ \ \phi(\mathbf{X})=\bar{\phi}\right\}\\
\mathcal{V}_0&=\left\{\delta\phi(\mathbf{X})|\delta\phi(\mathbf{X})\in\mathcal{H}^1(\Omega\setminus\Gamma_c),\ \ \delta\phi(\mathbf{X})=0\right\}
\end{aligned} \tag{3.207}$$

3.6.3.2 XFEM formulation for piezoelectric materials

In XFEM, the approximation of the displacement and electric potential field for fracture in a piezoelectric material is given by,

$$\begin{aligned}
\mathbf{u}^h(x)&=\sum_{i\in I}N_i(X)\boldsymbol{u_i}+\sum_{N=1}^{n_c}\sum_{j\in J}N_j(X)\boldsymbol{a}_j^{(N)}F_I^{(N)}\\
&+\sum_{M=1}^{m_t}\sum_{k\in K}N_k(X)\left(\sum_{i=1}^{4}G_i^{(M)}(r,\theta)\boldsymbol{b}_k^i\right)
\end{aligned} \tag{3.208}$$

$$\begin{aligned}
\phi^h(x)&=\sum_{i\in I}N_i(X)\boldsymbol{\phi_i}+\sum_{N=1}^{n_c}\sum_{j\in J}N_j(X)\boldsymbol{\alpha}_j^{(N)}F_I^{(N)}\\
&+\sum_{M=1}^{m_t}\sum_{k\in K}N_k(X)\left(\sum_{i=1}^{4}G_i^{(E)}(r,\theta)\boldsymbol{\beta}_k^i\right),
\end{aligned} \tag{3.209}$$

where J is the set of all nodes whose support is cut by the strong discontinuity or void boundary. The set K contains all the nodes that lies within a fixed region around the crack tip [6], n_c denotes the number of cracks/voids, m_t is the number of crack tips and l is the number of additional degrees of freedom of crack tip enriched nodes; $\boldsymbol{a_j}$, $\boldsymbol{b_k}$, $\boldsymbol{\alpha_j}$ and $\boldsymbol{\beta_k}$ are the additional degrees of freedom to be solved for. For cracks, the step function in J ensures the jump in displacement and electric potential field:

$$F_I^{(N)}=sign\left[f^{(N)}(\mathbf{X})\right]-sign\left[f^{(N)}(\mathbf{X}_I)\right] \tag{3.210}$$

$$f^{(N)}(\mathbf{X})=sign\left[\mathbf{n}\cdot(\mathbf{X}-\mathbf{X}^{(N)})\right]\min(\mathbf{X}-\mathbf{X}^{(N)});\ \mathbf{X}^{(N)}\in\Gamma^{(N)}, \tag{3.211}$$

where $\mathbf{n}$ is the outward unit normal of the crack face. For voids, $F_I^{(N)}=0$ and 1, for nodes that lie inside and outside the voids, respectively. The nodes that lie exactly over the void boundary are not enriched [59]. The

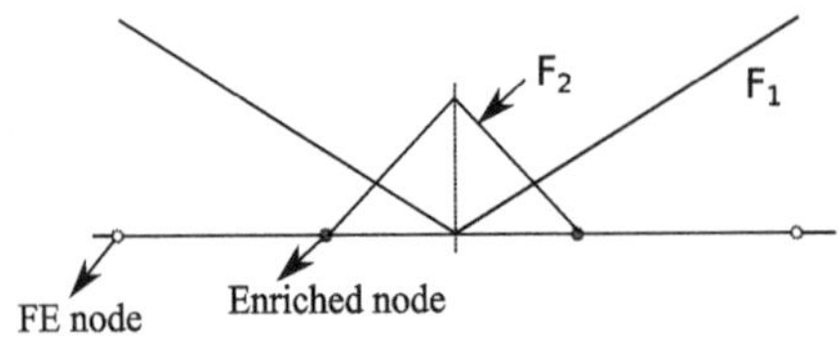

Figure 3.30 Absolute signed distance enrichment function [40].

last term in Eq. (3.208) and (3.209) will vanish for voids. The asymptotic crack tip enrichment function for the mechanical field is identical to the pure mechanical problem and defined as

$$G^M(r,\theta) = \left\{ \sqrt{r}sin\left(\frac{\theta}{2}\right) , \sqrt{r}cos\left(\frac{\theta}{2}\right) , \sqrt{r}sin\left(\frac{\theta}{2}sin\theta\right) , \sqrt{r}cos\left(\frac{\theta}{2}sin\theta\right) \right\} \tag{3.212}$$

where (r, θ) are the local polar coordinates at the crack tip. Bechet et al. [7] derived a crack tip enrichment for the electrical potential. However, comparing results obtained using this specifically designed six-fold enrichment for electromechanical problems with the standard four-fold enrichment of the isotropic elasticity problem, they found almost no differences.

For problems involving multi-materials, different enrichment functions are employed which ensure discontinuities in the derivatives of the displacement and electric potential field. As pointed out earlier, this can be achieved by using the absolute value of the signed distance function as the enrichment function:

$$F_1 = |\sum_{i \in I} N_i \Phi_i| \tag{3.213}$$

This enrichment function is shifted as proposed in Moes et al. [40] to avoid convergence issue due to blending elements:

$$F_2 = \sum_{i \in I} N_i |\Phi_i| - |\sum_{i \in I} N_i \Phi_i| \tag{3.214}$$

The enrichment function F_2 has zero value on the elements which are not crossed by the interface as illustrated in Fig. 3.30.

Substituting the displacement field from Eq. (3.208) and electric potential field from Eq. (3.209) into the weak form leads to the following system of equations for a fracture problem:

$$\begin{pmatrix} \mathbf{K}_{ij}^{uu} & \mathbf{K}_{ij}^{ua} & \mathbf{K}_{ij}^{ub} \\ \mathbf{K}_{ij}^{au} & \mathbf{K}_{ij}^{aa} & \mathbf{K}_{ij}^{ab} \\ \mathbf{K}_{ij}^{bu} & \mathbf{K}_{ij}^{ba} & \mathbf{K}_{ij}^{bb} \end{pmatrix} \begin{Bmatrix} \mathbf{u} \\ \mathbf{a} \\ \mathbf{b} \end{Bmatrix} + \begin{pmatrix} \mathbf{K}_{ij}^{u\phi} & \mathbf{K}_{ij}^{u\alpha} & \mathbf{K}_{ij}^{u\beta} \\ \mathbf{K}_{ij}^{a\phi} & \mathbf{K}_{ij}^{a\alpha} & \mathbf{K}_{ij}^{a\beta} \\ \mathbf{K}_{ij}^{b\phi} & \mathbf{K}_{ij}^{b\alpha} & \mathbf{K}_{ij}^{b\beta} \end{pmatrix} \begin{Bmatrix} \boldsymbol{\phi} \\ \boldsymbol{\alpha} \\ \boldsymbol{\beta} \end{Bmatrix} = \begin{Bmatrix} \mathbf{f} \\ \mathbf{f}^a \\ \mathbf{f}^b \end{Bmatrix} \tag{3.215}$$

$$\begin{pmatrix} \mathbf{K}_{ij}^{\phi u} & \mathbf{K}_{ij}^{\phi a} & \mathbf{K}_{ij}^{\phi b} \\ \mathbf{K}_{ij}^{\alpha u} & \mathbf{K}_{ij}^{\alpha a} & \mathbf{K}_{ij}^{\alpha b} \\ \mathbf{K}_{ij}^{\beta u} & \mathbf{K}_{ij}^{\beta a} & \mathbf{K}_{ij}^{\beta b} \end{pmatrix} \begin{Bmatrix} \mathbf{u} \\ \mathbf{a} \\ \mathbf{b} \end{Bmatrix} - \begin{pmatrix} \mathbf{K}_{ij}^{\phi\phi} & \mathbf{K}_{ij}^{\phi\alpha} & \mathbf{K}_{ij}^{\phi\beta} \\ \mathbf{K}_{ij}^{\alpha\phi} & \mathbf{K}_{ij}^{\alpha\alpha} & \mathbf{K}_{ij}^{\alpha\beta} \\ \mathbf{K}_{ij}^{\beta\phi} & \mathbf{K}_{ij}^{\beta\alpha} & \mathbf{K}_{ij}^{\beta\beta} \end{pmatrix} \begin{Bmatrix} \boldsymbol{\phi} \\ \boldsymbol{\alpha} \\ \boldsymbol{\beta} \end{Bmatrix} = \begin{Bmatrix} \mathbf{g} \\ \mathbf{g}^\alpha \\ \mathbf{g}^\beta \end{Bmatrix} \tag{3.216}$$

$$\mathbf{K}_{ij}^{uu} = \int_\Omega \mathbf{B}_{u_i}{}^T\mathbf{C}\mathbf{B}_{uj} d\Omega; \ \mathbf{K}_{ij}^{ua} = \int_\Omega \mathbf{B}_{ui}^T\mathbf{C}\mathbf{B}_{uj}F_j^{(N)} d\Omega = (\mathbf{K}_{ji}^{au})^T;$$

$$\mathbf{K}_{ij}^{ub} = \int_\Omega \mathbf{B}_{ui}^T\mathbf{C}\mathbf{B}_{uj}G_j^{(N)} d\Omega = (\mathbf{K}_{ji}^{bu})^T; \ \mathbf{K}_{ij}^{aa} = \int_\Omega F_i^{(N)}\mathbf{B}_{ui}^T\mathbf{C}\mathbf{B}_{uj}F_j^{(N)} d\Omega;$$

$$\mathbf{K}_{ij}^{ab} = \int_\Omega F_i^{(N)}\mathbf{B}_{ui}^T\mathbf{C}\mathbf{B}_{uj}G_j^{(N)} d\Omega = (\mathbf{K}_{ji}^{ba})^T; \ \mathbf{K}_{ij}^{bb} = \int_\Omega G_i^{(N)}\mathbf{B}_{ui}^T\mathbf{C}\mathbf{B}_{uj}G_j^{(N)} d\Omega;$$

$$\mathbf{K}_{ij}^{u\phi} = \int_\Omega \mathbf{B}_{ui}^T\mathbf{e}\mathbf{B}_{\phi j} d\Omega = (\mathbf{K}_{ji}^{\phi u})^T; \ \mathbf{K}_{ij}^{u\alpha} = \int_\Omega \mathbf{B}_{ui}^T\mathbf{e}\mathbf{B}_{\phi j}F_j^{(N)} d\Omega = (\mathbf{K}_{ji}^{\alpha u})^T;$$

$$\mathbf{K}_{ij}^{u\beta} = \int_\Omega \mathbf{B}_{ui}^T\mathbf{e}\mathbf{B}_{\phi j}G_j^{(N)} d\Omega = (\mathbf{K}_{ji}^{\beta u})^T; \ \mathbf{K}_{ij}^{a\phi} = \int_\Omega F_i^{(N)}\mathbf{B}_{ui}^T\mathbf{e}\mathbf{B}_{\phi j} d\Omega = (\mathbf{K}_{ji}^{\phi a})^T;$$

$$\mathbf{K}_{ij}^{a\alpha} = \int_\Omega G_i^{(N)}\mathbf{B}_{ui}^T\mathbf{e}\mathbf{B}_{\phi j}F_j^{(N)} d\Omega = (\mathbf{K}_{ji}^{\alpha a})^T;$$

$$\mathbf{K}_{ij}^{a\beta} = \int_\Omega F_i^{(N)}\mathbf{B}_{ui}^T\mathbf{e}\mathbf{B}_{\phi j}G_j^{(N)} d\Omega = (\mathbf{K}_{ji}^{\beta a})^T;$$

$$\mathbf{K}_{ij}^{b\phi} = \int_\Omega G_i^{(N)}\mathbf{B}_{ui}^T\mathbf{e}\mathbf{B}_{\phi j} d\Omega = (\mathbf{K}_{ji}^{\phi b})^T;$$

$$\mathbf{K}_{ij}^{b\alpha} = \int_\Omega G_i^{(N)}\mathbf{B}_{ui}^T\mathbf{e}\mathbf{B}_{\phi j}F_j^{(N)} d\Omega = (\mathbf{K}_{ji}^{b\alpha})^T;$$

$$\mathbf{K}_{ij}^{b\beta} = \int_\Omega G_i^{(N)}\mathbf{B}_{ui}^T\mathbf{e}\mathbf{B}_{\phi j}G_j^{(N)} d\Omega = (\mathbf{K}_{ji}^{b\beta})^T;$$

$$\mathbf{K}_{ij}^{\phi\phi} = \int_\Omega \mathbf{B}_{\phi i}^T\boldsymbol{\kappa}\mathbf{B}_{\phi j} d\Omega; \ \mathbf{K}_{ij}^{\phi\alpha} = \int_\Omega \mathbf{B}_{\phi i}^T\boldsymbol{\kappa}\mathbf{B}_{\phi j}F_j^{(N)} d\Omega = (\mathbf{K}_{ji}^{\alpha\phi})^T;$$

$$\mathbf{K}_{ij}^{\phi\beta} = \int_\Omega \mathbf{B}_{\phi i}^T\boldsymbol{\kappa}\mathbf{B}_{\phi j}G_j^{(N)} d\Omega = (\mathbf{K}_{ji}^{\beta\phi})^T; \ \mathbf{K}_{ij}^{\alpha\alpha} = \int_\Omega F_i^{(N)}\mathbf{B}_{\phi i}^T\boldsymbol{\kappa}\mathbf{B}_{\phi j}F_j^{(N)} d\Omega;$$

$$\mathbf{K}_{ij}^{\alpha\beta} = \int_\Omega F_i^{(N)}\mathbf{B}_{\phi i}^T\boldsymbol{\kappa}\mathbf{B}_{\phi j}\Phi_j^{(N)} d\Omega = (\mathbf{K}_{ji}^{\beta\alpha})^T; \ \mathbf{K}_{ij}^{\beta\beta} = \int_\Omega G_i^{(N)}\mathbf{B}_{\phi i}^T\boldsymbol{\kappa}\mathbf{B}_{\phi j}G_j^{(N)} d\Omega;$$

A similar final discrete system will be obtained for multi-materials. The integration can be done as discussed in Section 3.2.6.

3.6.4 Flexoelectricity

Flexoelectricity in solids was introduced by Mashkevich and Tolpygo [37] in the 1950s but received comparatively (to piezoelectricity) little attention in academia. Recent developments in nanotechnology have been attracting again new interests on flexoelectricity, probably due to the size dependent phenomenon where large strain gradients are obtainable at small length scales. There are two main differences between flexoelectricity and piezoelectricity: Firstly, the flexoelectric effect is generically available in all dielectrics including those with centrosymmetric crystalline structures while piezoelectricity only exists in materials with non-centrosymmetric crystals. This important feature makes flexoelectricity a more universal phenomenon than piezoelectricity. Secondly, in contrast to piezoelectricity (in the absence of surface effects at small length scales), flexoelectricity is a size dependence phenomenon (even in the absence of surface effects). The flexoelectric effect is typically insignificant relative to piezoelectricity at macroscopic scales, while at the nanoscale, flexoelectricity is dominant.

3.6.4.1 Governing equations and weak form

For a linear dielectric solid possessing only piezoelectric effects, the electric enthalpy density $\bar{\mathcal{H}}$ is a function of the linear strain tensor $\boldsymbol{\epsilon}$ and the electric field $\boldsymbol{E}$, i.e. $\bar{\mathcal{H}}(\epsilon_{ij}, E_i)$. When flexoelectric effects are accounted for, the enthalpy density $\mathcal{H}(\epsilon_{ij}, E_i)$ also becomes a function of the strain gradient and electric field gradient. Thus

$$\begin{aligned} \mathcal{H}(\epsilon_{ij}, E_i, \epsilon_{jk,l}, E_{i,j}) = {} & \frac{1}{2} C_{ijkl}\epsilon_{ij}\epsilon_{kl} - e_{ikl}E_i\epsilon_{kl} \\ & + (d_{ijkl}E_{i,j}\epsilon_{kl} + f_{ijkl}E_i\epsilon_{jk,l}) - \frac{1}{2}\kappa_{ij}E_iE_j \end{aligned} \tag{3.217}$$

where $E_i = -\theta_{,i}$ and $\boldsymbol{\theta}$ is the electric potential; $\boldsymbol{e}$ is the third-order tensor of piezoelectricity, $\boldsymbol{\kappa}$ is the second-order dielectric tensor, $\boldsymbol{C}$ is the fourth-order elasticity tensor, $\boldsymbol{f}$ is the fourth-order direct flexoelectric tensor and $\boldsymbol{d}$ is the fourth-order converse flexoelectric tensor. Let us consider the terms in the brackets on the RHS of Eq. (3.217)[9] containing the direct and reverse flexoelectric effects. Integrating these terms over the volume, using integration by parts and the Gauss divergence theorem on the first term

[9] This term is not present in the case of piezoelectricity.

yields

$$
\begin{aligned}
&\int_{\Omega}(d_{ijkl}E_{i,j}\epsilon_{kl}+f_{ijkl}E_i\epsilon_{jk,l})d\Omega=\int_{\Omega}d_{ijkl}E_{i,j}\epsilon_{kl}d\Omega+\int_{\Omega}f_{ijkl}E_i\epsilon_{jk,l}d\Omega\\
&=\int_{\partial\Omega}d_{ijkl}E_i\epsilon_{kl}ds-\int_{\Omega}d_{ijkl}E_i\epsilon_{kl,j}d\Omega+\int_{\Omega}f_{ijkl}E_i\epsilon_{jk,l}d\Omega\\
&=\int_{\Omega}(f_{ijkl}E_i\epsilon_{jk,l}-d_{ijkl}E_i\epsilon_{kl,j})d\Omega+\int_{\partial\Omega}d_{ijkl}E_i\epsilon_{kl}ds\\
&=-\int_{\Omega}(d_{iljk}-f_{ijkl})E_i\epsilon_{jk,l}d\Omega+\int_{\partial\Omega}d_{ijkl}E_i\epsilon_{kl}ds\\
&=-\int_{\Omega}\mu_{ijkl}E_i\epsilon_{jk,l}d\Omega+\int_{\partial\Omega}d_{ijkl}E_i\epsilon_{kl}ds
\end{aligned}
\tag{3.218}
$$

which is expressed in terms of only one material tensor, $\boldsymbol{\mu}$ with $\mu_{ijkl}=d_{iljk}-f_{ijkl}$. Therefore, one can rewrite Eq. (3.217) as

$$
\mathcal{H}(\epsilon_{ij},E_i,\epsilon_{jk,l})=\frac{1}{2}C_{ijkl}\epsilon_{ij}\epsilon_{kl}-e_{ikl}E_i\epsilon_{kl}-\mu_{ijkl}E_i\epsilon_{jk,l}-\frac{1}{2}\kappa_{ij}E_iE_j \tag{3.219}
$$

For a pure piezoelectric material, the electromechanical stress and electric displacement field is given by

$$
\sigma_{ij}=\frac{\partial\bar{\mathcal{H}}}{\partial\epsilon_{ij}}\quad,\quad D_i=-\frac{\partial\bar{\mathcal{H}}}{\partial E_i} \tag{3.220}
$$

while in the presence of flexoelectricity, besides the usual stress $\widehat{\sigma}_{ij}$ and electric displacement field $\widehat{D}_i$ for flexoelectricity, also higher order stress $\widetilde{\sigma}_{ijk}$ and electric displacement fields $\widetilde{D}_{ij}$ occur:

$$
\widehat{\sigma}_{ij}=\frac{\partial\mathcal{H}}{\partial\epsilon_{ij}}\quad,\quad\widehat{D}_i=-\frac{\partial\mathcal{H}}{\partial E_i}\quad,\quad\widetilde{\sigma}_{ijk}=\frac{\partial\mathcal{H}}{\partial\epsilon_{ij,k}}\quad,\quad\widetilde{D}_{ij}=-\frac{\partial\mathcal{H}}{\partial E_{i,j}} \tag{3.221}
$$

These flexoelectric quantities can be related to the piezoelectric quantities by

$$
\sigma_{ij}=\widehat{\sigma}_{ij}-\widetilde{\sigma}_{ijk,k}\quad,\quad D_i=\widehat{D}_i-\widetilde{D}_{ij,j} \tag{3.222}
$$

and hence

$$
\sigma_{ij}=\widehat{\sigma}_{ij}-\widetilde{\sigma}_{ijk,k}=C_{ijkl}\epsilon_{kl}-e_{kij}E_k+\mu_{lijk}E_{l,k} \tag{3.223}
$$

$$
D_i=\widehat{D}_i-\widetilde{D}_{ij,j}=e_{ikl}\epsilon_{kl}+\kappa_{ij}E_j+\mu_{ijkl}\epsilon_{jk,l} \tag{3.224}
$$

Rewriting Eq. (3.221)

$$\partial\mathcal{H} = \widehat{\sigma}_{ij}\partial\epsilon_{ij} \quad , \quad \partial\mathcal{H} = \widetilde{\sigma}_{ijk}\partial\epsilon_{ij,k} \quad , \quad \partial\mathcal{H} = -\widehat{D}_i\partial E_i \tag{3.225}$$

and integrating over Ω finally yields the total electrical enthalpy

$$H = \frac{1}{2}\int_{\Omega}(\widehat{\sigma}_{ij}\epsilon_{ij} + \widetilde{\sigma}_{ijk}\epsilon_{ij,k} - \widehat{D}_i E_i)d\Omega \tag{3.226}$$

The essential and natural electrical boundary conditions for Eq. (3.219) are given by

$$\theta = \overline{\theta} \text{ on } \Gamma_\theta, \quad D_i n_i = -\omega \text{ on } \Gamma_D, \quad \Gamma_\theta \cup \Gamma_D = \partial\Omega \text{ and } \Gamma_\theta \cap \Gamma_D = \emptyset \tag{3.227}$$

where $\overline{\theta}$ and ω are the prescribed electric potential and surface charge density; n_i is the unit normal to the boundary $\partial\Omega$ of the domain Ω. The mechanical boundary conditions are given by

$$u_i = \overline{u}_i \quad \text{on} \quad \Gamma_u,$$

$$t_k = n_j(\widehat{\sigma}_{jk} - \widetilde{\sigma}_{ijk,l}) - D_j(n_i\widetilde{\sigma}_{ijk}) - (D_p n_p)n_i n_j\widetilde{\sigma}_{ijk} = \overline{t}_k \quad \text{on} \quad \Gamma_t \tag{3.228}$$

$$\Gamma_u \cup \Gamma_t = \partial\Omega \text{ and } \Gamma_u \cap \Gamma_t = \emptyset \tag{3.229}$$

where $\overline{u}_i$ and $\overline{t}_k$ are the prescribed mechanical displacements and tractions; D_j is the surface gradient operator. In addition to these, the strain gradients result in other types of boundary conditions as follows

$$u_{i,j}n_j = \overline{v}_i \text{ on } \Gamma_v, \quad n_i n_j\widetilde{\sigma}_{ijk} = \overline{r}_k \text{ on } \Gamma_r,$$

$$\Gamma_v \cup \Gamma_r = \partial\Omega \text{ and } \Gamma_v \cap \Gamma_r = \emptyset \tag{3.230}$$

where $\overline{v}_i$ and $\overline{r}_k$ are the prescribed normal derivative of displacement and the higher order traction, respectively. The kinetic energy for the system K_E and the work W_{ext} done by the external surface mechanical and electrical forces can be written as

$$K_E = \frac{1}{2}\int_{\Omega}\varrho\dot{u}_i\dot{u}_i d\Omega, \quad W_{ext} = \int_{\Gamma_t}\overline{t}_i u_i dS - \int_{\Gamma_D}\omega\theta\, dS \tag{3.231}$$

where ϱ denotes the density and the superimposed dot indicate time derivative. Using Hamilton principle without the damping term we obtain

$$\delta\int_{t_1}^{t_2}(K_E - H + W_{ext})dt = 0 \tag{3.232}$$

and

$$\delta\int_{t_1}^{t_2}\left(\frac{1}{2}\int_\Omega \varrho\dot{u}_i\dot{u}_i d\Omega-\frac{1}{2}\int_\Omega(\widehat{\sigma}_{ij}\epsilon_{ij}+\tilde{\sigma}_{ijk}\epsilon_{ij,k}-\widehat{D}_iE_i)d\Omega\right.$$
$$\left.+\int_{\Gamma_t}\bar{t}_iu_idS-\int_{\Gamma_D}\omega\theta\,dS\right)dt=0 \tag{3.233}$$

Moving the variation operation into the integral operations leads to

$$\int_{t_1}^{t_2}\left(\frac{1}{2}\int_\Omega \delta(\varrho\dot{u}_i\dot{u}_i)d\Omega-\frac{1}{2}\int_\Omega\delta(\widehat{\sigma}_{ij}\epsilon_{ij}+\tilde{\sigma}_{ijk}\epsilon_{ij,k}-\widehat{D}_iE_i)d\Omega\right.$$
$$\left.+\int_{\Gamma_t}\bar{t}_i\delta u_idS-\int_{\Gamma_D}\omega\delta\theta\,dS\right)dt=0 \tag{3.234}$$

Changing the order of operations and using the chain rule of variation we have

$$\int_{t_1}^{t_2}\left[\frac{1}{2}\int_\Omega\delta(\varrho\dot{u}_i\dot{u}_i)d\Omega\right]dt=-\int_{t_1}^{t_2}\left[\int_\Omega\varrho(\delta u_i\ddot{u}_i)d\Omega\right]dt \tag{3.235}$$

$$\int_{t_1}^{t_2}\left[\frac{1}{2}\int_\Omega\delta(\widehat{\sigma}_{ij}\epsilon_{ij}+\tilde{\sigma}_{ijk}\epsilon_{ij,k}-\widehat{D}_iE_i)d\Omega\right]dt$$
$$=\int_{t_1}^{t_2}\left[\int_\Omega(\widehat{\sigma}_{ij}\delta\epsilon_{ij}+\tilde{\sigma}_{ijk}\delta\epsilon_{ij,k}-\widehat{D}_i\delta E_i)d\Omega\right]dt \tag{3.236}$$

Eq. (3.234) now becomes

$$\int_{t_1}^{t_2}\left(-\int_\Omega\varrho(\delta u_i\ddot{u}_i)d\Omega-\int_\Omega(\widehat{\sigma}_{ij}\delta\epsilon_{ij}+\tilde{\sigma}_{ijk}\delta\epsilon_{ij,k}-\widehat{D}_i\delta E_i)d\Omega\right.$$
$$\left.+\int_{\Gamma_t}\bar{t}_i\delta u_idS-\int_{\Gamma_D}\omega\delta\theta\,dS\right)dt=0 \tag{3.237}$$

To satisfy Eq. (3.237) for all possible choices of $\boldsymbol{u}$, the integrand of the time integration has to vanish, which leads to

$$\int_\Omega\varrho(\delta u_i\ddot{u}_i)d\Omega+\int_\Omega(\widehat{\sigma}_{ij}\delta\epsilon_{ij}+\tilde{\sigma}_{ijk}\delta\epsilon_{ij,k}-\widehat{D}_i\delta E_i)d\Omega$$
$$-\int_{\Gamma_t}\bar{t}_i\delta u_idS+\int_{\Gamma_D}\omega\delta\theta\,dS=0 \tag{3.238}$$

For a static problem, the inertia term is neglecting yielding

$$\int_\Omega(\widehat{\sigma}_{ij}\delta\epsilon_{ij}+\tilde{\sigma}_{ijk}\delta\epsilon_{ij,k}-\widehat{D}_i\delta E_i)d\Omega-\int_{\Gamma_t}\bar{t}_i\delta u_idS+\int_{\Gamma_D}\omega\delta\theta\,dS=0 \tag{3.239}$$

Substituting Eqs. (3.220)-(3.224) into Eq. (3.239) leads to the final weak form

$$\int_{\Omega}\left(C_{ijkl}\delta\epsilon_{ij}\epsilon_{kl} - e_{kij}E_k\delta\epsilon_{ij} - \mu_{lijk}E_l\delta\epsilon_{ij,k} - \kappa_{ij}\delta E_i E_j\right)d\Omega$$
$$- \int_{\Gamma_t}\bar{t}_i\delta u_i dS + \int_{\Gamma_D}\omega\delta\theta\, dS = 0 \qquad (3.240)$$

3.6.4.2 XIGA discretization

For weak discontinuities, the displacement field $\mathbf{u}^h$ and electric potential field, θ^h for a flexoelectric material in the XFEM formulation are expressed as

$$\boldsymbol{u}^h(\mathrm{X}) = \sum_{i\in I} N_i(\mathrm{X})\boldsymbol{u}_i + \sum_{N=1}^{n_c}\sum_{l\in L} N_l(\mathrm{X})\boldsymbol{a}_l^{(N)} f_l^{(N)} \qquad (3.241)$$

$$\theta^h(\mathrm{X}) = \sum_{i\in I} N_i(\mathrm{X})\theta_i + \sum_{N=1}^{n_c}\sum_{l\in L} N_l(\mathrm{X})\alpha_l^{(N)} f_l^{(N)} \qquad (3.242)$$

where $\boldsymbol{a}_j$ and α_j are the additional degrees of freedom that account for the jump in displacement and electric potential field, respectively, n_c denotes the number of inclusion interfaces, J is the set of all nodes whose support is cut by the material interface; $f^{(N)}$ in Eqs. (3.241) and (3.242) are the absolute signed distance function values from the interfaces. For impermeable cracks, a discretization for strong discontinuities would yield a similar structure as for the piezoelectric formulation. Due to the C^1 continuity requirement, a 'modified' weak form is required if standard Lagrange polynomials are employed which can guarantee only C^0 continuity globally. Such a formulation will be presented in the next subsection. However, when higher-order 'isogeometric' shape functions $N_i(\mathrm{X})$ such as NURBS or PHT-splines are employed, only the displacement and electrical potential field needs to be discretized as the IGA discretization ensure globally C^1 continuity. A NURBS geometry is represented in terms of knot vectors and control points and weights associated to each basis function. More details about the isogeometric concept will be provided in the course of this book. Substituting the discretization of the displacement field and electric potential field, the associated test functions and spatial derivatives into the weak form finally leads to the following system of algebraic equations

$$\underbrace{\begin{bmatrix} \boldsymbol{A}_{uu} & \boldsymbol{A}_{u\theta} \\ \boldsymbol{A}_{\theta u} & \boldsymbol{A}_{\theta\theta} \end{bmatrix}}_{\tilde{\mathbf{A}}} \underbrace{\begin{bmatrix} \boldsymbol{U} \\ \boldsymbol{\theta} \end{bmatrix}}_{\tilde{\mathbf{D}}} = \underbrace{\begin{bmatrix} \boldsymbol{f}_u \\ \boldsymbol{f}_\theta \end{bmatrix}}_{\tilde{\mathbf{f}}} \tag{3.243}$$

with

$$\boldsymbol{A}_{uu} = \sum_e \int_{\Omega_e} (\boldsymbol{B}_u)^T \boldsymbol{C}(\boldsymbol{B}_u) d\Omega_e,$$

$$\boldsymbol{A}_{u\theta} = \sum_e \int_{\Omega_e} \left[(\boldsymbol{B}_u)^T \boldsymbol{e}^T (\boldsymbol{B}_\theta) + (\boldsymbol{H}_u)^T \boldsymbol{\mu}^T (\boldsymbol{B}_\theta) \right] d\Omega_e \tag{3.244}$$

$$\boldsymbol{A}_{\theta u} = \sum_e \int_{\Omega_e} \left[(\boldsymbol{B}_\theta)^T \boldsymbol{e}(\boldsymbol{B}_u) + (\boldsymbol{B}_\theta)^T \boldsymbol{\mu}(\boldsymbol{H}_u) \right] d\Omega_e,$$

$$\boldsymbol{A}_{\theta\theta} = -\sum_e \int_{\Omega_e} (\boldsymbol{B}_\theta)^T \boldsymbol{\kappa}(\boldsymbol{B}_\theta) d\Omega_e \tag{3.245}$$

$$\boldsymbol{f}_u = \sum_e \int_{\Gamma_{t_e}} \boldsymbol{N}_u^T \boldsymbol{t}_\Gamma d\Gamma_{t_e}, \qquad \boldsymbol{f}_\theta = -\sum_e \int_{\Gamma_{D_e}} \boldsymbol{N}_\theta^T \omega d\Gamma_{D_e} \tag{3.246}$$

where the matrices $\boldsymbol{B}_i$, $i = u, \theta$ contain the spatial derivatives of the IGA shape functions and $\boldsymbol{H}_u$ contains the second derivatives of the shape functions; the subscript u and θ refers to the mechanical an electrical field, respectively. Note that the matrices $\boldsymbol{N}_i$, $i = u, \theta$, $\boldsymbol{B}_i$, $i = u, \theta$ and $\boldsymbol{H}_u$ contain the 'standard'.[10] Correspondingly, the vectors $\boldsymbol{U}$ and $\boldsymbol{\theta}$ contain also both, the standard degrees of freedom (DOFs) $\boldsymbol{u_i}$ and $\boldsymbol{\theta_i}$ as well as the DOFs of the enrichment $\boldsymbol{a}_I^{(N)}$ and $\boldsymbol{\alpha}_I^{(N)}$, respectively. In Eqs. (3.244)–(3.246), the subscript, e, in Ω_e, Γ_{t_e} and Γ_{D_e} denotes the eth element where $\Omega = \cup_e \Omega_e$.

3.7. XFEM for inverse analysis and topology optimization

3.7.1 Inverse problem

A direct problem is one in which the input and system parameters are known while output of the model is to be determined. A reconstruction problem is one in which the system parameters and output are known and input that has led to this output is to be determined. In an identification problem, the input and output are known and the system parameters which

[10] Related to the first term on the RHS of Eqs. (3.241) and (3.242) as well as the 'enriched' parts.

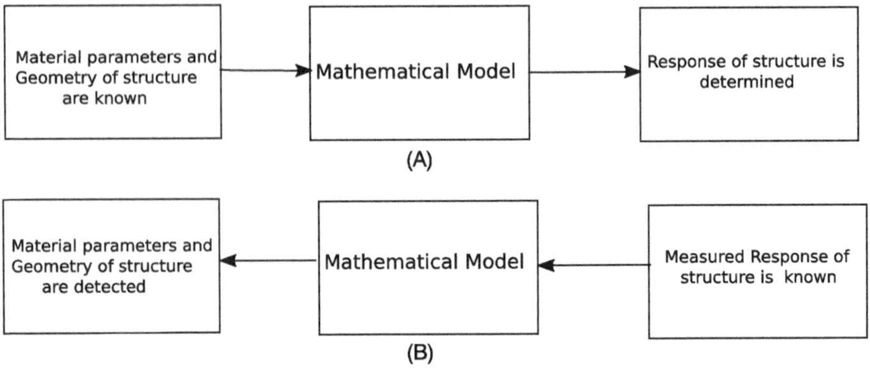

Figure 3.31 (A) Forward problem, (B) Inverse problem.

are in agreement with the relation between input and output is to be determined. The direct problem is a forward problem, in which the cause is known and effect is determined. The reconstruction and identification problems are called inverse problems because they involve finding out unknown causes of known consequences. The difference between forward and inverse problem in the context of structural mechanics is as shown in Fig. 3.31. The following list of inverse problems gives a good impression of the wide variety of applications [5]:

- the inverse problem of geomagnetic induction,
- X-ray tomography, ultrasound tomography, laser tomography,
- acoustic scattering, scattering in quantum mechanics,
- radio-astronomical imaging, image analysis,
- locating cracks or mines by electrical prospecting,
- seismic exploration, seismic tomography,
- the use of electrocardiography and magneto-cardiography,
- evolution backwards in time, inverse heat conduction,
- the inverse problem of potential theory,
- "can you hear the shape of a drum/manifold?"
- deconvolution, reconstruction of truncated signals,
- compartmental analysis, parameter identification,
- data assimilation,
- determining the volatility in models for financial markets,
- discrete tomography, shape from probing.

The French mathematician Jacques Hadamard in 1902 introduced the term well-posed for a mathematical problem. A well-posed problem has these features:

- A solution always exists.
- The solution is unique.
- A small change in the initial conditions leads to a small change in the solution (i.e.) the behavior of solution changes continuously with the initial conditions.

The opposite of a well-posed problem is an ill-posed problem, where:

- A solution may not exist.
- There may be more than one solution.
- A small change in the initial conditions leads to a big change in the solution.

Inverse problems are usually ill posed. The inverse problems described and solved in this book are to detect the following flaws in a piezoelectric domain:

- Detection of an edge or an interior crack.
- Detection of an elliptical void.
- Detection of an irregular shaped void.
- Detection of multiple voids.
- Detection of multiple cracks.
- Detection of location of both cracks and voids.
- Detection of interface of multiple inclusions, all of them made of same material.
- Detection of interface of three inclusions, two of them made of material A and the rest made of material B.

The ill-posedness of these inverse problems is negotiated by increasing the number of experimental setups from which the measurement or target data is obtained. The measured data may not be exact and may contain noise. The presence of noise in target data leads to poor performance of the detection algorithm. Regularization of inverse problem is performed in order to overcome the influence of noise and obtain a better detection quality. The commonly adopted regularization methods are:

- Tikhonov regularization.
- Truncated Singular Value Decomposition (TSVD).
- Total Variation (TV) regularization.

The total variation regularization is widely adopted in solving inverse problems using level set based schemes. Total variation is a regularization technique that does takes in to consideration that the data set is discontinuous. Most of the regularization methods assume the data sets to be smooth and continuous, but total variation does not assume the same. The common

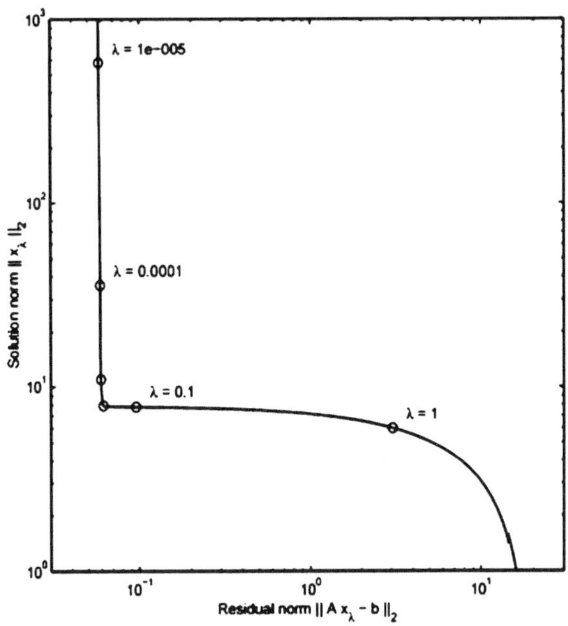

Figure 3.32 A standard L-Curve [24].

form of total variation regularization is as shown in the following equation,

$$T = \frac{1}{2}||Ax_\beta - b|| + \beta \int_\Omega \sqrt{|\nabla x_\beta|^2} d\Omega \tag{3.247}$$

The second term containing the total variation norm of the solution x_β regularizes the inverse problem and improves the quality of the solution. The term, β in the above equation is termed as regularization parameter. A value of this regularization parameter beyond a tolerance smoothens the solution while a value lesser than the tolerance leads to lack of regularization. The most popular method to determine the suitable value of regularization parameter β is the L-curve method. The L-curve is a log-log plot between the squared norm of the regularized solution and the squared norm of the regularized residual for a range of values of regularization parameter. A standard L curve for determining the regularization parameter is shown in Fig. 3.32 obtained from Hansen et al. [24]. The regularization parameter is denoted by λ in the figure.

Let us consider the specific task of identifying the size and location of cracks, voids and inclusions in a piezoelectric material. Such an inverse problem is commonly modified into an iterative optimization problem which in turn requires the solution of many forward problems for different void/inclusion configurations until convergence is obtained; meaning until the error between the measured and computational data reaches a certain minimum. The forward problem is often solved with the finite element method which requires remeshing of the domain in each iteration as the crack or void configuration varies with iterations. XFEM avoids remeshing and the forward problem can be solved on a fixed Eulerian mesh which does not change with iterations. Only the portion of the mesh affected by the voids or inclusions need to be 'adjusted'. Note that such a procedure can also be accomplished without XFEM by using a triangulation procedure without any enrichment. However, when the voids and inclusions are located inside an element, optimal convergence is lost in the forward problem. The way to describe the interfaces between the void/inclusion and the matrix material can be classified in two categories: (1) Smeared interface models and (2) sharp interface models. In the latter category one can further distinguish between explicit and implicit interface representations. Explicit interface representations align the discretization to the interface while implicit representations allow the interfaces being located inside an element for instance. One very powerful and popular implicit method is the level set method which fits well in the context of XFEM. While smeared approaches such as the Solid Isotropic Material with Penalization (SIMP) technique have been shown powerful in many applications, they have shown stability problems in multi-field and multi-constraint problems. Furthermore, in multiphysics problems the different sets of penalization parameters will directly and noticeably impact the final results in terms of the stability of the solution and the distinct void-solid representation. Subsequently, we present an efficient inverse analysis for a specific multi-field problem (piezoelectricity) which is based on XFEM and level sets.

3.7.1.1 Void and inclusion detection in piezoelectric materials

The interface of inclusions which is represented implicitly by the level set function, Φ is to be determined using the responses measured on the boundary of the piezoelectric structure. Mechanical displacements and electric potentials are the measurements taken on the boundary. The inverse problem is solved iteratively as an optimization problem with the following objective function,

$$J = \frac{1}{2}\int_{\Gamma} |\chi_i^{EXP} - \chi_i^{NUM}(\Phi)|^2 d\Gamma + \beta R(r) \tag{3.248}$$

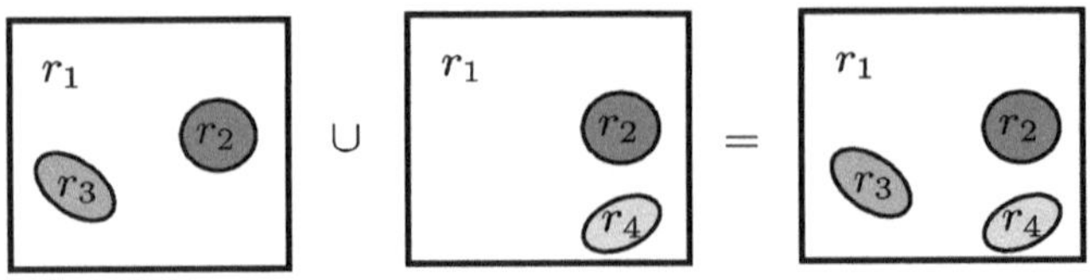

Figure 3.33 Multiple level sets representation: The union of two level set functions, ϕ_1 and ϕ_2, gives the actual domain with inclusions.

The level set function which minimizes the L^2 norm of the difference between numerical, χ^{NUM} and experimental responses, χ^{EXP} gives the actual inclusion interface. In the total variation norm of r, the material ratio is taken as the regularization, $R(r) = \int_\Omega |\nabla r| d\Omega$. The iterative methods of solving inverse problem are expensive and time consuming as the direct problem is solved at each step. Numerical method like FEM requires remeshing in each iteration so that the element edges align with the updated inclusion interface. In contrast, the XFEM offers the advantage of maintaining a fixed mesh irrespective of the varying interface in each iteration. Therefore in XFEM only the stiffness coefficients corresponding to enriched DOFS and the DOFS of those nodes which lie within the inclusions vary in each iteration whereas the FEM portion of global stiffness matrix, which comprises the bulk of matrix, remains the same.

The interface of inclusions can be implicitly represented by using the level sets, see Section 1.4. The level set function Φ can be in turn used to construct a function, r which can represent distinct material properties in different sub-domains. If the ratio of material constants with respect to the background material is r_1 in the subdomain with negative level set values and if the ratio of material constants is r_2 in the subdomain with positive level set values, then the function r can be written as,

$$r = \frac{1}{2}[r_1(1 + sign(\Phi)) + r_2(1 - sign(\Phi))]. \tag{3.249}$$

Multiple level sets can be used to represent more than two sub-domains as shown in Fig. 3.33. For example, the function r for representing four regions each of different material ratios can be written as,

$$r = \frac{1}{4}[r_1(1 + S_1)(1 + S_2) + r_2(1 + S_1)(1 - S_2) + r_3(1 - S_1)(1 + S_2) + r_4(1 - S_1)(1 - S_2)] \tag{3.250}$$

S_1 and S_2 correspond to the sign of the level set functions Φ_1 and Φ_2 respectively. r_1, r_2, r_3 and r_4 are material ratios in the four regions. The ratios can also be replaced directly by the material tensors of the corresponding sub regions.

The evolution of level set functions and the detection of different material regions can then be obtained by the solution of the stabilized Hamilton-Jacobi equation, Eq. (1.7) in Section 1.4. Determining Φ by solving Hamilton-Jacobi equation is equivalent to moving the level set isolines in the descent direction. Note that the Hamilton-Jacobi equation is posed not only along the interface but on the entire domain. The velocity V is related to the sensitivity of the objective function to variation in the material properties over the domain. The change in the objective function due to perturbation of inclusion interface is given by $\frac{\partial J}{\partial r}$. The derivative $\frac{\partial J}{\partial r}$ is commonly obtained by solving an adjoint problem.

3.7.1.2 Measurement techniques

The mechanical displacement in piezoelectric ceramics can be measured using fiber optics probes, laser interferometry [9] and capacitance gauge. These methods are capable of measuring displacement of less than 1 nm. In Burianova et al. [9], the three-dimensional deformed configuration of a bulk ceramic sample subjected an AC driving field is obtained using laser interferometry. The boundary displacements data required can be obtained by adopting these experimental methods. Similarly there are works [36] that present ways of measuring electric potential distribution on the surface of a piezoelectric ceramic.

3.7.1.3 Selection of the regularization parameter β

The regularization parameter can be determined by a variant of the L-curve criterion. Among the methods available in literature like discrepancy principle, Generalized Cross Validation (GCV), Unbiased Predictive Risk Estimator (UPRE) and so on [63], the L-curve criterion and GCV offer the advantage of not requiring prior knowledge about noise level.

An L-curve [24] is a log-log plot of residual norm and solution norm. There are two flat regions in the curve representing under regularization and over regularization. The corner point which corresponds to the transition between the two regions and the associated value of β gives the optimal regularization parameter.

New variants of the L-curve criterion are proposed in literature like the residual L-curve [51] and solution L-curve [52]. The residual L-curve is given by plot between regularization parameter and regularized residual norm.

3.7.1.4 The forward and adjoint problem

In [3], the adjoint problem for a general elliptic partial differential equation is defined, which is here extended for a piezoelectric case. The piezoelectric problem has already been stated in Section 3.6.3.1 and we therefore summarize the weak form given by

$$\begin{aligned}\int_{\Omega}(\boldsymbol{\epsilon}(\boldsymbol{u})^{T}:\boldsymbol{C}:\boldsymbol{\epsilon}(\boldsymbol{w})-\boldsymbol{\epsilon}(\boldsymbol{u})^{T}:\boldsymbol{e}:\boldsymbol{E}(\boldsymbol{\psi})\\ -\boldsymbol{E}(\boldsymbol{\phi})^{T}:\boldsymbol{e}:\boldsymbol{\epsilon}(\boldsymbol{w})-\boldsymbol{E}(\boldsymbol{\phi})^{T}:\boldsymbol{\kappa}:\boldsymbol{E}(\boldsymbol{\psi}))\,d\Omega\\ -\int_{\Gamma_N}\boldsymbol{p}.\,\boldsymbol{w}\,d\Gamma-\int_{\Gamma_N}\boldsymbol{q}.\,\boldsymbol{\psi}\,d\Gamma=0\end{aligned} \tag{3.251}$$

The objective function (without regularization term) to be minimized is,

$$J(\mathbf{C}(x),\boldsymbol{e}(x),\boldsymbol{\kappa}(x))=\int_{\Gamma^m}\frac{1}{2}(u-u^{meas})^2d\Gamma+\int_{\Gamma^m}\frac{1}{2}(\phi-\phi^{meas})^2d\Gamma \tag{3.252}$$

Let, $J_1=\frac{1}{2}(u-u^{meas})^2$ and $J_2=\frac{1}{2}(\phi-\phi^{meas})^2$. The material distribution in Ω is given by, $\mathbf{C}(x)=r(x)\mathbf{C_b}$, $\boldsymbol{\kappa}(x)=r(x)\boldsymbol{\kappa_b}$, $\mathbf{e}(x)=r(x)\mathbf{e_b}$. For the sake of simplicity it is assumed that the displacements and electrical potentials are measured along the same boundary, Γ^m.

The first variation of $J(\boldsymbol{C}(\boldsymbol{x}),\boldsymbol{e}(\boldsymbol{x}),\boldsymbol{\kappa}(\boldsymbol{x}))$ is given by,

$$\delta J(\mathbf{C}(x),\boldsymbol{e}(x),\boldsymbol{\kappa}(x))=\int_{\Gamma^m}\frac{\partial J_1}{\partial u}\delta u\,d\Gamma+\int_{\Gamma^m}\frac{\partial J_2}{\partial\phi}\delta\phi\,d\Gamma \tag{3.253}$$

The governing problem for the latter can be determined by differentiating the weak formulation, Eq. (3.251) as shown below,

$$\begin{aligned}\int_{\Omega}(\boldsymbol{\epsilon}(\boldsymbol{\delta u})^{T}:\boldsymbol{C}:\boldsymbol{\epsilon}(\boldsymbol{w})+\boldsymbol{\epsilon}(\boldsymbol{u})^{T}:\boldsymbol{\delta C}:\boldsymbol{\epsilon}(\boldsymbol{w})-\boldsymbol{\epsilon}(\boldsymbol{\delta u})^{T}:\boldsymbol{e}:\boldsymbol{E}(\boldsymbol{\psi})\\ -\boldsymbol{\epsilon}(\boldsymbol{u})^{T}:\boldsymbol{\delta e}:\boldsymbol{E}(\boldsymbol{\psi})-\boldsymbol{E}(\boldsymbol{\delta\phi})^{T}:\boldsymbol{e}:\boldsymbol{\epsilon}(\boldsymbol{w})-\boldsymbol{E}(\boldsymbol{\phi})^{T}:\boldsymbol{\delta e}:\boldsymbol{\epsilon}(\boldsymbol{w})\\ -\boldsymbol{E}(\boldsymbol{\delta\phi^{e}})^{T}:\boldsymbol{\kappa}:\boldsymbol{E}(\boldsymbol{\psi})-\boldsymbol{E}(\boldsymbol{\phi})^{T}:\boldsymbol{\delta\kappa}:\boldsymbol{E}(\boldsymbol{\psi}))\,d\Omega=0\end{aligned} \tag{3.254}$$

Considering $\mathbf{w}$ and $\boldsymbol{\psi}$ as the trial functions, the virtual work principle with $\boldsymbol{\delta}\mathbf{u}$ and $\boldsymbol{\delta\phi}$ as the virtual fields yields,

$$\int_{\Omega}(\boldsymbol{\epsilon}(\boldsymbol{\delta u})^{T}:\boldsymbol{C}:\boldsymbol{\epsilon}(\boldsymbol{w})-\boldsymbol{\epsilon}(\boldsymbol{\delta u})^{T}:\boldsymbol{e}:\boldsymbol{E}(\boldsymbol{\psi})-\boldsymbol{E}(\boldsymbol{\delta\phi})^{T}:\boldsymbol{e}:\boldsymbol{\epsilon}(\boldsymbol{w})$$

$$- \boldsymbol{E}(\delta\boldsymbol{\phi})^T : \boldsymbol{\kappa} : \boldsymbol{E}(\boldsymbol{\psi})) \, d\Omega - \int_{\Gamma_N} \boldsymbol{p}. \delta\boldsymbol{u} \, d\Gamma - \int_{\Gamma_N} \boldsymbol{q}. \delta\boldsymbol{\phi} \, d\Gamma = 0 \quad (3.255)$$

having set $p = \boldsymbol{\sigma}(\mathbf{w}, \boldsymbol{\psi}).\mathbf{n} = \frac{\partial J_1}{\partial u}$ and $q = \mathbf{D}(\mathbf{w}, \boldsymbol{\psi}).\mathbf{n} = \frac{\partial J_2}{\partial \phi}$.

Comparing Eqs. (3.254) and (3.255) we have,

$$\begin{aligned} &- \int_{\Omega} (\boldsymbol{\epsilon}(\boldsymbol{u})^T : \delta\boldsymbol{C} : \boldsymbol{\epsilon}(\boldsymbol{w}) - \boldsymbol{\epsilon}(\boldsymbol{u})^T : \delta\boldsymbol{e} : \boldsymbol{E}(\boldsymbol{\psi}) - \boldsymbol{E}(\boldsymbol{\phi})^T : \delta\boldsymbol{e} : \boldsymbol{\epsilon}(\boldsymbol{w}) \\ &- \boldsymbol{E}(\boldsymbol{\phi})^T : \delta\boldsymbol{\kappa} : \boldsymbol{E}(\boldsymbol{\psi}) d\Omega = \int_{\Gamma_N} \boldsymbol{p}. \delta\boldsymbol{u} \, d\Gamma + \int_{\Gamma_N} \boldsymbol{q}. \delta\boldsymbol{\phi} \, d\Gamma. \end{aligned} \quad (3.256)$$

Comparing the above identity with Eq. (3.253), the variation of $J(\boldsymbol{C}(\boldsymbol{x}), \boldsymbol{e}(\boldsymbol{x}), \boldsymbol{\kappa}(\boldsymbol{x}))$ is,

$$\begin{aligned} \delta J(\boldsymbol{C}(x), \boldsymbol{e}(x), \boldsymbol{\kappa}(x)) &= \boldsymbol{\epsilon}(\boldsymbol{u})^T : \delta\boldsymbol{C} : \boldsymbol{\epsilon}(\boldsymbol{w}) - \boldsymbol{\epsilon}(\boldsymbol{u})^T : \delta\boldsymbol{e} : \boldsymbol{E}(\boldsymbol{\psi}) \\ &- \boldsymbol{E}(\boldsymbol{\phi})^T : \delta\boldsymbol{e} : \boldsymbol{\epsilon}(\boldsymbol{w}) - \boldsymbol{E}(\boldsymbol{\phi})^T : \delta\boldsymbol{\kappa} : \boldsymbol{E}(\boldsymbol{\psi}) \end{aligned} \quad (3.257)$$

where $\mathbf{w}$ and $\boldsymbol{\psi}$ are solutions of the adjoint problem. The governing equations for the adjoint problem can be obtained from Eq. (3.255) as follows,

$$\begin{aligned} &- \int_{\Omega} \delta\boldsymbol{u}^T \, \nabla_u \cdot (\boldsymbol{C} : \boldsymbol{\epsilon}(\boldsymbol{w})) \, d\Omega + \int_{\Gamma} \delta\boldsymbol{u}^T \, \boldsymbol{C} : \boldsymbol{\epsilon}(\boldsymbol{w}) \, \boldsymbol{n} \, d\Gamma \\ &+ \int_{\Omega} \delta\boldsymbol{u}^T \, \nabla_e(\boldsymbol{e} : \boldsymbol{E}(\boldsymbol{\psi})) \, d\Omega - \int_{\Gamma} \delta\boldsymbol{u}^T \, \boldsymbol{e} : \boldsymbol{E}(\boldsymbol{\psi}) \, \boldsymbol{n} \, d\Gamma \\ &+ \int_{\Omega} \delta\boldsymbol{\phi}^T \, \nabla_u(\boldsymbol{e} : \boldsymbol{\epsilon}(\boldsymbol{w})) \, d\Omega - \int_{\Gamma} \delta\boldsymbol{\phi}^T \, \boldsymbol{e} : \boldsymbol{\epsilon}(\boldsymbol{w}) \, \boldsymbol{n} \, d\Gamma \\ &+ \int_{\Omega} \delta\boldsymbol{\phi}^{eT} \, \nabla_e(\boldsymbol{\kappa} : \, \boldsymbol{E}(\boldsymbol{\psi})) \, d\Omega - \int_{\Gamma} \delta\boldsymbol{\phi}^T \, \boldsymbol{\kappa} : \, \boldsymbol{E}(\boldsymbol{\psi}) \, \boldsymbol{n} \, d\Gamma \\ &- \int_{\Gamma_N} \boldsymbol{p}. \delta\boldsymbol{u} \, d\Gamma - \int_{\Gamma_N} \boldsymbol{q}. \delta\boldsymbol{\phi} \, d\Gamma = 0. \end{aligned} \quad (3.258)$$

Imposing $\int_{\Gamma_D} \delta\boldsymbol{u}^T \, (\boldsymbol{C} : \boldsymbol{\epsilon}(\boldsymbol{w}) - \boldsymbol{e} : \boldsymbol{E}(\boldsymbol{\psi})) \cdot \boldsymbol{n} \, d\Gamma = 0$ and $\int_{\Gamma_D} \delta\boldsymbol{\phi}^T \, (\boldsymbol{\kappa} : \, \boldsymbol{E}(\boldsymbol{\psi}) + \boldsymbol{e} : \boldsymbol{\epsilon}(\boldsymbol{w})) \cdot \boldsymbol{n} \, d\Gamma = 0$, leads to,

$$\begin{aligned} &\int_{\Omega} \delta\boldsymbol{u}^T \, (\nabla_u(-\boldsymbol{C} : \boldsymbol{\epsilon}(\boldsymbol{w}) + \boldsymbol{e}^T : \boldsymbol{E}(\boldsymbol{\psi})) \, d\Omega \\ &+ \int_{\Gamma_N} \delta\boldsymbol{u}^T \, ((\boldsymbol{C} : \boldsymbol{\epsilon}(\boldsymbol{w}) - \boldsymbol{e} : \boldsymbol{E}(\boldsymbol{\psi})) \cdot \boldsymbol{n} - \boldsymbol{p}) \, d\Gamma \\ &+ \int_{\Omega} \delta\boldsymbol{\phi}^T \, (\nabla_e(\boldsymbol{\kappa} : \, \boldsymbol{E}(\boldsymbol{\psi}) + \boldsymbol{e} : \boldsymbol{\epsilon}(\boldsymbol{w})) \, d\Omega \\ &+ \int_{\Gamma_N} \delta\boldsymbol{\phi}^T \, ((\boldsymbol{\kappa} : \, \boldsymbol{E}(\boldsymbol{\psi}) + \boldsymbol{e} : \boldsymbol{\epsilon}(\boldsymbol{w})) \cdot \boldsymbol{n} - \boldsymbol{q}) \, d\Gamma = 0. \end{aligned} \quad (3.259)$$

The strong form of the adjoint problem is,

$$\begin{aligned}\nabla_u \cdot (\boldsymbol{C} : \boldsymbol{\epsilon}(\boldsymbol{w}) - \boldsymbol{e}^T : \boldsymbol{E}(\boldsymbol{\psi})) = 0 \ \text{ in } \ \Omega \\ \nabla_e(\boldsymbol{\kappa} : \boldsymbol{E}(\boldsymbol{\psi}) + \boldsymbol{e} : \boldsymbol{\epsilon}(\boldsymbol{w})) = 0 \ \text{ in } \ \Omega \\ \boldsymbol{w} = 0 \ \text{ and } \ \boldsymbol{\psi} = 0 \ \text{ on } \ \Gamma_D \\ \boldsymbol{\sigma}(\boldsymbol{w}, \boldsymbol{\psi}).\boldsymbol{n} = \frac{\partial J_1}{\partial u} \ \text{ and } \ \boldsymbol{D}(\boldsymbol{w}, \boldsymbol{\psi}).\boldsymbol{n} = \frac{\partial J_2}{\partial \phi} \ \text{ on } \ \Gamma_N\end{aligned} \tag{3.260}$$

where, $\mathbf{C}(x) = r(x)\mathbf{C_b}$, $\boldsymbol{\kappa}(x) = r(x)\boldsymbol{\kappa}_\mathbf{b}$, $\mathbf{e}(x) = r(x)\mathbf{e_b}$. This is the simplest case in which the ratio remains the same for $\mathbf{C}$, $\boldsymbol{\kappa}$ and $\mathbf{e}$. $\mathbf{C_b}$, $\boldsymbol{\kappa}_\mathbf{b}$ and $\mathbf{e_b}$ correspond to elastic, permittivity and piezoelectric constants of the background material (PZT-4). From Eq. (3.257), the derivative of objective function with respect to material ratio can be written as,

$$\begin{aligned}\frac{\partial J}{\partial r} &= \boldsymbol{\epsilon}(\boldsymbol{u}) : \frac{\partial \boldsymbol{C}}{\partial r} : \boldsymbol{\epsilon}(\boldsymbol{w}) - \boldsymbol{\epsilon}(\boldsymbol{u}) : \frac{\partial \boldsymbol{e}}{\partial r} : \boldsymbol{E}(\boldsymbol{\psi}) - \boldsymbol{E}(\boldsymbol{\phi}) : \frac{\partial \boldsymbol{e}}{\partial r} : \boldsymbol{\epsilon}(\boldsymbol{w}) \\ &\quad - \boldsymbol{E}(\boldsymbol{\phi}) : \frac{\partial \boldsymbol{\kappa}}{\partial r} : \boldsymbol{E}(\boldsymbol{\psi}) \\ &= \boldsymbol{\epsilon}(\boldsymbol{u}) : \boldsymbol{C}_b : \boldsymbol{\epsilon}(\boldsymbol{w}) - \boldsymbol{\epsilon}(\boldsymbol{u}) : \boldsymbol{e}_b : \boldsymbol{E}(\boldsymbol{\psi}) - \boldsymbol{E}(\boldsymbol{\phi}) : \boldsymbol{e}_b : \boldsymbol{\epsilon}(\boldsymbol{w}) \\ &\quad - \boldsymbol{E}(\boldsymbol{\phi}) : \boldsymbol{\kappa}_b : \boldsymbol{E}(\boldsymbol{\psi}).\end{aligned} \tag{3.261}$$

The extended velocity of the zero level set at node i, V_i is as shown below,

$$-V_i = \left(\frac{\partial J}{\partial \Phi}\right)_i \tag{3.262}$$

$$\frac{\partial J}{\partial \Phi} = \left(\frac{\partial J}{\partial r} + \beta \frac{\partial R}{\partial r}\right) \cdot \left(\frac{\partial r}{\partial \Phi}\right) \tag{3.263}$$

$$\frac{\partial R}{\partial r} = \nabla \cdot \left(\frac{\nabla r}{|\nabla r|}\right) \tag{3.264}$$

The Hamilton-Jacobi equations usually do not admit smooth solutions. Existence and uniqueness are achieved in the framework of viscosity solutions which help in convenient definition of generalized shape motion. The discrete solution of H-J equation is obtained by an explicit first-order upwind scheme [2].

The steps involved in this inclusion interface detection algorithm are as follows:

- Initialization of level set function Φ_0. In order to avoid local optima, circular inclusions are distributed all over the domain. In the case of

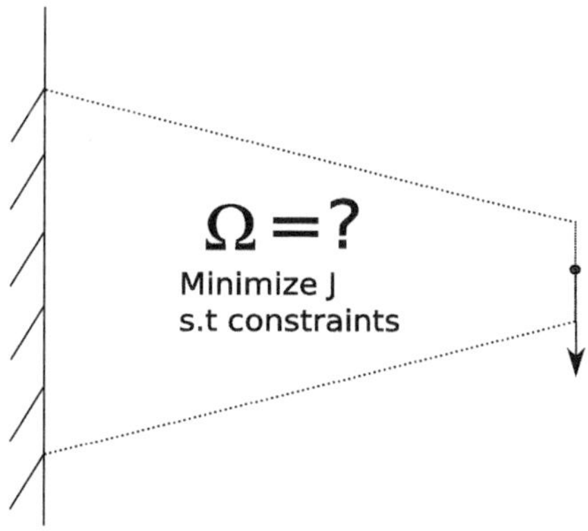

Figure 3.34 Structural optimization – Geometry of domain Ω to be determined such that objective function, *J* is minimized and the constraints are satisfied.

multiple level sets, this is the initial assumption for all the level set functions. In the case of three-dimensional structures, spherical inclusions are assumed to be distributed throughout the domain.

- In each iteration, n the actual state u_n and adjoint state p_n are determined by performing XFEM analysis. This involves solving Eqs. (3.251) and (3.260) for actual and adjoint states respectively. The shape derivative is obtained from Eq. (3.263).
- The updated inclusion interface is given by the level set function Φ_{n+1} obtained by solving the H-J Eq. (3.302) using upwind finite difference scheme, with time step Δt_n and velocity V_n, starting from the initial interface Φ_n.
- In case of multiple level sets, each of them is updated by solving H-J equation with its corresponding velocities. For example, in case of two level sets formulation expressed in Eq. (3.250), $(V_n)_1 = \frac{\partial J}{\partial r} \cdot \frac{\partial r}{\partial \Phi_1}$ and $(V_n)_2 = \frac{\partial J}{\partial r} \cdot \frac{\partial r}{\partial \Phi_2}$.
- The algorithm is regarded as converged when the gradient of the objective function is less than a fixed tolerance, which can also be seen from no significant change in the geometry of the inclusion interface between successive iterations.

For detailed numerical examples involving detection of material interfaces in piezoelectric structures we refer to [42].

3.7.2 Optimization problems

The structural optimization is the subject of making an assemblage of materials sustain loads in the best way. As shown in Fig. 3.34, the process of determining the shape, size or topology of the domain in question, Ω such that the minimum weight or minimum compliance or any other objective of this sort is achieved and the constraints imposed are satisfied. Based on the geometric feature which is to be optimized, structural optimization problems can be classified as,

- Sizing optimization.
- Shape optimization.
- Topology optimization.

Sizing optimization: Especially in case of trusses, the cross-sectional area of the truss members can be varied so that the intended objective function is minimized.

Shape optimization: In this case, some parts of the boundary of the structural domain is optimized to minimize the objective function. Considering a solid body, the behavior of which is described by a set of partial differential equations, shape optimization consists in choosing the domain for integrating the differential equations in an optimal manner. The important aspect of this type of optimization is that connectivity of the structure is not changed during the optimization process. The shape of the boundary is modified but new surfaces cannot be formed in this optimization.

Topology optimization: The most general form of structural optimization is topology optimization. For example, in case of a truss, when the cross-sectional areas of truss members is considered as the design variable, a bar is removed from the truss when the value of the design variable corresponding to this truss member is set as zero. During the optimization process, the connectivity of nodes is varied such that the topology of the truss gets modified. On the other hand if a discrete structure like truss is replaced by a continuum structure like a beam or plate, then topology changes can be achieved by letting the density of material to zero in certain subdomains of the structure. The number of subdomains in which density can be reduced to zero is restricted by a volume or a weight constraint. Ideally, shape optimization is a subclass of topology optimization, but practical implementations are based on very different techniques, so the two types are usually treated separately. When the state problem is a differential equation, we can say that shape optimization concerns control of domain of the equation, while sizing and topology optimization concern control of its parameters.

3.7.3 Mathematical form of a structural optimization problem

The function and variables that are basic components of a structural optimization problem are:

- Objective function (J): The function which is to be minimized. In most of the optimization problems the objective function remains as the measure of quality or efficiency of the design. Generally J is such that a small value is better than a large one (i.e.) a minimization problem. In literature and in practical applications the most commonly used objective functions are minimum weight, minimum compliance or minimum least square error compared to a target value.
- Design variable (Ω): The objective function varies in each iteration with change in the value of design variable. In the optimization problems solved in this book, the geometry of the domain, Ω is taken as the design variable. The geometry of the domain has to be defined in terms of parameters. Parametrization may be either explicit or implicit. Explicit parametrization corresponds to parametrizing the geometry of the design domain in terms of polynomials or splines. Implicit parametrization corresponds to level set representation of the geometry.
- State variable (U): The vector U corresponds to the response of the structure which are required to determine the value of objective function in each iteration. For an electro-mechanical structure, response means mechanical displacement, stress, strain, electric displacement, field or potential.

3.7.4 Solid isotropic material with penalization (SIMP)

The principle idea is an ersatz material approach where the pseudo density parameter is applied to each cell of the finite element mesh. By varying the parameters arbitrary structures can be modeled on a fixed finite element discretization. The SIMP method is very efficient in solving the resulting optimization problems which typically comprise several design variables. In solid isotropic material with penalization (SIMP) the intermediate designs are penalized by using the following constitutive matrix in Hooke's law,

$$D = \frac{\rho^p E}{1-\nu^2}\begin{pmatrix} 1 & \nu & 0 \\ \nu & 1 & 0 \\ 0 & 0 & \frac{1-\nu}{2} \end{pmatrix} \tag{3.265}$$

SIMP provides large regions with $\rho = 0$ or $\rho = 1$. When the value of ρ is zero, there is no material in the subdomain while when $\rho = 1$, then effective Young's modulus is E.

3.7.5 Level set based optimization

Level set methods first devised by Osher and Sethian [44] have become popular recently for tracking, modeling and simulating the motion of dynamic interfaces (moving free boundaries) and they are naturally coupled in the framework of XFEM. In the optimization problem, the interface is closed, nonintersecting and Lipschitz-continuous and represented implicitly through a Lipschitz-continuous level set function $\Phi(x)$, and the interface itself is the zero isosurface or zero level set $\{x \in R^d | \Phi(x) = 0\}$ (d = 2 or 3). The level set function may be utilized to define different regions in a domain as,

$$\begin{aligned} \text{Solid}: \quad & \boldsymbol{\Phi}(\mathbf{x}) < 0 \quad \forall \mathbf{x} \in \boldsymbol{\Omega} \setminus \partial \boldsymbol{\Omega} \\ \text{Boundary}: \quad & \boldsymbol{\Phi}(\mathbf{x}) = 0 \quad \forall \mathbf{x} \in \partial \boldsymbol{\Omega} \cap \mathbf{D} \\ \text{Void}: \quad & \boldsymbol{\Phi}(\mathbf{x}) > 0 \quad \forall \mathbf{x} \in \mathbf{D} \setminus \boldsymbol{\Omega} \end{aligned} \tag{3.266}$$

In the conventional level set methods, the Hamilton-Jacobi PDE is solved to evolve the interface using an Eulerian approach. The solving procedure requires appropriate choice of the upwind schemes, reinitialization algorithms and extension velocity methods. Fig. 3.35 shows the optimal topology obtained for minimum compliance problem for a short cantilever beam of size 32 × 20 units, subjected to point load at the bottom of the free end, obtained both using SIMP and level set based optimization. Subsequently, we will describe a level set based optimization method for two problems: Nanoelasticity and nanopiezoelectricity.

3.7.6 Nanoelasticity

The commonly synthesized structures in the order of a nanometer are nanotubes, oxide nanobelts and semiconductor nanowires. Though these materials have technological potential in applications such as nanoelectronic and photonic circuits, nano-sensors and electromechanical nanodevices, these applications require knowledge about the mechanical behavior of the nano materials. For example, application of piezoelectric nano wire in a nano generator requires understanding about their elastic, electric and coupling piezoelectric behavior. As a first step as the energy conversion of a nano wire depends on its mechanical strain, the study of the elastic properties of nanostructures is essential. The study of behavior of nano elastic structures is challenging both from an experimental and theoretical point of view. In the experimental side, the difficulties include the lack of reliable methods to quantitatively measure the elasticity and sometimes the friction

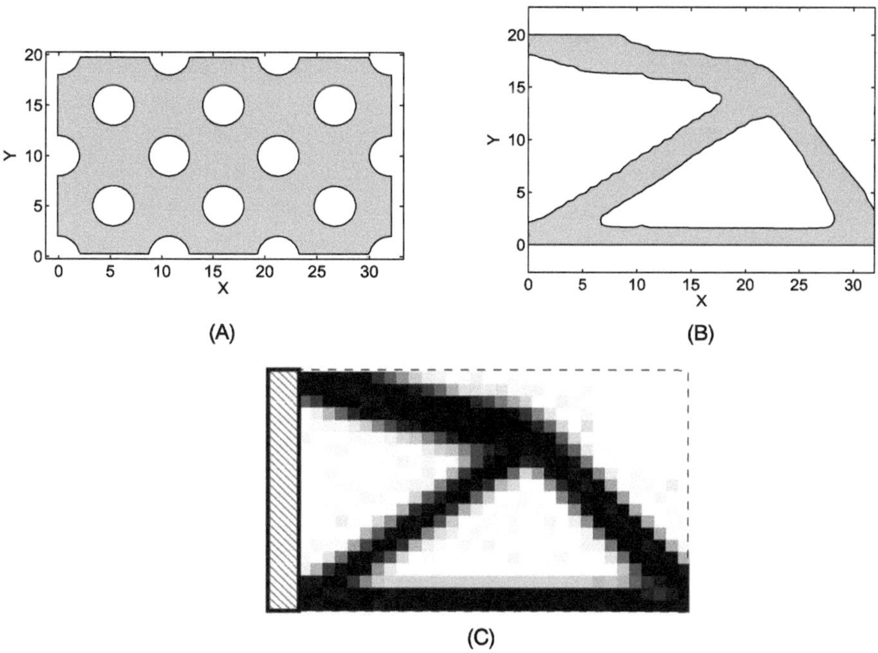

Figure 3.35 (A) Initialization, (B) Optimum topology for a short cantilever beam subjected to a point load at free end by Level set method, (C) by SIMP [55].

at the nanoscale. The problems are related to spatial and force resolution, instrument calibration as well as not well defined surface shape and chemistry because at this scale each atom makes a difference. From the theoretical side, developing a theory of elasticity at the nanoscale is an intriguing theoretical challenge, which lies at the cross-over between the atomic level and the continuum.

The physical origin of the surface effects is that atoms at the surfaces of a material have fewer bonding neighbors than atoms that lie within the material bulk [11]. This so-called undercoordination of the surface atoms causes them to exhibit different elastic properties than atoms in the bulk, which can lead to either stiffening or softening of the nanostructure These unique mechanical properties have motivated researchers to develop computational approaches that capture these surface effects based on either linear or nonlinear continuum theories. It is critical to consider surface effects when discussing the mechanical behavior and properties of nanomaterials, particularly when any characteristic dimension of the nanostructure is smaller than about 100 nm [46].

The computational approaches are based on the well-known Gurtin-Murdoch linear surface elasticity theory [22], which considers the surface to be an entity of zero thickness that has its own elastic properties that are distinct from the bulk. Other approaches have considered a bulk plus surface ansatz of various forms incorporating finite deformation kinematics.

Let us consider an elastic solid $\boldsymbol{\Omega}$ with a material surface $\boldsymbol{\partial\Omega}$. According to continuum theory of elastic material surfaces [22], the equilibrium equations for a nanostructure can be written as:

$$\nabla \cdot \boldsymbol{\sigma} + \boldsymbol{b} = 0 \qquad \text{in } \boldsymbol{\Omega} \tag{3.267}$$

$$\nabla_s \cdot \boldsymbol{\sigma}_s + [\boldsymbol{\sigma} \cdot \mathbf{n}] = 0 \qquad \text{on } \boldsymbol{\Gamma} \tag{3.268}$$

where the first equation refers to bulk equilibrium and the second equation refers to the generalized Young-Laplace equation [22] resulting from mechanical equilibrium on the surface. In the above equations, $\boldsymbol{\sigma}$ represents bulk Cauchy stress tensor, $\boldsymbol{b}$ represents the body force vector, $\boldsymbol{\sigma}_s$ denotes the surface stress tensor, $\mathbf{n}$ is the outward unit normal vector to $\boldsymbol{\Gamma}$, and $\nabla_s \cdot \boldsymbol{\sigma}_s = \nabla \boldsymbol{\sigma}_s : \mathbf{P}$. Here $\mathbf{P}$ is the tangential projection tensor to $\boldsymbol{\Gamma}$ at $\mathbf{x} \in \boldsymbol{\Gamma}$ which is defined as $\mathbf{P}(\mathbf{x}) = \mathbf{I} - \mathbf{n}(\mathbf{x}) \otimes \mathbf{n}(\mathbf{x})$, $\mathbf{I}$ is the second order unit tensor. Γ is the boundary of the domain Ω. Furthermore, the boundary conditions are given by

$$\begin{aligned} \boldsymbol{\sigma} \cdot \mathbf{n} &= \bar{\mathbf{t}} \qquad \text{on } \boldsymbol{\Gamma}_N \\ \boldsymbol{u} &= \bar{\boldsymbol{u}} \qquad \text{on } \boldsymbol{\Gamma}_D \end{aligned} \tag{3.269}$$

where $\bar{\mathbf{t}}$ and $\bar{\boldsymbol{u}}$ are the prescribed traction and displacement, respectively, and $\boldsymbol{\Gamma}_N$ and $\boldsymbol{\Gamma}_D$ are the Neumann and Dirichlet boundaries. The bulk strain tensor $\boldsymbol{\epsilon}$ and surface strain tensor $\boldsymbol{\epsilon}_s$ are written as

$$\boldsymbol{\epsilon} = \frac{1}{2}(\nabla \mathbf{u} + (\nabla \mathbf{u})^T) \tag{3.270}$$

$$\boldsymbol{\epsilon}_s = \mathbf{P} \cdot \boldsymbol{\epsilon} \cdot \mathbf{P} \tag{3.271}$$

where $\mathbf{u}$ is the displacement vector. Assuming a linear elastic bulk material and an isotropic linear elastic surface, the constitutive equations for the bulk and surface can be written as

$$\boldsymbol{\sigma} = \mathbb{C}^{bulk} : \boldsymbol{\epsilon} \tag{3.272}$$

$$\boldsymbol{\sigma}_s = \frac{\partial \gamma}{\partial \boldsymbol{\epsilon}_s} \tag{3.273}$$

where γ is the surface energy density given by

$$\gamma = \gamma_0 + \boldsymbol{\tau}_s : \boldsymbol{\epsilon}_s + \frac{1}{2}\boldsymbol{\epsilon}_s : \mathbb{C}^s : \boldsymbol{\epsilon}_s \tag{3.274}$$

where γ_0 is the surface free energy density that exists even when $\boldsymbol{\epsilon}_s = 0$, and $\boldsymbol{\tau}_s = \tau_s \mathbf{P}$ is the surface residual stress tensor. By substituting Eq. (3.274) into Eq. (3.273), $\boldsymbol{\sigma}_s$ can be obtained by

$$\boldsymbol{\sigma}_s = \boldsymbol{\tau}_s + \mathbb{C}^s : \boldsymbol{\epsilon}_s \tag{3.275}$$

In the above equations, $\mathbb{C}^{bulk}$ and $\mathbb{C}^s$ are the fourth-order elastic stiffness tensors associated with the bulk and surface, respectively, and are defined as

$$\mathbb{C}^{bulk}_{ijkl} = \lambda \delta_{ij}\delta_{kl} + \mu(\delta_{ik}\delta_{jl} + \delta_{il}\delta_{jk}) \tag{3.276}$$

$$\mathbb{C}^{s}_{ijkl} = \lambda_s P_{ij}P_{kl} + \mu_s(P_{ik}P_{jl} + P_{il}P_{jk}) \tag{3.277}$$

where λ and μ, and λ_s and μ_s are the Lamé constants of the bulk and surface, respectively.

It should be noted that the surface is considered as a special case of a coherent imperfect interface between two materials when one of them exists in a vacuum phase [69]. It is assumed that the surface adheres to the bulk and therefore we have:

$$[\![\mathbf{u}]\!] = 0 \quad \text{on } \boldsymbol{\Gamma} \tag{3.278}$$

where $[\![.]\!]$ denotes the jump across the interface.

Having defined the constitutive and field equations, we derive the weak form of the boundary value problem based on the principle of stationary potential energy. The total potential energy Π of the system is given by

$$\Pi = \Pi_{bulk} + \Pi_s - \Pi_{ext} \tag{3.279}$$

where Π_{bulk}, Π_s, and Π_{ext} represent the bulk elastic strain energy, surface elastic energy and the work of external forces, respectively, which are given by

$$\Pi_{bulk} = \frac{1}{2}\int_{\Omega} \boldsymbol{\epsilon} : \mathbb{C}^{bulk} : \boldsymbol{\epsilon} \, d\Omega \tag{3.280}$$

$$\Pi_s = \int_{\Gamma} \gamma \, d\Gamma \tag{3.281}$$

$$\Pi_{ext} = \int_{\Gamma_N} \mathbf{u} \cdot \bar{\mathbf{t}}\, d\Gamma + \int_{\Omega} \mathbf{u} \cdot \mathbf{b}\, d\Omega \tag{3.282}$$

The stationary condition of Eq. (3.279) is given by

$$D_{\delta\mathbf{u}}\Pi = 0 \tag{3.283}$$

where $D_{\mathbf{m}}\Upsilon$ is the directional derivative (or Gâteaux derivative) of the functional Υ in the direction $\mathbf{m}$. Applying the stationary condition, the weak form of the equilibrium equations can be obtained by finding $\mathbf{u} \in \{\mathbf{u} = \bar{\mathbf{u}} \quad \text{on } \mathbf{\Gamma}_D, \mathbf{u} \in H^1(\Omega)\}$ such that

$$\begin{aligned} &\int_{\Omega} \boldsymbol{\epsilon}(\mathbf{u}) : \mathbb{C}^{bulk} : \boldsymbol{\epsilon}(\delta\mathbf{u})\, d\Omega + \int_{\Gamma} \boldsymbol{\epsilon}_s(\mathbf{u}) : \mathbb{C}^s : \boldsymbol{\epsilon}_s(\delta\mathbf{u})\, d\Gamma \\ &= -\int_{\Gamma} \boldsymbol{\tau}_s : \boldsymbol{\epsilon}_s(\delta\mathbf{u})\, d\Gamma + \int_{\Gamma_N} \delta\mathbf{u} \cdot \bar{\mathbf{t}}\, d\Gamma + \int_{\Omega} \delta\mathbf{u} \cdot \mathbf{b}\, d\Omega \end{aligned} \tag{3.284}$$

for all $\delta\mathbf{u} \in \{\delta\mathbf{u} = 0 \quad \text{on } \mathbf{\Gamma}_D, \delta\mathbf{u} \in H^1(\Omega)\}$. This weak form can be written in a simplified form as

$$a(\mathbf{u}, \delta\mathbf{u}) + a_s(\mathbf{u}, \delta\mathbf{u}) = -l_s(\delta\mathbf{u}) + l(\delta\mathbf{u}) \tag{3.285}$$

where the bilinear functionals $a(\mathbf{u}, \boldsymbol{\delta}\mathbf{u})$ and $a_s(\mathbf{u}, \boldsymbol{\delta}\mathbf{u})$, and linear functionals $l(\boldsymbol{\delta}\mathbf{u})$ and $l_s(\boldsymbol{\delta}\mathbf{u})$ are defined as

$$\begin{aligned} a(\mathbf{u}, \boldsymbol{\delta}\mathbf{u}) &= \int_{\Omega} c(\mathbf{u}, \boldsymbol{\delta}\mathbf{u})\, d\Omega = \int_{\Omega} \boldsymbol{\epsilon}(\mathbf{u}) : \mathbb{C}^{bulk} : \boldsymbol{\epsilon}(\boldsymbol{\delta}\mathbf{u})\, d\Omega \\ a_s(\mathbf{u}, \boldsymbol{\delta}\mathbf{u}) &= \int_{\partial\Omega} c_s(\mathbf{u}, \boldsymbol{\delta}\mathbf{u})\, d\Gamma = \int_{\partial\Omega} \boldsymbol{\epsilon}_s(\mathbf{u}) : \mathbb{C}^s : \boldsymbol{\epsilon}_s(\boldsymbol{\delta}\mathbf{u})\, d\Gamma \\ l(\boldsymbol{\delta}\mathbf{u}) &= \int_{\partial\Omega_N} \boldsymbol{\delta}\mathbf{u} \cdot \bar{\mathbf{t}}\, d\Gamma + \int_{\Omega} \boldsymbol{\delta}\mathbf{u} \cdot \mathbf{b}\, d\Omega \\ l_s(\boldsymbol{\delta}\mathbf{u}) &= \int_{\partial\Omega} \boldsymbol{\tau}_s : \boldsymbol{\epsilon}_s(\boldsymbol{\delta}\mathbf{u})\, d\Gamma \end{aligned} \tag{3.286}$$

3.7.6.1 Discretization using XFEM

Let us recall the XFEM approximation of the displacement field in a material with several material subdomains given by

$$\mathbf{u}^{\mathbf{h}}(\mathbf{X}) = \sum_{i \in I} N_i(\mathbf{X})\mathbf{u_i} + \mathbf{u}^{\mathbf{enr}} \tag{3.287}$$

$$\mathbf{u}^{\mathbf{enr}} = \sum_{N=1}^{n_c} \sum_{j \in J} N_j(\mathbf{X})\mathbf{a}_{\mathbf{j}}^{(\mathbf{N})} F(\mathbf{X}) \tag{3.288}$$

where $\mathbf{a_j}$ is the additional degrees of freedom (DOF) that accounts for the jump in the strain field, n_c denotes the number of material interfaces, and J is the set of all nodes whose support is cut by the material interface. As discussed in the previous sections, the absolute enrichment function $F(x)$ [40]

$$\mathbf{F}(\mathbf{X}) = \sum_{I} N_i(\mathbf{X})|\phi_i(\mathbf{X})| - |N_i(x)\phi_i(\mathbf{X})| \tag{3.289}$$

is used in order to account for the discontinuous strain field along Γ. The voids are assumed to be filled with a material that is 1000 times softer than the stiffness of the nanostructure; this assumption is more stable than assuming voids which are filled with air. The usage of a softer material enables the traction and displacement boundary to intersect with the void boundary. The stiffness coefficients are determined by numerical integration performed over sub triangles on either side of the inclusion interface. Substituting the displacement field in Eq. (3.287) to the weak formulation, Eq. (3.284), the algebraic finite element equations can be obtained. The expressions for $a(\mathbf{u}, \delta\mathbf{u})$, $a_s(\mathbf{u}, \delta\mathbf{u})$, $l_s(\delta\mathbf{u})$ and $l(\delta\mathbf{u})$ for an element can be rewritten using the FE approximation as,

$$a^e(\mathbf{u}, \delta\mathbf{u}) = \delta\mathbf{u^e}^T\left(\int_{\Omega^e} \mathbf{B^T}\{\mathbb{C}^{bulk}\}B\, d\Omega^e\right)\mathbf{u^e} \tag{3.290}$$

$$\begin{aligned} a_s^e(\mathbf{u}, \delta\mathbf{u}) + l_s^e(\delta\mathbf{u}) = & \\ & \int_{\Gamma^e} (\mathbf{P}\epsilon(\mathbf{u})\mathbf{P})\,\{\mathbb{C}^s\}\,(\mathbf{P}\epsilon(\delta\mathbf{u})\mathbf{P})d\Gamma^e \\ & + \int_{\Gamma^e} \boldsymbol{\tau}_s\,(\mathbf{P}\epsilon(\mathbf{u})\mathbf{P})d\Gamma^e \\ & = \delta\mathbf{u^e}^T\left(\int_{\Gamma^e} \mathbf{B^T}\mathbf{M_p}^\mathbf{T}\{\mathbb{C}^s\}\mathbf{M_p}\mathbf{B}\, d\Gamma^e\right)\mathbf{u^e} \\ & + \delta\mathbf{u^e}^T\int_{\Gamma^e} \mathbf{B^T}\mathbf{M_p}^\mathbf{T}\boldsymbol{\tau}_s\, d\Gamma^e \end{aligned} \tag{3.291}$$

$$l^e(\delta\mathbf{u}) = \delta\mathbf{u^{eT}}\left(\int_{\Gamma_N^e} \mathbf{N^T}\bar{\mathbf{t}}\, d\Gamma^e + \int_{\Omega}^{e} \mathbf{N^T}\mathbf{b}\, d\Omega^e\right) \tag{3.292}$$

where $\mathbf{u} \in H^1(\Omega)$ and $\delta\mathbf{u} \in H^1(\Omega)$.

The final system of discrete algebraic XFEM equations is,

$$(\mathbf{K_b} + \mathbf{K_s})\mathbf{u} = -\mathbf{f_s} + \mathbf{f_{ext}} \tag{3.293}$$

$$
\begin{aligned}
\mathbf{K_b} &= \int_{\Omega} \mathbf{B^T}\{\mathbb{C}^{bulk}\}\mathbf{B}d\Omega \\
\mathbf{K_s} &= \int_{\Gamma} \mathbf{B^T M_p}^{\mathbf{T}} \cdot \{\mathbb{C}^s\} \cdot \mathbf{M_p B}\, d\Gamma \\
\mathbf{f_s} &= \int_{\Gamma} \mathbf{B^T} \cdot \mathbf{M_p}^{\mathbf{T}} \boldsymbol{\tau}_s \, d\Gamma \\
\mathbf{f_{ext}} &= \int_{\Gamma_N} N^T \bar{\mathbf{t}}\, d\Gamma + \int_{\Omega} N^T \mathbf{b}\, d\Omega
\end{aligned}
\tag{3.294}
$$

where K_s is the surface stiffness matrix, while f_s is the surface residual. M_P and C^s are defined as [20],

$$
M_P = \begin{pmatrix} P_{11}^2 & P_{12}^2 & P_{11}P_{12} \\ P_{12}^2 & P_{22}^2 & P_{12}P_{22} \\ 2P_{11}P_{12} & 2P_{12}P_{22} & P_{12}^2 + P_{11}P_{22} \end{pmatrix} \tag{3.295}
$$

$$
C^s = M_p^T S^s M_p \tag{3.296}
$$

$$
S^s = \begin{pmatrix} S_{1111} & S_{1122} & 0 \\ S_{1122} & S_{2222} & 0 \\ 0 & 0 & S_{1212} \end{pmatrix} \tag{3.297}
$$

We assume $\mathbf{D} \subset \mathbb{R}^d$ ($d = 2$ or 3) as the whole structural shape and topology design domain including all admissible shapes $\boldsymbol{\Omega}$, i.e. $\boldsymbol{\Omega} \subset \mathbf{D}$. A level set function $\boldsymbol{\Phi}(\mathbf{x})$ which partitions the design domain $\mathbf{D}$ into three parts, i.e. the solid, void and the boundary which are defined as

$$
\begin{aligned}
\text{Solid}: &\quad \boldsymbol{\Phi}(\mathbf{x}) < 0 \quad \forall \mathbf{x} \in \boldsymbol{\Omega} \setminus \partial\boldsymbol{\Omega} \\
\text{Boundary}: &\quad \boldsymbol{\Phi}(\mathbf{x}) = 0 \quad \forall \mathbf{x} \in \partial\boldsymbol{\Omega} \cap \mathbf{D} \\
\text{Void}: &\quad \boldsymbol{\Phi}(\mathbf{x}) > 0 \quad \forall \mathbf{x} \in \mathbf{D} \setminus \boldsymbol{\Omega}
\end{aligned}
\tag{3.298}
$$

The basic idea of the level set method for structural optimization is to describe the structural design boundary $\boldsymbol{\Gamma}(\mathbf{x})$ implicitly by the zero level set of a higher dimensional level set function (see Fig. 3.36):

$$
\boldsymbol{\Gamma}(\mathbf{x}) = \{\mathbf{x} \in \mathbb{R}^d \,|\, \boldsymbol{\Phi}(\mathbf{x}) = 0\} \tag{3.299}
$$

To allow the design boundary for a dynamic evolution in the optimization process, we introduce t as a fictitious time. Thus the dynamic design boundary is defined as

$$
\boldsymbol{\Gamma}(t) = \{\mathbf{x}(t) \in \mathbb{R}^d \,|\, \boldsymbol{\Phi}(\mathbf{x}(t), t) = 0\} \tag{3.300}
$$

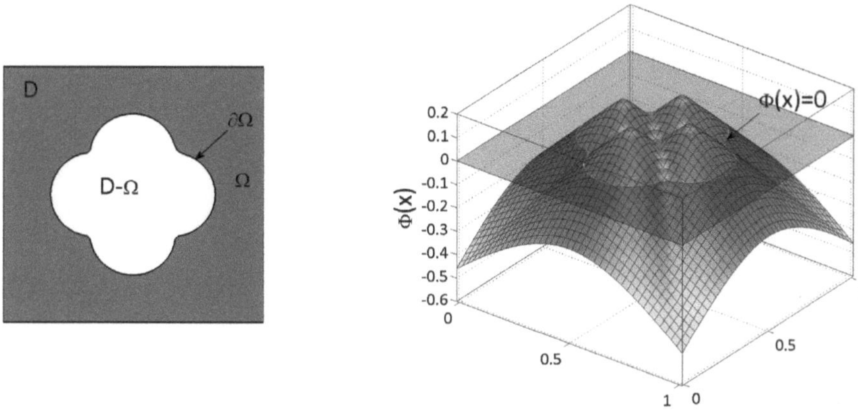

Figure 3.36 Level set description of a plate with a hole [41]. *(Left)* Design domain, *(right)* level set function.

By differentiating $\{\mathbf{\Phi}(\mathbf{x}(t), t) = 0\}$ with respect to time,[11] we obtain the well-known Hamilton-Jacobi partial differential equation

$$\frac{\partial \mathbf{\Phi}(\mathbf{x}(t), t)}{\partial t} + \nabla \mathbf{\Phi}(\mathbf{x}(t), t) \cdot \mathbf{V} = 0 \tag{3.301}$$

where $\mathbf{V} = \frac{\mathrm{d}\mathbf{x}}{\mathrm{d}t}$ denotes the velocity vector of the design boundary. This equation can be further written considering the unit outward normal $\mathbf{n} = \frac{\nabla \mathbf{\Phi}}{|\nabla \mathbf{\Phi}|}$ to the boundary and normal component of velocity vector $V_n = \mathbf{V} \cdot \mathbf{n}$,

$$\frac{\partial \mathbf{\Phi}}{\partial t} + V_n |\nabla \mathbf{\Phi}| = 0 \tag{3.302}$$

By solving this Hamilton-Jacobi equation, the level set function and consequently the structural design boundary is updated during the optimization process. It should be noted that here V_n is a quantity that links the level set method to the shape design sensitivity analysis [65]. The Hamilton-Jacobi equations usually do not admit smooth solutions. Existence and uniqueness are achieved in the framework of viscosity solutions which provide a convenient definition of the generalized shape motion. The discrete solution of the H-J equation is obtained by an explicit first-order upwind scheme. The level set function is regularized periodically by solving

$$\frac{\partial \mathbf{\Phi}}{\partial t} + sign\,(\mathbf{\Phi_0})\,(\|\nabla \mathbf{\Phi}\| - 1) = 0. \tag{3.303}$$

[11] This is the same as taking the material derivative of $\{\mathbf{\Phi}(\mathbf{x}(t), t) = 0\}$.

Solving this equation gives a signed distance function with respect to an initial isoline, $\mathbf{\Phi}_0$. This ensures smoother interfaces and also that the signed distance from the interface can be used as enrichment values for the nodes whose support is cut by the zero level sets, for the XFEM analysis performed in each iteration.

3.7.6.2 Material derivative

Consider an initial structural domain $\mathbf{\Omega}$ which is transformed into a deformed (or perturbed) structural domain $\mathbf{\Omega}_\tau$ in a fictitious time τ. This transformation can be viewed as a mapping $\mathbf{T}: x \to x_\tau(x),\ x \in \mathbf{\Omega}$ such that

$$\begin{aligned} x_\tau &\equiv \mathbf{T}(x, \tau) \\ \mathbf{\Omega}_\tau &\equiv \mathbf{T}(\mathbf{\Omega}, \tau) \end{aligned} \tag{3.304}$$

A design velocity field can be defined as

$$\mathbf{V}(x_\tau, \tau) \equiv \frac{\mathrm{d}x_\tau}{\mathrm{d}\tau} = \frac{\mathrm{d}\mathbf{T}(x, \tau)}{\mathrm{d}\tau} = \frac{\partial \mathbf{T}(x, \tau)}{\partial \tau} \tag{3.305}$$

Based on the linear Taylor's series expansion of $\mathbf{T}(x, \tau)$ around $\tau = 0$ [16], any material point in the initial domain $x \in \mathbf{\Omega}$ can be mapped onto a new material point in the perturbed domain $x_\tau \in \mathbf{\Omega}_\tau$ as

$$x_\tau(x) = \mathbf{T}(x, \tau) = x + \tau \mathbf{V}(x) \tag{3.306}$$

The material derivative of quantity z is defined as

$$\dot{z}(x) = \frac{\mathrm{d}}{\mathrm{d}\tau} z_\tau(x + \tau \mathbf{V}(x)) \bigg|_{\tau=0} = z'(x) + \nabla z(x) \cdot \mathbf{V}(x) \tag{3.307}$$

where the over dot represents the material derivative and the prime denotes a local derivative.

Lemma 3.1. *Let* Ψ_1 *be a domain functional defined as*

$$\Psi_1 = \int_{\Omega_\tau} f_\tau(x_\tau)\, d\Omega_\tau$$

with f_τ *being a regular function defined in* Ω_τ*, then the material derivative of* Ψ_1 *is given by*

$$\begin{aligned}\dot{\Psi}_1 &= \int_\Omega [\dot{f}(x) + f(x)(\nabla \cdot \mathbf{V}(x))]\, d\Omega \\ &= \int_\Omega [f'(x) + \nabla \cdot (f(x)\mathbf{V}(x))]\, d\Omega \\ &= \int_\Omega f'(x)\, d\Omega + \int_\Gamma f(x)(\mathbf{V}(x).\mathbf{n})\, d\Gamma\end{aligned}$$

Lemma 3.2. *Let Ψ_2 be a boundary functional defined as*

$$\Psi_2 = \int_{\Gamma_\tau} g_\tau(x_\tau)\, d\Gamma_\tau$$

with g_τ being a regular function defined on Γ_τ, then the material derivative of Ψ_2 is given by

$$\begin{aligned}\dot{\Psi}_2 &= \int_\Gamma [\dot{g}(x) + \kappa g(x)(\mathbf{V}(x).\mathbf{n})]\, d\Gamma \\ &= \int_\Gamma [g'(x) + (\nabla g(x) \cdot \mathbf{n} + \kappa g(x))(\mathbf{V}(x).\mathbf{n})]\, d\Gamma\end{aligned}$$

where $\kappa = div\mathbf{n} = \nabla \cdot \mathbf{n}$ is the curvature of Γ in $\mathbb{R}^2$ and twice the mean curvature of Γ in $\mathbb{R}^3$.

With these lemmas at hand, the material derivative of the objective functionals can be obtained as (see Appendix A for details),

$$\dot{J}_1 = \int_\Omega u'.b\, d\Omega + \int_\Gamma u.b\, V_n d\Gamma + \int_{\Gamma_N} (\nabla u.t.n + \kappa u.t)\, V_n d\Gamma \tag{3.308}$$

$$\begin{aligned}\dot{J}_2 = c_0 \cdot (&\int_\Gamma 2|u - u_0|u' d\Gamma + \int_\Gamma (\nabla(|u - u_0|^2)) \cdot n + \\ &\kappa|u - u_0|^2)\, V_n d\Gamma\end{aligned} \tag{3.309}$$

$$c_0 = \frac{1}{2}(\int_\Gamma |u - u_0|^2 d\Gamma)^{-\frac{1}{2}} \tag{3.310}$$

3.7.6.3 Sensitivity analysis

In order to convert the constrained optimization problem to an unconstrained problem, an augmented objective functional L is constructed as

$$\begin{aligned}L &= J(\mathbf{u}, \boldsymbol{\Omega}) + \chi(\boldsymbol{\Omega}) \\ \chi(\boldsymbol{\Omega}) &= \lambda(\int_\Omega d\Omega - \bar{V}) + \frac{1}{2\Lambda}(\int_\Omega d\Omega - \bar{V})^2\end{aligned} \tag{3.311}$$

in which λ is the Lagrange multiplier and Λ is a penalization parameter. These parameters are updated at each iteration k of the optimization process by the following rule

$$\begin{aligned} \lambda^{k+1} &= \lambda^k + \frac{1}{\Lambda^k}\left(\int_\Omega d\Omega - \bar{V}\right) \\ \Lambda^{k+1} &= \zeta \Lambda^k \end{aligned} \tag{3.312}$$

where $\zeta \in (0, 1)$ is a constant parameter.

The shape derivative of augmented Lagrangian L is defined as

$$L' = J'(\mathbf{u}, \boldsymbol{\Omega}) + \chi'(\boldsymbol{\Omega}) \tag{3.313}$$

$$J' = \int_\Gamma G.V_n d\Gamma \tag{3.314}$$

$$\begin{aligned} G = &- \int_\Gamma \boldsymbol{\epsilon}(\mathbf{u}) : \mathbb{C}^{bulk} : \boldsymbol{\epsilon}(\mathbf{w})\, d\Gamma \\ &- \int_\Gamma \kappa(\mathbf{P}\boldsymbol{\epsilon}(\mathbf{w})\mathbf{P} : \boldsymbol{\tau}_\mathbf{s}) d\Gamma \\ &- \int_\Gamma \kappa(\mathbf{P}\boldsymbol{\epsilon}(\mathbf{u})\mathbf{P} : \mathbb{C}^s : \mathbf{P}\boldsymbol{\epsilon}(\mathbf{w})\mathbf{P}) d\Gamma \end{aligned} \tag{3.315}$$

$$\chi'(\boldsymbol{\Omega}) = \int_{\partial\Omega} \max\{0, \lambda + \frac{1}{\Lambda}\left(\int_\Omega d\Omega - \bar{V}\right)\} V_n\, d\Gamma \tag{3.316}$$

Based on the steepest descent direction,

$$\begin{aligned} V_n = &- \int_\Gamma \boldsymbol{\epsilon}(\mathbf{u}) : \mathbb{C}^{bulk} : \boldsymbol{\epsilon}(\mathbf{w})\, d\Gamma \\ &- \int_\Gamma \kappa(\mathbf{P}\boldsymbol{\epsilon}(\mathbf{w})\mathbf{P} : \boldsymbol{\tau}_\mathbf{s}) d\Gamma \\ &- \int_\Gamma \kappa(\mathbf{P}\boldsymbol{\epsilon}(\mathbf{u})\mathbf{P} : \mathbb{C}^s : \mathbf{P}\boldsymbol{\epsilon}(\mathbf{w})\mathbf{P}) d\Gamma \end{aligned} \tag{3.317}$$

$$J' = -\int_\Gamma V_n^2\, d\Gamma \leq 0 \tag{3.318}$$

Velocity extension

The normal velocity of the front V_n is to be extended from the front to the whole design domain in order to solve the H-J Eq. (3.302). Different techniques for velocity extension have been proposed in the literature e.g. the normal, natural, Hilbertian and Helmholtz velocity extension methods (see [62] for a review on different velocity extension strategies). It is obvious

Table 3.2 *E* (bulk Young's modulus), ν (Poisson ratio) and S_{ijkl} (surface stiffness) for Gold (Au) from atomistic calculations [39].

E (GPa)	ν	$S_{1111} = S_{2222}$ (J/m^2)	$S_{1122} = S_{2211}$ (J/m^2)	S_{1212} (J/m^2)	τ^0 (J/m^2)
36	0.44	5.26	2.53	3.95	1.57

from Eq. (3.317) that the velocity comprises two parts, the bulk, V_b and surface terms, V_s. The bulk part of velocity, V_b can be obtained at each node whereas the surface part, V_s can be determined only along the surface. In order to solve H-J Eq. (3.302) that is posed throughout the domain, the surface part of velocity V_s is extended by extrapolation to the nodes that belong to the cut elements. The value of the speed function at the closest point on the surface is assigned as the extension velocity to the nodal point [34], such that the condition $V_{ext} = V_n$ at $\phi = 0$ is satisfied.

3.7.6.4 Numerical example

Let us consider the topology optimization of a cantilever beam at different length scales. The objective is to minimize the least square error compared to a target displacement. The nanobeam is assumed to be made of gold, where the bulk and surface properties are given in Table 3.2. The domain is discretized by bilinear quadrilateral (Q4) (XFEM) elements.

A cantilever nanobeam of size 40 × 10 nm is subjected to a point load of 3.6 nN at the free (40,0) nm end. The target displacement at the load location is 16 nm. The optimum topology obtained is shown in Fig. 3.37A, where the volume ratio of the optimum topology is 0.59 and the stiffness ratio of the 40 × 10 nm beam is 8.6%.

Now the aspect ratio is maintained as 4 and the thickness of the beam is gradually increased. Thus, Fig. 3.37 also shows the optimum shapes obtained for beams of size 80 × 20, 160 × 40 and 320 × 80 nm, which have stiffness ratios of 4.25%, 2.2% and 1.05%, respectively, and volume ratios of 0.65, 0.695, and 0.71, respectively.

The optimum topology obtained for the 160 × 40 and 320 × 80 nm beams appear similar, which suggests that for this particular aspect ratio and objective function, surface effects lose their effect once the nanobeam thickness is larger than about 20 nm. However, the 10 and 20 nm thickness nanobeams have different optimal designs, which are driven by the fact that the smaller nanostructures are stiffer as demonstrated by the stiffness ratios, and thus require less material to conform to the maximum displacement constraint.

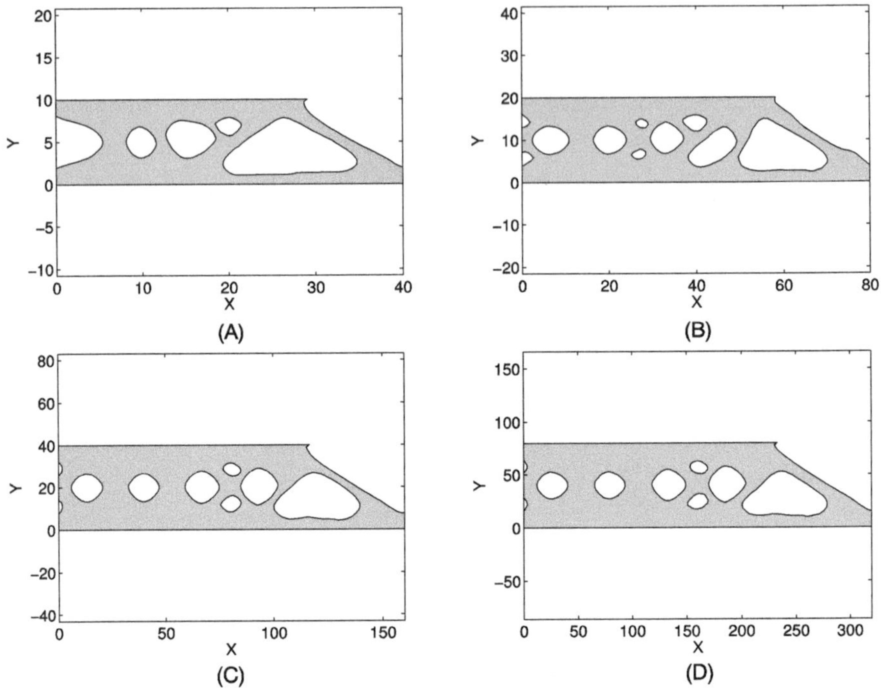

Figure 3.37 The optimal topology obtained for J_2 objective function for a (A) 40 × 10 nm, (B) 80 × 20 nm, (C) 160 × 40 nm and (D) 320 × 80 nm cantilever beams including surface effects.

To further clarify the effects of surfaces on the optimal nanodesign, Fig. 3.38 shows the optimum topologies for the same nanobeams, i.e. 40 × 10, 80 × 20, 160 × 40 and 320 × 80 nm, though when surface effects are not considered. As expected, the optimum topologies remain the same irrespective of the dimensions, which demonstrates the size-independence that would be expected when nanoscale surface effects are not considered.

3.7.7 Nanopiezoelectricity

Energy harvesting which means capturing minute amounts of energy from one or more of the surrounding energy sources where a remote application is deployed, and where such energy source is inexhaustible, is an increasingly attractive alternative to conventional batteries. The recent advances in micro-electromechanical systems has led to considerable increase in portable electronics and wireless sensors. Powering such devices with batteries of finite life span is a problematic task. If the ambient energy can

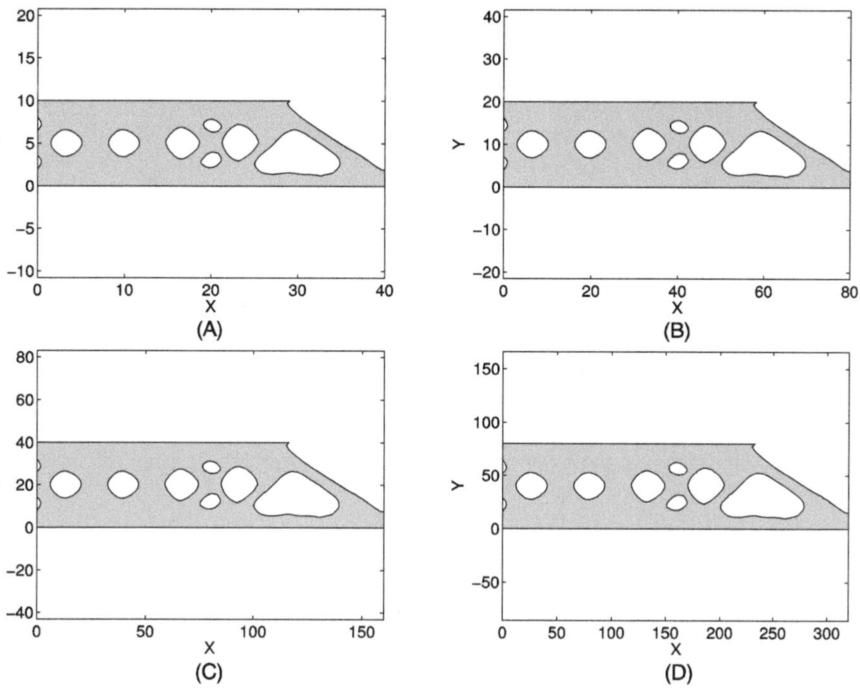

Figure 3.38 The optimal topology obtained for J_2 objective function for (A) 40 × 10 nm, (B) 80 × 20 nm, (C) 160 × 40 nm and (D) 320 × 80 nm cantilever beams where surface effects are not considered.

be captured as useful energy then ideally the device can be powered for infinite span. Several energy harvesting approaches are proposed using solar, electromagnetic, thermoelectric and piezoelectric materials at micro and nano scales. Marin et al. [35] concluded that the output power of piezoelectric mechanism is proportional to $v^{9/4}$ at mm^3 dimensions, where v is effective volume ratio. The main advantages of piezoelectric materials in energy harvesting compared to other transduction mechanisms are their large power densities and ease of application. Ambient vibrations provide energy to the system through base excitation. This excitation is converted to cyclic oscillation, which is then converted to cyclic electrical energy through piezoelectric effect. Electromechanical coupling coefficient, k defines the efficiency of energy conversion. A proof mass is attached at the free end of cantilever to modify the resonant frequency to match with ambient frequency as shown in Fig. 3.39. In d_{31} operating mode stress along direction 1 leads to an induced electric field in direction 3, that is the

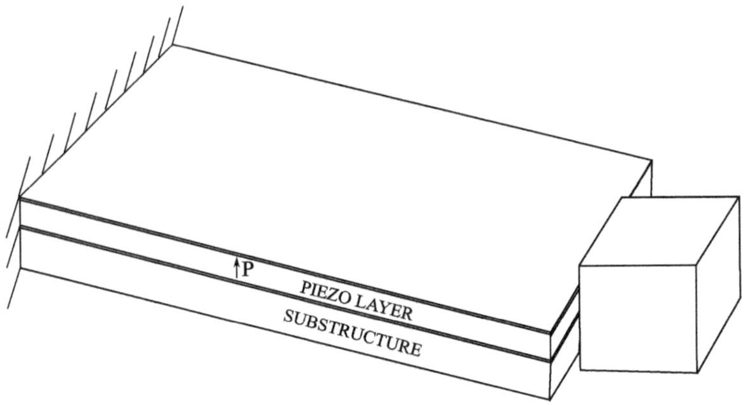

Figure 3.39 A typical cantilever energy harvester with piezoelectric layer, substrate and proof mass [43].

poling and mechanical load direction are perpendicular to each other. In d_{33} mode, the stress and electric field are in the same direction, while d_{15} mode indicates shear stress harvesters. Commonly adopted resonance based piezoelectric generators are of d_{31} mode. The d_{33} mode based vibration generators are less effective than d_{31} mode because of the percentage of piezoelectric material that does not contribute to energy conversion [10].

However, most piezoelectric energy harvesters have been used in bulk material systems. The exciting possibility of using nanoscale piezoelectric energy harvesters emerged in 2006 with the discovery of piezoelectricity from ZnO nanowires by Wang et al. [66]. Many researchers have since extended the original seminal work, including the development of self-powered nano generators that can provide gate voltage to effectively control charge transport [66], lateral and vertical integration of ZnO nanowires into arrays that are capable of producing sufficient power to operate real devices as presented in Xu et al. [67]. In this book, the energy conversion ability of nano piezoelectric beams and plates is studied. Possibility of performing topology optimization of such nano structures to further increase their energy conversion is explored.

3.7.7.1 Strong and weak form of surface piezoelectricity

Let us consider a piezoelectric domain $\boldsymbol{\Omega}$ with a material surface $\boldsymbol{\Gamma}$. Based on the continuum theory of surface piezoelectricity, the equilibrium equations are

$$\nabla \cdot \boldsymbol{\sigma} + \boldsymbol{b} = 0 \qquad \text{in } \boldsymbol{\Omega} \tag{3.319}$$

$$\nabla \cdot \mathbf{D} - q = 0 \qquad \text{in } \boldsymbol{\Omega} \tag{3.320}$$

$$\nabla_s \cdot \boldsymbol{\sigma}_s = 0 \qquad \text{on } \boldsymbol{\Gamma} \tag{3.321}$$

$$\nabla_s \cdot \boldsymbol{D}_s = 0 \qquad \text{on } \boldsymbol{\Gamma} \tag{3.322}$$

where $\boldsymbol{\sigma}$ and $\mathbf{D}$ are mechanical stress and electric displacement, respectively, while $\boldsymbol{\sigma}_s$ and $\mathbf{D_s}$ are the surface stress and the surface electric displacement, respectively. In the above equation, $\nabla_s \cdot \boldsymbol{\sigma}_s = \nabla\boldsymbol{\sigma} : \boldsymbol{P}$ and $\nabla_s \cdot \boldsymbol{D}_s = \nabla\boldsymbol{D} : \boldsymbol{P}$ where : is the double tensor contraction and where $\boldsymbol{P}$ is a second order tensor defined as $\boldsymbol{P} = \boldsymbol{I} - \boldsymbol{n} \otimes \boldsymbol{n}$.

The linear piezoelectric constitutive relations for the bulk and surface are,

$$\boldsymbol{\sigma} = \mathbb{C} : \boldsymbol{\epsilon} - \boldsymbol{e} \cdot \boldsymbol{E} \tag{3.323}$$

$$\boldsymbol{D} = \boldsymbol{e} : \boldsymbol{\epsilon} + \boldsymbol{\kappa} \cdot \boldsymbol{E} \tag{3.324}$$

$$\boldsymbol{\sigma}_s = \boldsymbol{\tau}_s + \mathbb{C}_s : \boldsymbol{\epsilon}_s - \boldsymbol{e}_s \cdot \boldsymbol{E}_s \tag{3.325}$$

$$\boldsymbol{D}_s = \boldsymbol{\omega}_s + \boldsymbol{e}_s^T : \boldsymbol{\epsilon}_s + \boldsymbol{\kappa}_s \cdot \boldsymbol{E}_s \tag{3.326}$$

where $\mathbb{C}$ and $\mathbb{C}_s$ are the fourth-order elastic stiffness tensors associated with the bulk and surface, respectively, while $\mathbf{e}$ and $\mathbf{e_s}$ correspond to the bulk and surface piezoelectric third order tensors, respectively. κ and κ_s are the bulk and surface second order permittivity tensors. $\boldsymbol{\tau}_s$ and $\boldsymbol{\omega}_s$ give the residual surface stresses and electric displacement, which are related to the residual surface strain and electric field, respectively. $\boldsymbol{\epsilon}$ and $\boldsymbol{E}$ are the bulk strain tensor and bulk electric field vector while $\boldsymbol{\epsilon}_s$ and $\boldsymbol{E}_s$ are their corresponding surface counterparts.

The surface energy density γ is given as,

$$\gamma = \gamma_0 + \boldsymbol{\tau}_s : \boldsymbol{\epsilon}_s + \boldsymbol{\omega}_s \cdot \mathbf{E_s} + \frac{1}{2}\boldsymbol{\epsilon}_s : \mathbb{C}_s : \boldsymbol{\epsilon}_s + \frac{1}{2}\boldsymbol{E}_s \cdot \boldsymbol{\kappa}_s \cdot \boldsymbol{E}_s + \mathbf{E_s} \cdot \mathbf{e_s} : \boldsymbol{\epsilon}_s \tag{3.327}$$

where γ_0 is surface free energy density. Having defined the constitutive and field equations, we derive the weak form of the boundary value problem based on the principle of stationary potential energy. The total potential energy Π of the system is given by

$$\Pi = \Pi_{bulk} + \Pi_s - \Pi_{ext} \tag{3.328}$$

where Π_{bulk}, Π_s, and Π_{ext} represent the bulk internal energy, surface internal energy and the work of external forces, respectively, which are given by

$$\Pi_{bulk} = \frac{1}{2}\boldsymbol{\epsilon} : \mathbb{C} : \boldsymbol{\epsilon} - \boldsymbol{E} \cdot \mathbf{e} : \boldsymbol{\epsilon} - \frac{1}{2}\boldsymbol{E} \cdot \boldsymbol{\kappa} \cdot \boldsymbol{E} \tag{3.329}$$

$$\Pi_s = \int_\Gamma \gamma \, d\Gamma \tag{3.330}$$

The stationary condition of (3.328) is given by

$$D_{\delta \mathbf{u}} \Pi = 0 \tag{3.331}$$

where $D_{\delta \mathbf{u}} \Pi$ is the directional (or Gâteaux) derivative of the functional Π in the direction $\boldsymbol{\delta}\mathbf{u}$. Applying the stationary condition, the weak form of the equilibrium equations can be obtained by finding $\mathbf{u} \in \{\mathbf{u} = \bar{\mathbf{u}} \quad \text{on } \boldsymbol{\Gamma}_u, \mathbf{u} \in H^1(\Omega)\}$ and $\boldsymbol{\phi} \in \{\boldsymbol{\phi} = \bar{\boldsymbol{\phi}} \quad \text{on } \boldsymbol{\Gamma}_\phi, \boldsymbol{\phi} \in H^1(\Omega)\}$ such that

$$\begin{aligned}
&\int_\Omega \boldsymbol{\epsilon}(\delta \mathbf{u}) : \mathbb{C} : \boldsymbol{\epsilon}(\mathbf{u}) \, d\Omega - \int_\Omega \mathbf{E}(\delta \boldsymbol{\phi}) \cdot \mathbf{e} : \boldsymbol{\epsilon}(\mathbf{u}) \, d\Omega \\
&- \int_\Omega \boldsymbol{\epsilon}(\delta \mathbf{u}) : \mathbf{e} \cdot \mathbf{E}(\boldsymbol{\phi}) \, d\Omega - \int_\Omega \mathbf{E}(\delta \boldsymbol{\phi}) \cdot \boldsymbol{\kappa} \cdot \mathbf{E}(\boldsymbol{\phi}) \, d\Omega \\
&+ \int_\Gamma \boldsymbol{\epsilon}_s(\delta \mathbf{u}) : \mathbb{C}_s : \boldsymbol{\epsilon}_s(\mathbf{u}) \, d\Gamma - \int_\Gamma \mathbf{E}_s(\delta \boldsymbol{\phi}) \cdot \mathbf{e}_s : \boldsymbol{\epsilon}_s(\mathbf{u}) \, d\Gamma \\
&- \int_\Gamma \boldsymbol{\epsilon}_s(\delta \mathbf{u}) : \mathbf{e}_s \cdot \mathbf{E}_s(\boldsymbol{\phi}) \, d\Gamma - \int_\Gamma \mathbf{E}_s(\delta \boldsymbol{\phi}) \cdot \boldsymbol{\kappa}_s \cdot \mathbf{E}_s(\boldsymbol{\phi}) \, d\Gamma \\
&= - \int_\Gamma \boldsymbol{\epsilon}_s(\delta \mathbf{u}) : \boldsymbol{\tau}_s \, d\Gamma - \int_\Gamma \mathbf{E}_s(\delta \boldsymbol{\phi}) \cdot \boldsymbol{\omega}_s \, d\Gamma + \int_{\Gamma_N} \delta \mathbf{u} \cdot \bar{\mathbf{t}} \, d\Gamma \\
&+ \int_\Omega \delta \mathbf{u} \cdot \mathbf{b} \, d\Omega - \int_\Omega \delta \boldsymbol{\phi} \, q \, d\Omega
\end{aligned} \tag{3.332}$$

for all $\delta \mathbf{u} \in \{\delta \mathbf{u} = 0 \quad \text{on } \boldsymbol{\Gamma}_u, \delta \mathbf{u} \in H^1(\Omega)\}$ and $\delta \boldsymbol{\phi} \in \{\delta \boldsymbol{\phi} = 0 \quad \text{on } \boldsymbol{\Gamma}_\phi, \delta \boldsymbol{\phi} \in H^1(\Omega)\}$.

This weak formulation can be written in simplified form as,

$$a(\mathbf{u}, \boldsymbol{\phi}, \delta \mathbf{u}, \boldsymbol{\delta \phi}) + a_s(\mathbf{u}, \boldsymbol{\phi}, \delta \mathbf{u}, \boldsymbol{\delta \phi}) = -l_s(\delta \mathbf{u}, \boldsymbol{\delta \phi}) + l(\delta \mathbf{u}, \boldsymbol{\delta \phi}) \tag{3.333}$$

3.7.7.2 XFEM formulation for surface piezoelectricity

As in the nanoelasticity problem, the displacement field $\boldsymbol{u}^h$ and electric potential field ϕ^h for a piezoelectric material in the XFEM formulation are expressed as:

$$u^h(\mathbf{X}) = \sum_{i \in I} N_i(\mathbf{X}) u_i + \sum_{N=1}^{n_c} \sum_{l \in L} N_l(\mathbf{X}) a_l^{(N)} f_l^{(N)} \tag{3.334}$$

$$\phi^h(\mathbf{X}) = \sum_{i \in I} N_i(\mathbf{X}) \phi_i + \sum_{N=1}^{n_c} \sum_{l \in L} N_l(\mathbf{X}) \alpha_l^{(N)} f_l^{(N)} \tag{3.335}$$

where a_j and α_j are the additional degrees of freedom that account for the jump in the derivatives of the displacement and electric potential field, respectively, n_c denotes the number of inclusion interfaces, L is the set of all nodes whose support is cut by the material interface and $f^{(N)}$ in Eqs. (3.334) and (3.335) is the absolute signed distance function values from the interfaces. The interface enrichment is required because in the optimization process, the voids are assumed to be filled with a material that is 1000 times softer than the stiffness of the nano structure. As mentioned before, this enables the optimization algorithm to recover from trial topologies in which the void boundary intersects with the force boundary.

The terms in the weak formulation shown in the previous section can be written using the finite element (FE) approximation as follows,

$$\begin{aligned} &a(\mathbf{u}, \boldsymbol{\phi}, \delta\mathbf{u}, \delta\boldsymbol{\phi}) \\ &= \delta\mathbf{u}^T \left(\int_\Omega \mathbf{B}_\mathbf{u}^\mathbf{T} \{\mathbb{C}\} \mathbf{B}_\mathbf{u} \, d\Omega \right) \mathbf{u} + \delta\mathbf{u}^T \left(\int_\Omega \mathbf{B}_\mathbf{u}^\mathbf{T} \{\mathbf{e}\}^T \boldsymbol{B}_\phi \, d\Omega \right) \boldsymbol{\phi} \\ &+ \delta\boldsymbol{\phi}^T \left(\int_\Omega \boldsymbol{B}_\phi^T \{\mathbf{e}\} \boldsymbol{B}_u \, d\Omega \right) \boldsymbol{u} - \delta\boldsymbol{\phi}^T \left(\int_\Omega \boldsymbol{B}_\phi^T \{\kappa\} \boldsymbol{B}_\phi \, d\Omega \right) \boldsymbol{\phi} \end{aligned} \tag{3.336}$$

$$\begin{aligned} &a_s(\mathbf{u}, \boldsymbol{\phi}, \delta\mathbf{u}, \delta\boldsymbol{\phi}) \\ &= \int_\Gamma (\boldsymbol{P}\boldsymbol{\varepsilon}(\delta\mathbf{u})\boldsymbol{P})^T : \{\mathbb{C}_s\} : (\boldsymbol{P}\boldsymbol{\varepsilon}(\mathbf{u})\boldsymbol{P}) d\Gamma + \int_\Gamma (\boldsymbol{P}\boldsymbol{\varepsilon}(\delta\mathbf{u})\boldsymbol{P})^T : \{\mathbf{e}_s\} \cdot (\boldsymbol{P}_\phi \boldsymbol{E}(\boldsymbol{\phi})) d\Gamma \\ &+ \int_\Gamma (\boldsymbol{P}_\phi \boldsymbol{E}(\delta\boldsymbol{\phi}))^T \cdot \{\mathbf{e}_s\} : (\boldsymbol{P}\boldsymbol{\varepsilon}(\mathbf{u})\boldsymbol{P}) d\Gamma - \int_\Gamma (\boldsymbol{P}_\phi \boldsymbol{E}(\delta\boldsymbol{\phi}))^T \cdot \{\kappa_s\} \cdot (\boldsymbol{P}_\phi \boldsymbol{E}(\boldsymbol{\phi})) d\Gamma \\ &= \delta\mathbf{u}^T \left(\int_\Gamma \boldsymbol{B}_u^T \boldsymbol{M}_p{}^T \{\mathbb{C}_s\} \boldsymbol{M}_p \boldsymbol{B}_u \, d\Gamma \right) \mathbf{u} + \delta\mathbf{u}^T \left(\int_\Gamma \boldsymbol{B}_u^T \boldsymbol{M}_p{}^T \{\mathbf{e}_s\}^T \boldsymbol{P}_\phi \boldsymbol{B}_\phi \, d\Gamma \right) \boldsymbol{\phi} \\ &+ \delta\boldsymbol{\phi}^T \left(\int_\Gamma \boldsymbol{B}_\phi^T \boldsymbol{P}_\phi{}^T \{\mathbf{e}_s\} \boldsymbol{M}_P \boldsymbol{B}_u \, d\Gamma \right) \boldsymbol{u} - \delta\boldsymbol{\phi}^T \left(\int_\Gamma \boldsymbol{B}_\phi^T \boldsymbol{P}_\phi{}^T \{\kappa_s\} \boldsymbol{P}_\phi \boldsymbol{B}_\phi \, d\Gamma \right) \boldsymbol{\phi} \end{aligned} \tag{3.337}$$

where $\mathbb{C}$ and $\mathbb{C}_s$ are the matrix forms of the fourth order elastic bulk and surface stiffness tensors, $\boldsymbol{e}$ and $\boldsymbol{e}_s$ are the matrix forms of the third order elastic bulk and surface coupling tensors, $\boldsymbol{\kappa}$ and $\boldsymbol{\kappa}_s$ are the matrix forms of the second order elastic bulk and surface stiffness tensors and $\boldsymbol{\varepsilon}$ and $\boldsymbol{\varepsilon}_s$ are the vector forms of the bulk and surface strain tensors.

The matrix $\boldsymbol{M}_P$ gives the relationship between surface strain $\boldsymbol{\epsilon}_s$ and the bulk strain $\boldsymbol{\epsilon}$ so that $\boldsymbol{\epsilon}_s = \boldsymbol{M}_P\ \boldsymbol{\epsilon}$ and takes the form:

$$\boldsymbol{M_P} = \begin{pmatrix} P_{11}^2 & P_{12}^2 & P_{11}P_{12} \\ P_{12}^2 & P_{22}^2 & P_{12}P_{22} \\ 2P_{11}P_{12} & 2P_{12}P_{22} & P_{12}^2 + P_{11}P_{22} \end{pmatrix} \tag{3.338}$$

$$\begin{aligned} l_s(\delta\mathbf{u}, \delta\boldsymbol{\phi}) &= \int_\Gamma (\boldsymbol{P}\boldsymbol{\varepsilon}(\delta\mathbf{u})\boldsymbol{P})^T : \boldsymbol{\tau}_s\, d\Gamma + \int_\Gamma (\mathbf{P}_\phi \boldsymbol{E}(\delta\boldsymbol{\phi}))^T \cdot \boldsymbol{\omega}_s\, d\Gamma \\ &= \delta\mathbf{u}^T \int_\Gamma \boldsymbol{B}_u^T \boldsymbol{M}_p{}^T \boldsymbol{\tau}_s\, d\Gamma + \delta\boldsymbol{\phi}^T \int_\Gamma \boldsymbol{B}_\phi^T \boldsymbol{P}_\phi{}^T \boldsymbol{\omega}_s\, d\Gamma \end{aligned} \tag{3.339}$$

$$l(\delta\mathbf{u}, \delta\boldsymbol{\phi}) = \delta\mathbf{u}^{\mathbf{T}} \left(\int_{\Gamma_N} \boldsymbol{N}^T \bar{\boldsymbol{t}} d\Gamma + \int_\Omega \boldsymbol{N}^T \boldsymbol{b} d\Omega \right) - \delta\boldsymbol{\phi}^T \int_\Omega \boldsymbol{N}^T q d\Omega \tag{3.340}$$

where $\boldsymbol{B}_u$ and $\boldsymbol{B}_\phi$ are the strain-displacement and electric field-potential matrices, respectively. From Eqs. (3.336)-(3.340) and the weak form (3.332), the system of algebraic XFEM equations to solve is,

$$(\boldsymbol{K}_{uu}^b + \boldsymbol{K}_{uu}^s)\boldsymbol{u} + (\boldsymbol{K}_{\phi u}^b + \boldsymbol{K}_{\phi u}^s)\boldsymbol{\phi} = -(\boldsymbol{f}_u^s + \boldsymbol{f}_u^{ext}) \tag{3.341}$$

$$(\boldsymbol{K}_{u\phi}^b + \boldsymbol{K}_{u\phi}^s)\boldsymbol{u} + (\boldsymbol{K}_{\phi\phi}^b + \boldsymbol{K}_{\phi\phi}^s)\boldsymbol{\phi} = -(\boldsymbol{g}_u^s + \boldsymbol{g}_u^{ext}) \tag{3.342}$$

The vectors $\boldsymbol{u}$ and $\boldsymbol{\phi}$ comprise the corresponding FE and enriched DOFs. The bulk and surface stiffness matrices corresponding to the discrete finite element equations in Eq. (3.342) are defined as

$$\begin{aligned} \boldsymbol{K}_{uu}^b &= \int_\Omega \boldsymbol{B}_u^T \{\mathbb{C}\} \boldsymbol{B}_u d\Omega \\ \boldsymbol{K}_{uu}^s &= \int_\Gamma \boldsymbol{B}_u^T \boldsymbol{M}_p{}^T \{\mathbb{C}^s\} \boldsymbol{M}_p \boldsymbol{B}_u\, d\Gamma \end{aligned} \tag{3.343}$$

$$\begin{aligned} \boldsymbol{K}_{\phi u}^b &= \int_\Omega \boldsymbol{B}_\phi^T \{\mathbf{e}\} \boldsymbol{B}_u d\Omega \\ \boldsymbol{K}_{\phi u}^s &= \int_\Gamma \boldsymbol{B}_\phi^T \boldsymbol{P}_\phi{}^T \{\mathbf{e}_s\} \boldsymbol{M}_p \boldsymbol{B}_u d\Gamma \end{aligned} \tag{3.344}$$

$$\begin{aligned} \boldsymbol{K}_{u\phi}^b &= \int_\Omega \boldsymbol{B}_u^T \{\mathbf{e}\}^T \boldsymbol{B}_\phi d\Omega \\ \boldsymbol{K}_{u\phi}^s &= \int_\Gamma \boldsymbol{B}_u^T \boldsymbol{M}_P{}^T \{\mathbf{e}_s\}^T \boldsymbol{P}_\phi \boldsymbol{B}_\phi d\Gamma \end{aligned} \tag{3.345}$$

$$
\begin{aligned}
\boldsymbol{K}_{\phi\phi}^{b} &= -\int_{\Omega} \boldsymbol{B}_{\phi}^{T}\{\kappa\}\boldsymbol{B}_{\phi}\, d\Omega \\
\boldsymbol{K}_{\phi\phi}^{s} &= -\int_{\Gamma} \boldsymbol{B}_{\phi}^{T}\boldsymbol{P}^{T}\{\kappa_s\}\boldsymbol{P}\boldsymbol{B}_{\phi}\, d\Gamma
\end{aligned}
\tag{3.346}
$$

$$
\begin{aligned}
\boldsymbol{f}_{u}^{s} &= \int_{\Gamma} \boldsymbol{B}_{u}^{T}\boldsymbol{M}_{p}{}^{T}\boldsymbol{\tau}_{s}\, d\Gamma \\
\boldsymbol{f}_{u}^{ext} &= \int_{\Gamma_N} \boldsymbol{N}^{T}\,\bar{\boldsymbol{t}} d\Gamma + \int_{\Omega} \boldsymbol{N}^{T}\,\boldsymbol{b} d\Omega \\
\boldsymbol{g}_{u}^{s} &= \int_{\Gamma} \boldsymbol{B}_{\phi}^{T}\boldsymbol{P}^{T}\boldsymbol{\omega}_{s}\, d\Gamma \\
\boldsymbol{g}_{u}^{ext} &= -\int_{\Omega} \boldsymbol{N}^{T}\,\boldsymbol{q} d\Omega
\end{aligned}
\tag{3.347}
$$

Here $\mathbf{P}$ is the tangential projection tensor to $\boldsymbol{\Gamma}$ at $\mathbf{x} \in \boldsymbol{\Gamma}$ which is defined as $\mathbf{P}(\mathbf{x}) = \mathbf{I} - \mathbf{n}(\mathbf{x}) \otimes \mathbf{n}(\mathbf{x})$, $\mathbf{I}$ is the second order unit tensor. $\boldsymbol{M}_p$ is defined as,

$$
\boldsymbol{M}_{P} = \begin{pmatrix} P_{11}^{2} & P_{12}^{2} & P_{11}P_{12} \\ P_{12}^{2} & P_{22}^{2} & P_{12}P_{22} \\ 2P_{11}P_{12} & 2P_{12}P_{22} & P_{12}^{2} + P_{11}P_{22} \end{pmatrix}
\tag{3.348}
$$

The matrices $\boldsymbol{B}_u$ and $\boldsymbol{B}_{\phi}$ comprise standard FEM and enrichment parts. The $\boldsymbol{B}_u$ and $\boldsymbol{B}_{\phi}$ matrices are,

$$
\begin{aligned}
\boldsymbol{B}_{u} &= \left[\boldsymbol{B}_{u,std}\;\; \boldsymbol{B}_{u,enr}\right] \\
\boldsymbol{B}_{u,std} &= \begin{pmatrix} N_x & 0 \\ 0 & N_y \\ N_y & N_x \end{pmatrix} \\
\boldsymbol{B}_{u,enr} &= \begin{pmatrix} fN_x + f,xN & 0 \\ 0 & fN_y + f,yN \\ fN_y + f,yN & fN_x + f,xN \end{pmatrix} \\
\boldsymbol{B}_{\phi} &= \left[\boldsymbol{B}_{\phi,std}\;\; \boldsymbol{B}_{\phi,enr}\right] \\
\boldsymbol{B}_{\phi,std} &= \begin{pmatrix} N_x \\ N_y \end{pmatrix} \\
\boldsymbol{B}_{\phi,enr} &= \begin{pmatrix} fN_x + f,xN \\ fN_y + f,yN \end{pmatrix}
\end{aligned}
\tag{3.349}
$$

The enrichment f used for modeling weak discontinuity is [40]

$$
f = \sum_{i \in I} N_i |\Psi_i| - |\sum_{i \in I} N_i \Psi_i|
\tag{3.350}
$$

3.7.7.3 Topology optimization of nanoscale piezoelectric energy harvesters

Objective function

A common objective function in topology optimization of piezoelectric energy harvesters is to maximize the energy conversion. The electromechanical coupling coefficient (EMCC), k is defined as [61],

$$k^2 = \frac{{\Pi_m}^2}{\Pi_e \Pi_d} \tag{3.351}$$

where Π_m is the elasto-dielectric energy and Π_e and Π_d are the stored elastic and dielectric energy respectively.

$$\Pi_m = \int_\Omega \boldsymbol{\varepsilon}(\boldsymbol{u})^T \boldsymbol{e}^T \boldsymbol{E}(\boldsymbol{\phi}) d\Omega \tag{3.352}$$

$$\Pi_e = \int_\Omega \boldsymbol{\varepsilon}(\boldsymbol{u})^T \boldsymbol{C}\boldsymbol{\epsilon}(\boldsymbol{u}) d\Omega \tag{3.353}$$

$$\Pi_d = \int_\Omega \boldsymbol{E}(\boldsymbol{\phi})^T \boldsymbol{\kappa} \boldsymbol{E}(\boldsymbol{\phi}) d\Omega \tag{3.354}$$

The energy harvesting device is assumed to be subjected only to static mechanical loads and so the EMCC may be rewritten as

$$k^2 = \frac{\Pi_d}{\Pi_e} \tag{3.355}$$

Clearly, the higher the EMCC the better from a piezoelectric energy harvesting point of view, which thus leads to the best performance of the energy harvesting device.

Within the XFEM numerical formulation, the shape and topology of the piezoelectric nanostructure is modified to maximize the energy conversion. This is done by defining the geometry of the nanostructure using the level set function. Doing so makes the design variable in the optimization problem the level set function Ψ. The corresponding objective function is:

$$\text{Minimize } J(\Psi) = \frac{1}{k^2} = \frac{\Pi_e}{\Pi_d} \tag{3.356}$$

$$\text{Subject to } \int_\Omega d\Omega - \bar{V} = 0 \tag{3.357}$$

$$a(\mathbf{u}, \boldsymbol{\phi}, \delta\mathbf{u}, \boldsymbol{\delta\phi}) + a_s(\mathbf{u}, \boldsymbol{\phi}, \delta\mathbf{u}, \boldsymbol{\delta\phi}) = -l_s(\delta\mathbf{u}, \boldsymbol{\delta\phi}) + l(\delta\mathbf{u}, \boldsymbol{\delta\phi}) \tag{3.358}$$

The optimum configuration is obtained by using the level set based topology optimization method described in the following section.

Level set method and sensitivity analysis

As for the inverse analysis, the level set method is ideally suited for optimization methods. Hence, the level set function is dynamically updated at each time step by solving the Hamilton-Jacobi (H-J) partial differential equation to minimize the objective function.

$$\frac{\partial \psi}{\partial t} + V_n |\nabla \psi| = 0 \tag{3.359}$$

and the velocity to update the level set function, V_n is obtained by performing sensitivity analysis which requires the material time derivative of the objective function (see Appendix B for details), i.e.

$$\dot{J} = \frac{\dot{\Pi}_e(\boldsymbol{u}, \boldsymbol{u})}{\Pi_d(\boldsymbol{\phi}, \boldsymbol{\phi})} - \frac{\Pi_e(\boldsymbol{u}, \boldsymbol{u})\dot{\Pi}_d(\boldsymbol{\phi}, \boldsymbol{\phi})}{\Pi_d(\boldsymbol{\phi}, \boldsymbol{\phi})^2} \tag{3.360}$$

$$= C_1\, \dot{\Pi}_e(\boldsymbol{u}, \boldsymbol{u}) + C_2\, \dot{\Pi}_d(\boldsymbol{\phi}, \boldsymbol{\phi}) \tag{3.361}$$

$$\begin{aligned}\dot{\Pi}_e(\boldsymbol{u}, \boldsymbol{u}) = &\int_\Omega 2\,\boldsymbol{\varepsilon}(\boldsymbol{u}')^T : \mathbb{C}^{bulk} : \boldsymbol{\varepsilon}(\boldsymbol{u})\, d\Omega + \int_\Gamma \boldsymbol{\varepsilon}(\boldsymbol{u})^T : \mathbb{C}^{bulk} : \boldsymbol{\varepsilon}(\boldsymbol{u})\, V_n\, d\Gamma \\ &+ \int_\Gamma 2\,\boldsymbol{\epsilon}_s(\boldsymbol{u}')^T : \mathbb{C}^s : \boldsymbol{\varepsilon}_s(\boldsymbol{u})\, d\Gamma \\ &+ \int_\Gamma [\nabla(\boldsymbol{\varepsilon}_s(\boldsymbol{u})^T : \mathbb{C}^s : \boldsymbol{\varepsilon}_s(\boldsymbol{u})) \cdot n + (\boldsymbol{\varepsilon}_s(\boldsymbol{u})^T : \mathbb{C}^s : \boldsymbol{\varepsilon}_s(u))\eta)]\, V_n\, d\Gamma\end{aligned} \tag{3.362}$$

$$\begin{aligned}\dot{\Pi}_d(\boldsymbol{\phi}, \boldsymbol{\phi}) = &\int_\Omega 2\,\boldsymbol{E}(\boldsymbol{\phi}')^T \cdot \boldsymbol{\kappa}^{bulk} \cdot \boldsymbol{E}(\boldsymbol{\phi})\, d\Omega + \int_\Gamma \boldsymbol{E}(\boldsymbol{\phi})^T \cdot \boldsymbol{\kappa}^{bulk} \cdot \boldsymbol{E}(\boldsymbol{\phi})\, V_n\, d\Gamma \\ &+ \int_\Gamma 2\,\boldsymbol{E}_s(\boldsymbol{\phi}')^T \cdot \boldsymbol{\kappa}^s \cdot \boldsymbol{E}_s(\boldsymbol{\phi})\, d\Gamma \\ &+ \int_\Gamma [\nabla(\boldsymbol{E}(\boldsymbol{\phi})^T : \boldsymbol{\kappa}^s \cdot \boldsymbol{E}_s(\boldsymbol{\phi})) \cdot n + (\boldsymbol{E}_s(\boldsymbol{u})^T \cdot \boldsymbol{\kappa}^s \cdot \boldsymbol{E}_s(\boldsymbol{\phi}))\eta)]\, V_n\, d\Gamma\end{aligned} \tag{3.363}$$

$$C_1 = \frac{1}{\Pi_d(\boldsymbol{\phi}, \boldsymbol{\phi})} \tag{3.364}$$

$$C_2 = -\frac{\Pi_e(\boldsymbol{u}, \boldsymbol{u})}{\Pi_d(\boldsymbol{\phi}, \boldsymbol{\phi})^2} \tag{3.365}$$

The augmented Lagrangian L defining the unconstrained optimization problem is

$$L = J(\mathbf{u}, \boldsymbol{\phi}, \boldsymbol{\Omega}) + \chi(\boldsymbol{\Omega}) \tag{3.366}$$

The shape derivative of augmented Lagrangian L is defined as

$$L' = J'(\mathbf{u}, \boldsymbol{\phi}, \boldsymbol{\Omega}) + \chi'(\boldsymbol{\Omega}) \tag{3.367}$$

$$J' = \int_{\Gamma} \mathbf{G} \cdot \boldsymbol{V}_n \, d\Gamma \tag{3.368}$$

where

$$\begin{aligned}
\mathbf{G} = & \int_{\Omega} C_1 \, \boldsymbol{\varepsilon}(\boldsymbol{u}')^T : \mathbb{C}^{bulk} : \boldsymbol{\varepsilon}(\boldsymbol{u}) \, d\Omega + \int_{\Gamma} C_1 \, \boldsymbol{\varepsilon}_s(\boldsymbol{u})^T : \mathbb{C}^s : \boldsymbol{\varepsilon}_s(\boldsymbol{u}) \, d\Gamma \\
& + \int_{\Omega} C_2 \, \mathbf{E}(\boldsymbol{\phi}')^T \cdot \boldsymbol{\kappa}^{bulk} \cdot \mathbf{E}(\boldsymbol{\phi}) \, d\Omega + \int_{\Gamma} C_2 \, \mathbf{E}_s(\boldsymbol{\phi})^T \cdot \boldsymbol{\kappa}^s \cdot \mathbf{E}_s(\boldsymbol{\phi}) \, d\Gamma \\
& - \int_{\Gamma} \boldsymbol{\varepsilon}_s(\boldsymbol{w}) \boldsymbol{\tau}_s \cdot \eta \, d\Gamma - \int_{\Gamma} \boldsymbol{\varepsilon}_s(\boldsymbol{u})^T : \mathbb{C}^s : \boldsymbol{\varepsilon}_s(\boldsymbol{w}) \cdot \eta \, d\Gamma \\
& + \int_{\Gamma} \boldsymbol{\varepsilon}_s(\boldsymbol{u})^T : \mathbf{e}^{sT} \cdot \boldsymbol{E}_s(\boldsymbol{\psi}) \cdot \eta \, d\Gamma + \int_{\Gamma} \mathbf{E}_s(\boldsymbol{\phi})^T \cdot e^s : \boldsymbol{\varepsilon}_s(\boldsymbol{w}) \, \eta \, d\Gamma \\
& + \int_{\Gamma} \mathbf{E}_s(\boldsymbol{\phi})^T \cdot \boldsymbol{\kappa}^s \cdot \boldsymbol{E}_s(\boldsymbol{\psi}) \, \eta \, d\Gamma - \int_{\Omega} \boldsymbol{\varepsilon}(\boldsymbol{u})^T : \mathbb{C}^{bulk} : \boldsymbol{\varepsilon}(\boldsymbol{w}) \, d\Omega \\
& + \int_{\Omega} \boldsymbol{\varepsilon}(\boldsymbol{u})^T : \mathbf{e}^{bulk^T} \cdot \mathbf{E}(\boldsymbol{\psi}) \, d\Gamma + \int_{\Omega} \mathbf{E}(\boldsymbol{\phi})^T \cdot e^{bulk} : \boldsymbol{\epsilon}(\boldsymbol{w}) \, d\Gamma \\
& + \int_{\Omega} \mathbf{E}(\boldsymbol{\phi})^T \cdot \boldsymbol{\kappa}^{bulk} \cdot \mathbf{E}(\boldsymbol{\psi}) \, d\Omega
\end{aligned} \tag{3.369}$$

Based on the steepest descent direction

$$\begin{aligned}
\mathbf{G} &= -\boldsymbol{V}_n \\
J' &= -\int_{\Gamma} \boldsymbol{V}_n^2 \, d\Gamma \leq 0
\end{aligned} \tag{3.370}$$

In Eq. (3.369), $\boldsymbol{u}$ and $\boldsymbol{\phi}$ are the actual displacement and potential variables respectively, while $\boldsymbol{w}$ and $\boldsymbol{\psi}$ are the adjoint displacement and potential variables. The displacement $\boldsymbol{u}$ and voltage $\boldsymbol{\phi}$ are obtained by solving Eqs. (3.341) and (3.342). The weak formulation for the adjoint equation is derived in Appendix B. The system of algebraic equations of the adjoint problem is as follows,

$$(\boldsymbol{K}_{uu}^b + \boldsymbol{K}_{uu}^s)\boldsymbol{w} + (\boldsymbol{K}_{\phi u}^b + \boldsymbol{K}_{\phi u}^s)\boldsymbol{\psi} = 2C_1(\boldsymbol{K}_{uu}^b + \boldsymbol{K}_{uu}^s)\boldsymbol{u} \tag{3.371}$$

$$(\boldsymbol{K}_{u\phi}^b + \boldsymbol{K}_{u\phi}^s)\boldsymbol{w} + (\boldsymbol{K}_{\phi\phi}^b + \boldsymbol{K}_{\phi\phi}^s)\boldsymbol{\psi} = 2C_2(\boldsymbol{K}_{\phi\phi}^b + \boldsymbol{K}_{\phi\phi}^s)\boldsymbol{\phi} \tag{3.372}$$

3.7.7.4 Numerical examples

In this section we study the energy harvesting capability of a piezoelectric nanobeam, nanoplate and a cantilever energy harvester with nanoscale

Table 3.3 Electromechanical properties of bulk ZnO.

Elastic constants	Piezoelectric constants	Dielectric constants
$C_{11} = 206$ GPa	$e_{31} = -0.58$ C/m^2	$\kappa_{11} = 8.11$ C/(GV m)
$C_{12} = 117$ GPa	$e_{33} = 1.55$ C/m^2	$\kappa_{33} = 11.2$ C/(GV m)
$C_{13} = 118$ GPa	$e_{15} = -0.48$ C/m^2	
$C_{33} = 211$ GPa		
$C_{44} = 44.3$ GPa		

Table 3.4 Electromechanical properties of ZnO surfaces.

Elastic constants	Piezoelectric constants
$C_{11}^s = 44.2$ N/m	$e_{31}^s = -0.216$ nC/m
$C_{12}^s = 14.2$ N/m	$e_{33}^s = 0.451$ nC/m
$C_{13}^s = 14.2$ N/m	$e_{15}^s = -0.253$ nC/m
$C_{33}^s = 35$ N/m	
$C_{44}^s = 11.7$ N/m	

piezoelectric layers. We also perform topology optimization of these piezoelectric structures. We consider examples both with and without surface elasticity and piezoelectricity to examine the effects that surface effects have on the energy harvesting ability. In all examples, ZnO is the piezoelectric material of choice, where the surface elastic and piezoelectric properties of ZnO are shown in Tables 3.3 and 3.4 [26].

Two-dimensional piezoelectric nanobeam

Optimization of two-dimensional piezoelectric nanobeam

We now perform topology optimization of the two-dimensional nanobeams under both open and closed circuit boundary conditions to examine not only the enhancements in energy conversion that are possible, but also to delineate the relative effects of surface piezoelectricity and elasticity on the energy conversion. To do so, we again subject the cantilever nanobeam to a mechanical point load acting vertically downwards at its free end, while the nanobeam is poled along the thickness direction.

The nanobeam is placed over a substrate made of material with a Young's modulus of E = 150 GPa, where the substrate dimensions are the same as the nanobeam, though surface effects on the substrate are neglected. As shown in Fig. 3.40, electrodes are placed above and below the nanobeam. Open circuit conditions are achieved by grounding the bottom

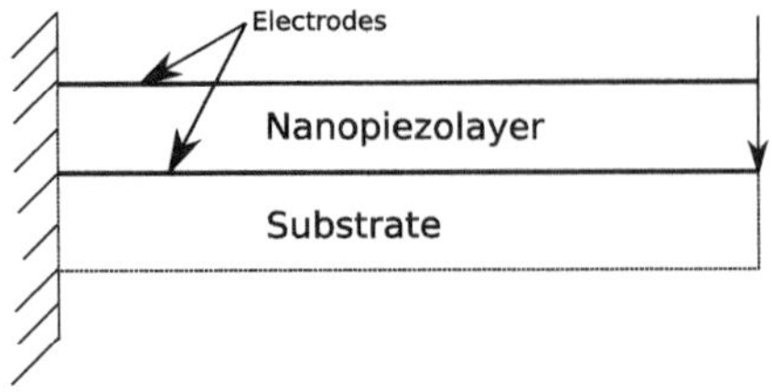

Figure 3.40 A cantilever beam energy harvester with nano piezoelectric layer and a substrate subject to a point load at the free end.

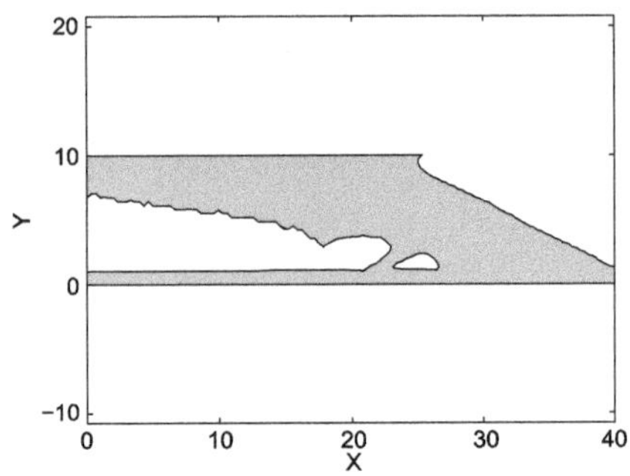

Figure 3.41 Optimal topology for maximizing the ECF of a ZnO nanobeam of dimensions 40 × 10 nm. Note that the substrate is not shown.

electrode while the top electrode is free to take any potential value, while in closed circuit both the top and bottom electrodes are grounded.

To give an example of an illustrative optimal topology, we show in Fig. 3.41 the optimized nanobeam topology for a 40 × 10 nm nanobeam for both open and closed circuit boundary conditions. The 40 × 10 nm beam was meshed with 120 × 30 Q4 elements, respectively. We note that a significant amount of material is removed from the region where the mechanical force is applied, along with the region where the beam is clamped, as shown in Fig. 3.41.

To explain the optimized structure that is obtained in Fig. 3.41, we first note that the piezoelectric layer lies above the neutral axis of the piezoelectric layer/substrate composite beam. Material is removed from the interior of the beam (i.e. between $x = 0$ and $x = 20$ nm in Fig. 3.41) rather than at the surface because the stresses and strains in the beam are largest at the surfaces, and not in the beam interior. Similarly, material close to the free end at $x = 40$ nm are relatively unstrained and thus can also be removed in the optimization.

Table 3.5 Nominal EMCC of optimized piezoelectric nanobeam under open circuit conditions.

Nanobeam dimensions (nm)	Nominal EMCC ($C^s \neq 0, e^s \neq 0$)
Aspect ratio = 4	
40 × 10	1.18
80 × 20	1.12
160 × 40	1.09
Aspect ratio = 8	
80 × 10	1.15
160 × 20	1.085
240 × 30	1.06

Table 3.6 Nominal EMCC of optimized piezoelectric nanobeam under closed circuit condition.

Nanobeam dimensions (nm)	Nominal EMCC ($C^s \neq 0, e^s \neq 0$)
Aspect ratio = 4	
40 × 10	2.3
80 × 20	2.05
160 × 40	1.9
Aspect ratio = 8	
80 × 10	2.35
160 × 20	2.1
240 × 30	2.0

The distribution of the material in the optimized beams also have a direct connection to the energy harvesting potential. Specifically, the nanobeam in Fig. 3.41 is only subject to mechanical loads, and so any electrical energy that is generated can only be due to the piezoelectric coupling constants. Material that is relatively unstrained is not needed for the mechanical integrity, and also contributes little to the generation of electrical energy. As the remaining material in the optimized structure is needed to resist mechanical loading, and is more highly strained, these regions will have larger electrical displacements, and thus conversion of mechanical to electrical energy, leading to increased EMCC.

Tables 3.5 and 3.6 show the nominal EMCC obtained for various geometries and aspect ratios considered for both open and closed circuit

Table 3.7 The mechanical and electrical energy values under open circuit boundary conditions for the 40 × 10 nm optimized beam shown in Fig. 3.41.

	Electrical energy (nJ)	Mechanical energy (nJ)	EMCC	Nominal EMCC
$C_s = 0,\ e_s \neq 0$	0.67	6.69	0.091	1.39
$C_s \neq 0,\ e_s \neq 0$	0.534	6.3	0.0781	1.19
$C_s \neq 0,\ e_s = 0$	0.442	6.47	0.064	0.97
$C_s = 0,\ e_s = 0$	0.543	6.924	0.0727	1.1

electrical boundary conditions, where the nominal EMCC is the ratio of the EMCC of the optimal topology with surface effects and the EMCC of the solid beam with surface effects. Different thicknesses were chosen to illustrate the size-dependent nature of the surface effects. There are several noticeable and interesting trends, which we now discuss.

The first effect is that, for both open and closed circuit boundary conditions as shown in Tables 3.5 and 3.6, the nominal EMCC decreases with increasing nanobeam thickness. In all cases the optimized EMCC is greater than one. This is because the inclusion of surface piezoelectric effects can only lead to an increase in the energy conversion for a solid ZnO nanobeam because the bulk and surface coupling constants e_{31} and e^s_{31} have the same sign [26], though the increase in energy conversion decreases with increasing nanobeam thickness.

In the optimal topology obtained, both surface piezoelectric and surface elastic effects compete to influence the energy conversion, as illustrated in Table 3.7 for the 40 × 10 nm nanobeam under the open circuit boundary condition, where we note that the following findings are also valid for the closed circuit boundary condition. To delineate these effects, we considered different combinations of including or neglecting surface piezoelectric and elastic effects via the corresponding surface elastic or piezoelectric stiffness tensors. As shown in Table 3.7, when surface elastic constants are ignored ($C_s = 0$), the nominal EMCC increases. If only surface elasticity is considered, both the mechanical and electrical energies are smaller than the no surface effects case ($C_s = 0,\ e_s = 0$), leading to a nominal EMCC of less than one. This is because the presence of surface elasticity increases the beam stiffness, which results in a decrease in strain. This decrease in strain reduces the resulting electrical energy to 0.442 nJ, which is the smallest among all cases shown in Table 3.7. For the purely surface piezoelectric case, the electrical energy is the largest among all the cases leading to the

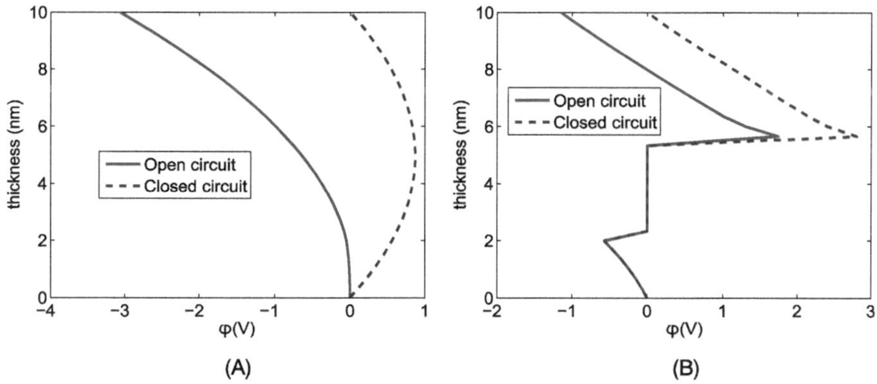

Figure 3.42 Electrical potential distribution across the thickness of the 40 × 10 nm, (A) Solid beam, (B) Optimal beam from Fig. 3.41, at $x = 10$ nm.

highest nominal EMCC. This parametric study demonstrates that surface elastic effects in ZnO act to reduce the energy conversion ability of the nanobeams while the surface piezoelectric effects enhance the energy conversion ability.

The second effect is that the nominal EMCC is substantially higher for the closed circuit boundary condition, which corresponds to the flow of current across the nanobeam cross section, as compared to the open circuit boundary condition, which corresponds to a build up of voltage across the nanobeam cross section. In the closed circuit condition, the potentials at the top and bottom surfaces of the piezoelectric layer are constrained to zero. However, during the optimization process, new surfaces are formed inside the beam. These new surfaces are not constrained to zero potential, which enables the EMCC for the closed circuit boundary condition to approach the open circuit value after optimization. This can also be observed by looking at the distribution of potential across the solid 40 × 10 nm beam at $x = 10$ nm as shown in Fig. 3.42A under open and closed circuit conditions. As can be seen in Fig. 3.42A for the solid beam with surface effects, the potential difference between the electrodes on the top and bottom surfaces is about 3 V for open circuit, but zero for closed circuit.

In contrast, the potential distribution is similar for both open and closed circuit conditions after optimization, as shown in Fig. 3.42B. In Fig. 3.41, it can be seen at $x = 10$ nm, there is material present from $y = 0$ to 1.3 nm, and then no material between $y = 1.3$ and 5.35 nm, which is the location of two newly formed free surfaces at $x = 10$. In Fig. 3.42B, it is shown that the potential difference between $y = 0$ and 1.3 nm is 0.34 V while the potential difference between $y = 5.35$ nm and $y = 10$ nm is around 2.85 V for both open and closed circuit conditions. Furthermore, the EMCC of the optimal 40×10 nm beam under closed circuit condition is 0.078, while under open circuit condition the EMCC is 0.0785, thus showing that as the potential distribution of open and closed circuit becomes similar for the optimal topology, the EMCC under closed circuit approaches the EMCC under open circuit condition. The EMCC of the solid 40×10 nm beam under closed circuit condition is 0.034 while under open circuit condition the EMCC is 0.067, which yields a nominal EMCC under open circuit condition of $\frac{0.0785}{0.067} = 1.18$, while under closed circuit condition, the nominal EMCC is $\frac{0.078}{0.034} = 2.3$. Therefore due to the appearance of new free surfaces in the optimal topology, the nominal EMCC under closed circuit condition is higher compared to the one under open circuit condition.

3.8. Conditioning and solution of ill-conditioned systems

The linear dependence of the XFEM shape functions and the enrichment leads to an ill-conditioned system matrix. The conditioning can be improved by orthogonalization of the enrichment functions. An alternative approach to improve the conditioning is to reduce the additional degrees of freedom as in the intrinsic XFEM [21], the Hansbo-Hansbo XFEM [23] or the phantom-node method [49,56]. The SXFEM [8] avoids an isoparametric mapping and also improves the conditioning of the system matrix. [50] proposed a multi-level approach to improve the conditioning of the system matrix.

An effective solution strategy to solve the ill-conditioned systems was proposed by [58]. Therefore, let us consider a linear system of equations of the form $\mathbf{AD} = \mathbf{F}$. The solution strategy was originally developed by [53] who showed that the matrix

$$\mathbf{A}_\epsilon := \mathbf{A} + \epsilon \mathbf{I} \tag{3.373}$$

is better conditioned than the matrix $\mathbf{A}$, ϵ being of the order of 10^{-12} and assuming that $\mathbf{A}_\epsilon$ is a positive definite and hence non-singular matrix.[12] Instead of solving the 'original' system of equations, the 'perturbed' system

[12] We assume that the values on the main diagonal of $\mathbf{A}$ are of the order 10^0 that can be achieved through scaling.

of equations is solved:

$$\mathbf{D}_0 = \mathbf{A}_\epsilon^{-1}\mathbf{F}, \quad \mathbf{r}_0 = \mathbf{F} - \mathbf{A}\mathbf{D}_0, \tag{3.374}$$

and subsequently

$$\mathbf{z}_i = \mathbf{A}_\epsilon^{-1}\mathbf{r}_i, \tag{3.375}$$

$$\mathbf{v}_i = \mathbf{A}\mathbf{z}_i, \tag{3.376}$$

$$\mathbf{D}_i = \mathbf{D}_0 + \sum_{j=0}^{i-1}\mathbf{z}_j, \tag{3.377}$$

$$\mathbf{r}_i = r_0 - \sum_{j=0}^{i-1}\mathbf{v}_j, \tag{3.378}$$

until

$$\frac{|\mathbf{z}_i^T\mathbf{A}\mathbf{z}_i|}{|\mathbf{D}_i^T\mathbf{A}\mathbf{D}_i|}$$

is 'sufficiently' small. Commonly, only a few iterations are needed until $\mathbf{D}_i$ converges to a solution of $\mathbf{AD} = \mathbf{F}$. The results are comparable to the results of a singular value decomposition but the proposed solution procedure is computationally less expensive.

References

[1] M.-A. Abellan, R. de Borst, Wave propagation and localisation in a softening two-phase medium, Computer Methods in Applied Mechanics and Engineering 195 (37) (2006) 5011–5019, https://doi.org/10.1016/j.cma.2005.05.056.
[2] G. Allaire, F. Jouve, A.M. Toader, Structural optimization using sensitivity analysis and a level-set method, Journal of Computational Physics 194 (2004) 363–393.
[3] D. Àlvarez, O. Dorn, N. Irishina, M. Moscoso, Crack reconstruction using a level-set strategy, Journal of Computational Physics 228 (16) (2009) 5710–5721.
[4] G. Bachelor, An Introduction to Fluid Mechanics, Cambridge University Press, Cambridge, UK, 1967.
[5] J. Baumeister, A. Leitão, Topics in inverse problems, in: Lecture Notes: 25th CBM, IMPA, July 2005.
[6] E. Béchet, H. Minnebo, N. Möes, B. Burgardt, Improved implementation and robustness study of the X-FEM for stress analysis around cracks, International Journal for Numerical Methods in Engineering 64 (8) (2005) 1033–1056.
[7] E. Béchet, M. Scherzer, M. Kuna, Application of the X-FEM to the fracture of piezoelectric materials, International Journal for Numerical Methods in Engineering 77 (2009) 1535–1565.
[8] S. Bordas, T. Rabczuk, H. Nguyen-Xuan, S. Natarajan, T. Bog, V. Nguyen, Q.D. Minh, H.N. Vinh, Strain smoothing in FEM and XFEM, Computers & Structures 88 (2010) 1419–1443.

[9] L. Burianova, C.R. Bowen, M. Prokopova, M. Sulc, Laser interferometric displacement measurements of multi-layer actuators and PZT ceramics, Ferroelectrics 320 (2005) 161–169.

[10] R. Calio, U.B. Rongala, D. Camboni, M. Milazzo, C. Stefanini, G. Petris, C.M. Oddo, Piezoelectric energy harvesting solutions, Sensors 14 (2014) 4755–4790.

[11] R.C. Cammarata, Surface and interface stress effects in thin films, Progress in Surface Science 46 (1) (1994) 1–38.

[12] J. Chen, C. Wu, S. Yoon, Y. You, A stabilized conforming nodal integration for Galerkin meshfree-methods, International Journal for Numerical Methods in Engineering 50 (2001) 435–466.

[13] L. Chen, T. Rabczuk, G. Liu, K. Zeng, P. Kerfriden, S. Bordas, Extended finite element method with edge-based strain smoothing (ESm-XFEM) for linear elastic crack growth, Computer Methods in Applied Mechanics and Engineering 209–212 (2012) 250–265.

[14] K. Cheng, T. Fries, Higher-order XFEM for curved strong and weak discontinuities, International Journal for Numerical Methods in Engineering 82 (5) (2010) 564–590, https://doi.org/10.1002/nme.2768.

[15] J. Chessa, H. Wang, T. Belytschko, On the construction of blending elements for local partition of unity enriched finite elements, International Journal for Numerical Methods in Engineering 57 (7) (2003) 1015–1038.

[16] K.K. Choi, N.H. Kim, Structural Sensitivity Analysis and Optimization, Springer, New York, 2005.

[17] H. Darcy, Les fontaines publiques de la ville de Dijon: exposition et application, Victor Dalmont, 1856.

[18] M. Duflot, The extended finite element method in thermoelastic fracture mechanics, International Journal for Numerical Methods in Engineering 74 (2007) 827–847.

[19] M. Enderlein, A. Ricoeur, M. Kuna, Finite element techniques for dynamic crack analysis in piezoelectrics, International Journal of Fracture 134 (2005) 191–208.

[20] M. Farsad, F.J. Vernerey, H.S. Park, An extended finite element/level set method to study surface effects on the mechanical behavior and properties of nanomaterials, International Journal for Numerical Methods in Engineering 84 (2010) 1466–1489.

[21] T. Fries, T. Belytschko, The intrinsic XFEM: a method for arbitrary discontinuities without additional unknowns, International Journal for Numerical Methods in Engineering 68 (2006) 1358–1385.

[22] M.E. Gurtin, A. Murdoch, A continuum theory of elastic material surfaces, Archive for Rational Mechanics and Analysis 57 (1975) 291–323.

[23] A. Hansbo, P. Hansbo, A finite element method for the simulation of strong and weak discontinuities in solid mechanics, Computer Methods in Applied Mechanics and Engineering 193 (33–35) (2004) 3523–3540.

[24] P. Hansen, D. OĹeary, The use of the L-curve in the regularization of discrete ill-posed problems, SIAM Journal on Scientific Computing 14 (1993) 1487–1503.

[25] S.M. Hassanizadeh, W.G. Gray, Mechanics and thermodynamics of multiphase flow in porous media including interphase boundaries, Advances in Water Resources 13 (4) (1990) 169–186, https://doi.org/10.1016/0309-1708(90)90040-B.

[26] M.T. Hoang, J. Yvonnet, A. Mitrushchenkov, G. Chambaud, First-principles based multiscale model of piezoelectric nanowires with surface effects, Journal of Applied Physics 113 (1) (2013) 014309.

[27] M. Kaltenbacher, Numerical simulation of mechatronic sensors and actuators, International Journal of Fracture 134 (2005) 191–208.

[28] M. Kuna, Finite element analyses of cracks in piezoelectric structures – a survey, Computer Aided Design 76 (2006) 725–745.

[29] M. Kuna, Fracture mechanics of piezoelectric materials? Where are we right now? Engineering Fracture Mechanics 77 (2010) 3635–3647.

[30] R.W. Lewis, B.A. Schrefler, The Finite Element Method in the Deformation and Consolidation of Porous Media, John Wiley and Sons Inc., New York, NY, 1987.

[31] M. Li, J.X. Yuan, D. Guan, W. Chen, Application of piezoelectric fiber composite actuator to aircraft wing for aerodynamic performance improvement, Science China. Technological Sciences 54 (2011) 394–408.

[32] G. Liu, N.-T. Trung, Smoothed Finite Element Methods, CRC Press, 2016.

[33] G. Liu, K. Dai, T. Nguyen, A smoothed finite element for mechanics problems, Computational Mechanics 39 (2007) 859–877.

[34] R. Malladi, J.A. Sethian, B.C. Vemuri, Shape modeling with front propagation: a level set approach, IEEE Transactions on Pattern Analysis and Machine Intelligence 17 (1995) 158–175.

[35] A. Marin, S. Bressers, S. Priya, Multiple cell configuration electromagnetic vibration energy harvester, Journal of Physics. D, Applied Physics 44 (2011) 295501.

[36] T. Martin, F. Pigache, S. Martin, Measurement of the electric potential distribution on piezoelectric ceramic surface, in: IEEE 11th International Workshop of Electronics, Control, Measurement, Signals and Their Application to Mechatronics (ECMSM), June 2013, pp. 1–5.

[37] V.S. Mashkevich, K. Tolpygo, Electrical, optical and elastic properties of diamond crystals, Soviet Physics, JETP-USSR 5 (1957) 435–439.

[38] J. Mergheim, E. Kuhl, P. Steinmann, A finite element method for the computational modelling of cohesive cracks, International Journal for Numerical Methods in Engineering 63 (2005) 276–289.

[39] C. Mi, S. Jun, D.A. Kouris, S.Y. Kim, Atomistic calculations of interface elastic properties in noncoherent metallic bilayers, Physical Review. B 77 (2008) 075425.

[40] N. Moes, M. Cloirec, P. Cartraud, J.F. Remacle, A computational approach to handle complex microstructure geometries, Computer Methods in Applied Mechanics and Engineering 192 (2003) 3163–3177.

[41] S.S. Nanthakumar, N. Valizadeh, H.S. Park, Surface effects on shape and topology optimization of nanostructures, Computational Mechanics 56 (1) (2015) 97–112.

[42] S.S. Nanthakumar, T. Lahmer, X. Zhuang, G. Zi, T. Rabczuk, Detection of material interfaces using a regularized level set method in piezoelectric structures, Inverse Problems in Science and Engineering 24 (1) (2016) 153–176.

[43] S.S. Nanthakumar, T. Lahmer, X. Zhuang, H.S. Park, T. Rabczuk, Topology optimization of piezoelectric nanostructures, Journal of the Mechanics and Physics of Solids 94 (2016) 153–176.

[44] S. Osher, J.A. Sethian, Front propagating with curvature dependent speed: algorithms based on Hamilton-Jacobi formulations, Journal of Computational Physics 78 (1988) 12–49.

[45] Y.E. Pak, Linear electro-elastic fracture mechanics of piezoelectric materials, International Journal of Fracture 54 (1992) 79–100.

[46] H.S. Park, W. Cai, H.D. Espinosa, H. Huang, Mechanics of crystalline nanowires, MRS Bulletin 34 (3) (2009) 178–183.

[47] S.W. Park, J.W. Kim, C.S. Lee, Effect of injector type on fuel-air mixture formulation of high-speed diesel sprays, Proceedings of the Institution of Mechanical Engineers. Part D, Journal of Automobile Engineering 220 (2006) 647.
[48] Q.H. Qin, Fracture Mechanics of Piezoelectric Materials, WIT Press, Southampton, Boston, 2001.
[49] T. Rabczuk, G. Zi, A. Gerstenberger, W. Wall, A new crack tip element for the phantom node method with arbitrary cohesive cracks, International Journal for Numerical Methods in Engineering 75 (2008) 577–599.
[50] J. Rannou, A. Gravouil, M. Baietto-Dubourg, A local multigrid X-FEM strategy for 3-d crack propagation, International Journal for Numerical Methods in Engineering 77 (2009) 581–600.
[51] L. Reichela, H. Sadok, A new L-curve for ill-posed problems, Journal of Computational and Applied Mathematics 219 (2008) 493–508.
[52] M. Rezghi, S.M. Hosseini, A new variant of L-curve for Tikhonov regularization, Journal of Computational and Applied Mathematics 231 (2009) 914–924.
[53] J. Riley, Solving systems of linear equations with a positive definite, symmetric, but possibly ill-conditioned matrix, Mathematical Tables and Other Aids To Computation 9 (1955) 96–101.
[54] A. Safari, E.K. Akdogan, Piezoelectric and Acoustic Materials for Transducer Applications, Springer, 2007.
[55] O. Sigmund, A 99 line topology optimization code written in Matlab, Structural and Multidisciplinary Optimization 21 (2001) 120–127.
[56] J.-H. Song, P. Areias, T. Belytschko, A method for dynamic crack and shear band propagation with phantom nodes, International Journal for Numerical Methods in Engineering 67 (6) (2006) 868–893.
[57] H. Sosa, Plane problems in piezoelectric media with defects, International Journal of Solids and Structures 28 (4) (1991) 491–505.
[58] T. Strouboulis, I. Babuška, K. Copps, The design and analysis of the generalized finite element method, International Journal for Numerical Methods in Engineering 181 (2000) 43–69.
[59] N. Sukumar, D.L. Chopp, N. Moes, T. Belytschko, Modeling holes and inclusions by level sets in the extended finite-element method, Computer Methods in Applied Mechanics and Engineering 190 (2001) 6183–6200.
[60] Z. Suo, C.M. Kuo, D.M. Barnett, J.R. Willis, Fracture mechanics for piezoelectric ceramics, Journal of the Mechanics and Physics of Solids 40 (4) (1992) 739–765.
[61] M.A. Trindade, A. Benjeddou, Effective electromechanical coupling coefficients of piezoelectric adaptive structures: critical evaluation and optimization, Mechanics of Advanced Materials and Structures 16 (3) (2009) 210–223.
[62] N.P. van Dijk, K. Maute, M. Langelaar, F. van Keulen, Level-set methods for structural topology optimization: a review, Structural and Multidisciplinary Optimization 48 (2013) 437–472.
[63] C. Vogel, Computational Methods for Inverse Problems, Society for Industrial and Applied Mathematics, 2002.
[64] N. Vu Bac, H. Nguyen-Xuan, L.C.S. Bordas, P. Kerfriden, R. Simpson, G. Liu, T. Rabczuk, A node-based smoothed extended finite element method (NS-XFEM) for fracture analysis, Computer Modeling in Engineering & Sciences 1898 (2011) 1–25.
[65] M.Y. Wang, X.M. Wang, D.M. Guo, A level set method for structural topology optimization, Computer Methods in Applied Mechanics and Engineering 192 (2003) 217–224.

[66] Z.L. Wang, J. Song, Piezoelectric nanogenerators based on zinc oxide nanowire arrays, Science 312 (2006) 242–246.
[67] S. Xu, Y. Qin, C. Xu, Y. Wei, R. Yang, Z.L. Wang, Self-powered nanowire devices, Nature Nanotechnology 5 (2010) 366–373.
[68] X.L. Xu, R.K.N.D. Rajapakse, Analytical solution for an arbitrarily oriented void/crack and fracture of piezoceramics, Acta Materialia 47 (6) (1999) 1735–1747.
[69] J. Yvonnet, A. Mitrushchenkov, G. Chambaud, Q.-C. He, Finite element model of ionic nanowires with size-dependent mechanical properties determined by ab initio calculations, Computer Methods in Applied Mechanics and Engineering 200 (2011) 614–625.
[70] G. Zi, J.-H. Song, E. Budyn, S.-H. Lee, T. Belytschko, A method for growing multiple cracks without remeshing and its application to fatigue crack growth, Modelling and Simulation in Materials Science and Engineering 12 (1) (2004) 901–915.

CHAPTER FOUR

Phantom node method

4.1. Formulation and concepts

As mentioned already in Section 3.1.2, an alternative to the standard XFEM for strong discontinuities was proposed by Hansbo and Hansbo [6]. This approach was later implemented for complex fracture problems in dynamics and named the phantom-node method Song et al. [12]. The basic idea in one dimension was already explained earlier and we will now focus on higher dimensions. Therefore, consider a body that is cracked as shown in Fig. 4.1 and the corresponding finite element discretization. Because of the crack, there are cracked elements cut by the crack. To have a set of full interpolation bases, the part of the cracked elements which belongs in the *real domain* Ω_0 are extended to the *phantom domain* Ω_{P}. Then the displacement in the real domain Ω_0 can be interpolated by using the degrees of freedom for the nodes in the phantom domain Ω_{P}. The nodes are called the *phantom nodes* and marked by empty circles in Fig. 4.1. The approximation of the displacement field is then given by [12]:

$$\mathbf{u}^h(\mathbf{X}, t) = \sum_{I \in \{\mathcal{W}_0^+,\, \mathcal{W}_{\mathrm{P}}^-\}} \mathbf{u}_I(t)\; N_I(\mathbf{X}) H(f(\mathbf{X})) + \sum_{J \in \{\mathcal{W}_0^-,\, \mathcal{W}_{\mathrm{P}}^+\}} \mathbf{u}_J(t)\; N_J(\mathbf{X}) H(-f(\mathbf{X})) \tag{4.1}$$

where $f(\mathbf{X})$ is the signed distance measured from the crack, $\mathcal{W}_0^+$, $\mathcal{W}_0^-$, $\mathcal{W}_{\mathrm{P}}^+$ and $\mathcal{W}_{\mathrm{P}}^-$ are nodes belonging to Ω_0^+, Ω_0^-, Ω_{P}^+ and Ω_{P}^-, respectively. $H(x)$ is the Heaviside function. As can be seen from Fig. 4.1, cracked elements have both real nodes and phantom nodes. The jump in the displacement field is realized by simply integrating only over the area from the side of the real nodes up to the crack, i.e. Ω_0^+ and Ω_0^-. We note that there are certain similarities to the visibility method used originally in meshfree methods, [2–4,10].

The key advantages over standard XFEM are:

- The phantom node method avoids the 'mixed' terms $\mathbf{K}^{ua}$ and $\mathbf{K}^{au}$ (or $\mathbf{M}^{ua}$ and $\mathbf{M}^{au}$).
- The phantom node method leads to a better conditioned system matrix.

Extended Finite Element and Meshfree Methods
https://doi.org/10.1016/B978-0-12-814106-9.00010-3
Copyright © 2020 Elsevier Inc.
All rights reserved.

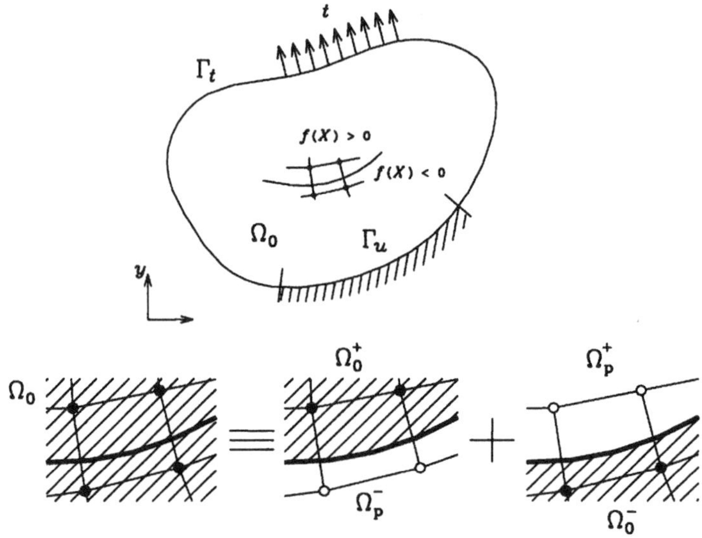

Figure 4.1 The principle of the phantom node method in which the hatched area is integrated to build the discrete momentum equation; the solid circles represent real nodes and the empty ones phantom nodes.

- Standard mass lumping schemes such as row-sum technique can be used. Though mass lumping strategies were developed for XFEM [7, 8], they depend on assumptions such as conservation of *kinetic* energy that is not guaranteed for general applications.
- The implementation of the phantom node method is simpler.

The key drawback is that the phantom node method is less flexible than standard XFEM. For example, the crack needs to be cross the entire element due to the absence of a crack tip enrichment. Therefore, Rabczuk et al. [11] proposed a crack tip phantom node element for triangular and quadrilateral elements that is explained in the next section.

4.2. A crack tip element for the phantom node methods

4.2.1 Three-node triangular element

The crack opening displacement $[\![\mathbf{u}]\!]$ should vanish at the crack tip. In the classical extended finite element method, the condition was fulfilled by using a nonsingular branch enrichment [9] or by a special element as proposed by Zi and Belytschko [13]; see also Section 3.3.1.

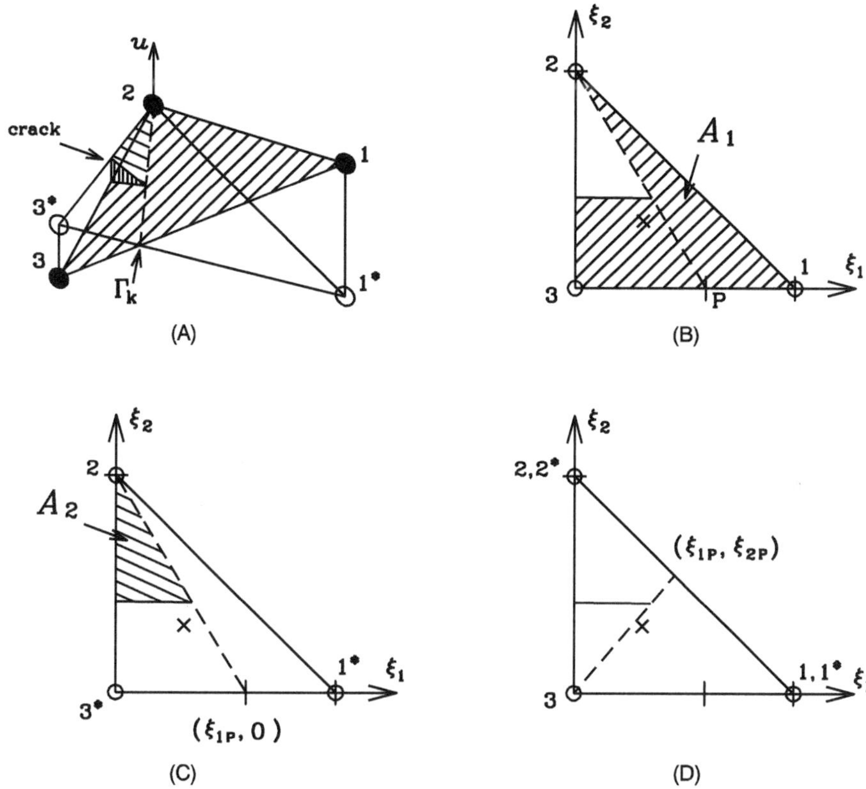

Figure 4.2 (A) A three-node tip element for the phantom node method in which solid circles represent physical nodes and empty symbols the extra degrees of freedom to be determined by a kinematical constraint, (B) and (C) the areas to be integrated and (D) another case of crack development.

A simple crack tip element for the phantom node method can be devised easily in the extension line of Zi and Belytschko [13]'s method. First, expand the small triangle representing the enriched displacement in Fig. 4.2 to obtain a new triangle. The new triangular element 1*2 3* is shown in Fig. 4.2A. Node 2 is shared by the two elements. The difference in the displacement of the two elements vanishes at Γ_k.

Integrating area A_1 in element 1 2 3 and A_2 in element 1*2 3*, we then have the same displacement field as in Fig. 4.2. It is trivial that the sum of A_1 and A_2 is equal to the area of element 1 2 3. Note that there are only three physical nodes which are the solid circles in Fig. 4.2A. The nodal values of the extra degrees of freedom, marked by the empty circles in

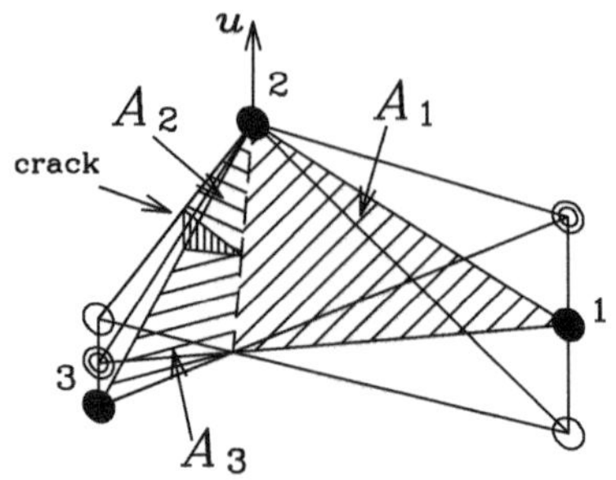

Figure 4.3 An alternative three-node tip element for the phantom node method in which solid circles represent physical nodes and empty symbols, the extra degrees of freedom to be determined by a kinematical constraint.

Fig. 4.2A should be determined from a kinematical constraint. For the case illustrated in Fig. 4.2A, B, C, displacement $\mathbf{u}^*$ is given by

$$\xi_{1P}(\mathbf{u}_1^* - \mathbf{u}_1) + (1 - \xi_{1P})(\mathbf{u}_3^* - \mathbf{u}_3) = 0 \tag{4.2}$$

$$(\mathbf{u}_2^* - \mathbf{u}_2) = 0 \tag{4.3}$$

in which $(\xi_{1P}, 0)$ is the intersection point of Γ_k and edge 31 in the parent coordinate of the element; see Fig. 4.2B, C.

There can be another case as in Fig. 4.2D such that node 3 is shared. The displacement $\mathbf{u}^*$ is given by

$$\xi_{1P}(\mathbf{u}_1^* - \mathbf{u}_1) + \xi_{2P}(\mathbf{u}_2^* - \mathbf{u}_2) = 0 \tag{4.4}$$

$$(\mathbf{u}_3^* - \mathbf{u}_3) = 0 \tag{4.5}$$

in which (ξ_{1P}, ξ_{2P}) is the intersection point of Γ_k and edge 12 in the parent coordinate of the element; see Fig. 4.2D. When the position of the crack tip is close to edge 12, this would produce a well-suited Γ_k. If the crack is too close to an element side or node, there is no need to close the crack within an element and the crack is closed as in the conventional phantom node method.

The approach mentioned above is not the only way. There is an alternative where three elements may be used; see Fig. 4.3. Again, the difference in the displacement of those elements vanishes at Γ_k. The sum of A_1, A_2 and A_3 is equal to the area of the triangle. The value of the extra degrees of freedom can be determined in a similar way as Eqs. (4.2) to (4.5).

To accurately integrate the hatched area in Fig. 4.2 or 4.3, it is inevitable to project internal variables stored in a set of quadrature points to another for general quadrature rules. The subtriangulation of a cracked element is commonly used for the numerical integration [9]. The standard quadrature scheme without the subtriangulation is also possible, in which there is no projection of internal variables. In that case, the use of a sufficiently high order quadrature is necessary for the reduction of the integration error due to the crack. If the one point quadrature is used, there is no need for the projection. The integration of the hatched area is carried out by, simply, the product of the area fraction A_1/A or A_2/A and the integration for the original triangle, in which A is the area of the original triangle 1 2 3.

The conventional phantom node method is used if the crack tip is too close to an edge or too close to a node, which is employed in the standard XFEM techniques [5].

4.2.2 Four-node quadrilateral element

The method developed with three-node triangular elements can be applied for four-node quadrilateral elements with a slight difference. The standard displacement field of a four-node quadrilateral element is bilinear. Therefore if we overlap two quadrilateral elements together, similar to the case with the three-node triangular elements, the intersection Γ_k is not always a line. The only case in which Γ_k is a line is when the intersection is parallel to one of the axes of the parent coordinates.

Let a crack always grow to the element from edge 41; see Fig. 4.4. Consider four phantom nodes 3*, 4*, 1** and 2**, and three rectangles 1 2 3 4, 1 2 3*4* and 1**2**3 4. The only case that there is a single line intersection Γ_k is that Γ_k is parallel to the axis of ξ_2. The intersection between Γ_k and edge 34 or 12 is point $(\xi_{1P}, -1)$ or $(\xi_{1P}, 1)$.

From the condition that the difference of the displacement must be equal to zero along Γ_k, we can devise a kinematical constraint corresponding to the configuration discussed above is given by

$$\mathbf{u}_3^* = \mathbf{u}_3 - k_c(\mathbf{u}_4^* - \mathbf{u}_4) \tag{4.6}$$

$$\mathbf{u}_2^{**} = \mathbf{u}_2 - k_c(\mathbf{u}_1^{**} - \mathbf{u}_1) \tag{4.7}$$

in which $k_c = (1 + \xi_{1P})/(1 - \xi_{1P})$. Note $\mathbf{u}_4^*$ and $\mathbf{u}_1^{**}$ are determined from the continuity condition with the adjacent elements. If Belytschko and Bindemann [1]'s element is used as the base element, the one point quadrature can be adopted for the four-node quadrilateral element.

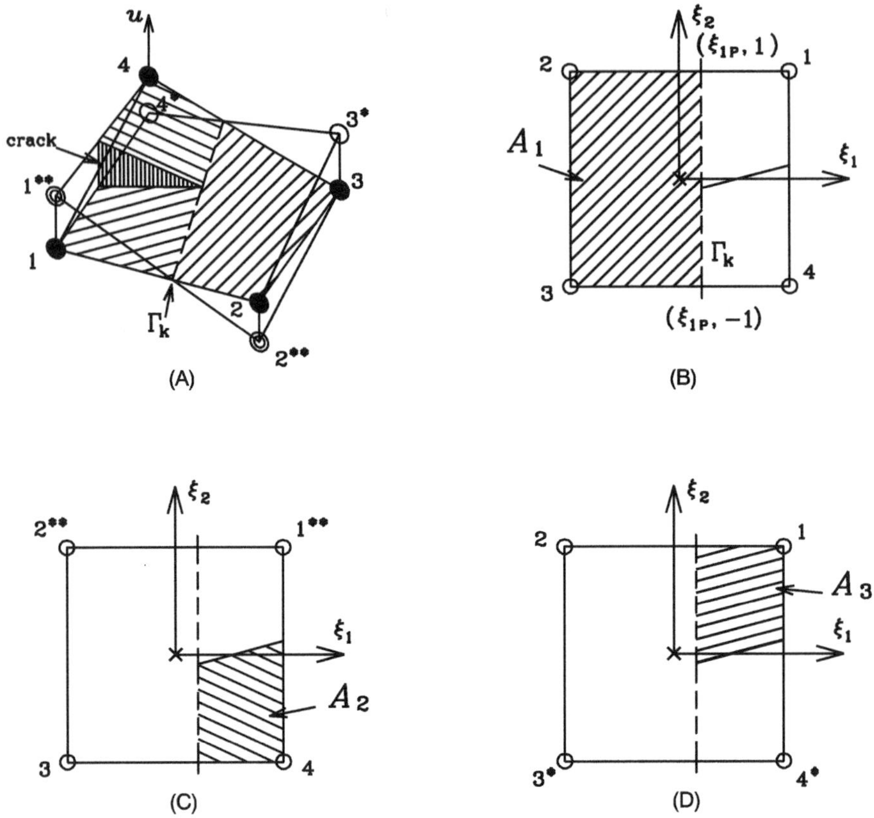

Figure 4.4 (A) A four-node tip element for the phantom node method in which solid circles represent physical nodes and empty symbols, the extra degrees of freedom to be determined by a kinematical constraint, and (B), (C) and (D) the areas to be integrated.

4.3. Multiple crack modeling

The concept of the overlapping element method can be easily extended to crack branch modeling. When the original crack, *crack 1*, branches into *crack 1* and *crack 2*, as shown in Fig. 4.5, the element in which the crack branches is replaced by three overlapping elements.

Let $f^1(\boldsymbol{X}) = 0$ describe the original crack and one branch, and let $f^2(\boldsymbol{X}) = 0$ describe the second branch. The discontinuous velocity field is then given by

$$\boldsymbol{u}(\boldsymbol{X}, t) = \boldsymbol{u}^{e_1}(\boldsymbol{X}, t) + \boldsymbol{u}^{e_2}(\boldsymbol{X}, t) + \boldsymbol{u}^{e_3}(\boldsymbol{X}, t) \tag{4.8}$$

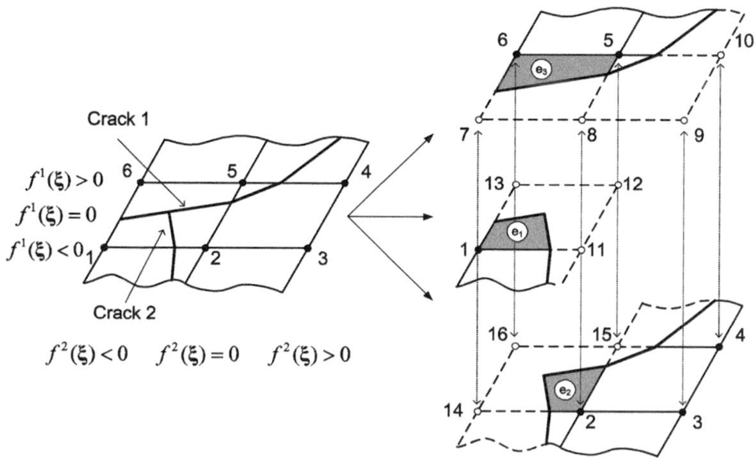

Figure 4.5 The decomposition of an element into three elements e_1, e_2 and e_3 to model crack branching; solid and hollow circles denote the original nodes and the added phantom nodes, respectively.

$$
\begin{aligned}
&= \sum_{I\in S_1} N_I(\mathbf{X})H(-f^1(\mathbf{X}))H(-f^2(\mathbf{X}))\boldsymbol{u}_I^{e_1}(t) \\
&+ \sum_{I\in S_2} N_I(\mathbf{X})H(-f^1(\mathbf{X}))H(f^2(\mathbf{X}))\boldsymbol{u}_I^{e_2}(t) \\
&+ \sum_{I\in S_3} N_I(\mathbf{X})H(f^1(\mathbf{X}))H(-f^2(\mathbf{X}))\boldsymbol{u}_I^{e_3}(t)
\end{aligned}
$$

For each of the overlapped elements modeling a crack, the nodal forces are given by the sum of the element internal force vector calculated with the activated area of the corresponding overlapping elements.

References

[1] T. Belytschko, L. Bindemann, Assumed strain stabilization of the 4-node quadrilateral with 1-point quadrature for nonlinear problems, Computer Methods in Applied Mechanics and Engineering 88 (1991) 311–340.

[2] T. Belytschko, Y. Lu, Element-free Galerkin methods for static and dynamic fracture, International Journal of Solids and Structures 32 (1995) 2547–2570.

[3] T. Belytschko, Y. Lu, L. Gu, Crack propagation by element-free Galerkin methods, Engineering Fracture Mechanics 51 (2) (1995) 295–315.

[4] T. Belytschko, D. Organ, M. Tabbara, Numerical simulations of mixed mode dynamic fracture in concrete using element-free Galerkin methods, in: ICES Conference Proceedings, 1995.

[5] J. Dolbow, N. Moes, T. Belytschko, Discontinuous enrichment in finite elements with a partition of unity method, Finite Elements in Analysis and Design 36 (3) (2000) 235–260.

[6] A. Hansbo, P. Hansbo, A finite element method for the simulation of strong and weak discontinuities in solid mechanics, Computer Methods in Applied Mechanics and Engineering 193 (33–35) (2004) 3523–3540.

[7] T. Menouillard, J. Rethore, A. Combescure, H. Bung, Efficient explicit time stepping for the extended finite element method (X-FEM), International Journal for Numerical Methods in Engineering 68 (9) (2006) 911–939.

[8] T. Menouillard, J. Rethore, N. Moes, A. Combescure, H. Bung, Mass lumping strategies for X-FEM explicit dynamics: application to crack propagation, International Journal for Numerical Methods in Engineering 74 (3) (2008) 447–474.

[9] N. Moes, J. Dolbow, T. Belytschko, A finite element method for crack growth without remeshing, International Journal for Numerical Methods in Engineering 46 (1) (1999) 133–150.

[10] D. Organ, M. Fleming, T. Terry, T. Belytschko, Continuous meshless approximations for nonconvex bodies by diffraction and transparency, Computational Mechanics 18 (1996) 225–235.

[11] T. Rabczuk, G. Zi, A. Gerstenberger, W. Wall, A new crack tip element for the phantom node method with arbitrary cohesive cracks, International Journal for Numerical Methods in Engineering 75 (2008) 577–599.

[12] J.-H. Song, P. Areias, T. Belytschko, A method for dynamic crack and shear band propagation with phantom nodes, International Journal for Numerical Methods in Engineering 67 (6) (2006) 868–893.

[13] G. Zi, T. Belytschko, New crack-tip elements for XFEM and applications to cohesive cracks, International Journal for Numerical Methods in Engineering 57 (15) (2003) 2221–2240.

CHAPTER FIVE

Extended meshfree methods

5.1. Introduction to meshfree methods

5.1.1 Basic approximation

Meshfree approximations for a scalar function u in terms of the material (Lagrangian) coordinates can be written as

$$u(\mathbf{X}, t) = \sum_{J \in \mathcal{S}} \Phi_J(\mathbf{X})\ u_J(t) \tag{5.1}$$

where $\Phi_J(\mathbf{X})$ are the shape functions and u_J is the value at the particle at the position $\mathbf{X}_J$ and $\mathcal{S}$ is the set of nodes J for which $\Phi_J(\mathbf{X}) \neq 0$. Note, that the above form is identical to the form of an FEM approximation. However, in contrast to an FEM, Eq. (5.1) is only an approximant and not an interpolant, since $u_I \neq u(\mathbf{X}_I)$. Therefore special techniques are needed to treat displacement boundary conditions.

The approximation can also be formulated in terms of the spatial (Eulerian) coordinates. In that case, it is conventional to approximate the velocities, instead of the displacements, so

$$\dot{u}(\mathbf{x}, t) = \sum_{J \in \mathcal{S}} \Phi_J(\mathbf{x})\ \dot{u}_J(t) \tag{5.2}$$

Eqs. (5.1) and (5.2) are the Lagrangian and Eulerian descriptions of motion, respectively. Note, that (5.2) does not correspond to the standard description of motion for a solid as given in (2.15). Furthermore, if all dependent variables such as the stress, strain and state variables are expressed in terms of the spatial coordinates $\mathbf{x}$ and time t, then transport terms must be included in the updated equations. The shape functions $\Phi_J(\mathbf{X})$ or $\Phi_J(\mathbf{x})$, are obtained from the weighting or kernel functions, which are denoted by $W_J(\mathbf{X})$ or $W_J(\mathbf{x})$. Usually radial kernel functions are chosen. For Lagrangian radial kernels $W_J(\mathbf{X}) = W(r_0)$ where $r_0 = \|\mathbf{X} - \mathbf{X}_J\|$ while for Eulerian kernels $W_J(\mathbf{x}, t) = W(r)$ where $r = \|\mathbf{x} - \mathbf{x}_J(t)\|$. As noted later, the time dependence of Eulerian kernels is always neglected, see [46,89,102]. The kernel functions are chosen to have compact support, i.e. $\Phi_J(\mathbf{X}) = 0$ if $\|\mathbf{X} - \mathbf{X}_J\| > h_J$ where h_J is the "interpolation" radius often called dilation

Extended Finite Element and Meshfree Methods
https://doi.org/10.1016/B978-0-12-814106-9.00011-5
Copyright © 2020 Elsevier Inc.
All rights reserved.

parameter. A very popular kernel function is the cubic B-spline given by:

$$W(r) = \begin{cases} \frac{C}{h^D}\left(1 - 1.5z^2 + 0.75z^3\right) & 0 \le z < 1 \\ \frac{C}{4\,h^D}\,(2-z)^3 & 1 \le z \le 2 \\ 0 & z > 2 \end{cases} \tag{5.3}$$

where D is the dimension, $z = r/h$ and C is a constant depending on the dimension

$$C = \begin{cases} 2/3 & D=1 \\ 10/(7\,\pi) & D=2 \\ 1/\pi & D=3 \end{cases} \tag{5.4}$$

The quartic spline is given by:

$$W(r) = \begin{cases} 1 - 6z^2 + 8z^3 - 3z^4 & 0 \le z < 1 \\ 0 & 1 \le z \end{cases} \tag{5.5}$$

Other popular kernels are:

$$W(r) = \begin{cases} \|\mathbf{x} - \mathbf{x}_I\| \equiv r & \text{linear} \\ z^2 \log z & \text{thin plate spline} \\ e^{-z^2/c^2} & \text{Gaussian} \\ \left(z^2 + R^2\right)^q & \text{multipolar} \end{cases} \tag{5.6}$$

where c, R and q are shape parameters.

The support of the kernel function is also called the domain of influence and corresponds to $\mathcal{S}$. A typical kernel function and its first spatial derivative in one dimension is shown in Fig. 5.1. A typical shape function in two dimensions with corresponding first and second spatial derivatives is shown in Fig. 5.2. As can be seen, the shape functions are very smooth. How the kernel function is related to the shape function depends on the approximation technique and is explained later for some specific methods.

5.1.2 Completeness and conservation

Completeness in Galerkin methods plays the same role as consistency in finite difference methods, [16]. While in finite differences, consistency describes how good the difference scheme approximates the differential operator, completeness is expressed in terms of the order of the polynomial which must be represented exactly. An approximation that is able to reproduce constant functions exactly is called zeroth-order complete, an

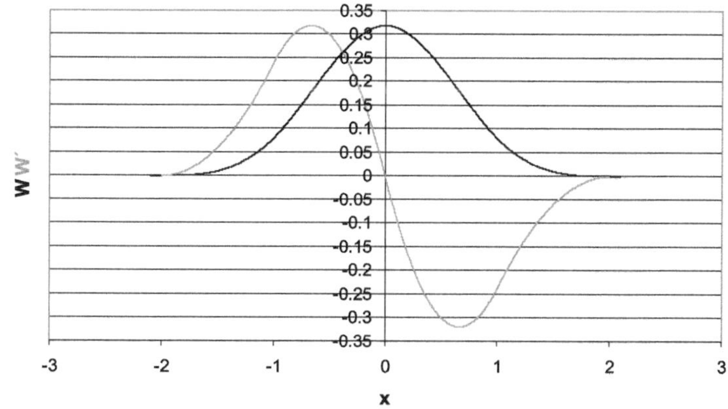

Figure 5.1 Typical kernel function and its first derivative in one dimension.

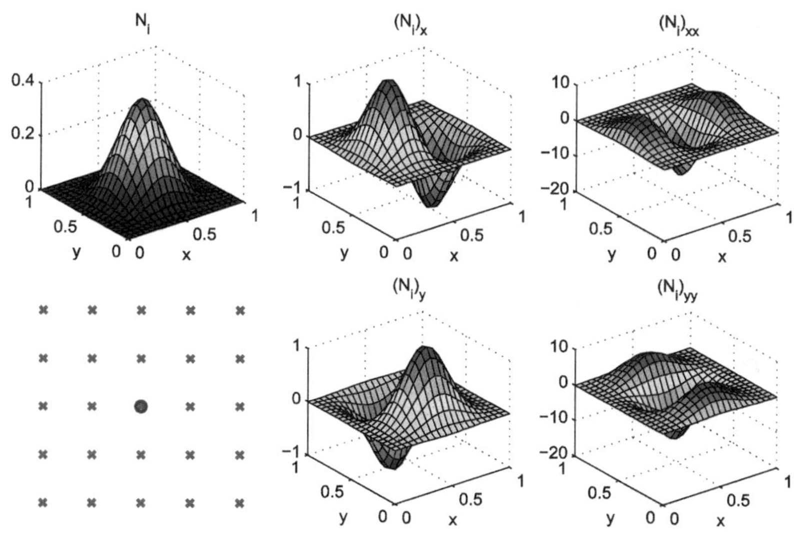

Figure 5.2 Distribution of particles, EFG approximation function and derivatives, with $\rho/h \simeq 2.2$ with circular supported cubic spline and linear consistency, from [56].

approximation that is able to reproduce linear functions exactly is called linear complete and so on. [14] have shown that a discretization has to be zeroth-order complete to guarantee conservation of linear momentum and linear complete to guarantee conservation of angular momentum. [20] have also demonstrated this property.

Conservation of linear momentum requires that the rate of change of linear momentum equals the total applied force such that the total change of linear momentum due to internal forces is zero. Thus, in the absence of external forces and body forces, conservation of linear momentum requires that

$$\frac{D}{Dt}\left(\sum_{I\in\mathcal{S}}\mathbf{m}_I\mathbf{v}_I\right)=\sum_{I\in\mathcal{S}}\mathbf{m}_I\cdot\dot{\mathbf{v}}_I=0 \tag{5.7}$$

where $\mathbf{m}_I$ is the diagonalized mass matrix and $\mathbf{v}$ the velocity field. We know that

$$\mathbf{m}_I\dot{\mathbf{v}}_I=-\sum_{J\in\mathcal{S}}\nabla\Phi_I(\mathbf{X}_J)\cdot\boldsymbol{\sigma}(\mathbf{X}_J)\ w \tag{5.8}$$

where $\Phi_I(\mathbf{X}_J)$ are shape functions and w are the quadrature weights. Substituting the RHS of Eq. (5.8) into Eq. (5.7) gives

$$\begin{aligned}\sum_{I\in\mathcal{S}}\mathbf{m}_I\cdot\dot{\mathbf{v}}_I&=-\sum_{I\in\mathcal{S}}\sum_{J\in\mathcal{S}}\nabla\Phi_I(\mathbf{X}_J)\cdot\boldsymbol{\sigma}(\mathbf{X}_J)\ w\\&=-\sum_{J\in\mathcal{S}}\sum_{I\in\mathcal{S}}\nabla\Phi_I(\mathbf{X}_J)\cdot\boldsymbol{\sigma}(\mathbf{X}_J)\ w=0\end{aligned} \tag{5.9}$$

that requires that $\sum_{I\in\mathcal{S}}\nabla\Phi_I(\mathbf{X}_J)=0$ for arbitrary stress states that in turn requires zero-order complete shape functions. Similarly, it can be shown that first order completeness is necessary for conservation of angular momentum. Since we consider neither mass distributions of polar momenta nor magnetized media, we refer the interested reader to [14].

5.1.3 Consistency, stability and convergence

A method is convergent of order k $(k>0)$ if

$$\underbrace{\max}_{i}|u(x_i)-u_i|\le Ch^k \tag{5.10}$$

where C is a constant and h is the dilation parameter in meshfree methods, see [56] (h would denote the element size in finite elements). According to the Lax-Richtmeyr equivalence theorem, a method is convergent if it is consistent and stable. Stability guarantees that a small defect in the discretization stays small. As discussed in the previous section, completeness takes the role of consistency in a Galerkin method.

[38] showed convergence for the hp-cloud method and that the convergence rate is given by

$$|u(\mathbf{X})-u^h(\mathbf{X})|_{m,\Omega}\le Ch^{k+1-m}|u|_{k+1,\Omega} \tag{5.11}$$

where h is the dilation parameter, k is the polynomial degree of the basis, m is the order of the governing equation and the seminorms are the usual L_2 generated norms on Sobolev spaces. For more details, the interested reader is referred to the book by [73].

[74] showed convergence of the RKPM and EFG method. The a priori error bound is very similar to the bound in finite elements. The meshfree dilation parameter h_0 plays the role of h in finite elements, and m (the order of consistency) plays the role of the degree of the approximation polynomials in the finite element mesh. Convergence properties depend on m and h_0. They do not depend on the distance between particles because usually this distance is proportional to h_0, i.e. the ratio between the particle distance over the dilation parameter is of order one, see [74].

5.1.4 Continuity

An approximation is considered to be C^n if its shape functions are n times continuous differentiable. For all the meshfree methods explained in this book, it can easily be shown that the continuity of the approximation depends on the continuity of the kernel or weighting function. Most kernel functions are at least C^1 and it is easily possible to construct even C^∞ kernel functions. Therefore, even meshfree methods that are not even zeroth-order complete can be C^∞. This is a huge advantage over finite elements where the increase in continuity will also lead to an increase in completeness, see Fig. 5.3. This higher order continuity of meshfree methods can be exploited in many areas as mentioned in the Introduction. Note, that when the ratio between dilation parameter h and particle separation x goes to one, the linear finite element shape functions are recovered as might become obvious in Fig. 5.3.

5.1.5 Partition of unity

As was shown by [75,114], all ansatz-methods such as finite elements and most meshfree methods can be derived from a partition of Unity concept. A partition of unity of n-th order (completeness) can be constructed in two different ways in meshfree methods, either *intrinsically* or *extrinsically*. In the intrinsic formulation, the meshfree shape functions in Eq. (5.1) are chosen so that they are n-th order complete. For example, to obtain a one-dimensional second-order complete approximation, the polynomial basis $\mathbf{p}$ is $\mathbf{p} = [1\ X\ X^2]$, see also Section 5.2. It is also possible, to increase the order

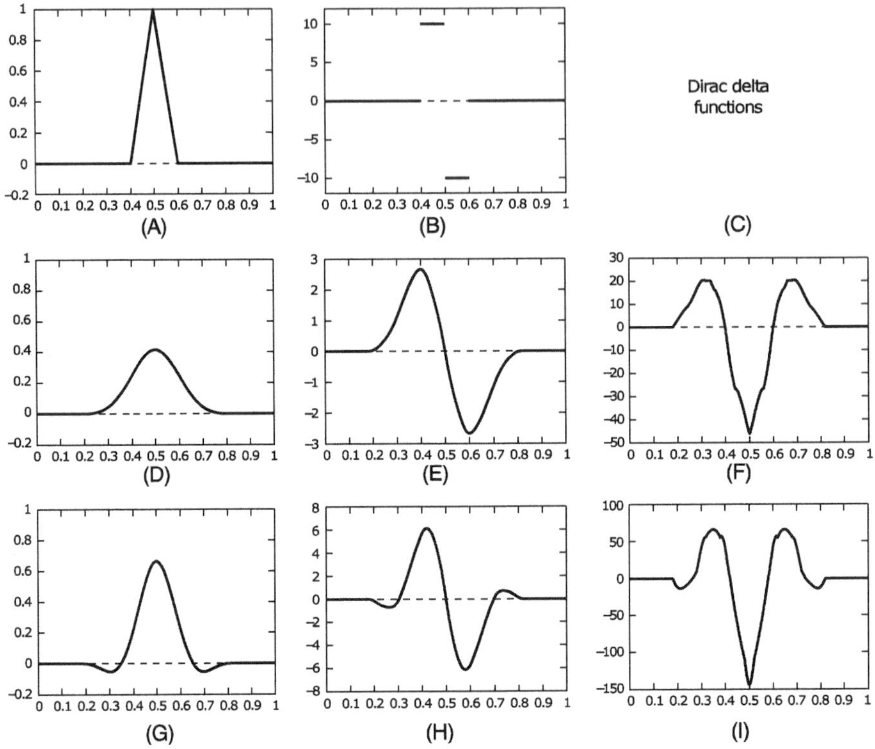

Figure 5.3 Shape function and derivatives for linear finite elements and the EFG approximation; (A) linear finite element shape function, (B) first spatial derivative of the finite element shape function, (C) second spatial derivative of the finite element shape function, (D) EFG shape function with linear polynomial basis $\mathbf{p} = (1\ X)^T$ and a relation $h_0/dx = 3.2$ with the (E) first spatial derivative and (F) second spatial derivative, (G) EFG shape function with quadratic polynomial basis $\mathbf{p} = (1\ X\ X^2)^T$ and a relation $h_0/dx = 3.2$ with the (H) first spatial derivative and (I) second spatial derivative, from [56].

of completeness by an extrinsic basic:

$$u^h(\mathbf{X}, t) = \sum_{J \in \mathcal{S}} \Phi_J(\mathbf{X}) u_J(t) + \sum_{J \in \mathcal{S}} \sum_{I=1}^{n} \tilde{\Phi}_J(\mathbf{X}) p_I(\mathbf{X}) q_{JI} \tag{5.12}$$

where q_{JI} are additional unknowns and p are n polynomials to be reproduces exactly that can be chosen arbitrarily. Note that the polynomials p can differ from particle to particle facilitating the incorporation of hp-adaptivity. It should be noted that the extrinsic concept is essential for p-adaptivity. An intrinsic basis cannot be varied from particle to particle without introducing a discontinuity, [56]. This extrinsic (hp) concept was first proposed by [37] to increase the order of completeness of the approximations. This idea was later generalized by [75] in the PUFEM (partition of unity finite element method) and by [114] in the GFEM (generalized finite element method). It is worth to mention that XFEM (extended finite element method) is also based on an extrinsic "basis". However, the extrinsic basis is not used to increase the order of completeness as e.g. in [37], but to fit the approximation to the kinematics of the crack. Thus, we do not talk about an extrinsic *basis* but instead of an extrinsic *enrichment*. Note, that in contrast to the hp-clouds or PUFEM, XFEM is based on a *local* partition-of-unity since an extrinsic enrichment is only used in a certain domain. At this point, we would like to mention that in meshfree methods, it is also possible to capture the correct crack kinematics with an intrinsic enrichment. This will be discussed in Section 5.5.1.

5.1.6 Kernel functions

5.1.6.1 Construction of the kernel function

The kernel function plays an important role in meshfree methods. Other names for the kernel function are window and weighting function. The kernel function may be defined in various manners. Typically, three requirements are imposed on the kernel function:

$$\underbrace{\lim}_{h_0 \to 0} W(\mathbf{X}_I - \mathbf{X}_J, h_0) = \delta(\mathbf{X}_I - \mathbf{X}_J) \tag{5.13}$$

$$\int_{\Omega_0} W(\mathbf{X}_I - \mathbf{X}_J, h_0)\, d\Omega_0 = 1 \tag{5.14}$$

$$W(\mathbf{X}_I - \mathbf{X}_J, h_0) = 0 \quad \forall\ \mathbf{X}_I - \mathbf{X}_J \geq R \tag{5.15}$$

where δ denotes the Dirac delta function and R is a shape parameter often related to h_0. The last condition is called compact support of the shape function.

Usually, also the following symmetry conditions are imposed on the kernel function:

$$W(\mathbf{X}_I - \mathbf{X}_J, h_0) = W(\mathbf{X}_J - \mathbf{X}_I, h_0) \tag{5.16}$$

$$\nabla_0 W(\mathbf{X}_I - \mathbf{X}_J, h_0) = -\nabla_0 W(\mathbf{X}_J - \mathbf{X}_I, h_0) \tag{5.17}$$

A huge variety of kernel functions were constructed. Some of them are already given in Section 5.1.1; for more details see e.g. [73]. There are basically two ways to construct the kernel function to higher order dimensions [56] that lead to different sizes and shapes of the domain of influence. Either, the kernel function with spherical support[1] is constructed

$$W(\mathbf{x}) = W_{1D}(\|\mathbf{x}\|), \tag{5.18}$$

or a kernel function with rectangular support is constructed by multiplying the one-dimensional kernel functions

$$W(\mathbf{x}) = W_{1D}(|x_1|)\, W_{1D}(|x_2|)\, W_{1D}(|x_3|) \tag{5.19}$$

where, as usual, $\mathbf{x} = (x_1, x_2, x_3)$ and $\|\mathbf{x}\| = \sqrt{x_1^2 + x_2^2 + x_3^2}$.

In contrast to finite elements, meshfree methods have one more "numerical" parameter to fit, the dilation parameter h_0, see e.g. Eq. (5.13). This nonlocal approximation character can be advantageous in certain applications, e.g. it improves the accuracy around the crack tip. However, if not (physically) motivated otherwise, it can be shown that there is an optimal value for the ratio between the dilation parameter h and the distance between particles x. Fig. 5.4 shows that for a fixed distribution of particles, x constant, the dilation parameter must be large enough to avoid aliasing (spurious short waves in the approximated solution). It also shows that an excessively large value for h will lead to excessive smoothing. For this reason, it is common to maintain a constant ratio between the dilation parameter h and the distance between particles x. It has been shown, e.g. by [56,92] that the optimal ratio between dilation parameter h and particle separation x lies in the range between 2.8 and 3.2.

5.1.6.2 Eulerian and Lagrangian kernels

In most large deformation meshfree methods such as SPH procedures (see e.g. [89], [33,34]), the shape function is directly expressed in terms of an Eulerian kernel:

$$W_J(\mathbf{x}) = W(\mathbf{x} - \mathbf{x}_J(t), h(\mathbf{x}, t)) \tag{5.20}$$

Note that the Eulerian kernel is expressed in terms of spatial coordinates. The radius h of the support depends on the spatial coordinates and can change in time. Simple approaches to update h are given by:

$$\begin{aligned} h^{t+\Delta t} &= h^t + \dot{h}\,\Delta t \\ \dot{h} &= 1/3\ \varrho\ \nabla \cdot \mathbf{v} \end{aligned} \tag{5.21}$$

[1] Though it's possible to create anisotropic kernel functions, e.g. with elliptical support.

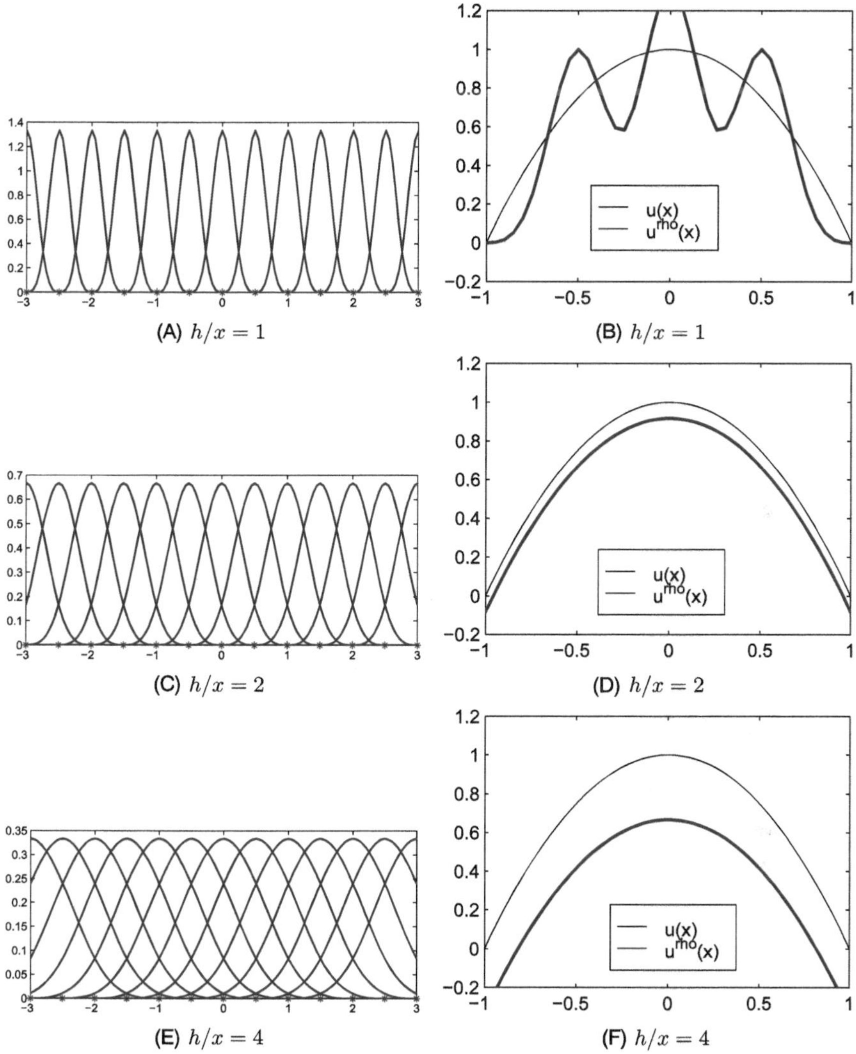

Figure 5.4 SPH approximation functions and approximation of $u(x) = 1 - x^2$ with cubic spline window function, distance between particles $x = 0.5$ and quadrature weights $\omega_i = x$, for $h/x = 1, 2, 4$, from [56].

where $\mathbf{v}$ indicates the particle velocity and ϱ is the density. However, the shape of the domain of influence of a given particle still remains a sphere. More realistic tensorial support sizes $\mathbf{h}$ can be used which can be updated by use of the deformation gradient:

$$\mathbf{h} = h_0\, \mathbf{F} \tag{5.22}$$

For radial supported shape functions, $\mathbf{h}$ spans a sphere in the initial configuration with h_0 as radius. In the current configuration, the domain of influence becomes an ellipsoid. The eigenvalues and eigenvectors of $\mathbf{h}$ have to be computed which give the axis of the ellipsoid. Kernels with rectangular support are easier to implement since one-dimensional kernel functions can be combined. It was shown [95] that for strain softening brittle materials, there are only minor differences in the final results between methods based on tensorial and scalar support sizes, respectively, since the stresses decay to zero before large strains occur.

In [15], it was shown that particle discretizations of solids with an Eulerian kernel lead to a distortion of the stable domain of the material in stress space; the tensile instability analyzed in [119] is one manifestion of this distortion.

The Lagrangian kernel is expressed in terms of material coordinates, so:

$$W_J(\mathbf{X}) = W(\mathbf{X} - \mathbf{X}_J, h_0) \tag{5.23}$$

For Lagrangian kernels, the neighbors of influence do not change during the course of the simulation but the domain of influence in the current configuration changes with time. For radial kernel functions, the domain of influence in the initial configuration is a circle in two dimensions and a sphere in three dimensions. The Lagrangian kernel eliminates the tensile instability. Therefore, the representation of the kernel in terms of material coordinates provides a more consistent procedure when simulating material fracture or other material instabilities because instabilities will not occur due to numerical artifacts. This issue will be addressed in detail in Section 5.3.

In the current configuration the domain of influence can be extremely distorted. This is a disadvantage in simulating fluid flow problems and other problems with large distortions.

In [95], a mixed Lagrangian/Eulerian kernel formulation for large deformation problems is proposed. In that method, a Lagrangian kernel is used until the fracture process is completed, i.e. the cohesive tractions decayed to zero. Afterwards, an Eulerian kernel is used. Another opportunity is the use of a so-called updated Lagrangian kernel formulation [91] where the Lagrangian kernel is re-initialized, i.e. referred to a different reference configuration, every n-th time step.

In the following sections, the approximation techniques are described in terms of a Lagrangian kernel. The equations for an Eulerian kernel can be derived similarly.

5.2. Some specific methods

We won't give a comprehensive overview about meshfree methods, but instead will focus on the methods we used in our examples. A large variety of methods can be created depending on the choice of approximating the test and trial functions in combination with different integration schemes.

5.2.1 Approximation of the displacement field

5.2.1.1 The SPH-method

In the SPH method with a Lagrangian kernel, the shape functions are given by a product of the particle volume and the weighting function:

$$\Phi_J(\mathbf{X}) = W(\mathbf{X} - \mathbf{X}_J, h_0)\, V_J^0 \tag{5.24}$$

where V_J^0 is the volume associated with the node J in the initial configuration. For an Eulerian kernel the shape functions are

$$\Phi_J(\mathbf{x}) = W(\mathbf{x} - \mathbf{x}_J, h)\, V_J \tag{5.25}$$

where V_J is the current volume associated with the particle at J.

In the application to PDE's, an approximation of a function gradient is needed. In SPH, an approximation of the gradient of a function in terms of the Lagrangian kernel is given by

$$\nabla u_h = -\sum_{J\in\mathcal{S}} u_J \nabla_0 \Phi_J \quad \text{with} \quad \nabla_0 \Phi_J = \nabla_0 W(\mathbf{X} - \mathbf{X}_J, h_0)\, V_J^0 \tag{5.26}$$

where the ∇_0 indicates the gradient in terms of the material coordinates. The minus sign on the right hand side of Eq. (5.26) results from the integration by parts, that is standard in SPH ([46]).

One drawback of the standard SPH-method is its inability to accurately approximate a function when particles are unevenly spaced and sized. This is evidenced by its inability to reproduce linear fields, which is necessary for convergence, see Section 5.1.2 and 5.1.3. The momentum equations with stress rates depending on the velocity gradient are second order PDEs, so linear completeness, i.e. the ability to reproduce linear functions, is desirable for convergence. These are among the properties checked in the well known patch test. Also for boundary particles, the constant reproducing conditions are violated even for uniformly spaced nodes.

As mentioned in Section 5.1.2, the reproducing conditions (or completeness) of an approximation corresponds to the order of the polynomial which can be represented exactly. The following are the conditions for zeroth and first-order completeness:

$$\sum_{J\in\mathcal{S}}\Phi_J(X)=1 \quad \sum_{J\in\mathcal{S}}\nabla_0\Phi_J(X)=0 \tag{5.27}$$

$$\sum_{J\in\mathcal{S}}\Phi_J(X)\ X_J=X \quad \sum_{J\in\mathcal{S}}\Phi_J(X)\ Y_J=Y \quad \sum_{J\in\mathcal{S}}\Phi_J(X)\ Z_J\ =Z \tag{5.28}$$

Approximations that satisfy (5.27) have zeroth order completeness. Since Eq. (5.26) does not fulfill zeroth order completeness on the boundary even for a regular particle configuration, [46] introduced a so-called symmetrization. In this procedure they assume that

$$\left(\sum_{J\in\mathcal{S}}\nabla_0 W(\mathbf{X}-\mathbf{X}_J,h_0)\ V_J^0\right)\ u_I\equiv 0 \tag{5.29}$$

although this is only true for a uniform distribution of particles away from any boundary. Note that Eq. (5.29) is equivalent to assuming zeroth order completeness (note the quantity inside the parenthesis in (5.29)). Adding Eq. (5.29) to Eq. (5.26) gives

$$\nabla u_h(\mathbf{X})=-\sum_{J\in\mathcal{S}}(u_J-u_I)\,\nabla_0 W(\mathbf{X}_I-\mathbf{X}_J,h_0)\ V_J^0 \tag{5.30}$$

A remarkable feature of the symmetrization procedure is that it yields zeroth-order completeness for the derivatives of a function for an irregular particle arrangement.

5.2.1.2 Krongauz-Belytschko correction

[64] noted that errors in the extensional strains due to lack of linear completeness could be corrected by simple scaling, thus improving accuracy. A correction that enables the derivatives of constant or linear fields to be reproduced exactly was developed by [102] and [66]. The corrected derivatives are approximated by

$$\nabla u_h(\mathbf{X},t)=\sum_{I\in\mathcal{S}}\mathbf{G}_I(\mathbf{X})\ \mathbf{u}_I(t) \tag{5.31}$$

where $\mathbf{G}_I$ is a linear combination of the Shepard functions. The approximation functions for the derivatives $\mathbf{G}_I$ are defined as linear combinations of the exact derivatives by a linear transformation

$$\mathbf{G}_I(\mathbf{X}) = \mathbf{a}(\mathbf{X}) \cdot \nabla_0 W_I^S(\mathbf{X}) \tag{5.32}$$

where $\mathbf{a}(\mathbf{X})$ are arbitrary parameters and $W_I^S(\mathbf{X})$ are the Shepard functions given by

$$W_I^S(\mathbf{X}) = \frac{W_I(\mathbf{X})}{\sum_{I \in \mathcal{S}} W_I(\mathbf{X})} \tag{5.33}$$

Note that $\mathbf{a}(\mathbf{X})$ is invariant with respect to time for a Lagrangian kernel, so they need only be computed once and stored. The following reproducing condition for the derivative of a linear function must be fulfilled:

$$\sum_{I \in \mathcal{S}} \mathbf{G}_I(\mathbf{X}) \otimes \mathbf{X}_I = \delta_{ij} \tag{5.34}$$

Let $\mathbf{A}$ be the matrix of the cross product between the derivatives of the Shepard function and the Lagrangian coordinate vector. Then the parameters $\mathbf{a}$ can be easily determined from

$$\mathbf{A}\,\mathbf{a}^T = \mathbf{I} \tag{5.35}$$

where $\mathbf{I}$ is the identity matrix and

$$\mathbf{A} = \begin{bmatrix} W_{I,X}^S\, X_I & W_{I,Y}^S\, X_I & W_{I,Z}^S\, X_I \\ W_{I,X}^S\, Y_I & W_{I,Y}^S\, Y_I & W_{I,Z}^S\, Y_I \\ W_{I,X}^S\, Z_I & W_{I,Y}^S\, Z_I & W_{I,Z}^S\, Z_I \end{bmatrix}$$

$$\mathbf{a} = \begin{bmatrix} a_{XX} & a_{XY} & a_{XZ} \\ a_{YX} & a_{YY} & a_{YZ} \\ a_{ZX} & a_{ZY} & a_{ZZ} \end{bmatrix}$$

Finally, we obtain the approximation for the derivatives of a function from (5.32) and (5.34) as

$$\nabla_0 u_h(\mathbf{X}, t) = \sum_{I \in \mathcal{S}} \mathbf{a}(\mathbf{X}) \cdot \nabla_0 W_I^S(\mathbf{X})\, u_I(t) \tag{5.36}$$

Since only the derivatives of the approximation are modified they are usually not integrable. [66] showed that the correction of the derivatives for both the test functions and the trial functions lead to a violation of the patch test. However, in a Petrov-Galerkin method they showed that the patch test can be satisfied if Shepard functions are used as the test functions. This means that the test functions do not meet linear completeness, so that global angular momentum is not conserved. However, in linear problems they observed excellent convergence.

5.2.1.3 Randles-Libersky correction

[102] developed a similar correction which they called normalization (NSPH). We develop it here for a Lagrangian kernel, which is a straightforward extension of their work. To fulfill the first-order completeness, they modified the SPH approximation for the gradient of the function u with a matrix $\mathbf{B}$:

$$\nabla u_h(\mathbf{X}, t) = \left(-\sum_{J \in \mathcal{S}} \left(u_J(t) - u_I(t)\right) \ \nabla_0 W(\mathbf{X}_J - \mathbf{X}, h_0) \ V_J^0 \right) \cdot \mathbf{B}(\mathbf{X}) \quad (5.37)$$

with

$$\mathbf{B}(\mathbf{X}) = \left(-\sum_{J \in \mathcal{S}} (\mathbf{X}_J - \mathbf{X}) \otimes \nabla_0 W(\mathbf{X}_J - \mathbf{X}, h_0) \ V_J^0 \right)^{-1} \quad (5.38)$$

If the SPH shape functions $W(\mathbf{X} - \mathbf{X}_J, h) \ V_J^0$ are replaced by the Shepard functions, the expression for $\mathbf{B}$ becomes:

$$\mathbf{B}(\mathbf{X}) = \left(-\sum_{J \in \mathcal{S}} \mathbf{X}_J \otimes \nabla_0 W^S(\mathbf{X}_J - \mathbf{X}, h_0) \right)^{-1} \quad (5.39)$$

which is similar to the Krongauz-Belytschko correction; subtle differences arise from the symmetrization in (5.37) and (5.38). Note, that the SPH shape functions in Eq. (5.37) have to be replaced by the Shepard functions when using Eq. (5.39) for the computation of $\mathbf{B}$. The approximation for the gradient of the function u is then formulated in the unsymmetrized form because the Shepard functions are zeroth order complete by construction:

$$\nabla u_h(\mathbf{X}, t) = \left(-\sum_{J \in \mathcal{S}} u_J(t) \ \nabla_0 W^S(\mathbf{X}_J - \mathbf{X}, h_0) \ V_J^0 \right) \cdot \mathbf{B}(\mathbf{X}) \quad (5.40)$$

A Petrov-Galerkin procedure is probably preferable here as in the Krongauz-Belytschko correction.

5.2.1.4 The MLS-approximation

The MLS-approximation fulfills the reproducing conditions by construction, so no corrections are needed. The moving least square approximation was introduced in the EFG-method by [11], and was first applied in a SPH setting by [33]. To satisfy the linear reproducing conditions linear base functions $\mathbf{p}$ are chosen to be

$$\mathbf{p}(\mathbf{X}) = \begin{pmatrix} 1 & X & Y & Z \end{pmatrix} \quad \forall \mathbf{X} \in \Re^3 \tag{5.41}$$

The MLS approximation is

$$u_h(\mathbf{X}, t) = \sum_{J \in \mathcal{S}} \mathbf{p}(\mathbf{X}_J)\, \mathbf{a}(\mathbf{X}, t) \tag{5.42}$$

with $\mathbf{a}$ chosen to minimize the quadratic form

$$\mathfrak{J} = \sum_{J \in \mathcal{S}} \left(\mathbf{p}(\mathbf{X}_J)^T \mathbf{a}(\mathbf{X}, t) - u_J(t)\right)^2 W(\mathbf{X} - \mathbf{X}_J, h_0) \tag{5.43}$$

Minimizing Eq. (5.43) with respect to $\mathbf{a}$ leads to the approximation

$$u_h(\mathbf{X}, t) = \sum_{J \in \mathcal{S}} u_J(t)\ \Phi_J(\mathbf{X}) \tag{5.44}$$

with

$$\Phi_J(\mathbf{X}) = \mathbf{p}(\mathbf{X}_J)^T \cdot \mathbf{A}(\mathbf{X})^{-1} \cdot \mathbf{p}(\mathbf{X}_J)\, W(\mathbf{X} - \mathbf{X}_J, h_0) \tag{5.45}$$

$$\mathbf{A}(\mathbf{X}) = \sum_{J \in \mathcal{S}} \mathbf{p}(\mathbf{X}_J)\ \mathbf{p}^T(\mathbf{X}_J)\, W(\mathbf{X} - \mathbf{X}_J, h_0) \tag{5.46}$$

For zeroth-order completeness ($p(\mathbf{X}) = [1]$) the shape function Φ is

$$\Phi_J(\mathbf{X}) = \frac{W_{JI}}{\sum_{J \in \mathcal{S}} W_{JI}} \tag{5.47}$$

where $W_{JI} = W(\mathbf{X}_J - \mathbf{X}_I, h_0)$ and which is known as Shepard function. A fast evaluation procedure for the gradient of the shape function Φ_J can be derived by rewriting Eq. (5.45):

$$\Phi_J(\mathbf{X}) = \boldsymbol{\gamma}(\mathbf{X}) \cdot \mathbf{p}(\mathbf{X}_J)\, W(\mathbf{X} - \mathbf{X}_J, h_0) \tag{5.48}$$

with

$$\mathbf{A}(\mathbf{X}) \cdot \boldsymbol{\gamma}(\mathbf{X}) = \mathbf{p}(\mathbf{X}_J) \tag{5.49}$$

such that the coefficients $\boldsymbol{\gamma}$ can be obtained by an LU decomposition and backsubstitution that requires fewer computations than inverting the matrix $\mathbf{A}$. Then the derivatives of the shape functions can be written as

$$\nabla_0 \mathbf{A}(\mathbf{X}) \cdot \boldsymbol{\gamma}(\mathbf{X}) + \mathbf{A}(\mathbf{X}) \cdot \nabla_0 \boldsymbol{\gamma}(\mathbf{X}) = \nabla_0 \mathbf{p}(\mathbf{X}_J) \tag{5.50}$$

Rearranging this equation, $\nabla_0 \boldsymbol{\gamma}(\mathbf{X})$ is obtained. More details can be found e.g. in [13].

In summary, the evaluation of the derivatives of the shape functions requires little extra computer cost and, moreover, higher order derivatives can also be computed repeating the same process. The same development can be done for the centered and scaled approach defined subsequently.

For computational purposes, it is usual and preferable to center in $\mathbf{X}_J$ and scale with h_0 also the polynomials involved in the definition of the meshfree approximation functions, see [74] or [55]. Thus, another expression for the EFG shape functions is employed:

$$\Phi_I(\mathbf{X}) = W(\mathbf{X}_I, \mathbf{X})\, \mathbf{P}^T \left(\frac{\mathbf{X}_I - \mathbf{X}}{h_0} \right) \boldsymbol{\gamma}(\mathbf{X}), \tag{5.51}$$

which is similar to (5.45). Recall also that typical expressions for the window function are of the following type: $W(\mathbf{Y}, \mathbf{X}) = W\left((\mathbf{Y} - \mathbf{X})/h_0\right)$. The consistency condition becomes in this case:

$$\mathbf{P}(0) = \sum_{I \in \mathcal{S}} \Phi_I(\mathbf{X})\, \mathbf{P}\left(\frac{\mathbf{X}_I - \mathbf{X}}{h_0} \right) \tag{5.52}$$

After substitution of (5.51) in (5.52) the linear system of equations that determines $\boldsymbol{\gamma}(\mathbf{X})$ is obtained:

$$\mathbf{M}(\mathbf{X})\, \boldsymbol{\gamma}(\mathbf{X}) = \mathbf{P}(0) \tag{5.53}$$

where

$$\mathbf{M}(\mathbf{X}) = \sum_{J \in \mathcal{S}} W(\mathbf{X}_J, \mathbf{X})\, \mathbf{P}^T \left(\frac{\mathbf{X}_J - \mathbf{X}}{h_0} \right) \mathbf{P}\left(\frac{\mathbf{X}_J - \mathbf{X}}{h_0} \right) \tag{5.54}$$

For a varying dilation parameter h_{0I}, h_{0I} associated to particle $\mathbf{X}_I$, is embedded in the definition of the weighting function:

$$W(\mathbf{X}_I, \mathbf{X}) = W\left(\frac{\mathbf{X}_I - \mathbf{X}}{h_{0I}} \right) \tag{5.55}$$

Note that a constant h_0 is employed in the scaling of the polynomials $\mathbf{P}$. The constant value h_0 is typically chosen as the mean value of all the h_{0I}. The consistency condition in this case is also (5.52). It also imposes the reproducibility of the polynomials in $\mathbf{P}$.

This centered expression for the EFG shape functions can also be obtained with a discrete MLS development with the discrete centered scalar product

$$\langle f, g\rangle_{\mathbf{X}} = \sum_{J\in\mathcal{S}} W(\mathbf{X}_J, \mathbf{X}) f\left(\frac{\mathbf{X}_J - \mathbf{X}}{h_0}\right) g\left(\frac{\mathbf{X}_J - \mathbf{X}}{h_0}\right) \tag{5.56}$$

The MLS development in this case is as follows: for fixed $\mathbf{X}$, and for $\mathbf{Z}$ near $\mathbf{X}$, u is approximated as

$$u(\mathbf{Z}) \simeq u^h(\mathbf{Z}, \mathbf{X}) = \mathbf{P}^T\left(\frac{\mathbf{Z} - \mathbf{X}}{h_0}\right) \mathbf{c}(\mathbf{X}) \tag{5.57}$$

where $\mathbf{c}$ is obtained, as usual, through a least-squares fitting with the discrete centered scalar product (5.56).

Note that the MLS shape functions do not allow for the space $\mathcal{H}^1$ to be represented exactly by the polynomial basis because the MLS shape functions don't vanish along the boundaries. Techniques like coupling MLS with finite element shape functions can be as a modification of the approximation space to account for essential boundary conditions such that $\mathcal{H}^1$ is represented exactly, [67].

5.2.2 Spatial integration

Numerical integration is the final step to obtain the discrete equations from the weak form. Since the shape functions in meshfree methods are often not polynomial (as e.g. for the EFG method), exact integration of the weak form is difficult to impossible.

Integration techniques in meshfree methods can basically be classified into three categories. The easiest way to obtain the discrete equations is a *nodal integration*, based on the trapezoid rule. Such methods are closely related to collocation methods based on the strong form. [14] reinterpreted strong form based methods, such as SPH, as nodally integrated Galerkin methods and provides a clear framework for integration techniques in meshfree methods. It can be shown that meshfree methods based on nodal integration lead to numerical instabilities due to underintegration of the weak form. This instability due to rank deficiency can be removed by adding stress points to the nodes that leads to the notation of *stress-point* integration. This concept was proposed first by [41] in one-dimensions. It was later extended to higher order dimensions by [15] and [103]. There is a subtle difference between the stress point integration of [15] and the one of [103] since in [103], stresses are *only* evaluated at the stress points and *not* at the particles. [103] applied their stress-point integration technique in the context of strong form based SPH versions, typically a symmetric version. This might remove the instability due to rank deficiency. However, we were not able to capture material instabilities correctly with this method as shown in Section 5.3 for some selected strain softening constitutive models. Another alternative is Gauss quadrature. Therefore, a background mesh has to be constructed. In this context, [4] introduced the notation of *truly* meshfree methods. These are methods that do not need a background mesh for integration. However, for a wide range of applications in solid mechanics, it is quite simple to create a background mesh.

5.2.2.1 Nodal integration

An efficient and simple discretization in a Galerkin method is nodal integration. In most computations of transient phenomena spatial integration by background cells is quite expensive. To obtain the disrcete governing equations from the weak form, a nodal integration is given by:

$$\int_{\Omega_0} f(\mathbf{X})\ d\Omega_0 = \sum_{J \in \mathcal{S}} f(\mathbf{X}_J)\ V_J^0 \tag{5.58}$$

where the weights V_J^0 represent tributary volumes associated with particle J. Applying this technique to the computation of the internal forces gives:

$$\mathbf{f}_I^{int} = \sum_{J \in \mathcal{S}} V_J^0\ \nabla_0 \Phi_I(\mathbf{X}_J) \cdot \mathbf{P}_J \tag{5.59}$$

The momentum equation is then

$$\begin{aligned} m_I\ \frac{d\dot{\mathbf{u}}_I}{dt} &= \sum_{J \in \mathcal{S}} -\nabla_0 \Phi_I(\mathbf{X}_J) \cdot \mathbf{P}(\mathbf{X}_J)\ V_J^0 \\ &+ \Phi_I(\mathbf{X}_J)\ \bar{\mathbf{t}}_0(\mathbf{X}_J)\ \Gamma_J^0 + \Phi_I(\mathbf{X}_J)\ \mathbf{b}(\mathbf{X}_J) \varrho(\mathbf{X}_J)\ V_J^0 \end{aligned} \tag{5.60}$$

where Γ_J^0 are the boundary weights. In Eulerian coordinates the momentum equation is:

$$m_I \frac{d\mathbf{v}_I}{dt} = \sum_{J\in\mathcal{S}} -\nabla\Phi_I(\mathbf{x}_J)\cdot\boldsymbol{\sigma}(\mathbf{x}_J)\ V_J + \Phi_I(\mathbf{x}_J)\ \bar{\mathbf{t}}(\mathbf{x}_J)\ \Gamma_J + \Phi_I(\mathbf{x}_J)\ \mathbf{b}(\mathbf{x}_J)\varrho(\mathbf{x}_J)\ V_J \tag{5.61}$$

The nodal masses m_I are often obtained by a Voronoi procedure, i.e. after placement of the nodes, a triangulation is performed and the intersections of the perpendicular bisectors of the side on the triangles form the Voronoi cell, see also Fig. 5.7. The extension of this procedure to three dimensions is straightforward. The particle masses m_I are then computed just by multiplying the volume of the Voronoi cell V_J^0 with the density. Since the physical mass m_I remains constant during the simulation, the volume V_J changes in time for methods based on Eulerian kernels. The update of the volume is obtained by the simple algebraic relation (2.17) or (2.28). The volume update can also be realized via the continuity equation. Note that in general the physical mass m_I varies from the numerical mass m_{IJ}. However, it can be shown that the physical mass is recovered by diagonalizing the consistent mass matrix m_{IJ} with a standard row-sum technique:

$$m_I = \sum_{J\in\mathcal{S}} m_{IJ} = \sum_{J\in\mathcal{S}} \int_{\Omega_0} \varrho_0\ \Phi_I(\mathbf{X})\ \Psi_J(\mathbf{X})\ d\Omega_0 = \int_{\Omega_0} \varrho_0\ \Phi_I(\mathbf{X}) \left(\sum_{J\in\mathcal{S}} \Psi_J(\mathbf{X})\right)\ d\Omega_0 \tag{5.62}$$

With an at least zeroth-order complete approximation, Eq. (5.62) finally becomes

$$m_I = \int_{\Omega_0} \varrho_0\ \Phi_I(\mathbf{X})\ d\Omega_0 \tag{5.63}$$

After nodal integration we obtain

$$m_I = \sum_{J\in\mathcal{S}} \varrho_J\ \Phi_I(\mathbf{X})\ V_J^0 \tag{5.64}$$

When the test functions are also at least zero-order complete, it is easy to verify that after adding up all lumped masses, the total physical mass M of

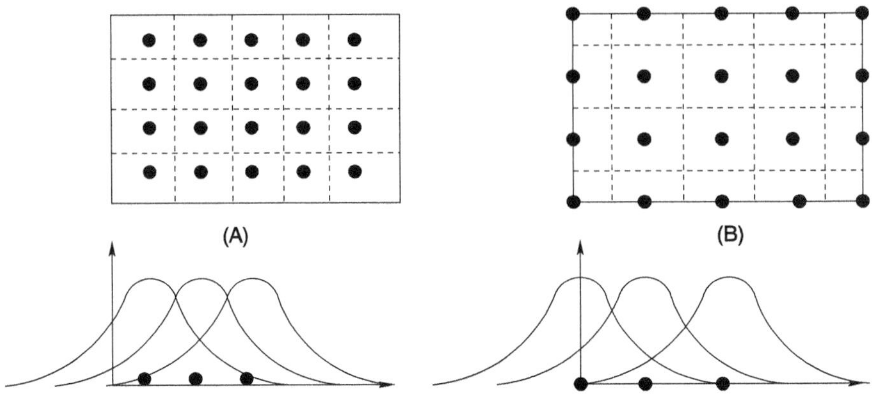

Figure 5.5 Value of the shape function for boundary particles lying directly on the boundaries or inside the boundaries.

the body is retained:

$$\sum_{I=1}^{N_{tot}} m_I = \sum_{I=1}^{N_{tot}} \sum_{J \in \mathcal{S}} \varrho_J \; \Phi_I(\mathbf{X}) \; V_J^0 = \sum_{J \in \mathcal{S}} \varrho_J \left(\sum_{I=1}^{N_{tot}} \Phi_I(\mathbf{X}) \right) V_J^0 = \sum_{J \in \mathcal{S}} \varrho_J \; V_J^0 = M \tag{5.65}$$

At this point, we would already like to mention that for cracking problems when additional unknowns are introduced into the variational formulation, a row-sum technique will not recover the physical mass any more. Thus, a consistent mass matrix has to be used, at least for parts of the system. This is clearly a drawback for explicit time integration schemes. This issue is addressed at a later point in this book.

Crucial is also how to consider and discretize the 'real' geometric volume of a body. It can be discretized with particles arranged as in Fig. 5.5A where the particles are inside the volume. No particles lie on the boundaries of the body. The body can also be discretized as illustrated on Fig. 5.5B where particles are placed exactly on the boundaries. For Fig. 5.5A, all masses are equal; this leads to poor reflection conditions on boundaries. For the discretization in Fig. 5.5B, the boundary particles have only half of the mass of the interior particles for a Voronoi method. The masses of the particles at the corners are only a quarter of the masses of the interior particles. [96] have shown that this assignment of nodal masses leads to more accurate wave reflections in wave propagation problems.

The boundary integrals for natural boundary conditions (external forces) also differs for the two models as shown in Fig. 5.5. When the particles are located directly on the boundary, the value of the shape function differs from when the particles are at a certain distance from the boundaries (see Fig. 5.5).

In SPH, the discrete equations are obtained from the strong form by collocation, i.e. the governing equations are enforced at each particle. Neglecting body forces and boundary conditions, the momentum equation with the Lagrangian kernel is

$$m_I\,\ddot{\mathbf{u}} = V_I^0 \sum_{J\in\mathcal{S}} V_J^0\;\nabla_0\;W_{JI}\cdot\mathbf{P}_J \tag{5.66}$$

and in Eulerian coordinates the momentum equation reads:

$$m_I\,\dot{\mathbf{v}} = V_I \sum_{J\in\mathcal{S}} V_J\;\nabla\;W_{JI}\cdot\boldsymbol{\sigma}_J \tag{5.67}$$

Usually a symmetrization procedure is performed, see Section 5.2.1.1. When the supports of all particles are of the same size, the discrete collocation equations are equivalent to the discrete equations obtained from a nodal integration of the Galerkin form, except at the boundaries, compare Eqs. (5.60) and (5.66).

5.2.2.2 Stabilized conforming nodal integration

Nodal integration often leads to instabilities due to rank deficiency and to low convergence rates. [26] showed that the vanishing derivatives of the meshfree shape functions at the particles are the cause of the instabilities. They notice that linear complete meshfree methods have to satisfy the following integration constraint

$$\int_{\Omega_0} \nabla\Phi_I(\mathbf{X})\;d\Omega_0 = \int_{\Gamma_0} \mathbf{n}_0\Phi_I(\mathbf{X})\;d\Gamma_0 \tag{5.68}$$

This integration constraint comes from the equilibrium of the internal and external forces of the linear complete Galerkin approximation and is similar to the linear consistency in the constant stress patch test in finite elements. [26] proposes a stabilized conforming nodal integration using strain smoothing. In the strain smoothing procedure, the nodal strains are computed as the divergence of a spatial average of the strain field. The strain smoothing avoids evaluating derivatives of the shape functions at the nodes and hence eliminates defective modes. The smooth strain field $\tilde{\epsilon}$ at a material point $\mathbf{X}_M$ can be expressed as

$$\tilde{\epsilon}(\mathbf{X}_M) = \int_{\Omega_0} \epsilon \; \Psi(\mathbf{X} - \mathbf{X}_M) \; d\Omega_0 \tag{5.69}$$

where Ω_0 is the domain of the Voronoi cell and $\Psi(\mathbf{X} - \mathbf{X}_M)$ is the smoothing function that has to fulfill the following requirements:

$$\begin{aligned} &\Psi(\mathbf{X} - \mathbf{X}_M) \geq 0 \\ &\int_{\Omega_0} \Psi(\mathbf{X} - \mathbf{X}_M) \; d\Omega_0 = 1 \end{aligned} \tag{5.70}$$

[26] chose

$$\Psi(\mathbf{X} - \mathbf{X}_M) = \frac{1}{A_M} \quad \forall \mathbf{X}_M \in \Omega_0, \;\; \text{otherwise } \Psi(\mathbf{X} - \mathbf{X}_M) = 0 \tag{5.71}$$

where A_M is the area of the smoothing (=Voronoi) cell. By expanding ϵ in a Taylor series

$$\epsilon(\mathbf{X}) = \epsilon(\mathbf{X}_M) + \nabla_0 \epsilon(\mathbf{X}) \cdot (\mathbf{X} - \mathbf{X}_M) + \ldots \tag{5.72}$$

and substituting Eq. (5.72) into (5.69), we obtain with (5.71)

$$\begin{aligned} \tilde{\epsilon}(\mathbf{X}_M) &= \frac{\nabla_0 \epsilon(\mathbf{X}_M)}{A_M} \int_{\Omega_0} (\mathbf{X} - \mathbf{X}_M) \; d\Omega_0 \\ \tilde{\epsilon}(\mathbf{X}_M) &= \frac{1}{2A_M} \int_{\Omega_0} \left(u_{i,j} + u_{j,i}\right) \; d\Omega_0 = \frac{1}{2A_M} \int_{\Gamma_0} \left(u_i \; n_j + u_j \; n_i\right) \; d\Gamma_0 \end{aligned} \tag{5.73}$$

As can be seen, integration is performed along a surface (the surface of the Voronoi cell) instead of a volume. It should be noted that the smoothed strain field does not satisfy the compatibility relations with the displacement field at all points in the discretized domain. However, the smooth, nonlocal strain and the local strain can be considered as two independent fields – the nonlocal strain field can be viewed as an assumed strain field. Thus, a two-field variational principle is suitable for this approximation.

5.2.2.3 Stress-point integration

Stress point integration was proposed by [41] in one dimension to stabilize the SPH method. [103] extended stress point integration to higher

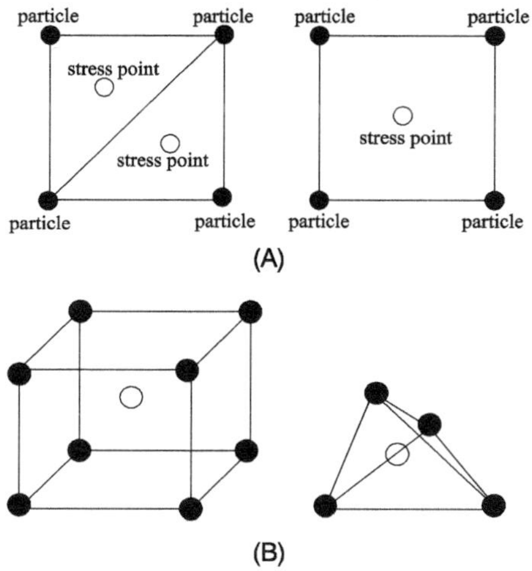

Figure 5.6 Arrangements of stress-points.

dimensions to stabilize the normalized form of SPH. Stress point integration eliminates instabilities due to rank deficiency.

In stress point integration methods, stress points are interspersed between the particles and the contributions of the stresses are added to the integration of (5.58) as described later. There are different methods for arranging the stress points between the original particles, see [15]. In most of our applications, stress points are added as shown in Fig. 5.6A for the 2D case and in Fig. 5.6B for the 3D case. Note that all kinematic values such as displacements and velocities are obtained via interpolation from the original particles:

$$\mathbf{u}_I^S = \sum_{J\in\mathcal{S}} \Phi_J(\mathbf{X}_I^S)\,\mathbf{u}_J^P, \qquad \mathbf{v}_I^S = \sum_{J\in\mathcal{S}} \Phi_J(\mathbf{X}_I^S)\,\mathbf{v}_J^P, \qquad \nabla_0\,\mathbf{v}_I^S = \sum_{J\in\mathcal{S}} \nabla\Phi_J(\mathbf{X}_I^S)\,\mathbf{v}_J \tag{5.74}$$

where the superscripts S indicate stress points and the P the original particles. $\Phi_J(\mathbf{X}_I^S)$ is the shape function of the supporting master node J at $\mathbf{X}_I^S$.

The internal forces for examples are calculated by

$$\mathbf{f}_I^{int} = \sum_{J\in\mathcal{N}_P} V_J^{0P}\,\nabla_0\Phi_I(\mathbf{X}_J^P)\cdot\mathbf{P}_J^P + \sum_{J\in\mathcal{N}_S} V_J^{0S}\,\nabla_0\Phi_I(\mathbf{X}_J^S)\cdot\mathbf{P}_J^S \tag{5.75}$$

where $\mathcal{N}_P$ and $\mathcal{N}_S$ indicate the supporting original particles and stress point nodes, respectively, to the original particle $\mathbf{X}_I^P$. The volumes V_J^{0P} and V_J^{0S}

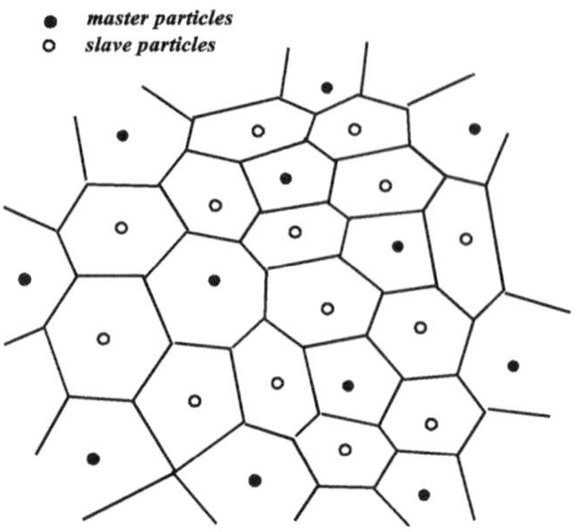

Figure 5.7 Scheme of stress point integration.

are computed from the Voronoi diagram (see Fig. 5.7) so that their sums result in the total geometric initial volume:

$$V^0 = \sum_{J\in\mathcal{N}_P} V_J^{0P} + \sum_{J\in\mathcal{N}_S} V_J^{0S} \tag{5.76}$$

In Eulerian coordinates, Eq. (5.75) reads:

$$\mathbf{f}_I^{int} = \sum_{J\in\mathcal{N}_P} V_J^P\, \nabla\Phi_I(\mathbf{x}_J^P)\cdot\boldsymbol{\sigma}_J^P + \sum_{J\in\mathcal{N}_S} V_J^S\, \nabla\Phi_I(\mathbf{X}_J^S)\cdot\boldsymbol{\sigma}_J^S \tag{5.77}$$

and

$$V = \sum_{J\in\mathcal{N}_P} V_J^P + \sum_{J\in\mathcal{N}_S} V_J^S \tag{5.78}$$

5.2.2.4 Cell integration

In EFG the integrals are usually evaluated over background cells based on an octree structure [11]. Note that the background integration mesh does not need to be conforming thus facilitating h-adaptivity. In each integration cell Gauss quadrature is performed. Note that in Gauss quadrature a polynomial of degree $2n_Q - 1$ can be exactly integrated with n_Q quadrature points.[2] However, as mentioned earlier, exact integration is difficult

[2] **Example**. For a 2D quadrilateral with bilinear shape functions, we have a maximum of quadratic terms for the shape functions plus linear terms for the Jacobian, i.e. maximum cubic terms; so we need $n_Q = 2$ quadrature points. An 8-node quadrilateral will require $n_Q = 3$ quadrature points since the highest polynomial is of order 5 (order 3 for the shape function and order 2 for the Jacobian).

to perform in meshfree methods since meshfree shape functions are often not polynomial. It is common to increase the number of quadrature points to reduce integration errors. The number of Gauss points depends on the number of nodes in a cell. Here, at least $n_Q \times n_Q$ Gauss points are needed. The number of Gauss points can be given by numerical experiments according to

$$n_Q = \sqrt{m} + 2 \tag{5.79}$$

where m denotes the number of nodes in one cell.

In 2D, the integral of a function is given by

$$\int_{\Omega_0} f(\mathbf{X})\, d\Omega_0 = \int_{-1}^{+1} \int_{-1}^{+1} f(\xi, \eta)\, det\, \mathbf{J}^{\xi}(\xi, \eta)\, d\xi\, d\eta = \sum_{J=1}^{m} w_J\, f(\boldsymbol{\xi}_J)\, det\, \mathbf{J}^{\xi}(\boldsymbol{\xi}_J) \tag{5.80}$$

where $\boldsymbol{\xi} = (\xi, \eta)$ are scaled local coordinates, m is the total number of quadrature points, $w_J = w(\xi_J)\, w(\eta_J)$ are the quadrature weights which is the product of the weight at the corresponding Gauss point in ξ- and η-direction and $det\, \mathbf{J}^{\xi}$ is the Jacobian determinant given by

$$\mathbf{J}^{\xi} = \frac{\partial \mathbf{X}}{\partial \boldsymbol{\xi}} \tag{5.81}$$

The internal forces in 2D are then

$$\mathbf{f}_{int} = \sum_{J=1}^{m} w_J\, det \mathbf{J}^{\xi}(\boldsymbol{\xi}_J)\, \nabla_0 \Phi(\mathbf{X}(\boldsymbol{\xi}_J) - \mathbf{X}^P)\, \mathbf{P}(\boldsymbol{\xi}_J) \tag{5.82}$$

where the superscript P indicates the particle position.

5.2.3 Essential boundary conditions

Many specific techniques have been developed in the recent years in order to impose essential boundary conditions in meshfree methods. Some possibilities are: (1) Lagrange multipliers ([11]), (2) modified variational principles ([11]), (3) penalty methods ([19,133]), (4) perturbed Lagrangian ([29]), (5) coupling to finite elements ([12,55,97,124]), or (6) modified shape functions ([48,51,123]) among others.

The first attempts to define shape functions with the "delta property" along the boundary (see [48]), namely $\Phi_I(\mathbf{X}_J) = \delta_{IJ}$ for all $\mathbf{X}_J$ in Γ_D, have serious difficulties for complex domains and for the integration of the weak forms.

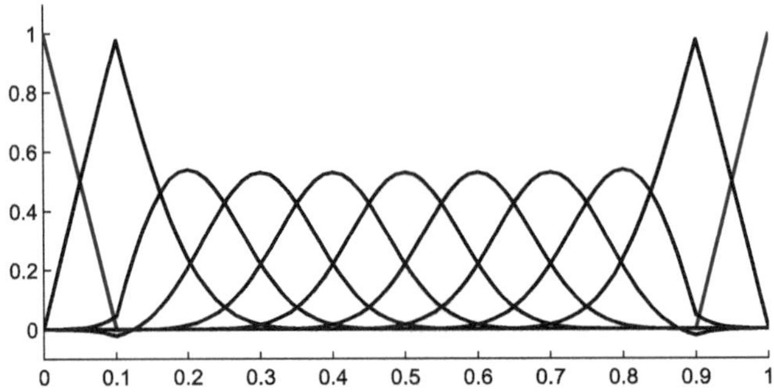

Figure 5.8 Mixed interpolation with linear finite element nodes near the boundary and particles in the interior of the domain, with $\rho/h = 3.2$, cubic spline and linear consistency in all the domain, from [56].

In the recent years, mixed interpolations that combine finite elements with meshfree methods have been developed. Mixed interpolations can be quite effective for imposing essential boundary conditions. The idea is to use one or two layers of finite elements next to the Dirichlet boundary and use a meshfree approximation in the rest of the domain. Thus, the essential boundary conditions can be imposed as in standard finite elements. In [12] a mixed interpolation is defined in the transition area (from the finite elements region to the particles region). This mixed interpolation requires the substitution of finite element nodes by particles and the definition of ramp functions. Thus, the transition is of the size of one element and the interpolation is linear. Following this idea, [55] propose a more general mixed interpolation, for any order of interpolation with no need for ramp functions and no substitution of nodes by particles. This is done preserving consistency and continuity of the solution. Fig. 5.8 shows an example of this mixed interpolation in 1D: two finite element nodes are considered at the boundary of the domain, with their corresponding shape functions in blue (gray in print version), and the meshfree shape functions are modified in order to preserve consistency, in black. A detailed overview about imposing essential boundary conditions in meshfree methods is given in [56].

5.2.4 Comparison of different methods

In this section, for two simple benchmark problems in statics and dynamics, we will compare five selected methods:

1. The standard SPH-procedure with nodal integration (SPH).
2. The Randles-Libersky correction with nodal integration (RLM ni).

Table 5.1 Discrete internal forces and deformation gradient for a Lagrangian kernel.

	Internal forces	Deformation gradient
SPH	$\mathbf{f}_{int,I} = \sum_{J\in S} (\mathbf{P}_J + \mathbf{P}_I) \cdot \nabla_0 W_J V_J^0$	$\mathbf{F}_I = -\sum_{J\in S} (\mathbf{u}_J - \mathbf{u}_I) \otimes \nabla_0 W_J V_J^0$
RLM (ni)	$\mathbf{f}_{int,I} = \left(-\sum_{J\in S}(\mathbf{P}_J + \mathbf{P}_I) \otimes \nabla_0 W_{JI} V_J^0\right) : \mathbf{B}$	$\mathbf{F}_I = \left(-\sum_{J\in S}(\mathbf{u}_J - \mathbf{u}_I) \otimes \nabla_0 W_J V_J^0\right) \cdot \mathbf{B}$
KBM (si)	$\mathbf{f}_{int} = \sum_{J\in N_P} \nabla_0 W^S(\mathbf{X}_J^P) \cdot \mathbf{P}(\mathbf{X}_J^P) V_J^{P0}$ $+ \sum_{J\in N_S} \nabla_0 W^S(\mathbf{X}_J^S) \cdot \mathbf{P}(\mathbf{X}_J^S) V_J^{S0}$	$\mathbf{F}^P = \sum_{J\in N_P} \mathbf{u}_J \otimes \mathbf{G}(\mathbf{X}_J^P)$ $\mathbf{F}^S = \sum_{J\in N_S} \mathbf{u}_J \otimes \mathbf{G}(\mathbf{X}_J^S)$
EFG (ni)	$\mathbf{f}_{int} = \sum_{J\in S} \mathbf{P}_J \cdot \nabla_0 \Phi(\mathbf{X}_J)\ V_J^0$	$\mathbf{F}_I = -\sum_{J\in S}(\mathbf{u}_J - \mathbf{u}_I) \otimes \nabla_0 \Phi(\mathbf{X}_J)$
EFG (ci)	$\mathbf{f}_{int} = \sum_{J=1}^{m} w_J\ det\mathbf{J}_J\ \mathbf{P}_J \cdot \nabla_0 \Phi(\mathbf{X}_J - \mathbf{X}_P)$	$\mathbf{F} = \sum_{J=1}^{N_P} \mathbf{u}_J \otimes \nabla_0 \Phi(\mathbf{X}_{GP} - \mathbf{X}_J)$

SPH-SPH with collocation, RLM (ni)-Randles-Libersky correction with nodal integration.
KBM (si)-Krongauz-Belytschko correction with stress point integration.
EFG (ni)-EFG with nodal integration, EFG (ci)-EFG with cell integration.

Table 5.2 Discrete internal forces and deformation gradient for an Eulerian kernel.

	Internal forces	Velocity gradient
SPH	$\mathbf{f}_{int,I} = \sum_{J\in S} (\boldsymbol{\sigma}_J + \boldsymbol{\sigma}_I) \cdot \nabla W_J V_J$	$\mathbf{L}_I = -\sum_{J\in S} (\mathbf{v}_J - \mathbf{v}_I) \otimes \nabla W_J V_J$
RLM (ni)	$\mathbf{f}_{int,I} = \left(-\sum_{J\in S}(\boldsymbol{\sigma}_J + \boldsymbol{\sigma}_I) \otimes \nabla W_J V_J\right) : \mathbf{B}$	$\mathbf{L}_I = \left(-\sum_{J\in S}(\mathbf{v}_J - \mathbf{v}_I) \otimes \nabla W_J V_J\right) \cdot \mathbf{B}$
KBM (si)	$\mathbf{f}_{int} = \sum_{J\in N_P} \nabla W^S(\mathbf{x}_J^P) \cdot \boldsymbol{\sigma}(\mathbf{x}_J^P) V_J^P$ $+ \sum_{J\in N_S} \nabla W^S(\mathbf{x}_J^S) \cdot \boldsymbol{\sigma}(\mathbf{x}_J^S) V_J^S$	$\mathbf{L}^P = \nabla \otimes \mathbf{v}^P = \sum_{J\in N_P} \mathbf{v}_J \otimes \mathbf{G}(\mathbf{x}_J^P)$ $\mathbf{L}^S = \nabla \otimes \mathbf{v}^S = \sum_{J\in N_S} \mathbf{v}_J \otimes \mathbf{G}(\mathbf{x}_J^S)$
EFG (ni)	$\mathbf{f}_{int} = \sum_{J\in S} \boldsymbol{\sigma}_J \cdot \nabla \Phi(\mathbf{x}_J)\ V_J$	$\mathbf{L}_I = -\sum_{J\in S}(\mathbf{v}_J - \mathbf{v}_I) \otimes \nabla \Phi(\mathbf{x}_J)$
EFG (ci)	$\mathbf{f}_{int} = \sum_{J=1}^{m} w_J\ det\mathbf{J}_J\ \boldsymbol{\sigma}_J \cdot \nabla \Phi(\mathbf{x}_J - \mathbf{x}_P)$	$\mathbf{L} = \sum_{J=1}^{N_P} \mathbf{v}_J \otimes \nabla \Phi(\mathbf{x}_{GP} - \mathbf{x}_J)$

SPH-SPH with collocation, RLM (ni)-Randles-Libersky correction with nodal integration.
KBM (si)-Krongauz-Belytschko correction with stress point integration.
EFG (ni)-EFG with nodal integration, EFG (ci)-EFG with cell integration.

3. The Krongauz-Belytschko correction and Shepard functions in a Petrov-Galerkin method with stress-point integration (KBM si).
4. The EFG method with nodal integration (EFG ni).
5. The EFG method with cell integration (EFG ci).

The equations for the internal forces and the incremental deformation gradient are summarized in Table 5.1 and 5.3 for a Lagrangian kernel and in Table 5.2 and 5.4 for an Eulerian kernel. For the constitutive model in terms of the Eulerian kernel the velocity gradient $\mathbf{L}$ has to be determined instead of the deformation gradient. Note that $\mathbf{L}$ can directly obtained via $\mathbf{F}$ by the following relation: $\mathbf{L} = \dot{\mathbf{F}} \cdot \mathbf{F}^{-1}$. For SPH and the Randles-Libersky correction method (RLM), a symmetrized version is used as shown in Table 5.2 and 5.1, where $\mathbf{P}_I$ and $\boldsymbol{\sigma}_I$, respectively, are the stresses at the central particle. It should be mentioned, that in Randles' original paper, the stress term is subtracted in the momentum equation. This leads to problems when

Table 5.3 Discrete internal forces and deformation gradient for a Lagrangian kernel in indicial notation.

	Internal forces	Deformation gradient
SPH	$f_{iI}^{int} = \sum_{J\in S} \left(P_{ij}(\mathbf{X}_J) + P_{ij}(\mathbf{X}_I)\right)\ \nabla_{j0} W_J V_J^0$	$F_{ijI} = -\sum_{J\in S} \left(u_{iJ} - u_{iI}\right)\ \nabla_{j0} W_J V_J^0$
RLM (ni)	$f_{iI}^{int} = \left(-\sum_{J\in S}(P_{ij}(\mathbf{X})_J + P_{ij}(\mathbf{X})_I)\nabla_{k0} W_{JI} V_J^0\right) B_{kj}$	$F_{ijI} = \left(-\sum_{J\in S}(u_{iJ} - u_{iI})\nabla_{k0} W_J V_J^0\right) B_{kj}$
KBM (si)	$f_{iI}^{int} = \sum_{J\in N_P} \nabla_{j0} W^S(\mathbf{X}_J^P)\ P_{ji}(\mathbf{X}_J^P) V_J^{P0}$ $+\sum_{J\in N_S} \nabla_{j0} W^S(\mathbf{X}_J^S)\ P_{ji}(\mathbf{X}_J^S) V_J^{S0}$	$F_{ij}^P = \sum_{J\in N_P} u_{iJ} G_j(\mathbf{X}_J^P)$ $F_{ij}^S = \sum_{J\in N_S} u_{iJ}\ G_j(\mathbf{X}_J^S)$
EFG (ni)	$f_i^{int} = \sum_{J\in S} P_{ij}(\mathbf{X}_J)\ \nabla_{j0}\Phi(\mathbf{X}_J)\ V_J^0$	$F_{ij} = -\sum_{J\in S}(u_{iJ} - u_{iI})\ \nabla_{j0}\Phi(\mathbf{X}_J)$
EFG (ci)	$f_i^{int} = \sum_{J=1}^{m} w_J\ det\mathbf{J}_J\ P_{ij}(\mathbf{X}_J)\ \nabla_{j0}\Phi(\mathbf{X}_J - \mathbf{X}_P)$	$F_{ij} = \sum_{J=1}^{N_P} u_{iJ}\ \nabla_{j0}\Phi(\mathbf{X}_{GP} - \mathbf{X}_J)$

SPH-SPH with collocation, RLM (ni)-Randles-Libersky correction with nodal integration.
KBM (si)-Krongauz-Belytschko correction with stress point integration.
EFG (ni)-EFG with nodal integration, EFG (ci)-EFG with cell integration.

Table 5.4 Discrete internal forces and deformation gradient for an Eulerian kernel in indicial notation.

	Internal forces	Velocity gradient
SPH	$f_{iI}^{int} = \sum_{J\in S} \left(\sigma_{ij}(\mathbf{x}_J) + \sigma_{ij}(\mathbf{x}_I)\right)\ \nabla_j W_J V_J$	$L_{ijI} = -\sum_{J\in S} \left(v_{iJ} - v_{iI}\right)\ \nabla_j W_J V_J$
RLM (ni)	$f_{iI}^{int} = \left(-\sum_{J\in S}(\sigma_{ij}(\mathbf{x})_J + \sigma_{ij}(\mathbf{x})_I)\nabla_k W_{JI} V_J\right)\ B_{kj}$	$L_{ijI} = \left(-\sum_{J\in S}(v_{iJ} - v_{iI})\nabla_k W_J V_J\right)\ B_{kj}$
KBM (si)	$f_{iI}^{int} = \sum_{J\in N_P} \nabla_j W^S(\mathbf{x}_J^P)\ \sigma_{ji}(\mathbf{x}_J^P) V_J^P$ $+\sum_{J\in N_S} \nabla_j W^S(\mathbf{x}_J^S)\ \sigma_{ji}(\mathbf{X}_J^S) V_J^{S0}$	$L_{ij}^P = \sum_{J\in N_P} v_{iJ} G_j(\mathbf{x}_J^P)$ $L_{ij}^S = \sum_{J\in N_S} v_{iJ}\ G_j(\mathbf{X}_J^S)$
EFG (ni)	$f_i^{int} = \sum_{J\in S} \sigma_{ij}(\mathbf{x}_J)\ \nabla_j\Phi(\mathbf{x}_J)\ V_J$	$L_{ij} = -\sum_{J\in S}(v_{iJ} - v_{iI})\ \nabla_j\Phi(\mathbf{X}_J)$
EFG (ci)	$f_i^{int} = \sum_{J=1}^{m} w_J\ det\mathbf{J}_J\ \sigma_{ij}(\mathbf{x}_J)\ \nabla_j\Phi(\mathbf{x}_J - \mathbf{x}_P)$	$L_{ij} = \sum_{J=1}^{N_P} v_{iJ}\ \nabla_{j0}\Phi(\mathbf{x}_{GP} - \mathbf{x}_J)$

SPH-SPH with collocation, RLM (ni)-Randles-Libersky correction with nodal integration.
KBM (si)-Krongauz-Belytschko correction with stress point integration.
EFG (ni)-EFG with nodal integration, EFG (ci)-EFG with cell integration.

treating stress-free boundaries. Therefore, [102] recommended to rotate the stress tensor at the boundaries, set the traction perpendicular to the boundary to zero and rotate the stress back.

5.2.4.1 Cantilever beam

The first example compares the 5 methods for a problem in static. Therefore, consider the cantilever beam as illustrated in Fig. 5.9. Linear elastic material behavior is assumed. The beam is loaded with a parabolic traction at the end of the beam as shown in Fig. 5.9. The analytic solution for this problem can be found in [47]. The L_2 error in the displacement as well as the error in the energy is checked. The L_2 error in the displacement is computed by

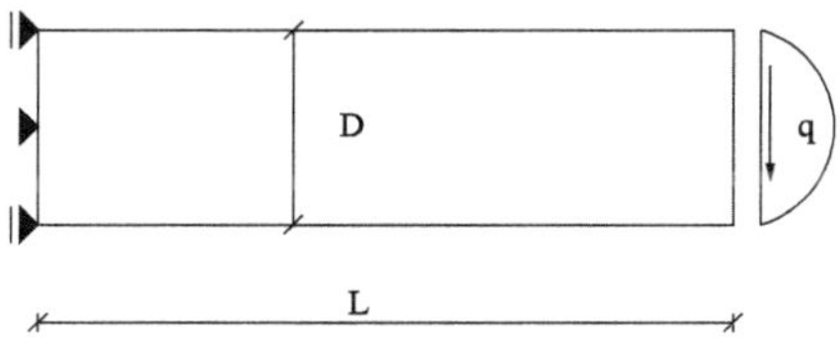

Figure 5.9 Model description of the cantilever beam.

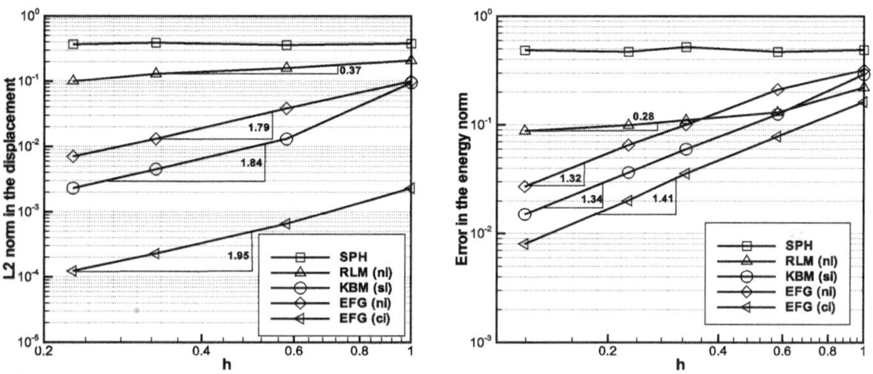

Figure 5.10 L_2 norm of error in displacement and error in energy for different particle methods.

$$\|\mathbf{u}\|_{L_2} = \frac{\|\mathbf{u}_h - \mathbf{u}_{analytic}\|_{L_2}}{\|\mathbf{u}_{analytic}\|_{L_2}} \tag{5.83}$$

with

$$\|\mathbf{u}\|_{L_2} = \sqrt{\left(\int_{\Omega_0} \mathbf{u} \cdot \mathbf{u} \; d\Omega_0\right)} \tag{5.84}$$

The error in the energy is obtained by

$$\|\mathbf{u}\|_{energy} = \frac{\|\mathbf{u}_h - \mathbf{u}_{analytic}\|_{energy}}{\|\mathbf{u}_{analytic}\|_{energy}} \tag{5.85}$$

with

$$\|\mathbf{u}\|_{energy} = \left(\int_{\Omega_0} \mathbf{E}^T(\mathbf{u}) : \mathbf{C} : \mathbf{E}(\mathbf{u}) \; d\Omega_0\right)^{1/2} \tag{5.86}$$

where $\mathbf{E}$ is the linear strain tensor. The results are presented in Fig. 5.10. As expected the best results are obtained with EFG with cell integration. The Randles-Libersky correction combined with a nodal integration yields poor results. SPH does not converge at all.

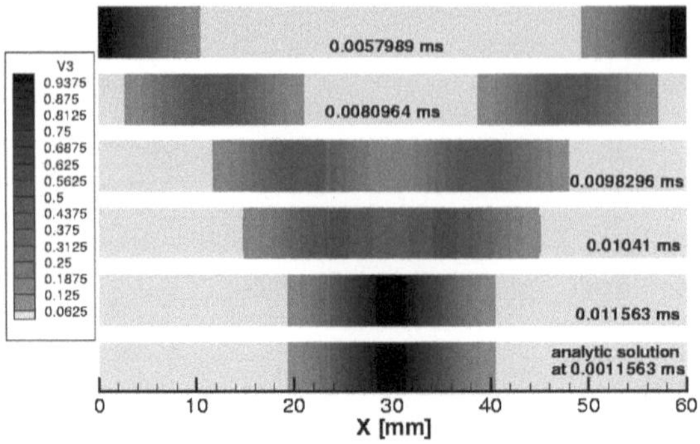

Figure 5.11 Velocity profile in the rod for an initial condition (Gauss distribution of the velocities) and EFG with cell integration at different times.

5.2.4.2 Linear elastic rod with initial displacement

To study the performance of the methods in dynamics, we investigate these methods for wave propagation and reflection in a linear elastic rod. The rod is discretized in 2D (plane strain). It is 60 mm long and its height is 5 mm. An initial velocity condition of $v = e^{(-\alpha\ (x-30)^2)}$, $\alpha = 0.025$ was prescribed. With the sound speed of $c = \sqrt{E/\varrho}$ with $E = 210{,}000$ MPa and $\varrho = 0.0078$ g/mm^3, the wave needs approximately $t = 0.011563$ ms to reach its original position. Fig. 5.11 shows the velocity field in the rod at different time steps for EFG with cell integration. It can be seen how the wave propagates to the left and right hand end, is reflected and after 0.011563 ms reaches its original position again. As can be seen, the curves of the EFG simulation and the analytic solution matches very well. The wave dispersion is higher for SPH. Also, the maximum velocity is not reached for the SPH computation. In Table 5.5, the error in the velocity norm is shown for the different methods at $t = 0.011562$ ms and $t = 0.023126$ ms. At $t = 0.011562$ ms, the wave reaches its first time its original position, at $t = 0.023126$ ms it reaches its original position for the second time. As expected, the best results are obtained with EFG and a cell integration but even SPH can reproduce the velocities quite well where the error is approximately 5% (see Table 5.5).

5.3. Numerical instabilities

Certain meshfree methods show unstable behavior under certain circumstances. The reason of the instabilities have different causes. In the

Table 5.5 L_2 error in the velocities for an initial condition (Gauss distribution of the velocities) for different particle methods before and after the wave reflection.

	Error at 0.0116 ms	Error at 0.0231 ms	Number of particles	Number of str. points	Number of cells	CPU time [minutes]
SPH (ni)	0.05026	0.04999	20,000	–	–	6.2
RLM (ni)	0.01301	0.01297	20,000	–	–	7.8
KBM (si)	0.00813	0.00842	10,500	10,200	–	8.4
EFG (ni)	0.008322	0.008232	20,000	–	–	7.6
EFG (ci)	0.006817	0.006834	4,100	–	15,700	20.1

so-called *tensile instability*, detected first by [119], the method becomes unstable when tensile stresses occur. This spurious instability is caused by the use of an Eulerian kernel. [10,15,129] have shown that by simply using a Lagrangian kernel, this instability can be avoided.

The second instability is an *instability due to rank deficiency* caused by underintegration of the weak form comparable to reduced integrated finite elements. This instability occurs only for methods that use a nodal integration. By adding stress points to the nodes, the instability due to rank deficiency can be avoided; also background cell integrated meshfree methods don't show this unstable behavior.

The third instability is the *material instability* and is physically motivated. All materials with strain softening show this unstable behavior as discussed in the Introduction. Also non-associated hardening plasticity models show this material instability. The material instability depends on the constitutive model and the chosen material parameters. It is therefore not possible to perform a generalized material stability analysis. The results will change with changing the constitutive model. A well accepted material stability analysis is the so-called von Neumann stability analysis and will be described in Section 5.3.4.1. As mentioned earlier, [10,96] have shown that particle discretizations of solids with an Eulerian kernel cannot capture the material stability correctly. Fracture will occur due to numerical artifacts instead of physical conditions. For all material models in this book,[3] a Lagrangian kernel was able to capture the material instability appropriately. This will be shown and discussed in Section 5.3.4.2 exemplarily for two material models in two dimensions.

[3] For some models we also tested different sets of material parameters.

5.3.1 Instability due to rank deficiency

The instability due to rank deficiency is inherent for all particle discretizations based on nodal integration. It might be comparable to the hourglass modes in reduced integrated finite element methods. The instability occurs in the absence of stresses. This is called a singular mode since for a frequency equal to zero, a mode other than the rigid body mode is spurious. The instability due to rank deficiency is a weak instability and grows linear in time. It is called rank deficient instability since it reflects a rank deficiency in the stiffness matrix **K** in the linearized equilibrium equation. The proper rank of the matrix is given by [16]

$$\text{proper rank}(\mathbf{K}) = \text{order}(\mathbf{K}) - n_{RB} \tag{5.87}$$

where **K** denotes the stiffness matrix of the element or particle and n_{RB} is the number of rigid body modes. We know from linear algebra that the rank of the product of two matrices is always less or equal to the rank of either matrices:

$$\text{rank}(\mathbf{K}) = \min\left(\text{rank}(\tilde{\mathbf{B}}), \text{rank}(\tilde{\mathbf{C}})\right) \tag{5.88}$$

where $\tilde{\mathbf{B}} = [\mathbf{B}(\boldsymbol{\xi}_1), \mathbf{B}(\boldsymbol{\xi}_2), ..., \mathbf{B}(\boldsymbol{\xi}_{n_Q})]^T$ is the B-operator that contains the spatial derivatives of the shape functions and $\tilde{\mathbf{C}}$ is the tangent modulus where n_Q is the number of quadrature points:

$$\tilde{\mathbf{C}} = \begin{bmatrix} \bar{w}_1 J_\xi[\mathbf{C}] & \mathbf{0} & \dots & \mathbf{0} \\ \mathbf{0} & \bar{w}_2 J_\xi[\mathbf{C}] & \dots & \mathbf{0} \\ \dots & \dots & \dots & \dots \\ \mathbf{0} & \mathbf{0} & \dots & \bar{w}_{n_Q} J_\xi[\mathbf{C}] \end{bmatrix}$$

If we consider a quadrilateral bilinear finite element with four nodes and two degrees of freedom the order of $\mathbf{K} = 8$. There are two translational and one rotational rigid body modes so the proper rank of **K** is 5. The rank of **B** is 3 since the number of rows are 3. If we assume that the rank of **B** is less than the rank of **C**, the rank of **K** is equal to $\text{rank}(\tilde{\mathbf{B}}) = 12$ where we have assumed that $n_Q = 4$. This exceeds the rank of **K**. However, it can be shown that only 5 rows in $\tilde{\mathbf{B}}$ are linearly dependent. For a reduced integrated linear element, i.e. $n_Q = 1$, the rank **K** is only three that is less than the required rank of 5.

The instability due to rank deficiency can be avoided by added stress points (stress-point integration) to the nodes; also integration based on a background mesh avoids the instability due to rank deficiency.

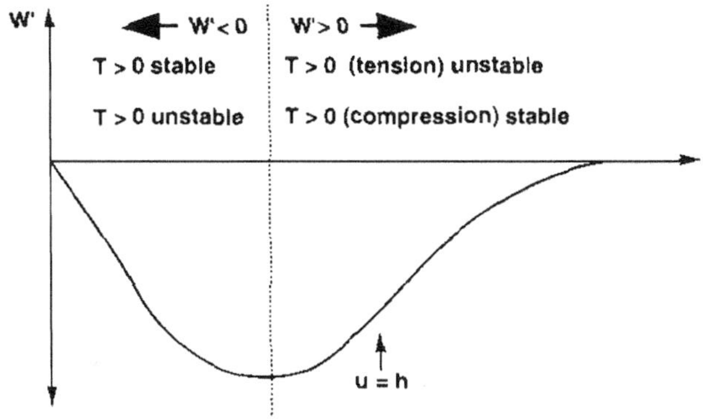

Figure 5.12 Unstable regions for the original SPH method using the cubic B-spline in one dimension, from [119].

5.3.2 Tensile instability

The so-called tensile instability was first detected by [119] in the SPH-method. They carried out a one-dimensional von Neumann material stability analysis (with linear elastic material behavior), that will be explained in the next section.

The main result is that the discretization becomes unstable when the product of the second spatial derivative of the kernel function $\frac{\partial W^2}{\partial^2 x}$ and the stress is smaller than 0. Thus, the tensile instability can theoretically also occur under compression. They further showed that the instability occurred independent from the form of the kernel and the peak stress or peak strain. This contradicts the observations of [18] who reported stable behavior for materials with small tensile strength. For the cubic B-spline in 1D, the instable regions in compression and tension are shown in Fig. 5.12.

As pointed out at the beginning of this section, it can be shown that the tensile instability is caused by the use of an Eulerian kernel. Lagrangian kernels don't show this kind of instable behavior.

5.3.3 Attempts to remove instabilities

For sake of completeness, we will briefly mention some attempts to improve particle methods, especially SPH, with respect to their instable behavior. [34] has argued that the tensile instability can be alleviated by the use of MLS approximations for the dependent variables. [82] suggested the use of higher order spline functions. [118] eliminated the tensile instability with a conservative smoothing scheme. [53] suggested to shift the shape functions such that the instability condition – the product of second derivative of the kernel with the stress – does no longer occur. The instability due to rank efficiency was successfully removed by a least-square stabilization, see [15]. [26] suggested a stabilized conforming nodal integration to eliminate the instability due to rank deficiency. [68] proposed an umbrella spherical integration as stabilization. There are probably much more approaches to remove instabilities in particle methods. Finally, we would like to remark that –in our opinion – the best choice for a stable particle method in solid mechanics is a method based on a Lagrangian kernel (with stress point, stabilized nodal integration or cell integration), also with respect to modeling material failure. This will be discussed in the following section.

5.3.4 Material instability in meshfree methods

5.3.4.1 Material stability for continua

Another physical instability is the material instability which occurs for all materials with strain softening. In order to avoid numerical fracture, it is important that the physical instability is captured exactly. As mentioned earlier, the material instability depends on the constitutive model and the chosen material parameters. Therefore, it is not possible to perform a general material stability analysis for any constitutive model. A well accepted linearized material stability analysis for continuum is the von Neumann stability analysis of an infinite slab of the material that can be traced back to Jean Baptist Joseph Fourier. Such analyses are conventionally made since material behavior in this model problem is relevant to arbitrary shaped bodies in more complex states of stress. The slab is initially at rest and in a state of uniform stress and strain. Let us assume a perturbation in the displacement field given by

$$\bar{\mathbf{u}} = \mathbf{u} + \tilde{\mathbf{u}} \tag{5.89}$$

where the superimposed *tilde* denotes the perturbation. The perturbed linearized momentum equation in the absence of body forces is given by

$$\varrho_0 \ddot{\tilde{\mathbf{u}}} = \nabla_0 \tilde{\mathbf{P}} \tag{5.90}$$

where $\mathbf{P}$ is the nominal stress tensor. With the relation $\mathbf{P} = \mathbf{S}\,\mathbf{F}^T$ where $\mathbf{S}$ is the second Piola Kirchhoff (2PK) tensor and $\mathbf{F}$ is the deformation gradient, one obtains

$$\tilde{\mathbf{P}} = \tilde{\mathbf{S}}\,\mathbf{F}^T + \mathbf{S}\,\tilde{\mathbf{F}}^T \tag{5.91}$$

The perturbed rate of the 2PK tensor is given by

$$\dot{\tilde{\mathbf{S}}} = \mathbf{C}^{SE}\dot{\tilde{\mathbf{E}}} \tag{5.92}$$

where $\mathbf{C}^{SE}$ is the tangent modulus tensor and with the perturbed Green-Lagrange strain tensor

$$\tilde{\mathbf{E}} = \frac{1}{2}\left(\tilde{\mathbf{C}} - \mathbf{I}\right) \tag{5.93}$$

with the perturbed right Cauchy Green tensor $\tilde{\mathbf{C}}$

$$\tilde{\mathbf{C}} = \tilde{\mathbf{F}}^T\ \mathbf{F} + \mathbf{F}^T\ \tilde{\mathbf{F}} \tag{5.94}$$

Assuming minor symmetry of $\mathbf{C}^{SE}$, with $\tilde{\mathbf{C}} = \mathbf{F}^T\ \tilde{\mathbf{F}}$ and using the rate form of the constitutive equations ($\tilde{\mathbf{S}}$ is considered as stress rate), Eq. (5.91) can be rewritten as

$$\tilde{P}_{ij} = C^{SE}_{ikab}F_ar^T\tilde{F}_{rb}F^T_{kj} + S_{ib}F^T_{bj} = F_{ra}F_{jk}C^{SE}_{ikab}\tilde{F}_{rb} = A^0_{ijrb}\tilde{F}_{rb} \tag{5.95}$$

where we have used indicial notation for better clarity. The above defines the first elasticity tensor

$$A^0_{ijrb} = F_{ra}F_{jk}C^{SE}_{ikab} + S_{ib}\delta_{jr} \tag{5.96}$$

where δ_{jr} is the Kronecker delta. The linearized momentum equation for the disturbed displacement finally reads:

$$\varrho_0\ddot{\tilde{\mathbf{u}}}(\mathbf{X},t) = \frac{\partial A^0_{jirb}(\mathbf{X},t)\tilde{F}_{rb}(\mathbf{X},t)}{\partial X_j} \tag{5.97}$$

Let us assume a perturbation of displacement in the form of a plane wave

$$\tilde{\mathbf{u}} = \mathbf{h}\ e^{\omega t + ik\mathbf{n}^0\cdot\mathbf{X}} \tag{5.98}$$

where ω is a complex frequency, k is the wave number, $\mathbf{n}^0$ is the direction of the wave front with respect to the initial configuration and $\mathbf{h}$ is the polarization of the wave. Defining $\alpha = \omega t + ik\mathbf{n}^0\cdot\mathbf{X}$, we obtain

$$\ddot{\tilde{\mathbf{u}}} = \omega^2\mathbf{h}\ e^{\alpha} \tag{5.99}$$

and

$$\frac{\partial A^0_{jirb}(\mathbf{X},t)\tilde{F}_{rb}(\mathbf{X},t)}{\partial X_j} = -k^2A^0_{jirb}n^0_bn^0_jh_re^{\alpha} \tag{5.100}$$

leads the final perturbed equation

$$\varrho_0 \omega^2 h_i - k^2 A^0_{jirb} n^0_b n^0_j h_r = 0 \tag{5.101}$$

or

$$\left(\frac{\omega^2}{k^2} \delta_{ir} + \frac{1}{\varrho_0} A^0_{jirb} n^0_b n^0_j \right) h_r = 0 \tag{5.102}$$

Thus, the characteristic equation for the continuum is

$$\det \left(\frac{\omega^2}{k^2} \delta_{ir} + \frac{1}{\varrho_0} A^0_{jirb} n^0_b n^0_j \right) = 0 \tag{5.103}$$

If the matrix $A_{ir} = A^0_{jirb} n^0_b n^0_j$ is positive definite, then the roots of ω must be imaginary and the continuum stable, otherwise unstable. Note that A^0_{ir} can loose positive definiteness when $\mathbf{C}^{SE}$ looses positive definiteness that occurs in strain softening or with the absence of major symmetry $C^{SE}_{ijab} \neq C^{SE}_{abij}$, i.e. with a non-associative law. The condition

$$\mathbf{h} \otimes \mathbf{n}^0 : \mathbf{A}^0 : \mathbf{n} \otimes \mathbf{h} > 0 \ \forall \ \mathbf{n}^0 \text{ and } \mathbf{h} \tag{5.104}$$

is often also called strong ellipticity condition since an elliptic PDE looses ellipticity if condition (5.104) is violated. For hyperbolic PDEs, we talk about loss of hyperbolicity. For a rate-dependent constitutive model, the condition is called material stability condition. Note that the underlying PDE based on strong nonlocal formulations will not loose ellipticity/hyperbolicity/stability. The tensor $A^0_{ir} = A^0_{jirb} n^0_b n^0_j$ is often called *acoustic tensor.*

Eq. (5.103) can be used to obtain the stability condition in terms of Eulerian variables by letting the current configuration be the reference configuration. This yields $\mathbf{F} = \mathbf{I}$, $\mathbf{S} = \boldsymbol{\sigma}$ (the Cauchy stress), $\dot{\mathbf{S}} = \boldsymbol{\sigma}^{\nabla \mathfrak{T}}$ (the Truesdell rate) and $\mathbf{C}^{SE} = \mathbf{C}^{\sigma \mathfrak{T}}$ where $\mathbf{C}^{\sigma \mathfrak{T}}$ is the modulus relating the Truesdell rate of the Cauchy stress to the rate of deformation. More details can be found in [16]. The final equation reads

$$\det \left(\frac{\omega^2}{k^2} \delta_{ir} + \frac{1}{\varrho} Q_{jirb} n_b n_j \right) = 0 \tag{5.105}$$

5.3.4.2 Material stability analyses of meshfree methods

In this section, we will study Eulerian and Lagrangian kernel based meshfree methods with respect to their ability to predict the onset of fracture. Exemplarily, we will show the results obtained with the element-free Galerkin method and a Petrov Galerkin method with corrected derivatives.

Hyperelastic material model with strain softening

In this section, we will perform a material stability analysis for a hyperelastic constitutive model. The analysis was done by [121] in the context of a M.Sc.-thesis and is based on the results presented in [15]. The hyperelastic material is characterized by the existence of a stored (or strain) energy function that is the potential for the stress

$$\mathbf{S} = 2\frac{\partial \psi}{\partial \mathbf{C}} = \frac{\partial \mathbf{w}(\mathbf{E})}{\partial \mathbf{E}} \tag{5.106}$$

Here an *isotropic hyperelastic material* model will be considered whose potential is given by

$$\psi = \frac{1}{2}c_1 I_1 + \frac{1}{2}c_2 I_2 - [\frac{3}{2}c_1 I_3^{\frac{1}{3}} + \frac{3}{2}c_2 I_3^{\frac{2}{3}} - \frac{\lambda}{4}(lnI_3)^2] \tag{5.107}$$

where I_1, I_2 and I_3 are the principal invariants of the right Cauchy-Green deformation tensor $\mathbf{C}$. Since $\mathbf{C}$ is a symmetric matrix, these principal invariants are given by

$$\begin{aligned} I_1(\mathbf{C}) &= trace(\mathbf{C}) = \lambda_1^2 + \lambda_2^2 + 1 \\ I_2(\mathbf{C}) &= \frac{1}{2}[(trace(\mathbf{C}))^2 - trace(\mathbf{C}^2)] = 0 \\ I_3(\mathbf{C}) &= det(\mathbf{C}) = \lambda_1^2\lambda_2^2 \end{aligned} \tag{5.108}$$

This model is similar to the Mooney-Rivlin material model except that the Mooney-Rivlin material is incompressible, i.e. $I_3 = 1$. We use this hyperelastic material to analyze stability because it develops material instability more readily than a Mooney-Rivlin material model. The first derivatives of the principal invariants are:

$$\frac{\partial I_1}{\partial \mathbf{C}} = \mathbf{I}, \quad \frac{\partial I_2}{\partial \mathbf{C}} = I_1\mathbf{I} - \mathbf{C}^T, \quad \frac{\partial I_3}{\partial \mathbf{C}} = I_3\mathbf{C}^{-T} \tag{5.109}$$

The second Piola-Kirchhoff stress tensor $\mathbf{S}$ is given by

$$\mathbf{S} = 2\frac{\partial \psi}{\partial \mathbf{C}} = 2(\frac{\partial \psi}{\partial I_1}\frac{\partial I_1}{\partial \mathbf{C}} + \frac{\partial \psi}{\partial I_2}\frac{\partial I_2}{\partial \mathbf{C}} + \frac{\partial \psi}{\partial I_3}\frac{\partial I_3}{\partial \mathbf{C}}) \tag{5.110}$$

or

$$\mathbf{S} = (c_1 + c_2 I_1)\mathbf{I} - c_2\mathbf{C} - (c_1 I_3^{1/3} + 2c_2 I_3^{2/3} - \lambda lnI_3)\mathbf{C}^{-1} \tag{5.111}$$

The tangent modulus $\mathbf{C}^{SE}$, which is also called the second elasticity tensor relates the rate of the second Piola-Kirchhoff stress to the Green strain rate by $\dot{\mathbf{S}} = \mathbf{C}^{SE} : \dot{\mathbf{E}}$. It is given by

$$\mathbf{C}^{SE} = 2\frac{\partial \mathbf{S}(\mathbf{C})}{\mathbf{C}} = 2c_2\mathbf{I} \otimes \mathbf{I} - 2c_2\mathbf{I} - 2s_0\frac{\partial \mathbf{C}^{-1}}{\partial \mathbf{C}} - 2s_1\mathbf{C}^{-1} \otimes \mathbf{C}^{-1} \tag{5.112}$$

where $s_0 = c_1 I_3^{1/3} + 2c_2 I_3^{2/3}$ and $s_1 = (\frac{1}{3}c_1 I_3^{-2/3} + \frac{4}{3}c_2 I_3^{-1/3} - \frac{\lambda}{I_3})I_3$. Rewriting Eq. (5.112) in indicial form, the tangent modulus tensor is given by

$$C_{ijkl}^{SE} = 2c_2\delta_{ij}\delta_{kl} - c_2(\delta_{jk}\delta_{il} + \delta_{il}\delta_{jk}) + s_0(C_{ik}^{-1}C_{jl}^{-1} + C_{il}^{-1}C_{jk}^{-1}) - 2s_1 C_{ij}^{-1}C_{kl}^{-1} \tag{5.113}$$

In the spatial form, the second elastic tensor is the fourth elasticity tensor $\mathbf{C}^{\tau}$ or is referred to as the spatial tangent moduli. It relates the convected rate of Kirchhoff to the rate-of-deformation, i.e. $\boldsymbol{\tau}^{\nabla c} = \mathbf{C}^{\tau} : \mathbf{D}$. This fourth elasticity tensor in indicial form is given by

$$C_{ijkl}^{\tau} = F_{im}F_{jn}F_{kp}F_{lq}C_{mnpq}^{SE} \tag{5.114}$$

The relation between the Truesdell rate of the Cauchy stress and rate-of-deformation is

$$\boldsymbol{\sigma}^{\nabla T} = J^{-1}\mathbf{C}^{\tau} : \mathbf{D} = \mathbf{C}^{\sigma T} : \mathbf{D} \tag{5.115}$$

where $\mathbf{C}^{\sigma T} = J^{-1}\mathbf{C}^{\tau}$ and the Truesdell rate of Cauchy stress is given by

$$\boldsymbol{\sigma}^{\nabla T} = \dot{\boldsymbol{\sigma}} + div(\mathbf{v})\boldsymbol{\sigma} - \mathbf{L}\cdot\boldsymbol{\sigma} - \boldsymbol{\sigma}\cdot\mathbf{L}^T \tag{5.116}$$

and

$$\boldsymbol{\sigma} = \frac{1}{J}\mathbf{F}\cdot\mathbf{S}\cdot\mathbf{F}^T, \quad \mathbf{L} = (\nabla\mathbf{v})^T \tag{5.117}$$

The elasticity tensors defined above play an important role in stability analysis and uniqueness of solutions in finite strain elasticity.

The characteristic equation that determines stability is obtained by setting the determinant of the characteristic matrix equal to zero. The eigenvalue problem corresponding to Eq. (5.105) is

$$\mathbf{A}\mathbf{g} = \boldsymbol{\lambda}\mathbf{g} \tag{5.118}$$

where

$$\boldsymbol{\lambda} = -\frac{\rho_0\omega^2}{\kappa^2} \tag{5.119}$$

From the characteristic equation (5.98), a simple sufficient condition for stability can be deduced. A material is stable when its eigenvalues of the acoustic tensor are real and positive ($\lambda > 0$) $\forall \mathbf{n}^0, \mathbf{h}$. Therefore, all values of ω that we derive from Eq. (5.119) will be imaginary without real parts. As mentioned above, no growth of perturbation appears, i.e. the response will be stable. The deformation gradient whose diagonal show states of deformation is the symmetric matrix, so the right Cauchy-Green deformation tensor is

$$\mathbf{C} = \mathbf{F}^{\mathrm{T}} \cdot \mathbf{F} = \begin{bmatrix} \lambda_1^2 & 0 \\ 0 & \lambda_2^2 \end{bmatrix} \tag{5.120}$$

Substituting Eq. (5.120) into Eq. (5.111) we obtain the second Piola-Kirchhoff stress $\mathbf{S}$:

$$\mathbf{S} = (c_1 + c_2 I_1)\begin{bmatrix} 1 & 0 \\ 0 & 1 \end{bmatrix} - c_2 \begin{bmatrix} \lambda_1^2 & 0 \\ 0 & \lambda_2^2 \end{bmatrix} - s_0 \begin{bmatrix} \lambda_1^{-2} & 0 \\ 0 & \lambda_2^{-2} \end{bmatrix}$$

Then, the components of the second elasticity tensor are computed from Eq. (5.112) and note that this tangent modulus tensor $\mathbf{C}^{SE}$ is major symmetric matrix i.e. $C_{ijkl}^{SE} = C_{klij}^{SE}$ with

$$\begin{aligned}
C_{1111}^{SE} &= \frac{2s_0 - 2s_1}{\lambda_1^4} \\
C_{2222}^{SE} &= \frac{2s_0 - 2s_1}{\lambda_2^4} \\
C_{1122}^{SE} &= C_{2211}^{SE} = 2c_2 - \frac{2s_1}{\lambda_1^2 \lambda_2^2} \\
C_{2121}^{SE} &= C_{1212}^{SE} = C_{1221}^{SE} = C_{2121}^{SE} = -c_2 + \frac{s_0}{\lambda_1^2 \lambda_2^2}
\end{aligned} \tag{5.121}$$

The effective tangent modulus tensor C_{ijkl}^0 will then be given by Eq. (5.96)

$$\begin{aligned}
C_{1111}^0 &= C_{1111}^{SE} \lambda_1^2 + (c_1 + c_2 I_1) - c_2 \lambda_1^2 - s_0 \lambda_1^{-2} \\
C_{2112}^0 &= C_{2112}^{SE} \lambda_1^2 + (c_1 + c_2 I_1) - c_2 \lambda_2^2 - s_0 \lambda_2^{-2} \\
C_{1122}^0 &= C_{1122}^{SE} \lambda_1 \lambda_2 \\
C_{2121}^0 &= C_{2121}^{SE} \lambda_1 \lambda_2 \\
C_{2211}^0 &= C_{2211}^{SE} \lambda_1 \lambda_2 \\
C_{1212}^0 &= C_{1212}^{SE} \lambda_1 \lambda_2 \\
C_{1221}^0 &= C_{1221}^{SE} \lambda_2^2 + (c_1 + c_2 I_1) - c_2 \lambda_1^2 - s_0 \lambda_1^{-2}
\end{aligned}$$

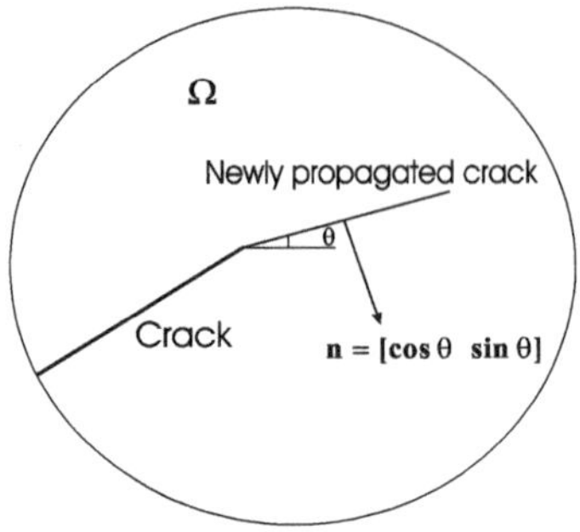

Figure 5.13 Crack with its new propagation and the definition of θ.

$$C^0_{2222} = C^{SE}_{2222}\lambda_2^2 + (c_1 + c_2 I_1) - c_2\lambda_2^2 - s_0\lambda_2^{-2} \tag{5.122}$$

where c_1 =0.1256 MPa, c_2 =0.01012 MPa and λ =10.12 MPa are the material constants. In two dimensions, $\mathbf{n} = [\cos\theta \;\; \sin\theta]$ and θ represents the wave front direction. See Fig. 5.13.

The acoustic tensor $\mathbf{A}$ is computed from the effective tangent modulus tensor $\mathbf{C}^0$ as follows:

$$\begin{aligned}
A_{11} &= C^0_{1111}\cos^2\theta + C^0_{2112}\sin^2\theta \\
A_{12} &= C^0_{1122}\cos\theta\sin\theta + C^0_{2121}\cos\theta\sin\theta \\
A_{21} &= C^0_{2211}\cos\theta\sin\theta + C^0_{1212}\cos\theta\sin\theta \\
A_{22} &= C^0_{1221}\cos^2\theta + C_{222}\sin^2\theta
\end{aligned} \tag{5.123}$$

The stable domain of this hyperelastic material is given by Eq. (5.104). If the minimum eigenvalues of the acoustic tensor $\mathbf{A}$ are positive values, the perturbation in Eq. (5.98) will decay in time and the response is stable. Note that instabilities can occur when $\mathbf{A}$ loses positive definiteness, which can occur when $\mathbf{C^0}$ loses positive definiteness in $\mathbf{C^0}$ due to *strain-softening* or with absence of the major symmetric $C^0_{jirb} = C^0_{rbji}$.

The stable domain for the hyperelastic material considered is shown in Fig. 5.14. Note that set of asterisks '.' in the figure denote the stable domain of material in tension while the set of circle 'o' indicate the stable domain under compression ($0 < \lambda_i \leq 1$). From this stable domain, we get the curve bound of domain that corresponds to the *onset of the instability* in Fig. 5.15. In the stable domain of material, the minimum eigenvalues of the acoustic tensor $\mathbf{A}$ are positive for all angles θ. On the other hand, states of deformation gradient like $\lambda_1 = 2.0$, $\lambda_2 = 1.5$, $\lambda_1 = 2.0$, $\lambda_2 = 2.0$, ... for

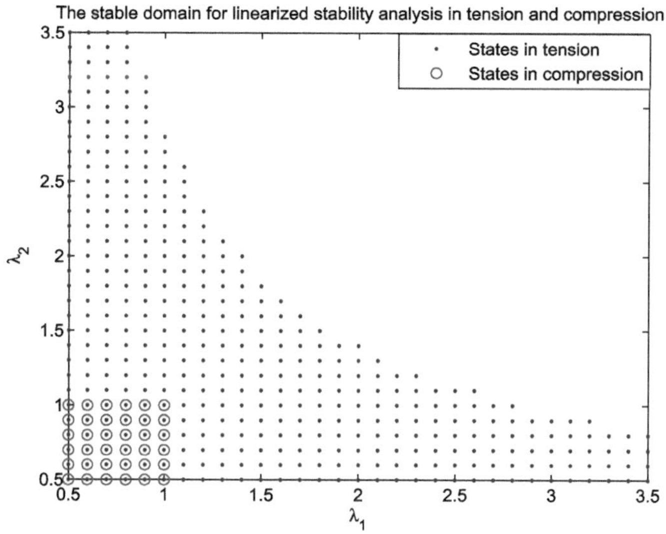

Figure 5.14 Stable domain is made by states of deformation gradient for hyperelastic material model.

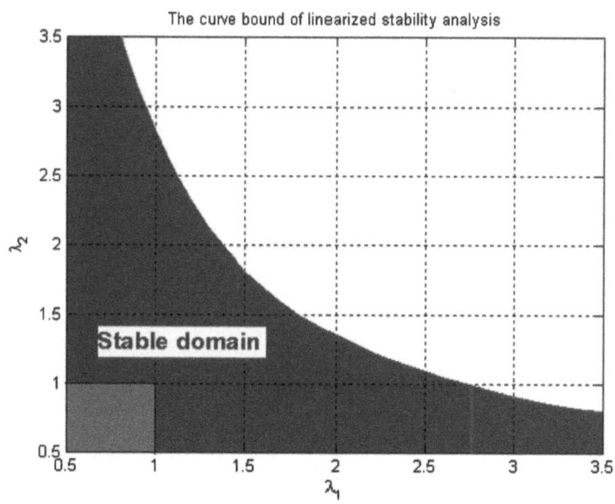

Figure 5.15 Stable domain with curve bound for hyperelastic material model.

which the minimum eigenvalues of $\mathbf{A}$ are negative for all θ, this leads that the values of frequency ω are positive then the perturbation grows with time and its response will be unstable (see Fig. 5.16).

For a Lagrangian kernel and using nodal integration, it can be easily shown that the internal nodal forces for the discrete perturbed momentum

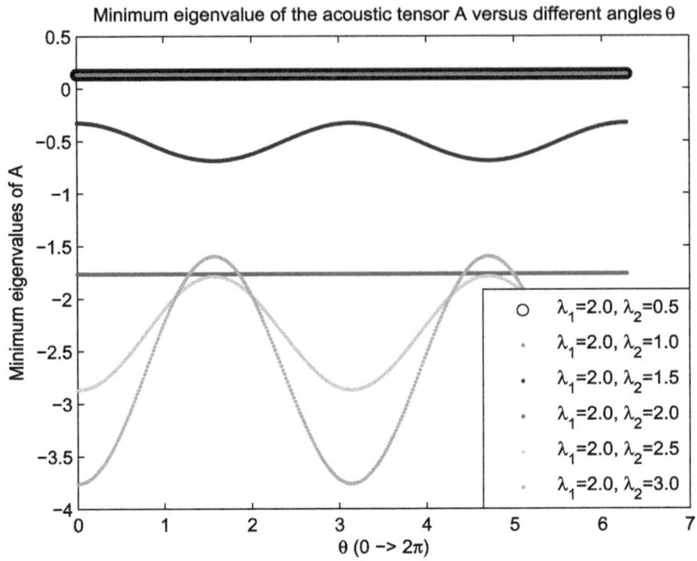

Figure 5.16 Relation between the minimum eigenvalues of A and different angles for some states of deformation gradient.

equation

$$m_I \ddot{\tilde{u}}_{iI}(\mathbf{X}_I) = -\tilde{f}_{iI}^{int} = -\sum_{J\in N} V_J^0 \frac{\partial \Phi_I(\mathbf{X}_J)}{\partial \mathbf{X}_j} \tilde{P}_{ji}(\mathbf{X}_J) \tag{5.124}$$

is given by

$$f_{iI}^{int} = \sum_{J\in N} V_J^0 \frac{\partial \Phi_I(\mathbf{X}_j)}{\partial \mathbf{X}_j} P_{ji}(\mathbf{X}_J) \tag{5.125}$$

In compact form, the perturbed momentum equation reads

$$m_I \ddot{\tilde{\mathbf{u}}}_I = -\tilde{\mathbf{f}}_I^{int} = -\sum_{J\in N} V_J^0 \nabla_0 \Phi_I(\mathbf{X}_J) \tilde{\mathbf{P}}_J \tag{5.126}$$

where $m_I = V_I \rho_I$ is mass of one particle and $\ddot{\tilde{\mathbf{u}}}_I = \omega^2 g e^{\omega t + i\kappa \mathbf{n}^0 \mathbf{X}_I}$ so that the LHS of Eq. (5.126) becomes

$$m_I \ddot{\tilde{\mathbf{u}}}_I = V_I \rho_I \omega^2 \mathbf{g} e^{\omega t + i\kappa \mathbf{n}^0 \mathbf{X}_I} \tag{5.127}$$

The perturbed nominal stress tensor then is given by

$$\tilde{\mathbf{P}}_J = \mathbf{C}_J^0 \tilde{\mathbf{F}}_J \quad \text{or} \quad \tilde{P}_{ij} = C_{ijkl}^0 \tilde{F}_{kl} \tag{5.128}$$

where the effective tangent modulus tensor C^0_{ijkl} is defined by Eq. (5.96) as follows

$$\mathbf{C}^0_J = (\mathbf{C}^{SE}\mathbf{F}^2 + \mathbf{S})_J \quad \text{or} \quad C^0_{ijkl} = C^{SE}_{iral} F_{jr} F_{ka} + S_{il}\delta_{jk} \tag{5.129}$$

The perturbed deformation gradient in two dimension is given by

$$\widetilde{\mathbf{F}}_J = \sum_{K \in N'} \tilde{\mathbf{u}}_K \otimes \nabla_0 \Phi_J(\mathbf{X}_K) \quad \text{or} \quad \widetilde{F}_{ijJ} = \sum_{K \in N'} \tilde{u}_{iK} \frac{\partial \Phi_J(\mathbf{X}_K)}{\partial \mathbf{X}_j} \tag{5.130}$$

Then the internal nodal force becomes

$$\mathbf{f}^{int}_I = \sum_{J \in N} V^0_J \nabla_0 \Phi_I(\mathbf{X}_J) \mathbf{C}^0_J \widetilde{\mathbf{F}}_J \quad \text{or} \quad f^{int}_{iI} = \sum_{J \in N} V^0_J \frac{\partial \Phi_I(X_J)}{\partial X_j} C^0_{ijkl} \widetilde{F}_{kl} \tag{5.131}$$

Now the perturbed equation (5.126) reads

$$\begin{aligned} &V_I \rho_I \omega^2 \mathbf{g} e^{\omega t + i\kappa \mathbf{n}^0 \mathbf{X}_I} \\ &= -\sum_{J \in N} V^0_J \nabla_0 \Phi_I(\mathbf{X}_J) \mathbf{C}^0_J \left[\sum_{K \in N'} \left[\mathbf{g} e^{\omega t + i\kappa \mathbf{n}^0 \mathbf{X}_K} \right] \otimes \nabla_0 \Phi_J(\mathbf{X}_K) \right] \end{aligned} \tag{5.132}$$

Note that N is set of particle J in the domain influence of particle I and N' is set of particle K in the domain influence of particle J. The normal to the crack is $\mathbf{n}^0 = [cos\theta \quad sin\theta]$. The shape functions Φ_I and Φ_J are computed based on the MLS approximation in the EFG method. Besides we also have

$$e^{i\kappa \mathbf{n}^0 \mathbf{X}_I} = cos(\kappa \mathbf{n}^0 X_I) + isin(\kappa \mathbf{n}^0 \mathbf{X}_I) \tag{5.133}$$

Combining Eq. (5.133) with Eq. (5.132), we obtain the dispersion equations

$$\omega^2_i = \frac{1}{V_I \rho_I g_i} \sum_{J \in N} V^0_J \Phi_{I,j}(X_J) C^0_{jikl} g_k \left[\sum_{K \in N'} \Phi_{J,l}(X_K) cos\beta \right] \tag{5.134}$$

where $cos\beta = cos[\kappa(X_K cos\theta + Y_K sin\theta) - \kappa(X_I cos\theta + Y_I sin\theta)]$. The dispersion equations give the wave speed for a polarization $\mathbf{h}$. As discussed before, the material is stable when all frequencies ω are imaginary. This means that the material is stable if the square of frequencies are negative for any direction of the wave front θ under a given deformation gradient in Eq. (5.129)

$$\mathbf{F} = \begin{bmatrix} \lambda_1 & 0 \\ 0 & \lambda_2 \end{bmatrix}$$

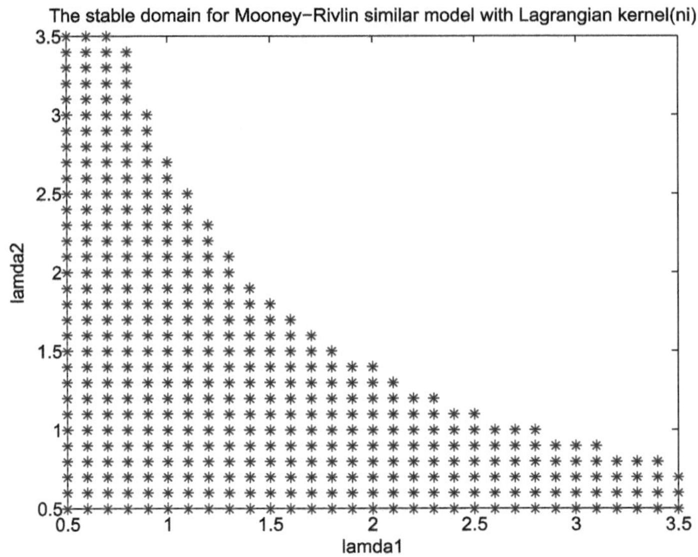

Figure 5.17 The stable domain of Lagrangian kernel method with nodal integration.

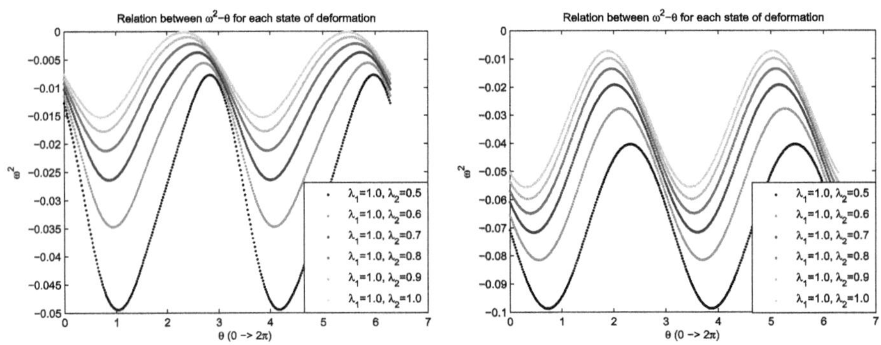

Figure 5.18 The frequencies $\omega^2 - \theta$ relations by EFG with nodal integration and Lagrangian kernel for Dmax=2 and Dmax=3.

The results are shown in Figs. 5.17 and 5.19. As can be seen, the onset of material stability is captured exactly; note some irregularities that occur only due to post-processing issues.[4] Fig. 5.18 shows the values for ω^2.

For Eulerian kernels, the stability analysis depends on the stress and the second derivative of the shape function. The Cauchy stress for the initial

[4] The stable domain is calculated at discrete values.

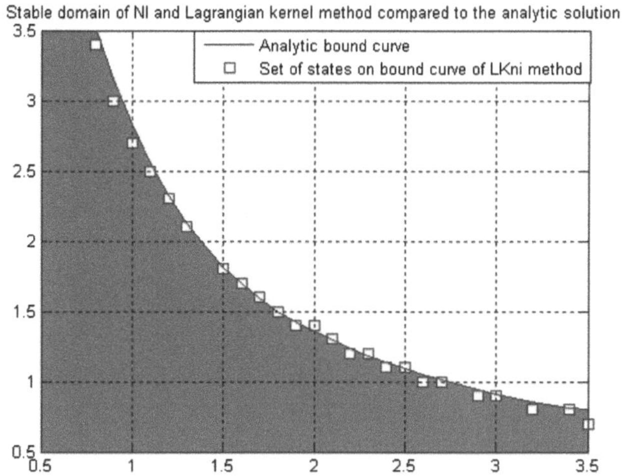

Figure 5.19 Stable domain for MLS particle method with nodal integration and Lagrangian kernel compared to the stable domain of linearized stability analysis.

state is given by

$$\boldsymbol{\sigma} = \frac{1}{J}\mathbf{F}\cdot\mathbf{S}\cdot\mathbf{F}^T = \frac{1}{\lambda_1\lambda_2}\begin{bmatrix} S_{11}\lambda_1^2 & 0 \\ 0 & S_{22}\lambda_2^2 \end{bmatrix} \tag{5.135}$$

In EFG with an Eulerian kernel, the internal force is computed in terms of the current domain and Cauchy stress as follows

$$\mathbf{f}_I^{int} = \int_\Omega \nabla\Phi_I(x)\boldsymbol{\sigma}(\mathbf{x})d\Omega \tag{5.136}$$

For nodal integration, the above integral is evaluated by summing the integrand at the nodes. The internal nodal force is then

$$\mathbf{f}_I^{int} = \sum_J V_J\Phi_I'(\mathbf{x}_j)\boldsymbol{\sigma}_J, \qquad V_J = \frac{m_J\mathbf{F}_J}{\rho_{0J}} \tag{5.137}$$

Thus the discrete equations are

$$m_I\ddot{\mathbf{u}}_I = -\sum_J V_J^0\Phi_I'(\mathbf{x}_j)\boldsymbol{\sigma}_J\mathbf{F}_J \tag{5.138}$$

We also consider a perturbation in displacement and identify all perturbation by a superposed $\tilde{}$. The perturbed equation is given by

$$m_I\tilde{\ddot{\mathbf{u}}}_I = -\sum_J V_J^0\left[\Phi_I'(\mathbf{x}_j)(\tilde{\boldsymbol{\sigma}}_J\mathbf{F}_J + \boldsymbol{\sigma}_J\tilde{\mathbf{F}}_J) + \boldsymbol{\sigma}_J\mathbf{F}_J\tilde{\Phi}_I'(\mathbf{x}_j)\right] \tag{5.139}$$

where

$$\widetilde{\Phi}_I'(\mathbf{x}_j) = \Phi_I''(\mathbf{x}_J)(\tilde{\mathbf{u}}_J - \tilde{\mathbf{u}}_I) \tag{5.140}$$

$$\tilde{\boldsymbol{\sigma}}_I(\mathbf{x}_J) = \bar{\mathbf{C}}_I^0 \sum_K \Phi_K'(\mathbf{x}_J)\tilde{\mathbf{u}}_K(\mathbf{x}_J) \tag{5.141}$$

The effective tangent modulus tensor $\bar{\mathbf{C}}_\mathbf{I}^\mathbf{0}$ is written in indicial form by

$$\bar{C}_{ijkl}^0 = \frac{1}{2J}(C_{ijkl}^\tau + C_{ijlk}^\tau) - \sigma_{ij}\delta_{kl} + \sigma_{il}\delta_{jk} + \sigma_{jl}\delta_{ik} \tag{5.142}$$

where $C_{ijkl}^\tau = F_{ip}F_{jq}F_{ks}F_{lt}C_{pqst}^{SE}$. Letting the current configuration be the reference configuration will give $\mathbf{F} = \mathbf{I}$, $\boldsymbol{\sigma} = \mathbf{S}$ and we get consequently that $\mathbf{C}^\tau = \mathbf{C}^{\mathbf{SE}}$. Substituting Eqs. (5.142) and (5.141) into (5.139) and using the perturbation of the displacement given by $\ddot{\mathbf{u}}_I = \omega^2 \mathbf{g} e^{\omega t + i\kappa \mathbf{n}^0 \mathbf{x}_I}$, the dispersion equations for nodal integration with an Eulerian kernel become

$$\begin{aligned}\omega_i^2 = \frac{1}{V_I \rho_I g_i} \sum_{J \in N} V_J \Bigg[& \Phi_{I,j} \bar{C}_{jirl}^0 g_r \sum_{K \in N'} \Phi_{K,l}(\mathbf{x}_J)\cos\beta \\ & + \sigma_{ji}\Phi_{I,j} g_t \sum_{K \in N'} \Phi_{K,i}(\mathbf{x}_J)\cos\beta + \sigma_{ji}\Phi_{I,jt} g_t(\cos\gamma - 1)\Bigg]\end{aligned} \tag{5.143}$$

where

$$\begin{aligned}\cos\beta &= \cos[\kappa(x_K\cos\theta + y_K\sin\theta) - \kappa(x_I\cos\theta + y_I\sin\theta)] \\ \cos\gamma &= \cos[\kappa(x_J\cos\theta + y_J\sin\theta) - \kappa(x_I\cos\theta + y_I\sin\theta)]\end{aligned} \tag{5.144}$$

The results of the stability analysis are shown in Figs. 5.20 and 5.21 The squares of frequency are positive in the domain of $(0 < \lambda_1 < 1) \cap (0 < \lambda_2 < 1)$ are stable. However, when tensile stresses occur, the material is unstable.

The same observations hold when stress-point integration is used: Lagrangian kernel formulations exactly replicate the onset of material instabilities while Eulerian kernel formulations become unstable in tension.

Example of an instability for hyperelastic material

This example shows the onset of a tensile instability for an Eulerian kernel in a 2D problem. We use a hyperelastic material [65]:

$$\boldsymbol{\sigma} = \frac{1}{\sqrt{I_3}}[(c_1 + c_2 I_1)\mathbf{B} - c_2\mathbf{B}^2 - (c_1 I_3^{\frac{1}{3}} + 2c_2 I_3^{\frac{2}{3}} - \lambda ln I_3)\mathbf{I}] \tag{5.145}$$

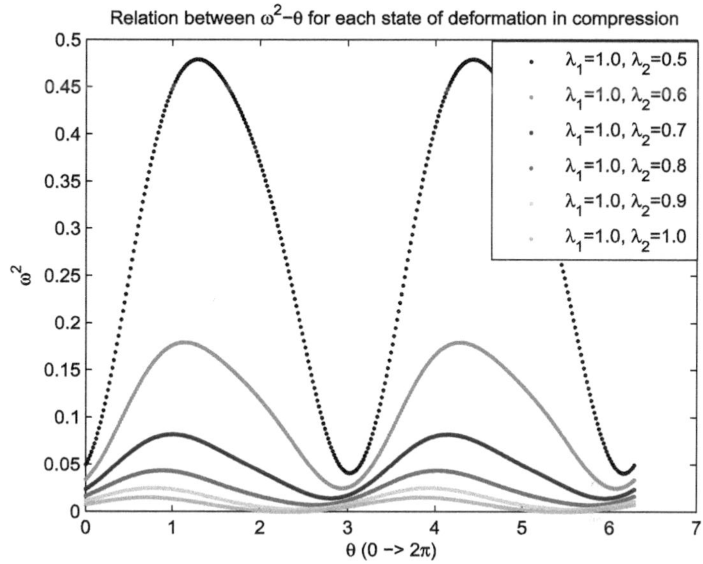

Figure 5.20 The frequencies $\omega^2 - \theta$ relations by EFG with nodal integration and Eulerian kernel in compression.

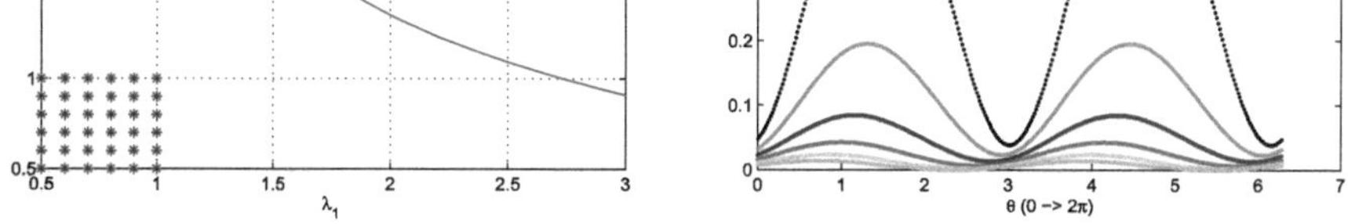

Figure 5.21 The stable domain and the frequencies $\omega^2 - \theta$ relations with different support size for nodal integration and Eulerian kernel for hyperelastic material in compression.

where $\mathbf{B} = \mathbf{F}\mathbf{F}^T$. The material constants are $c_1 = 1.265e5$ N/m^2, $c_2 = 1.012e4$ N/m^2 and $\lambda = 1.012e7$ N/m^2. $I_1 = tr(\mathbf{B})$ and $I_3 = det(\mathbf{B})$. The material density is $\rho = 125.4$ kg/m^3. The stress point integration with Lagrangian kernels can reproduce the material instability but Eulerian kernels distort it severely.

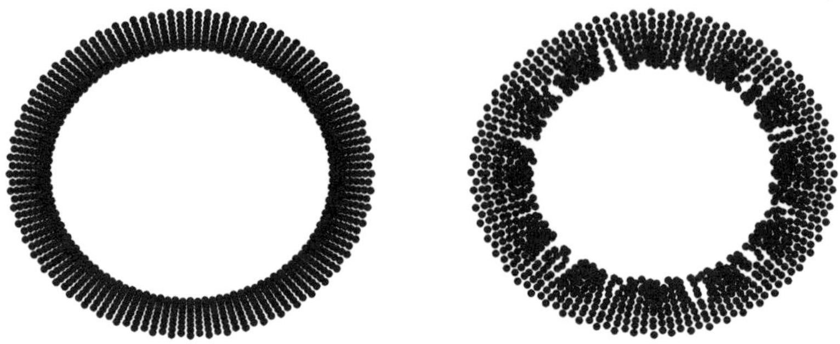

(A) Lagrangian kernel (B) Eulerian kernel

Figure 5.22 Deformed rubber ring by stress points ($t = 0.12$ ms).

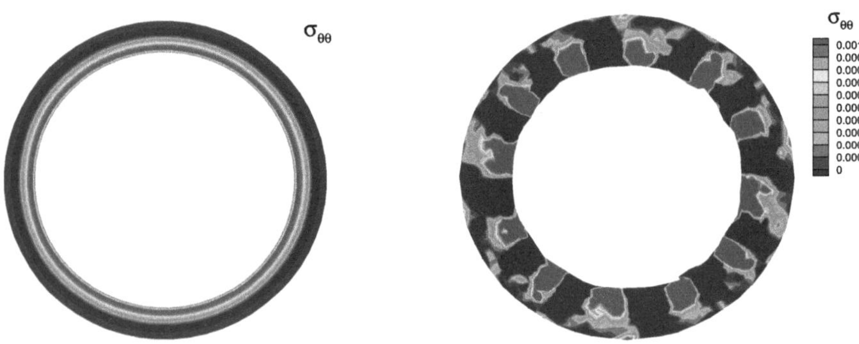

(A) Lagrangian kernel (B) Eulerian kernel

Figure 5.23 $\sigma_{\theta\theta}$ distribution at $t = 0.12$ ms.

A pressure of $6.2e7$ N/m^2 is applied for 0.1 ms on the inner surface of the rubber ring. The discrete model with stress points consists of 540 particles and 480 stress points. Fig. 5.22 shows the deformed rubber ring at the same time by a Lagrangian kernel and an Eulerian kernel with stress point integration. Fig. 5.22B shows the clustering of particles in the Eulerian kernel which is typical for the tensile instability. Fig. 5.23 shows the distribution of $\sigma_{\theta\theta}$ on the deformed configuration as shown in Fig. 5.22B. As can be seen, the stress is concentrated at the discrete locations along the circumference of the ring. If the ring is getting more and more deformations, the instability will occur due to its own material instability even when stress points with Lagrangian kernels are used.

5.4. Fracture modeling in meshfree methods

5.4.1 The visibility method

The visibility method is the first approach that introduces a discrete crack into the meshfree discretization. As all the methods in this section, the visibility method modifies the original shape function.

In the visibility method, the crack boundary is considered to be opaque. Thus, the displacement discontinuity is modeled by excluding the particles on the opposite side of the crack in the approximation of the displacement field, see Fig. 5.24:

$$\mathbf{u}^h = \sum_{I \in \mathcal{S}^+} \Phi_I(\mathbf{X})\ \mathbf{u}_I(t) \tag{5.146}$$

This is identical to setting the shape function across the crack to zero as shown in Fig. 5.24. The jump in the displacement is then computed by

$$[\![\mathbf{u}]\!] = \left(\sum_{I \in \mathcal{S}^+} \Phi_I(\mathbf{X})\ \mathbf{u}_I(t) \right) - \left(\sum_{I \in \mathcal{S}^-} \Phi_I(\mathbf{X})\ \mathbf{u}_I(t) \right) \tag{5.147}$$

where sampling points can be generated across the crack line. Note that these sampling points are coincident at crack initiation where the two crack surfaces are located at the same position.

Difficulties arise for particles close to the crack tip since undesired interior discontinuities occur since the shape function is cut abruptly, see Fig. 5.24 and 5.25A. Nevertheless, [67] showed convergence for the visibility method. For linear complete EFG shape functions, they even showed that the convergence rate is not affected by the discontinuity. The length and size of the (undesired) discontinuities depends on the nodal spacing near the crack boundary. If the nodal spacing approaches zero, the length of the discontinuities tends to zero. With this argument and the theory of non-conforming finite elements, convergence of the discontinuous displacement field can be shown.

It should also be noted that the visibility criterion leads to discontinuities in shape functions near non-convex boundaries such as kinks, crack edges and holes, as shown in Fig. 5.25 in two dimensions. Methods that can also handle non-convex discontinuities will be described in the next sections. For convex discontinuities, the visibility criterion seems to be suitable.

There are many ways to implement the visibility method. We will briefly describe two possibilities how to implement the visibility criterion

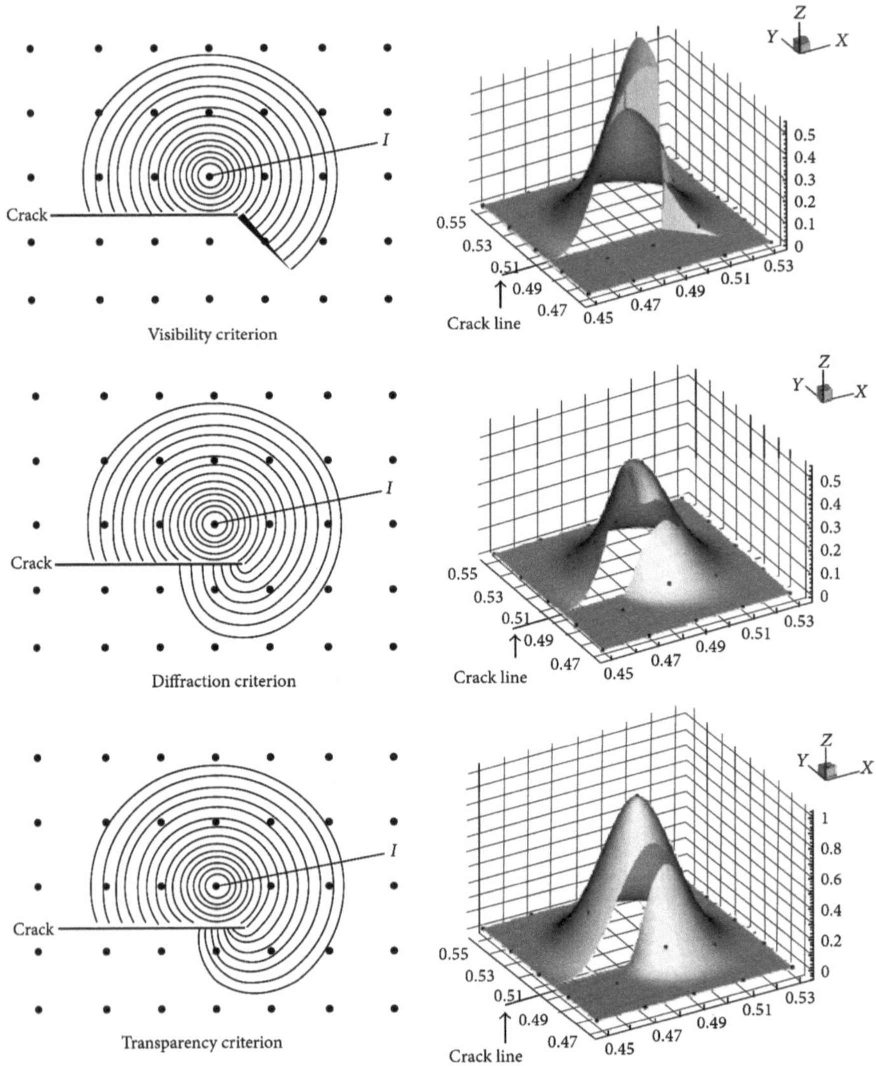

Figure 5.24 Principle of the visibility, diffraction and transparency method with corresponding shape functions, from [14].

in 2D. Consider the vectors $\mathbf{g}$ from $\mathbf{b}$ to $\mathbf{e}$, $\bar{\mathbf{g}}$ from $\mathbf{x}$ to $\mathbf{b}$ and $\hat{\mathbf{g}}$ from $\mathbf{x}$ to $\hat{\mathbf{x}}$ as illustrated in Fig. 5.26. For the vectors $\tilde{\lambda}\mathbf{g}$, $\bar{\mathbf{g}}$ and $\hat{\lambda}\hat{\mathbf{g}}$, we can write (5.148):

$$\bar{\mathbf{g}} + \tilde{\lambda}\, \mathbf{g} = \hat{\lambda}\, \hat{\mathbf{g}} \tag{5.148}$$

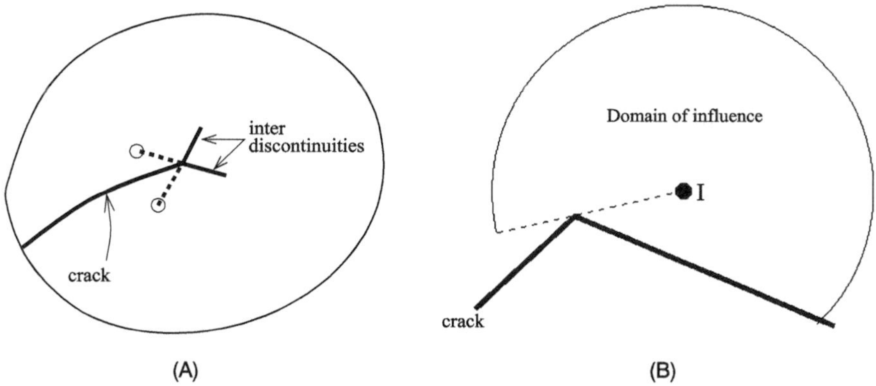

Figure 5.25 (A) Undesired introduced discontinuities by the visibility method; (B) Difficulties with the visibility method for concave boundaries and kinks.

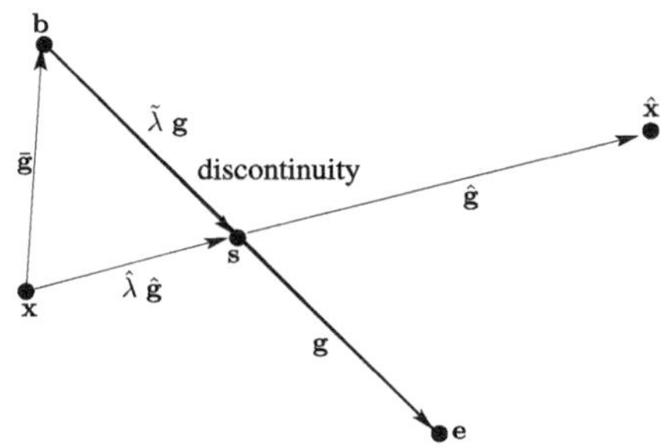

Figure 5.26 A crack modeled with the visibility criterion.

which can also be written as

$$\mathbf{G}\,\boldsymbol{\lambda} = \bar{\mathbf{g}} \tag{5.149}$$

with

$$\mathbf{G} = \begin{bmatrix} -g_x & \hat{g}_x \\ -g_y & \hat{g}_y \end{bmatrix} \qquad \boldsymbol{\lambda} = \begin{bmatrix} \tilde{\lambda} \\ \hat{\lambda} \end{bmatrix} \qquad \bar{\mathbf{g}} = \begin{bmatrix} \bar{g}_x \\ \bar{g}_y \end{bmatrix}$$

The straight lines $\mathbf{g}$ and $\hat{\mathbf{g}}$ have a common intersection $\mathbf{s}$, if $0 < \tilde{\lambda} < 1$ and $0 < \hat{\lambda} < 1$. If $\det \mathbf{G} = 0$, the vectors $\mathbf{g}$ and $\hat{\mathbf{g}}$ are parallel.

Another implementation of the visibility method is illustrated in Fig. 5.27 where the characteristic angle θ_c denotes the angle between two adjacent crack segments. To find if two particles are on the same side of the

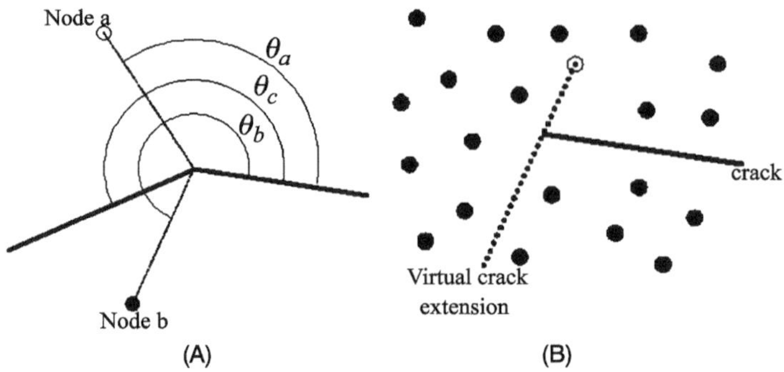

Figure 5.27 (A) Implementation of the visibility criterion; (B) Virtual crack extension around the crack tip for the visibility method shown in figure.

crack, let us define the following sign functions:

$$G(\theta) = \begin{cases} 1 & \forall \theta_a/\theta_c \geq 1 \\ -1 & \forall \theta_a/\theta_c < 1 \end{cases} \tag{5.150}$$

$$H(\theta) = \begin{cases} 1 & \forall \theta_b/\theta_c \geq 1 \\ -1 & \forall \theta_b/\theta_c < 1 \end{cases} \tag{5.151}$$

If the product $G(\theta)\ H(\theta)$ is positive, then the particles are on the same side of the crack, otherwise they are not and henceforth excluded in the summation. A special technique has to be used at the crack tip since the angle θ_c cannot be defined. Therefore, the crack is "virtually" extended for particles across the crack tip as shown in Fig. 5.27.

The extension of the visibility method in three dimensions is straightforward. However, the implementation, especially for general cases is cumbersome. Several particularities have to be considered as shown e.g. in Fig. 5.28 for nodes near or on the crack front. Usually, the crack is presented by planar triangles. Although it is possible[5] to represent the crack by smooth surfaces, [50,81],[6] there is according to our knowledge no publication that combines the visibility method with smooth crack surfaces in three dimensions.

5.4.2 The diffraction method

The diffraction method is an improvement of the visibility method. It removes the *undesired* interior discontinuities, see Fig. 5.29 (see also Fig. 5.24)

[5] It will cumbersome.

[6] [50,81] made use of level sets.

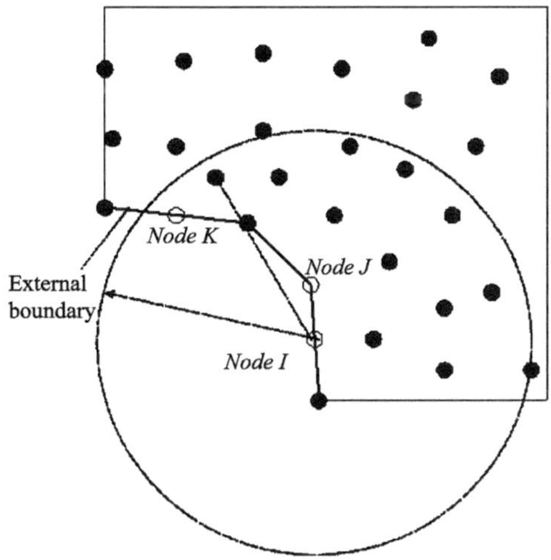

Figure 5.28 Ambiguities for modeling 3D cracks with the visibility criterion.

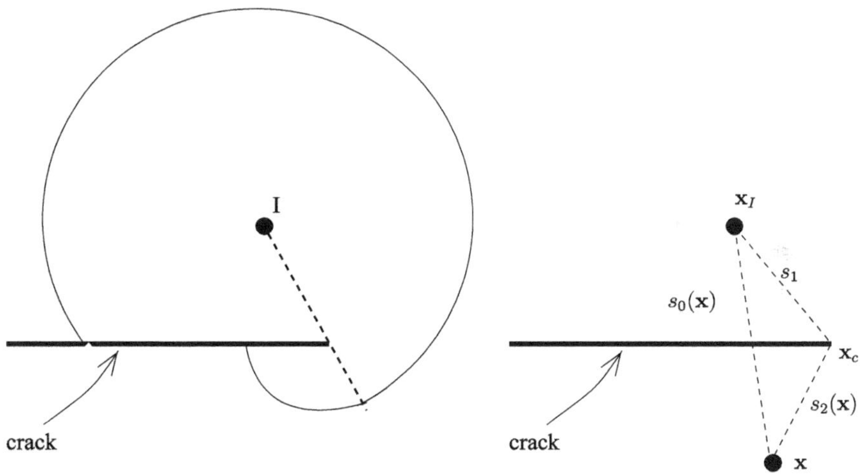

Figure 5.29 The diffraction method.

and the diffraction method is also suitable for non-convex crack boundaries. The method is motivated by the way light diffracts around a sharp corner but the equations used in constructing the domain of influence and the weight function bear almost no relationship to the equation of diffraction. The method is only applicable to radial basis kernel functions with a single parameter h_{I0}.

The idea of the diffraction method is to treat the crack as opaque but to evaluate the length of the ray h_0 by a path which passes around the corner of the discontinuity. This removes the abrupt cut of the shape function to zero across the *undesired* interior discontinuity. A typical weight function is shown in Fig. 5.24. The weight parameter h_{I0} is computed by

$$h_{I0}(\mathbf{X}) = \left(\frac{s_1 + s_2(\mathbf{X})}{s_0(\mathbf{X})} \right)^{\lambda} s_0(\mathbf{X}) \tag{5.152}$$

where

$$\begin{aligned} s_0(\mathbf{X}) &= \|\mathbf{X} - \mathbf{X}_I\| \\ s_1 &= \|\mathbf{X}_c - \mathbf{X}_I\| \\ s_2(\mathbf{X}) &= \|\mathbf{X} - \mathbf{X}_c\| \end{aligned} \tag{5.153}$$

The parameter λ is usually set between one and two and adjusts the distance of the support on the opposite side of the crack. For a better illustration, see also Fig. 5.29.

It should be noted that the shape function of the diffraction method is quite complex with several areas of rapidly varying derivatives that complicates quadrature of the discrete Galerkin form. Moreover, the extension of the diffraction method into three dimensions is complex.

The corrected shape functions are obtained by substituting the modified dilation parameter h_{I0} into the corresponding kernel function. For the use in a Galerkin formulation, the spatial derivatives of the shape functions are needed. This requires the derivatives of the kernel function:

$$\frac{\partial W}{\partial X_i} = \frac{\partial W}{\partial h_{0I}} \frac{\partial h_{0I}}{\partial X_i} \tag{5.154}$$

The first term on the right hand side is unchanged and the second term is

$$\frac{\partial h_{0I}}{\partial X_i} = \lambda \left(\frac{s_1 + s_2(\mathbf{X})}{s_0(\mathbf{X})} \right)^{\lambda - 1} \frac{\partial s_2}{\partial X_i} + (1 - \lambda) \left(\frac{s_1 + s_2(\mathbf{X})}{s_0(\mathbf{X})} \right)^{\lambda} \frac{\partial s_0}{\partial X_i} \tag{5.155}$$

with

$$\begin{aligned} \frac{\partial s_2}{\partial \mathbf{X}} &= \frac{\mathbf{X} - \mathbf{X}_c}{s_2(\mathbf{X})} \\ \frac{\partial s_0}{\partial \mathbf{X}} &= \frac{\mathbf{X} - \mathbf{X}_I}{s_0(\mathbf{X})} \end{aligned} \tag{5.156}$$

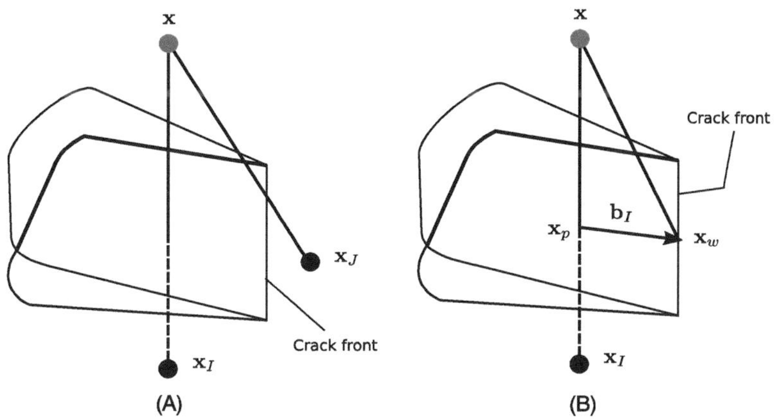

Figure 5.30 (A) The principle of the visibility method in 3D; (B) the principle of the diffraction method in 3D.

Note that the derivatives of the shape functions are often evaluated numerically in three dimensions due to the high complexity, [14].

A typical algorithm for 3D crack propagation with the diffraction method in three dimensions is given by [39,117]:

1. Find $\mathbf{x}_p$, the intersection of the line from $\mathbf{x}$ to $\mathbf{x}_I$ with the crack surface, see Fig. 5.30.
2. Choose the vector $\mathbf{p}_I$ in the direction from $\mathbf{x}_p$ to the closest point on the crack surface Γ_c.
3. The vectors $\mathbf{p}_I$ and $(\mathbf{x}-\mathbf{x}_I)$ define a plane containing $\mathbf{x}_p$. The wrap-around point $\mathbf{x}_w$ is a point on the crack front in this plane which is closest to $\mathbf{x}_p$. The determination of $\mathbf{x}_w$ for a straight crack front is illustrated in Fig. 5.30B.
4. Compute s_0, s_1, s_2, h_{0I}, Eqs. (5.152), (5.153).

5.4.3 The transparency method

The transparency method was developed as an alternative to the diffraction method by [87]. The transparency method is easier extendable into three dimensions than the diffraction method. In the transparency method, the crack is made transparent near the crack tip. The degree of transparency is related to the distance from the crack tip to the point of intersection, see Fig. 5.31.

Consider a ray from the evaluation point $\mathbf{X}$ to the node $\mathbf{X}_I$, Fig. 5.31. The dilation parameter h_{0I} is modified as follows when the ray intersects

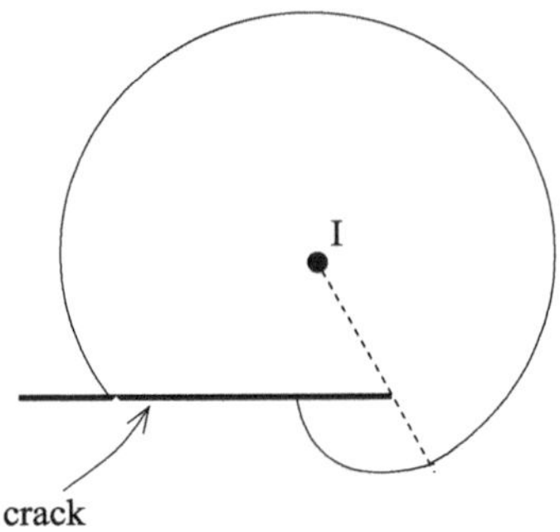

Figure 5.31 The transparency method.

the crack:

$$h_{0I} = s_0(\mathbf{X}) + h_{mI}\left(\frac{s_c(\mathbf{X})}{\bar{s}_c}\right)^{\lambda}, \quad \lambda \geq 2 \tag{5.157}$$

where $s_0(\mathbf{X})$ is defined in Eq. (5.153), h_{mI} is the radius of the nodal support $\mathcal{S}_I$ and $s_c(\mathbf{X})$ is the distance from the crack tip to the intersection point, Fig. 5.31. The parameter $\bar{s}_c = \kappa h$ sets the intersection distance at which the crack segment is completely opaque where κ is used to vary the opacity and h is a measure of nodal spacing. Note that the additional term in Eq. (5.157) is at least quadratic so that the weight function derivatives will be continuous.

An additional requirement is usually imposed for particles close to the crack. Since the angle between the crack and the ray from the node to the crack tip is small, a sharp gradient in the weight function across the line ahead of the crack is introduced. In order to reduce this effect, [87] imposed that all nodes have a minimum distance from the crack surface, i.e. the normal distance to the crack surface must be larger than γh with $0 < \gamma < 1$. [14] suggested $\gamma = 1/4$. Note that this will also have a positive effect on the critical time steps if an explicit scheme is used.

The spatial derivatives are obtained by formal differentiation:

$$\frac{\partial h_{0I}}{\partial \mathbf{X}} = \frac{\partial s_0}{\mathbf{X}} + \lambda h_{mI} \frac{s_c^{\lambda-1}}{\bar{s}_c^{\lambda}} \frac{\partial s_c}{\partial \mathbf{X}} \tag{5.158}$$

where we can write in two dimensions:

$$\frac{\partial s_0}{\partial \mathbf{X}} = \frac{\mathbf{X} - \mathbf{X}_I}{s_0(\mathbf{X})}$$

$$\frac{\partial s_c}{\partial X_1} = -\cos(\theta) = \frac{X_b - X_c}{s_c(\mathbf{X})}$$
$$\frac{\partial s_c}{\partial X_2} = -\sin(\theta) = \frac{Y_b - Y_c}{s_2(\mathbf{X})} \quad (5.159)$$

where θ is the angle between the crack and the x-axis. The weight function of the transparency method looks similar as the one for the diffraction method and is illustrated in Fig. 5.24.

5.4.4 The "see through" and "continuous line" method

The "see-through" method was proposed by [120] for constructing continuous approximations near non-convex boundaries. Therefore, the boundary was considered as completely transparent such that the discontinuity is removed. Though the "see-through" methods works well for capturing features such as interior holes, it is not well suited to model cracks.

In the continuous line method from [67] and [37], the crack is completely transparent at the crack tip. In other words, particles whose domain of influence are partially cut by the crack, can see through the crack. This drastically shortens the crack. If no special techniques are introduced, the crack also does not close at the crack tip that leads to inaccurate solutions. A possible solution is to enforce crack closure at the crack tip with Lagrange multiplier or to decrease the domain of influence of nodes close to the crack tip. This will be discussed later for a (Partition of Unity) PU-based cracking technique, Section 5.6.2.

[9] suggested to combine different methods depending on the convexity of the crack boundary e.g. to use the visibility for convex boundaries and other methods for non-convex crack boundaries. They suggest a criterion based on the angle of the wedge that can be written in terms of the surface normal, see Fig. 5.32. When $\mathbf{n}^A \cdot \mathbf{n}^B \leq \beta$ with $\beta = 0°$ as cutoff value, the boundary can be considered as convex, otherwise non-convex. The cutoff value of $\beta = 0°$ corresponds to the wedge angle of $\omega = 90°$ in Fig. 5.32.

5.5. The concept of enrichment

There are basically two concepts of enrichment, an intrinsic enrichment and an extrinsic enrichment. The motivation of enrichment is to increase the accuracy of the solution by introducing new, more accurate information into the approximation. In case of a cohesive crack, we know that the displacement field is discontinuous across the crack and that the crack

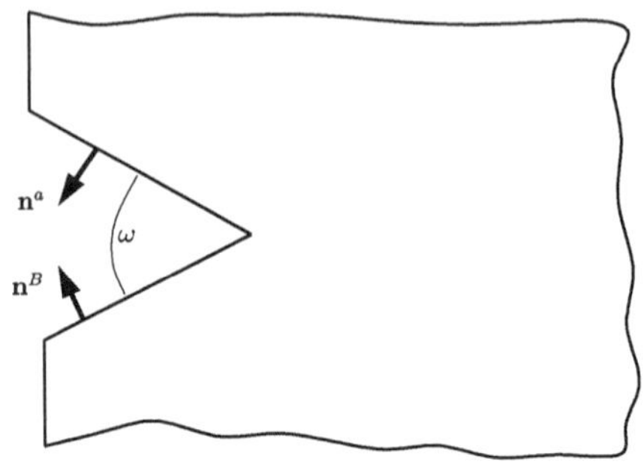

Figure 5.32 Domain of influence near a wedge-shaped non-convex boundary. The boundary is enforced if $\mathbf{n}^A \cdot \mathbf{n}^B \leq \beta$.

has to close at the crack tip. In linear elastic fracture mechanics (LEFM) where the analytical solutions are known (e.g. the stress at the crack tip is singular), the accuracy of the method can be increased enormously by an enrichment. In an intrinsic enrichment, the enrichment function is included in the polynomial basis of the EFG shape function. In an extrinsic enrichment, additional functions are introduced externally to the EFG basis. In addition, the extrinsic enrichment can be classified in an extrinsic *MLS* enrichment and an extrinsic *PU* enrichment as will be outlined in Section 5.5.2. We will show the concept of enrichment in the following two Sections 5.5.1 and 5.5.2 (mainly) for crack tip fields in LEFM. In Section 5.6, we will modify the enrichment for cohesive cracks and focus on some features with respect to the propagation of the cracks.

In LEFM, the asymptotic near crack tip displacement field in two dimensions is given by

$$
\begin{aligned}
u_1 &= \frac{1}{G}\sqrt{\frac{r}{2G}}\left(K_I Q_I^1(\theta) + K_{II} Q_{II}^1(\theta)\right) \\
u_2 &= \frac{1}{G}\sqrt{\frac{r}{2G}}\left(K_I Q_I^2(\theta) + K_{II} Q_{II}^2(\theta)\right)
\end{aligned} \tag{5.160}
$$

where G is the shear modulus, r and θ are explained in Fig. 5.33 and

$$
\begin{aligned}
Q_I^1(\theta) &= \kappa - \cos\frac{\theta}{2} + \sin\theta \,\sin\frac{\theta}{2} \\
Q_I^2(\theta) &= \kappa + \sin\frac{\theta}{2} + \sin\theta \,\cos\frac{\theta}{2} \\
Q_{II}^1(\theta) &= \kappa + \sin\frac{\theta}{2} + \sin\theta \,\cos\frac{\theta}{2}
\end{aligned}
$$

$$Q_{II}^2(\theta) = \kappa - \cos\frac{\theta}{2} - \sin\theta \, \sin\frac{\theta}{2} \tag{5.161}$$

are the well known angular functions for LEFM, K_I and K_{II} are the mode-I and mode-II stress intensity factors (SIF) where $\kappa = (3 - \nu)/(1 + \nu)$ for plane stress and $\kappa = (3 - 4\nu)$ for plane strain is the Kolosov constant. Using trigonometric identities, it can be shown that the basis, given by

$$\mathbf{p}^T(\mathbf{X}) = \left[\sqrt{r}\sin(\theta/2), \sqrt{r}\cos(\theta/2), \sqrt{r}\sin(\theta/2)\sin(\theta), \sqrt{r}\cos(\theta/2)\sin(\theta)\right] \tag{5.162}$$

spans the LEFM crack-tip displacement field in Eq. (5.160).

Though we don't examine mode-III problems, we will give the full asymptotic near crack tip displacement field for 3D mixed mode failure for sake of completeness:

$$\mathbf{u} = K_I \mathbf{Q}_I(\mathbf{X}) + K_{II}\mathbf{Q}_{II}(\mathbf{X}) + K_{III}\mathbf{Q}_{III}(\mathbf{X}) \tag{5.163}$$

with Q_I^1, Q_I^2, Q_{II}^1, Q_{II}^2 from (5.161), $Q_I^3 = Q_{II}^3 = Q_{III}^1 = Q_{III}^2 = 0$ and

$$Q_{III}^3 = \frac{1}{G}\sqrt{\frac{2r}{\pi}}\sin\frac{\theta}{2} \tag{5.164}$$

While there are many publications on LEFM (e.g. [21,39,104]), there are only a few on elastic-plastic fracture mechanics (EPFM). [105] modified the intrinsic basis for EPFM. They fitted the enrichment to the asymptotic solution of the (HRR) [61,108] crack-tip field under EPFM conditions. [130] employed the EFG-method to EPFM as well but without discussing the enrichment.

5.5.1 Intrinsic enrichment

In the EFG method, the approximation can be enriched intrinsically by extending the MLS basis function $\mathbf{p}$. In LEFM, the accuracy of the approximation is drastically improved by adding the asymptotic near-tip displacement field of the Westergaard solution into $\mathbf{p}$:

$$\mathbf{p}^T(\mathbf{X}) = \left[1, X, Y, \sqrt{r}\sin\frac{\theta}{2}, \sqrt{r}\cos\frac{\theta}{2}, \sqrt{r}\sin\frac{\theta}{2}\sin(\theta), \sqrt{r}\cos\frac{\theta}{2}\sin(\theta)\right] \tag{5.165}$$

where r is the radial distance to the crack tip and θ the angle to the crack (or its virtual extension), see Fig. 5.33. Note that the linear terms are not

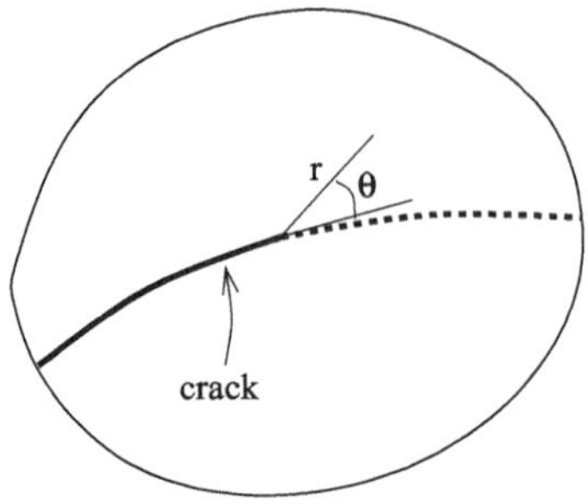

Figure 5.33 Definitions for the crack tip enrichment [86].

related to the asymptotic near crack tip displacement field but the usual polynomial basis. We found that the accuracy is barely affected by omitted the last three terms in (5.165).

Recalling the EFG shape functions (5.45)

$$\Phi_J(\mathbf{X}) = \mathbf{p}(\mathbf{X})^T \cdot \mathbf{A}(\mathbf{X})^{-1} \cdot \mathbf{p}_J(\mathbf{X}) \, W(\mathbf{X} - \mathbf{X}_J, h)$$
$$\mathbf{A}(\mathbf{X}) = \sum_{J \in S} \mathbf{p}_J(\mathbf{X}) \, \mathbf{p}_J^T(\mathbf{X}) \, W(\mathbf{X} - \mathbf{X}_J, h)$$

the additional computational cost that occurs due to the intrinsic enrichment becomes obvious. Moreover, the moment matrix $\mathbf{A}$ becomes ill conditioned. The size of the domain of influence has to be enlarged to guarantee the regularity of $\mathbf{A}$. Though this ill-conditioning does not affect the final solution, it is quite troublesome. By diagonalizing $\mathbf{A}$ with a Gram-Schmidt orthogonalization (a procedure as proposed in Eqs. (5.48) to (5.50), Section 5.2.1.4), the regularity of the moment matrix can be drastically improved but leads to more complex shape functions. Moreover, interior discontinuities will be introduced unless the intrinsic basis is employed globally. Therefore, [43] proposed a procedure to blend nodes with different basis. The procedure is similar to the coupling procedure of meshfree methods and finite elements, [12,14], and is realized by a transition region, see Fig. 5.34. The approximation is written by

$$\mathbf{u}^h(\mathbf{X}) = R \, \mathbf{u}^{enr}(\mathbf{X}) + (1 - R) \, \mathbf{u}^{lin}(\mathbf{X}) \tag{5.166}$$

where $\mathbf{u}^{enr}(\mathbf{X})$ is the enriched approximation, $\mathbf{u}$ is the linear approximation and R is a ramp function that is one on the enriched boundary of the coupling domain and zero on the linear boundary of the coupling domain. This is the only requirement imposed on R. [9] suggested to choose $R =$

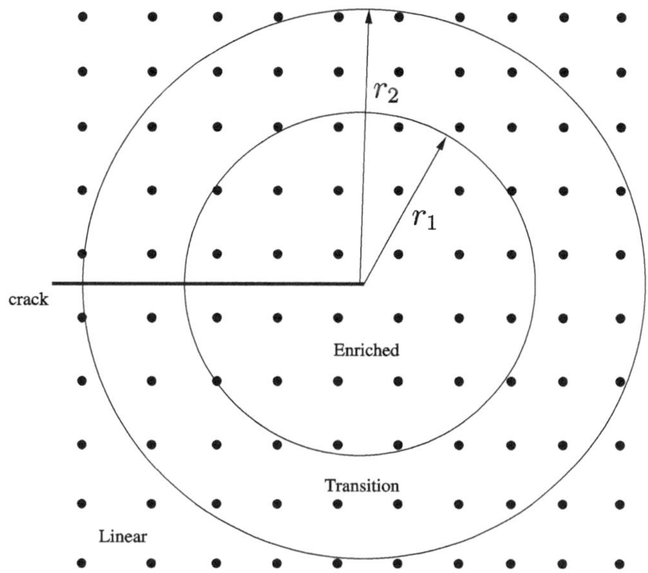

Figure 5.34 Coupling between enriched and linear approximations.

$1-\xi$ or $R = 1 - 10\xi^3 + 15\xi^4 - 6\xi^5$ with $\xi = (r - r_1)(r_2 - r_1)$ (see Fig. 5.34) and reported excellent results.

The final approximation (5.44) then reads

$$\mathbf{u}^h(\mathbf{X}, t) = \sum_{J \in S} u_J(t)\ \bar{\Phi}_J(\mathbf{X}) \tag{5.167}$$

with

$$\bar{\Phi}_J(\mathbf{X}) = R\ \Phi_J^{enr}(\mathbf{X}) + (1 - R)\Phi_J^{lin}(\mathbf{X}) \tag{5.168}$$

where the shape functions $\Phi_J^{enr}(\mathbf{X})$ and $\Phi_J^{lin}(\mathbf{X})$ are formed by the enriched and linear basis, respectively. This blending ensures continuity in the displacement field. However, in case for linear blending, i.e. $R = 1 - \xi$, discontinuities in the strain field occur. In [43], it was reported that good results were obtained by just switching to a linear basis if this switching is done far away from the crack tip such that the computation of the stress intensity factors are not influenced. In this case, the formulation is not compatible any more.

Another popular alternative to the tip enrichment (5.165) is the radial enrichment

$$\mathbf{p}^T(\mathbf{X}) = \left[1, X, Y, \sqrt{r}\right] \tag{5.169}$$

The advantage of this enrichment is that the intrinsic basis is only extended by one term and thus the computation of the shape functions is cheaper. In LEFM, it is a good alternative to the methods mentioned in Section 5.4 since no smoothing technique is necessary. The crack closes automatically at the crack tip and the stress is singular at the crack tip. However, since this enrichment does not contain the discontinuity behind the crack tip, its convergence is much slower.

In [40], an alternative intrinsic enrichment by enriched kernel functions was proposed. In particular, the kernel functions are modified such that they behave in a similar to Eq. (5.161). The following kernel functions are given:

$$\begin{aligned} W_c(\mathbf{X}) &= \alpha\sqrt{r}\cos\frac{\theta}{2}\, W_4(\mathbf{X}) \\ W_p(\mathbf{X}) &= \alpha\sqrt{r}\left(1+\sin\frac{\theta}{2}\right) W_4(\mathbf{X}) \\ W_p(\mathbf{X}) &= \alpha\sqrt{r}\left(1-\sin\frac{\theta}{2}\right) W_4(\mathbf{X}) \end{aligned} \tag{5.170}$$

where $W_4(\mathbf{X})$ is the quartic spline and the factor α controls the amplitude of the enriched kernel function compared with the amplitude of the regular nodes. In [40], the authors suggest a value of $\alpha = 1$. The indices c, m and p stand for cos, plus sin and minus sin, respectively. An advantage of this method is that no blending domain needs to be introduced. This enriched method was employed in combination with the diffraction method in [40].

For cohesive cracks, that we are actually interested in, the stress singularity is generally not allowed. It is well known from experiments that for cohesive cracks, a crack tip enrichment of the form

$$\mathbf{p}^T(\mathbf{X}) = \left[1, X, Y, r^n \sin(\theta/2)\right] \tag{5.171}$$

with $n = 1, 2, 3$ reflects the reality pretty well. For $n = 2, 3...$, nonlinear crack openings can be modeled while for $n = 1$, the crack opening will be linear. Note that there is no paper (according to the knowledge of the author) that tries to model cohesive cracks with such an intrinsic enrichment probably due to the advantages of cohesive crack models based on extrinsic enrichments.

5.5.2 Extrinsic enrichment

5.5.2.1 Extrinsic MLS enrichment

In the MLS extrinsic enrichment, the near crack tip asymptotic field of LEFM is as follows:

$$\mathbf{u}^h(\mathbf{X},t)=\sum_{J\in S}\mathbf{p}(\mathbf{X}_J)^T\,\mathbf{a}(\mathbf{X},t)+\sum_{K=1}^{n_c}\left(k_I^K\,\mathbf{Q}_I^K(\mathbf{X}_I)+k_{II}^K\,\mathbf{Q}_{II}^K(\mathbf{X}_I)\right)\tag{5.172}$$

where n_c is the number of cracks in the model, $\mathbf{u}^h$ is the approximation of $\mathbf{u}$, $\mathbf{p}$ is the usual polynomial basis of nth order completeness, Section 5.2.1.4, and k_I and k_{II} are additional degrees of freedom associated with mode-I fracture and mode-II fracture. It was reported in [43] that k_I and k_{II} physically indicate the stress intensity factors (SIF) so that the approximate SIFs that can be obtained by directly solving the system of equations without considering the J-integral. However, it was shown by [131] that the linear term in (5.172) pollutes the solution and it is more accurate to obtain the SIF via the J-integral. They also showed that by omitting the linear term and adding terms of the series, the accuracy of the results can be greatly improved.

In [71], a local augmentation using four auxiliary supports was suggested and showed an excellent and improved accuracy though the approach is only local. The functions Q_{Ii} and Q_{IIi}, $i=1,2$ describe the near-tip displacement field and are given by

$$Q_{I1}(\mathbf{X})=\frac{1}{2G}\sqrt{\frac{r}{2\pi}}\cos(0.5\theta)\left(\kappa-1+2\sin^2(0.5\theta)\right)\tag{5.173}$$

$$Q_{I2}(\mathbf{X})=\frac{1}{2G}\sqrt{\frac{r}{2\pi}}\sin(0.5\theta)\left(\kappa+1-2\cos^2(0.5\theta)\right)\tag{5.174}$$

$$Q_{II1}(\mathbf{X})=\frac{1}{2G}\sqrt{\frac{r}{2\pi}}\sin(0.5\theta)\left(\kappa+1+2\cos^2(0.5\theta)\right)\tag{5.175}$$

$$Q_{II2}(\mathbf{X})=-\frac{1}{2G}\sqrt{\frac{r}{2\pi}}\cos(0.5\theta)\left(\kappa-1-2\sin^2(0.5\theta)\right)\tag{5.176}$$

where G is the shear modulus and κ is the Kolosov constant defined as $\kappa=3-4\nu$ for plane strain and $\kappa=(3-\nu)/(1+\nu)$ for plane stress conditions where ν is the Poisson ratio.

As for the standard MLS approximation, the coefficients $\mathbf{a}$ are determined by minimizing the weighted discrete L_2 error norm

$$\mathcal{J}=\sum_{J\in S}\frac{1}{2}\left(\mathbf{p}(\mathbf{X}_J)^T\,\mathbf{a}(\mathbf{X},t)+\sum_{K=1}^{n_c}\left[k_I^K\mathbf{Q}_I^K+k_{II}^K\mathbf{Q}_{II}^K\right]-\mathbf{u}_J(t)\right)^2 W(\mathbf{X}-\mathbf{X}_J,h_0)\tag{5.177}$$

that leads with the stationarity of $\mathfrak{J}$ to

$$\mathbf{A}(\mathbf{X})\mathbf{a}(\mathbf{X}) = \sum_{J\in\mathcal{S}} \mathbf{P}_J(\mathbf{X}) \left(\mathbf{u}_J - \sum_{K=1}^{n_c} \left[k_I^K \mathbf{Q}_I^K + k_{II}^K \mathbf{Q}_{II}^K\right]\right) \tag{5.178}$$

with

$$\mathbf{A}(\mathbf{X}) = \sum_{J\in S} \mathbf{p}(\mathbf{X}_J)\, \mathbf{p}^T(\mathbf{X}_J)\, W(\mathbf{X} - \mathbf{X}_J, h_0) \tag{5.179}$$

and

$$\mathbf{P}_J(\mathbf{X}) = \left[W(\mathbf{X} - \mathbf{X}_1, h_0)\mathbf{p}(\mathbf{X}_1), \ldots, W(\mathbf{X} - \mathbf{X}_n, h_0)\mathbf{p}(\mathbf{X}_n)\right] \tag{5.180}$$

with n numbers of nodes. Solving Eq. (5.178) with respect to **a** gives

$$\mathbf{a}(\mathbf{X}) = \sum_{J\in\mathcal{S}} \mathbf{A}^{-1}(\mathbf{X})\mathbf{P}_J(\mathbf{X}) \left(\mathbf{u}_J - \sum_{K=1}^{n_c} \left[k_I^K \mathbf{Q}_I^K + k_{II}^K \mathbf{Q}_{II}^K\right]\right) \tag{5.181}$$

After some algebra, we obtain the final approximation in terms of the nodal parameters:

$$\begin{aligned}\mathbf{u}^h(\mathbf{X}) = &\sum_{J\in\mathcal{S}} \mathbf{p}^T(\mathbf{X})\mathbf{A}^{-1}(\mathbf{X})\mathbf{P}_J(\mathbf{X}) \left(\mathbf{u}_J - \sum_{K=1}^{n_c} \left[k_I^K \mathbf{Q}_I^K + k_{II}^K \mathbf{Q}_{II}^K\right]\right)\\ &+ \sum_{K=1}^{n_c} \left[k_I^K \mathbf{Q}_I^K + k_{II}^K \mathbf{Q}_{II}^K\right]\end{aligned} \tag{5.182}$$

with the shape functions

$$\Phi_J(\mathbf{X}) = \mathbf{p}^T(\mathbf{X})\mathbf{A}^{-1}(\mathbf{X})\mathbf{P}_J(\mathbf{X}) \tag{5.183}$$

that can be written simplified as

$$\mathbf{u}^h(\mathbf{X}) = \sum_{J\in\mathcal{S}} \Phi_J(\mathbf{X})\, \tilde{\mathbf{u}}_J + \sum_{K=1}^{n_c} \left[k_I^K \mathbf{Q}_I^K + k_{II}^K \mathbf{Q}_{II}^K\right] \tag{5.184}$$

with

$$\tilde{\mathbf{u}}_J = \mathbf{u}_J - \sum_{K=1}^{n_c} \left[k_I^K \mathbf{Q}_I^K + k_{II}^K \mathbf{Q}_{II}^K\right] \tag{5.185}$$

Note that the parameters k_I and k_{II} are global parameters and no spatial derivatives are taken with respect to them.

5.5.2.2 Extrinsic PU enrichment

A simpler extrinsic enrichment is based on the partition of unity concept. In this extrinsic enrichment, the asymptotic near-tip displacement field of the Westergaard solution, Eq. (5.165), is added externally to the approximation:

$$\mathbf{u}^h(\mathbf{X}) = \sum_{I \in \mathcal{S}} \Phi_I(\mathbf{X}) \left(\mathbf{u}_I + \sum_{J \in \mathcal{S}_c} \mathbf{b}_{IJ} \mathbf{p}_J(\mathbf{X}) \right) \tag{5.186}$$

where $\mathbf{b}_{IJ}$ are additional unknowns introduced into the variational formulation and $\mathcal{S}_c$ is the set of nodes whose domain of influence is cut by the crack tip. The approximation (5.186) is clearly a partition of unity. It was shown in [114] that partition of unity is even guaranteed if different shape functions for the standard and the enriched part is employed:

$$\mathbf{u}^h(\mathbf{X}) = \sum_{I \in \mathcal{S}} \Phi_I(\mathbf{X}) \mathbf{u}_I + \sum_{I \in \mathcal{S}} \sum_{J \in \mathcal{S}_c} \tilde{\Phi}_I(\mathbf{X}) \mathbf{b}_{IJ} \mathbf{p}_J(\mathbf{X}) \tag{5.187}$$

where the superimposed *tilde* denotes the shape function of the enrichment. As might be noted, a compatible approximation is guaranteed by a *local* partition of unity, meaning only nodes close to the crack tip have to be enriched. Finally, we would like to mention that the extrinsic PU concept is ideally suited for cohesive cracks. As already outlined for the intrinsic enrichment, just by changing the tip enrichment (5.165) to (5.171), cohesive cracks are realized. We will not go into too much detail now but come back to this concept in the next section.

5.6. (Extrinsically) enriched local PU meshfree methods

In this section, we will focus in more detail on extrinsically enriched meshfree methods based on the local partition of unity (PU) concept. These discretizations are used in methods that enforce crack path continuity. Even though we haven't discussed cracking criteria,[7] we will discuss methods, especially in three dimensions, to track the crack pathes.

[7] That will tell us when a crack is initiated/propagated and give the crack orientation and crack length.

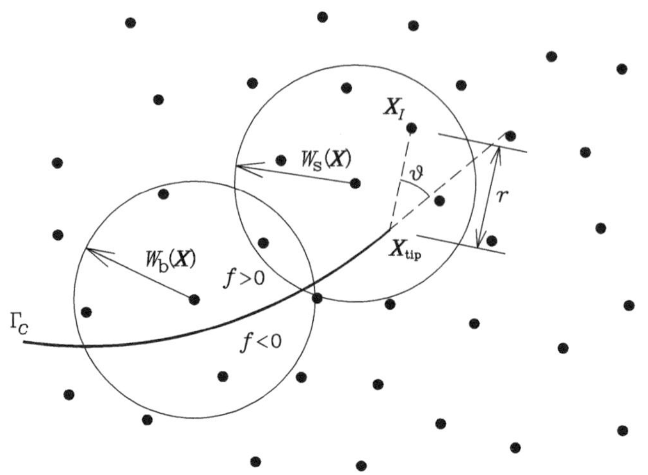

Figure 5.35 Crack with partial cut and complete cut domain of influence particles.

5.6.1 Enriched methods with crack tip enrichment

The main idea to capture the crack is to enrich the test and trial functions with additional unknowns so that the approximation is continuous in the whole domain but discontinuous along the crack as done in methods such as XFEM [80] and as already outlined in detail in the previous section. Since it is advantageous for tracking the crack path, the test and trial functions are written in terms of a signed distance function f or level set function[8] (see Fig. 5.35). Then, the test and trial functions can be decomposed into a usual part and an enriched part. The latter will give the jump in the displacement field:

$$\delta \mathbf{u} = \delta \mathbf{u}^0 + \delta \mathbf{u}^e \quad , \quad \mathbf{u} = \mathbf{u}^0 + \mathbf{u}^e \tag{5.188}$$

where the superimposed 0 denotes the usual part and the superimposed e denotes the enriched part so that the test and trial functions can be written as

[8] The level set function is the function where the signed distance function is zero; note that the level set does not necessarily be identical to the signed distance function; a vector level set for example stores in addition to the sign distance also the distance vector to the crack surface.

$$\delta\mathbf{u}(\mathbf{X}) = \underbrace{\sum_{I\in\mathcal{W}(\mathbf{X})} \Phi_I(\mathbf{X})\ \delta\mathbf{u}_I}_{\mathbf{u}^0} + \underbrace{\sum_{n=1}^{n_c}\sum_{I\in\mathcal{W}_b(\mathbf{X})} \Phi_I(\mathbf{X})\ S(f_I(\mathbf{X}))\ \delta\mathbf{a}_I + \sum_{m=1}^{m_t}\sum_{I\in\mathcal{W}_s(\mathbf{X})} \Phi_I(\mathbf{X}) \sum_K \mathbf{B}_K(\mathbf{X})\ \delta\mathbf{b}_{KI}}_{\mathbf{u}^e} \tag{5.189}$$

$$\mathbf{u}(\mathbf{X}) = \sum_{I\in\mathcal{W}(\mathbf{X})} \Phi_I(\mathbf{X})\ \mathbf{u}_I + \underbrace{\sum_{n=1}^{n_c}\sum_{I\in\mathcal{W}_b(\mathbf{X})} \Phi_I(\mathbf{X})\ S(f_I(\mathbf{X}))\ \mathbf{a}_I + \sum_{m=1}^{m_t}\sum_{I\in\mathcal{W}_s(\mathbf{X})} \Phi_I(\mathbf{X}) \sum_K \mathbf{B}_K(\mathbf{X})\ \mathbf{b}_{KI}}_{\mathbf{u}^e} \tag{5.190}$$

where $\mathcal{W}(\mathbf{X})$ is the entire domain, $\mathcal{W}_b(\mathbf{X})$ is the completely cut domain for n_c cracks, $\mathcal{W}_s(\mathbf{X})$ is the partial cut domain for m_t crack tips, and S and $\mathbf{B}$ are the enrichment functions explained later. The first term on the right hand side of Eq. (5.189) or (5.190), respectively, is the usual approximation where Φ_I are the shape functions, and $\mathbf{u}_I$ and $\delta\mathbf{u}_I$ are the parameters. The second and third term is the enrichment, in which the coefficient $\delta\mathbf{a}$ and $\delta\mathbf{b}$ or $\mathbf{a}$ and $\mathbf{b}$, respectively, are additional unknowns introduced for the crack.

$S(f(\mathbf{X}))$ depends on the signed distance function $f_I(\mathbf{X})$ and is defined as:

$$\begin{aligned} S(f_I(\mathbf{X})) &= 1 \quad \text{if } f_I(\mathbf{X}) > 0 \\ S(f_I(\mathbf{X})) &= -1 \quad \text{if } f_I(\mathbf{X}) < 0 \end{aligned} \tag{5.191}$$

with

$$f_I(\mathbf{X}) = \begin{cases} \operatorname{sign}[\mathbf{n}\cdot(\mathbf{X}_I - \mathbf{X})]\min\|\mathbf{X}_I - \mathbf{X}\|, & \text{for } \mathbf{X}_I \in W_b \\ \mathbf{n}\cdot(\mathbf{X}_{\text{tip}} - \mathbf{X}_I), & \text{for } \mathbf{X}_I \in W_s \end{cases} \tag{5.192}$$

where $\mathbf{X}_{\text{tip}}$ are the coordinates of the crack tip and $\mathbf{n}$ is the crack normal. Only nodes which are located in the domain $\mathcal{W}_b(\mathbf{X})$ are enriched with the additional unknowns $\delta\mathbf{a}$ and $\mathbf{a}$. The second term of Eq. (5.189) or (5.190) is called the step enrichment. Another popular alternative is to use the Heaviside function instead of the step function. In X-FEM, a Heaviside step

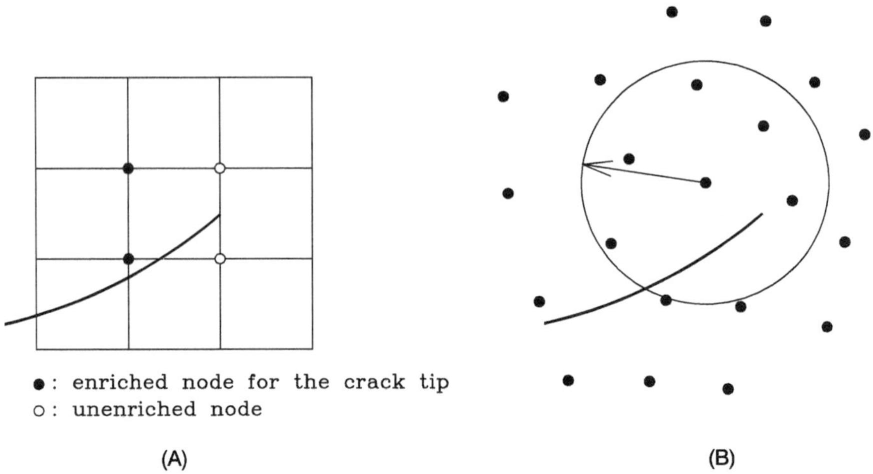

Figure 5.36 Crack with enriched nodes in (A) XFEM and (B) meshfree methods.

function or a so-called "modified" step function that is identical to (5.191) have been used. The sign function and modified step function are more symmetric than the Heaviside step function, but the results for all three are identical.

The third term of Eqs. (5.189) and (5.190) is applied around the crack tip $\mathcal{W}_s(\mathbf{X})$. In linear elastic fracture mechanics, $\mathbf{B}$ is chosen to be continuous in the whole domain $\mathcal{W}_s(\mathbf{X})$, but discontinuous at the crack line:

$$\mathbf{B} = \left(\sqrt{r}\,\sin\frac{\theta}{2}, \sqrt{r}\,\cos\frac{\theta}{2}, \sqrt{r}\,\sin\frac{\theta}{2}\,\sin\theta, \sqrt{r}\,\cos\frac{\theta}{2}\,\sin\theta\right) \qquad (5.193)$$

according to the analytical solution around the crack tip where r is the distance of $\mathbf{X}$ to the crack tip and $\theta(\mathbf{X}) = \sin^{-1}(f/r)$ is the angle between the tangent to the crack line and the segment $\mathbf{X} - \mathbf{X}_{\text{tip}}$, see Fig. 5.35. It is called the 'branch' enrichment.

For cohesive cracks, there is no crack tip singularity and the crack opening displacement, which the cohesive traction depends on, may be described by the additional unknown $\mathbf{a}$ only. In XFEM, this procedure is straightforward since it is easy to impose the appropriate boundary conditions, see Fig. 5.36A, i.e. the crack has to close at the end of the element edge. This can be accomplished e.g. by not enriching the nodes at the element edge where the crack tip is located as shown in Fig. 5.36A. However, in meshfree methods, this technique cannot be applied analogous to the way in XFEM, see Fig. 5.36B.

Therefore, we keep the branch enrichment for cohesive cracks, but without the crack tip singularity:

$$\mathbf{B} = \left(r^m \sin\frac{\theta}{2} \right) \quad m = 1, 2, 3 \tag{5.194}$$

For other applications, e.g. multi-field problems, shear bands or problems involving high gradients, *tanh*-functions or exponential functions are commonly used. A good overview is given in [45].

Furthermore, we shifted the function $S_I(f(\mathbf{X}))$ and $\mathbf{B}_K(\mathbf{X})$ by their values at the position of particle I, i.e. $S_I(f(\mathbf{X}_I))$ and $\mathbf{B}_K(\mathbf{X}_I)$, respectively:

$$\bar{S}_I^n(\mathbf{X}) = S^n\left(f^n(\mathbf{X})\right) - S^n\left(f^n(\mathbf{X}_I)\right) \tag{5.195}$$

$$\bar{\mathbf{B}}_K^m(\mathbf{X}) = \mathbf{B}_K^m(\mathbf{X}) - \mathbf{B}_K^m(\mathbf{X}_I) \tag{5.196}$$

which makes the enriched region narrower. To avoid having heavy notations, we drop $\bar{\cdot}$ in the following sections; unless mentioned otherwise, S and $\mathbf{B}$ stand for $\bar{S}$ and $\bar{\mathbf{B}}$ of Eqs. (5.195) and (5.196), respectively.

We would like to mention that for particles in the blending region, i.e. the particles whose domain of influence is not cut but influenced by the enriched particles, only the usual approximation (first term on the right hand side of Eqs. (5.189) and (5.190)) is considered in the approximation of the test and trial functions.

The jump in the displacement is governed only by the enrichment and is given by

$$\begin{aligned}
[\![\mathbf{u}(\mathbf{X})]\!] &= \sum_{n=1}^{n_c} \sum_{I \in \mathcal{W}_b(\mathbf{X})} \Phi_I(\mathbf{X}) [\![S(\mathbf{X})]\!] \mathbf{a}_I \\
&+ \sum_{m=1}^{m_t} \sum_{I \in \mathcal{W}_s(\mathbf{X})} \Phi_I(\mathbf{X}) \sum_K [\![\mathbf{B}_K(\mathbf{X})]\!]\, \mathbf{b}_{KI} \\
&= 2 \sum_{n=1}^{n_c} \sum_{I \in \mathcal{W}_b(\mathbf{X})} \Phi_I(\mathbf{X})\, \mathbf{a}_I \\
&+ \sum_{m=1}^{m_t} \sum_{I \in \mathcal{W}_s(\mathbf{X})} \Phi_I(\mathbf{X}) \sum_K [\![\mathbf{B}_K(\mathbf{X})]\!]\, \mathbf{b}_{KI}
\end{aligned} \tag{5.197}$$

The normal part δ_n, i.e. the crack opening and the tangential part δ_t, the crack sliding is given by

$$\delta_n = [\![\mathbf{u}(\mathbf{X})]\!]_n = \mathbf{n} \cdot [\![\mathbf{u}(\mathbf{X})]\!] \tag{5.198}$$

$$\delta_t = [\![\mathbf{u}(\mathbf{X})]\!]_\tau = \left\| [\![\mathbf{u}(\mathbf{X})]\!] - (\delta_n \mathbf{n}) \right\| \tag{5.199}$$

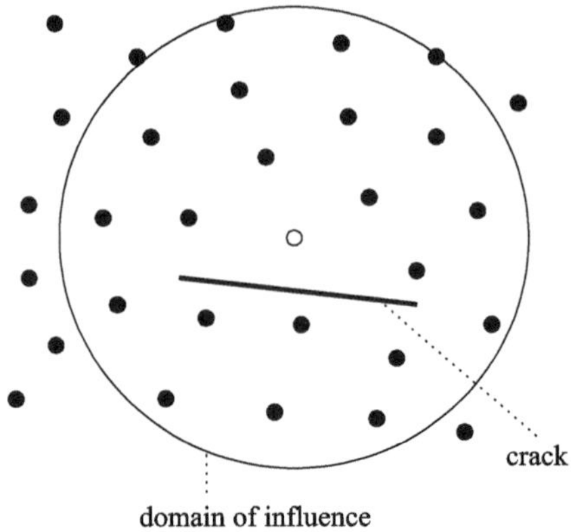

Figure 5.37 A crack that has two tips within the domain of influence of a single particle.

5.6.2 Enriched methods without crack tip enrichment

The approximation of the test and trial functions, Eqs. (5.189) and (5.190), including the crack tip enrichment have some drawbacks. First of all, they are computational more expensive and more cumbersome to implement, especially in 3D and for crack branching and junctions. Second, certain assumptions have to be made how the particles in front of the crack tip are enriched. The enriched crack tip domain spans in three dimensions a cylinder around the crack front that will have kinks if the crack front is not straight or smooth. This complicates the crack front enrichment, i.e. the local crack coordinate system is not defined at the kinks and hence, the crack tip angle θ cannot be computed uniquely. Third, at crack initiation, the initiated crack has to be related to the size of the domain of influence of a particle since the results are inaccurate when the entire crack front (two crack tips of the same crack) fall in only a single domain of influence, see Fig. 5.37 for the two-dimensional illustration. And fourth, crack tip functions for general nonlinear material and cohesive laws are usually not known.

Therefore, it can be advantageous to omit the crack tip enrichment. However, if crack path continuity is enforced, crack closure at the crack front has to be enforced as well that is difficult for particle methods as discussed in the previous section. In this section, we will propose two techniques how to remove the crack tip enrichment in a meshfree method. The formulation is based, as in the previous section, on the decomposition of the displacement into a usual and an enriched part:

$$\mathbf{u}(\mathbf{X}, t) = \mathbf{u}^0(\mathbf{X}, t) + \mathbf{u}^e(\mathbf{X}, t) \tag{5.200}$$

where $\mathbf{u}^0$ is the continuous displacement field defined by

$$\mathbf{u}(\mathbf{X}, t) = \sum_{I \in \mathcal{W}} \Phi_I(\mathbf{X})\ \mathbf{u}_I(t) \tag{5.201}$$

and $\mathbf{u}^e$ is the discontinuous (or the enriched) displacement field which is given by

$$\mathbf{u}^e(\mathbf{X}, t) = \sum_{J \in \mathcal{E}} \mathbf{u}^{e,J}(\mathbf{X}, t) \tag{5.202}$$

Here $\mathcal{E}$ is the set of all the cracks in the domain and $\mathbf{u}^{e,J}$ is the displacement enrichment by crack J. The displacement enrichment $\mathbf{u}^{e,J}$ is given by

$$\mathbf{u}^{e,J}(\mathbf{X}, t) = \sum_{I \in \mathcal{W}_J} \Phi_I(\mathbf{X})\ \Psi_I^J(\mathbf{X})\ \mathbf{a}_I^J(t) \tag{5.203}$$

where $\mathcal{W}_J$ is the set of the particles whose domain of influence is cut by crack J, Ψ_I^J is the enrichment function for particle I and crack J and $\mathbf{a}_I$ are the additional degrees of freedom for the enrichment Ψ_I^J. Note that the shape function Φ_I in Eq. (5.203) does not have to be the same as Φ_I in Eq. (5.201) [28,113,114].

If a domain of influence is completely cut by a crack, it is enriched by using the sign function or the Heaviside function, which is now classical. We will use again the sign function as the enrichment function rather than the Heaviside function because of its appealing symmetry. The enrichment is given by

$$\Psi_I^J(\mathbf{X}) = \text{sign}\left[f^J(\mathbf{X})\right] - \text{sign}\left[f^J(\mathbf{X}_I)\right] \tag{5.204}$$

Here $f^J(\mathbf{X})$ is the signed distance measured from $\mathbf{X}$ to crack J. It is defined as

$$f^J(\mathbf{X}) = \text{sign}\left[\mathbf{n} \cdot (\mathbf{X} - \mathbf{X}^J)\right] \min \|\mathbf{X} - \mathbf{X}^J\| \quad \text{with} \quad \mathbf{X}^J \in \Gamma_c^J \tag{5.205}$$

in which $\mathbf{n}$ is the crack normal and Γ_c^J represents crack J. The choice of the direction of the crack normal is completely arbitrary as long as it is consistent throughout the entire computation. The sign function is defined as

$$\text{sign}(x) = \begin{cases} 1 & \text{for} \quad x > 0 \\ -1 & \text{for} \quad x < 0 \end{cases} \tag{5.206}$$

If the domain of influence is partially cut such as a 'tip element' in which the crack tip is located, the enrichment of Eq. (5.204) is not working but

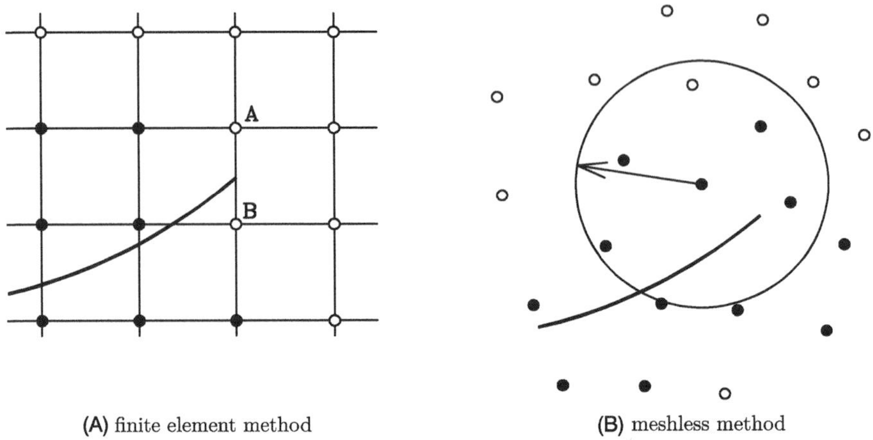

Figure 5.38 The enrichment for the crack tip by using the step function in (A) the finite element method and (B) meshless methods; solids are enriched nodes and circles un-enriched nodes [136].

the branch enrichment is needed [17,79]. If the cohesive crack is considered and a crack tip is located on an edge of a domain of influence, not inside of any domain of influence at the same time, a successful enrichment can be devised by using only the sign function or the Heaviside step function without other enrichment functions [35,134].

The above idea can be applied to the extended finite element method only. The shape function for a node in the standard finite element method with C^0 approximation is completely decoupled from others except those for the neighbor nodes. Imagine a crack tip in a two-dimensional finite element mesh; see Fig. 5.38A. The crack tip is positioned on the edge connecting nodes A and B. Because the crack tip must close at the tip, i.e. the crack opening displacement at crack tip must be zero, nodes A and B should not be enriched. Fig. 5.38B is the case for meshless methods. The domain of influence for a particle in meshless methods is heavily overlapped with others. Therefore it is difficult to apply the idea to meshless methods.

5.6.2.1 Domain decrease method

[134] proposed a simple idea to enrich triangular elements with only the sign function (5.206) without using the branch enrichment even when the crack tip is located inside of an element, not on the edge. They modified the shape function of the discontinuous displacement field $\mathbf{u}^{e,J}$ in Eq. (5.203) so that the crack tip is always placed on the edge of the shape

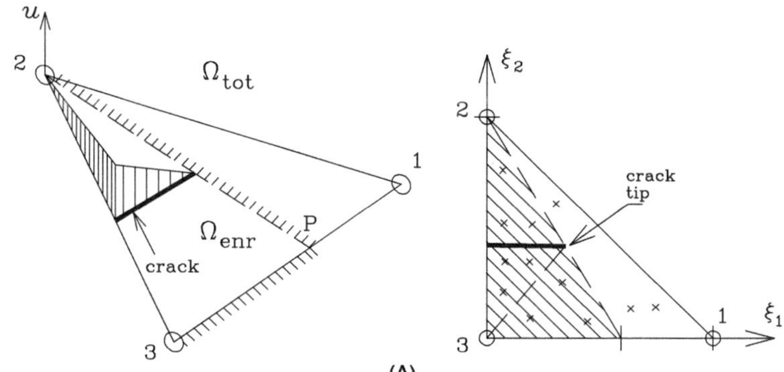

(A)

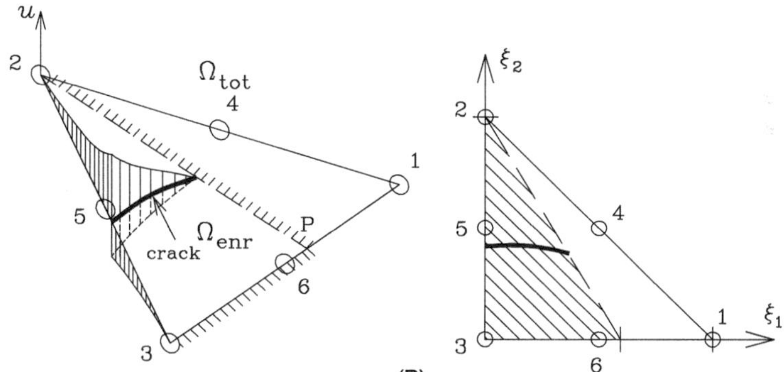

(B)

Figure 5.39 The discontinuous displacement field by [134]'s scheme in which the shape function for the enrichment is modified so that the crack tip is position at the edge; (A) three node triangular element and (B) six node triangular element.

function in (5.203) as the crack grows. The modification is only for the tip element in which the crack tip is located. Once the crack tip grows into an element which then becomes the tip element. The shape function of the discontinuous displacement field for the tip element is modified as shown in Fig. 5.39. Because the shape function vanishes at its edge, the discontinuity from Eq. (5.203) must be zero beyond the edge. So any branch enrichment is not needed. Of course the shape function of the continuous displacement $\mathbf{u}^0$ is completely independent of the crack growth.

We adopt their idea for meshless methods. A crack is shown in Fig. 5.40. The domain of influence of node 1 is completely cut by the crack and it is enriched by the sign function as in Eq. (5.204). Crack tip P is located inside the domain of influence of node 2 which is partially cut. As [134] modified the shape function of triangular elements, we scale down the shape function

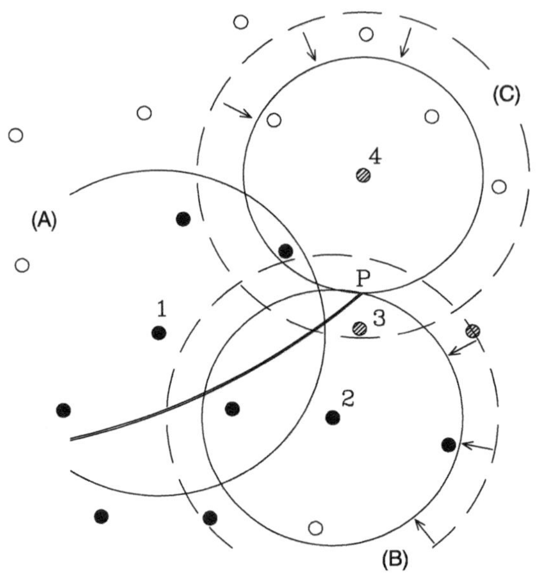

Figure 5.40 Domain of influences (A) completely cut by the crack and (B), (C) partially cut; (B) is decreased so that the crack tip is positioned at its edge and (C) is not needed as its size is decreased; circles and hatched circles are unenriched nodes, and solids enriched nodes.

Φ_I for the discontinuous displacement in Eq. (5.203) so that the crack tip is positioned at its edge as shown in the figure, i.e.

$$\Phi_J^*(\mathbf{X}) = \mathbf{p}^T(\mathbf{X}) \cdot \mathbf{A}^*(\mathbf{X})^{-1} \cdot \mathbf{D}^*(\mathbf{X}_J) \tag{5.207}$$

$$\mathbf{A}^*(\mathbf{X}) = \sum_J \mathbf{p}(\mathbf{X}_J)\, \mathbf{p}^T(\mathbf{X}_J)\, W(\bar{r}_J^*;\, h^*) \tag{5.208}$$

$$\mathbf{D}^*(\mathbf{X}_J) = \sum_J \mathbf{p}(\mathbf{X}_J)\, W(\bar{r}_J^*;\, h^*) \tag{5.209}$$

in which the asterisk denotes the modification for the crack tip and h^* is the modified size of the domain of influence. Note that the shape function for the continuous displacement field Eq. (5.201) is not changed.

The domain of influence of node 3 is also partially cut and the shape function may be shrunk, too. However, we do not enrich the node. Because the node is very close to the crack tip, the shape function becomes very small compared to others after it is modified and the approximation for the discontinuous displacement field becomes bumpy. Therefore, when the domain of influence of a node is partially cut, we enrich the node if the

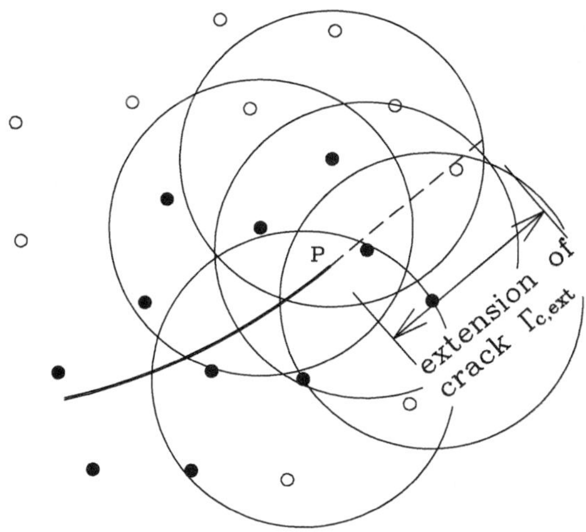

Figure 5.41 The discontinuity $\Gamma_{c,ext}$ beyond crack tip P when nodes are enriched by using only the sign function.

support of the shape function includes at least one enriched node after it is modified. Node 4 is not enriched because the shape function becomes not cut by the crack as it is modified.

One drawback of the method is that the crack appears to be shorter for particles close to the crack and the crack tip. However, we did not observe any severe difficulties compared to more accurate techniques as e.g. explained in the next section. As might become obvious, the method is well suited for adaptive procedures that will not only provide higher accuracy but also a better particle distribution around the crack tip.

The jump in the displacement is governed only by the enrichment and is given for the step function by

$$[\![\mathbf{u}(\mathbf{X})]\!] = 2\sum_{J\in\mathcal{E}}\sum_{I\in W_J(\mathbf{X})} \Phi_I^J(\mathbf{X})\ \mathbf{a}_I^J \qquad (5.210)$$

5.6.2.2 Lagrange multiplier method

Instead of modifying the shape function of the node of which domain of influence is partially cut, we may consider the use of Lagrange multiplier method. If only the sign function enrichment of Eq. (5.204) is used, there is the extension of discontinuity $\Gamma_{c,ext}$ beyond the crack tip; see Fig. 5.41. To model the crack, the discontinuity on $\Gamma_{c,ext}$ should vanish. Because the condition should be satisfied along a line, not at a point, the Lagrange

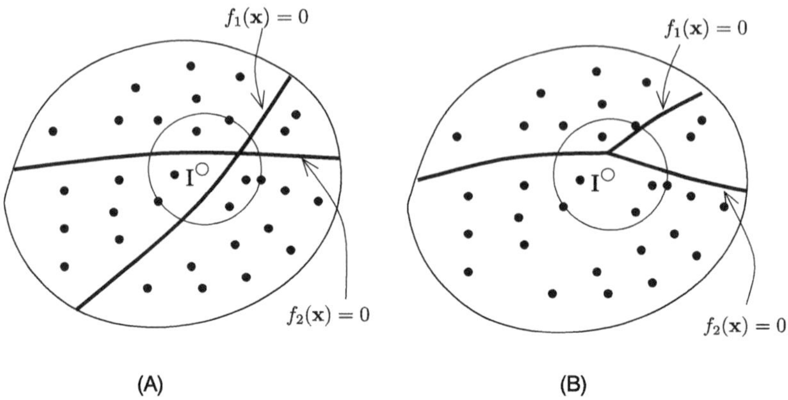

Figure 5.42 Support of node I with (A) intersecting discontinuities and (B) branching discontinuities.

multiplier must be discretized, too. To avoid introducing another nodes for the discretization, we use the same shape functions as those partially cut by the crack. The detailed formulation is given later.

5.6.3 Crack branching and crack junction

As already mentioned, the crack surface is defined by an implicit function f which is zero along the crack path and has the value of the minimum distance to the crack with plus or minus sign. The choice of the sign is completely arbitrary as long as it is consistent throughout the entire calculation.

In this section, we describe how we treat crack branching and crack intersection which is different from the approach in [32] and [17] in the sense that we do not use any special branch function in addition to the usual enrichment. Consider cracks shown in Fig. 5.42. Let $\mathcal{W}_b^1$ be the set of nodes whose domain of influence is completely cut by the discontinuity $f_1(\mathbf{X}) = 0$ and $\mathcal{W}_b^2$ the corresponding set for $f_2(\mathbf{X}) = 0$. $\mathcal{W}_b^3 = \mathcal{W}_b^1 \bigcap \mathcal{W}_b^2$. The same applies accordingly for nodes whose domain of influence is cut by the crack tip enrichment. We will denote this set of nodes with $\mathcal{W}_s^1$ and $\mathcal{W}_s^2$. Then the approximation of the displacement may be given by [32]

$$\mathbf{u}(\mathbf{X}) = \sum_{I \in \mathcal{W}(\mathbf{X})} \Phi_I(\mathbf{X})\ \mathbf{u}_I + \sum_{I \in \mathcal{W}_b^1(\mathbf{X})} \Phi_I(\mathbf{X})\ H(f_1(\mathbf{X}))\ \mathbf{a}_I^{(1)}$$
$$+ \sum_{I \in \mathcal{W}_b^2(\mathbf{X})} \Phi_I(\mathbf{X})\ H(f_2(\mathbf{X}))\ \mathbf{a}_I^{(2)}$$

$$+ \sum_{I \in \mathcal{W}_b^3(\mathbf{X})} \Phi_I(\mathbf{X})\, H(f_1(\mathbf{X}))\, H(f_2(\mathbf{X}))\, \mathbf{a}_I^{(3)}$$
$$+ \sum_{I \in \mathcal{W}_s^1(\mathbf{X})} \Phi_I(\mathbf{X}) \sum_K \mathbf{B}_K^{(1)}(\mathbf{X})\, \mathbf{b}_{KI}^{(1)}$$
$$+ \sum_{I \in \mathcal{W}_s^2(\mathbf{X})} \Phi_I(\mathbf{X}) \sum_K \mathbf{B}_K^{(2)}(\mathbf{X})\, \mathbf{b}_{KI}^{(2)} \qquad (5.211)$$

Principally, more than two branches can be included at one time and a branched crack can branch again. As can be easily seen by Eq. (5.211), additional complexity is then introduced. We would like to mention, that Eq. (5.211) looks worse than it is since only very few nodes are included in all sets $\mathcal{W}$. However, a crack branching requires the introduction of another level set which makes the computation cumbersome for many cracks.

[135] proposed a computationally more efficient approach than (5.211) by modifying the signed distance functions so that no cross terms are needed for junction or branch problems.

When two cracks are joining, the crack tip enrichment is removed. By using the signed distance functions of the pre-existing and approaching crack, the signed distance function of the approaching crack is modified. Consider Fig. 5.43, where three different subdomains have to be considered: $(f_1 < 0, f_2 < 0)$, $(f_1 > 0, f_2 > 0)$, $(f_1 > 0, f_2 < 0)$ as in Fig. 5.43B or $(f_1 > 0, f_2 < 0)$, $(f_1 > 0, f_2 > 0)$, $(f_1 < 0, f_2 < 0)$ as in Fig. 5.43D. The signed distance function of crack 1 of a point $\mathbf{X}$ is then obtained by:

$$f_1(\mathbf{X}) = \begin{cases} f_1^0(\mathbf{X}), & \text{if } f_2^0(\mathbf{X_1}) f_2^0(\mathbf{X}) > 0 \\ f_2^0(\mathbf{X}), & \text{if } f_2^0(\mathbf{X_1}) f_2^0(\mathbf{X}) < 0 \end{cases} \qquad (5.212)$$

where the superimposed 0 denotes the sign distance function before crack junction. Therefore, the final approximation without the cross term reads:

$$\mathbf{u}(\mathbf{X}) = \sum_{I \in \mathcal{W}(\mathbf{X})} \Phi_I(\mathbf{X})\, \mathbf{u}_I + \sum_{n=1}^{n_c} \sum_{I \in \mathcal{W}_b(\mathbf{X})} \Phi_I(\mathbf{X})\, H(f_I^{(n)}(\mathbf{X}))\, \mathbf{a}_I^{(n)}$$
$$+ \sum_{m=1}^{m_t} \sum_{I \in \mathcal{W}_s(\mathbf{X})} \Phi_I(\mathbf{X}) \sum_K \mathbf{B}_K^{(m)}(\mathbf{X})\, \mathbf{b}_{KI}^{(m)} \qquad (5.213)$$

where n_c and m_t are the number of cracks that completely or partially, respectively, cross the domain of influence of the corresponding particle.

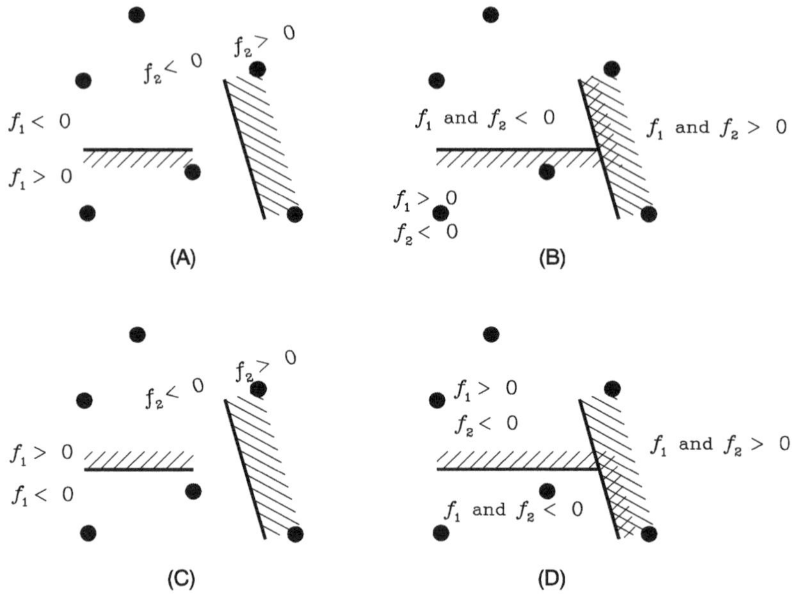

Figure 5.43 Sign functions for crack junction.

The test functions are expressed according to Eq. (5.213):

$$\delta \mathbf{u}(\mathbf{X}) = \sum_{I \in \mathcal{W}(\mathbf{X})} \Phi_I(\mathbf{X})\, \delta \mathbf{u}_I + \sum_{n=1}^{n_c} \sum_{I \in \mathcal{W}_b(\mathbf{X})} \Phi_I(\mathbf{X})\, H(f_I^{(n)}(\mathbf{X}))\delta\, \mathbf{a}_I^{(n)} + \sum_{m=1}^{m_t} \sum_{I \in \mathcal{W}_s(\mathbf{X})} \Phi_I(\mathbf{X}) \sum_K \mathbf{B}_K^{(m)}(\mathbf{X})\, \delta \mathbf{b}_{KI}^{(m)} \tag{5.214}$$

5.7. Extended local maximum entropy (XLME)

Maximum entropy shape functions are a relatively new class of approximation functions, as they were first introduced in [116] in the context of polygonal interpolation. The idea of these functions is to maximize the Shannon entropy [110] of the basis functions, which gives a measure of the uncertainty in the approximation scheme. The principle of maximum entropy (max-ent) was developed by Jaynes [62,63], who showed that there is a natural correspondence between statistical mechanics and information theory. In particular, max-ent offers the least-biased statistical inference when the shape functions are viewed as probability distributions subject to the approximation constraints (such as linear reproducing properties). However, without additional constraints, the basis functions are nonlocal, which due to increased overlapping makes them unsuitable for analysis using Galerkin methods. The increased overlapping of the basis functions generally leads to more expensive numerical integration due to the large number of evaluation points. It also produces a non-sparse stiffness matrix, resulting in a linear system that is much more expensive to solve.

The local maximum-entropy (LME) approximation schemes were developed in [3] using a framework similar to meshfree methods. Here the support of the basis functions is introduced as a thermalization (or penalty) parameter β in the constraint equations. When $\beta = 0$, then the max-ent principle is fully satisfied and the basis functions will be least biased. For example, if only zero-order consistency is required, the shape functions are Shepard approximants [111] with Gaussian weight function. When β is large, then the shape functions have minimal support. In particular, they become the usual linear finite element functions defined on a Delaunay triangulation of the domain associated with the given node set. In [3] it was shown that for some values of β, the approximation properties of the maximum-entropy basis functions are greatly superior to those of the finite element linear functions, even when the added computational cost due to larger support is taken into account.

Enriched maximum entropy shape functions can be used for the analysis of fracture problems. It was shown in [1] that this method is more accurate than standard XFEM and does not require the so-called blending elements (the elements near the crack tip). When compared to usual meshfree methods for crack propagation, such as Element Free Galerkin (EFG), the method presented here can more easily deal with essential boundary conditions, due to the fact that the shape functions satisfy a weak Kronecker delta property. The shape functions are also very smooth (C^∞), which results in an accurate numerical integration with a relatively low number of integration points, especially for Gauss-Legendre quadrature [3,31,88].

5.7.1 Local Maximum Entropy (LME) approximants

LME meshfree approximants, introduced in [3], are related to other convex approximation schemes, such as natural neighbor approximants [115], subdivision approximants [30], or B-spline and NURBS basis functions [58]. The LME basis functions will be denoted by $p_a(\mathbf{x})$, $a = 1, \ldots, N$ with $\mathbf{x} \in \mathbb{R}^d$, d is the dimension of the physical domain. They are non-negative and are required to satisfy the zeroth-order and first-order consistency conditions:

$$p_a(\mathbf{x}) \geq 0, \tag{5.215}$$

$$\sum_{a=1}^{N} p_a(\mathbf{x}) = 1, \tag{5.216}$$

$$\sum_{a=1}^{N} p_a(\mathbf{x})\mathbf{x}_a = \mathbf{x}. \tag{5.217}$$

In the last equation, the vector $\mathbf{x}_a$ identifies the positions of the nodes associated with each basis function. Consider a set of nodes $X = \{\mathbf{x}_a\}_{a=1,\dots,N}$, which we will call *the node set*. The convex hull of X is the set

$$\text{conv}X := \{\mathbf{x} \in \mathbb{R}^d | \mathbf{x} = \mathbf{X}\lambda,\ \lambda \in \mathbb{R}^N_+,\ \mathbf{1} \cdot \lambda = 1\} \tag{5.218}$$

Here $\mathbb{R}^N_+$ is the non-negative orthant, $\mathbf{1}$ denotes the vector in $\mathbb{R}^N$ whose entries are one, and $\mathbf{X}$ is the $d \times N$ matrix whose columns are the co-ordinates of the position vectors of the nodes in the node set X [3]. Convex approximants, which are in the span of convex basis functions, can only exist within the convex hull of X (or subsets of it) and satisfy a weak Kronecker delta property at the boundary of the convex hull of the nodes. This means that the shape functions corresponding to the interior nodes vanish on the boundary. With this property, the imposition of essential boundary conditions in the Galerkin method is straightforward.

The principle of maximum entropy comes from statistical physics and information theory, which consider the measure of uncertainty or information entropy [110]. Consider a random variable $\chi : I \to \mathbb{R}^d$, where I is the index set $I = \{1, \dots, N\}$ and $\chi(a) = \mathbf{x}_a$ gives to each index the position vector of its corresponding node. Since the shape functions of a convex approximation scheme are non-negative and add to one, we regard $\{p_1(\mathbf{x}), \dots, p_N(\mathbf{x})\}$ as the corresponding probabilities. The statistical expectation or average of this random variable, as regarding Eq. (5.217), is $\mathbf{x}$. According to this interpretation, the approximation of a function $u(\mathbf{x}) \approx \sum_{a=1}^{N} p_a(\mathbf{x}) u_a$ from the nodal values $\{u_a\}_{a=1,\dots,N}$ is understood as an expected value $u(\mathbf{x})$ of a random variable $\mu : I \to \mathbb{R}$ where $\mu(a) = u_a$.

The main idea of max-ent is to maximize the Shannon's entropy, $H(p_1, p_2, \dots, p_N)$, subject to the consistency constraints as follows:

$$\text{(ME)} \quad \text{For a fixed } \mathbf{x} \text{ maximize} \tag{5.219}$$

$$H(p_1, p_2, \ldots, p_N) = -\sum_{a=1}^{N} p_a \log(p_a)$$
$$\text{subject to} \quad p_a \geq 0, \quad a = 1, \ldots, N$$
$$\sum_{a=1}^{N} p_a = 1$$
$$\sum_{a=1}^{N} p_a \mathbf{x}_a = \mathbf{x}$$

Solving the (ME) problem produces the set of basis functions, $p_a := p_a(\mathbf{x})$, $a = 1, \ldots, N$. However, these basis functions are nonlocal, i.e. they have support in all of conv X, and are not suitable for use in a Galerkin approximation because it would lead to a full, non-banded matrix. Nevertheless, they have been used in [116] as basis functions for polygonal elements.

Another optimization problem which takes into account the locality of the shape functions is Rajan's form of the Delaunay triangulation [101]. This can be stated as the following linear program:

$$\text{(RAJ)} \quad \text{For a fixed } \mathbf{x} \text{ minimize} \tag{5.220}$$
$$U(\mathbf{x}, p_1, p_2, \ldots, p_N) = \sum_{a=1}^{N} p_a \left|\mathbf{x} - \mathbf{x}_a\right|^2$$
$$\text{subject to} \quad p_a \geq 0, \quad a = 1, \ldots, N$$
$$\sum_{a=1}^{N} p_a = 1$$
$$\sum_{a=1}^{N} p_a \mathbf{x}_a = \mathbf{x}$$

It is easy to see that $U(\mathbf{x}, p_1, p_2, \ldots, p_N)$ is minimized when the shape functions $p_1, \ldots, p_N$ decay rapidly as the distance from the corresponding nodes $\mathbf{x}_a$ increases. There, the shape functions that satisfy (RAJ) problem will have small supports, where the support can be defined up to a small tolerance ϵ by

$$\text{supp}(p_a) = \left\{\mathbf{x} : p_a(\mathbf{x}) > \epsilon\right\}$$

The main idea of LME approximants is to compromise between the (ME) problem and the (RAJ) problem by introducing parameters β_a that control the support of the p_a. Therefore we write:

$$\text{For a fixed } \mathbf{x} \text{ minimize} \quad \sum_{a=1}^{N} \beta_a p_a |\mathbf{x}-\mathbf{x}_a|^2 + \sum_{a=1}^{N} p_a \log(p_a) \quad \text{subject to} \quad p_a \geq 0, \quad a=1,\ldots,N; \quad \sum_{a=1}^{N} p_a = 1; \quad \sum_{a=1}^{N} p_a \mathbf{x}_a = \mathbf{x} \tag{5.221}$$

The non-negative parameters β_a can in general be functions of the position $\mathbf{x}$. This convex optimization problem is solved efficiently by a duality method as described in [3]. Finally, the shape functions are written in the form:

$$p_a(\mathbf{x}) = \frac{1}{Z(\mathbf{x}, \lambda^*(\mathbf{x}))} \exp[-\beta_a |\mathbf{x}-\mathbf{x}_a|^2 + \lambda^*(\mathbf{x}) \cdot (\mathbf{x}-\mathbf{x}_a)]$$

where

$$Z(\mathbf{x}, \lambda) = \sum_{b=1}^{N} \exp[-\beta_b |\mathbf{x}-\mathbf{x}_b|^2 + \lambda \cdot (\mathbf{x}-\mathbf{x}_b)]$$

is a function associated with the node set X and $\lambda^*(\mathbf{x})$ is defined by

$$\lambda^*(\mathbf{x}) = \arg \min_{\lambda \in \mathbb{R}^d} \log Z(\mathbf{x}, \lambda)$$

The local max-ent shape functions are as smooth as $\beta(\mathbf{x})$ and $p_a(\mathbf{x}, \beta_a)$ is a continuous function of $\beta \in [0, +\infty)$ [3]. For example LME shape functions are C^∞ if β is constant. Here we choose $\beta = \frac{\gamma}{h^2}$, where h is a measure of the nodal spacing and γ is constant over the domain. In this case the shape functions are smooth and their degree of locality is controlled by the parameter γ. A plot of the LME functions for $\gamma = 1.8$ and a particular choice of nodes is given in Fig. 5.44. In general, the optimal β is not obvious and this will be discussed later.

As we mentioned before, LME shape functions satisfy a weak Kronecker delta property at the boundary of the convex hull of the nodes. Therefore, the shape functions that correspond to interior nodes vanish on the boundary.

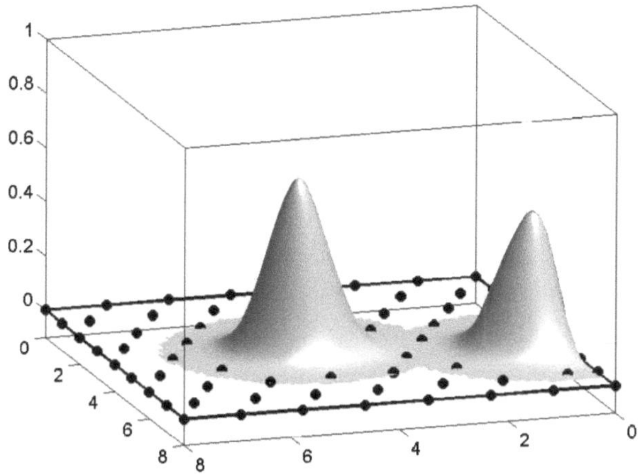

Figure 5.44 Local max-ent shape functions in 2D.

5.7.2 Numerical integration

5.7.2.1 Numerical integration for LME

The numerical integration of LME shape functions poses similar challenges as that of the shape functions used in meshless methods. In particular, the integrands used in the assembly of the stiffness matrix are non-polynomial and (depending on the values of the parameter γ) the supports of the shape functions overlap more than in standard finite elements. However, the shape functions are smooth so only a relatively small number of integration points are required.

In the examples we considered, we used quadrilateral background integration cells for integrating the shape functions whose support does not intersect the crack. For the values of γ between 4.8 and 1.8, and for uniformly spaced nodes and square we found that the 4×4 Gauss quadrature rule is sufficient to ensure optimal convergence. Moreover, a quadrature rule with 8×8 Gauss points provides close to exact integration (i.e. the results change by less than 10^{-6} when the number of Gauss points is further increased).

5.7.2.2 Numerical integration for enriched LME

The usual numerical integration methods, for example Gauss quadrature, are less accurate for PU-enriched methods for fracture. This happens due to the discontinuity along the crack, and the singularity at the crack tip. The usual rule is to use a simple splitting of integration cells crossed by the crack

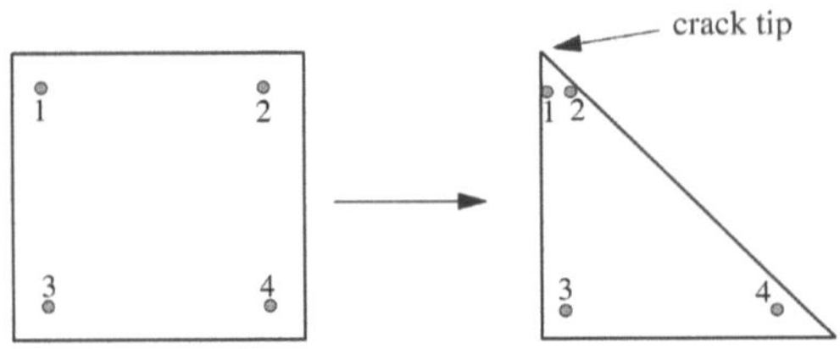

Figure 5.45 Transformation of an integration method on a square into an integration method on a triangle for crack tip functions.

[69]. In [84], a method was proposed in which each part of the elements that are cut or intersected by a discontinuity is mapped onto the unit disk using a conformal Schwarz-Christoffel map. However, for straight cracks, a triangulation of the elements cut by the crack which takes into account the location of the discontinuity is relatively easy to implement and was used in this work.

For the integration cells that contain the crack tip, special care has to be taken. These cells contain the discontinuity and a singularity together. So, simply refining the triangles that make up the integration cells leads to less accurate numerical results. A simple solution is to refine locally each split triangle, until an acceptable estimate of the integrands is achieved. Unfortunately, this method is expensive. To solve this problem, the *almost polar integration* was introduced in [69]. The main idea is to build a quadrature rule on a triangle from a quadrature rule on the unit square (see Fig. 5.45). The map is:

$$T : (x, y) \longrightarrow (xy, y)$$

which maps a square into a triangle. By looking at the integrands which contain the derivatives of the branch functions, we notice that the Jacobian of the transformation T, will cancel the $r^{-1/2}$ singularity. This integration method gives excellent results with a low number of integration points and is used on the sub-triangles having the crack tip as a vertex. In the other integration cells, we found it is sufficient to use standard Gauss quadrature over a background mesh (such as the Delaunay triangulation of the nodes that takes in to account the discontinuity for the cells cut by the crack).

An important distinction between meshless methods and standard finite elements is that, in the former, the numerical integration is almost never exact. Recent work [6] has shown that integration errors in meshless methods negatively impact the stability of the method when a large number of degrees of freedom is involved. In particular, as the value of the discretization parameter h decreases, the accuracy of the numerical integration should increase proportionally, so that optimal convergence can be obtained. We have conducted a detailed study on the effect of approximate integration for one of the numerical examples shown below.

5.7.3 Condition number

There are two ways to choose the enrichment area: *topological enrichment* in which the area of enrichment shrinks with the nodal spacing h, and *geometric enrichment* which uses a fixed enrichment area. In topological enrichment, the branch functions are multiplied by shape functions on a small set of nodes around the crack tip. These singular functions live on a compact support vanishing as h goes to zero. In the context of meshless methods, only topological enrichment has been studied, which leads to non-optimal convergence rate. However, the numerical results in Section 5.9 show that the enrichment area should have a size independent of the mesh parameter (i.e. it should be geometric) to obtain optimal convergence, as seen for standard XFEM in [7,69]. Unfortunately, adding singular functions on all the nodes within a fixed area around the crack tip leads to an increase in the number of degrees of freedom and an increase in the condition number (see Fig. 5.46).

Some methods were proposed to improve the condition number of the stiffness matrix, such as preconditioning schemes. Here we use a method introduced in [7] which relies on a Cholesky decomposition of the diagonal blocks of the stiffness matrix corresponding to enriched nodes. This method noticeably improves the condition number (see Fig. 5.46), but not the rate of increase as the mesh is refined. A robust preconditioning scheme for XFEM was proposed in [76], which is based on a domain decomposition and results in a condition number close to the finite element matrices without enrichment. Another promising development for improving the condition number of geometric enrichment has been developed in [5]. This improvement will be discussed in a future work.

5.8. Cracking particle methods

Cracking particle methods (CPM) [93–95,98,100] do not require a continuous crack path and greatly simplify the representation of the crack surface, in particular in 3D. Branching cracks and joining cracks are to some extend also a more natural outcome of the analysis. In contrast to

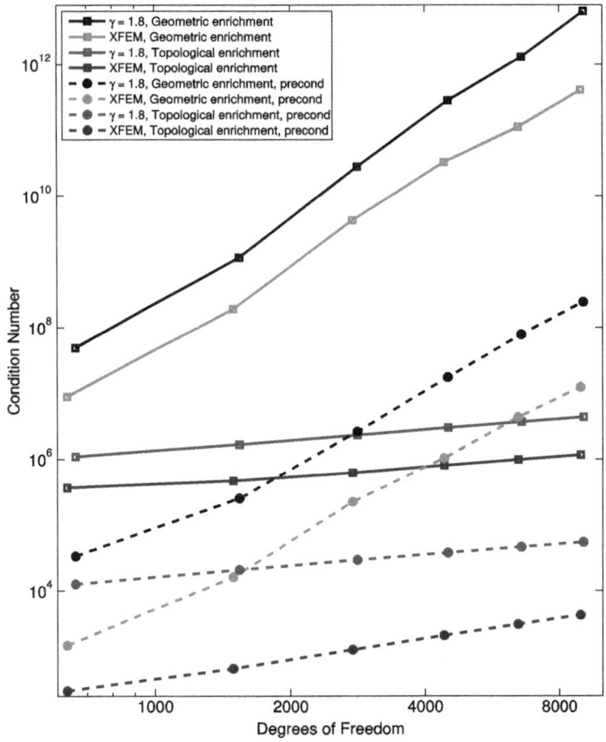

Figure 5.46 The condition number of geometric and topological enrichment for $\gamma = 1.8$ and $\gamma = 4.8$, using a direct solver and the preconditioning method.

finite element methods, that do not enforce crack path continuity, those methods do not suffer mesh orientation bias as shown e.g. by [52].

5.8.1 The enriched cracking particles method

Consider a displacement field which is continuous in the entire domain except at the cracks where a discontinuity occurs in the displacements. To describe this discontinuity, the displacement is (as usual) decomposed into continuous and discontinuous parts (see also e.g. references [8,36,80]):

$$\mathbf{u}(\mathbf{X}, t) = \mathbf{u}^{cont}(\mathbf{X}, t) + \mathbf{u}^{enr}(\mathbf{X}, t) \tag{5.222}$$

where $\mathbf{X}$ are the material coordinates, t is the time, $\mathbf{u}^{cont}$ denotes the continuous displacement and $\mathbf{u}^{enr}$ the discontinuous part, which is also called the enrichment.

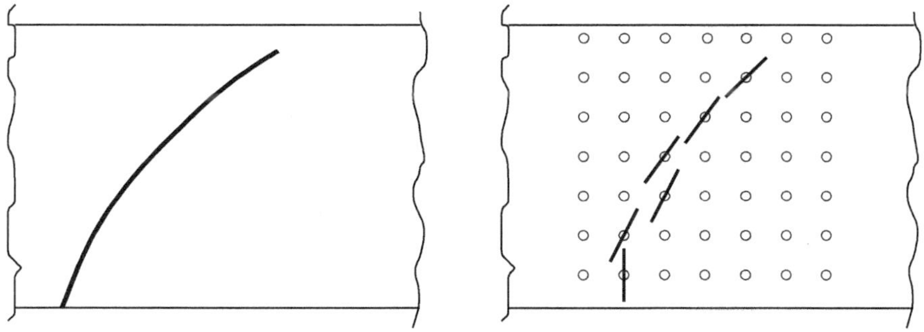

Figure 5.47 Schematic on the right shows a crack model for the crack on the left.

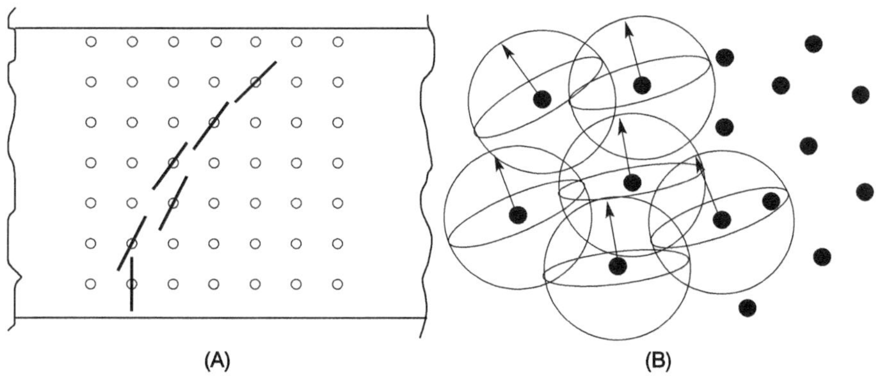

Figure 5.48 Schematic of the crack model for two (A) and three (B) dimensions.

The crack is modeled by a set of discrete cracks as shown in Fig. 5.47, 5.48 and 5.49. These discrete cracks are restricted to lie on the particles, i.e. each crack plane (or line in 2D) always passes through a particle. Since the crack geometry is described by the set of cracked particles, we do not have to provide a representation for the geometry of the crack. By contrast, in XFEM or in the Element free Galerkin method a crack does not have to be located at nodes and the crack surface is continuous. But as mentioned before, this entails additional complexity.

Let $\mathcal{N}$ be the total set of nodes in the model and $\mathcal{N}_c$ the set of cracked nodes. To model the discontinuous part of the displacement, the test and trial functions are enriched with sign functions which are parametrized by $\delta\mathbf{q}_I$ and $\mathbf{q}_I$, respectively.

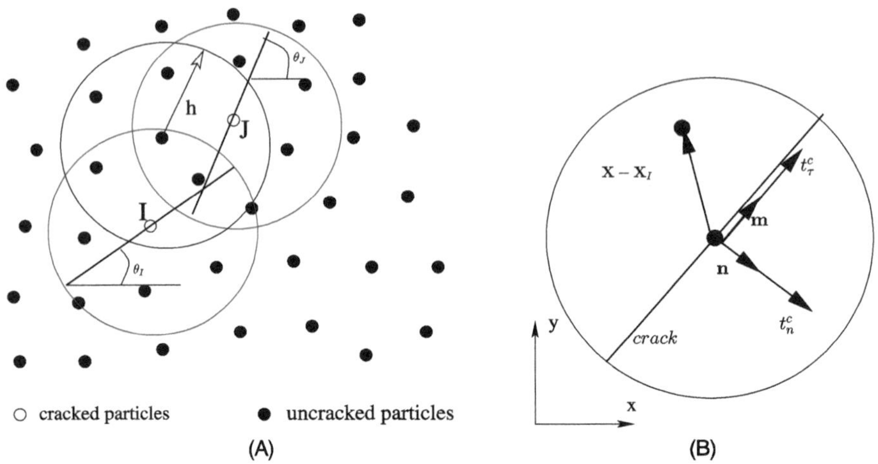

Figure 5.49 (A) Schematic of the crack when particles I and J are cracked; they are cracked at angles θ_I and θ_J, respectively. (B) Particle with a crack and its domain of influence with crack boundary.

The test and trial functions are

$$\mathbf{u}^h(\mathbf{X}, t) = \sum_{I \in \mathcal{N}} \hat{\Phi}_I(\mathbf{X})\ \mathbf{u}_I(t) + \sum_{I \in \mathcal{N}_c} \Psi_I(\mathbf{X})\ S(f_I(\mathbf{X}))\ \mathbf{q}_I(t) \tag{5.223}$$

$$\delta\mathbf{u}^h(\mathbf{X}) = \sum_{I \in \mathcal{N}} \Phi_I(\mathbf{X})\ \delta\mathbf{u}_I + \sum_{I \in \mathcal{N}_c} \Psi_I(\mathbf{X})\ S(f_I(\mathbf{X}))\ \delta\mathbf{q}_I \tag{5.224}$$

where $f_I(\mathbf{X})$ is given by

$$f_I(\mathbf{X}) = \mathbf{n}_0 \cdot (\mathbf{X} - \mathbf{X}_I) \tag{5.225}$$

where $\mathbf{n}_0$ is the normal to the crack in the reference configuration, $\mathbf{u}_I$ are the particle displacements and S is the step function, Eq. (5.191). The normal in the initial configuration $\mathbf{n}_0$ is found by $\mathbf{n}$ from Nanson's law

$$\mathbf{n}_0 = J^{-1}\mathbf{n}(t_{cr}) \cdot \mathbf{F}(\mathbf{X}_I, t_{cr}) \tag{5.226}$$

where t_{cr} is the time at which the particle cracks.

Note that in general different shape functions for the continuous part, $\Phi_I(\mathbf{X})$, and discontinuous part, $\Psi_I(\mathbf{X})$ can be used, see Eq. (5.224) and (5.223).

The deformation tensor $\mathbf{F}$ is also decomposed into a continuous and discontinuous part $\mathbf{F} = \hat{\mathbf{F}} + \tilde{\mathbf{F}}$ where $\hat{\mathbf{F}}$ is the continuous and $\tilde{\mathbf{F}}$ the discon-

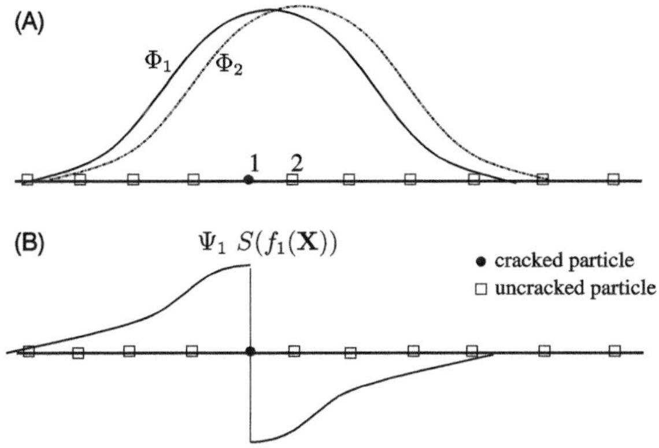

Figure 5.50 Shape functions for cracked and uncracked particles: (A) Shape function Φ for cracked particle 1 and uncracked particle 2, (B) $\Psi_I(\mathbf{X}))\ S(f_I(\mathbf{X}))$ for cracked particles 1.

tinuous part:

$$\nabla_0 \mathbf{u}(\mathbf{X}, t) = \underbrace{\sum_{I \in \mathcal{N}} \nabla_0 \hat{\Phi}_I(\mathbf{X}) \otimes \mathbf{u}_I(t)}_{\hat{\mathbf{F}}} + \underbrace{\sum_{I \in \mathcal{N}_c} \nabla_0 \Psi_I(\mathbf{X})\ S(f_I(\mathbf{X})) \otimes \mathbf{q}_I(t)}_{\tilde{\mathbf{F}}}$$
$$+ \underbrace{\sum_{I \in \mathcal{N}_c} \Psi_I(\mathbf{X})\ \nabla_0 S(f_I(\mathbf{X})) \otimes \mathbf{q}_I(t)}_{=0} \tag{5.227}$$

The first term on the RHS is the usual approximation, the third term vanishes in the open set $\Omega_0 \setminus \Gamma^c$; the second term are the discontinuous enrichment strains.

To illustrate the resulting approximation, consider a one-dimensional example. The continuous and discontinuous shape functions (in 1D the domain of influence is always cut completely) are illustrated in Fig. 5.50. Particle 1 is cracked while particle 2 is uncracked. In Fig. 5.50, the enrichment functions for particle 1 and the shape functions $\Phi_I(\mathbf{X})$ are illustrated. Fig. 5.50B shows the jump in $\Psi_I(\mathbf{X})\ S(f(\mathbf{X}))$ for the cracked particle.

The jump in the displacement across the crack depends only on the discontinuous part of the displacement field $\mathbf{u}^{enr}(\mathbf{X})$ and hence the enrichment parameters $\mathbf{q}_I$. It can be shown from (5.223) that

$$[\![\mathbf{u}(\mathbf{X})]\!] = 2 \sum_{I \in \mathcal{N}_c} \Psi_I(\mathbf{X})\ \mathbf{q}_I \tag{5.228}$$

5.8.2 Applications to large deformations

In large deformation theory, a mixed Lagrangian/Eulerian kernel formulation or an updated Lagrangian kernel formulation is used, see Section 5.1.6.2. When the cohesive tractions have decayed to zero, the enrichment is removed and the particle is split into two particles. Kinematic values such as velocity are assigned to the corresponding particle according to the jump in the velocity field. By using initially a Lagrangian kernel, we ensure that material fracture occurs physically and not due to numerical artifacts. The Eulerian kernel or the updated Lagrangian kernel ensures that the solution is stable and allows large separations.

In the blending domain between Lagrangian and Eulerian kernels, we proceed as follows: The kernels that are not cracked are kept Lagrangian but exclude particles on the opposite side of the crack from the domain of influence. In other words, we check if the opposite cracked particles are not in the domain of influence in the current configuration of the blended particles. If they do, they are removed from the domain of influence of the blended particles. This is a straightforward check since all neighbors and connectivities are known.

5.8.3 The cracking particles method without enrichment

[95] showed that the original CPM, where only cracked nodes were enriched, does not lead convergent results for problems in linear elastic fracture mechanics and proposed to modify the method through local visibility. In [100], a CPM without enrichment was proposed. The principle idea is the same as in [93] where the crack is modeled by a set of cracked particles. However, in contrast to [93], the displacement discontinuity is not introduced by additional unknowns in the variational formulation. Instead, a cohesive surface is introduced at the location where the cracking criteria is met and the corresponding cracked particle is split into two particles lying on associated cohesive crack segments as shown in Fig. 5.51. The size of the circular crack segment (or crack line in 2D) is determined by the support size of the cracked particle. In the initial configuration, the cohesive crack segments will coincide. Crack opening is assumed to be piece-wise constant, i.e. the movement of the corresponding cohesive crack segments are entirely described by the movement of the cracked (and split) particles, Fig. 5.51.

In order to account for the discontinuous displacement field, a local visibility method with respect to the corresponding crack segment is used. The jump in the displacement field (with respect to every cohesive crack

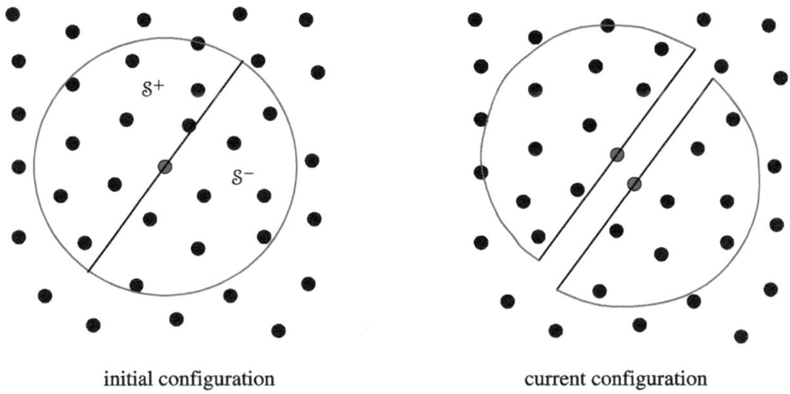

Figure 5.51 Crack model.

segment, see Fig. 5.51), necessary for the cohesive traction, is then obtained by

$$[\![\mathbf{u}(\mathbf{X})]\!] = \sum_{I \in \mathcal{S}^+} \Phi_I(\mathbf{X}^+)\ \mathbf{u}_I - \sum_{I \in \mathcal{S}^-} \Phi_I(\mathbf{X}^-)\ \mathbf{u}_I \tag{5.229}$$

where $\mathcal{S}^+$ denotes the set of nodes on one side of the crack and $\mathcal{S}^-$ the set of nodes on the opposite side of the crack, Fig. 5.51. The orientation of the cohesive crack segments that is governed by its normal $\mathbf{n}$ depends on the cracking criterion. It is simple to determine if a particle is lying in $\mathcal{S}^+$ or $\mathcal{S}^-$, Fig. 5.51, by checking the following condition:

$$H(\mathbf{x}) = \begin{cases} +1 & \text{if } \mathbf{n} \cdot (\mathbf{x} - \mathbf{x}_I) > 0 \\ -1 & \text{if } \mathbf{n} \cdot (\mathbf{x} - \mathbf{x}_I) < 0 \end{cases} \tag{5.230}$$

If $H = 1$, the particles are in $\mathcal{S}^+$, otherwise they are in $\mathcal{S}^-$. Note that in a Total Lagrangian description, this condition has to be checked only once with respect to the initial frame.

5.8.4 Cracking rules for cracking particle methods

Cracking Particle Methods that do not enforce crack path continuity might suffer from spurious crack patterns, see e.g. [27,132] and Fig. 5.52. However, they can be eliminated by a set of simple crack injection rules [112]. Firstly, one has to distinguish between crack propagation and crack initiation. Since the crack is not considered as continuous surface, crack propagation is assumed when no cracked node is detected in the vicinity

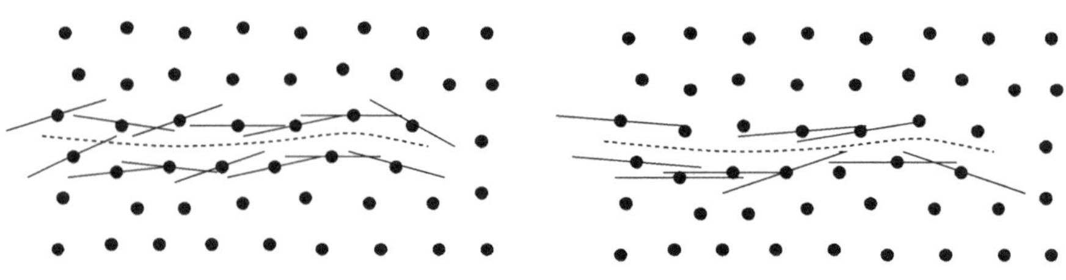

Figure 5.52 (A) Spurious cracking and (B) improved crack pattern.

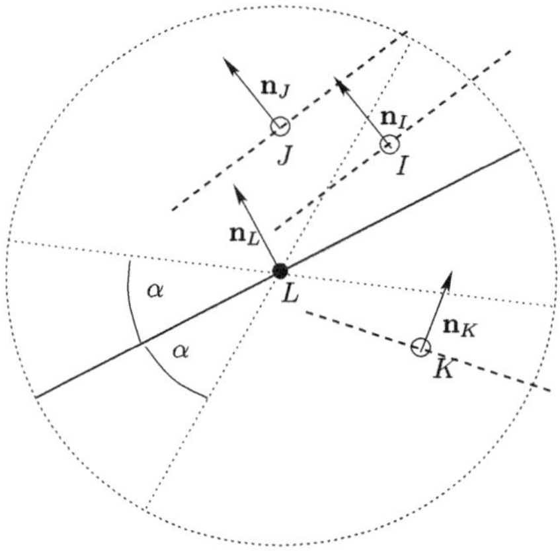

Figure 5.53 Crack prevention: nodes *J* and *K* are located in the crack prevention zone, node *I* is located in the crack propagation zone.

(in a circle [sphere in 3D] with radius $r = \alpha h$; for most calculations shown later, we choose $\alpha = 1.5$) of a sampling point where cracking is detected. For propagating cracks, spurious cracks as shown in Fig. 5.52 are avoided by introducing a prevention zone, Fig. 5.53. For example for particle L, the crack prevention zone is given through the angle α:

$$|\mathbf{n} \cdot \frac{\mathbf{x} - \mathbf{x}_L}{||\mathbf{x} - \mathbf{x}_L||}| < cos\left(\frac{\Pi}{2} - \alpha\right) \tag{5.231}$$

For branching cracks, it is often observed that the orientation of neighboring cracks severely differ, see node K in Fig. 5.53. Therefore, cracks are also allowed in the prevention zone when the following criterion is fulfilled:

$$\mathbf{n}_L \cdot \mathbf{n}_K < cos\beta \tag{5.232}$$

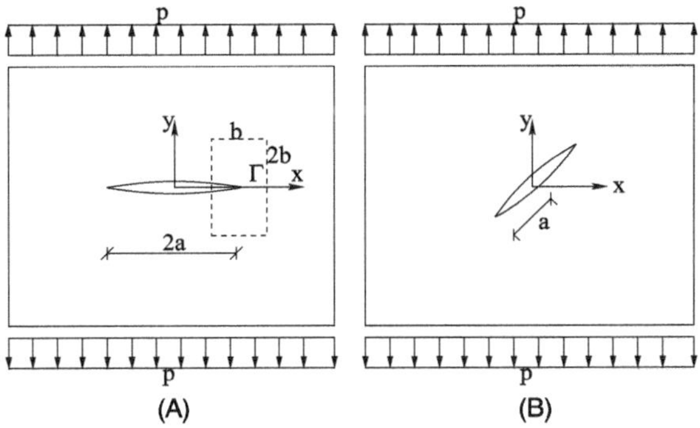

Figure 5.54 (A) Mode I problem, (B) (Mixed) Mode I-II problem.

with $\alpha = 2\Pi/9$ and $\beta = \Pi/3$. Similar criteria are used within the cracking particle method in [125]. [132] reported that he was avoided cracking rules by use of rotating crack segments.

5.9. Comparison of different methods

In this section, we will compare different meshfree methods for some standard examples where an analytical solution is available: a mode I and a mixed mode problems in 2D and a penny shaped crack problem in 3D. We will check local and global convergence. Global convergence is checked in terms of the error in the energy norm:

$$\|err\|_{energy} = \frac{\|\mathbf{u}_h - \mathbf{u}_{analytic}\|_{energy}}{\|\mathbf{u}_{analytic}\|_{energy}} \tag{5.233}$$

with

$$\|\mathbf{u}\|_{energy} = \left(\int_{\Omega_0} \mathbf{E}^T(\mathbf{u}) : \mathbf{C} : \mathbf{E}(\mathbf{u}) \; d\Omega_0 \right)^{1/2} \tag{5.234}$$

where $\mathbf{E}$ is the Green strain. For local convergence, we will compute the stress intensity factors (SIFs) K_I, K_{II} and K_{III}.

5.9.1 The mode I crack problem

Consider a mode I crack problem illustrated in Fig. 5.54A. Plane strain conditions and linear elastic material behavior are assumed. For an infinite

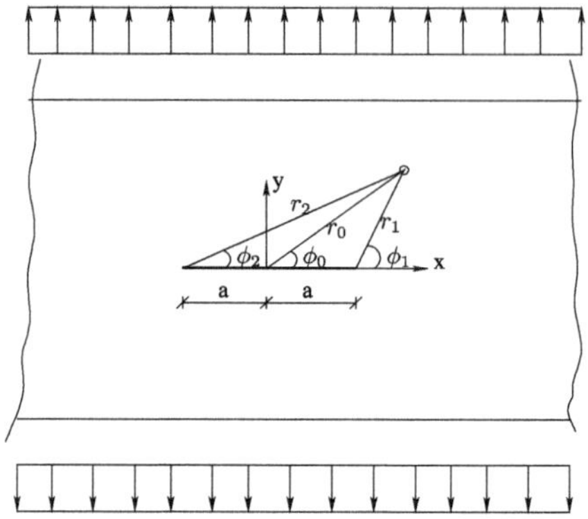

Figure 5.55 The Griffith problem.

plate, [126] provided a solution for this problem:

$$\sigma_{xx} = p\left(\frac{r_0}{\sqrt{r_1\ r_2}}\cos\left(\phi_0 - \frac{\phi_1+\phi_2}{2}\right) - \frac{a^2\ r_0}{(r_1\ r_2)^{1.5}}\sin\phi_0\ \sin\left[1.5\ (\phi_1+\phi_2)\right] - 1\right) \tag{5.235}$$

$$\sigma_{yy} = p\left(\frac{r_0}{\sqrt{r_1\ r_2}}\cos\left(\phi_0 - \frac{\phi_1+\phi_2}{2}\right) + \frac{a^2\ r_0}{(r_1\ r_2)^{1.5}}\sin\phi_0\ \sin\left[1.5\ (\phi_1+\phi_2)\right]\right) \tag{5.236}$$

$$\tau_{xy} = p\frac{a^2\ r_0}{(r_1\ r_2)^{1.5}}\sin\phi_0\ \cos\left[1.5(\phi_1+\phi_2)\right] \tag{5.237}$$

where the parameters r_0, r_1, r_2, ϕ_0, ϕ_1 and ϕ_2 are explained in Fig. 5.55. The exact tractions may be calculated from Eqs. (5.235) to (5.237) and are applied to the boundary shown in Fig. 5.54. The near tip stress field is given by

$$\sigma_{xx}(r,\phi) = \frac{K_I}{\sqrt{2\pi r}}\cos\frac{\phi}{2}\left(1 - \sin\frac{\phi}{2}\sin\frac{3\phi}{2}\right) \tag{5.238}$$

$$\sigma_{yy}(r,\phi) = \frac{K_I}{\sqrt{2\pi r}}\cos\frac{\phi}{2}\left(1 + \sin\frac{\phi}{2}\sin\frac{3\phi}{2}\right) \tag{5.239}$$

$$\tau_{xy}(r,\phi) = \frac{K_I}{\sqrt{2\pi r}}\sin\frac{\phi}{2}\cos\frac{\phi}{2}\cos\frac{3\phi}{2} \tag{5.240}$$

with polar coordinates r and ϕ.

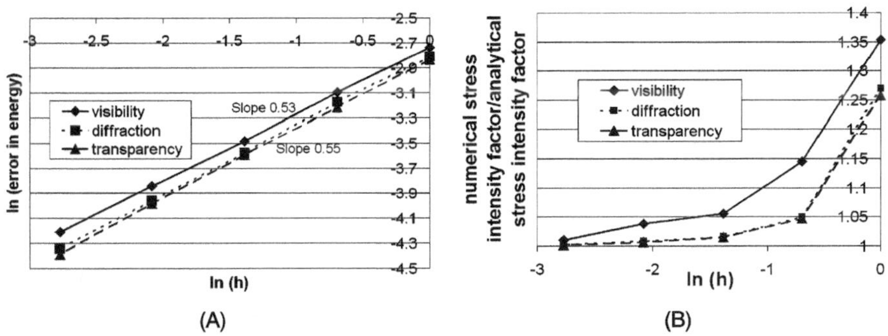

Figure 5.56 (A) Error in the energy for the *mode I* problem using the visibility, diffraction and transparency criterion; (B) normalized stress intensity factor versus *h*.

5.9.1.1 EFG with visibility, diffraction and transparency methods

EFG with linear complete shape functions and an integration order of 4 are used. The error in the energy norm is illustrated in Fig. 5.56A for the visibility, diffraction and transparency method. No crack tip enrichment is used. The convergence rate of the three methods are almost identical though the absolute error is smaller for the diffraction and the transparency method. Fig. 5.56B shows the computed SIF K_I. For a better illustration, the computed SIF K_I is divided by the analytical value. Also for local convergence, the diffraction and transparency methods perform better. There is almost no difference in the results for the transparency and diffraction method.

5.9.1.2 PU methods with intrinsic and extrinsic enrichments

The error in the energy norm is illustrated in Fig. 5.57A for the intrinsic PU enrichment, the extrinsic PU and extrinsic MLS enrichment. For all PU enrichments, we have used the complete crack tip enrichment (5.165). For the intrinsic enrichment, we have used the higher order blending away from the crack tip as described in Section 5.5.1. The extrinsic PU enrichment is also only applied around the crack tip. We have also included the results obtained with the visibility criterion in this figure. The methods that include the crack tip enrichment give more accurate results and a much better convergence rate than the crack approach with the "pure" visibility criterion. This is expected due to the crack tip enrichment with the Westergaard solution. The most accurate results and the highest convergence rate of 0.94 are obtained with the extrinsic MLS enrichment. However, the computational cost is higher since the enrichment is applied in the en-

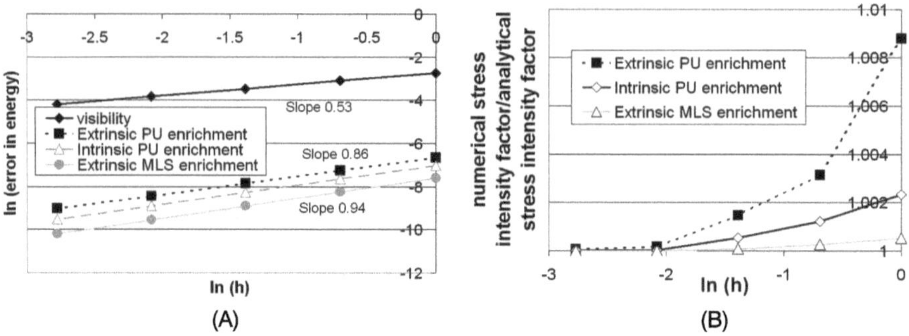

Figure 5.57 (A) Error in the energy for the *mode I* problem using the intrinsic and extrinsic PU enrichment, respectively and the extrinsic MLS enrichment; (B) normalized stress intensity factor versus h.

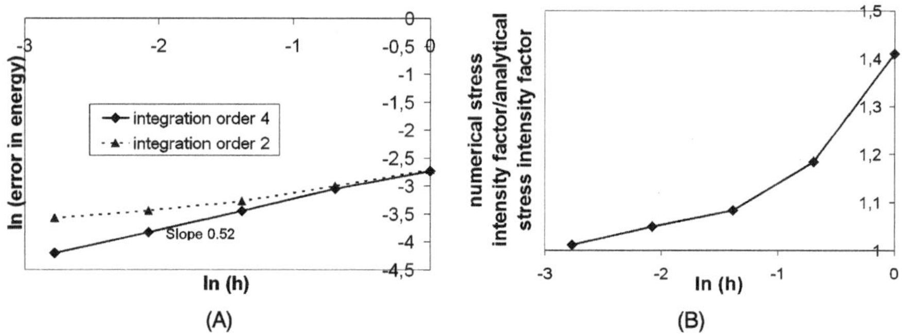

Figure 5.58 Convergence for mode I crack problem for the cracking particle method (A) convergence in energy norm; (B) convergence in K_I (integration order is order of Gauss quadrature in cells).

tire domain. The extrinsic PU enrichment gives a convergence rate of 0.86 and is only slightly less accurate than the extrinsic MLS enrichment. The intrinsic PU enrichment lies in between these two results. The same observation can be made for local convergence. The fact that the SIFs can be directly obtained is a major advantage of the extrinsic MLS enrichment and probably leads to more accurate results with respect to local convergence. Nevertheless, also the PU enrichment based method give excellent local convergence, see the scale of the y-axis in Fig. 5.56B and 5.57B.

5.9.1.3 Cracking particle method

The results obtained with the *cracking particle* method are shown in Fig. 5.58. We have shown the results of this method using Gauss quadrature in order

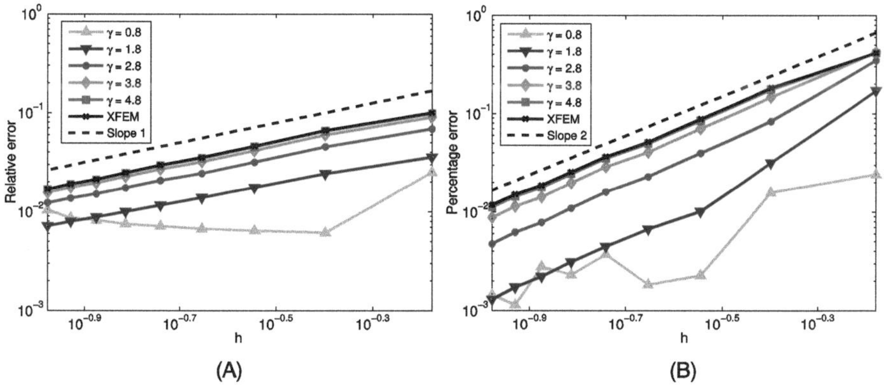

Figure 5.59 Convergence for mode I crack problem for the XLME method (A) convergence in energy norm; (B) convergence in K_I.

to compare them with the other methods. The results for an integration order of 2 is also shown in Fig. 5.58A. With decreasing integration order, the results become worse. The results for a stress point integration are not shown here. They are slightly worse and depend also on the stress points arrangement. The results are similar to the visibility method. The convergence rate is slightly worse and the results are a little less accurate, compare with Fig. 5.56.

5.9.1.4 XLME method

For computations with enriched LME shape functions, Dirichlet boundary conditions are used on the bottom, right and top edges and Neumann boundary conditions on the left edge which includes the crack. As we mentioned in Section 5.7, LME shape functions satisfy a weak Kronecker delta property. This property allows us to impose Dirichlet boundary conditions by computing a node-based interpolant or an L^2 projection of the boundary data. The latter can also be used for edges that contain enriched nodes. Numerical integration is performed on a background mesh of rectangular elements and the almost polar integration is used on the elements containing a crack tip.

Approximation errors in the energy norm are illustrated in Fig. 5.59A for different values of γ. Fig. 5.59B shows the percentage error for stress intensity factor (SIFs). It is obvious from these figures that in this case there is an optimal value for the parameter γ of around 1.8 for which accuracy is maximized. For very low values of γ, convergence is degraded. This is

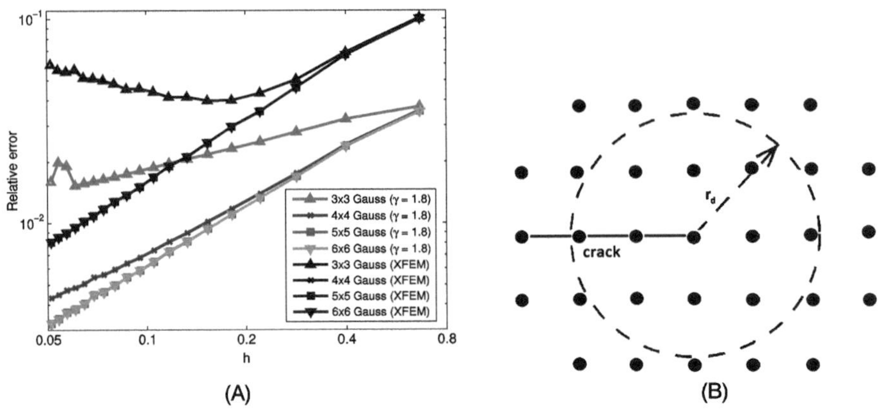

Figure 5.60 (A) Error in the energy norm for XLME and XFEM and different quadrature rules. (B) Region which contains a node within a ball of radius r_d around the crack tip.

due to numerical integration. With a higher number of Gauss points and $\gamma = 0.8$, the optimal rate of convergence for a plane elasticity problem was recovered in [3]. But in that case, the method is very expensive. The LME results converge to the standard XFEM results as γ increases.

As shown in Fig. 5.59A, the rate of convergence for different values of γ, the parameter that controls the support of the shape functions, is 1 for the energy norm. This agrees with the *a priori* error estimates for XFEM established in the literature (see [85]). For a fixed number of nodes, when γ decreases the error also decreases. However, as γ decreases, because the support of the LME shape functions becomes larger, we also need to consider a larger radius of influence (the distance of the neighbor search between the nodes), which leads to more function evaluations and increases the computational cost. In [1], it was found that choosing $\gamma = 1.8$, which corresponds to a radius of influence of 3 nodes, provides a reasonable balance between accuracy and computational cost.

For the problems studied, even in the cases of standard XFEM, the integration is not exact. This is because the Branch enrichment functions are non-polynomial in nature. To study the effect of approximate integration on the accuracy and stability of the solution, we have considered Gauss quadratures with a varying number of evaluation points. The relative errors in energy norm obtained for XFEM and for XLME with $\gamma = 1.8$ are shown in Fig. 5.60A. The figure shows the convergence study for a larger number of nodes (up to $n = 196 \times 196$). We observe that for both XFEM and XLME, a 3×3 Gauss quadrature is not sufficient for a stable solution and the results diverge in the case of XFEM, or become unstable in the case of XLME. However, with a Gauss quadrature of 4×4 or more points, the error in XFEM remains constant and optimal convergence is achieved (the lines corresponding to 4×4, 5×5 and 6×6 Gauss points overlap and have slope $m = 1.00$).

For XLME with $\gamma = 1.8$ and with a 4×4 Gauss quadrature, the convergence rate becomes sub-optimal as the number of degrees of freedom increases (the slope is $m = 0.83$). However, the lines corresponding to 5×5 and 6×6 Gauss points overlap almost completely, with only a slight difference that appears when the number of degrees of freedom is very large (greater than 100,000). The slope of the convergence line that best fits the data points is $m = 0.95$ in both cases. This indicates that the error due to numerical integration when 5×5 or more Gauss points are used is very small. It is possible that as the number of degrees of freedom increases, an even larger number of Gauss points per integration element will be needed, in line with the results obtained by [6]. In such cases, an adaptive numerical quadrature method may be needed. However, the LME shape functions are very smooth (C^∞), so in general the integration should be less problematic in comparison to other meshless methods.

We compute the stress intensity factors by the interaction integral method, where the domain form of the interaction integral is given by [80]

$$I^{(1,2)} = \int_A \left[\sigma_{ij}^{(1)} \frac{\partial u_i^2}{\partial x_1} - \sigma_{ij}^{(2)} \frac{\partial u_i^1}{\partial x_1} - W^{(1,2)} \delta_{1j} \right] \frac{\partial q}{\partial x_j} dA$$

The domain of integration, A, is set to be the union of all the elements which have a node within a ball of radius r_d around the crack tip (see Fig. 5.60B). Since we use a fixed area enrichment, r_d is also a fixed distance. We found that most accurate results are obtained when r_d is half of the modeled crack length. This results in a superconvergent ($O(h^2)$) rate for K_I, as also reported for XFEM in [69] and [7].

The weight function q is taken to have a value of unity for all nodes within the ball r_d, and zero on the outside of the ball. Hence, the bilinear shape functions are used as the weight functions. $W^{(1,2)}$ is the interaction strain energy density

$$W^{(1,2)} = \sigma_{ij}^{(1)} \epsilon_{ij}^{(2)} = \sigma_{ij}^{(2)} \epsilon_{ij}^{(1)}$$

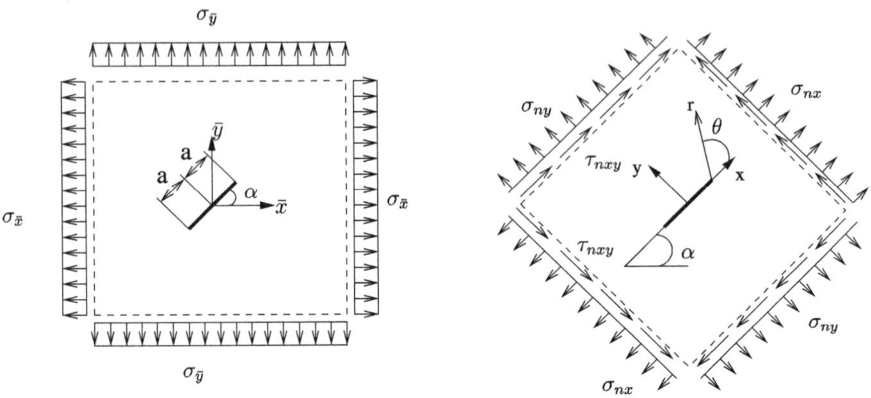

Figure 5.61 Mixed mode crack problem.

$\sigma_{ij}^{(1)}$ and $\epsilon_{ij}^{(1)}$ are computed stresses and strains and $\sigma_{ij}^{(2)}$ and $\epsilon_{ij}^{(2)}$ are auxiliary stresses and strains derived by Westergaard and Williams, corresponding to mode 1 and mode 2 as described in [80].

5.9.2 The mixed mode problem

Let us consider a (mixed) mode I-II problem. The analytical near-tip field solution for this problem is given by:

$$\sigma_r = \frac{1}{4\sqrt{2\pi r}}\left[K_I\left(5\cos\frac{\theta}{2} - \cos\frac{3\theta}{2}\right) + K_{II}\left(-5\sin\frac{\theta}{2} + 3\sin\frac{3\theta}{2}\right)\right] \tag{5.241}$$

$$\sigma_\theta = \frac{1}{4\sqrt{2\pi r}}\left[K_I\left(3\cos\frac{\theta}{2} - \cos\frac{3\theta}{2}\right) + K_{II}\left(-5\sin\frac{\theta}{2} + 3\sin\frac{3\theta}{2}\right)\right] \tag{5.242}$$

$$\tau_{r\theta} = \frac{1}{4\sqrt{2\pi r}}\left[K_I\left(\sin\frac{\theta}{2} + \sin\frac{3\theta}{2}\right) + K_{II}\left(\cos\frac{\theta}{2} + \cos\frac{3\theta}{2}\right)\right] \tag{5.243}$$

where r and θ are explained in Fig. 5.61 and with $K_I = \sigma_n\sqrt{\pi\ a}$ and $K_{II} = \tau_n\sqrt{\pi\ a}$, with the loading conditions σ_n and τ_n from Fig. 5.61 and where

$$\sigma_{ny} = \frac{\sigma_{\bar{x}} + \sigma_{\bar{y}}}{2} + \frac{\sigma_{\bar{x}} - \sigma_{\bar{y}}}{2}\cos 2\alpha \tag{5.244}$$

$$\sigma_{nx} = \frac{\sigma_{\bar{x}} + \sigma_{\bar{y}}}{2} - \frac{\sigma_{\bar{x}} - \sigma_{\bar{y}}}{2}\cos 2\alpha \tag{5.245}$$

$$\tau_{nxy} = \frac{\sigma_{\bar{x}} - \sigma_{\bar{y}}}{2}\sin 2\alpha \tag{5.246}$$

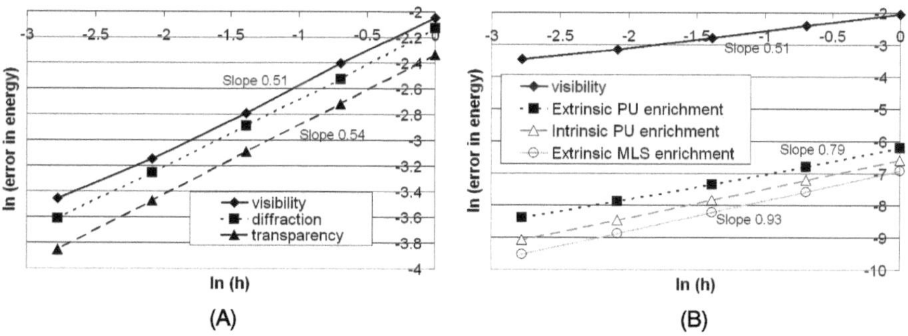

Figure 5.62 Error in the energy for the *mixed mode* problem using (A) the visibility, diffraction and transparency criterion; (B) the extrinsic PU, extrinsic MLS and intrinsic PU enrichment.

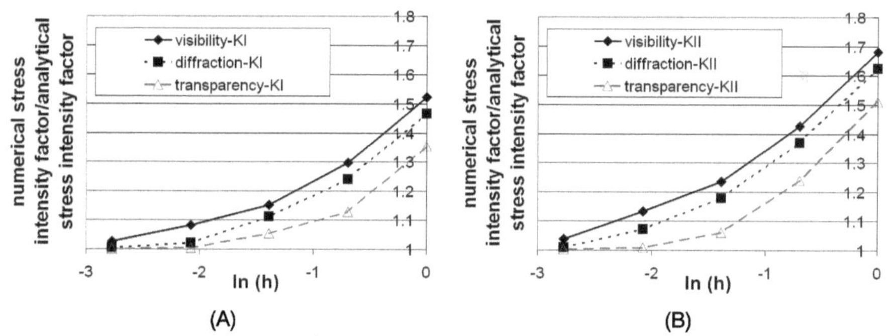

Figure 5.63 Normalized stress intensity factor versus *h* for the *mixed mode* problem using the visibility, diffraction and transparency method, (A) SIF K_I, (B) SIF K_{II}.

5.9.2.1 EFG with visibility, diffraction and transparency methods

We study again the error in the energy norm. The results for the visibility, diffraction and transparency method is shown in Fig. 5.62A and 5.63. The convergence rate of the three methods is similar to the convergence rate in the mode *I* crack problem. For the mixed mode problem, the transparency methods performs a little better than the diffraction method. The visibility method shows the worst results. This tendency is observed for global as well as local convergence.

5.9.2.2 PU methods with intrinsic and extrinsic enrichments

The results for the extrinsic PU, intrinsic PU and extrinsic MLS methods are shown in Fig. 5.62B and 5.64. We have again included the results for the visibility method in Fig. 5.62B. It can be seen that the enrichment im-

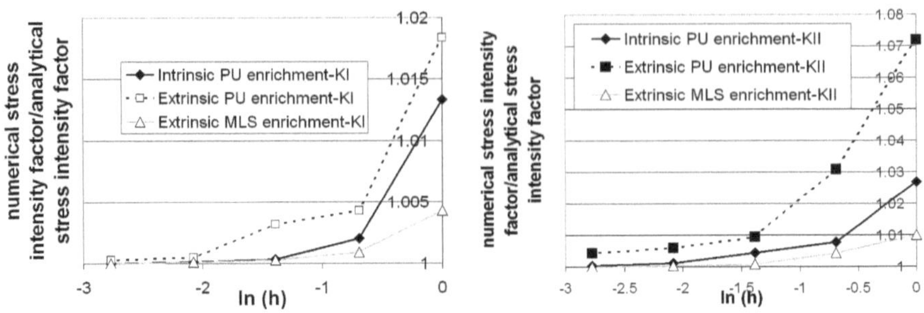

(A) (B)

Figure 5.64 Normalized stress intensity factor versus h for the *mixed mode* problem using the intrinsic PU, extrinsic PU and extrinsic MLS method, (A) SIF K_I, (B) SIF K_{II}.

proves the results significantly. The same observations made for the mode I problem apply here. The highest convergence rate of 0.93 and the most accurate results are obtained with the extrinsic MLS enrichment. The extrinsic PU enrichment gives with a convergence rate with 0.86 the worst result of the enriched methods. The same applies for local convergence. Note that no matter what type of enrichment is used, the results are all excellent and only minor differences occur.

5.9.2.3 XLME method

The same tendency as for the previous example is observed for this mixed mode problem. As before, LME shape functions with $\gamma = 1.8$ give the most accurate results and this method has a convergence rate of approximately $O(h^2)$.

We note that when γ decreases to the optimal value, in this case $\gamma = 1.8$, the error decreases, however the computational cost increases due to a larger radius of influence of the shape functions. Nevertheless, we note that the error is much smaller (almost an order of magnitude) compared to $\gamma = 4.8$, which is virtually the same as standard XFEM. It can be observed that there is only a very small difference between the $\alpha = 15°$ and $\alpha = 30°$. This can be explained by the fact that the discretization is identical, the only difference being the size of the forces applied to the boundaries, as can be seen from Fig. 5.61. The log-log plots indicating the convergence rates of K_I and K_{II} with $\alpha = 30°$ are shown in Fig. 5.65. The errors for K_I and K_{II} for angles $\alpha = 45°, 60°, 75°$ can be computed with similar results.

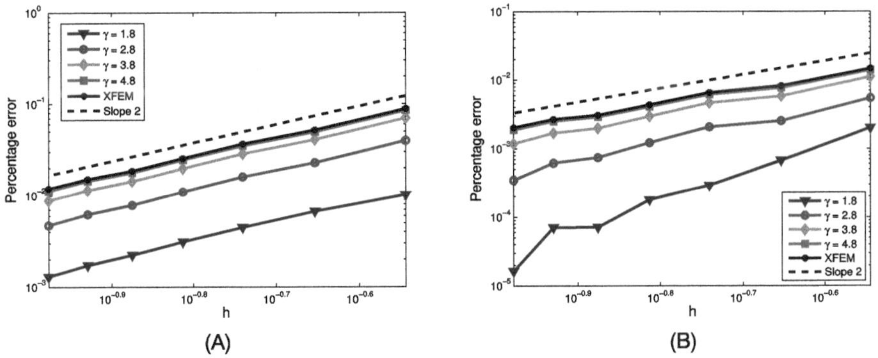

Figure 5.65 Normalized stress intensity factor versus *h* for the *mixed mode* problem using XLME method, (A) SIF K_I, (B) SIF K_{II}.

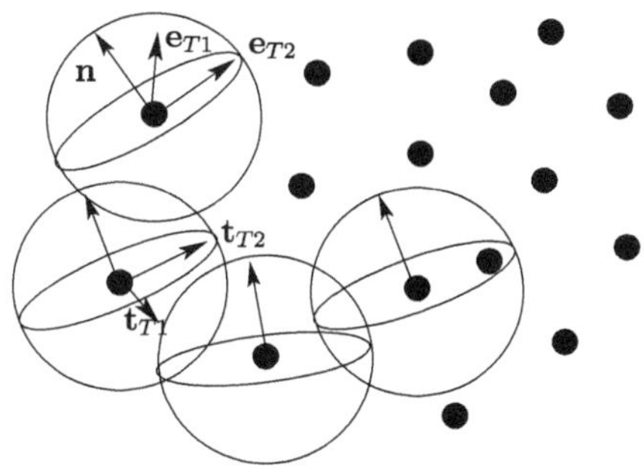

Figure 5.66 Nomenclature.

5.10. Extensions to mode II kinematics

In this section, we will modify and extend our cracking approach to simulate pure mode II failure. Mode II failure occurs in shear bands in metal and slip lines in geological structures. While the principal displacement kinematics is the same in both cases, shear bands in metallic structures are physically different from slip lines in geological materials. We will describe this extension only for the cracking particle method. The extension of the other methods is straightforward.

5.10.1 Enriching in the shear band plane

The shear band is modeled by a set of discrete shear bands as shown in Fig. 5.66, i.e. by *sheared* particles. The velocity field is additively partitioned into a continuous part and a discontinuous part by

$$\dot{\mathbf{u}}(\mathbf{X},t) = \dot{\mathbf{u}}^{cont}(\mathbf{X},t) + \dot{\mathbf{u}}^{disc}(\mathbf{X},t) \tag{5.247}$$

where $\mathbf{u} \in \Re^{n_{SD}}$ is the displacement, $\mathbf{X} \in \Re^{n_{SD}}$ are the material coordinates, t is the time and n_{SD} the number of space dimensions and the superimposed dots denote time derivatives.

Let the set of *sheared* particles be denoted by $\mathcal{N}_s$ and the total set of particles by $\mathcal{N}$. The *sheared* particles are determined by a stability criterion described later. For each sheared particle we assume that the normal to the shearing plane $\mathbf{n}$ is provided by a material stability analysis. This enables us to construct a set of tangent vectors $\mathbf{e}_{\alpha T}$, $\alpha = 1$ to $n_{SD} - 1$, such that $\mathbf{e}_{\alpha T} \cdot \mathbf{n} = 0$. The continuous and discontinuous displacement fields are then given by

$$\begin{aligned} \dot{\mathbf{u}}^{cont}(\mathbf{X},t) &= \sum_{I \in \mathcal{N}} \Phi_I(\mathbf{X})\ \dot{\mathbf{u}}_I(t) \\ \dot{\mathbf{u}}^{disc}(\mathbf{X},t) &= \sum_{I \in \mathcal{N}_s} \sum_{\alpha=1}^{n_{SD}-1} \Phi_I(\mathbf{X})\ S(f_I(\mathbf{x}))\ \dot{q}_{\alpha I}(t)\ \mathbf{e}_{\alpha T} \end{aligned} \tag{5.248}$$

where $\Phi_I(\mathbf{X})$ are the meshfree shape functions, $\mathbf{u}_I(t)$ are the nodal displacements, $S(f_I(\mathbf{x}))$ is the sign function defined by

$$S(\xi) = \begin{cases} 1 & \text{if } \xi > 0 \\ -1 & \text{if } \xi < 0 \end{cases} \tag{5.249}$$

and $f_I(\mathbf{x})$ is given by

$$f_I(\mathbf{x}) = \mathbf{n} \cdot (\mathbf{x} - \mathbf{x}_I) \tag{5.250}$$

where $\mathbf{x} = \mathbf{X} + \mathbf{u}$ and $\mathbf{x}_I$ is the current position of the node I. The nomenclature is illustrated in Fig. 5.66.

The role of the sheared particles is to mimic a displacement field with a jump in the tangential displacements. To represent a linear tangential jump, there must be a sufficient number of *sheared* particles so that the linear completeness of the shape function is inherited by the jump field. A linear basis was chosen for the shape functions employed for the discontinuous field.

Consider a solid Ω with material points $\mathbf{X}$, see Fig. 5.67. Its boundary Γ is partitioned into two subsets, Γ^t and Γ^u, upon which tractions and displacements are applied, the corresponding entities in the reference

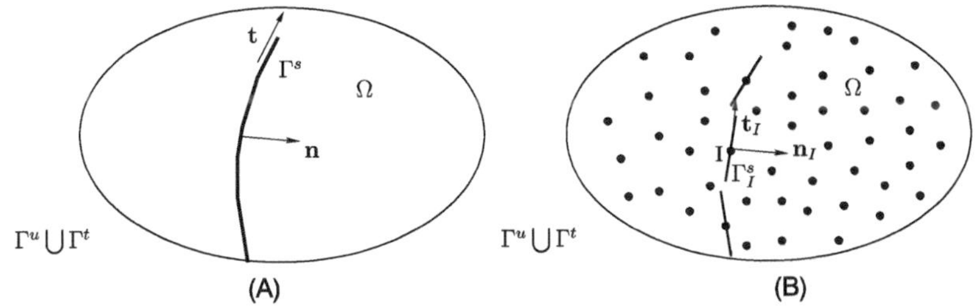

Figure 5.67 Body with a shear band Γ^S and its representation of Γ^S by the cracked particle method.

configuration are denoted by Ω_0, Γ_0^t and Γ_0^u. The internal surface of discontinuity Γ^s is approximated by local discontinuities Γ_I^s; in the following $\Gamma^s = \bigcup_I \Gamma_I^s$.

5.10.2 Enforcing mode II-kinematics with the penalty method

Instead of introducing a "local" enrichment, i.e. in 2D one enrichment variable is introduced and in 3D two enrichment variables are introduced, the enrichment in Section 5.8.1 is kept. To be able to capture a pure mode II crack, a penalty term is then used to prevent the crack from opening. This penalty term is introduced into the linearized momentum equation.

5.11. Discrete system of equations for pure mechanical problems

5.11.1 Methods without enrichment

The discrete system of equations for all methods without enrichment[9] are presented for linear problems only. The variational (or weak form) of the equilibrium equation is given by: find $\mathbf{u} \in \mathcal{V}$ such that

$$\delta W = \delta W_{int} - \delta W_{ext} - \delta W_u = 0 \quad \forall \delta \mathbf{u} \tag{5.251}$$

where

$$\delta W_{int} = \int_{\Omega_0 \backslash \Gamma_0^c} (\nabla \otimes \delta \mathbf{u})^T : \mathbf{P} \, d\Omega_0 \tag{5.252}$$

$$\delta W_{ext} = \int_{\Omega_0 \backslash \Gamma_0^c} \varrho_0 \, \delta \mathbf{u} \cdot \mathbf{b} \, d\Omega_0 + \int_{\Gamma_0^t} \delta \mathbf{u} \cdot \bar{\mathbf{t}}_0 \, d\Gamma_0 + \int_{\Gamma_0^c} \delta [\![\mathbf{u}]\!] \cdot \mathbf{t}_{c0} \, d\Gamma_0 \tag{5.253}$$

[9] Visibility method, transparency method, diffraction method.

where

$$\mathcal{V} = \left\{\mathbf{u}(\cdot,t)|\mathbf{u}(\cdot,t) \in \mathcal{H}^1,\ \mathbf{u}(\cdot,t) \text{ discontinuous on } \Gamma_0^c,\ \mathbf{u}(\cdot,t) = \bar{\mathbf{u}}(t) \text{ on } \Gamma_0^u\right\}$$
$$\mathcal{V}_0 = \left\{\delta\mathbf{u}|\delta\mathbf{u} \in \mathcal{H}^1,\ \delta\mathbf{u} \text{ discontinuous on } \Gamma_0^c,\ \delta\mathbf{u} = 0 \text{ on } \Gamma_0^u\right\} \tag{5.254}$$

The term δW_u in Eq. (5.251) is necessary for the imposition of Dirichlet boundary conditions. The discrete form of (5.251) is given by substituting the test and trial functions into Eq. (5.251). For elastostatics it leads to the following system of equations:

$$\mathbf{K}\,\mathbf{u} = \mathbf{f}^{ext} \tag{5.255}$$

where $\mathbf{K}$ is the stiffness matrix

$$\mathbf{K}_{IJ} = \int_{\Omega_0} \mathbf{B}_I\ \mathbf{C}^t\mathbf{B}_J\ d\Omega_0 \tag{5.256}$$

with the following $\mathbf{B}$-matrix in two dimensions

$$\mathbf{B}_I = \begin{bmatrix} \Phi_{I,X} & 0 \\ 0 & \Phi_{I,Y} \\ \Phi_{I,Y} & \Phi_{I,X} \end{bmatrix} \tag{5.257}$$

and in 3D

$$\mathbf{B}_I = \begin{bmatrix} \Phi_{I,X} & 0 & 0 \\ 0 & \Phi_{I,Y} & 0 \\ 0 & 0 & \Phi_{I,Z} \\ \Phi_{I,Y} & \Phi_{I,X} & 0 \\ 0 & \Phi_{I,Z} & \Phi_{I,Y} \\ \Phi_{I,Z} & 0 & \Phi_{I,X} \end{bmatrix} \tag{5.258}$$

The external and body forces are given by

$$\mathbf{f}_I^{ext} = \int_{\Gamma_0^t} \Phi_I(\mathbf{X})\ \mathbf{t}_0\ d\Gamma_0 + \int_{\Omega_0} \Phi_I(\mathbf{X})\ \mathbf{b}\ d\Omega_0 \tag{5.259}$$

The Dirichlet boundary conditions can be imposed by methods briefly explained in Section 5.2.3. With the Lagrange multiplier method, δW_u reads:

$$\delta W_u = \int_{\Gamma_u^0} \delta\boldsymbol{\lambda} \cdot (\mathbf{u} - \bar{\mathbf{u}})\ d\Gamma_0 + \int_{\Gamma_0} \delta\mathbf{u}\boldsymbol{\lambda}\ d\Gamma_0 \tag{5.260}$$

where $\boldsymbol{\lambda}$ are the Lagrange multipliers or with the penalty method:

$$\delta W_u = 0.5p \int_{\Gamma_u^0} \|\mathbf{u} - \bar{\mathbf{u}}\|^2 d\Gamma_0 \tag{5.261}$$

where p is the penalty parameter that has to be specified by the user. Note, that these equations are only valid for elastostatics. The equations for non-linear materials will be derived in the next section.

5.11.2 Enriched methods

5.11.2.1 The intrinsic PU-method

The discrete equations for the intrinsic MLS-method for LEFM are the same as the equations given in Section 5.11.1.

5.11.2.2 The extrinsic MLS-method

The final system of discrete equations has the same structure as before, i.e.

$$\bar{\mathbf{K}}\,\bar{\mathbf{d}} = \bar{\mathbf{f}}^{ext} \tag{5.262}$$

with

$$\bar{\mathbf{K}} = \begin{bmatrix} \mathbf{K} & K_{I1} & K_{I2} \\ K_{I1}^T & K_{11} & K_{12} \\ K_{I2}^T & K_{12}^T & K_{22} \end{bmatrix} \tag{5.263}$$

where $\mathbf{K}$ is the "usual" stiffness matrix

$$\mathbf{K}_{IJ} = \int_{\Omega_0} \mathbf{B}_I\ \mathbf{C}^t \mathbf{B}_J\ d\Omega_0 \tag{5.264}$$

and

$$K_{Ii} = \int_{\Omega_0} \mathbf{B}_I^T\ \mathbf{C}^t \hat{\mathbf{B}}_i\ d\Omega_0 \quad i = 1, 2 \tag{5.265}$$

where the subscript I indicates the node I and i is associated to the fracture mode and

$$K_{ij} = \int_{\Omega_0} \hat{\mathbf{B}}_i^T\ \mathbf{C}^t \hat{\mathbf{B}}_j\ d\Omega_0 \quad i, j = 1, 2 \tag{5.266}$$

with

$$\hat{\mathbf{B}}_1 = \begin{bmatrix} Q_{11,X}(\mathbf{X}) \\ Q_{12,Y}(\mathbf{X}) \\ Q_{11,Y}(\mathbf{X}) + Q_{12,X}(\mathbf{X}) \end{bmatrix} \tag{5.267}$$

$$\hat{\mathbf{B}}_2 = \begin{bmatrix} Q_{21,X}(\mathbf{X}) \\ Q_{12,Y}(\mathbf{X}) \\ Q_{21,Y}(\mathbf{X}) + Q_{22,X}(\mathbf{X}) \end{bmatrix} \tag{5.268}$$

The external and body force vector $\bar{\mathbf{f}}^{ext,T} = \left[\mathbf{f}^{ext}, f_1^1, f_2^1, \ldots, f_1^j, f_2^j\right]$ contains additional terms

$$\begin{aligned} f_1 &= \int_{\Gamma_0^t} \hat{\Phi}_1 \, \bar{\mathbf{t}} \, d\Gamma_0 + \int_{\Omega_0} \hat{\Phi}_1 \, \mathbf{b} \, d\Omega_0 \\ f_2 &= \int_{\Gamma_0^t} \hat{\Phi}_2 \, \bar{\mathbf{t}} \, d\Gamma_0 + \int_{\Omega_0} \hat{\Phi}_2 \, \mathbf{b} \, d\Omega_0 \end{aligned} \tag{5.269}$$

with

$$\begin{aligned} \hat{\Phi}_1^T &= \left[Q_{11}(\mathbf{X}), \, Q_{12}(\mathbf{X})\right] \\ \hat{\Phi}_2^T &= \left[Q_{21}(\mathbf{X}), \, Q_{22}(\mathbf{X})\right] \end{aligned} \tag{5.270}$$

Accordingly, we have the global displacement vector

$$\bar{\mathbf{d}} = \left[\tilde{\mathbf{u}}(\mathbf{X}), \, k_1^1, \, k_2^1, \ldots k_1^j, \, k_2^j\right] \tag{5.271}$$

5.11.2.3 The extrinsic PU-method

In this section, we will derive the discrete equations for material non-linearities and cohesive cracks. Let us consider Eq. (5.251). Minimizing Eq. (5.251) with respect to $\mathbf{u}$ leads to

$$\frac{\partial \delta W}{\partial \mathbf{u}} = \frac{\partial \delta W_{int}}{\partial \mathbf{u}} - \frac{\partial \delta W_{ext}}{\partial \mathbf{u}} = \mathbf{f}_{int} - \mathbf{f}_{ext} = 0 \tag{5.272}$$

The derivatives of δW_{int} and δW_{ext} with respect to $\mathbf{u}$ are the internal and external forces, respectively:

$$\mathbf{f}_{int} = \int_{\Omega_0 \backslash \Gamma_0^c} (\nabla_0 \otimes \delta \mathbf{u})^T : \mathbf{P} \, d\Omega_0 \tag{5.273}$$

$$\mathbf{f}_{ext} = \int_{\Omega_0 \backslash \Gamma_0^c} \delta \mathbf{u} \cdot \mathbf{b} \, d\Omega_0 + \int_{\Gamma_0^t} \delta \mathbf{u} \cdot \bar{\mathbf{t}}_0 \, d\Gamma_0 + \int_{\Gamma_0^c} \mathbf{t}_0^c \cdot \delta [\![\mathbf{u}]\!] \, d\Gamma_0 \tag{5.274}$$

To obtain the discrete system of nonlinear equations we will do a linearization as described e.g. in [16]. Therefore, we take a Taylor series expansion of Eq. (5.272) neglecting any higher order terms:

$$\mathbf{f}_{int} - \mathbf{f}_{ext} + \frac{\partial \mathbf{f}_{int}}{\partial \mathbf{u}} \Delta \mathbf{u} - \frac{\partial \mathbf{f}_{ext}}{\partial \mathbf{u}} \Delta \mathbf{u} = 0 \tag{5.275}$$

With the test and trial functions from Section 5.6.1, Eqs. (5.189) and (5.190), respectively, we obtain the final linearized system of equations:

$$\sum_{I} \bar{\mathbf{K}}_{IJ} \, \Delta \mathbf{d}_J = \mathbf{F} \tag{5.276}$$

or in matrix form

$$\begin{bmatrix} \mathbf{K}_{IJ}^{uu} & \mathbf{K}_{IJ}^{ua} & \mathbf{K}_{IJ}^{ubK} \\ \mathbf{K}_{IJ}^{au} & \mathbf{K}_{IJ}^{aa} & \mathbf{K}_{IJ}^{abK} \\ \mathbf{K}_{IJ}^{buK} & \mathbf{K}_{IJ}^{baK} & \mathbf{K}_{IJ}^{bbK} \end{bmatrix} \cdot \begin{bmatrix} \Delta \mathbf{u}_J \\ \Delta \mathbf{a}_J \\ \Delta \mathbf{b}_J^K \end{bmatrix} = \begin{bmatrix} \mathbf{f}_{ext}^u - \mathbf{f}_{int}^u \\ \mathbf{f}_{ext}^a - \mathbf{f}_{int}^a \\ \mathbf{f}_{ext}^b - \mathbf{f}_{int}^b \end{bmatrix} \tag{5.277}$$

with

$$\begin{aligned}
\mathbf{K}_{IJ}^{uu} &= \int_{\Omega_0 \backslash \Gamma_0^c} \left(\mathbf{B}_I^u\right)^T \mathbf{C}\, \mathbf{B}_J^u \, d\Omega_0 \\
\mathbf{K}_{IJ}^{ua} &= \int_{\Omega_0 \backslash \Gamma_0^c} \left(\mathbf{B}_I^u\right)^T \mathbf{C}\, \mathbf{B}_J^a \, S(f_I(\mathbf{X})) \, d\Omega_0 \\
\mathbf{K}_{IJ}^{ubK} &= \int_{\Omega_0 \backslash \Gamma_0^c} \left(\mathbf{B}_I^u\right)^T \mathbf{C}\, \mathbf{B}_J^{bK} \, d\Omega_0 \\
\mathbf{K}_{IJ}^{aa} &= \int_{\Omega_0 \backslash \Gamma_0^c} \left(S(f_I(\mathbf{X}))\, \mathbf{B}_I^a\right)^T \mathbf{C}\, \mathbf{B}_J^a \, S(f_J(\mathbf{X})) \, d\Omega_0 \\
\mathbf{K}_{IJ}^{abK} &= \int_{\Omega_0 \backslash \Gamma_0^c} \left(\mathbf{B}_I^a\right)^T \mathbf{C}\, \mathbf{B}_J^{bK} \, d\Omega_0 \\
\mathbf{K}_{IJ}^{bbK} &= \int_{\Omega_0 \backslash \Gamma_0^c} \left(\mathbf{B}_I^{bK}\right)^T \mathbf{C}\, \mathbf{B}_J^{bK} \, d\Omega_0 \\
\mathbf{K}^{au} &= \mathbf{K}^{ua} \\
\mathbf{K}^{ab} &= \mathbf{K}^{ba} \\
\mathbf{K}^{bu} &= \mathbf{K}^{ub}
\end{aligned} \tag{5.278}$$

and

$$\mathbf{B}_I^u = \begin{bmatrix} \Phi_{I,X} & 0 \\ 0 & \Phi_{I,Y} \\ \Phi_{I,Y} & \Phi_{I,X} \end{bmatrix} \tag{5.279}$$

$$\mathbf{B}_I^a = \begin{bmatrix} \Phi_{I,X} S(f_I(\mathbf{X})) & 0 \\ 0 & \Phi_{I,Y} S(f_I(\mathbf{X})) \\ \Phi_{I,Y} S(f_I(\mathbf{X})) & \Phi_{I,X} S(f_I(\mathbf{X})) \end{bmatrix} \tag{5.280}$$

$$\mathbf{B}_I^{bK} = \begin{bmatrix} \Phi_{I,X}\mathbf{B}_I^K + \Phi_I \mathbf{B}_{I,X}^K & 0 \\ 0 & \Phi_{I,Y}\mathbf{B}_I^K + \Phi_I \mathbf{B}_{I,Y}^K \\ \Phi_{I,Y}\mathbf{B}_I^K + \Phi_I \mathbf{B}_{I,Y}^K & \Phi_{I,X}\mathbf{B}_I^K + \Phi_I \mathbf{B}_{I,X}^K \end{bmatrix} \tag{5.281}$$

The three-dimensional **B**-*operators* are obtained in the same manner, see also Eq. (5.258). Essential boundary conditions can be imposed with one of the methods described in Section 5.2.3, see also Section 5.11.1.

5.11.3 Extension to dynamics

5.11.3.1 The PU-method with crack tip enrichment

The weak form of the momentum equation is given by

$$\delta W = \delta W_{\text{int}} + \delta W_{\text{kin}} - \delta W_{\text{ext}} - \delta W_{\text{coh}} \tag{5.282}$$

in which δW_{int}, δW_{kin}, δW_{ext}, δW_{coh} are the virtual works by the internal stress, the inertia force, the external traction and the cohesive traction, respectively. They are given by

$$\delta W_{\text{int}} = \int_{\Omega_0 \backslash \Gamma_0^c} (\nabla_0 \otimes \delta \mathbf{u})^T : \mathbf{P} \, \mathrm{d}\Omega_0 \tag{5.283}$$

$$\delta W_{\text{kin}} = \int_{\Omega_0 \backslash \Gamma_0^c} \varrho_0 \, \delta \mathbf{u} \cdot \ddot{\mathbf{u}} \, \mathrm{d}\Omega_0 \tag{5.284}$$

$$\delta W_{\text{ext}} = \int_{\Omega_0 \backslash \Gamma_0^c} \varrho_0 \, \delta \mathbf{u} \cdot \mathbf{b} \, \mathrm{d}\Omega_0 + \int_{\Gamma_0^t} \delta \mathbf{u} \cdot \bar{\mathbf{t}}_0 \, \mathrm{d}\Gamma_0 \tag{5.285}$$

$$\delta W_{\text{coh}} = \int_{\Gamma_0^c} [\![\delta \mathbf{u}]\!] \cdot \boldsymbol{\tau} \, \mathrm{d}\Gamma_0 \tag{5.286}$$

Substituting the test and trial functions from Section 5.6.3 (Eqs. (5.213) and (5.214), respectively) into Eqs. (5.282) to (5.286), we obtain

$$\delta W_{kin} = \sum_I \delta \mathbf{u}_I \sum_J \int_{\Omega \backslash \Gamma^c} \varrho \; \Phi_I(\mathbf{X}) \cdot \Phi_J(\mathbf{X}) \; d\Omega \; \ddot{\mathbf{u}}_J$$

$$
\begin{aligned}
&+\sum_I \delta \mathbf{u}_I \sum_J \sum_n \int_{\Omega\backslash\Gamma^c} \varrho\ \Phi_I(\mathbf{X}) \cdot \Phi_J(\mathbf{X})\ \bar{H}(f_J^{(n)}(\mathbf{X}))\ d\Omega\ \ddot{\mathbf{a}}_J^{(n)} \\
&+\sum_I \delta \mathbf{u}_I \sum_J \sum_m \int_{\Omega\backslash\Gamma^c} \varrho\ \Phi_I(\mathbf{X}) \cdot \Phi_J(\mathbf{X})\ \bar{\mathbf{B}}_K^{(m)}\ d\Omega\ \ddot{\mathbf{b}}_{KJ}^{(m)} \\
&+\sum_I \sum_n \delta \mathbf{a}_I^{(n)} \sum_J \int_{\Omega\backslash\Gamma^c} \varrho\ \left[\Phi_I(\mathbf{X})\ \bar{H}(f_I^{(n)}(\mathbf{X}))\right] \cdot \Phi_J(\mathbf{X})\ d\Omega\ \ddot{\mathbf{u}}_J \\
&+\sum_I \sum_n \delta \mathbf{a}_I^{(n)} \sum_J \sum_N \int_{\Omega\backslash\Gamma^c} \varrho\ \left[\Phi_I(\mathbf{X})\ \bar{H}(f_I^{(n)}(\mathbf{X}))\right] \\
&\qquad \cdot \left[\Phi_J(\mathbf{X})\ \bar{H}(f_J^{(N)}(\mathbf{X}))\right]\ d\Omega\ \ddot{\mathbf{a}}_J^{(N)} \\
&+\sum_I \sum_n \delta \mathbf{a}_I^{(n)} \sum_J \sum_m \int_{\Omega\backslash\Gamma^c} \varrho\ \left[\Phi_I(\mathbf{X})\ \bar{H}(f_I^{(n)}(\mathbf{X}))\right] \\
&\qquad \cdot \left[\Phi_J(\mathbf{X})\ \bar{\mathbf{B}}_K^{(m)}\right]\ d\Omega\ \ddot{\mathbf{b}}_{KJ}^{(m)} \\
&+\sum_I \sum_m \delta \mathbf{b}_{RI}^{(m)} \sum_J \int_{\Omega\backslash\Gamma^c} \varrho\ \left[\Phi_I(\mathbf{X})\ \bar{B}_R^{(m)}\right] \cdot \Phi_J(\mathbf{X})\ d\Omega\ \ddot{\mathbf{u}}_J \\
&+\sum_I \sum_m \delta \mathbf{b}_{RI}^{(m)} \sum_J \sum_n \int_{\Omega\backslash\Gamma^c} \varrho\ \left[\Phi_I(\mathbf{X})\ \bar{B}_R^{(m)}\right] \\
&\qquad \cdot \left[\Phi_J(\mathbf{X})\ \bar{H}(f_J^{(n)}(\mathbf{X}))\right]\ d\Omega\ \ddot{\mathbf{a}}_J^{(n)} \\
&+\sum_I \sum_m \delta \mathbf{b}_{RI}^{(m)} \sum_J \sum_M \int_{\Omega\backslash\Gamma^c} \varrho\ \left[\Phi_I(\mathbf{X})\ \bar{B}_R^{(m)}\right] \cdot \left[\Phi_J(\mathbf{X})\ \bar{B}_K^{(M)}\right]\ d\Omega\ \ddot{\mathbf{b}}_{KJ}^{(M)}
\end{aligned} \tag{5.287}
$$

$$
\begin{aligned}
\delta W_{int} = &\sum_I \delta \mathbf{u}_I \int_{\Omega\backslash\Gamma^c} \nabla\Phi_I(\mathbf{X}) \cdot \boldsymbol{\sigma}\ d\Omega \\
&+\sum_I \sum_n \delta \mathbf{a}_I^{(n)} \int_{\Omega\backslash\Gamma^c} \left[\nabla\Phi_I(\mathbf{X})\ \bar{H}(f_I^{(n)}(\mathbf{X})) + \nabla\bar{H}(f_I^{(n)}(\mathbf{X}))\ \Phi_I(\mathbf{X})\right] \cdot \boldsymbol{\sigma}\ d\Omega \\
&+\sum_I \sum_m \delta \mathbf{b}_{KI}^{(m)} \int_{\Omega\backslash\Gamma^c} \left[\nabla\Phi_I(\mathbf{X})\ \bar{B}_K^{(m)} + \nabla\bar{B}_K^{(m)}\ \Phi_I(\mathbf{X})\right] \cdot \boldsymbol{\sigma}\ d\Omega
\end{aligned} \tag{5.288}
$$

$$
\begin{aligned}
\delta W_{ext} = &\sum_I \delta \mathbf{u}_I \int_{\Omega\backslash\Gamma^c} \varrho\ \Phi_I(\mathbf{X}) \cdot \mathbf{b}\ d\Omega \\
&+\sum_I \delta \mathbf{u}_I \int_{\Gamma^t} \Phi_I(\mathbf{X}) \cdot \bar{\mathbf{t}}\ d\Gamma \\
&+\sum_I \sum_n \delta \mathbf{a}_I^{(n)} \int_{\Omega\backslash\Gamma^c} \varrho \left[\Phi_I(\mathbf{X})\ \bar{H}(f_I^{(n)}(\mathbf{X}))\right] \cdot \mathbf{b}\ d\Omega
\end{aligned}
$$

$$+\sum_I\sum_n \delta\mathbf{a}_I^{(n)} \int_{\Gamma^t} \left[\Phi_I(\mathbf{X})\ \bar{H}(f_I^{(n)}(\mathbf{X}))\right]\cdot\bar{\mathbf{t}}\ d\Gamma$$

$$+\sum_I\sum_m \delta\mathbf{b}_{KI}^{(m)} \int_{\Omega\backslash\Gamma^c} \varrho\left[\Phi_I(\mathbf{X})\ \bar{B}_K^{(m)}\right]\cdot\mathbf{b}\ d\Omega$$

$$+\sum_I\sum_m \delta\mathbf{b}_{KI}^{(m)} \int_{\Gamma^t} \tag{5.289}$$

$$\delta W_{coh} = \left[\Phi_I(\mathbf{X})\ \bar{B}_K^{(m)}\right]\cdot\bar{\mathbf{t}}\ d\Gamma$$

$$+\sum_I\sum_n \int_{\Gamma^c} [\![\delta\mathbf{a}_I^{(n)}\left[\Phi_I(\mathbf{X})\ \bar{H}(f_I^{(n)}(\mathbf{X}))\right]]\!]\cdot\mathbf{t}_c\ d\Gamma$$

$$+\sum_I\sum_m \int_{\Gamma^c} [\![\delta\mathbf{b}_{KI}^{(m)}\left[\Phi_I(\mathbf{X})\ \bar{B}_K^{(m)}\right]]\!]\cdot\mathbf{t}_c\ d\Gamma \tag{5.290}$$

where Eqs. (5.289) to (5.290) are for δW_{ext}. After some algebraic operations the final form of the equation of motion is obtained by

$$\mathbf{M}_{IJ}\cdot\ddot{\mathbf{u}}_I = \mathbf{F}_I^{ext} - \mathbf{F}_I^{int} \tag{5.291}$$

with

$$\mathbf{M}_{IJ} = \begin{bmatrix} \mathbf{m}_{IJ}^{uu} & \mathbf{m}_{IJ}^{ua} & \mathbf{m}_{IJ}^{ub} \\ \mathbf{m}_{IJ}^{au} & \mathbf{m}_{IJ}^{aa} & \mathbf{m}_{IJ}^{ab} \\ \mathbf{m}_{IJ}^{bu} & \mathbf{m}_{IJ}^{ba} & \mathbf{m}_{IJ}^{bb} \end{bmatrix} \tag{5.292}$$

$$\ddot{\mathbf{u}}_I = \begin{bmatrix} \ddot{\mathbf{u}}_I^u \\ \ddot{\mathbf{a}}_I \\ \ddot{\mathbf{b}}_{IK} \end{bmatrix} \tag{5.293}$$

$$\mathbf{F}_I^{ext} = \begin{bmatrix} \mathbf{f}_I^{u,ext} \\ \mathbf{f}_I^{a,ext} \\ \mathbf{f}_{IK}^{b,ext} \end{bmatrix} \tag{5.294}$$

$$\mathbf{F}_I^{int} = \begin{bmatrix} \mathbf{f}_I^{u,int} \\ \mathbf{f}_I^{a,int} \\ \mathbf{f}_{IK}^{b,int} \end{bmatrix} \tag{5.295}$$

with

$$\mathbf{m}_{IJ}^{uu} = \int_{\Omega_0\backslash\Gamma_0^c} \varrho_0\ \Phi_I(\mathbf{X})\ \Phi_J(\mathbf{X})\ d\Omega_0$$

$$\mathbf{m}_{IJ}^{ua} = \int_{\Omega_0 \setminus \Gamma_0^c} \varrho_0 \; \Phi_I(\mathbf{X}) \; \Phi_J(\mathbf{X}) \; S(f_I^{(n)}(\mathbf{X})) \; d\Omega_0 \, , \; \mathbf{m}_{IJ}^{ua} = \mathbf{m}_{IJ}^{au}$$
$$\mathbf{m}_{IJ}^{ub} = \int_{\Omega_0 \setminus \Gamma_0^c} \varrho_0 \; \Phi_I(\mathbf{X}) \; \Phi_J(\mathbf{X}) \; \mathbf{B}_K^{(m)} \; d\Omega_0 \, , \; \mathbf{m}_{IJ}^{ub} = \mathbf{m}_{IJ}^{bu}$$
$$\mathbf{m}_{IJ}^{aa} = \int_{\Omega_0 \setminus \Gamma_0^c} \varrho_0 \; \Phi_I(\mathbf{X}) \; S(f_I^{(n)}(\mathbf{X})) \; \Phi_J(\mathbf{X}) \; S(f_I^{(n)}(\mathbf{X})) \; d\Omega_0$$
$$\mathbf{m}_{IJ}^{ab} = \int_{\Omega_0 \setminus \Gamma_0^c} \varrho_0 \; \Phi_I(\mathbf{X}) \; S(f_I^{(n)}(\mathbf{X})) \; \Phi_J(\mathbf{X}) \; \mathbf{B}_K^{(m)} \; d\Omega_0$$
$$\mathbf{m}_{IJ}^{ab} = \mathbf{m}_{IJ}^{ba}$$
$$\mathbf{m}_{IJ}^{bb} = \int_{\Omega_0 \setminus \Gamma_0^c} \varrho_0 \; \Phi_I(\mathbf{X}) \; \mathbf{B}_K^{(m)} \; \Phi_J(\mathbf{X}) \; \mathbf{B}_K^{(m)} \; d\Omega_0 \tag{5.296}$$

$$\mathbf{f}_I^{u,ext} = \int_{\Omega_0 \setminus \Gamma_0^c} \varrho_0 \; \mathbf{b} \; \Phi_I(\mathbf{X}) \; d\Omega_0 + \int_{\Gamma_0^t} \bar{\mathbf{t}}_0 \; \Phi_I(\mathbf{X}) \; d\Gamma_0 + \mathbf{f}_I^{u,cr}$$
$$\mathbf{f}_I^{a,ext} = \int_{\Omega_0 \setminus \Gamma_0^c} \varrho_0 \; \mathbf{b} \; \Phi_I(\mathbf{X}) \; S(f_I^{(n)}(\mathbf{X})) \; d\Omega_0 + \int_{\Gamma_0^t} \bar{\mathbf{t}}_0 \; \Phi_I(\mathbf{X}) \; S(f_I^{(n)}(\mathbf{X})) \; d\Gamma_0 + \mathbf{f}_I^{a,cr}$$
$$\mathbf{f}_I^{b,ext} = \int_{\Omega_0 \setminus \Gamma_0^c} \varrho_0 \; \mathbf{b} \; \Phi_I(\mathbf{X}) \; \mathbf{B}_K^{(m)} \; d\Omega_0 + \int_{\Gamma_0^t} \bar{\mathbf{t}}_0 \; \Phi_I(\mathbf{X}) \; \mathbf{B}_K^{(m)} \; d\Gamma_0 + \mathbf{f}_I^{b,cr} \tag{5.297}$$

where

$$\mathbf{f}_I^{a,cr} = \int_{\Gamma_0^c} \Phi_I(\mathbf{X}) [\![S(f_I^{(n)}(\mathbf{X}))]\!] \; \mathbf{t}_c0 \; d\Gamma_0$$
$$\mathbf{f}_I^{b,cr} = \int_{\Gamma_0^c} \Phi_I(\mathbf{X}) [\![\mathbf{B}_K^{(m)}]\!] \; \mathbf{t}_c0 \; d\Gamma_0 \tag{5.298}$$

$$\mathbf{f}_I^{u,int} = \int_{\Omega_0 \setminus \Gamma_0^c} \nabla_0 \Phi_I(\mathbf{X}) \cdot \mathbf{P} \; d\Omega_0$$
$$\mathbf{f}_I^{a,int} = \int_{\Omega_0 \setminus \Gamma_0^c} \left(\left(\nabla_0 \Phi_I(\mathbf{X}) \; S(f_I^{(n)}(\mathbf{X})) + \Phi_I(\mathbf{X}) \; \nabla_0 S(f_I^{(n)}(\mathbf{X})) \right) \cdot \mathbf{P} \right) d\Omega_0$$
$$\mathbf{f}_{IK}^{b,int} = \int_{\Omega_0 \setminus \Gamma_0^c} \left(\left(\nabla_0 \Phi_I(\mathbf{X}) \; \mathbf{B}_K^{(m)} + \Phi_I(\mathbf{X}) \; \nabla_0 \mathbf{B}_K^{(m)} \right) \cdot \mathbf{P} \right) d\Omega_0 \tag{5.299}$$

Eq. (5.292) is the consistent mass matrix. In Eq. (5.299), the spatial derivatives of S vanish since the domain is considered as an open set. The cohesive forces are taken into account in the external forces, Eq. (5.297). Usually, we employed Gauss quadrature to obtain the discrete equations in this method. More details are given in Section 5.12.

5.11.3.2 The PU-method without crack tip enrichment

The weak form of the momentum equation is given by Eq. (5.282). Substituting the continuous and discontinuous displacement fields $\mathbf{u}^0$ and $\mathbf{u}^e$ from Section 5.6.2, Eqs. (5.201) and (5.202), and the crack opening displacement $[\![\mathbf{u}]\!]$ in Eq. (5.210) to the weak form, we obtain

$$\begin{aligned}\delta W_{\text{int}} &= \sum_{I\in\mathcal{W}} \delta\mathbf{u}_I^T \int_{\Omega_0\setminus\Gamma_0^c} \nabla_0\Phi_I(\mathbf{X})^T : \mathbf{P}\,\mathrm{d}\Omega \\ &+ \sum_{J\in\mathcal{E}}\sum_{K\in\mathcal{W}_J} \delta\mathbf{a}_K^{J\,T} \int_{\Omega_0\setminus\Gamma_0^c} \nabla_0\left[\Phi_K(\mathbf{X})\,\Psi_K^J(\mathbf{X})\right]^T : \mathbf{P}\,\mathrm{d}\Omega\end{aligned} \tag{5.300}$$

$$\begin{aligned}\delta W_{\text{kin}} &= \sum_{I\in\mathcal{W}} \delta\mathbf{u}_I^T \sum_{J\in\mathcal{W}} \int_{\Omega_0\setminus\Gamma_0^c} \varrho_0\,\Phi_I(\mathbf{X})^T\cdot\Phi_J(\mathbf{X})\,\mathrm{d}\Omega\,\ddot{\mathbf{u}}_J \\ &+ \sum_{I\in\mathcal{W}} \delta\mathbf{u}_I^T \sum_{J\in\mathcal{E}}\sum_{K\in\mathcal{W}_J} \int_{\Omega_0\setminus\Gamma_0^c} \varrho_0\,\Phi_I(\mathbf{X})^T\cdot\Phi_K(\mathbf{X})\Psi_K^J(\mathbf{X})\,\mathrm{d}\Omega\,\ddot{\mathbf{a}}_J^K \\ &+ \sum_{J\in\mathcal{E}}\sum_{K\in\mathcal{W}_J} \delta\mathbf{a}_K^{J\,T} \sum_{I\in\mathcal{W}} \int_{\Omega_0\setminus\Gamma_0^c} \varrho_0\left[\Phi_K(\mathbf{X})\Psi_K^J(\mathbf{X})\right]^T\cdot\Phi_I(\mathbf{X})\,\mathrm{d}\Omega\,\ddot{\mathbf{u}}_I \\ &+ \sum_{J\in\mathcal{E}}\sum_{K\in\mathcal{W}_J} \delta\mathbf{a}_I^{K\,T} \sum_{L\in\mathcal{E}}\sum_{M\in\mathcal{W}_L} \int_{\Omega_0\setminus\Gamma_0^c} \varrho_0\left[\Phi_K(\mathbf{X})\Psi_K^J(\mathbf{X})\right]^T\cdot \\ &\quad \Phi_M(\mathbf{X})\Psi_M^L(\mathbf{X})\,\mathrm{d}\Omega\,\ddot{\mathbf{a}}_M^L\end{aligned} \tag{5.301}$$

$$\begin{aligned}\delta W_{\text{ext}} &= \sum_{I\in\mathcal{W}} \delta\mathbf{u}_I^T \int_{\Omega_0\setminus\Gamma_0^c} \varrho_0\,\Phi_I(\mathbf{X})^T\cdot\mathbf{b}\,\mathrm{d}\Omega \\ &+ \sum_{I\in\mathcal{W}} \delta\mathbf{u}_I^T \int_{\Gamma_0^t} \Phi_I(\mathbf{X})^T\cdot\bar{\mathbf{t}}_0\,\mathrm{d}\Gamma\end{aligned} \tag{5.302}$$

$$\delta W_{\text{coh}} = 2\sum_{J\in\mathcal{E}}\sum_{K\in\mathcal{W}_J} \delta\mathbf{a}_K^T \int_{\Gamma_0^c} \Phi_K(\mathbf{X})^T\cdot\bar{\mathbf{t}}_c\,\mathrm{d}\Gamma \tag{5.303}$$

Using the fundamental lemma of the variational principle, one obtains the discretized equation, i.e.

$$\mathbf{M}\,\ddot{\mathbf{q}} = \mathbf{f}_{\text{ext}} + \mathbf{f}_{\text{coh}} - \mathbf{f}_{\text{int}} \tag{5.304}$$

where $\mathbf{M}$ is the consistent mass matrix, $\mathbf{q}$ is the generalized parameters, $\mathbf{f}_{\text{ext}}$, $\mathbf{f}_{\text{int}}$, $\mathbf{f}_{\text{coh}}$ are the discrete external, internal and cohesive force vectors, respectively. The expressions for $\mathbf{M}$, $\mathbf{q}$, $\mathbf{f}_{\text{ext}}$, $\mathbf{f}_{\text{coh}}$ and $\mathbf{f}_{\text{coh}}$ are given by

$$\mathbf{M}=\int_{\Omega_0\backslash\Gamma_0^c}\varrho_0\begin{bmatrix}\boldsymbol{\Phi}^{0\,T}\boldsymbol{\Phi}^0 & \boldsymbol{\Phi}^{0\,T}\boldsymbol{\Phi}^e\\ \boldsymbol{\Phi}^{e\,T}\boldsymbol{\Phi}^0 & \boldsymbol{\Phi}^{e\,T}\boldsymbol{\Phi}^0\end{bmatrix}\mathrm{d}\Omega \tag{5.305}$$

$$\mathbf{f}_{\text{int}}=\int_{\Omega_0\backslash\Gamma_0^c}\mathbf{B}^{0\,T}\mathbf{P}\,\mathrm{d}\Omega+\int_{\Omega_0\backslash\Gamma_0^c}\mathbf{B}^{e\,T}\mathbf{P}\,\mathrm{d}\Omega \tag{5.306}$$

$$\mathbf{f}_{\text{ext}}=\int_{\Omega_0\backslash\Gamma_0^c}\varrho_0\,\boldsymbol{\Phi}^T\mathbf{b}\,\mathrm{d}\Omega+\int_{\Gamma_0^t}\boldsymbol{\Phi}^T\bar{\mathbf{t}}_0\,\mathrm{d}\Gamma \tag{5.307}$$

$$\mathbf{f}_{\text{coh}}=2\int_{\Gamma_0^c}\boldsymbol{\Phi}^T\mathbf{t}_{c0}\,\mathrm{d}\Gamma \tag{5.308}$$

$$\mathbf{q}=\begin{Bmatrix}\mathbf{u}\\ \mathbf{a}\end{Bmatrix} \tag{5.309}$$

$$\mathbf{u}=[\mathbf{u}_I]^T\quad\forall I\in\mathcal{W}\ \text{ and }\ \mathbf{a}=\left[\mathbf{a}_K^J\right]^T\quad\forall K\in\mathcal{W}_J,\ \forall J\in\mathcal{E} \tag{5.310}$$

$$\boldsymbol{\Phi}^0=[\boldsymbol{\Phi}_I]\quad\forall I\in\mathcal{W}\ \text{ and }\ \boldsymbol{\Phi}^e=\left[\boldsymbol{\Phi}_J\boldsymbol{\Psi}_J^K\right]\quad\forall K\in\mathcal{W}_J,\ \forall J\in\mathcal{E} \tag{5.311}$$

$$\mathbf{B}^0=\nabla_0\boldsymbol{\Phi}^0\ \text{ and }\ \mathbf{B}^e=\nabla_0\boldsymbol{\Phi}^e \tag{5.312}$$

If the Lagrange multiplier is used, Eq. (5.282) should be modified as follows:

$$\delta W_{\mathrm{L}}=\delta W+\delta(\boldsymbol{\Lambda}\cdot[\![\mathbf{u}]\!]) \tag{5.313}$$

in which δW_{L} is the general variation with constraint and $\boldsymbol{\Lambda}$ is the Lagrange multiplier vector. As mentioned in Section 5.6.2.2, the Lagrange multiplier is defined for $\Gamma_{\text{c,ext}}$ and is interpolated using the shape function of the meshless method. Therefore the discretized Lagrange multiplier is given by

$$\boldsymbol{\Lambda}=\boldsymbol{\Phi}^0\boldsymbol{\lambda} \tag{5.314}$$

where $\boldsymbol{\lambda}$ are the vector parameters to interpolate the Lagrange multiplier Λ. Through the standard procedure for deriving the discrete equations (e.g., [16]), we obtain

$$\mathbf{M}\,\ddot{\mathbf{q}}=\mathbf{f}_{\text{ext}}+\mathbf{f}_{\text{coh}}-\mathbf{f}_{\text{int}}-\mathbf{f}_{\text{con}} \tag{5.315}$$

$$\mathbf{G}\,\mathbf{a}=\mathbf{0} \tag{5.316}$$

Here $\mathbf{f}_{\text{con}}$ is the extra force term due to the constraint and is given by

$$\mathbf{f}_{\text{con}}=\boldsymbol{\lambda}^T\mathbf{G} \tag{5.317}$$

in which $\mathbf{G}$ is

$$\mathbf{G} = \int_{\Gamma_0^c} 2\, \mathbf{\Phi}^{0^T} \mathbf{\Phi}^e \, d\Gamma_0^c \tag{5.318}$$

5.11.3.3 The cracking particles method

The discrete equations are obtained by substituting the test and trial functions, see Section 5.8.1, into the linear momentum equation:

$$\begin{aligned} \delta W_{ext} = & \sum_I \int_{\Omega_0 \backslash \Gamma_0^c} \varrho_0 \mathbf{b} \cdot \left(\Phi_I(\mathbf{X})\, \delta \mathbf{u}_I + \Psi_I(\mathbf{X})\, S(f_I(\mathbf{X}))\, \delta \mathbf{q}_I \right) d\Gamma_0 \\ & + \sum_I \int_{\Gamma_0^t} \bar{\mathbf{t}}_0 \cdot \left(\Phi_I(\mathbf{X})\, \delta \mathbf{u}_I + \Psi_I(\mathbf{X})\, S(f_I(\mathbf{X}))\, \delta \mathbf{q}_I \right) d\Gamma_0 \\ & + \sum_I \int_{\Gamma^c} \delta \mathbf{q}_I \cdot \mathbf{t}_c \Psi_I(\mathbf{X})\, [\![S(f_I(\mathbf{X}))]\!]\, d\Gamma \end{aligned} \tag{5.319}$$

$$\begin{aligned} \delta W_{int} = & \sum_I \int_{\Omega_0 \backslash \Gamma_0^c} (\delta \mathbf{u}_I \otimes \nabla \Phi_I(\mathbf{X})) : \mathbf{P}(\mathbf{X})\, d\Omega_0 \\ & + \sum_I \int_{\Omega_0 \backslash \Gamma_0^c} \left(\delta \mathbf{q}_I \otimes \nabla \Psi_I(\mathbf{X}) \right) : \mathbf{P}(\mathbf{X})\, S(f_I(\mathbf{X}))\, d\Omega_0 \end{aligned} \tag{5.320}$$

$$\begin{aligned} \delta W_{kin} = & \sum_I \int_{\Omega_0 \backslash \Gamma_0^c} \varrho_0 \left(\Phi_I(\mathbf{X}) \delta \mathbf{u}_I + \Psi_I(\mathbf{X})\, S(f_I(\mathbf{X}))\, \delta \mathbf{q}_I \right) \\ & \cdot \left(\hat{\Phi}_J(\mathbf{X}) \ddot{\mathbf{u}}_J + \Psi_J(\mathbf{X})\, S(f_I(\mathbf{X}))\, \ddot{\mathbf{q}}_J \right) d\Omega_0 \end{aligned} \tag{5.321}$$

The cohesive tractions across the crack are treated as external nodal forces. From the fact that $\delta \mathbf{u}_I$ and $\delta \mathbf{q}_I$ are arbitrary, the well known equations of motion can be obtained from Eqs. (5.319) to (5.321)

$$\mathbf{M}_{IJ} \cdot \ddot{\mathbf{u}}_J = \mathbf{f}_I^{ext} - \mathbf{f}_I^{int} \tag{5.322}$$

In the above $\mathbf{M}_{IJ}$ is the mass matrix

$$\mathbf{M}_{IJ} = \begin{bmatrix} \mathbf{m}_{IJ}^{uu} & \mathbf{m}_{IJ}^{uq} \\ \mathbf{m}_{IJ}^{qu} & \mathbf{m}_{IJ}^{qq} \end{bmatrix} \tag{5.323}$$

$$\ddot{\mathbf{u}}_I = \left\{ \begin{array}{c} \ddot{\mathbf{u}}_I^P \\ \ddot{\mathbf{q}}_I \end{array} \right\} \tag{5.324}$$

where

$$\begin{aligned}
\mathbf{m}_{IJ}^{uu} &= \int_{\Omega_0 \backslash \Gamma_0^c} \varrho_0 \ \Phi_I(\mathbf{X}) \ \hat{\Phi}_J(\mathbf{X}) \ \mathbf{I} \ d\Omega_0 \\
\mathbf{m}_{IJ}^{uq} &= \int_{\Omega_0 \backslash \Gamma_0^c} \varrho_0 \ \Phi_I(\mathbf{X}) \ \Psi_J(\mathbf{X}) \ S(f_J(\mathbf{X})) \ \mathbf{I} \ d\Omega_0 \\
\mathbf{m}_{IJ}^{qu} &= \int_{\Omega_0 \backslash \Gamma_0^c} \varrho_0 \ \Psi_I(\mathbf{X}) \ S(f_J(\mathbf{X})) \ \hat{\Phi}_J(\mathbf{X}) \ \mathbf{I} \ d\Omega_0 \\
\mathbf{m}_{IJ}^{qq} &= \int_{\Omega_0 \backslash \Gamma_0^c} \varrho_0 \ \Psi_I(\mathbf{X}) \ S(f_I(\mathbf{X})) \ \Psi_J(\mathbf{X}) \ S(f_J(\mathbf{X})) \ \mathbf{I} \ d\Omega_0 \\
&= \int_{\Omega_0 \backslash \Gamma_0^c} \varrho_0 \ \Psi_I(\mathbf{X}) \ \hat{\Psi}_J(\mathbf{X}) \ \mathbf{I} \ d\Omega_0
\end{aligned} \tag{5.325}$$

where $\mathbf{I}$ is the identity matrix. The matrix $\mathbf{f}_I^{ext}$ is the external nodal force vector and given by

$$\mathbf{f}_I^{ext} = \left\{ \begin{array}{c} \mathbf{f}_I^{u,ext} \\ \mathbf{f}_I^{q,ext} \end{array} \right\} \tag{5.326}$$

$$\begin{aligned}
\mathbf{f}_I^{u,ext} &= \int_{\Omega_0 \backslash \Gamma_0^c} \varrho_0 \mathbf{b} \ \Phi_I(\mathbf{X}) \ d\Omega_0 + \int_{\Gamma_0^t} \bar{\mathbf{t}}_0 \ \Phi_I(\mathbf{X}) \ d\Gamma_0 \\
\mathbf{f}_I^{q,ext} &= \int_{\Omega_0 \backslash \Gamma_0^c} \varrho_0 \mathbf{b} \ \Psi_I(\mathbf{X}) \ S(f_I(\mathbf{X})) \ d\Omega_0 + \int_{\Gamma^c} \mathbf{t}_c \Psi_I(\mathbf{X}) \ [\![S(f_I(\mathbf{X}))]\!] \ d\Gamma \\
&+ \int_{\Gamma_0^t} \bar{\mathbf{t}}_0 \ \Psi_I(\mathbf{X}) \ S(f_I(\mathbf{X})) \ d\Gamma_0
\end{aligned} \tag{5.327}$$

The matrix $\mathbf{f}_I^{int}$ is the internal nodal force vector and given by

$$\mathbf{f}_I^{int} = \left\{ \begin{array}{c} \mathbf{f}_I^{u,int} \\ \mathbf{f}_I^{q,int} \end{array} \right\} \tag{5.328}$$

$$\begin{aligned}
\mathbf{f}_I^{u,int} &= \int_{\Omega_0 \backslash \Gamma_0^c} \nabla_0 \Phi_I(\mathbf{X}) \cdot \mathbf{P}(\mathbf{X}) \ d\Omega_0 \\
\mathbf{f}_I^{q,int} &= \int_{\Omega_0 \backslash \Gamma_0^c} \nabla_0 \Psi_I(\mathbf{X}) \cdot \mathbf{P}(\mathbf{X}) \ S(f_I(\mathbf{X})) \ d\Omega_0 \\
&+ \int_{\Omega_0 \backslash \Gamma_0^c} \Psi_I(\mathbf{X}) \ \nabla_0 S(f_I(\mathbf{X})) \ \mathbf{P}(\mathbf{X}) \ d\Omega_0
\end{aligned} \tag{5.329}$$

Since the domain Ω_0 is considered an open set with the crack excluded and $\nabla_0 S = 0$ except at the discontinuity, the second term on the RHS of the second line of Eq. (5.329) vanishes. As for the enriched methods in the previous sections, the introduction of additional degrees of freedom complicate the solution of the system of equations.

We mainly used the CPM combined with a stress point integration due to the good performance of the method with respect to dynamic applications with large deformations. For a stress point integration, the discrete equations are given below. Note, that the discrete momentum equation is not imposed on the stress points since the test function does not depend on these particle values, see (5.223). All kinematic values at the stress points are approximated via the shape functions from the original particles. The masses and the particle volumes are computed by the Voronoi diagram as shown in Fig. 5.69. The condition that the sum of particle volumes and stress point volumes equal the total volume must be met:

$$V^0 = \sum_{I \in \mathcal{N}_I^P} V_I^{0P} + \sum_{I \in \mathcal{N}_I^S} V_I^{0S} \tag{5.330}$$

The surface area associated with stress point or particle I is computed as the intersection of the Voronoi cell I and the surface Γ_0. So if the domain of Voronoi cell I is denoted by Ω_{0I}^V, then $A_I^0 = \Omega_{0I}^V \bigcap \Gamma_0$. From the definition, it follows that

$$\sum_{I \in \mathcal{N}_P \bigcup \mathcal{N}_S} A_I^0 = \int_{\Gamma_0} \Gamma_0 \tag{5.331}$$

The areas of the crack surface are obtained similarly.

The internal forces are computed by numerical integration of (5.329) with the particles and the stress points serving as quadrature points, which gives

$$\begin{aligned}
\mathbf{f}_I^{u,int} &= \sum_{J \in \mathcal{N}_I^P} V_J^{0P}\, \nabla_0 \Phi_I(\mathbf{X}_J^P) \cdot \mathbf{P}(\mathbf{X}_J^P) + \sum_{J \in \mathcal{N}_I^S} V_J^{0S}\, \nabla_0 \Phi_I(\mathbf{X}_J^S)\, \mathbf{P}(\mathbf{X}_J^S) \\
\mathbf{f}_I^{q,int} &= \sum_{J \in \mathcal{N}_I^P} \left(V_J^{0P}\, \nabla_0 \Psi_I(\mathbf{X}_J^P)\, S(f_I(\mathbf{X}_J^P)) \cdot \mathbf{P}(\mathbf{X}_J^P) \right) \\
&+ \sum_{J \in \mathcal{N}_I^S} \left(V_J^{0S}\, \nabla_0 \Psi_I(\mathbf{X}_J^S)\, S(f_I(\mathbf{X}_J^S)) \cdot \mathbf{P}(\mathbf{X}_J^S) \right)
\end{aligned} \tag{5.332}$$

where $\mathcal{N}_I^P = \mathcal{N}_P(\mathbf{X}_I)$ and $\mathcal{N}_I^S = \mathcal{N}_S(\mathbf{X}_I)$ are the particles whose support is non-zero at $\mathbf{X}_I$, i.e.

$$J \in \mathcal{N}_I^P \ \ if \ \ \|\mathbf{X}_J - \mathbf{X}_I\| \leq h_I \tag{5.333}$$

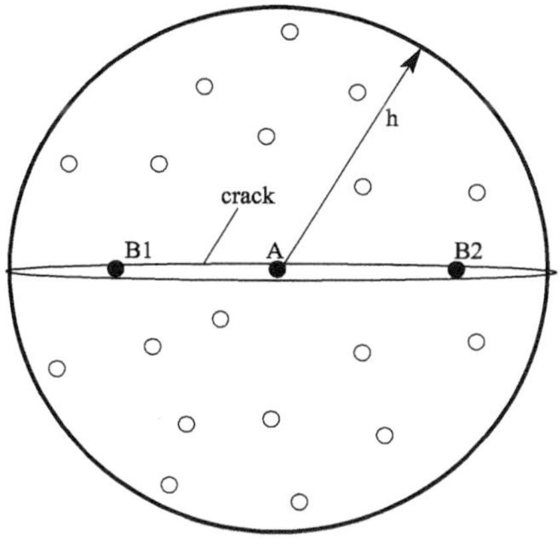

Figure 5.68 Quadrature points across the crack (point *A* is used here, 3 point quadrature scheme using *B*1, *A* and *B*2 could be used).

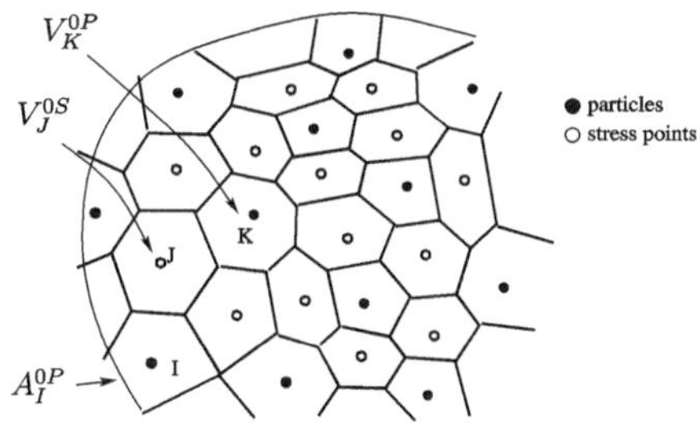

Figure 5.69 Voronoi cells for stress point integration.

with similar definition for $\mathcal{N}_I^S$; any particles J on a crack are omitted. Note, that the cracked particles do not contribute to the internal nodal forces. Furthermore, in this formulation the derivatives of the shape functions for the central particle are zero, since $\nabla_0\, w_J(\mathbf{X}_J) = 0$.

Let $\mathbf{b}_J = \mathbf{b}(\mathbf{X}_J)$, $\mathbf{t}_{0J} = \mathbf{t}_0(\mathbf{X}_J)$ and $\mathbf{t}_{cJ} = \mathbf{t}_c(\mathbf{X}_J)$. The external forces are similarly obtained from (5.327), giving

$$
\begin{aligned}
\mathbf{f}_I^{u,ext} &= \sum_{J\in\mathcal{N}_I^P} V_J^{0P}\, \Phi_I(\mathbf{X}_J^P)\, \varrho_0\, \mathbf{b}_J + \sum_{J\in\mathcal{N}_I^S} V_J^{0S}\, \Phi_I(\mathbf{X}_J^S)\, \varrho_0\, \mathbf{b}_J \\
&\quad + \sum_{J\in\mathcal{N}_I^P} A_J^{0P}\, \Phi_I(\mathbf{X}_J^P)\, \bar{\mathbf{t}}_{0J} + \sum_{J\in\mathcal{N}_I^S} A_J^{0S}\, \Phi_I(\mathbf{X}_J^S)\, \bar{\mathbf{t}}_{0J} \\
\mathbf{f}_I^{q,ext} &= \sum_{J\in\mathcal{N}_I^P} V_J^{0P}\, \Psi_I(\mathbf{X}_J^P)\, S(f_I(\mathbf{X}_J^P))\, \varrho_0\, \mathbf{b}_J
\end{aligned}
$$

$$+\sum_{J\in\mathcal{N}_I^S} V_J^{0S}\ \Psi_I(\mathbf{X}_J^S)\ S(f_I(\mathbf{X}_J^S))\ \varrho_0\ \mathbf{b}_J$$
$$+\sum_{J\in\mathcal{N}_I^P} A_J^{0P}\ \Psi_I(\mathbf{X}_J^P)\ S(f_I(\mathbf{X}_J^P))\ \bar{\mathbf{t}}_{0J}$$
$$+\sum_{J\in\mathcal{N}_I^S} A_J^{0S}\ \Psi_I(\mathbf{X}_J^S)\ S(f_I(\mathbf{X}_J^S))\ \bar{\mathbf{t}}_{0J}$$
$$+2\,c_I\left(\sum_{J\in\mathcal{N}_I^P} A_J^{cP}\ \Psi_I(\mathbf{X}_J^P)\ \mathbf{t}_{cJ}+\sum_{J\in\mathcal{N}_I^S} A_J^{cS}\ \Psi_I(\mathbf{X}_J^S)\ \mathbf{t}_{cJ}\right) \tag{5.334}$$

where A_J^{cP} and A_J^{cS} is the crack surface at node J and $c_I = 1$ if particle I is cracked, $c_I = 0$ otherwise. Note that we compute the cohesive nodal forces in the current configuration. For this purpose, we obtain the current normal from Nanson's law from $\mathbf{n}_0$ by (5.226). This procedure avoids the need to update the current normal by a rate relation, which could engender integration errors.

We remark that the one point quadrature of the cohesive force used here underestimates the cohesive force $\mathbf{t}_c$ since $[\![\mathbf{u}(\mathbf{X})]\!]$ is greatest at the quadrature point (point A in Fig. 5.68), and decreases with increasing $[\![\mathbf{u}(\mathbf{X})]\!]$. Three point quadrature across the crack with additional points, such as $B1$ and $B2$ in Fig. 5.68, would increase the discrete force corresponding to the cohesive traction.

5.11.3.4 The cracking particles method for shear bands

We will propose a B-bar formulation to model shear bands within the CPM. This principle is easily extendable to any other extrinsically enriched method described in this book.

Therefore, the deformation gradient is decomposed into a deviatoric and volumetric part, see [44]:

$$\mathbf{F} = \mathbf{F}^{dev}\ \mathbf{F}^{vol} \tag{5.335}$$

The approximations for $\mathbf{F}$ and $\bar{\mathbf{F}}$ are given by

$$\mathbf{F} = \sum_{I\in\mathcal{N}} \nabla\Phi_I(\mathbf{X})\ \mathbf{u}_I(t) + \sum_{I\in\mathcal{N}_c} \nabla\Phi_I(\mathbf{X})\ S(f_I(\mathbf{X}))\ q_{\alpha I}(t)\ \mathbf{e}_{\alpha T}(\mathbf{X}) + \mathbf{I}$$
$$\bar{\mathbf{F}} = \sum_{I\in\mathcal{N}} \nabla\bar{\Phi}_I(\mathbf{X})\ \mathbf{u}_I(t)$$

$$+\sum_{I\in\mathcal{N}_c}\nabla\bar{\Phi}_I(\mathbf{X})\ S(f_I(\mathbf{X}))\ q_{\alpha I}(t)\ \mathbf{e}_{\alpha T}(\mathbf{X})+\mathbf{I} \tag{5.336}$$

where $\Phi_I(\mathbf{X})$ are shape functions based on a bilinear or quadratic basis, respectively, and the shape functions $\bar{\Phi}_I(\mathbf{X})$ use a constant or linear basis, respectively. The facts that $\nabla S(f_I(\mathbf{X}))$ and $\nabla e_{\alpha T}$ vanish are already incorporated into Eq. (5.336). The former term vanishes since the shear band is considered as an open set and the latter since the discontinuity is piecewise constant. The volumetric part of the deformation gradient can then be computed in terms of the constant-basis deformation gradient $\bar{\mathbf{F}}$ by

$$\bar{\mathbf{F}}^{vol}=\frac{1}{3}\mathrm{tr}(\bar{\mathbf{F}})\ \mathbf{I} \tag{5.337}$$

The pressure $\bar{p}$ is computed using the following constitutive law

$$\bar{p}=K(\theta-1) \tag{5.338}$$

with

$$\theta=\det(\bar{\mathbf{F}}) \tag{5.339}$$

The deviatoric deformation gradient is computed by

$$\mathbf{F}^{dev}=\mathbf{F}\left(\mathbf{F}^{vol}\right)^{-1} \tag{5.340}$$

For meshfree shape functions $\Phi_I(\mathbf{X})$ with bilinear basis and shape functions $\bar{\Phi}_I(\mathbf{X})$ with constant basis, instabilities can occur. They can be eliminated by a GLS (Generalized Least Square) or SUPG stabilization procedure, [23, 60]. The instabilities can be also prevented by using higher order shape functions.

Substituting the approximations of the test and trial functions and invoking the arbitrariness of $\delta\mathbf{u}_I$ and $\delta\mathbf{q}_{\alpha I}$ into the linear momentum equation we obtain

$$\begin{aligned}
&\sum_{I\in\mathcal{N}}\delta\mathbf{u}_I\cdot\left(\mathbf{f}_I^{int}+\mathbf{M}_{IK}^{uu}\cdot\ddot{\mathbf{u}}_K+\mathbf{M}_{IK}^{uq_1}\ \ddot{q}_{1K}+\mathbf{M}_{IK}^{uq_2}\ \ddot{q}_{2K}-\mathbf{f}_I^{ext}\right)+\\
&\sum_{I\in\mathcal{N}_c}\delta q_{1I}\ \left(Q_{1I}^{int}+\mathbf{M}_{IK}^{uq_1}\cdot\ddot{\mathbf{u}}_K+\mathbf{M}_{IK}^{q_1q_1}\ \ddot{q}_{1K}+\mathbf{M}_{IK}^{q_1q_2}\ \ddot{q}_{2K}-Q_{1I}^{ext}\right)+\\
&\sum_{I\in\mathcal{N}_c}\delta q_{2I}\ \left(Q_{2I}^{int}+\mathbf{M}_{IK}^{uq_2}\cdot\ddot{\mathbf{u}}_K+\mathbf{M}_{IK}^{q_1q_2}\ \ddot{q}_{1K}+\mathbf{M}_{IK}^{q_2q_2}\ \ddot{q}_{2K}-Q_{2I}^{ext}\right)=0
\end{aligned} \tag{5.341}$$

where

$$
\begin{aligned}
\mathbf{f}_I^{int} &= \int_{\Omega\backslash\Gamma^s} \nabla\Phi_I(\mathbf{X}) \cdot \text{dev}(\boldsymbol{\sigma}(\mathbf{X}))\ d\Omega \\
&+ \int_{\Omega\backslash\Gamma^s} \nabla\bar{\Phi}_I(\mathbf{X}) \cdot \mathbf{I}\ \bar{p}(\mathbf{X})\ d\Omega
\end{aligned} \tag{5.342}
$$

$$
\mathbf{f}_I^{ext} = \int_{\Omega\backslash\Gamma^s} \varrho\mathbf{b}\ \Phi_I(\mathbf{X})\ d\Omega + \int_{\Gamma^t} \bar{\mathbf{t}}\ \Phi_I(\mathbf{X})\ d\Gamma \tag{5.343}
$$

$$
\begin{aligned}
Q_{\alpha I}^{int} &= \int_{\Omega\backslash\Gamma^s} S(f_I(\mathbf{X}))\,(\nabla\Phi_I(\mathbf{X}) \otimes \mathbf{e}_{\alpha T})^S : \text{dev}(\boldsymbol{\sigma}(\mathbf{X}))\ d\Omega \\
&\underbrace{+ \int_{\Omega\backslash\Gamma^s} S(f_I(\mathbf{X}))\Phi_I(\mathbf{X})\,(\nabla \otimes \mathbf{e}_{\alpha T})^S : \text{dev}(\boldsymbol{\sigma}(\mathbf{X}))\ d\Omega}_{=0} \\
&+ \int_{\Omega\backslash\Gamma^s} S(f_I(\mathbf{X}))\left(\nabla\bar{\Phi}_I(\mathbf{X}) \otimes \mathbf{e}_{\alpha T}\right)^S : \mathbf{I}\ \bar{p}(\mathbf{X})\ d\Omega \\
&\underbrace{+ \int_{\Omega\backslash\Gamma^s} S(f_I(\mathbf{X}))\bar{\Phi}_I(\mathbf{X})\,(\nabla \otimes \mathbf{e}_{\alpha T})^S : \mathbf{I}\ \bar{p}(\mathbf{X})\ d\Omega}_{=0}
\end{aligned} \tag{5.344}
$$

$$
\begin{aligned}
Q_{\alpha I}^{ext} &= \int_{\Omega\backslash\Gamma^s} \varrho\left(\mathbf{b} \cdot \mathbf{e}_{\alpha T}\right)\ S(f_I(\mathbf{X}))\ \Phi_I(\mathbf{X})\ d\Omega \\
&+ \int_{\Gamma^t} \left(\bar{\mathbf{t}} \cdot \mathbf{e}_{\alpha T}\right)\ S(f_I(\mathbf{X}))\ \Phi_I(\mathbf{X})\ d\Gamma \\
&- \int_{\Gamma^s} \mathbf{t}_c\Phi_I(\mathbf{X})\ [\![S(f_I(\mathbf{X}))]\!]\ d\Gamma
\end{aligned} \tag{5.345}
$$

where the spatial derivatives with respect to $\mathbf{e}_{\alpha T}$, $\alpha = 1, 2$ vanish as indicated above since the discontinuity is piecewise constant. Each of the parenthesis in Eq. (5.341) gives a discrete equation of motion. The mass matrix is given by

$$
\mathbf{M}_{IJ} = \begin{bmatrix} \mathbf{m}_{IJ}^{uu} & \mathbf{m}_{IJ}^{uq_1} & \mathbf{m}_{IJ}^{uq_2} \\ \mathbf{m}_{IJ}^{uq_1} & \mathbf{m}_{IJ}^{q_1q_1} & \mathbf{m}_{IJ}^{q_1q_2} \\ \mathbf{m}_{IJ}^{uq_2} & \mathbf{m}_{IJ}^{q_1q_2} & \mathbf{m}_{IJ}^{q_2q_2} \end{bmatrix} \tag{5.346}
$$

with

$$
\mathbf{m}_{IJ}^{uu} = \int_{\Omega\backslash\Gamma^s} \varrho_0\ \Phi_I(\mathbf{X})\ \hat{\Phi}_J(\mathbf{X})\ \mathbf{I}\ d\Omega
$$

$$\mathbf{m}_{IJ}^{uq_\alpha} = \int_{\Omega \backslash \Gamma^s} \varrho_0 \; \Phi_I(\mathbf{X}) \; \Phi_J(\mathbf{X}) \; S(f_J(\mathbf{X})) \; \mathbf{e}_{\alpha T} \; d\Omega$$

$$\mathbf{m}_{IJ}^{q_\alpha q_\alpha} = \int_{\Omega \backslash \Gamma^s} \varrho_0 \; \Phi_I(\mathbf{X}) \; S(f_I(\mathbf{X})) \; \Phi_J(\mathbf{X}) \; S(f_J(\mathbf{X})) \; d\Omega \qquad (5.347)$$

In addition to the B-bar approach, we have used a selective reduced integration to obtain the discrete equations. For example, when the nodal internal forces that depends on the pressure, the second term in Eqs. (5.342) and (5.344), are obtained by nodal integration:

$$\int_{\Omega_0} \Phi_I(\mathbf{X}) g(\mathbf{X}) d\Omega_0 = \sum_{J \in \mathcal{N}_I} \Phi_I(\mathbf{X}_J) g(\mathbf{X}_J) \, V_J^0 \qquad (5.348)$$

where V_J^0 is the Voronoi volume of node J and $\mathcal{N}_I$ is the set of nodes in the support of $\Phi_I(\mathbf{X})$, then a stress point integration is used for the deviatoric terms.

5.12. Spatial integration

Spatial integration techniques for meshfree methods were already discussed in Section 5.2.2. However, in particular the integration of non-polynomial functions such as the crack tip enrichment causes difficulties. The simples way to reduce integration errors is to increase the number of Gauss points [42,70,90]. For LEFM, [122] proposed a strategy that prevents the integration of the singularity. Similar procedures were proposed by [49] and [22] in the context of the smoothed extended finite element method. The basic idea is to transform domain integrals into line integration. We note that the method in [22] is based on the SCNI also used in meshfree methods. The almost polar integration of [83] also avoids the integration of a singularity and was used in [7,70] in the context of XFEM.

Also the integration of step-enriched shape functions require special attention. The most common integration strategy is based on sub-dividing an existing background cell into several smaller cells, see Fig. 5.70 for the 2D case and Fig. 5.71 for the 3D case. Integration is particularly difficult for curved crack geometries. If a crack curves as shown for the two-dimensional case in Fig. 5.72, an additional node is introduced to maintain integration as accurate as possible. Adding one more point, the error is reduced second order small.

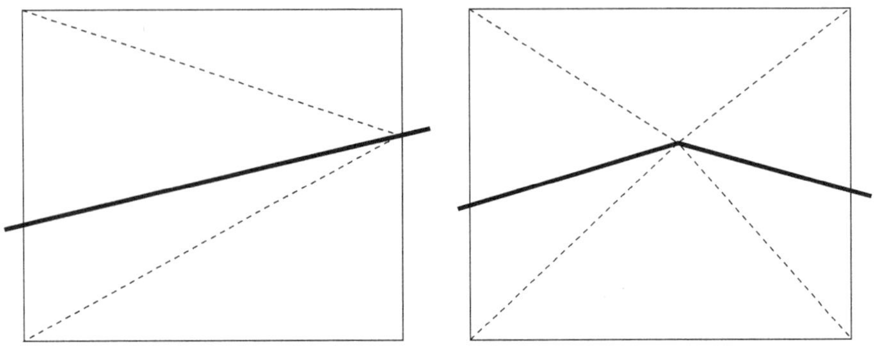

Figure 5.70 Subtriangulation for integration purposes.

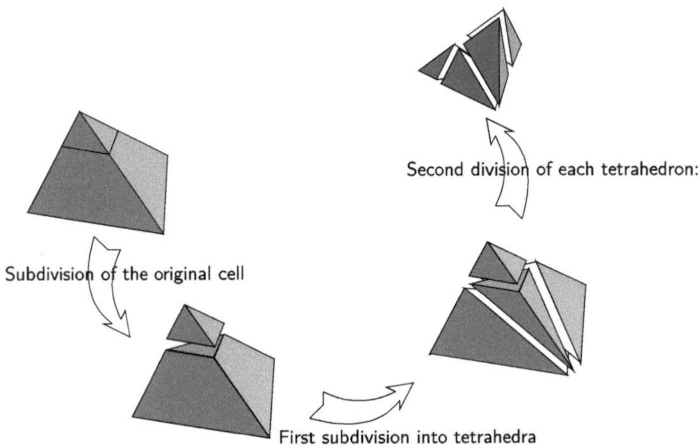

Figure 5.71 Decomposition of tetrahedra split by a discontinuity for accurate integration of the weak form [99].

Usually, the Gauss or Voronoi cell is sub-triangulated (in 3D, tetrahedra are created). The integration of a function $\mathcal{F}$ can be accomplished by

$$\begin{aligned}\mathcal{F} &= \int_{\Omega^-} \mathcal{F}(\mathbf{X}) d\Omega + \int_{\Omega^+} \mathcal{F}(\mathbf{X}) \\ &= \int_{\bar{\Omega}^-} \mathcal{F}(\mathbf{X}(\boldsymbol{\xi})) \, \det\mathbf{J}^-(\boldsymbol{\xi}) d\Omega + \int_{\bar{\Omega}^+} \mathcal{F}(\mathbf{X}(\boldsymbol{\xi})) \, \det\mathbf{J}^+(\boldsymbol{\xi}) \, d\Omega \qquad (5.349)\end{aligned}$$

where the Jacobian maps the domains $\bar{\Omega}^\pm$ onto $\Omega^\pm$. With the mapping $\tilde{\Omega}^\pm \rightarrow \bar{\Omega}^\pm$, we finally obtain:

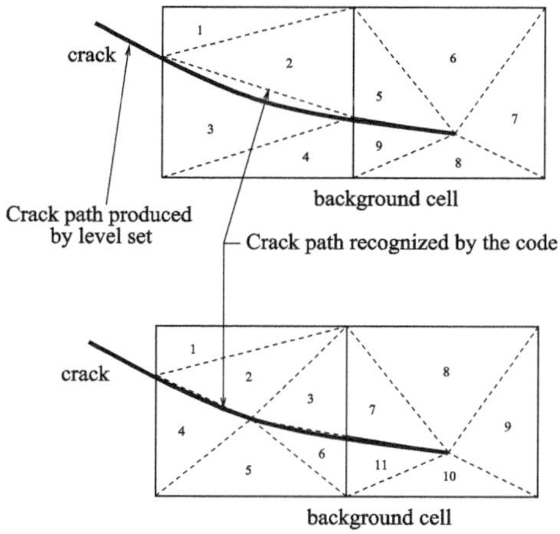

Figure 5.72 Sub-triangulation of background cells.

$$\mathcal{F} = \int_{\tilde{\Omega}^-} \mathcal{F}(\mathbf{X}(\boldsymbol{\xi})(\boldsymbol{\eta}))\ \det\mathbf{J}^-(\boldsymbol{\xi}(\boldsymbol{\eta}))\ \det\bar{\mathbf{J}}^-(\boldsymbol{\eta})d\Omega + \int_{\tilde{\Omega}^+} \mathcal{F}(\mathbf{X}(\boldsymbol{\xi})(\boldsymbol{\eta}))\ \det\mathbf{J}^+(\boldsymbol{\xi}(\boldsymbol{\eta}))\ \det\bar{\mathbf{J}}^+(\boldsymbol{\eta})d\Omega \tag{5.350}$$

and in its discrete form using Gauss quadrature:

$$\mathcal{F} = \sum_{I=1}^{n_{GP}^-} \mathcal{F}(\boldsymbol{\eta}_I)\ \det\mathbf{J}^-(\boldsymbol{\eta})\ \det\bar{\mathbf{J}}^-(\boldsymbol{\eta})w_I + \sum_{I=1}^{n_{GP}^+} \mathcal{F}(\boldsymbol{\eta}_I)\ \det\mathbf{J}^+(\boldsymbol{\eta})\ \det\bar{\mathbf{J}}^+(\boldsymbol{\eta})w_I \tag{5.351}$$

where n_{GP}^- and n_{GP}^+ are the Gauss points in Ω^- and Ω^+, respectively, and $\boldsymbol{\eta}$ are the local coordinates of the Gauss points and w_I are their quadrature weights.

A simpler approach that leads to the same results is the modification of the quadrature weights, that is illustrated Fig. 5.73 simplified for a nodal or stress point integration in two dimensions though we used this method also for Gauss quadrature. This approach was used in the context of cracks in shells using XFEM, [2] and is here proposed in a meshfree context. In that method, the quadrature weights crossed by the crack are computed according to their areas (or volumes in 3D) V^+ and V^-:

$$w_I^+ = w\frac{V_I^+}{V_I}$$
$$w_I^- = w\frac{V_I^-}{V_I} \tag{5.352}$$

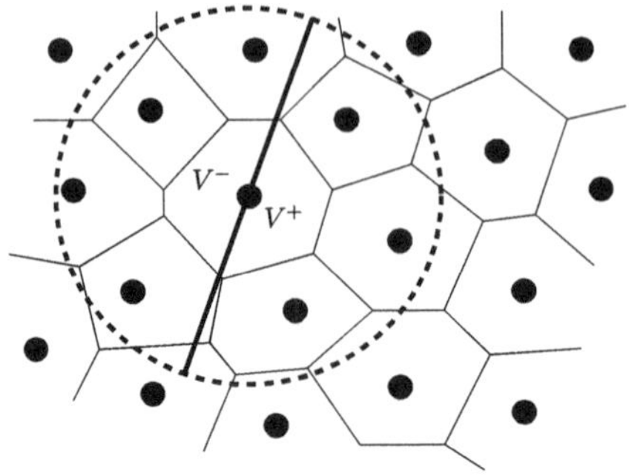

Figure 5.73 Integration.

The method requires subdivision of the Voronoi or quadrature cells; in case of a nodal or stress point integration, the quadrature weights are the volumes of the Voronoi cells, simplified shown for the 2D case in Fig. 5.73 though our method is three-dimensional. The intersection between the crack area with the Voronoi cells are easily computed since algorithms of the initial discretization generation can be adopted.

Higher order integration schemes were inter alia proposed by [70] and [128]. The cohesive tractions across the crack are integrated by the standard surface integration.

5.13. Time integration

5.13.1 Explicit-implicit time integration

For most of our dynamic computations we employed the central difference method that is of second order accuracy. While methods that are not based on an enrichment use a row sum technique to diagonalize the consistent mass matrix, lumping the consistent mass matrix of enriched methods is not straightforward and can result in unphysical (negative) masses. Since formulations based on a consistent mass matrix require the solution of a system of equations are therefore computationally expensive, we suggest an explicit-implicit integration scheme [59] to reduce computational cost. For the standard degrees of freedom, the mass matrix can still be lumped and the explicit central difference method used. For all particles influenced by the crack, the consistent mass matrix is retained. After solving for the regular part, we solve for the enrichment with the Newmark-beta scheme:

$$\Delta \mathbf{q} = -\mathbf{A}^{-1} \cdot \mathbf{r} \tag{5.353}$$

with

$$\mathbf{r} = \frac{1}{\Delta t^2 \beta}\mathbf{M}\cdot\left(\Delta\mathbf{q}^{n+1} - \mathbf{u}^{n+1}\right) + \mathbf{f}_{imp}^{int,n+1} + \mathbf{f}_{exp,n+1}^{int} - \mathbf{f}^{ext,n+1} \tag{5.354}$$

where the indices *exp* and *imp* denote explicit and implicit time integration and

$$\mathbf{A} = \frac{\partial \mathbf{r}}{\partial \Delta \mathbf{q}^{n+1}} = \frac{1}{\Delta t^2 \beta}\mathbf{M} + \left(\frac{\partial \mathbf{d}_{imp}^{int}}{\partial \mathbf{q}}\right)^{n+1} \tag{5.355}$$

The method has to be used with care since spurious wave reflections may occur. Moreover, the critical time step is drastically decreased when the crack approaches a node. In XFEM, the critical time step tends to zero while that's not the case for enriched meshfree methods as will be shown in the next section. Methods were developed to diagonalize the consistent mass matrix. In the context of the XFEM, [77,78] proposed a lumping technique of the consistent mass matrix (see also [109]):

$$m_{diag} = \frac{m}{n_{nodes}}\frac{1}{mes(\Omega)}\int_\Omega S^2 \, d\Omega \tag{5.356}$$

where Ω is the element being considered, m is its mass, n_{nodes} the number of nodes in Ω and S is the enrichment function. [77] used the step function as enrichment function and shows that the mass matrix can be diagonalized by one of these two procedures:

$$\mathbf{M}_{II}^{lumped} = \sum_J \mathbf{M}_{IJ}^{consistent}, \quad \text{or}$$
$$\mathbf{M}_{II}^{lumped} = m\frac{\mathbf{M}_{II}^{consistent}}{\sum_J \mathbf{M}_{IJ}^{consistent}} \tag{5.357}$$

Moreover, they showed that the critical time step is not diminished drastically when the discontinuity, i.e. the crack, approaches a node; drastically means by a factor around 2 (compared to the CFL criterion of the element without discontinuity). Note that the lumping procedure is slightly different when the Heaviside function is used as enrichment.

5.13.2 Explicit time integration, critical time step and mass lumping

In this section, we propose two strategies to diagonalize the consistent mass matrix. We will also show that the critical time step for enriched meshfree

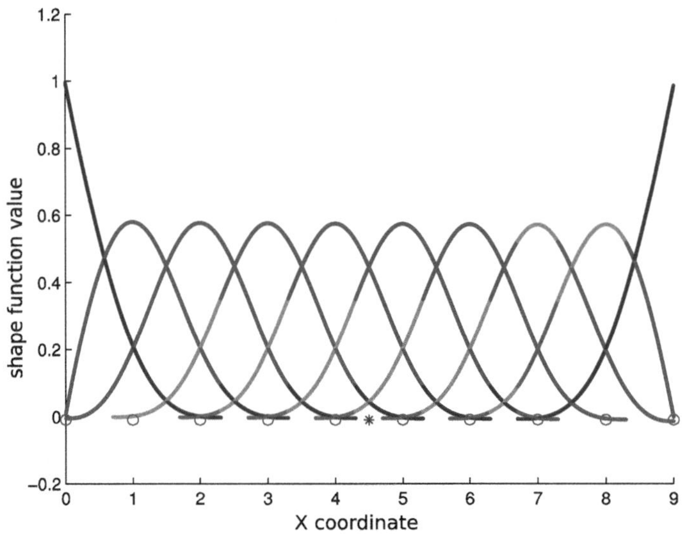

Figure 5.74 Meshfree shape functions for a set of ten nodes and DMAX = 2.3. The point of discontinuity is located at $X_{cr} = 4.5$.

methods based on the central difference time integration scheme[10] and using a lumped mass matrix will be of the same order as the critical time step without enrichment. We will also show that in contrast to the XFEM, even for a consistent mass matrix, the critical time step won't tend to zero even when the crack directly crosses a node. For the examples in the following subsections, Gauss quadrature is used though it is well known that nodally integrated methods will lead to much larger (factor of 100) critical time steps [91].

5.13.2.1 Critical time step and consistent mass matrix

Let us consider a bar in one dimension discretized with 10 particles as shown in Fig. 5.74. The Young's modulus of the bar is $E = 200$ GPa and the density is 7850 kg/m^3. Assuming that the domain of influence ($DMAX$) is 2.3, the meshfree shape functions are plotted for all the nodes in Fig. 5.74. The particle spacing is uniform and the distance between two neighboring particles is 1 m. We assume a point of discontinuity in the middle of the bar shown as a star in Fig. 5.74. The enriched nodes are those whose domain of influence is cut by the crack (point of discontinuity here). Therefore,

[10] In order to integrate the linear momentum equation in time.

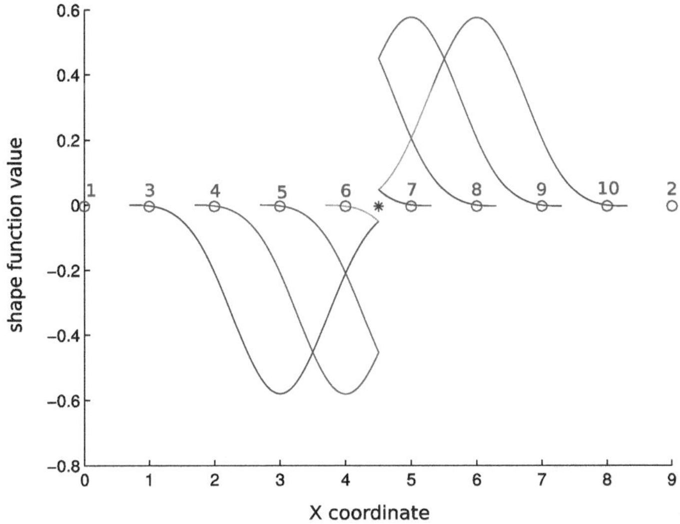

Figure 5.75 Enriched shape functions for a set of ten nodes with the point of discontinuity at $X_{cr} = 4.5$.

if for example $DMAX = 2.3$, then the nodes 5, 6, 7, 8 are affected by the crack and therefore should be enriched. Every enriched node will have one extra degree of freedom. Using a step function to model the jump in the displacement field, we build the enriched shape functions that are shown in Fig. 5.75. The node numbering is also shown in the same figure.

The critical time step, Δt, for the above consistent mass matrix is dependent on the position of the crack; however unlike XFEM, due to the smoothed nature of meshfree methods Δt does not vanish when the location of the discontinuity approaches the node (Fig. 5.76). Moreover [15] showed that the critical time step depends on the size of domain of influence. Thus, for a patch of nodes with nodal spacing of 1 m different size of domain of influence ($DMAX$), the normalized critical time step $\Delta t_{normalized} = \dfrac{\Delta t_{discon}}{\Delta t_{con}}$ versus the relative position of the crack to the neighboring nodes is plotted in Fig. 5.76 for 1D and Fig. 5.77 for 2D. The size of the domain of influence affects the critical time step. Also, the critical time step is maximum when the crack is located in the middle of the nodes.

5.13.2.2 Mass lumping strategy 1 (MLS1)

The mass lumping technique introduced by [54], see also [57,127], is a strategy that always leads to positive masses at the nodes. For finite elements,

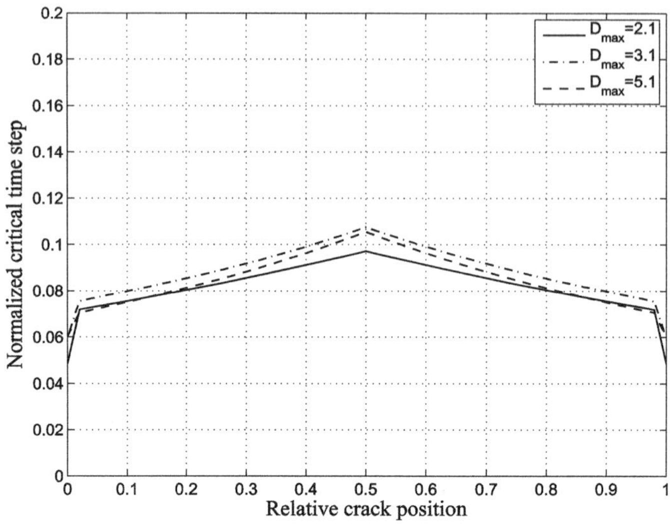

Figure 5.76 Normalized critical time step versus the crack position for the consistent mass matrix in 1D.

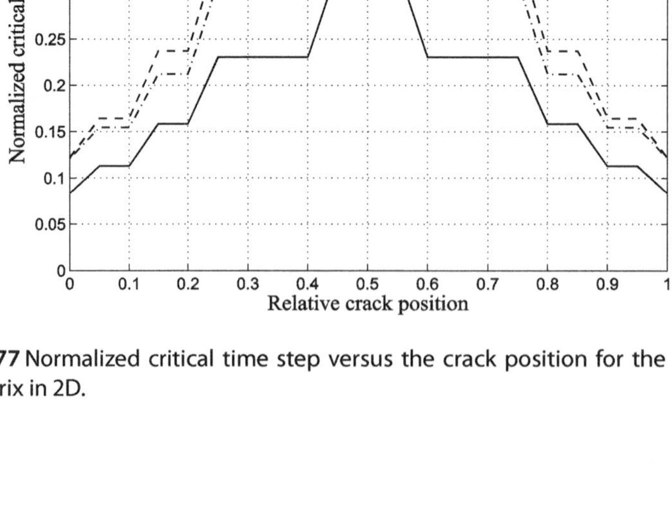

Figure 5.77 Normalized critical time step versus the crack position for the consistent mass matrix in 2D.

the idea is to start from a consistent mass matrix and to scale the diagonal terms in such a way that the mass is constant within the element. With little change this procedure can be used for meshfree methods where instead of the elements we have the background cells [72]:

$$m_I^{MM} = \vartheta_h M_{II} \mathbf{I} \tag{5.358}$$

where

$$M_{II} = \int_{\Omega_h} \varrho_0 \tilde{\Phi}_I^2 d\Omega \tag{5.359}$$

and the scaling factor

$$\vartheta_h = \frac{M_h}{\sum_{I=1}^{N_h} M_{II}}, \qquad M_h = \int_{\Omega_h} \varrho_0 d\Omega \tag{5.360}$$

where m_I^{MM} is the lumped mass value for the particle, N_h is the number of particles that are in the domain of influence of the integration point $\mathbf{X}$ and M_h is the mass of the integration cell. Also, $\tilde{\Phi}_I$ refers to any shape function including enriched and non-enriched ones. Using Eq. (5.358) leads to a lumped mass matrix that conserves the mass of the whole system even when the enrichment functions exist.

The justification of Hinton's special lumping technique is that it retains the diagonal part of the consistent mass matrix, and assumes that the diagonal part of the consistent mass matrix covers the correct frequency range of the dynamic response, whereas the non-diagonal part of the consistent mass matrix is not essential for the final results. This technique ensures the positive definiteness of the mass matrix, and eliminates the singular mode. A possible setback could be that it cuts of the connection, or interaction between the neighboring material particles. However, this setback may be compensated by the nonlocal nature of mesh-free methods, because each material point in mesh-free methods is covered by more than one shape function; therefore the interaction between the adjacent particles is always present. This special lumping technique was proven to produce high-quality, detailed resolution shear-band solutions in numerical simulations using the RKPM [72].

5.13.2.3 Mass lumping strategy 2 (MLS2)

The second lumping strategy is based on the idea proposed by Menouillard et al. [42,78]. The goal is that the discretization preserves the value

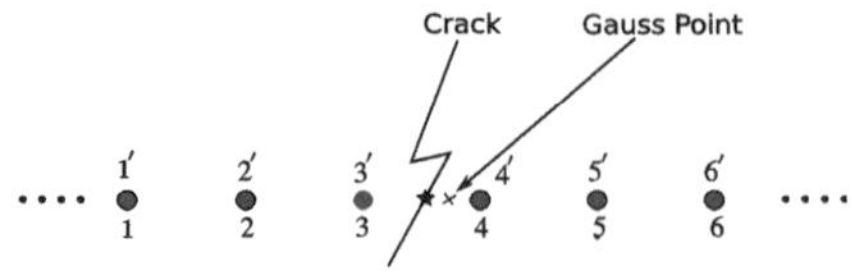

Figure 5.78 A sample Gauss point where the neighboring nodes are enriched.

of the kinetic energy with respect to rigid body translations. Adapting this idea to meshfree methods introduces some complications, since MMs are generally not interpolatory. Thus, we use some simplifying assumptions in order to derive a workable expression for the lumped mass matrix. Numerical experiments will show later the good performance of the obtained expression.

Let us consider an equally spaced sample patch of N nodes, for which there is a point of discontinuity in the middle. See Fig. 5.78. The patch of nodes resembles a finite element or an enriched element when there is a discontinuity. Having this in mind and following Menouillard et al. [78], we assume that all the nodes in the patch are enriched; therefore we will not consider the blending region. Thus we have N extra degrees of freedom, the same as number of nodes in the patch. As noted in [91], for any meshfree or finite element method, the lumped mass of the ordinary degrees of freedom can be calculated by:

$$M_I = \sum_{J=1}^{N} \rho\left(X_J\right) \Phi_I\left(X_J\right) W_J \quad \text{and} \quad M_I = \rho\left(X_I\right) W_I \quad \text{if } \Phi\left(X_J\right) = \delta_{IJ} \tag{5.361}$$

where W_I is the nodal weighting, which is the length in 1D, the area in 2D and the volume in 3D. Eq. (5.361) is the same as the result of the row-sum technique to compute the lumped masses. The lumped mass obtained from the expressions at the left and at the right in Eq. (5.361) are close if the support is tight.

Now we need a mass matrix in the format below:

$$\mathbf{M} = \begin{bmatrix} M_1 & & & & & \\ & M_2 & & & 0 & \\ & & \cdot & & & \\ & & & M_N & & \\ & & & & M_1' & \\ & 0 & & & \cdot & \\ & & & & & M_N' \end{bmatrix} \tag{5.362}$$

where M_I are the mass for ordinary degrees of freedom and the M_I' stand for the mass of the extra degrees of freedom. As noted by [78] we can still use the row-sum technique for the usual degrees of freedom. To obtain the mass for the extra degrees of freedom, we use the concept of preserving the value of the exact kinetic energy with respect to rigid body motions. In order to do this, the arbitrary enrichment function ψ is imposed as an initial velocity condition everywhere within the patch of nodes by using the velocity array $\hat{\mathbf{V}} = [\ 0 \quad \ldots \quad 0 \quad \psi \quad \ldots \quad \psi\ ]^T$, where $\hat{\mathbf{V}}$ denotes the value of the velocity approximation evaluated at the nodes, as opposed to the value of the approximation parameters related to these nodes, which are arranged in $\mathbf{V}$. Arrays $\hat{\mathbf{V}}$ and $\mathbf{V}$ can be related by the transformation method in [25]. With the assumptions made above, we can consider that $\hat{\mathbf{V}} = \mathbf{V}$, which, in a general case, does not hold (remember that meshfree shape functions do not satisfy the Kronecker delta properties).

We impose that the corresponding discrete and exact kinetic energies are equal in order to preserve the value of the kinetic energy, which reads

$$\frac{1}{2}\int_{\Omega} \rho V^2 d\Omega = \frac{1}{2}\mathbf{V}^T\mathbf{M}\mathbf{V} \tag{5.363}$$

Since the meshfree shape functions we are considering satisfy the partition of unity and velocity is imposed equally to all the additional degrees of freedom, the nodal values of the initial velocity are equal to the real values. Therefore, we will have

$$\frac{1}{2}\int_{\Omega} \rho V^2 d\Omega = \frac{1}{2}\int_{\Omega} \psi^2 \rho d\Omega = \frac{1}{2}\mathbf{V}^T\mathbf{M}\mathbf{V} = \frac{1}{2}\hat{\mathbf{V}}^{\mathbf{T}}\hat{\mathbf{M}}\hat{\mathbf{V}} \tag{5.364}$$

and $\hat{M}_I = \hat{M}_J$, as well as $M_I = \hat{M}_I$. The symbol $\hat{M}_I$ stands for the nodal masses calculated from the shape functions, with $\Phi_I(X_J) = \delta_{IJ}$. Of course, this is not true for the nodes located on the boundary. However, we also assume our patch of nodes is away enough from the boundary. Notice that $\hat{\mathbf{M}}$ and $\mathbf{M}$ are both diagonal and equal to each other, with the assumptions made here. However, in general, we would have to consider a non-diagonal $\mathbf{M}$, which is of no use for our purposes, or $M_I \neq M_J$, which will complicate the expressions of the resulting lumped mass matrix.

With all the above considerations, from Eq. (5.364) we have:

$$\hat{M}_I' = \frac{1}{\sum_{I=1}^{N_Q} \psi^2}\int_{\Omega} \psi^2 \rho d\Omega \tag{5.365}$$

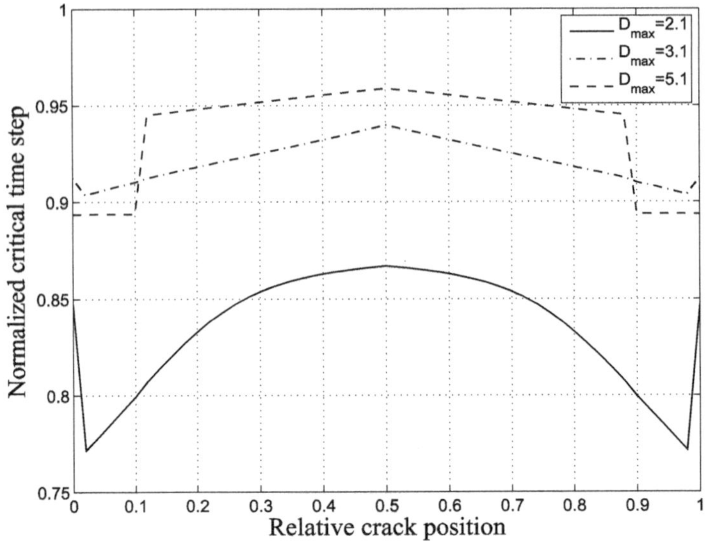

Figure 5.79 Normalized critical time step versus the crack position for mass lumping strategy 1 (MLS1) in 1D.

This expression is similar to the one obtained in [42]. For the step enrichment we obtain:

$$\hat{M}_I' = \frac{1}{\sum_{I=1}^{N} \psi^2} \int_\Omega \psi^2 \rho d\Omega = \frac{1}{\sum_{I=1}^{N} 1^2} \int_\Omega 1^2 \rho d\Omega = \frac{1}{N} \int_\Omega \rho d\Omega = \hat{M}_I \tag{5.366}$$

5.13.2.4 Critical time step analysis

MLS1

Let us consider again the one-dimensional bar. Fig. 5.79 shows the normalized Δt for the lumped mass matrix versus the crack position. As the figure clearly illustrates, Δt_{discon} is of order of Δt_{con}. As can be seen in Fig. 5.79 there is a jump in the graph which belongs to the point that the enriched nodes are changed. This is directly related to the value of $DMAX$. The jumps in Fig. 5.79 occur due to the change in the nodes to be enriched. In other words when the position of the discontinuity is altered, the enrichment of the nodes change as well. Fig. 5.80 also shows the values of Δt versus crack position in 1D. It is obvious that with increasing size of domain of influence (DMAX) the time step size increases; however, a bigger DMAX implies excessive smoothing of the fields and more neighboring nodes which in turn means more computational time.

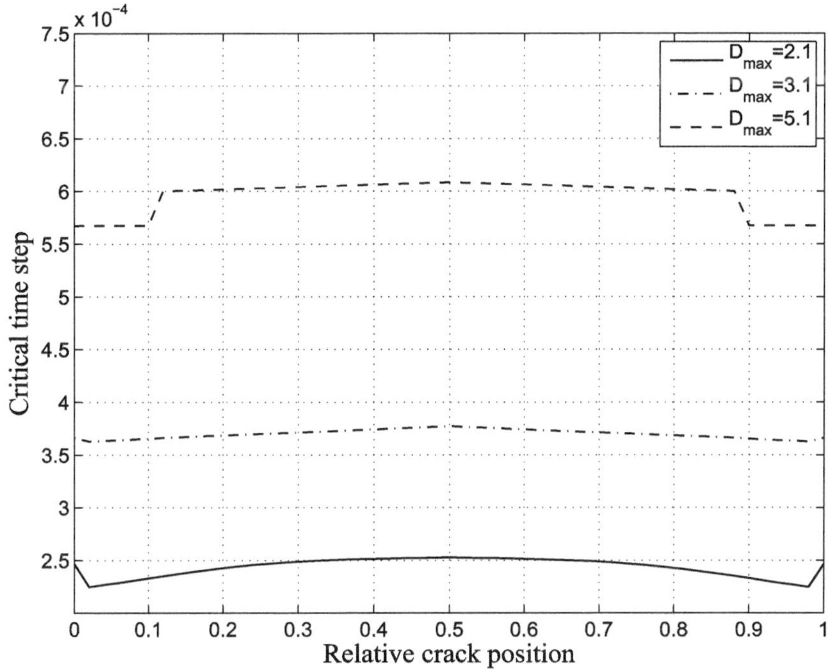

Figure 5.80 Critical time step versus the crack position for mass lumping strategy 1 (MLS1) in 1D.

In 2D, the crack position is varied as illustrated in Fig. 5.81 and Fig. 5.82. The dimensions of the patch are 19 by 19 and 20 particles in every direction. The density, Young's modulus and the Poisson's ratio are 7.8, 210000 and 0.3 respectively. We first consider a pure step enrichment, Fig. 5.79, where we vary the crack position in the x direction while the crack completely cuts the patch. Then, the tip is placed at the relative potion of $y = 0.9$ and x is varying. This resembles a rather difficult case since the crack tip will be very close to the neighboring nodes. Fig. 5.83 and Fig. 5.85 show the normalized Δt results for mass lumping strategy 1 in 2D with and without tip enrichment. Fig. 5.84 also shows the computed Δt for different DMAX. As it is realized from Fig. 5.85, the order of Δt_{discon} to Δt_{con} for the first lumping strategy drops down to 30 percent.

MLS2

First, we consider again the one-dimensional example, Fig. 5.86. Fig. 5.86 shows the normalized Δt and Fig. 5.87 shows the computed values of Δt with different DMAX. The critical time step of MLS2 (Δt_{discon}) is simi-

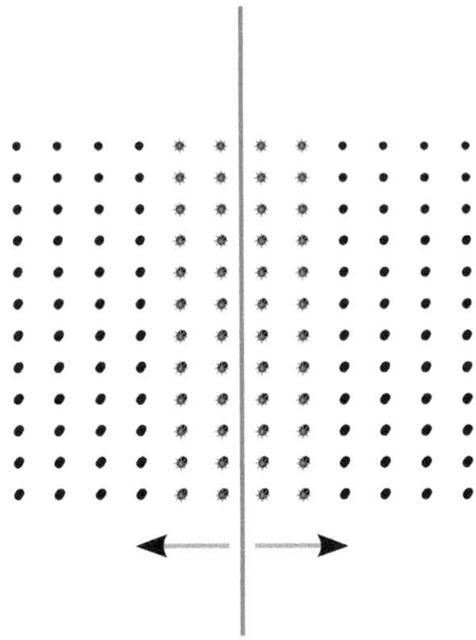

Figure 5.81 The 2D patch of particles to compute the critical time step versus crack position with step enrichment only.

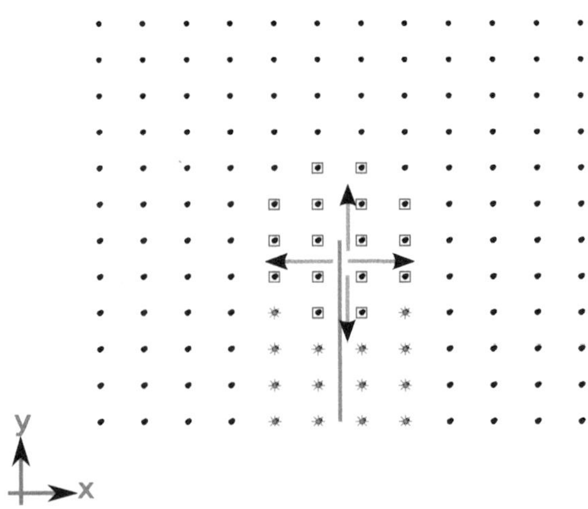

Figure 5.82 The 2D patch of particles to compute the critical time step versus crack position with tip enrichment.

lar to Δt_{con}. Fig. 5.88, Fig. 5.89, Fig. 5.90 and Fig. 5.91 show the results for Δt in 2D. The time step for the enrichment drops in the worst case to 85% compared to the critical time step without discontinuity. Moreover, the changes in Δt with respect to DMAX complies with the results obtained by Belytschko et al. [15] in the continuous problem. While the

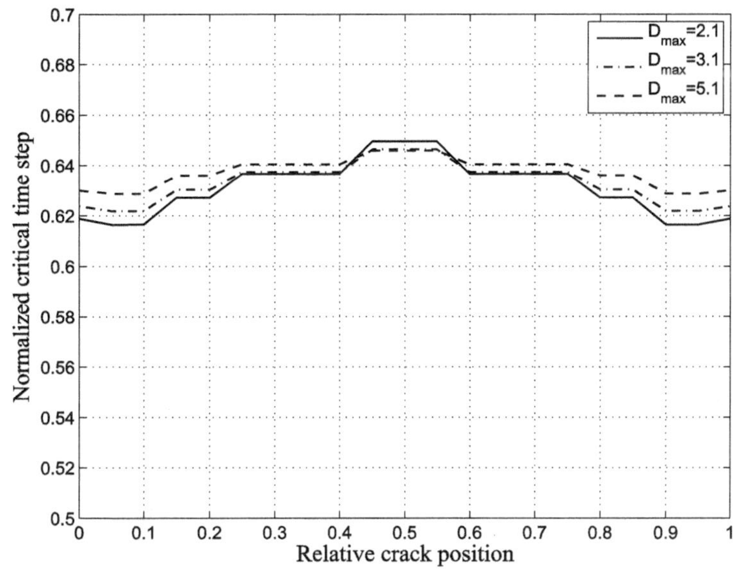

Figure 5.83 Normalized critical time step versus the crack position for mass lumping strategy 1 (MLS1) in 2D.

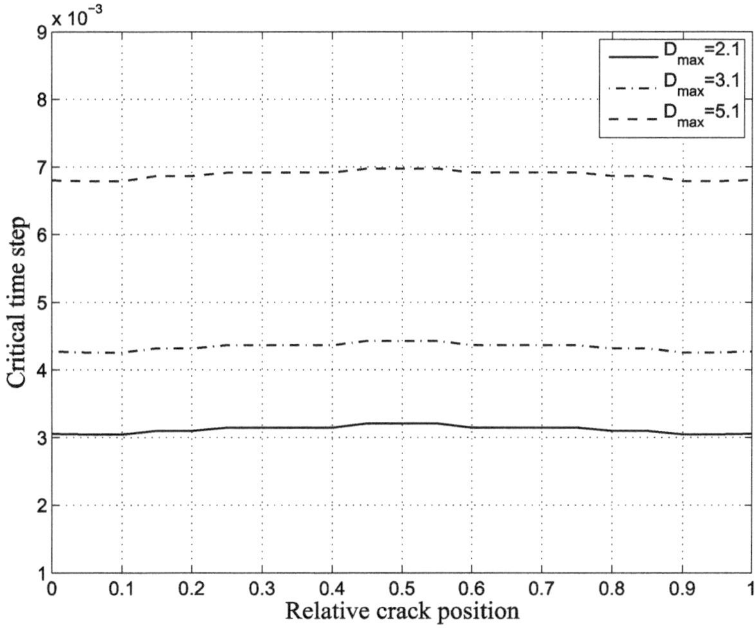

Figure 5.84 Critical time step versus the crack position for mass lumping strategy 1 (MLS1) in 2D.

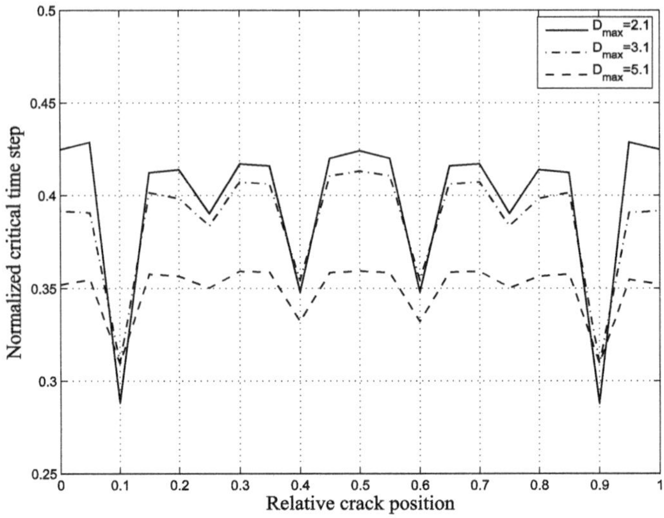

Figure 5.85 Normalized critical time step versus the crack position for mass lumping strategy 1 (MLS1) in 2D with tip enrichment.

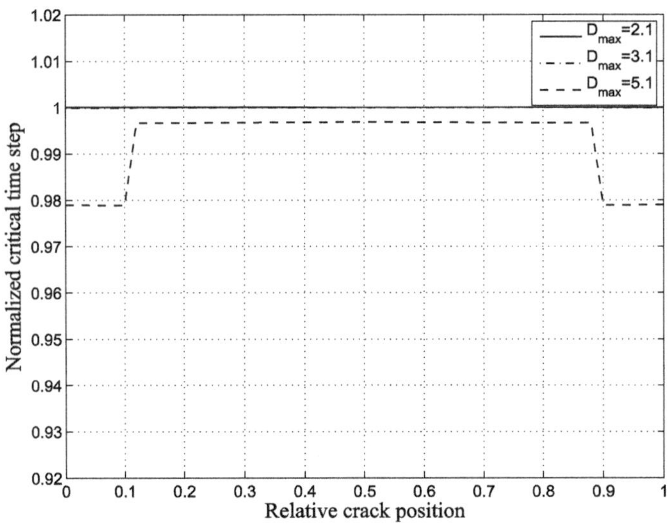

Figure 5.86 Normalized critical time step versus the crack position for mass lumping strategy 2 (MLS2) in 1D.

critical time step in MLS2 drops to 80% of Δt_{con} the MLS1 allows only 30% of Δt_{con}. We attribute these difference to the different mass values. While MLS1 'conserves' the mass of the whole system which implies smaller mass

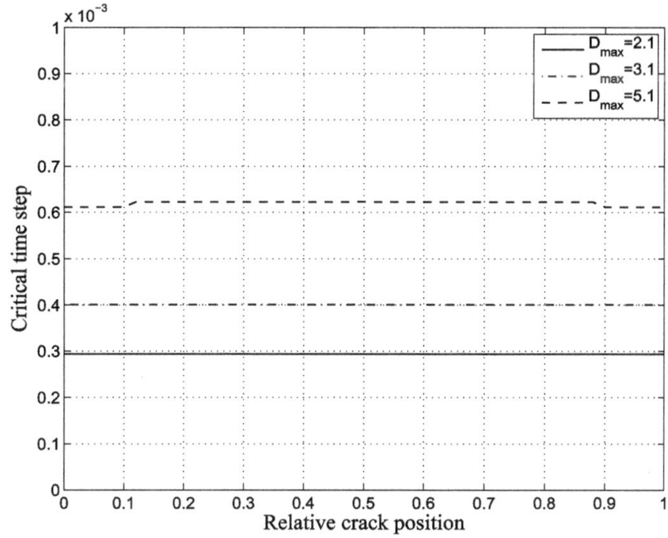

Figure 5.87 Critical time step versus the crack position for mass lumping strategy 2 (MLS2) in 2D without tip enrichment.

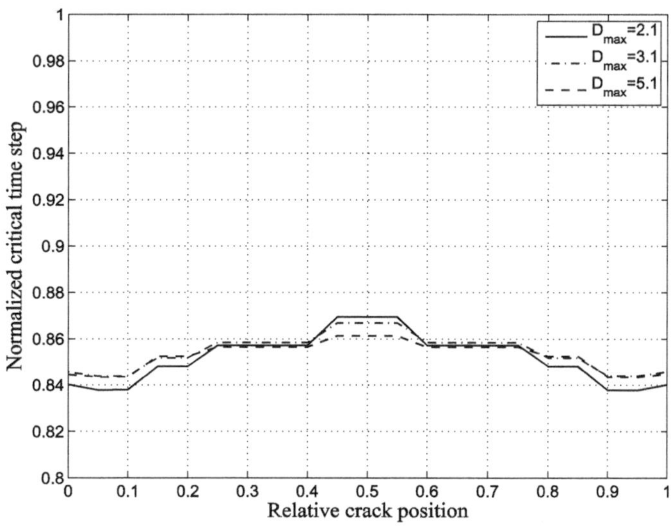

Figure 5.88 Normalized critical time step versus the crack position for mass lumping strategy 2 (MLS2) in 2D without tip enrichment.

for some degrees of freedom which leads to smaller time step, the MLS2 increases the mass or in other words: $\sum_{I=1}^{N_{total}} M_{II}^{MLS1} < \sum_{I=1}^{N_{total}} M_{II}^{MLS2}$. For a pure rigid body motion of two crack surfaces, with $\boldsymbol{u} = \mathbf{0}$, MLS2 seems

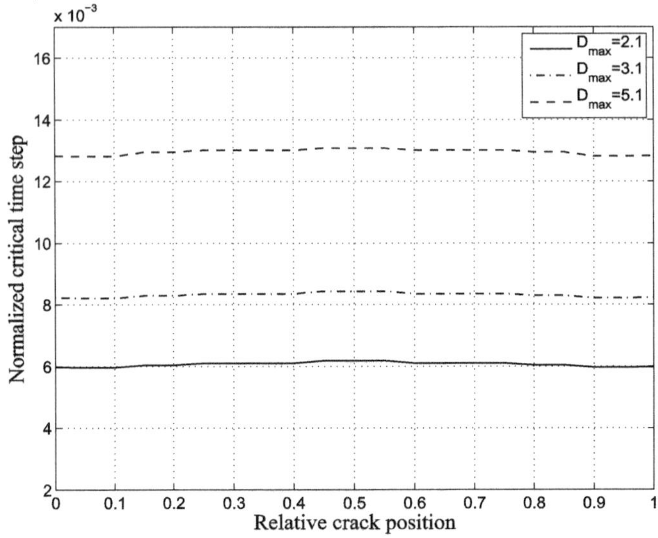

Figure 5.89 Critical time step versus the crack position for mass lumping strategy 2 (MLS2) in 2D without tip enrichment.

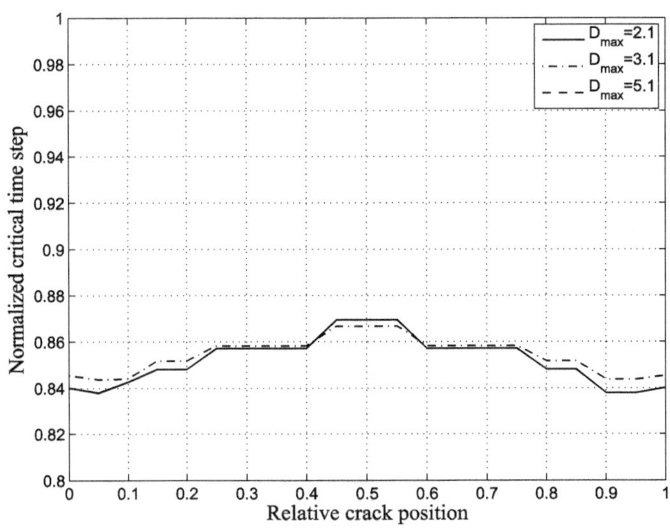

Figure 5.90 Normalized critical time step versus the crack position for mass lumping strategy 2 (MLS2) in 2D with tip enrichment.

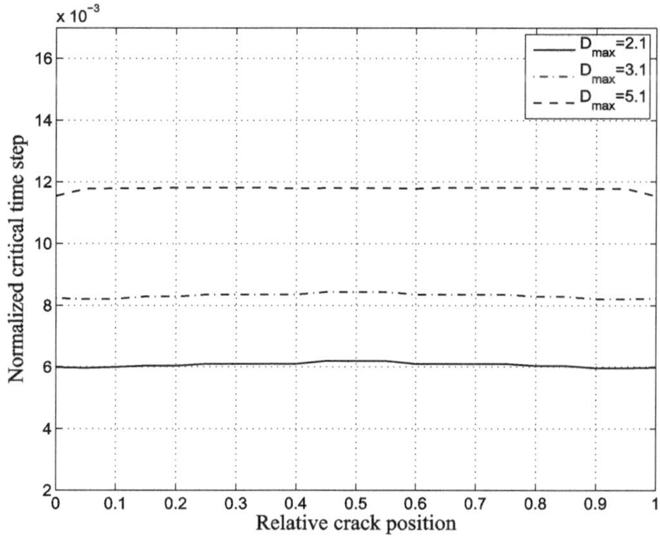

Figure 5.91 Critical time step versus the crack position for mass lumping strategy 2 (MLS2) in 2D with tip enrichment.

more appropriate and MLS1 will produce too small time steps; however, when the kinetic is not conserved,[11] MLS2 will provide a mass value that is too large.

5.13.2.5 Analytical critical time step estimates

By using the fact that, for both mass lumping techniques studied here, the values of the critical time steps for the case with and without discontinuities are similar, we get some critical time step estimates for the latter case. Note that the second mass lumping technique reduces to the row-sum method when no extra degrees of freedom are considered.

Now, let us consider a problem in one dimension with equidistant particle arrangement and constant density in the entire domain so that

$$\boldsymbol{M}_{\text{cell}} = \varrho \int_{\Omega^c} \boldsymbol{\Phi}^T(x)\boldsymbol{\Phi}(x)d\Omega^c \tag{5.367}$$

and

$$\boldsymbol{K}_{\text{cell}} = \mathrm{E} \int_{\Omega^c} \boldsymbol{\Phi}_x(x)\boldsymbol{\Phi}_x(x)d\Omega^c \tag{5.368}$$

[11] This will not be the case for generic problems.

where E is the Young's modulus,

$$\boldsymbol{\Phi} = \begin{bmatrix} \Phi_1(x) & \Phi_2(x) & \ldots & \Phi_n(x) \end{bmatrix} \tag{5.369}$$

are the shape functions, n the number of nodes that are different from zero in the integration cell and

$$\boldsymbol{\Phi}_{,x} = \begin{bmatrix} \Phi_{1,x}(x) & \Phi_{2,x}(x) & \ldots & \Phi_{n,x}(x) \end{bmatrix} \tag{5.370}$$

are the derivatives of the shape functions. Solving the integrals defined in mass and stiffness matrices, we will obtain

$$\boldsymbol{M}_{\text{cell}} = \rho\Delta x \begin{bmatrix} m_{11} & m_{12} & \ldots & m_{1n} \\ m_{21} & m_{22} & \ldots & m_{2n} \\ \vdots & \vdots & & \vdots \\ m_{n1} & m_{n2} & \ldots & m_{nn} \end{bmatrix} = \rho\Delta x \boldsymbol{M}_{\text{dmax}} \tag{5.371}$$

and

$$\boldsymbol{K}_{\text{cell}} = \frac{\mathrm{E}}{\Delta x} \begin{bmatrix} k_{11} & k_{12} & \ldots & k_{1n} \\ k_{21} & k_{22} & \ldots & k_{2n} \\ \vdots & \vdots & & \vdots \\ k_{n1} & k_{n2} & \ldots & k_{nn} \end{bmatrix} = \mathrm{E}\frac{1}{\Delta x}\boldsymbol{K}_{\text{dmax}} \tag{5.372}$$

where Δx is the length of the integration cell and m_{ij} and k_{ij} depend on D_{max}. The matrices $\boldsymbol{M}_{\text{dmax}}$ and $\boldsymbol{K}_{\text{dmax}}$ are introduced for the sake of convenience in further derivations.

Solving the generalized eigenvalue problem, it can be shown that

$$\lambda\left(\boldsymbol{M}_{\text{cell}}, \boldsymbol{K}_{\text{cell}}\right) = \frac{\mathrm{E}}{\rho\Delta x^2}\lambda\left(\boldsymbol{M}_{\text{dmax}}, \boldsymbol{K}_{\text{dmax}}\right) \tag{5.373}$$

where λ is a generalized eigenvalue. The above expression is valid for both the consistent and the lumped mass matrices.

For finding an estimate for the critical time step, valid for various values of the parameters involved, we will solve the eigenvalue problem for different values of D_{max} and then obtain an analytic expression by fitting techniques.

In [15] the largest eigenvalue obtained for the row-sum technique for the same one-dimensional problem is

$$\lambda = \alpha \frac{E}{\varrho \Delta x^2 D_{max}^2} \tag{5.374}$$

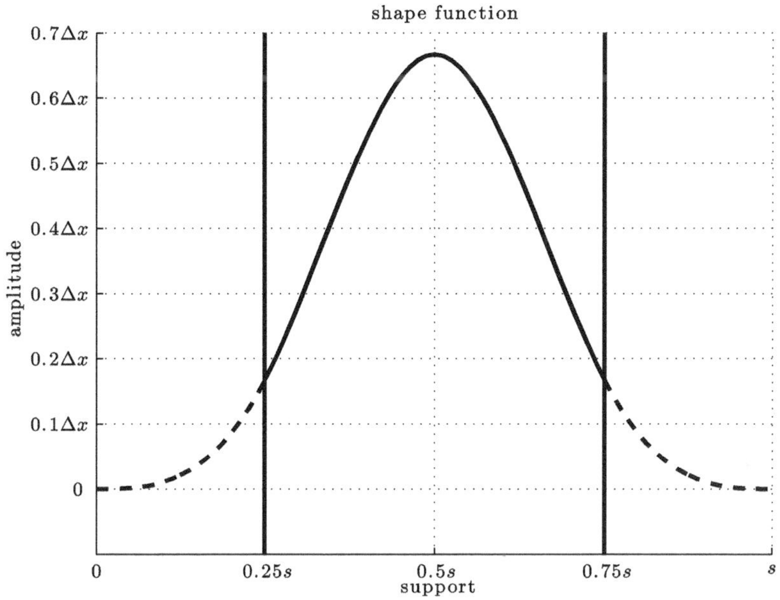

Figure 5.92 Shape function. The dotted lines are the values of the function that are not considered.

where $\alpha = 13.30$. As shown in [15] this result can be generalized for two-dimensional problems. We will adopt the form of this expression in order to perform the above mentioned fitting. With our framework, for the same mass lumping strategy we get $\alpha = 12.46$.

However, the expression obtained for λ with the first mass lumping is not very sharp. In practice, the critical time step estimates obtained seem to be excessively large. For these reasons, we propose an alternative that allows to obtain better estimations for λ.

In general, in the mass matrix integration cell there exist very small contributions of some nodes compared with other nodes. This effect enlarges the value of λ. However, in the global system, these big differences between the nodal contributions are not observed. Moreover, in practice, it seems that these small contributions decrease the critical time step estimation with respect to the real critical time step of the global system.

We propose the following heuristic solution. Let us denote the length of the shape function support as s, and consider a local coordinate whose origin coincides with the initial point of the cell. In order to avoid small contributions, we will only consider the support of any shape function from 0.25 s to 0.75 s. In Fig. 5.92, the dotted line shows the values that are

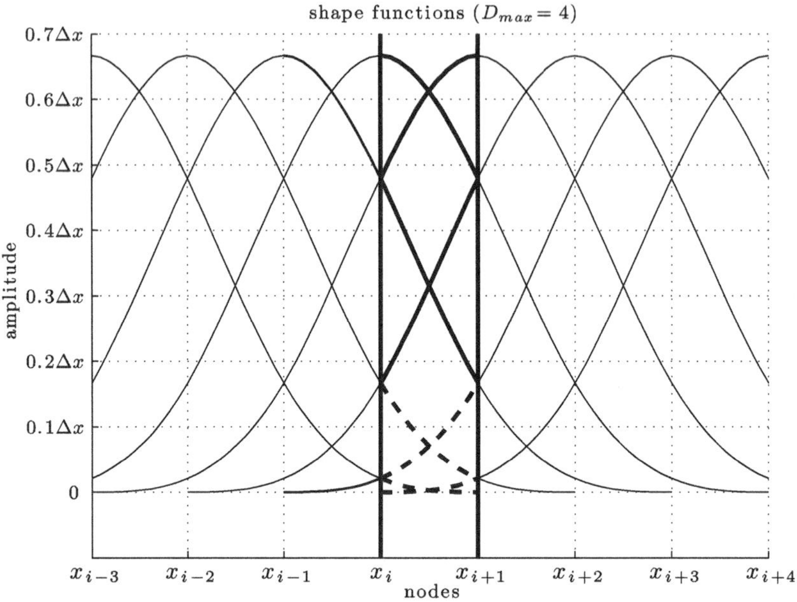

Figure 5.93 Nodal shape functions in an integration cell. This cell is defined in the interval between the node x_i and the node x_{i+1}. The dotted line indicates the values of the shape functions that are not considered.

excluded from the shape function. Also, we will only use even values for the parameter D_{max}. It is easy to see that this avoids very little contribution of some nodes in mass matrix.

As an example, see Fig. 5.93. We have an arbitrary integration cell defined between the nodes x_i and x_{i+1}. The value of D_{max} is equal to 4. The dotted line shows the values of the nodal shape functions that we exclude. That is, we are not considering the contribution of the shape functions associated with nodes x_{i-3}, x_{i-2}, x_{i+3} and x_{i+4}. Notice that these contributions are the smallest in the integration cell considered.

Finally, solving the eigenvalue problem for different even values of D_{max}, we obtain $\alpha = 7.1$ in Eq. (5.374) for the row-sum technique (relevant for mass lumping 2) and $\alpha = 10$ for the first mass lumping technique.

5.13.3 Crack propagation in time

In [106], it was reported that inaccuracies can occur in dynamic applications due to the change of the solution space over time when the cracks grow, i.e. the energy balance is not guaranteed if the additional unknowns are not

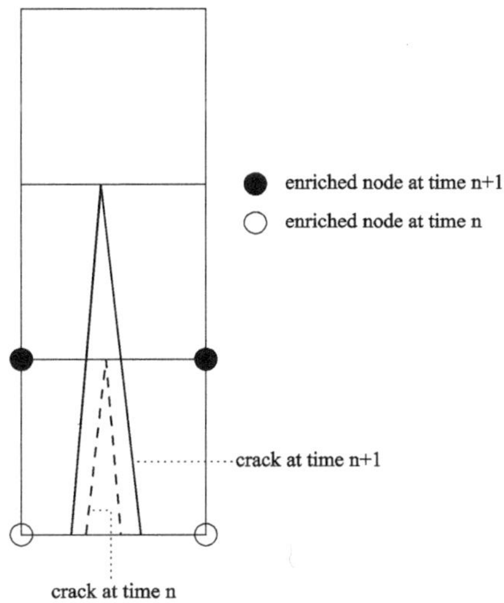

Figure 5.94 Crack configuration at different time steps.

'initialized' correctly. Therefore, we consider a propagation crack as shown in Fig. 5.94. The crack configuration is illustrated at two different time steps. Usually, the test and trial functions are evaluated at time $n+1$ for the crack configuration at $n+1$ as illustrated in Fig. 5.94 and when the crack evolves in time, they are evaluated at time n also for the crack configuration $n+1$. This does not take into account the change in the solution space that arises from time n at crack configuration n that will lead to an error in the energy balance. Let us consider the crack configurations in Fig. 5.94. To avoid that error, the two black nodes that will be enriched at time $n+1$ due to the propagation of the crack, have to be enriched at time n as well but initialized properly such that the crack has to close at the element edge at time n. Otherwise, an error is introduced in the energy balance due to the change of the solution space. In most of our problems, we did not find large errors in the energy balance even without this "correct initialization". However, the change in the solution space might explain the oscillations in the stress field in front of the crack tip that are observed in dynamic XFEM computation. The source of these oscillations might be the level set function that changes direction when the crack changes direction, which in turn changes the shape functions. These problems might be avoided by using a space-time method, [24,107]. Another source of the stress oscillations around the crack tip that occur in finite elements is due to the stress jumps across the element edges. These jumps do not occur in meshfree methods due to the higher order continuous shape functions.

References

[1] F. Amiri, C. Anitescu, M. Arroyo, S. Bordas, T. Rabczuk, XLME interpolants, a seamless bridge between XFEM and enriched meshless methods, Computational Mechanics 53 (1) (2014) 45–57.

[2] P. Areias, J. Song, T. Belytschko, Analysis of fracture in thin shells by overlapping paired elements, International Journal for Numerical Methods in Engineering 195 (2006) 5343–5360.

[3] M. Arroyo, M. Ortiz, Local maximum-entropy approximation schemes: a seamless bridge between finite elements and meshfree methods, International Journal for Numerical Methods in Engineering 65 (2006) 2167–2202.

[4] S. Atluri, S. Shen, The meshless local Petrov-Galerkin (MLPG) method: a simple and less-costly alternative to the finite element and boundary element methods, Computer Modeling in Engineering & Sciences 3 (2002) 11–51.

[5] I. Babuška, U. Banerjee, Stable generalized finite element method (SGFEM), Computer Methods in Applied Mechanics and Engineering 201 (Jan. 2012) 91–111.

[6] I. Babuška, U. Banerjee, J.E. Osborn, Q. Zhang, Effect of numerical integration on meshless methods, Computer Methods in Applied Mechanics and Engineering 198 (37) (2009) 2886–2897.

[7] E. Bechet, H. Minnebo, N. Moës, B. Burgardt, Improved implementation and robustness study of the X-FEM for stress analysis around cracks, International Journal for Numerical Methods in Engineering 64 (2005) 1033–1056.

[8] T. Belytschko, T. Black, Elastic crack growth in finite elements with minimal remeshing, International Journal for Numerical Methods in Engineering 45 (5) (1999) 601–620.

[9] T. Belytschko, M. Fleming, Smoothing, enrichment and contact in the element-free Galerkin method, Computers & Structures 71 (1999) 173–195.

[10] T. Belytschko, S. Xiao, Stability analysis of particle methods with corrected derivatives, Computers & Mathematics with Applications 43 (2002) 329–350.

[11] T. Belytschko, Y. Lu, L. Gu, Element-free Galerkin methods, International Journal for Numerical Methods in Engineering 37 (1994) 229–256.

[12] T. Belytschko, D. Organ, Y. Krongauz, A coupled finite element–element-free Galerkin method, Computational Mechanics 17 (3) (1995) 186–195.

[13] T. Belytschko, M. Fleming, D. Organ, Y. Krongauz, W. Liu, Smoothing and accelerated computations in the element free Galerkin method, Journal of Computational and Applied Mathematics 74 (1996) 111–126.

[14] T. Belytschko, Y. Krongauz, D. Organ, M. Fleming, P. Krysl, Meshless methods: an overview and recent developments, Computer Methods in Applied Mechanics and Engineering 139 (1996) 3–47.

[15] T. Belytschko, Y. Guo, W. Liu, S. Xiao, A unified stability analysis of meshfree particle methods, International Journal for Numerical Methods in Engineering 48 (2000) 1359–1400.

[16] T. Belytschko, W.K. Liu, B. Moran, Nonlinear Finite Elements for Continua and Structures, John Wiley and Sons, Chichester, 2000.

[17] T. Belytschko, N. Moes, S. Usui, C. Parimi, Arbitrary discontinuities in finite elements, International Journal for Numerical Methods in Engineering 50 (4) (2001) 993–1013.
[18] W. Benz, E. Asphaug, Simulations of brittle solids using smooth particle hydrodynamics, Computer Physics Communications 87 (1995) 253–265.
[19] J. Bonet, S. Kulasegaram, Correction and stabilization of smooth particle hydrodynamics methods with applications in metal forming simulations, International Journal for Numerical Methods in Engineering 47 (6) (2000) 1189–1214.
[20] J. Bonet, T. Lok, Variational and momentum preservation aspects of smooth particle hydrodynamic formulations, Computer Methods in Applied Mechanics and Engineering 180 (1–2) (1999) 97–115.
[21] S. Bordas, Extended Finite Element and Level Set Methods With Applications to Growth of Cracks and Biofilms, Phd thesis, Northwestern University, December 2003.
[22] S. Bordas, T. Rabczuk, H. Nguyen-Xuan, S. Natarajan, T. Bog, N. Phu, Q.D. Minh, H.N. Vinh, Strain smoothing in FEM and XFEM, Computers & Structures 88 (23–24) (2010) 1419–1443, https://doi.org/10.1016/j.compstruc.2008.07.006.
[23] A. Brooks, T. Hughes, Streamline upwind/Petrov-Galerkin formulations for convective dominated flows with particular emphasis on the incompressible Navier-Stokes equations, Computer Methods in Applied Mechanics and Engineering 32 (1982) 199–259.
[24] P. Cavin, A. Gravouil, A. Lubrecht, A. Combescure, Automatic energy conserving space-time refinement for linear dynamic structural problems, International Journal for Numerical Methods in Engineering 64 (3) (2005) 304–321.
[25] J. Chen, H. Wang, New boundary condition treatments in meshfree computation of contact problems, Computer Methods in Applied Mechanics and Engineering 187 (2000) 441–468.
[26] J. Chen, C. Wu, S. Yoon, Y. You, A stabilized conforming nodal integration for Galerkin meshfree-methods, International Journal for Numerical Methods in Engineering 50 (2001) 435–466.
[27] L. Chen, Y. Zhang, Dynamic fracture analysis using discrete cohesive crack method, International Journal for Numerical Methods in Biomedical Engineering 26 (11) (2010) 1493–1502.
[28] J. Chessa, H. Wang, T. Belytschko, On the construction of blending elements for local partition of unity enriched finite elements, International Journal for Numerical Methods in Engineering 57 (7) (2003) 1015–1038.
[29] Y.A. Chu, B. Moran, A computational model for nucleation of solid-solid phase transformations, Modelling and Simulation in Materials Science and Engineering 3 (1995) 455–471.
[30] F. Cirak, M. Ortiz, P. Schröder, Subdivision surfaces: a new paradigm for thin-shell finite-element analysis, International Journal for Numerical Methods in Engineering 47 (12) (2000) 2039–2072.
[31] C.J. Cyron, M. Arroyo, M. Ortiz, Smooth, second order, non-negative meshfree approximants selected by maximum entropy, International Journal for Numerical Methods in Engineering 79 (Sept. 2009) 1605–1632.
[32] C. Daux, N. Moes, J. Dolbow, N. Sukumar, T. Belytschko, Arbitrary branched and intersection cracks with the extended finite element method, International Journal for Numerical Methods in Engineering 48 (2000) 1731–1760.

[33] G. Dilts, Moving least square particle hydrodynamics I: consistency and stability, International Journal for Numerical Methods in Engineering 44 (2000) 1115–1155.
[34] G. Dilts, Moving least square particle hydrodynamics II: conservation and boundaries, International Journal for Numerical Methods in Engineering 48 (10) (2000) 1503–1524.
[35] J. Dolbow, A. Devan, Enrichment of enhanced assumed strain approximation for representing strong discontinuities: addressing volumetric incompressibility and the discontinuous path test, International Journal for Numerical Methods in Engineering 59 (2003) 47–67.
[36] J. Dolbow, N. Moes, T. Belytschko, Discontinuous enrichment in finite elements with a partition of unity method, Finite Elements in Analysis and Design 36 (3) (2000) 235–260.
[37] C. Duarte, J. Oden, An h-p adaptive method using clouds, Computer Methods in Applied Mechanics and Engineering 139 (1–4) (1996) 237–262.
[38] C. Duarte, J. Oden, H-p clouds- an h-p meshless method, Numerical Methods for Partial Differential Equations 12 (6) (1996) 673–705.
[39] M. Duflot, A meshless method with enriched weight functions for three-dimensional crack propagation, International Journal for Numerical Methods in Engineering 65 (12) (2006) 1970–2006.
[40] M. Duflot, N. Hung, A meshless method with enriched weight functions for fatigue crack growth, International Journal for Numerical Methods in Engineering 59 (2004) 1945–1961.
[41] C. Dyka, R. Ingel, An approach for tensile instability in smoothed particle hydrodynamics, Computers & Structures 57 (1995) 573–580.
[42] T. Elguedj, A. Gravouil, A. Combescure, Appropriate extended functions for XFEM simulation of plastic fracture mechanics, Computer Methods in Applied Mechanics and Engineering 195 (7–8) (2006) 501–515.
[43] M. Fleming, Y. Chu, B. Moran, T. Belytschko, Enriched element-free Galerkin methods for crack tip fields, International Journal for Numerical Methods in Engineering 40 (8) (1997) 1483–1504.
[44] R. Flory, Thermodynamic relations for highly elastic materials, Transactions of the Faraday Society 57 (1961) 829–838.
[45] T. Fries, T. Belytschko, The extended/generalized finite element method: an overview of the method and its applications, International Journal for Numerical Methods in Engineering 84 (3) (2010) 253–304, https://doi.org/10.1002/nme.2914.
[46] R. Gingold, J. Monaghan, Smoothed particle hydrodynamics: theory and applications to non-spherical stars, Monthly Notices of the Royal Astronomical Society 181 (1977) 375–389.
[47] K. Girkmann, Flaechentragwerke. 6. Auflage, Springer Verlag, Wiley, New York, 1974.
[48] J. Gosz, W.K. Liu, Admissible approximations for essential boundary conditions in the reproducing kernel particle method, Computational Mechanics 19 (2) (1996) 120–135.
[49] R. Gracy, H. Wang, T. Belytschko, Blending in the extended finite element method by discontinuous Galerkin and assumed strain methods, International Journal for Numerical Methods in Engineering 74 (2008) 1645–1669.
[50] A. Gravouil, N. Moes, T. Belytschko, Non-planar 3D crack growth by the extended finite element and level sets – part II: level set update, International Journal for Numerical Methods in Engineering 53 (2002) 2569–2586.

[51] F.C. Günter, W.K. Liu, Implementation of boundary conditions for meshless methods, Computer Methods in Applied Mechanics and Engineering 163 (1–4) (1998) 205–230.
[52] V. Gupta, S. Rajagopal, N. Gupta, A comparative study of meshfree methods for fracture, International Journal of Damage Mechanics 20 (5) (2010) 729–751, https://doi.org/10.1177/1056789510371222.
[53] D. Hicks, L. Liebrock, Sph hydrocodes can be stabilized with shape-shifting, Computers & Mathematics with Applications 38 (1999) 1–16.
[54] E. Hinton, T. Rock, O. Zienkiewicz, A note on mass lumping and related processes in finite element method, Earthquake Engineering and Structural Dynamics 4 (3) (1976) 245–249.
[55] A. Huerta, S. Fernández-Méndez, Enrichment and coupling of the finite element and meshless methods, International Journal for Numerical Methods in Engineering 48 (11) (2000) 1615–1636.
[56] A. Huerta, T. Belytschko, S. Fernández-Méndez, T. Rabczuk, Meshfree methods, in: Encyclopedia of Computational Mechanics, John Wiley and Sons, 2004.
[57] T. Hughes, The Finite Element Method: Linear Static and Dynamic Finite Element Analysis, Dover Publications, New York, 1987.
[58] T. Hughes, J. Cottrell, Y. Bazilevs, Isogeometric analysis: CAD, finite elements, NURBS, exact geometry and mesh refinement, Computer Methods in Applied Mechanics and Engineering 194 (39) (2005) 4135–4195.
[59] T.J.R. Hughes, W.K. Liu, Implicit-explicit finite elements in transient analysis: implementation and numerical examples, Journal of Applied Mechanics 45 (1978) 375–378.
[60] T.J.R. Hughes, L. Franca, G. Hulber, A new finite element formulation for computational fluid-dynamics: VIII. The Galerkin/least squares method for advective diffusive equations, Computer Methods in Applied Mechanics and Engineering 73 (1989) 173–189.
[61] J. Hutchinson, Singular behavior at the end of a tensile crack in a hardening material, Journal of the Mechanics and Physics of Solids 16 (1) (1968) 13–31.
[62] E.T. Jaynes, Information theory and statistical mechanics, Physical Review 106 (May 1957) 620–630.
[63] E.T. Jaynes, Information theory and statistical mechanics. II, Physical Review 108 (Oct 1957) 171–190.
[64] G. Johnson, S. Beissel, Normalized smoothing functions for SPH impact computations, International Journal for Numerical Methods in Engineering 39 (1996) 2725–2741.
[65] L.M. Kachanov, Time of the rupture process under creep conditions, Izvestiâ Akademii Nauk SSR. Otdelenie Tehničeskih Nauk 8 (1958) 26–31.
[66] Y. Krongauz, T. Belytschko, Consistent pseudo derivatives in meshless methods, Computer Methods in Applied Mechanics and Engineering 146 (1997) 371–386.
[67] P. Krysl, T. Belytschko, Element-free Galerkin method: convergence of the continuous and discontinuous shape functions, Computer Methods in Applied Mechanics and Engineering 148 (1997) 257–277.
[68] L. Kucherov, E. Tadmor, R. Miller, Umbrella spherical integration: a stable meshless method for nonlinear solids, International Journal for Numerical Methods in Engineering 69 (13) (2007) 2807–2847.
[69] P. Laborde, J. Pommier, Y. Renard, M. Salaun, High-order extended finite element method for cracked domains, International Journal for Numerical Methods in Engineering 64 (2005) 354–381.

[70] P. Laborde, J. Pommier, Y. Renard, M. Solau, Higher-order extended finite element method for cracked domains, International Journal for Numerical Methods in Engineering 64 (3) (2005) 354–381.
[71] S.-H. Lee, Y.-C. Yoon, Y. Kim, New local near-tip functions for the element-free Galerkin method, Communications in Numerical Methods in Engineering 21 (3) (2005) 133–148.
[72] S. Li, W. Liu, Numerical simulations of strain localization in inelastic solids using mesh-free methods, International Journal for Numerical Methods in Engineering 48 (13) (2000) 1285–1309.
[73] S. Li, W. Liu, Meshfree Particle Methods, Springer, 2004.
[74] W.K. Liu, S. Li, T. Belytschko, Moving least square reproducing kernel methods. Part I: methodology and convergence, Computer Methods in Applied Mechanics and Engineering 143 (1–2) (1997) 113–154.
[75] J.M. Melenk, I. Babuska, The partition of unity finite element method: basic theory and applications, Computer Methods in Applied Mechanics and Engineering 139 (1996) 289–314.
[76] A. Menk, S.P.A. Bordas, A robust preconditioning technique for the extended finite element method, International Journal for Numerical Methods in Engineering 85 (13) (2011) 1609–1632.
[77] T. Menouillard, J. Rethore, A. Combescure, H. Bung, Efficient explicit time stepping for the eXtended Finite Element Method (X-FEM), International Journal for Numerical Methods in Engineering 68 (9) (2006) 911–939.
[78] T. Menouillard, J. Rethore, N. Moes, A. Combescure, H. Bung, Mass lumping strategies for X-FEM explicit dynamics: application to crack propagation, International Journal for Numerical Methods in Engineering 74 (3) (2008) 447–474.
[79] N. Moes, T. Belytschko, Extended finite element method for cohesive crack growth, Engineering Fracture Mechanics 69 (2002) 813–834.
[80] N. Moes, J. Dolbow, T. Belytschko, A finite element method for crack growth without remeshing, International Journal for Numerical Methods in Engineering 46 (1999) 131–150.
[81] N. Moes, A. Gravouil, T. Belytschko, Non-planar 3-D crack growth by the extended finite element method and level sets, part I: mechanical model, International Journal for Numerical Methods in Engineering 53 (11) (2002) 2549–2568.
[82] J. Morris, A study of the stability properties of smoothed particle hydrodynamics, Publications of the Astronomical Society of Australia 13 (1) (1996) 97–102.
[83] A. Nagarajan, S. Mukherjee, A mapping method for numerical evaluation of two-dimensional integrals with 1/r singularity, Computational Mechanics 12 (1993) 19–26.
[84] S. Natarajan, D.R. Mahapatra, S.P.A. Bordas, Integrating strong and weak discontinuities without integration subcells and example applications in an XFEM/GFEM framework, International Journal for Numerical Methods in Engineering 83 (3) (2010) 269–294.
[85] S. Nicaise, Y. Renard, E. Chahine, Optimal convergence analysis for the extended finite element method, International Journal for Numerical Methods in Engineering 86 (Apr. 2011) 528–548.
[86] H. Nguyen-Vinh, I. Bakar, M.A. Msekh, J.-H. Song, J. Muthu, G. Zi, P. Le, S.P.A. Bordas, R. Simpson, S. Natarajan, T. Lahmer, T. Rabczuk, Extended finite element method for dynamic fracture of piezo-electric materials, Engineering Fracture Mechanics 92 (2012) 19–31.

[87] D. Organ, M. Fleming, T. Terry, T. Belytschko, Continuous meshless approximations for nonconvex bodies by diffraction and transparency, Computational Mechanics 18 (1996) 225–235.
[88] A. Ortiz, M. Puso, N. Sukumar, Maximum-entropy meshfree method for compressible and near-incompressible elasticity, Computer Methods in Applied Mechanics and Engineering 199 (25–28) (2010) 1859–1871.
[89] A. Petschek, L. Libersky, Cylindrical smoothed particle hydrodynamics, Journal of Computational Physics 109 (1) (1993) 76–83.
[90] B. Prabel, A. Combescure, A. Gravouil, S. Marie, Level set XFEM non-matching meshes: application to dynamic crack propagation in elasto-plastic media, International Journal for Numerical Methods in Engineering 69 (8) (2007) 1553–1569.
[91] M. Puso, J. Chen, E. Zywicz, W. Elmer, Meshfree and finite element nodal integration methods, International Journal for Numerical Methods in Engineering 74 (3) (2008) 416–446.
[92] T. Rabczuk, Numerische Untersuchungen zum Fragmentierungsverhalten von Beton, PhD thesis, Institut fuer Massivbau und Baustofftechnologie, Universitaet Karlsruhe, 2002.
[93] T. Rabczuk, T. Belytschko, Cracking particles: a simplified meshfree method for arbitrary evolving cracks, International Journal for Numerical Methods in Engineering 61 (13) (2004) 2316–2343.
[94] T. Rabczuk, T. Belytschko, Application of meshfree particle methods to static fracture of reinforced concrete structures, International Journal of Fracture 137 (1–4) (2006) 19–49.
[95] T. Rabczuk, T. Belytschko, A three dimensional large deformation meshfree method for arbitrary evolving cracks, Computer Methods in Applied Mechanics and Engineering 196 (29–30) (2007) 2777–2799.
[96] T. Rabczuk, T. Belytschko, S. Xiao, Stable particle methods based on Lagrangian kernels, Computer Methods in Applied Mechanics and Engineering 193 (2004) 1035–1063.
[97] T. Rabczuk, S. Xiao, M. Sauer, Coupling of meshfree methods with finite elements: basic concepts and test results, Communications in Numerical Methods in Engineering 22 (10) (2006) 1031–1065.
[98] T. Rabczuk, P. Areias, T. Belytschko, A simplified meshfree method for shear bands with cohesive surfaces, International Journal for Numerical Methods in Engineering 69 (5) (2007) 993–1021.
[99] T. Rabczuk, S. Bordas, G. Zi, A three-dimensional meshfree method for continuous multiple-crack initiation, propagation and junction in statics and dynamics, Computational Mechanics 40 (3) (2007) 473–495.
[100] T. Rabczuk, G. Zi, S. Bordas, H. Nguyen-Xuan, A simple and robust three dimensional cracking particle method without enrichment, Computer Methods in Applied Mechanics and Engineering 30 (2010) 1159–1172.
[101] V. Rajan, Optimality of the Delaunay triangulation in R^d, Discrete & Computational Geometry 12 (2) (1994) 189–202.
[102] P. Randles, L. Libersky, Recent improvements in SPH modeling of hypervelocity impact, International Journal of Impact Engineering 20 (1997) 525–532.
[103] P. Randles, L. Libersky, Normalized SPH with stress points, International Journal for Numerical Methods in Engineering 48 (2000) 1445–1462.

[104] B.N. Rao, S. Rahman, An efficient meshless method for fracture analysis of cracks, Computational Mechanics 29 (2000) 398–408.
[105] B.N. Rao, S. Rahman, An enriched meshless method for non-linear fracture mechanics, International Journal for Numerical Methods in Engineering 59 (2004) 197–223.
[106] J. Rethore, A. Gravouil, A. Combescure, An energy-conserving scheme for dynamic crack growth using the extended finite element method, International Journal for Numerical Methods in Engineering 63 (5) (2005) 631–659.
[107] J. Rethore, A. Gravouil, A. Combescure, A combined space-time extended finite element method, International Journal for Numerical Methods in Engineering 64 (2) (2005) 260–284.
[108] J.R. Rice, G. Rosengreen, Plain strain deformation near a crack tip in a power hardening material, Journal of the Mechanics and Physics of Solids 16 (1968) 1–13.
[109] P. Rozycki, N. Moes, E. Bechet, C. Dubois, X-FEM explicit dynamics for constant strain elements to alleviate mesh constraints on internal or external boundaries, Computer Methods in Applied Mechanics and Engineering 197 (5) (2008) 349–363.
[110] C. Shannon, A mathematical theory of communication, The Bell System Technical Journal 27 (1948) 623–656.
[111] D. Shepard, A two-dimensional interpolation function for irregularly spaced data, in: Proc. ACM Nat. Conf., 1968, pp. 517–524.
[112] J.-H. Song, T. Belytschko, Cracking node method for dynamic fracture with finite elements, International Journal for Numerical Methods in Engineering 77 (3) (2009) 360–385.
[113] F. Stazi, E. Budyn, J. Chessa, T. Belytschko, XFEM for fracture mechanics with quadratic elements, Computational Mechanics 31 (1–2) (2003) 38–48.
[114] T. Strouboulis, K. Copps, I. Babuška, The generalized finite element method: an example of its implementation and illustration of its performance, International Journal for Numerical Methods in Engineering 47 (8) (2000) 1401–1417.
[115] N. Sukumar, The Natural Element Method in Solid Mechanics, PhD thesis, Northwestern University, 1998.
[116] N. Sukumar, Construction of polygonal interpolants: a maximum entropy approach, International Journal for Numerical Methods in Engineering 61 (12) (2004) 2159–2181.
[117] N. Sukumar, B. Moran, T. Black, T. Belytschko, An element-free Galerkin method for three-dimensional fracture mechanics, Computational Mechanics 20 (1997) 170–175.
[118] L. Swegle, S. Attaway, M. Heinstein, F. Mello, D. Hicks, An Analysis of Smooth Particle Hydrodynamics, Report No. SAND93-2513-UC-705, Sandia National Laboratories, Albuquerque, New Mexico, 1994.
[119] L. Swegle, D. Hicks, S. Attaway, Smooth particle hydrodynamics stability analysis, Journal of Computational Physics 116 (1995) 123–134.
[120] T. Terry, Fatigue Crack Propagation Modeling Using the Element Free Galerkin Method, Master thesis, Northwestern University, 1994.
[121] P. Tung, Material Stability Analysis of Meshfree Methods, Master thesis, Technical University of Munich, 2007.
[122] G. Ventura, R. Gracie, T. Belytschko, Fast integration and weight function blending in the extended finite element method, International Journal for Numerical Methods in Engineering 77 (2009) 1–29.
[123] G.J. Wagner, W.K. Liu, Application of essential boundary conditions in mesh-free methods: a corrected collocation method, International Journal for Numerical Methods in Engineering 47 (8) (2000) 1367–1379.

[124] G.J. Wagner, W.K. Liu, Hierarchical enrichment for bridging scales and mesh-free boundary conditions, International Journal for Numerical Methods in Engineering 50 (3) (2001) 507–524.

[125] H. Wang, S. Wang, Analysis of dynamic fracture with cohesive crack segment method, Computer Modeling in Engineering & Sciences 35 (2008) 253–274.

[126] H. Westergaard, Bearing pressures and cracks, Journal of Applied Mechanics 6 (1939) A49–A53.

[127] P. Wriggers, Nonlinear Finite Element Methods, Springer, 2008.

[128] B. Xiao, B. Karihaloo, Improving the accuracy of XFEM crack tip fields using higher order quadrature and statically admissible stress recovery, International Journal for Numerical Methods in Engineering 66 (9) (2006) 1378–1410.

[129] S. Xiao, T. Belytschko, Material stability analysis of particle methods, Advances in Computational Mathematics 23 (1–2) (2005) 171–190.

[130] Y. Xu, S. Saigal, Element-free Galerkin study of steady quasi-static crack growth in plane strain tension in elastic-plastic materials, Computational Mechanics 22 (1998) 255–265.

[131] A. Zamani, R. Gracie, M. Eslami, Higher order tip enrichment of extended finite element method in thermoelasticity, Computational Mechanics 46 (2010) 851–866.

[132] Y. Zhang, Meshless modelling of crack growth with discrete rotating crack segments, International Journal of Mechanics and Materials in Design 4 (2008) 71–77.

[133] T. Zhu, S.N. Atluri, A modified collocation method and a penalty formulation for enforcing the essential boundary conditions in the element free Galerkin method, Computational Mechanics 21 (3) (1998) 211–222.

[134] G. Zi, T. Belytschko, New crack-tip elements for XFEM and applications to cohesive cracks, International Journal for Numerical Methods in Engineering 57 (15) (2003) 2221–2240.

[135] G. Zi, J.-H. Song, E. Budyn, S.-H. Lee, T. Belytschko, A method for grawing multiple cracks without remeshing and its application to fatigue crack growth, Modelling and Simulation in Materials Science and Engineering 12 (1) (2004) 901–915.

[136] G. Zi, T. Rabczuk, W. Wolfgang, Extended meshfree methods without branch enrichment for cohesive cracks, Computational Mechanics 40 (2) (2007) 367–382.

CHAPTER SIX

Extended isogeometric analysis

6.1. Formulation and concepts

6.1.1 B-splines and NURBS

Given a knot vector of $n+1$ knots $U = \{u_0, u_1, \cdots, u_n\}$ in the parametric domain, the set of basis functions $N_{i,p}(u)$ of polynomial degree p are defined recursively as follows:

$$
\begin{aligned}
N_{i,0}(u) &= \begin{cases} 1 & : \text{if} \quad u_i \leq u \leq u_{i+1} \\ 0 & : \text{otherwise} \end{cases} \\
N_{i,p}(u) &= \frac{u - u_i}{u_{i+p} - u_i} N_{i,p-1}(u) + \frac{u_{i+p+1} - u}{u_{i+p+1} - u_{i+1}} N_{i+1,p-1}(u),
\end{aligned}
\tag{6.1}
$$

where $u_i \in \mathbb{R}$ is the ith knot. The basis functions are completely determined by the knot-vector and the polynomial degree. For a knot-vector with no repeated values, the continuity of the basis is C^{p-1}, i.e. the first $p-1$ derivatives are continuous at all points in the domain. The knot values in a knot vector may be repeated, in which case the continuity order at that point is reduced by 1 for each repetition.

A B-spline curve $T : [u_0, u_{n+p+2}] \to \mathbb{R}^d$ of degree p is defined in terms of basis functions as:

$$
\begin{aligned}
T(u) &= \sum_{i=0}^{p} P_i N_{i,p}(u) \\
&= \mathbf{P}^T \mathbf{N}(u) \,.
\end{aligned}
\tag{6.2}
$$

The points on the curve corresponding to $T(u_i)$ are called the knot points. An example of a B-spline curve of degree $p = 3$ defined over the knot vector $U = \{0, 0, 0, 0, 1/3, 2/3, 1, 1, 1, 1\}$ is shown in Fig. 6.1. In Fig. 6.1A, the spline is represented by the red curve, the control polygon is represented by the blue lines, and the black dots are the control points. The corresponding basis functions are shown in Fig. 6.1B, where the red dots represent the knots. B-spline have several appealing properties including geometry invariance under translation and rotation. In addition, the basis functions form a partition of unity. Moreover, the h-/p-/k-refinement properties of

Extended Finite Element and Meshfree Methods
https://doi.org/10.1016/B978-0-12-814106-9.00012-7
Copyright © 2020 Elsevier Inc.
All rights reserved.

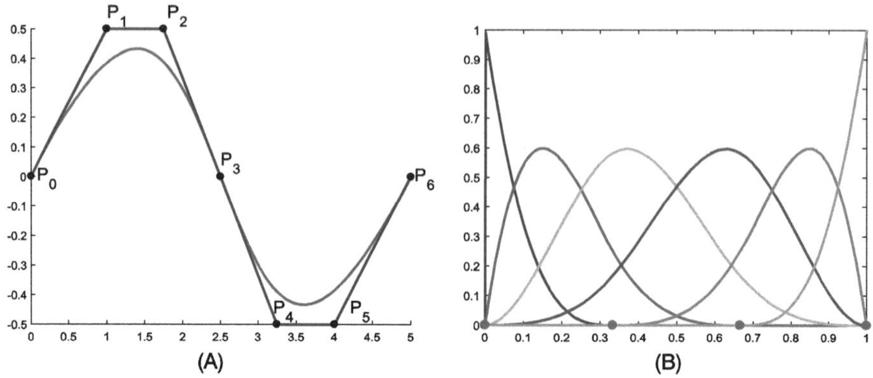

Figure 6.1 B-spline example: (A) B-spline, and (B) basis functions.

the basis function makes them suitable for adaptive analysis with B-splines. The approximation space S_p in parametric coordinates is given by:

$$S_p(U) = \text{span}\{N_{i,p}(\cdot)\}_{i=1,\ldots,n} \tag{6.3}$$

The surface and volume B-splines can be represented by using tensor products of the 1D B-spline basis functions. For 2D, it is defined as:

$$\begin{aligned} T(u, v) &= \sum_{i=0}^{p}\sum_{j=0}^{p} P_i N_{i,p}(u) N_{j,p}(v) \\ &= \mathbf{P}^T \mathbf{N}(u, v), \end{aligned} \tag{6.4}$$

where $P_i = \begin{bmatrix} x_i \\ y_i \end{bmatrix}$ are the 2D control points corresponding to the basis functions. It can be extended directly to 3D as:

$$\begin{aligned} T(u, v, w) &= \sum_{i=0}^{p}\sum_{j=0}^{p}\sum_{k=0}^{p} P_i N_{i,p}(u) N_{j,p}(v) N_{k,p}(w) \\ &= \mathbf{P}^T \mathbf{N}(u, v, w), \end{aligned} \tag{6.5}$$

and the control points in 3D is given by $P_i = \begin{bmatrix} x_i \\ y_i \\ z_i \end{bmatrix}$.

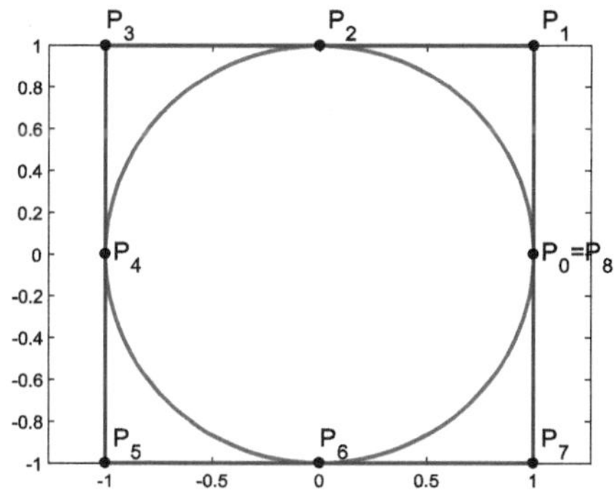

Figure 6.2 Representation of circle with NURBS.

A NURBS curve $C(u) \in \mathbb{R}^2$ is the projection of a B-spline in $\mathbb{R}^3$. It is defined by the basis functions and points with weights as:

$$
\begin{aligned}
C(u) &= \sum_{i=0}^{p} P_i \frac{w_i N_{i,p}(u)}{\sum_{\hat{i}=1}^{n} w_{\hat{i}} N_{\hat{i},p}(u)} \\
&= \mathbf{P}^T \mathbf{R}(u),
\end{aligned}
\tag{6.6}
$$

where w_i represent the weights. The advantage of NURBS is that it allows exact geometry representation which leads to more accurate result in modeling and analysis. Fig. 6.2 shows the example of the boundary of a circle formed by NURBS of degree 2. In Fig. 6.2 the red curve is the NURBS, the black dots denote the control points, and the control polygon is represented by the blue lines. The knot vector used to generate the circle is given by $U = \{0, 0, 0, 1/4, 1/4, 1/2, 1/2, 3/4, 3/4, 1, 1, 1\}$, while the control points and the corresponding weight are listed in Table 6.1.

6.1.2 Bézier extraction

Since the B-splines are nonlocal (i.e. they span up to $p + 1$ elements) or knot-spans, their implementation in standard finite element codes may be cumbersome. Therefore, it is advantageous to obtain a local representation of the values of the shape functions corresponding to each element independently. This provides an element structure for the isogeometric analysis that is compatible with existing finite element computations [5]. We note this only applies for $p > 1$, as linear B-splines ($p = 1$) are equivalent to the standard linear finite element basis in 1D or on tensor-product quadrilateral or hexahedral meshes.

Table 6.1 Control points and weight of NURBS in Fig. 6.2.

x	y	Weight
1	0	1
1	1	$\frac{1}{\sqrt{2}}$
0	1	1
−1	1	$\frac{1}{\sqrt{2}}$
−1	0	1
−1	−1	$\frac{1}{\sqrt{2}}$
0	−1	1
1	−1	$\frac{1}{\sqrt{2}}$
1	0	1

A Bézier curve $C : [0, 1] \rightarrow \mathbb{R}^d$ of polynomial degree p is defined as follows:

$$\begin{aligned} C(u) &= \sum_{i=0}^{p} P_i B_{i,p}(u) \\ &= \mathbf{P}^T \mathbf{B}(u), \quad u \in [0, 1]. \end{aligned} \tag{6.7}$$

Here P_i are the control points in d spatial dimensions, and $\mathbf{P} = \{P_i\}_{i=0}^{p}$ is a $(p+1) \times d$ matrix given by:

$$\mathbf{P} = \begin{bmatrix} P_0^1 & P_0^2 & \dots & P_0^d \\ P_1^1 & P_1^2 & \dots & P_1^d \\ \vdots & \vdots & \vdots & \vdots \\ P_p^1 & P_p^2 & \dots & P_p^d \end{bmatrix},$$

where $\mathbf{B}(u) = \{B_{i,p}(u)\}_{i=0}^{p}$, and $B_{i,p}(u)$ are the Bernstein basis polynomials of degree p defined as:

$$B_{i,p}(u) = \binom{p}{i} u^i (1-u)^{p-i}. \tag{6.8}$$

The binomial coefficients $\binom{p}{i}$ are given by:

$$\binom{p}{i} = \frac{p!}{i!(p-i)!}. \tag{6.9}$$

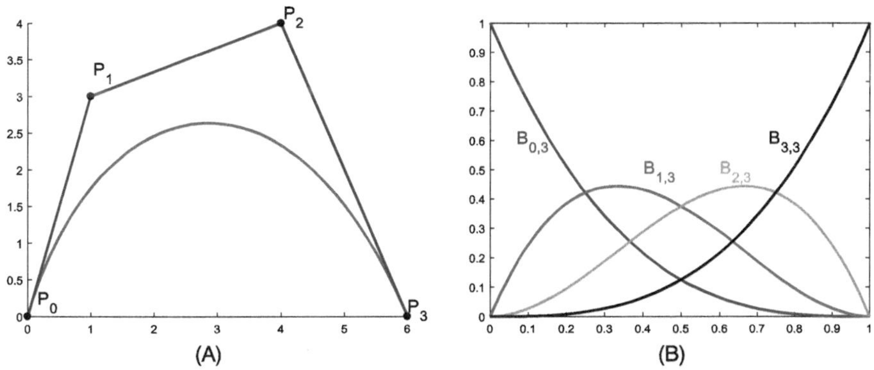

Figure 6.3 Bézier curve example: (A) Bézier curve, and (B) Bernstein polynomial.

The shape of the curve is defined by the *Bézier polygon* which is formed by connecting the control points (from P_0 to P_p) with lines. An example of a cubic Bézier curve with four control points and its *Bézier polygon* is shown in Fig. 6.3A. In Fig. 6.3A, the red curve is the Bézier curve, while the blue lines represent the *Bézier polygon*, and the black dots are the control points labeled accordingly. Fig. 6.3B shows the corresponding Bernstein polynomials for one Bézier element.

A B-spline curve can be decomposed into Bézier curves by defining the basis functions as a linear combination of Bézier extraction operator $\boldsymbol{C}$ and Bernstein polynomials $\mathbf{B}(u)$ as follows:

$$\mathbf{N}(u) = \mathbf{CB}(u). \tag{6.10}$$

Thus, (6.2) can be rewritten as:

$$T(u) = \boldsymbol{P}^T \boldsymbol{C}\boldsymbol{B}(u). \tag{6.11}$$

The computation of $\boldsymbol{C}$ is described in [5]. Note that non-zero elements in the matrix $\boldsymbol{C}$ are in fact the Bézier control points of the corresponding Bernstein polynomials, they are also known as the Bézier ordinates. In 2D, $\boldsymbol{C}$ is defined as:

$$\boldsymbol{C} = \boldsymbol{C}_\eta^{\hat{i}} \otimes \boldsymbol{C}_\xi^{\hat{j}},$$

and in 3D it is given by:

$$\boldsymbol{C} = \boldsymbol{C}_\eta^{\hat{i}} \otimes \boldsymbol{C}_\xi^{\hat{j}} \otimes \boldsymbol{C}_\zeta^{\hat{k}},$$

where $\mathbf{C}_\eta^{\hat{i}}$, $\mathbf{C}_\xi^{\hat{j}}$ and $\mathbf{C}_\zeta^{\hat{k}}$ are the $\hat{i}$th, $\hat{j}$th, and $\hat{k}$th extraction operators in η, ξ and ζ directions respectively.

6.2. Hierarchical refinement with PHT-splines

A PHT-spline is the generalization of B-splines over a hierarchical T-mesh [13]. A T-mesh refers to a partition of a given domain with rectangular grid which includes T-junctions (hanging nodes) [32]. A grid point on the T-mesh is also known as a vertex. A PHT-spline $N_{i,p}$ is defined by

- The corresponding control point P_i.
- A local knot vector which defines the support of the basis function.
- The corresponding Bézier extraction operator $\mathbf{C}$.

A PHT-spline mesh can be described by a recursive pattern, from a lower level to a higher level. We denote a PHT-spline parametrization at level ℓ as $F^\ell(u)$, and the basis function as $\mathbf{N}^\ell(u)$. The initial PHT-spline parametrization ($\ell = 0$) is given by:

$$\begin{aligned} F^0(u) &= \sum_{i=1}^{n} P_i N_{i,p}^0(u) \\ &= \mathbf{P}^T \mathbf{N}_p^0(u). \end{aligned} \tag{6.12}$$

We are particularly interested odd polynomial p and PHT-splines of continuity $\alpha = (p-1)/2$. For $p = 1$, this reduces to a standard FEM discretization on quadrilaterals or hexahedra with hanging nodes, where the control points coincide with the boundary or the interior (non-hanging) nodes. In general, each interior knot u_i is supported by $(\alpha + 1)^d$ basis functions, which are known as the *associated basis functions*. Thus, when a new basis node is inserted, it will introduce $(\alpha + 1)^d$ new basis functions and $(\alpha + 1)^d$ control points to the PHT-spline discretization.

In the following, we will consider the case for cubic splines ($p = 3$). Suppose we have a PHT-spline at level l, when a new knot is inserted into a selected domain, the next level PHT-spline will be:

$$\begin{aligned} F^{l+1}(u) &= \sum_{i=1}^{n} P_i \hat{N}_{i,3}^l(u) + \sum_{j=1}^{2^d} \hat{P}_j N_{j,3}^{l+1}(u) \\ &= \mathbf{P}^T \hat{\mathbf{N}}^l(u) + \hat{\mathbf{P}}^T \mathbf{N}^{l+1}(u) \end{aligned} \tag{6.13}$$

where $\mathbf{P}^T$ are the control points of level l, and $\hat{\mathbf{N}}^l(u)$ are the basis functions modified from $\mathbf{N}^l(u)$. While $\hat{\mathbf{P}}^T$ and $\mathbf{N}^{l+1}(u)$ are the new control

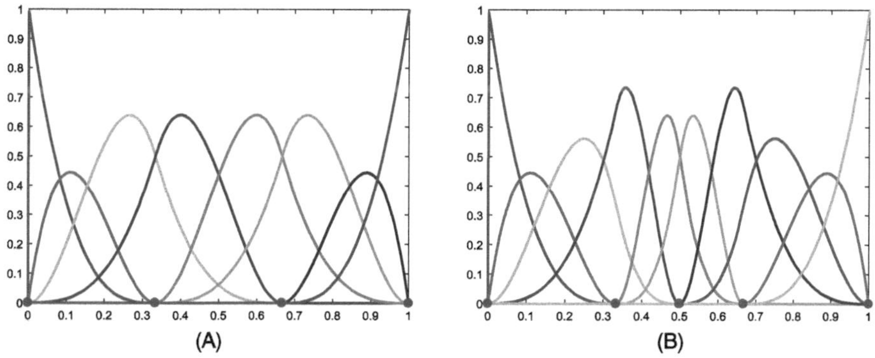

Figure 6.4 Example of PHT-spline basis functions: (A) PHT-spline basis functions, and (B) modify and new basis functions when a knot at $\frac{1}{2}$ is inserted.

points and *basis functions* associate to the new knot. Fig. 6.4A illustrates the basis functions of a PHT-spline defined over the knot vector ($U = \{0, 0, 0, 0, 1/3, 1/3, 2/3, 2/3, 1, 1, 1, 1\}$). The new basis functions when a pair of repeated knots ($\bar{u} = \frac{1}{2}$) are inserted as shown in Fig. 6.4B. An advantage of the PHT-spline is that when a knot is inserted in a selected element, only the corresponding basis functions of that element are changed while the other parts remain unchanged. Thus, PHT-splines have a good local control property.

6.2.1 PHT-spline space

Given a set of elements $\mathcal{F}$ in a T-mesh $\mathcal{T}(\mathcal{F})$, and $C^{1,1}(\Omega)$ represent the space consisting of all the C^1 continuous functions in each spacial direction of Ω, the spline space is defined as:

$$\mathcal{T}(\mathcal{F}): = \{s \in C^{1,1}(\Omega) : s|_\phi \in \mathbb{P}_3 \text{ for any } \phi \in \mathcal{F}\}, \tag{6.14}$$

where $\mathbb{P}_3$ denotes the space of all cubic polynomials. Refinement of mesh is done by "cross-insertion", where a square element in 2D is split into four elements. The intersection of the four elements, and the boundaries and corner vertices are the basis vertices. In 3D, a hexahedral element is split into eight sub-hexahedra. The basis vertices included the intersections of the eight elements in the interior, four elements on the faces, two elements on the edges and corner vertices. The vertices at the T-junctions are known as the T-vertices, they have no basis function associated with them.

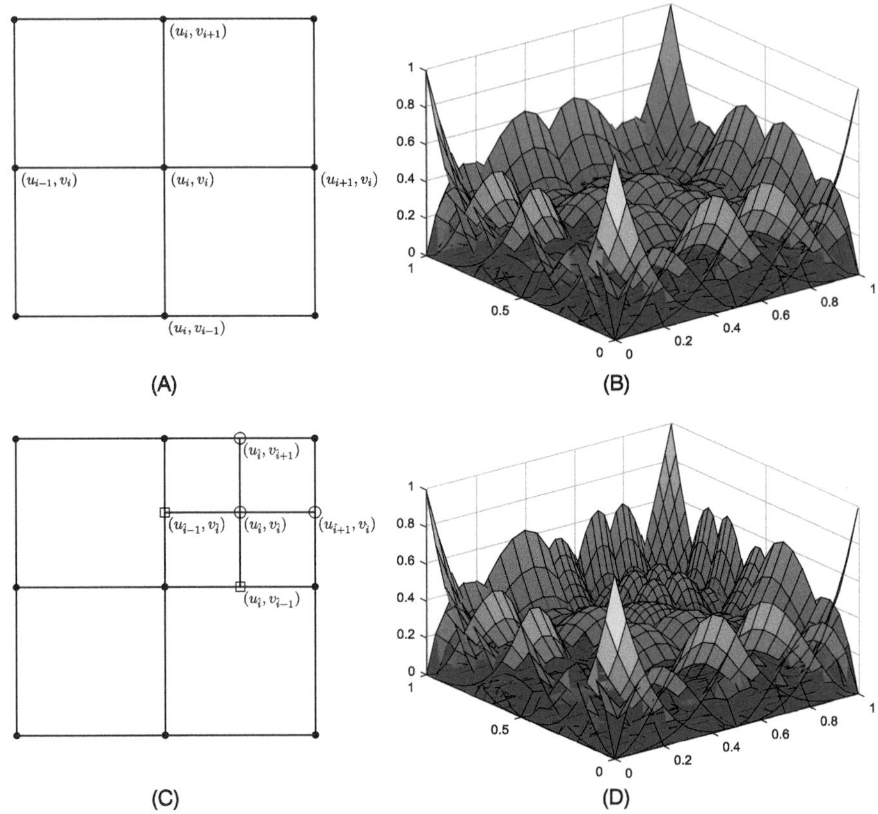

Figure 6.5 Modification of basis functions in 2D, (A) PHT-splines mesh at level ℓ, (B) basis functions at level ℓ, (C) PHT-splines mesh at level ℓ, and (D) basis functions at level $\ell + 1$.

For PHT-splines of continuity α and degree $p = 2\alpha + 1$, the dimension space over the mesh $\mathcal{F}$ is given by:

$$\dim(\mathcal{T}(\mathcal{F})) = (\alpha + 1)^d \cdot (\text{number of basis vertices}).$$

Suppose the spline space at levels ℓ are given by $\mathcal{T}(\mathcal{F}_{\ell+1})$. After refinement, the spline space at level $\ell + 1$, $\mathcal{T}(\mathcal{F}_{\ell+1})$ will have minimal changes because the basis functions for the elements which are not refined will be unmodified. Changes only occur at the elements that contain new basis vertices. Fig. 6.5A shows the 2D PHT-spline mesh at level ℓ and its basis functions are illustrated in Fig. 6.5B. When a vertex is inserted in one of the element as shown in Fig. 6.5C, we can see that only the basis functions in that particular element are changed (see Fig. 6.5D). The spline space of the neighbor of a refined element will only be changed when a T-vertex is converted to a basis vertex. This is an advantage of PHT-splines over T-splines which insertion of an edge will cause additional refinement further away from the inserted edge. The modification of the existing basis function and the computation of the new basis functions in 3D are similar with 2D. However, in 3D there are $(\alpha+1)^3$ basis functions supporting each vertex. Therefore, there will be $(\alpha+1)^3$ additional basis functions introduced to the approximation space for each new vertex.

6.2.2 Computing the control points

At the initial mesh $\ell=0$, there are linear mapping between the parameter space and physical space. Thus, the control points are set as the location of the Greville Abscissae. At level $\ell \geq 1$, we compute the control points by using the geometric information of the basis functions. Where the geometric information is given by the location of the basis vertex in the physical space and the derivatives values of the mapping at the basis vertex. For simplicity, we first consider the case $p=3$, $\alpha=1$. In 1D, the geometric information is defined as:

$$\mathcal{L}F(u) = \left(F(u), \frac{\partial F(u)}{\partial u}\right) = (F(u), F_u(u)).$$

Suppose the spline is given by:

$$F(u) = \sum_{i=1}^{n} B_i(u) \cdot P_i, \tag{6.15}$$

the geometric information of the spline at knot u_a is given by:

$$\begin{aligned} \mathcal{L}F(u_a) &= \sum_{j=1}^{2} \mathcal{L}(B_{i_j}(u_a) \cdot P_i) \\ &= \boldsymbol{P} \cdot \boldsymbol{B}, \end{aligned} \tag{6.16}$$

where i_1 and i_2 denote the two basis indices corresponding to the knot u_a. We can solve for $\boldsymbol{P}$ from:

$$\boldsymbol{P} = \mathcal{L}F(u_a) \cdot \boldsymbol{B}^{-1}. \tag{6.17}$$

In 2D, the geometric information is defined as:

$$\mathcal{L}F(u,v) = \left[F(u;v), \frac{\partial F(u,v)}{\partial u}, \frac{\partial F(u,v)}{\partial v}, \frac{\partial F^2(u,v)}{\partial u \partial v}\right] \tag{6.18}$$

Thus, for a basis vertex (u_a, v_a), the control points can be solved from the following relation:

$$\boldsymbol{P} = \mathcal{L}F(u_a, v_a) \cdot \boldsymbol{B}^{-1}. \tag{6.19}$$

Referring to [13], **B** is given by

$$\boldsymbol{B} = \begin{bmatrix} (1-\lambda)(1-\mu) & -\beta(1-\mu) & \gamma(1-\lambda) & \beta\gamma \\ \lambda(1-\mu) & \beta(1-\mu) & -\gamma\lambda & -\beta\gamma \\ (1-\lambda)\mu & -\beta\mu & \gamma(1-\lambda) & -\beta\gamma \\ \lambda\mu & \beta\mu & \gamma\lambda & \beta\gamma \end{bmatrix}, \tag{6.20}$$

where $\beta = \frac{1}{\Delta u_1 + \Delta u_2}$, $\gamma = \frac{1}{\Delta v_1 + \Delta v_2}$, $\lambda = \beta \Delta u_1$, and $\mu = \gamma \Delta v_1$. Note that Δu_i and Δv_i, $i = 1, 2$ are the differences between u_a and v_a, with the neighbor knots in the negative and positive direction respectively.

For 3D, the control points of a basis vertex (u_a, v_a, w_a) can be computed from:

$$\boldsymbol{P} = \mathcal{L}F(u_a, v_a, w_a) \cdot \boldsymbol{B}^{-1}. \tag{6.21}$$

B is given by the Kronecker product between **U**, **V** and **w** as follows:

$$\boldsymbol{B} = \boldsymbol{W} \otimes [\boldsymbol{V} \otimes \boldsymbol{U}], \tag{6.22}$$

where **U**, **V**, and **W** are given by:

$$\boldsymbol{U} = \begin{bmatrix} (1-\lambda) & \lambda \\ -\beta & \beta \end{bmatrix}, \boldsymbol{V} = \begin{bmatrix} (1-\mu) & \mu \\ -\gamma & \gamma \end{bmatrix}, \text{ and } \boldsymbol{W} = \begin{bmatrix} (1-\nu) & \nu \\ -\eta & \eta \end{bmatrix}.$$

α, β, λ and μ are defined as mentioned above, while $\eta = \frac{1}{\Delta w_1 + \Delta w_2}$ and Δw_i, $i = 1, 2$ are the differences between neighbor knot in the w direction. For the case when $\alpha > 1$, $p = 2\alpha + 1$ the control points can be computed similarly by considering the derivatives up to order α in each parameter direction.

6.3. Analysis using splines

In this section, we discuss solving boundary problem using spline. Suppose we would like to solve a boundary value problem as below:

$$\begin{aligned} \Delta u &= f \text{ in } \Omega \\ u &= g \text{ on } \partial\Omega \end{aligned} \tag{6.23}$$

The weak form of (6.23) is obtained by considering trial solutions ($u : \Omega \to \mathbb{R}$), multiplying both sides by the weighting function (w) and integration by parts:

$$\int_{\Omega} \Delta w \, \Delta u \, d\Omega = \int_{\Omega} wf \, d\Omega. \tag{6.24}$$

The collection of trial solutions ($\mathcal{S}$) is given by:

$$\mathcal{S} = \{u | u \in H^1(\Omega), u|_{\partial\Omega} = g\}. \tag{6.25}$$

The weighting functions are denoted by a set $\mathcal{V}$ as:

$$\mathcal{V} = \{w | w \in H^1(\Omega), w|_{\partial\Omega} = g\}, \tag{6.26}$$

where $H^1(\Omega)$ is the Sobolev space. The weak form can be rewritten as:

$$a(w, u) = L(w) \tag{6.27}$$

where

$$a(w, u) = \int_{\Omega} \Delta w \, \Delta u \, d\Omega \tag{6.28}$$

and

$$L(w) = \int_{\Omega} wf \, d\Omega \tag{6.29}$$

The solution to (6.24) and (6.27) is known as the weak solution. In the next section, we will discuss finding the solution using Galerkin method.

6.3.1 Galerkin method

The weak form of the problem can be converted into a system of algebraic equations using the Galerkin method. The main concept of Galerkin method is that it is used to construct a finite-dimensional approximation of $\mathcal{S}$ and $\mathcal{V}$, where the $\mathcal{S}^h$ and $\mathcal{V}^h$ are the corresponding subspaces. Suppose $g^h \in \mathcal{S}^h$ such that $g^h|_{\partial\Omega} = g$, then there is a unique $v^h \in \mathcal{V}^h$ for every $u^h \in \mathcal{S}^h$ such that

$$u^h = v^h + g^h \tag{6.30}$$

Assume that g^h exists, the Galerkin form of the problem can therefore be written as: given g^h, find $u^h = v^h + g^h$ such that for all w^h in $\mathcal{V}^h$,

$$a(w^h, u^h) = L(w^h) \tag{6.31}$$

Referring to (6.30), (6.31) can be rewritten as:

$$a(w^h, v^h) = L(w^h) - a(w^h, g^h) \tag{6.32}$$

In IGA, the solution space is formed by linear combination of a set of NURBS functions $N_A : \Omega \to \mathbb{R}$, $A = 1, \ldots, n_{nb}$ where for all $w^h \in \mathcal{V}^h$ there exists c_A, $A = 1, \ldots, n_{eq}$ such that

$$w^h = \sum_{A=1}^{n_{eq}} N_A c_A \tag{6.33}$$

The function g^h is given similarly as:

$$g^h = \sum_{A=n_{eq}+1}^{n_{nb}} N_A g_A \tag{6.34}$$

By using (6.33) and (6.34) the trial function can be express as:

$$u^h = \sum_{A=1}^{n_{eq}} N_A d_A + \sum_{A=n_{eq}+1}^{n_{nb}} N_A g_A \tag{6.35}$$

By substituting (6.33) and (6.35) into (6.32), we obtain:

$$\sum_{A=1}^{n_{eq}} c_A \left(\sum_{B=1}^{n_{eq}} a(N_A, N_B) d_B - L(N_A) + a(N_A, g^h) \right) = 0 \tag{6.36}$$

Since c_A are arbitrary, we have:

$$\sum_{B=1}^{n_{eq}} a(N_A, N_B) d_B = L(N_A) - a(N_A, g^h) \tag{6.37}$$

Let K_{AB} and F_A be

$$K_{AB} = a(N_A, N_B), \tag{6.38}$$

and

$$F_A = L(N_A) - a(N_A, g^h). \tag{6.39}$$

They can be written in matrix form as:

$$\mathbf{K} = [K_{ab}], \tag{6.40}$$

$$\mathbf{F} = \{F_a\}, \tag{6.41}$$

$$\mathbf{d} = \{d_a\}. \tag{6.42}$$

Therefore, for $A, B = 1, \ldots, n_{eq}$, the algebraic equation can be express as:

$$\mathbf{Kd} = \mathbf{F}, \tag{6.43}$$

where $\mathbf{K}$ is the stiffness matrix and $\mathbf{F}$ is the force vector. The displacement vector $\mathbf{d}$ can be solved from:

$$\mathbf{d} = \mathbf{K}^{-1}\mathbf{F}. \tag{6.44}$$

Finally, the solution can be obtained by substituting (6.44) into (6.35).

6.3.2 Linear elasticity

In this section, we show the examples of application of spline-based IGA on the problem of linear elasticity. Given a domain Ω, bounded by $\Gamma = \overline{\Gamma_t \cup \Gamma_u}$, $\Gamma_t \cap \Gamma_t = \emptyset$, where displacements and tractions $(\overline{t})$ are prescribed on Γ_u and Γ_t respectively. The weak form of a linear elastostatics problem is given by: find the displacement $\mathbf{u}$ in the trial space, such that for all test functions $\delta\mathbf{u}$ in the test place satisfy the following statement:

$$\int_\Omega \varepsilon(\mathbf{u}) : \mathbf{D} : \varepsilon(\delta\mathbf{u}) d\Omega = \int_{\Gamma_t} \overline{\mathbf{t}} \cdot \delta\mathbf{u} d\Gamma + \int_\Omega \mathbf{b} \cdot \delta\mathbf{u} d\Omega. \tag{6.45}$$

In (6.45), $\mathbf{D}$ is known as the elasticity matrix, while $\mathbf{b}$ and $\varepsilon = \frac{1}{2}(\nabla\mathbf{u} + \nabla^T\mathbf{u})$ denote the body force and displacement gradient respectively. By using the Galerkin method, $\mathbf{u}$ and $\delta\mathbf{u}$ can be expressed as:

$$\mathbf{u}(X) = \sum_I^{nn} R_I(\xi)\mathbf{u}_I, \tag{6.46}$$

and

$$\delta\mathbf{u}(X) = \sum_I^{nn} R_I(\xi)\delta\mathbf{u}_I. \tag{6.47}$$

Note that R_I is the NURBS basis functions, nn represents the number of control points, $\mathbf{u}_I = [u_{xI}, u_{yI}]^T$ is the nodal unknown vector, and $\delta\mathbf{u}_I$ denotes the nodal displacement variations.

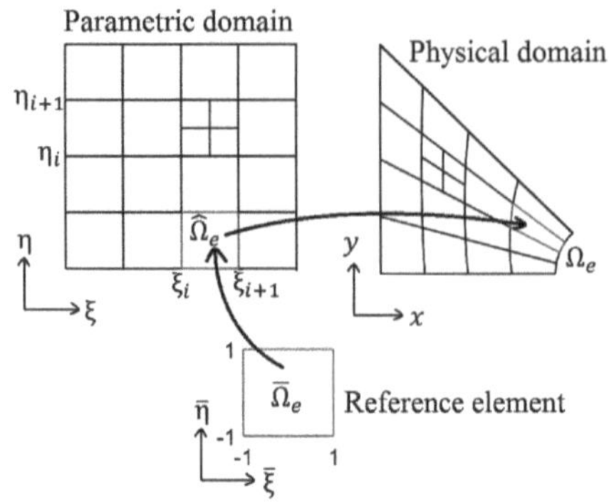

Figure 6.6 Domains in isogeometric analysis.

By substituting (6.46) and (6.47) into (6.45), we obtain the discrete equation:

$$\mathbf{K}\mathbf{u} = \mathbf{f} \tag{6.48}$$

where

$$\mathbf{K}_{IJ} = \int_{\Omega} \mathbf{B}_I^T \mathbf{D} \mathbf{B}_J d\Omega, \tag{6.49}$$

and

$$\mathbf{f}_I = \int_{\Gamma_t} R_I \bar{\mathbf{t}} d\Gamma + \int_{\Omega} R_I \mathbf{b} d\Omega. \tag{6.50}$$

The strain-displacement $\mathbf{B}_I$ in 2D is defined as:

$$\mathbf{B}_I = \begin{bmatrix} R_{I,x} & 0 \\ 0 & R_{I,y} \\ R_{I,y} & R_{I,x} \end{bmatrix}, \tag{6.51}$$

where $R_{I,x}$ represents the first order partial derivatives of R_I with respect to x. We can then obtain the solution $\mathbf{u}$ by solving the linear system (6.48).

Note that spline basis functions are defined in the parameter space, while the quadrature rule are defined in the parent domain. The relation between the domains is illustrated in Fig. 6.6, and the domain integral in (6.49) and (6.50) can be computed as follows:

$$\int_{\Omega} f(x, y) d\Omega = \bigcup_{e=1}^{nel} \int_{\Omega_e} f(x, y) d\Omega_e$$

$$= \bigcup_{e=1}^{nel} \int_{\hat{\Omega}_e} f(x(\xi), y(\eta)) |J_\xi| d\hat{\Omega}_e$$
$$= \bigcup_{e=1}^{nel} \int_{\overline{\Omega}_e} f(\overline{\xi}, \overline{\eta}) |J_\xi| |J_{\overline{\xi}}| d\overline{\Omega}_e, \tag{6.52}$$

where $\bigcup$ is the assembly operator, and *nel* represents the number of finite elements which are defined as non-zeros knot spans. Meanwhile, $|J_\xi|$ denotes the determinant of Jacobian of the transformation from the parametric domain to the physical domain, while $|J_{\overline{\xi}}|$ represents the determinant of Jacobian of the transformation from the reference element to the parametric domain. The reference element contains the Bernstein polynomials of degree p, as described in Section 6.1.2, which is scaled to $[-1, 1]^d$ from $[0, 1]^d$. The integration can then be computed using Gauss-Legendre quadrature.

6.4. Numerical examples

In this section, we show some 2D and 3D benchmark examples of spline-based isogeometric analysis for linear elasticity problem.

6.4.1 Infinite plate with circular hole

We first show the 2D example of an infinite plate with circular hole. The plate is subject to constant in-plane tension at infinity, and the exact solution of this problem is given by:

$$\sigma_{rr}(r\theta) = \frac{T_x}{2}\left(1 - \frac{R^2}{r^2}\right) + \frac{T_x}{2}\left(1 - 4\frac{R^2}{r^2} + 3\frac{R^4}{r^4}\right)\cos 2\theta \tag{6.53}$$

$$\sigma_{\theta\theta}(r\theta) = \frac{T_x}{2}\left(1 - \frac{R^2}{r^2}\right) + \frac{T_x}{2}\left(1 + 3\frac{R^4}{r^4}\right)\cos 2\theta \tag{6.54}$$

$$\sigma_{r\theta}(r\theta) = -\frac{T_x}{2}\left(1 + 2\frac{R^2}{r^2} - 3\frac{R^4}{r^4}\right)\cos 2\theta \tag{6.55}$$

where T_x is the magnitude of the applied stress, R is the radius of the circular hole, and L is the edge length of the plate. The domain of this problem is infinite, and the problem is symmetric about the origin. Therefore, we represent the model with one quadrant of a square plate with vicinity of the hole as shown in Fig. 6.7, and apply symmetric boundary condition on it. In this example, we set the isotropic material of the plate to be: Young modulus $E = 3 \times 10^5$ and Poisson ratio $\nu = 0.3$. The geometry of

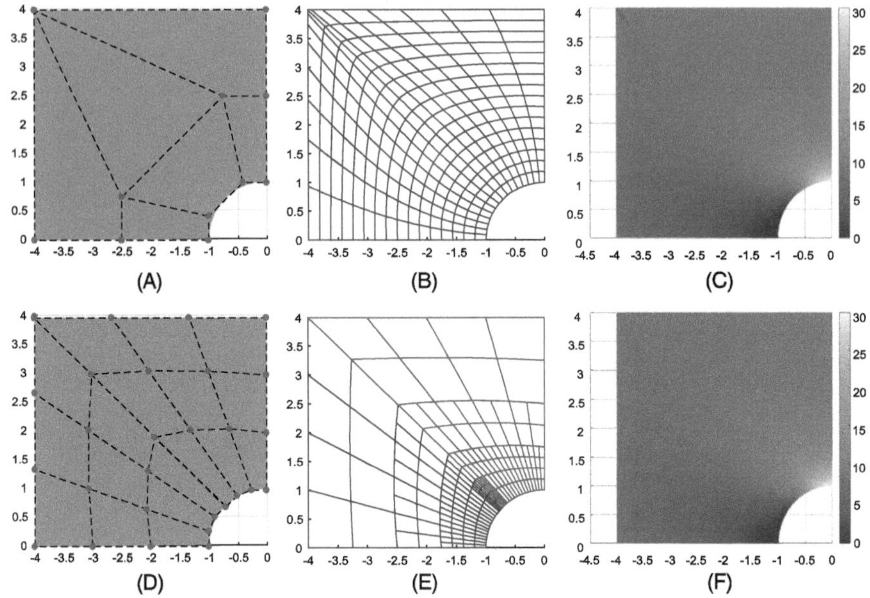

Figure 6.7 Infinite plate with circular hold example, (A) C^1 continuous NURBS mesh with control points, (B) refined mesh of (A), (C) σ_{xx} displacement of (B), (D) C^0 continuous NURBS mesh with control points, (E) refined mesh of (D), and (F) σ_{xx} displacement of (E).

the model can be represented by using a quadratic element that is C^1 continuous. Fig. 6.7A shows the control points of the quadratic element. Note that there are two overlapping points at the upper left corner of the model, which cause a singularity in parametrization. The refined mesh is shown in Fig. 6.7B, and the corresponding result is given in Fig. 6.7C. In the result, we can see that there is a small area with less accurate computed stresses at the upper left corner. Alternatively, the model can be represented by using two quadratic elements with C^0 continuity as shown in Fig. 6.7D. The solution will converge to exact solution (see Fig. 6.7F) as the mesh is refined as shown in Fig. 6.7E. A method to obtain a C^1 singularity-free parametrization for domains such as the one in this example has been discussed in [6].

6.4.2 Open spanner

Next we show a more complex 2D example – an open spanner, which is assumed to be made from linear elastic material with Young modulus $E = 3 \times 10^5$ and Poisson ratio $= 0.3$. Traction is applied at the end of the handle and the jaw is fixed. The spline mesh of the spanner is illustrated

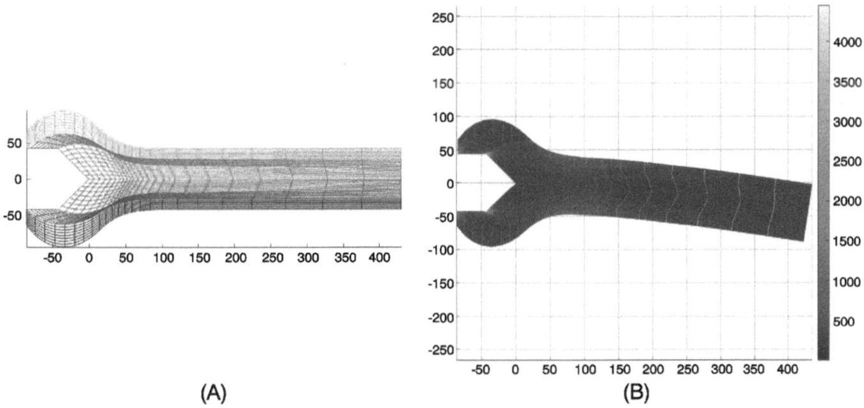

Figure 6.8 Spanner example, (A) splines mesh, (B) displacement and Von Mises stress.

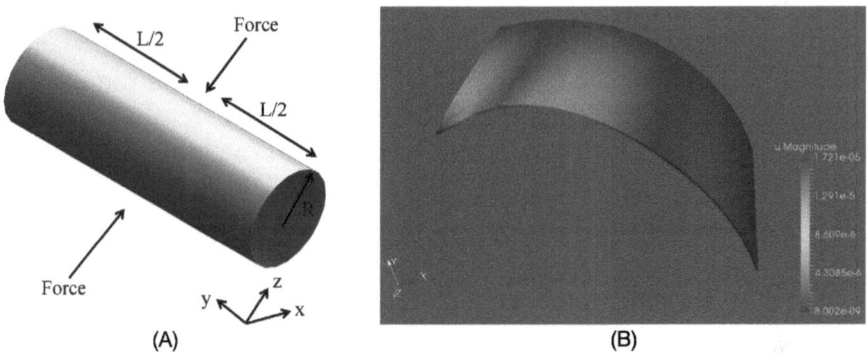

Figure 6.9 Pinched cylinder example, (A) geometry and boundary conditions, (B) displacement of pinched cylinder (one eighth of the model is shown).

in Fig. 6.8A, while the displacement and the Von Mises stress is shown in Fig. 6.8B. We note that some elements in the interior have a high aspect ratio, which may lead to less accurate results, and the continuity between the different regions of the domains is only C^0. A method for parametrizing geometries from the boundary description with at least C^1 continuity in the interior has been proposed in [7] and further refined in [8].

6.4.3 Pinched cylinder

Following is the thick-wall shell example of the pinched cylinder. The cylinder is subjected to two opposite loads at the middle of the cylinder as shown in Fig. 6.9A. Since the stress and boundary condition are sym-

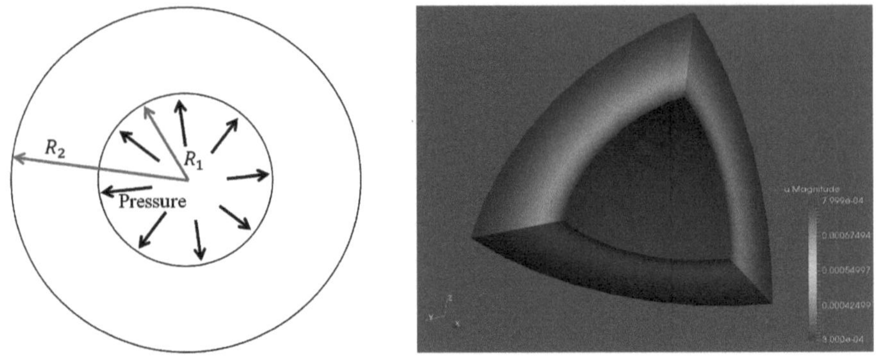

(A) (B)

Figure 6.10 Hollow sphere example, (A) pinched cylinder, (B) displacement of pinched cylinder.

metric, the result can be illustrated by using one eighth of the cylinder. In this example, the cylinder is modeled with one cubic element through the thickness. The Young modulus and Poission ration are set as 3×10^6 and 0.3 respectively. Fig. 6.9B shown the displacement of the cylinder.

6.4.4 Hollow sphere

Finally we show the 3D example of a hollow sphere. The elastic material of the sphere is set as: Young modulus $E = 1 \times 10^3$, and Poisson ration $\nu = 0.3$. In this example, pressure is applied on the interior surface. Fig. 6.10A shows the cross-section of the hollow sphere, where the black arrows represent the direction of the pressure, and the blue arrows labeled with R_1 and R_2 denote the interior and outer radius of the sphere. Fig. 6.10 shows one eight of the hollow sphere, illustrating the analysis result.

Note that for all the examples given above, the model is constructed by using one patch (parametric domain). For the pinched cylinder and hollow sphere examples with symmetric boundary condition, one eighth of the body which can be represented by using one patch is sufficient to illustrate the stress distribution and displacement. However, when working with more complex shapes, it is often necessary to use more patches, which involve additional parameter domains which are mapped to different regions of the physical domain. Normally, the continuity between the patches is at most C^0. It is often desirable to have higher order continuity in the geometry description, such that it can be used for solving fourth or higher order partial differential problems, which is the subject of ongoing research.

6.5. Adaptive analysis

For large-scale finite element computations, adaptivity is an important technique for reducing the computational cost of the simulation. The main observation is that certain regions of the domain require a more fine discretization to better capture singularities or other non-smooth features of the solution fields in order to obtain an acceptable approximation. A key factor in the efficient implementation of adaptivity is the availability of reliable and easy to compute *a posteriori* error estimators. Unlike *a priori* error estimators, which are based on the theoretical convergence rates for a given approximation and contain unknown constants, a posteriori error estimators are based on the numerical solution itself and typically can estimate the error within some given bounds.

The two main types of a posteriori error estimators are residual-based and recovery-based. In residual-based estimators, the errors related to the approximation of the governing equations (e.g. the balance equations or the boundary conditions) are evaluated. In the case of piecewise linear or C^0 approximation spaces, this requires an evaluation of the jump terms over the element boundaries, however this procedure can be simplified somewhat in the case of smoother approximations such as those resulting from IGA or mesh-free methods. In recovery-based error estimators, an "improved" or recovered approximation is obtaining by sampling the computed solution at a set of points and fitting them with a smoother or higher-order basis. In the case that where the sampled points have increased approximation property (or *superconvergence*), then the resulting recovered solution is closer to the true solution as well. Then the error can be simply estimated by taking the difference between the recovered and computed solution.

In the following presentation, we will focus mainly on the recovery-based error estimators applied to PHT-splines, although in principle the same ideas can be applied to different kinds of basis functions. We follow the derivations in [3], where we first determine the superconvergent point locations, then give an overview to the superconvergent patch recovery as applied to hierarchical splines. Finally, we include a brief discussion of different element marking strategies.

6.5.1 Determining the superconvergent point locations

To find the locations of the superconvergent points, we proceed similarly as in [2,37], by considering the Taylor expansion of degree $p+1$ of the exact solution. Suppose Ω is the computational domain (in the physical space) and $\Omega_1 \subset \Omega_0 \subset \Omega$ is a subdomain containing a point x_0. Let u be the exact solution, and let u_h be an H^1 projection in the approximation space $S^h(\Omega)$, i.e. u_h satisfies

$$B(u - u_h, \chi) = 0, \quad \forall \chi \in S^h(\Omega), \tag{6.56}$$

where

$$B(u, v) = \int_\Omega \nabla u \nabla v \, d\Omega, \quad \forall u, v \in H^1(\Omega). \tag{6.57}$$

Furthermore, let Q be a $p+1$ degree polynomial approximation to u (e.g. the Taylor expansion of u centered at x_0). A crucial observation is that for a uniform mesh and when the exact solution is a polynomial of degree $p+1$, i.e. $u = Q$, the spline approximation of degree p to u can be locally described (up to higher order terms) by a periodic function (see [37], Lemma 12.2.1). More precisely, we can write,

$$(u - u_h)(x) = \psi(x) + R(x), \text{ with } ||R||_{W^{i,\infty}(\Omega_1)} \le Ch^{p-i+1+\delta} \text{ for } x \in \Omega_1, \tag{6.58}$$

where ψ is a periodic function which can be computed on a reference knot-span, $\delta > 0$ and $W^{i,\infty}$ is the usual Sobolev norm i.e. $||u||_{W^{i,\infty}}(\Omega) = \max |D^i u|_{L^\infty}$.

To further describe the function ψ, we first consider the *periodic subspace* of S^h. Suppose $x_0, x_1, \ldots, x_n$ are uniformly spaced nodes (knots) in Ω_0. Then the periodic subspace of S^h consists of the functions which have the same values at the nodes, as well as the same derivatives up to order α, i.e.

$$S^h_{per}(\Omega_0) := \left\{ \chi : \frac{d^\beta \chi}{dx^\beta}(x_i) = \frac{d^\beta \chi}{dx^\beta}(x_{i+1}) \right\}, \quad \beta = 0, \ldots, \alpha, \text{ and } i = 0, \ldots, n-1. \tag{6.59}$$

We now consider $P_{per}\rho \in S^h_{per}(\Omega_0)$, the *periodic projection* of a function ρ over the space $S^h_{per}(\Omega_0)$, which satisfies

$$B(\rho - P_{per}\rho, \chi) = 0, \quad \forall \chi \in S^h_{per}(\Omega_0), \text{ and} \tag{6.60}$$

$$\int_\Omega (\rho - P_{per}\rho) \, d\Omega = 0. \tag{6.61}$$

Now we can define ψ as in terms a spline interpolant $I^h[Q]$ of Q, and the periodic projection of the difference between Q and its interpolant as follows:

$$\psi := (Q - I^h[Q]) - P_{per}(Q - I^h[Q]). \tag{6.62}$$

Here $I^h[Q]$ is an interpolant chosen such that $Q - I^h[Q]$ is a periodic function in Ω_0. For the space of C^α splines, $I^h[u]$ is chosen such that it matches u and the derivatives up to order α at the nodes:

$$\frac{d^\beta I^h[u]}{dx^\beta}(x_i) = \frac{d^\beta u}{dx^\beta}(x_i), \quad \beta = 0, \ldots, \alpha, \text{ and } i = 0, \ldots, n. \tag{6.63}$$

We emphasize that the superconvergent points, which are the roots of ψ or ψ', only need to be computed once for a reference element and the obtained coordinates can then be used using scaling and translations to any uniform mesh. A general procedure for obtaining the superconvergent points for spline approximations of degree p with continuity α, with $p \geq 2\alpha + 1$, on a reference knot-span $[-1, 1]$ is as follows:

1. Consider a knot vector of the form:

$$U := \{\underbrace{-3, \ldots, -3}_{p-\alpha \text{ times}}, \underbrace{-1, \ldots, -1}_{p-\alpha \text{ times}}, \underbrace{1, \ldots, 1}_{p-\alpha \text{ times}}, \underbrace{3, \ldots, 3}_{p-\alpha \text{ times}}\} \tag{6.64}$$

   From this knot vector, we only need to consider the $p + 1$ splines $\phi_1, \phi_2, \ldots, \phi_{p+1}$ that have support on $[-1, 1]$ and the local approximation space for this reference element:

$$S^{h,ref} := \text{span}\{\phi_1, \phi_2, \ldots, \phi_{p+1}\}. \tag{6.65}$$

2. We build an interpolant $I^h[Q] \in S^{h,ref}$ for $Q(x) = x^{p+1}$ which satisfies (6.63) at the points $x_0 := -1$ and $x_1 := 1$. We note that when $p = 2\alpha + 1$, we already have $p + 1$ constraints for the $p + 1$ unknown coefficients of ϕ_i to be solved. When $p > 2\alpha + 1$, we can interpolate $Q(x)$ at equally spaced points inside $[-1, 1]$ to impose additional constraints.
3. We compute a basis for the periodic subspace $S^{h,ref}_{per}$ of $S^{h,ref}$, which satisfies (6.59) for x_0 and x_1. It can easily be checked that

$$S^{h,ref}_{per} = \text{span}\Big\{\{\phi_1 + \phi_{p-\alpha}, \ldots, \phi_{\alpha+1} + \phi_{p+1}\} \cup \{\phi_{\alpha+2}, \ldots, \phi_{p-\alpha-1}\}\Big\}, \tag{6.66}$$

   which shows that the basis functions in the periodic space are the first $\alpha + 1$ sums of corresponding splines which have support at the two

Table 6.2 The superconvergent points for splines of degree p with continuity C^α on interval $[-1, 1]$.

p	α	Roots of ψ'
3	1	$\pm 1, 0$
4	1	$\pm\sqrt{(3/7) \pm (2/7)\sqrt{6/5}}$
5	2	$\pm 1, \pm\sqrt{1/3}, 0$
6	2	$\pm 0.790208564, \pm 0.2800702925$
7	3	$\pm 1, \pm 0.5294113738, 0$

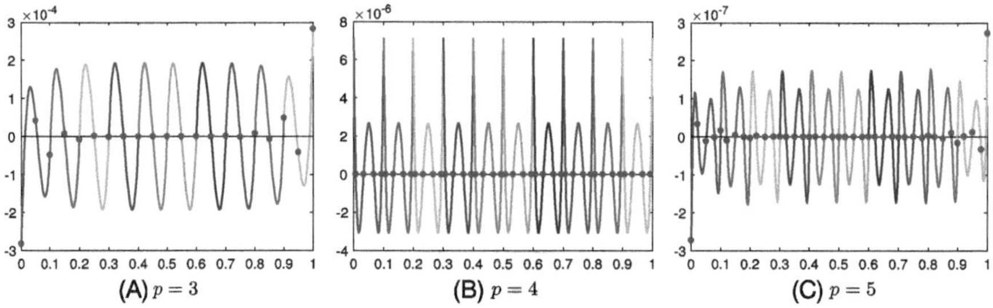

Figure 6.11 Plots of $(u - u_h)'(x)$ for $u = x^{p+1}$ and uniform meshes with knot-span length 1/10. The red dots correspond to the coordinates of the superconvergent points, i.e. the roots of ψ' computed on the reference interval and whose coordinates are scaled and translated to each knot-span in the discretization.

end-points, together with the "middle" $p - 2\alpha - 1$ splines that do not have support at the endpoints. Clearly, these latter basis functions only appear for $p > 2\alpha + 1$.

4. Now the function ψ can be computed according to (6.62). The superconvergent points for second order PDEs are then the roots of ψ or rather, for gradient recovery purposes, of ψ'. The coordinates of superconvergent points for several values of p and α are given in Table 6.2.

We note that the superconvergent points for $p = 4$ and $\alpha = 1$ are the same as the coordinates for the 4-point Gauss-Legendre quadrature rule. The superconvergent points for the other values of p and α are however different from the Gauss quadrature points.

It is instructive to check numerically that (6.58), or rather its derivative obtained by differentiating both sides, holds. In Fig. 6.11, we plot the error in the approximation of u' for $u(x) = x^{p+1}$, highlighting the values of the difference $e(x) := (u - u_h)'(x)$ at the (scaled and translated) coordinates of the superconvergence points. It can be seen that in the interior of the domain the values of $e(x)$ are much smaller compared to the maximum error. This superconvergence property holds for other functions besides x^{p+1} provided that it is smooth enough (the $p+1$ degree Taylor expansion exists and is a good approximation to u). However, near the boundaries of the uniform mesh, the superconvergence property is lost (except for B-splines of degree $p=4$). This will lead to a lesser effectivity of the error estimator developed in the next subsection, however in general, as it will be seen later, it has a small impact on the overall estimation and refinement scheme. The results of superconvergence can also be extended to tensor-product splines in higher dimensions, as in [26,37], where the superconvergent points are obtained using the tensor-product of the coordinates in one dimension. The points are mapped to the physical domain through the geometry mapping $\mathbf{F}$, though as is usual in IGA, all the computations are done in the parameter space.

6.5.2 Superconvergent patch recovery

Once the superconvergent points are determined, they can be used to build an enhanced "recovered" solution. In the following, we will particularly focus on constructing a recovered stress field $\boldsymbol{\sigma}_h^*$ obtained from the computed stresses $\boldsymbol{\sigma}_h$. The procedure is similar to that of the Zienkiewicz-Zhu (ZZ) patch recovery technique [39,40].

The task at hand is constructing a stress recovery operator which outputs a better (i.e. superconvergent) approximation to the solution. In particular, we want to compute $G[\sigma_h] = \sigma_h^*$, where the recovered stress σ_h^* is obtained by fitting a polynomial spline of higher degree through the superconvergent points on local subdomains. Let Ω_k, $k=1,\dots,n$ be a set of non-overlapping patches such that $\bigcup_{k=1}^{n}\Omega_k = \Omega$, where n is number of patches in the domain. Furthermore, let $\mathbf{x}_{i,k}^*$, with $i=1,\dots,N_k$ be the set of the N_k superconvergent points on the patch Ω_k. Then we choose the operator $G[\sigma_h]$ such that:

$$G[\sigma_h](\mathbf{x}) = \sum_{i=1}^{M_k} \phi_{i,k}^*(\mathbf{x})\mathbf{C}_{i,k}^*, \quad \text{for } \mathbf{x}\in\Omega_k \text{ with } M_k \geq N_k \text{ and} \tag{6.67}$$

$$G[\sigma_h](\mathbf{x}_{i,k}^*) = \sigma_h(\mathbf{x}_{i,k}^*), \quad \text{for } i=1,\dots,N_k,\, k=1,\dots,n. \tag{6.68}$$

Here $\phi_{i,k}^*$ are B-splines of degree $p^* \geq p$ and continuity $\alpha^* \geq \alpha$ with support on the local patches Ω_k. This gives rise to n linear systems of the form $\mathbf{A}^{(k)}\mathbf{C}_k^* = \mathbf{b_k}$, where

$$\mathbf{A}_{ij}^{(k)} = \phi_{j,k}^*(\mathbf{x}_{i,k}^*), \qquad i=1,\dots,N_k, j=1,\dots,M_k, \tag{6.69}$$

Table 6.3 Values of p^* and α^* for given p and α with the size of the resulting matrix $A^{(k)}$. The last column shows the location of the superconvergent points from Table 6.2 in one dimension ($d = 1$).

p	α	p^*	α^*	Size of $A^{(k)}$	Superconvergent points
3	1	3	2	$5^d \times 5^d$	
4	1	5	3	$8^d \times 8^d$	
5	2	7	6	$9^d \times 9^d$	
6	2	6	5	$8^d \times 8^d$	
7	3	7	6	$9^d \times 9^d$	

and $\mathbf{b}_k$ is an $N_k \times e$ matrix with $e = 3$ in two dimensions ($d = 2$) and $e = 6$ in three dimensions ($d = 3$) containing the stress components of σ_h evaluated at the points $\mathbf{x}_{i,k}$. These linear systems are overdetermined if $N_k > M_k$ and can be solved in a least-squares sense. Nevertheless, it is more computationally efficient to select Ω_k and $\phi^*_{i,k}$ such that $N_k = M_k$. A practical way to accomplish this in the context of hierarchical meshes is as follows:

i. We perform one level of cross-insertion on all the elements on the initial coarse mesh and let each subdomain Ω_k to be the set of 2^d elements which have a common parent. Subsequently we refine all the elements in a subdomain together, i.e. we mark all the elements in a given Ω_k for refinement and split it into 2^d subdomains which are used for patch recovery in the next step. This ensures that the number of elements in the mesh is divisible by 2^d and that every element belongs to a patch of 2^d elements which have the same refinement level.

ii. The polynomial degree and continuity of the recovery splines $\phi_{i,k}$ is chosen to match the number of superconvergent points in each recovery subdomain. This ensures that we have the same number of unknown coefficients as we have equations for the fitting of the recovered solution. The values of p^* and α^* for different values of p and α along with the size of the matrix $\mathbf{A}^{(k)}$ needed for computing $G[\sigma_h]$ on each recovery subdomain are given in Table 6.3. We note that for $p = 4$ and $p = 5$, other values of p^* and α^* are possible which result in a basis $\{\phi^*_{i,k}\}$ of the same dimension.

A recovery subdomain and the superconvergent points in the parameter and physical space are shown in Fig. 6.12. In general, we choose these subdomains such that the elements inside share the same parent element, which ensures that they are non-overlapping and the recovery cost is minimized. Since performing the recovery is completely local, the computational cost for computing $G[\sigma_h]$ is typically less than 10% than that of computing u_h

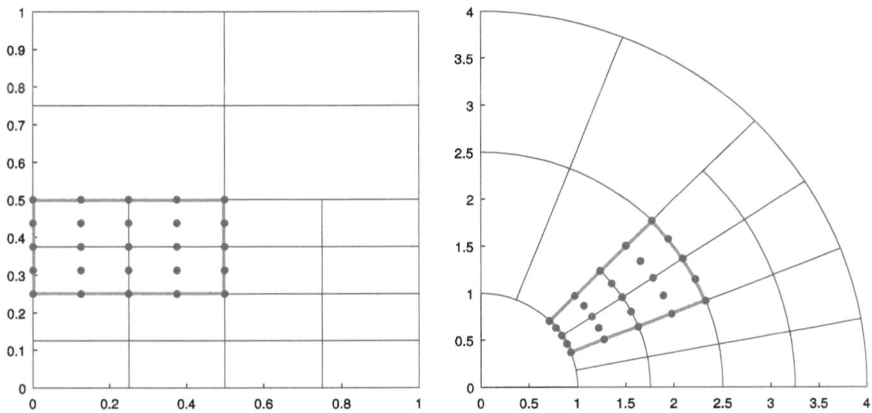

Figure 6.12 A recovery subdomain on an annular shaped domain together with the superconvergent points for $p = 3$.

and σ_h. Moreover, this operation can be trivially parallelized since the recovery on each subdomain is computed independently of the others.

Once $G[\sigma_h]$ has been computed, we estimate the error in energy norm by:

$$\eta(\sigma_h; \Omega) := \left(\int_\Omega (G[\sigma_h] - \sigma_h)^T D^{-1} (G[\sigma_h] - \sigma_h)\, d\Omega \right)^{1/2}, \tag{6.70}$$

where D^{-1} is the compliance matrix. We are now interested to analyze the asymptotic behavior of $G[\sigma_h]$. It is known [1,23] that if certain properties, which include consistency, locality, boundedness and linearity, are satisfied, then the recovery operator satisfies the *superapproximation property*:

$$e(G[\sigma_h]; \Omega) \leq \left(\int_\Omega (\sigma - G[\sigma_h])^T D^{-1} (\sigma - G[\sigma_h]]\, d\Omega \right)^{1/2} \leq Ch^{p+\delta}, \tag{6.71}$$

where δ is the same as in (6.58) and σ is the exact stress. If $\delta > 0$, then the error estimator $\eta(\sigma_h; \Omega)$ is asymptotically exact, i.e.

$$\theta(\sigma_h; \Omega) := \frac{\eta(\sigma_h; \Omega)}{e(\sigma_h; \Omega)} \to 1 \text{ as } h \to 0, \tag{6.72}$$

where $\theta(\sigma_h; \Omega)$ is the *effectivity index* of the error estimator $\eta(\sigma_h; \Omega)$. We note that asymptotic exactness can typically be achieved only under some rather stringent assumptions on the regularity of the solution and uniformity of the meshes. However, in many cases constructing a smoother, higher-order approximation to the stress field is enough to drive the adaptivity, though usually the error is over-estimated and finer meshes than needed are employed. This can be seen for example in [28,29] where over-refinement is observed in some of the numerical examples considered there.

6.5.3 Marking algorithm

It is important to consider for the hierarchical refinement process not just the accuracy and robustness of the error estimator, but also the choice of the marking algorithm. In general, there is a trade-off between the number of refinement steps required to reach a certain (estimated) accuracy and the number of elements in the final mesh. In other words, refining in small increments results in "optimal" meshes that have as few as possible elements while refining more elements at each refinement step results in fewer overall refinement steps but less optimal meshes. This is particularly true in case of problems in singularities, as for coarse meshes the presence of a singularity results in significant errors even some distance away in the domain (the so-called "polution errors"); however these errors become less significant as the area around the singularity is refined.

Given a particular error tolerance (in percentage or as a relative error), a simple refinement strategy is to mark for refinement all the elements where the error exceeds some threshold (which can be determined as a percentage of the maximum estimated error). This is referred to as the "absolute threshold" marking strategy in [18]. Since this method does not take into account the overall distribution of the error, it is preferable to use the "Dörfler marking" strategy [14], where the elements with the largest contribution to the total error are selected. In particular we want to choose the set of marked elements $\mathscr{M}$ of minimal cardinality such that given a parameter $\theta \in (0, 1]$:

$$\eta(\theta_h; \Omega_{\mathscr{M}}) \leq \theta\eta(\theta_h; \Omega), \tag{6.73}$$

where $\Omega_{\mathscr{M}} \subseteq \Omega$ is the region in the domain that is marked for refinement. We note that $\theta = 1$ results in uniform refinement, while $\theta \ll 1$ results in smaller refinement steps. In practice, $\theta = 0.75$ for problems with smooth solutions and $\theta = 0.5$ for problems with singularities provides a reasonable compromise between the number of refinement steps and the optimality of the meshes.

6.6. Multi-patch formulations for complex geometry

Finally, we mention that for complex domains it is generally desirable to consider multiple geometric patches, each of which are associated with a different parameter domain. These patches (which is the standard terminology in isogeometric analysis) are different from the recovery patches used in error estimation. Several techniques for weak coupling have been proposed (see for example [27,31]), but the simplest and most robust way to couple multiple conforming patches is by identifying the degrees of freedom at the patch interfaces. Due to the hierarchical structure of the meshes which allow local refinements, it is relatively easy to obtain conforming patches by matching the knot-vectors along the common boundaries. This procedure ensures that C^0 continuity is strongly enforced along the patch boundaries, with a minimal amount of extra refinement or computational cost. Other proposed methods to couple multi-patch geometries can also be used without additional difficulties.

In this chapter, we give an example of multipatch coupling of a three-dimensional L-Shaped bracket with inclusions. The model is obtained by joining 18 patches. We note that the geometry is represented exactly and that the different features of the bracket (e.g. the dimensions and radii of the inclusions) can be parametrized so that different designs can be obtained quickly. For numerical analysis, we consider a linear elastic model where the beam is fixed at one end and a displacement of -2 units is applied at the opposite end. The geometry description and the von Mises stresses obtained after 3 levels of non-uniform refinement with PHT splines are shown in Fig. 6.13. More detailed examples of multi-patch coupling and refinements for some benchmark problems are given in [3].

6.7. XIGA for interface problems

Problems with material interfaces where the solution is only C^0-continuous solutions are quite common in both computational and material engineering. The discontinuity in the gradient field occurs because of the different material properties on the two sides of the material interface. These are also known as weak discontinuities, which are not accurately modeled by the standard FEM without a conforming mesh. Matching the material geometry at the interface as close as possible can improve the accuracy of the solution. However, in the case of finite elements, creating an adequate conforming mesh is not feasible or even impossible for problems

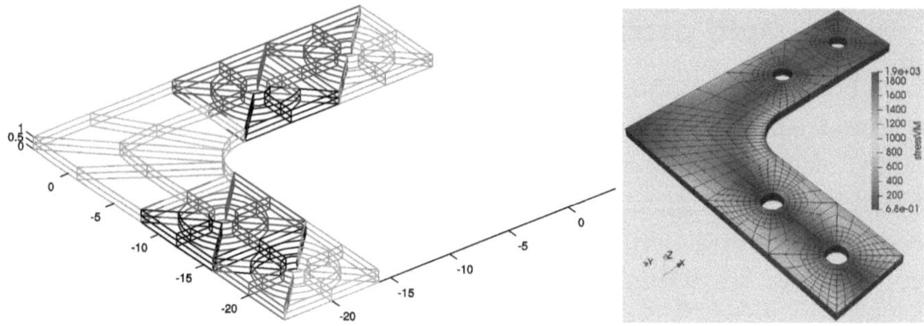

(A) Multi-patch geometry and the initial mesh

(B) von Mises stresses after 3 levels of refinement

Figure 6.13 L-Shape bracket geometry and analysis results.

with weak discontinuities or when the material interface is curved. The framework of isogeometric analysis does allow somewhat more flexibility for the explicit modeling of complex boundaries. However it is often desirable, particularly in the case of moving interfaces to have a mesh that is as independent as possible from the modeled discontinuities.

In this section, we follow the presentation in [22] an optimal XIGA method to solve material interface problems, by combining the advantages from XFEM and IGA. We show that the XIGA achieves optimal convergence rate while the IGA only converges at a lower rate for the Poisson's equation with weakly discontinuous solutions. The method is implemented for solving three numerical test problems, and is also compared with the traditional IGA.

6.7.1 Governing and weak form equations

For a two-dimensional Poisson's equation problem, the strong form of the boundary value problem is given by:

$$\begin{cases} -\nabla \cdot (\mathbf{K}\nabla u) = f & \text{in} \quad \Omega \\ u = \bar{u} & \text{on} \quad \Gamma_e \\ \frac{\partial u}{\partial \mathbf{n}} = \bar{q} & \text{on} \quad \Gamma_n \end{cases}, \tag{6.74}$$

where the two-dimensional domain Ω is an open set with Lipschitz continuous boundary $\partial\Omega$. $\mathbf{K} \subset \mathbb{R}^2 \times \mathbb{R}^2$ is a symmetric positive definitive matrix. When $\mathbf{K}$ is the identity matrix, (6.74) is identical to the classical Poisson's model problem. In the following numerical examples, $\mathbf{K}$ describes different material properties in different regions. $f : \Omega \rightarrow \mathbb{R}$ is a given function. $\bar{u}$ and

$\bar{q}$ denote known variables and flux boundary conditions. $\mathbf{n}$ is the outward normal unit vector at the boundary Γ_n. Γ_e and Γ_n are called Dirichlet/essential boundary and Neumann/natural boundary, respectively. $\Gamma_n \bigcap \Gamma_e = \emptyset$ and $\Gamma_n \bigcup \Gamma_e = \partial\Omega$.

Let us define two spaces $\mathcal{S}$ and $\mathcal{V}$:

$$\mathcal{S} = \{u \mid u \in H^1, u\,|_{\Gamma_e} = \bar{u}\} \tag{6.75}$$

$$\mathcal{V} = \{v \mid v \in H^1, v\,|_{\Gamma_e} = 0\}, \tag{6.76}$$

where H^1 is the Sobolev space. The functions in this space are in L^2 with square-integrable first derivatives. The functions in $\mathcal{S}$ need to satisfy the Dirichlet boundary condition. The homogeneous counterpart of the Dirichlet boundary condition is required in $\mathcal{V}$. We note that both $\mathcal{S}$ and $\mathcal{V}$ do not require the Neumann boundary condition in the definition.

The weak form of (6.74) is obtained by multiplying a test function v ($v \in \mathcal{V}$), and integrating over the region Ω,

$$-\int_\Omega v\nabla \cdot (\mathbf{K}\nabla u)\mathrm{d}\Omega = \int_\Omega vf\mathrm{d}\Omega \ , \tag{6.77}$$

$$-\oint_\Gamma v\frac{\partial \mathbf{K}u}{\partial \mathbf{n}}\mathrm{d}\Gamma + \int_\Omega \nabla v\mathbf{K}\nabla u\mathrm{d}\Omega = \int_\Omega vf\mathrm{d}\Omega \ . \tag{6.78}$$

The weak form reads: find $u \in \mathcal{S}$, for all $v \in \mathcal{V}$ such that

$$\int_\Omega \nabla v\mathbf{K}\nabla u\mathrm{d}\Omega = \int_\Omega vf\mathrm{d}\Omega + \int_{\Gamma_n} v\frac{\partial \mathbf{K}u}{\partial \mathbf{n}}\mathrm{d}\Gamma_n \ . \tag{6.79}$$

The effect of the Galerkin projection is to replace the above infinite-dimensional spaces $\mathcal{S}$ and $\mathcal{V}$ by their finite-dimensional subspaces $\mathcal{S}_h$ and $\mathcal{V}_h$, which contain finite independent basis functions. After applying the Galerkin projection, the problem reads: find $u_h \in \mathcal{S}_h$, for all $v_h \in \mathcal{V}_h$ such that

$$\int_\Omega \nabla v_h\mathbf{K}\nabla u_h\mathrm{d}\Omega = \int_\Omega v_hf\mathrm{d}\Omega + \int_{\Gamma_n} v_h\frac{\partial \mathbf{K}u_h}{\partial \mathbf{n}}\mathrm{d}\Gamma_n \ . \tag{6.80}$$

In (6.80), ∇ are applied respect to the physical coordinates (x, y). We denoted this by $\nabla_{(x,y)}$. The NURBS basis functions are defined in the parametric space. Using the chain rule, (6.81) can be written as:

$$\nabla_{(x,y)}u_h(x, y) = J(\xi, \eta)^{-T}\nabla_{(\xi,\eta)}u_h(\xi, \eta) \ , \tag{6.81}$$

where J is 2×2 Jacobian matrix,

$$J(\xi, \eta) = \begin{pmatrix} \frac{\partial F_1}{\partial \xi} & \frac{\partial F_1}{\partial \eta} \\ \frac{\partial F_2}{\partial \xi} & \frac{\partial F_2}{\partial \eta} \end{pmatrix}, \tag{6.82}$$

and $\mathbf{F}(\xi, \eta) = (F_1, F_2)$ is a transformation from the parametric coordinates (ξ, η) into the physical coordinates (x, y):

$$\mathbf{F}(\xi, \eta) = \sum_{i \in \mathbf{S}} N_i(\xi, \eta) \cdot \mathbf{C}_i \ , \tag{6.83}$$

where $\mathbf{C}_i$ are control points defined in the physical space.

Substituting (6.81) into both sides of (6.80), we obtain the following results:

$$\int_\Omega \nabla v_h \mathbf{K} \nabla u_h \mathrm{d}x\mathrm{d}y$$
$$= \int_{\Omega_0} (J^{-T}(\xi, \eta)\nabla v_h(\xi, \eta))\mathbf{K}(J^{-T}(\xi, \eta)\nabla u_h(\xi, \eta)) \mid \det J(\xi, \eta) \mid \mathrm{d}\xi \mathrm{d}\eta \ , \tag{6.84}$$

$$\int_\Omega v_h f \mathrm{d}x\mathrm{d}y = \int_{\Omega_0} (v_h f)(J(\xi, \eta)) \mid \det J(\xi, \eta) \mid \mathrm{d}\xi \mathrm{d}\eta \ , \tag{6.85}$$

$$\int_{\Gamma_n} v_h \frac{\partial \mathbf{K} u_h}{\partial \mathbf{n}} \mathrm{d}\Gamma_n = \int_{\Gamma_{n0}} (v_h \frac{\partial \mathbf{K} u_h}{\partial \mathbf{n}})(J(\xi, \eta)) \mid \det J(\xi, \eta) \mid \mathrm{d}\xi \mathrm{d}\eta \ . \tag{6.86}$$

For n-dimensional space with n basis functions $\mathbf{N} = \{N_1, N_2, \cdots, N_n\}$, we choose $v_h = N_i$, $i = 1 \ldots n$, and $u_h = \sum_{i=1}^n N_i u_i$, which are the fundamental of IGA method approximation. We are looking for n unknown coefficients u_i, with $i = 1 \ldots n$ for following linear problem:

$$\sum_{j=1}^n u_j a(N_j, N_i) = (f, N_i) + (N_i, \bar{q})_{\Gamma_n} \ , \tag{6.87}$$

where

$$a(N_j, N_i) = \int_{\Omega_0} \nabla N_j \mathbf{K} \nabla N_i \mathrm{d}\Omega \ , \tag{6.88}$$

$$(f, N_i) = \int_{\Omega_0} f N_i \mathrm{d}\Omega \ , \tag{6.89}$$

$$(N_i, \bar{q})_{\Gamma_n} = \int_{\Gamma_{n0}} N_i \bar{q} \mathrm{d}\Gamma_n \ . \tag{6.90}$$

In this paper, the error is estimated by either L^2 norm or the Energy norm. L^2 norm of the error is defined by:

$$\| u - u_h \|_{L^2}^2 = \int_\Omega (u - u_h)^2 \mathrm{d}\Omega \ .$$

In the classical FEM, an a priori error estimator is given by the formula:

$$\| u - u_h \|_m \leq Ch^\beta \| u \|_r \ , \tag{6.91}$$

where $\| \cdot \|_m$ and $\| \cdot \|_r$ are the norms defined in Sobolev spaces H_m and H_r [11]. u_h is the approximated solution of FEM, and u is the exact solution. h is a characteristic length scale related to the size of the elements in the mesh. $\beta = min(p + 1 - m \ , \ r - m)$ is the convergence rate. C is a constant, which is independent of u and h. This fundamental error estimate has been extended to IGA. More details can be found in [9].

The energy norm of the error is defined by:

$$\| u - u_h \|_E^2 = \int_\Omega \nabla(u - u_h) \, \mathbf{K} \nabla(u - u_h)^T \mathrm{d}\Omega \ . \tag{6.92}$$

According to (6.91) we give the bound and the optimal convergence rate by $\| u - u_h \|_{L2} \leq Ch^3 \| u \|_3$ for L^2 norm of the error and $\| u - u_h \|_E \leq Ch^2 \| u \|_3$ for energy norm of the error. However, it is based on studying the continuous problem. For the following discontinuous problem, we will show that this approximation needs to be improved. In some situations, we can reach the optimal convergence rate in numerical experiments.

6.7.2 Enriched basis functions selection

The enrichment means to expand the traditional continuous approximation space of IGA by adding particular functions which have non-smooth characteristics. There are two types of enrichments. The first one contains the discontinuities in the displacement field, which is suitable for modeling strong discontinuous problems. The second type includes the discontinuities in the gradient field, which is used for modeling weak discontinuous problems. Our study will focus on this later category.

In XIGA, only a subset of the original NURBS basis functions need to be enriched. The first step is thus to distinguish which functions should be enriched. We only enrich the basis functions that have support on the elements containing the discontinuity.

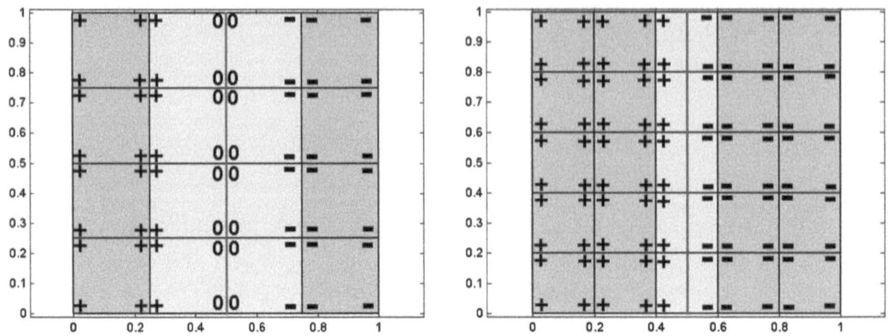

Figure 6.14 Applying the signed-distance function to distinguish the types of elements.

The coordinates of all nodes of the element are found and compared with the material interface, by using the Signed-Distance Function as follows

$$\xi(\mathbf{x}) = \underbrace{\min \| \mathbf{x} - \mathbf{x}_\Gamma \|}_{\mathbf{x}_\Gamma \in \Gamma} \operatorname{sign}(\mathbf{n} \cdot (\mathbf{x} - \mathbf{x}_\Gamma)) \ , \tag{6.93}$$

where $\mathbf{x}$ can be any point in the domain. Here $\mathbf{x}$ stands for one node of an element. $\mathbf{x}_\Gamma$ is the normal projection of $\mathbf{x}$ onto the material interface Γ. $\mathbf{n}$ is a normal unit vector. If all nodes of one element are on one side of the material interface, they are given the same sign. Otherwise some nodes will have the different signs with the others in the same element.

After applying the signed-distance function (6.93), there will be three possible situations as shown in Fig. 6.14. It shows that these points can have zero distance (0), positive distance (+) or negative distance (−).

Remark. The signed-distance function can be applied to both parametric and physical domains. In our study, we define the enriched basis functions in the parametric domain, because all the basis functions of IGA are originally defined in the parametric domain. The supports of each basis function are cut by straight lines within a square. The parameter space is also used for integration, where normally each integration cell is a quadrilateral. It is easier to map just the discontinuities in the parameter space and construct enrichment based on the image of the interface segment than to construct the enrichment functions in the physical space and map them back to the parameter space. The method needed is to find the parametric coordinates of the discontinuities is discussed in Section 6.7.6.

The XIGA method [12] [17] follows the PU framework of [4,24]. The main idea is to extend the classical solution space through multiplying the enrichment functions by the subset of these same basis functions which can ensure a conforming approximation. We define one set $\mathbf{N} = \{ N_1, N_2, \cdots, N_n \}$, N_i are the original IGA basis functions. The basis set of XIGA (represented by $\mathbf{N}^\star$) is the union of two sets, $\mathbf{N}^\star = \mathbf{N} \bigcup \mathbf{M}$, where $\mathbf{M}$ is the new enriched basis functions set, $\mathbf{M} = \{ M_1, M_2, \cdots, M_m \}$. M_j are constructed by multiplying N_i with a certain enrichment function ψ, i.e. $M_1 = N_{\hat{1}} \cdot \psi$, $M_2 = N_{\hat{2}} \cdot \psi$, $\cdots$, $M_m = N_{\hat{m}} \cdot \psi$. We use a different subscript notation of $N_{\hat{i}}$ from that of N_i, since $N_{\hat{i}}$ are the basis functions needed to be enriched. The approximation of XIGA can be expressed by:

$$u_h = \underbrace{\sum_{i \in \mathbf{N}} N_i u_i}_{\mathbf{u}^{st}} + \underbrace{\sum_{j \in \mathbf{M}} N_j \psi a_j}_{\mathbf{u}^{enr}} \ . \tag{6.94}$$

In the above, we have introduced the framework of IGA and XIGA. In the rest of this section, we will present some details and techniques applied in our XIGA method.

6.7.3 Enrichment functions

6.7.3.1 Ramp enrichment function

One kind of the enrichment functions is called Ramp enrichment functions. More details about the Ramp enrichment function can be found in [21]. We will apply it in our numerical Example I. The Ramp enrichment functions are formed by two independent parts ϕ_1 and ϕ_2:

$$\phi_1 = \begin{cases} x - 1/2 & \text{if} \quad x \geq 1/2 \\ 0 & \text{otherwise} \ , \end{cases} \tag{6.95}$$

$$\phi_2 = \begin{cases} -x + 1/2 & \text{if} \quad x < 1/2 \\ 0 & \text{otherwise} \ . \end{cases} \tag{6.96}$$

If N_i stand for the standard basis functions, then $N_i \cdot \phi_1$ and $N_i \cdot \phi_2$ are two new enriched basis functions. The extended approximation can be written as:

$$u_h = \underbrace{\sum_{i \in \mathbf{S}} N_i u_i}_{\mathbf{u}^{st}} + \underbrace{\sum_{j \in \mathbf{S}_{e1}} N_j \phi_1 a_j}_{\mathbf{u}^{enr1}} + \underbrace{\sum_{k \in \mathbf{S}_{e2}} N_k \phi_2 b_k}_{\mathbf{u}^{enr2}} \ ,$$

where $\mathbf{S}$ contains the indices of all global basis functions. $\mathbf{S}_{e1}$ and $\mathbf{S}_{e2}$ contain the indices of the enriched basis functions.

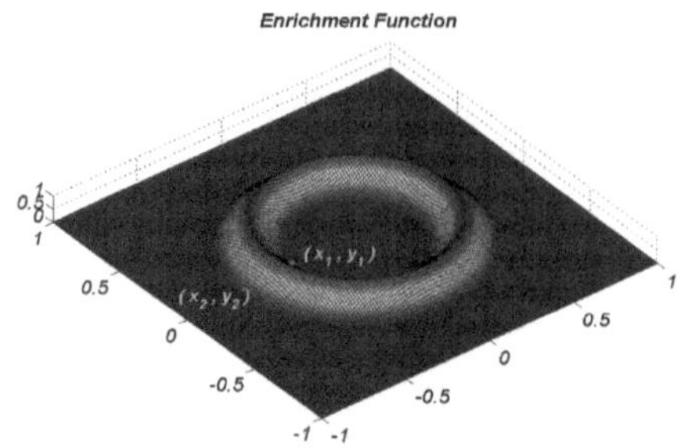

Figure 6.15 The enrichment function $\psi_1(x, y)$.

We define another two ramp enrichment functions φ_1 and φ_2 for Example III:

$$\varphi_1(x, y) = \begin{cases} \sqrt{x^2+y^2} - \frac{1}{2} & \text{if } \frac{1}{4} \le x^2 + y^2 \\ 0 & \text{otherwise} , \end{cases} \tag{6.97}$$

$$\varphi_2(x, y) = \begin{cases} -\sqrt{x^2+y^2} + \frac{1}{2} & \text{if } x^2 + y^2 \le \frac{1}{4} \\ 0 & \text{otherwise} . \end{cases} \tag{6.98}$$

6.7.3.2 *Moës enrichment function*

Another kind of enrichment function is called the Moës enrichment function. The detailed discussion about Moës enrichment functions can be seen in [25]. We will apply it for solving numerical Example II. It's defined by:

$$\psi_1(x, y) = \begin{cases} \frac{\sqrt{x^2+y^2} - \sqrt{{x_1}^2+{y_1}^2}}{1/2 - \sqrt{{x_1}^2+{y_1}^2}} & \text{if } \sqrt{{x_1}^2+{y_1}^2} \le \sqrt{x^2+y^2} \le \frac{1}{2} \\ \frac{\sqrt{{x_2}^2+{y_2}^2} - \sqrt{x^2+y^2}}{\sqrt{{x_2}^2+{y_2}^2} - 1/2} & \text{if } \frac{1}{2} < \sqrt{x^2+y^2} \le \sqrt{{x_2}^2+{y_2}^2} , \end{cases} \tag{6.99}$$

where (x_1, y_1) and (x_2, y_2) are two neighbor points of material interface shown in Fig. 6.15.

6.7.4 Greville Abscissae

In this subsection, we will introduce a way of selecting control points using Greville Abscissae. For quadratic and other higher orders, if control points are uniformly spaced, the mapping from the parameter space to the physical space is no longer identity. However, the Greville Abscissae can ensure this mapping to be an identity mapping. For more details, we refer to [15].

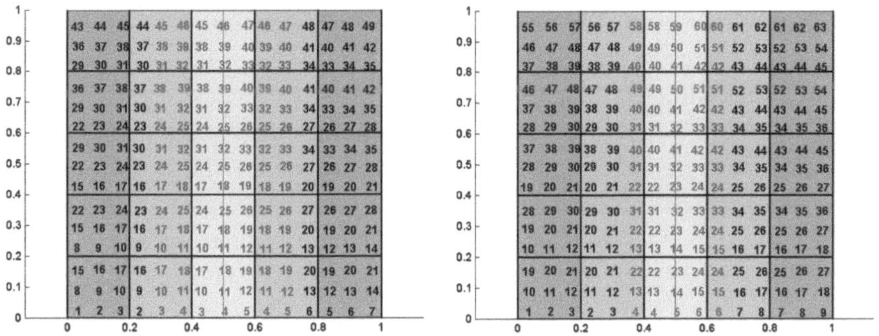

Figure 6.16 The basis functions distribution before repeating middle neighbor knots (left) after repeating middle neighbor knots (right).

In the Greville Abscissae, the control points $\mathbf{C} = (C_x, C_y)$ are calculated by:

$$C_x(i) = \frac{\sum_{j=i+1}^{i+p} \xi_j^u}{p}, \tag{6.100}$$

$$C_y(i) = \frac{\sum_{j=i+1}^{i+q} \xi_j^v}{q}, \tag{6.101}$$

where $\{ \xi_k^u \mid k = 1, \cdots, n_u \}$ and $\{ \xi_k^v \mid k = 1, \cdots, n_v \}$ are the knots in u and v directions respectively. n_u and n_v represent the numbers of NURBS basis functions in each direction.

6.7.5 Repeating middle neighbor knots

Repeating middle neighbor knots uses a unique property of NURBS function. Recall that the pth order NURBS function has support over $p + 1$ elements. We can repeat the neighbor knots of the material interface so that we can reduce the support of some NURBS functions.

Fig. 6.16 illustrates how the repeating neighbor knots works. On the left plot of Fig. 6.16, we want to enrich the elements in yellow (light gray in print version) color. All the basis functions that have supports in these elements will be enriched. A basis function has support in several elements, since each element is not isolated from others. As a result, some extra elements may be enriched, when these yellow (light gray in print version) elements are enriched. These elements are called blending elements in XFEM. In these blending elements, only some of the global basis functions will be enriched. The accuracy of the approximation is lower in these blending elements, as discussed in [16].

The distribution of the global basis functions is shown in the right panel of Fig. 6.16 after applying the repeating middle neighbor knots. The repeating middle neighbor knots decrease the spans for the basis functions in blending elements. However an obvious disadvantage is that a larger number of degrees-of-freedom is needed, which leads to more expensive calculation cost.

6.7.6 Inverse mapping

The basis functions are defined in the parametric space, and the numerical integration is also done in the parametric space. However usually the problems are described in the physical space. The material interface can have different shapes in the parametric and physical spaces. Hence, an inverse mapping needs to relate these two different shapes. The inverse mapping is set up by finding a pair of parameter coordinates (u, v) and satisfying the distance equation:

$$\mathbf{r}(u, v) = \mathbf{S}(u, v) - \mathbf{P} = 0 \ , \tag{6.102}$$

where $\mathbf{P}$ stands for the coordinates of the material interface. $\mathbf{S}$ stands for the coordinates of the NURBS surface point. Both points are defined in the physical space.

The Newton-Raphson method is applied to solve the nonlinear (6.102). Here the function $\mathbf{r}(u, v)$ is continuous and has a continuous derivative $\nabla\mathbf{r}(u, v)$. (6.102) can be expanded in the Taylor series:

$$\begin{aligned} &\mathbf{r}(u, v) + \delta\mathbf{u}\nabla\mathbf{r}(u, v) = 0 \ , \\ &\delta\mathbf{u}\nabla\mathbf{S}(u, v) = -(\mathbf{S}(u, v) - \mathbf{P}) \ . \end{aligned} \tag{6.103}$$

The solution of this linear system is of the form:

$$\begin{aligned} \delta\mathbf{u} &= \begin{pmatrix} u_{n+1} - u_n \\ v_{n+1} - v_n \end{pmatrix}^T , \\ u_{n+1} &= u_n + \delta u, \\ v_{n+1} &= v_n + \delta v. \end{aligned} \tag{6.104}$$

Remark. We note that it is important to choose proper initial values u_0 and v_0, the tolerance $\varepsilon > 0$, and the maximum number of iterations N.

The inverse mapping is done iteratively using the gradient values which gives good results in many applications. Finding good initial values usually depends on the scale and the shape of the problem domain studied. For example, in our third numerical example, we initially choose $u_0 = 0$ and $v_0 = 0$. Then in each iterative step, we check the stopping criteria $\| \mathbf{S}(u, v) - \mathbf{P} \| \leq \varepsilon$ with the tolerance $\varepsilon = 10^{-7}$ in our implementation.

6.7.7 Curve fitting

The inverse mapping is needed for finding the points in the parametric space corresponding to the material interface. These points are discrete, so they need to be recomputed for each curved material interface problem. Finding two intersection points in each enriched elements is very important, so it is better compute the path of this curve in the parametric space in the pre-processing stage. In [30], several methods for curve fitting are described. Here, we apply the least squares curve approximation method to calculate the control points of the B-spline curve of the material interface in the parametric space.

All weights of the control points are set to 1. In the following, p stands for the polynomial order of the basis functions and n is the number of the control points. $Q_0, Q_1, \ldots, Q_{m-1}, Q_m$ are data points from the inverse mapping. We look for a pth degree non-rational curve:

$$\mathbf{C}(\lambda) = \sum_{i=1}^{n} N_{i,p}(\lambda)\mathbf{P}_i \quad \lambda \in [0, 1] \ . \tag{6.105}$$

These data points $\mathbf{Q}_k$, with $k = 1, 2, \ldots, m-1, m$, are approximated in the least squares sense:

$$\begin{aligned}
f &= \sum_{k=1}^{m} \mid \mathbf{Q}_k - \mathbf{C}(\hat{u}_k) \mid^2 \\
&= \sum_{k=1}^{m} (\mathbf{Q}_k - \mathbf{C}(\hat{u}_k))(\mathbf{Q}_k - \mathbf{C}(\hat{u}_k)) \\
&= \sum_{k=1}^{m} [\mathbf{Q}_k\mathbf{Q}_k - 2\mathbf{Q}_k\mathbf{C}(\hat{u}_k) + \mathbf{C}(\hat{u}_k)\mathbf{C}(\hat{u}_k)] \\
&= \sum_{k=1}^{m} [\mathbf{Q}_k\mathbf{Q}_k - 2\sum_{i=1}^{n} N_{i,p}(\hat{u}_k)(\mathbf{Q}_k \cdot \mathbf{P}_i) + (\sum_{i=1}^{n} N_{i,p}(\hat{u}_k)\mathbf{P}_i)(\sum_{i=1}^{n} N_{i,p}(\hat{u}_k)\mathbf{P}_i)] \ .
\end{aligned} \tag{6.106}$$

The standard technique for least squares curve fitting is to minimize f by setting the derivatives of f with respect to the n points equal to zeros:

$$\frac{\partial f}{\partial \mathbf{P}_l} = \sum_{k=1}^{m} [-2N_{l,p}(\hat{u}_k)\mathbf{Q}_k + 2N_{l,p}(\hat{u}_k)\sum_{i=1}^{n} N_{i,p}(\hat{u}_k)\mathbf{P}_i] \ , \tag{6.107}$$

which yields

$$\sum_{i=1}^{n}(\sum_{k=1}^{m} N_{l,p}(\hat{u}_k)N_{i,p}(\hat{u}_k))\mathbf{P}_i = \sum_{k=1}^{m} N_{l,p}(\hat{u}_k)\mathbf{Q}_k \ , \tag{6.108}$$

$$(\mathbf{N}^T\mathbf{N})\mathbf{P} = \mathbf{N}^T\mathbf{Q} \ , \tag{6.109}$$

$\mathbf{N}$ is the $m \times n$ matrix

$$\mathbf{N} = \begin{pmatrix} N_{1,p}(\hat{u}_1) & N_{2,p}(\hat{u}_1) & \cdots & N_{n,p}(\hat{u}_1) \\ \vdots & \vdots & \ddots & \vdots \\ N_{1,p}(\hat{u}_m) & N_{2,p}(\hat{u}_m) & \cdots & N_{n,p}(\hat{u}_m) \end{pmatrix} \ , \tag{6.110}$$

$$\mathbf{Q} = [\mathbf{Q}_1 \ \mathbf{Q}_2 \ \cdots \ \mathbf{Q}_m]$$

$$\mathbf{P} = [\mathbf{P}_1 \ \mathbf{P}_2 \ \cdots \ \mathbf{P}_n] \ , \tag{6.111}$$

with $\mathbf{P} = (\mathbf{N}^T\mathbf{N})^{-1}\mathbf{N}^T\mathbf{Q}$.

Before solving the least square system, we should build a parametric coordinates consequence $\{\hat{u}_1, \hat{u}_2, \cdots, \hat{u}_m\}$ corresponding to the data points $\mathbf{Q}_k$, $k = 1, 2, \cdots m$. There are several ways to get this set. A good way is to calculate the chord length. d is the total chord length defined by

$$d = \sum_{k=2}^{m} \mid \mathbf{Q}_k - \mathbf{Q}_{k-1} \mid \ . \tag{6.112}$$

We have $\hat{u}_1 = 0$, $\hat{u}_m = 1$

$$\hat{u}_k = \hat{u}_{k-1} + \frac{\mid \mathbf{Q}_k - \mathbf{Q}_{k-1} \mid}{d} \ , \quad k = 1, 2, \cdots, m-1 \tag{6.113}$$

We now find the knots vector, which can be obtained by the equally spaced method or the averaging technique. We will use the averaging technique, which is recommended in [30]. Following this technique, we can calculate:

$$u_1 = \cdots = u_{p+1} = 0,$$

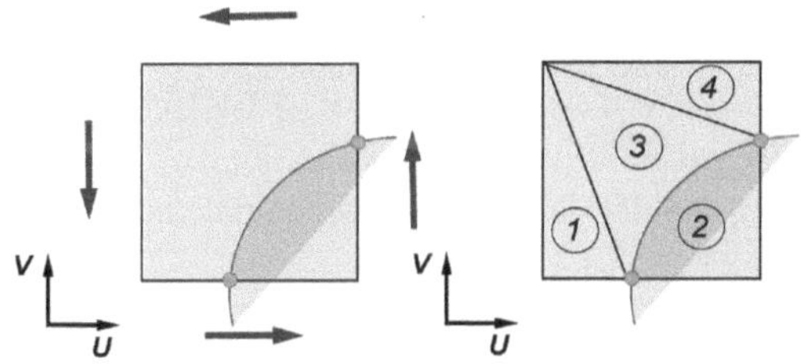

Figure 6.17 The element is crossed by the curved material interface (left). The enriched element is divided by triangulation (right).

$$u_{j+p} = \frac{1}{p}\sum_{i=j}^{j+p} \hat{u}_i \quad j = 1, 2, \cdots, n-p+1,$$

$$u_{m+1-p} = \cdots = u_{m+2} = 1 \ .$$

6.7.8 Intersection points

Now we can rebuild the material interface in the parametric space by using the results of the curve fitting. The curved material interface has its own knots vector and the control points, so the parametric space mesh is independent of the material interface. In this section, we are going to apply the Newton-Raphson method to find the intersection points between the material interface and the parametric mesh. We have,

$$r(\lambda) = C(\lambda) - \xi \ . \tag{6.114}$$

An illustration is shown in Fig. 6.17. We loop the edges of the element for each quadrilateral enriched element. For a horizontal edge, we take ξ to be the v coordinate value of the edge. We take ξ to be the u coordinate value of the edge for a vertical edge.

As mentioned in the section on inverse mappings, we always have to consider and pick some proper initial values for starting the iteration procedure. Here, we use the previous calculation of the fitted curve. We pick the first knot of the curve as the initial value of λ_0. We could also reinitialize the value of λ_0 by replacing with sequences of known knots inside the knots vector of the curve. After picking the values for λ_0, we iterate using the standard Newton-Raphson iteration:

$$\lambda_{n+1} = \lambda_n - \frac{r(\lambda_n)}{r'(\lambda_n)} \ . \tag{6.115}$$

Finally we find the parameter value corresponding to the intersection points on the material interface. In our case, after the loop for each enriched element, we should get two parameter value λ_1 and λ_2. We do not need to calculate λ for the special case, where the curve has only one intersection point in the corner of the element.

6.7.9 Triangular integration

In the case of curved interfaces, we need to consider more details for the numerical integration. Because of the shape of material interface, curved integration elements are needed to improve the quality of the domain discretization. The curved element development is derived from introducing isoparametric elements for handling curved boundaries in [38]. The triangular element with one curved edge is analyzed mathematically in [36,41, 42], etc. [19] also introduced transfinite elements which consist of a reference square mapped to a subdomain with curved boundaries. Our curved triangular elements construction is based on the study in [33–35]. We calculate the coordinates of the intersection points for each enriched element that is crossed by the material interface. We use the two intersection points together with the four nodes of the quadrilateral element for constructing a Delaunay triangulation. Since the material interface is curved in the parameter space, we have two triangles containing curved edges for matching it. For each quadrilateral element, there are two kinds of triangles, i.e. normal straight side triangles and curved edge triangles. For the normal triangle, we do the standard triangle transformation T_1 which is shown in Fig. 6.18.

$$T_1 : \xi \rightarrow (1 - \hat{\xi}^1 - \hat{\xi}^2)\xi_1 + \hat{\xi}^1\xi_2 + \hat{\xi}^2\xi_3 \ , \tag{6.116}$$

where ξ_1, ξ_2 and ξ_3 are the coordinates of the standard parent triangular element. $(\hat{\xi}^1, \hat{\xi}^2)$ are the coordinates of the Gaussian point. ξ is the new coordinates of the Gaussian point after transformation. For a curved triangle, we apply the transformation T_2 for constructing a curved edge triangle. This transformation is also shown in Fig. 6.18.

$$T_2 : \xi \rightarrow \frac{1 - \hat{\xi}^1 - \hat{\xi}^2}{1 - \hat{\xi}^1} C(\lambda(\hat{\xi}^1)) + \frac{\hat{\xi}^1\hat{\xi}^2}{1 - \hat{\xi}^1}\xi_2 + \hat{\xi}^2\xi_3 \ , \tag{6.117}$$

where $C(\lambda(\hat{\xi}^1))$ is the B-spline curve at point $\lambda(\hat{\xi}^1)$ and $\lambda(\hat{\xi}^1) = \lambda_1 + (\lambda_2 - \lambda_1)\hat{\xi}^1$, λ_1 and λ_2 are the parametric coordinates of the two intersection points ξ_1 and ξ_2. Using T_1 and T_2 allows us to consider the exact geometry in the integration process.

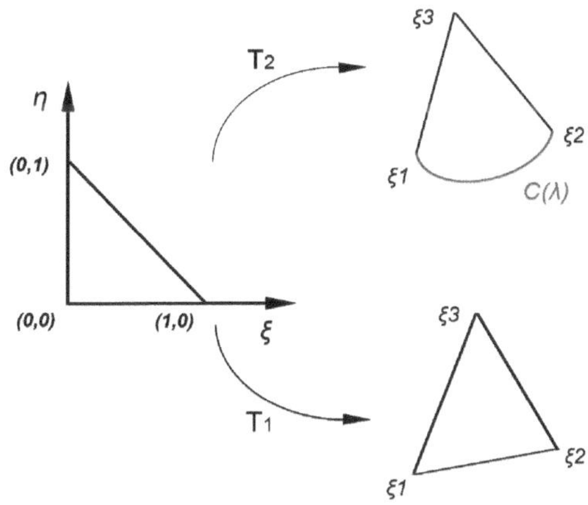

Figure 6.18 Triangle transformation.

Remark. Carrying out the integration efficiently and accurately needs particular attention, especially for the basis functions which contain discontinuities in the approximation space, since the Gaussian quadrature is not exact there. The first reason is that the curved triangle are needed to maintain all the Gaussian points which are inside the triangle on one side of the discontinuity. Once the curved edge matches the discontinuity, accurate integration is guaranteed. On the other hand, if we do not use curved triangles, the integration is not accurate anymore, and the optimal convergence rate will be lost. When the discontinuity passes through the integration triangle, the accuracy is completely lost. For more systematical curved elements studies with a plenty of numerical examples, we refer to the works of [34,35]. Numerical results illustrating these procedures are given in [22].

References

[1] M. Ainsworth, J. Oden, A posteriori error estimation in finite element analysis, Computer Methods in Applied Mechanics and Engineering 142 (1–2) (1997) 1–88.

[2] C. Anitescu, Y. Jia, Y.J. Zhang, T. Rabczuk, An isogeometric collocation method using superconvergent points, in: Isogeometric Analysis Special Issue, Computer Methods in Applied Mechanics and Engineering 284 (2015) 1073–1097.

[3] C. Anitescu, M.N. Hossain, T. Rabczuk, Recovery-based error estimation and adaptivity using high-order splines over hierarchical t-meshes, Computer Methods in Applied Mechanics and Engineering 328 (2018) 638–662.

[4] I. Babŭska, J. Melenk, The partition of unity method, International Journal for Numerical Methods in Engineering 40 (1997) 727–758.

[5] M.J. Borden, M.A. Scott, J.A. Evans, T.J.R. Hughes, Isogeometric finite element data structures based on Bézier extraction of NURBS, International Journal for Numerical Methods in Engineering 87 (1–5) (2011) 15–47.
[6] C.L. Chan, C. Anitescu, T. Rabczuk, Volumetric parametrization from a level set boundary representation with PHT-splines, Computer-Aided Design 82 (2017) 29–41.
[7] C.L. Chan, C. Anitescu, T. Rabczuk, Isogeometric analysis with strong multipatch C^1-coupling, Computer Aided Geometric Design 62 (2018) 294–310.
[8] C.L. Chan, C. Anitescu, T. Rabczuk, Strong multipatch C^1-coupling for isogeometric analysis on 2D and 3D domains, Computer Methods in Applied Mechanics and Engineering 357 (2019) 112599.
[9] J.A. Cottrell, T.J. Hughes, Y. Bazilers, Isogeometric Analysis, John Wiley & Sons, Ltd, 2009.
[10] J.A. Cottrell, T.J.R. Hughes, Y. Bazilevs, Isogeometric Analysis: Toward Integration of CAD and FEA, 1st edition, Wiley Publishing, 2009.
[11] R.F. Curtain, Functional Analysis in Modern Applied Mathematics, Academic Press INC. (London) Ltd, 1977.
[12] E. De Luycker, D.J. Benson, T. Belytschko, Y. Bazilevs, M.C. Hsu, X-FEM in isogeometric analysis for linear fracture mechanics, International Journal for Numerical Methods in Engineering 87 (6) (2011) 541–565.
[13] J. Deng, F. Chen, X. Li, C. Hu, W. Tong, Z. Yang, Y. Feng, Polynomial splines over hierarchical t-meshes, Graphical Models 70 (4) (2008) 76–86.
[14] W. Dörfler, A convergent adaptive algorithm for Poisson's equation, SIAM Journal on Numerical Analysis 33 (3) (1996) 1106–1124.
[15] G. Farin, Curves and Surfaces for CAGD, Morgan Kaufmann Publishers, 1999.
[16] T. Fries, A corrected XFEM approximation without problems in blending elements, International Journal for Numerical Methods in Engineering 75 (2008) 503–532.
[17] S.S. Ghorashi, N. Valizadeh, S. Mohammadi, Extended isogeometric analysis for simulation of stationary and propagating cracks, International Journal for Numerical Methods in Engineering 89 (2012) 1069–1101.
[18] C. Giannelli, B. Jüttler, S.K. Kleiss, A. Mantzaflaris, B. Simeon, J. Špeh, THB-splines: an effective mathematical technology for adaptive refinement in geometric design and isogeometric analysis, Computer Methods in Applied Mechanics and Engineering 299 (2016) 337–365.
[19] W.J. Gordon, C.A. Hall, Transfinite element methods: blending-function interpolation over arbitrary curved element domains, Numerische Mathematik 21 (1973) 109–129.
[20] T. Hughes, J. Cottrell, Y. Bazilevs, Isogeometric analysis: CAD, finite elements, NURBS, exact geometry and mesh refinement, Computer Methods in Applied Mechanics and Engineering 194 (39) (2005) 4135–4195.
[21] H. Ji, D. Chopp, J.E. Dolbow, A hybrid extended finite element/level set method for modeling phase transformations, International Journal for Numerical Methods in Engineering 54 (2002) 1209–1233.
[22] Y. Jia, C. Anitescu, S.S. Ghorashi, T. Rabczuk, Extended isogeometric analysis for material interface problems, IMA Journal of Applied Mathematics 80 (3) (June 2015) 608–633.
[23] M. Kumar, T. Kvamsdal, K.A. Johannessen, Superconvergent patch recovery and a posteriori error estimation technique in adaptive isogeometric analysis, in: Special Issue on Isogeometric Analysis: Progress and Challenges, Computer Methods in Applied Mechanics and Engineering 316 (2017) 1086–1156.

[24] J. Melenk, I. Babŭska, The partition of unity finite element method: basic theory and applications, Computer Methods in Applied Mechanics and Engineering 139 (1–4) (1996) 289–314.
[25] N. Moes, M. Cloirec, P. Cartraud, J.-F. Remacle, A computational approach to handle complex microstructure geometries, Computer Methods in Applied Mechanics and Engineering 192 (2003) 3163–3177.
[26] M. Montardini, G. Sangalli, L. Tamellini, Optimal-order isogeometric collocation at Galerkin superconvergent points, in: Special Issue on Isogeometric Analysis: Progress and Challenges, Computer Methods in Applied Mechanics and Engineering 316 (2017) 741–757.
[27] V.P. Nguyen, P. Kerfriden, M. Brino, S.P. Bordas, E. Bonisoli, Nitsche's method for two and three dimensional NURBS patch coupling, Computational Mechanics 53 (6) (2014) 1163–1182.
[28] N. Nguyen-Thanh, J. Muthu, X. Zhuang, T. Rabczuk, An adaptive three-dimensional RHT-splines formulation in linear elasto-statics and elasto-dynamics, Computational Mechanics 53 (2) (2014) 369–385.
[29] N. Nguyen-Thanh, K. Zhou, X. Zhuang, P. Areias, H. Nguyen-Xuan, Y. Bazilevs, T. Rabczuk, Isogeometric analysis of large-deformation thin shells using RHT-splines for multiple-patch coupling, in: Special Issue on Isogeometric Analysis: Progress and Challenges, Computer Methods in Applied Mechanics and Engineering 316 (2017) 1157–1178.
[30] L. Piegl, W. Tiller, The NURBS Book, Springer-Verlag, London, UK, 1995.
[31] M. Ruess, D. Schillinger, A.I. Özcan, E. Rank, Weak coupling for isogeometric analysis of non-matching and trimmed multi-patch geometries, Computer Methods in Applied Mechanics and Engineering 269 (2014) 46–71.
[32] T.W. Sederberg, J. Zheng, A. Bakenov, A. Nasri, T-splines and T-NURCCs, ACM Transactions on Graphics 22 (3) (July 2003) 477–484.
[33] R. Sevilla, S. Fernández-Méndez, NURBS-shaped domains with applications to NURBS-enhanced FEM, Finite Elements in Analysis and Design 47 (2011) 1209–1220.
[34] R. Sevilla, S. Fernández-Méndez, A. Huerta, NURBS-enhanced finite element method (NEFEM), International Journal for Numerical Methods in Engineering 76 (2008) 56–83.
[35] R. Sevilla, S. Fernández-Méndez, A. Huerta, NURBS-enhanced finite element method (NEFEM). A seamless bridge between CAD and FEM, Archives of Computational Methods in Engineering 18 (2011) 441–484.
[36] E.L. Wachspress, A rational basis for function approximation II. Curved sides, Journal of the Institute of Mathematics and Its Applications 11 (1973) 83–104.
[37] L. Wahlbin, Superconvergence in Galerkin Finite Element Methods, Springer-Verlag Berlin Heidelberg, 1995.
[38] O.C. Zienkiewicz, The Finite Element Method, McGraw-Hill, London, 1971.
[39] O.C. Zienkiewicz, J.Z. Zhu, The superconvergent patch recovery and a posteriori error estimates. Part 1: the recovery technique, International Journal for Numerical Methods in Engineering 33 (1992) 1331–1364.
[40] O.C. Zienkiewicz, J.Z. Zhu, The superconvergent patch recovery and a posteriori error estimates. Part 2: error estimates and adaptivity, International Journal for Numerical Methods in Engineering 33 (1992) 1365–1382.

[41] M. Zlamal, Curved elements in the finite element method I, SIAM Journal on Numerical Analysis 10 (1973) 229–240.
[42] M. Zlamal, The finite element method in domains with curved boundaries, International Journal for Numerical Methods in Engineering 5 (1973) 367–373.

CHAPTER SEVEN

Fracture in plates and shells

7.1. Fractures in shell and plates using XFEM

In the following, we first present a smoothed extended finite element method for Mindlin Reissner plates, following the exposition in [6].

7.1.1 Weak form

Let Ω be the domain of a flat isotropic homogeneous thick plate, Γ the boundary, and h the thickness. The midplane of the plate is taken as the reference plane, see Fig. 7.1.

The basic assumption for displacements is

$$\begin{Bmatrix} u(x,y,z) \\ v(x,y,z) \\ w(x,y,z) \end{Bmatrix} = \begin{Bmatrix} z\beta_x(x,y) \\ z\beta_y(x,y) \\ w(x,y) \end{Bmatrix}, \tag{7.1}$$

where $u(x,y,z)$, $v(x,y,z)$, and $w(x,y,z)$ are the components of displacement at a general point in Ω on the x, y, and z axes, respectively; $w(x,y)$ represents the transverse deflection and $\beta_x(x,y)$ and $\beta_y(x,y)$ are the rotations in the x and y directions of the middle surface, respectively. The bending and shear strains for Mindlin-Reissner plate theory are given by

$$\boldsymbol{\kappa} = \begin{bmatrix} \beta_{x,x} \\ \beta_{y,y} \\ \beta_{x,y} + \beta_{y,x} \end{bmatrix}, \boldsymbol{\gamma} = \begin{bmatrix} \beta_x + w_{,x} \\ \beta_y + w_{,y} \end{bmatrix}. \tag{7.2}$$

The total potential energy for a shear deformable plate subjected to in-plane prebuckling stresses ($\hat{\boldsymbol{\sigma}}_0$), in the absence of other external forces and neglecting terms with third and higher powers, can be written as

$$\Pi = \frac{1}{2}\iiint_V \boldsymbol{\kappa}^T \mathbf{Q}_b \boldsymbol{\kappa}\, dxdydz + \frac{1}{2}\iiint_V \boldsymbol{\gamma}^T \mathbf{Q}_s \boldsymbol{\gamma}\, dxdydz + \iiint_V \sigma_{xx}^0 \varepsilon_{xx}^{NL}\, dxdydz + \iiint_V \sigma_{yy}^0 \varepsilon_{yy}^{NL}\, dxdydz + \iiint_V \sigma_{xy}^0 \varepsilon_{xy}^{NL}\, dxdydz, \tag{7.3}$$

Extended Finite Element and Meshfree Methods
https://doi.org/10.1016/B978-0-12-814106-9.00013-9
Copyright © 2020 Elsevier Inc.
All rights reserved.

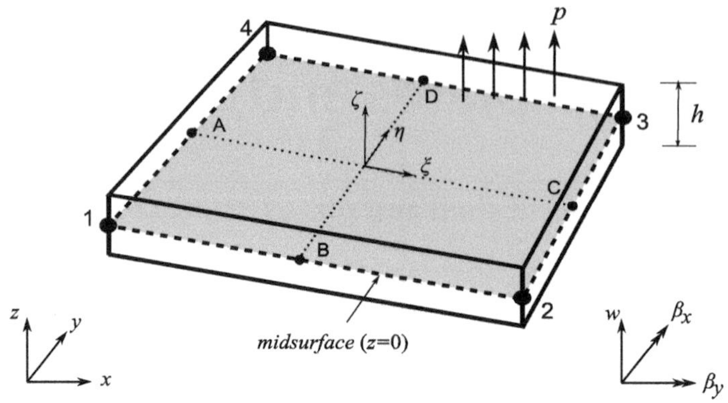

Figure 7.1 Quadrilateral shear deformable plate element.

where

$$\left\{ \begin{array}{c} \varepsilon_{xx}^{NL} \\ \varepsilon_{yy}^{NL} \\ \varepsilon_{xy}^{NL} \end{array} \right\} = \left\{ \begin{array}{c} \frac{1}{2}((u_{,x})^2 + (v_{,x})^2 + (w_{,x})^2) \\ \frac{1}{2}((u_{,y})^2 + (v_{,y})^2 + (w_{,y})^2) \\ (u_{,x}u_{,y} + v_{,x}v_{,y} + w_{,x}w_{,y}) \end{array} \right\}. \tag{7.4}$$

Combining (7.1)-(7.4) and integrating over the thickness (7.3), the total potential energy can be rewritten

$$\begin{aligned} \Pi = & \frac{1}{2}\iint_{\Omega} \boldsymbol{\kappa}^T \mathbf{D}_b \boldsymbol{\kappa}\, dxdy + \frac{1}{2}\iint_{\Omega} \boldsymbol{\gamma}^T \mathbf{D}_s \boldsymbol{\gamma}\, dxdy \\ & + \frac{1}{2}\iint_{\Omega} \left[\begin{array}{cc} w_{,x} & w_{,y} \end{array} \right] \hat{\boldsymbol{\sigma}}_0 \left\{ \begin{array}{c} w_{,x} \\ w_{,y} \end{array} \right\} h dxdy \\ & + \frac{1}{2}\iint_{\Omega} \left[\begin{array}{cc} \beta_{x,x} & \beta_{x,y} \end{array} \right] \hat{\boldsymbol{\sigma}}_0 \left\{ \begin{array}{c} \beta_{x,x} \\ \beta_{x,y} \end{array} \right\} \frac{h^3}{12} dxdy \\ & + \frac{1}{2}\iint_{\Omega} \left[\begin{array}{cc} \beta_{y,x} & \beta_{y,y} \end{array} \right] \hat{\boldsymbol{\sigma}}_0 \left\{ \begin{array}{c} \beta_{y,x} \\ \beta_{y,y} \end{array} \right\} \frac{h^3}{12} dxdy. \end{aligned} \tag{7.5}$$

The coefficient matrices in (7.3) are defined as

$$\mathbf{D}_b = \frac{Eh^3}{12(1-\upsilon^2)} \left[\begin{array}{ccc} 1 & \upsilon & 0 \\ \upsilon & 1 & 0 \\ 0 & 0 & \frac{1}{2}(1-\upsilon) \end{array} \right], \mathbf{D}_s = \frac{k_s Eh}{2(1+\upsilon)} \left[\begin{array}{cc} 1 & 0 \\ 0 & 1 \end{array} \right], \tag{7.6}$$

$$\hat{\boldsymbol{\sigma}}_0 = \left[\begin{array}{cc} \sigma_{xx}^0 & \sigma_{xy}^0 \\ \sigma_{yx}^0 & \sigma_{yy}^0 \end{array} \right], \tag{7.7}$$

where $\boldsymbol{D}_b$ is the bending stiffness matrix, $\boldsymbol{D}_s$ the shear stiffness matrix, E the Young's modulus, υ the Poisson ratio, and k_s the transverse shear correction factor (taken as $\frac{5}{6}$ in this work).

7.1.2 Implementation based on the Q4 element

The problem domain Ω will be discretized into a finite number of quadrilateral isoparametric elements N_e:

$$\Omega \approx \Omega^h = \bigcup_{e=1}^{Ne} \Omega^e.$$

Using the shape functions of the quadrilateral element shown in Fig. 7.1, lateral displacement and rotations can be expressed as

$$w = \sum_{i=1}^{4} N_i w_i,\ \beta_x = \sum_{i=1}^{4} N_i \beta_{xi},\ \beta_y = \sum_{i=1}^{4} N_i \beta_{y_i}, \tag{7.8}$$

where w_i, β_{xi}, and β_{y_i} denote nodal displacements and rotations and N_i the vector of bilinear shape functions. Using the approximation in (7.8) the following expansions can be written:

$$\boldsymbol{\kappa} = \boldsymbol{B}_b \boldsymbol{q},\ \boldsymbol{\gamma} = \boldsymbol{B}_s \boldsymbol{q}, \tag{7.9}$$

$$\left\{ \begin{array}{c} w_{,x} \\ w_{,y} \end{array} \right\} = \boldsymbol{G}_b \boldsymbol{q},\ \left\{ \begin{array}{c} \beta_{x,x} \\ \beta_{x,y} \end{array} \right\} = \boldsymbol{G}_{s1} \boldsymbol{q},\ \left\{ \begin{array}{c} \beta_{y,x} \\ \beta_{y,y} \end{array} \right\} = \boldsymbol{G}_{s2} \boldsymbol{q}, \tag{7.10}$$

where

$$\boldsymbol{B}_{bi} = \begin{bmatrix} 0 & N_{i,x} & 0 \\ 0 & 0 & N_{i,y} \\ 0 & N_{i,y} & N_{i,x} \end{bmatrix},\ \boldsymbol{B}_{si} = \begin{bmatrix} N_{i,x} & N_i & 0 \\ N_{i,y} & 0 & N_i \end{bmatrix},\ \boldsymbol{q}_i = \begin{bmatrix} w_i \\ \beta_{xi} \\ \beta_{y_i} \end{bmatrix}, \tag{7.11}$$

$$\boldsymbol{G}_{bi} = \begin{bmatrix} N_{i,x} & 0 & 0 \\ N_{i,y} & 0 & 0 \end{bmatrix},\ \boldsymbol{G}_{s1i} = \begin{bmatrix} 0 & N_{i,x} & 0 \\ 0 & N_{i,y} & 0 \end{bmatrix},$$

$$\boldsymbol{G}_{s2i} = \begin{bmatrix} 0 & 0 & N_{i,x} \\ 0 & 0 & N_{i,y} \end{bmatrix}. \tag{7.12}$$

Using (7.8)-(7.12), the stationary form of the total potential energy expression (7.3) can be expressed as

$$(\boldsymbol{K} - \lambda \boldsymbol{K}_G)\boldsymbol{q}^m = 0, \quad m = 1, 2, \ldots \text{ degrees of freedom}, \tag{7.13}$$

where $\boldsymbol{K}$ is the global stiffness matrix, $\boldsymbol{K}_G$ is the geometrical stiffness matrix, λ is a scalar by which the chosen in-plane loads must be multiplied in order to cause buckling, and vector $\boldsymbol{q}^m$ is the m-th buckling mode. The stiffness matrices in (7.13) can be explicitly written as

$$\boldsymbol{K} = \boldsymbol{K}_b + \boldsymbol{K}_s$$

that is,

$$\boldsymbol{K} = \int_{\Omega^e} \boldsymbol{B}_b^T \boldsymbol{D}_b \boldsymbol{B}_b d\Omega^e + \int_{\Omega^e} \boldsymbol{B}_s^T \boldsymbol{D}_s \boldsymbol{B}_s d\Omega^e \tag{7.14}$$

and $\boldsymbol{K}_G \boldsymbol{K}_{Gb} + \boldsymbol{K}_{Gs}$, with

$$\boldsymbol{K}_{Gb} = h \int_{\Omega^e} \mathbf{G}_b^T \hat{\boldsymbol{\sigma}}_0 \mathbf{G}_b d\Omega^e, \tag{7.15}$$

$$\boldsymbol{K}_{Gs} = \frac{h^3}{12} \int_{\Omega^e} \mathbf{G}_{s1}^T \hat{\boldsymbol{\sigma}}_0 \mathbf{G}_{s1} d\Omega^e + \frac{h^3}{12} \int_{\Omega^e} \mathbf{G}_{s2}^T \hat{\boldsymbol{\sigma}}_0 \mathbf{G}_{s2} d\Omega^e. \tag{7.16}$$

The shear contribution for the geometric stiffness matrix ($\boldsymbol{K}_{Gs}$) is negligible for thin plates, but its effect becomes more significant as the plate thickness increases. Most of the available literature dealing with buckling of cracked plates does not consider the effect of (7.16).

7.1.3 Shear locking

It is well known that the bilinear element described above exhibits shear locking as the plate thickness approaches the Kirchhoff limit. This is due to the fact that when using the bilinear interpolation for displacements and rotations the transverse shear strains cannot vanish at all points in the element when subjected to a constant bending moment. To overcome this deficiency, various remedies such as reduced integration have been proposed. In the present work, classical reduced integration and MITC (mixed interpolation of tensorial components) [7] approaches will be used to eliminate shear locking.

In the classical approach, the shear part of the stiffness matrix, $\boldsymbol{K}_s$ in (7.14), will be integrated using a 1×1 Gauss quadrature to avoid shear locking; this element will be referred to as Q4R.

In the approach proposed by [7] displacement and rotations are interpolated as usual, but for the transverse shear strains, the covariant components measured in the natural coordinate system are interpolated. Following [7] for the approximation of the shear strains, the second equation in (7.2) can be expressed as

$$\gamma = \begin{bmatrix} \gamma_x \\ \gamma_x \end{bmatrix} = J^{-1} \begin{bmatrix} \gamma_\xi \\ \gamma_\eta \end{bmatrix}, \tag{7.17}$$

with

$$\gamma_\xi = \frac{1}{2}[(1-\eta)\gamma_\xi^B + (1+\eta)\gamma_\xi^D], \tag{7.18}$$

$$\gamma_\eta = \frac{1}{2}[(1-\xi)\gamma_\eta^A + (1+\xi)\gamma_\eta^C], \tag{7.19}$$

where J is the Jacobian matrix and γ_η^A, γ_ξ^B, γ_η^C, and γ_ξ^D are the (physical) shear strains at the midside points A, B, C, and D, shown in Fig. 7.1 together with the global (x, y, z) and local (ξ, η, ζ) coordinate systems. Using (7.17) and (7.19) and following the description in [24], the shear part of the stiffness matrix, K_s in (7.14), can be rewritten as

$$\bar{B}_{si} = J^{-1} \begin{bmatrix} N_{i,\xi} & b_i^{11} N_{i,\xi} & b_i^{12} N_{i,\xi} \\ N_{i,\eta} & b_i^{21} N_{i,\eta} & b_i^{22} N_{i,\eta} \end{bmatrix}, \tag{7.20}$$

where

$$b_i^{11} = \xi_i x_{,\xi}^M,\ b_i^{12} = \xi_i y_{,\xi}^M,\ b_i^{21} = \eta_i x_{,\eta}^L,\ b_i^{22} = \eta_i y_{,\eta}^L. \tag{7.21}$$

The coordinates of the unit square are $\xi_i \in \{-1, 1, 1, -1\}$, $\eta_i \in \{-1, -1, 1, 1\}$ and the allocation of the midside nodes to the corner nodes is given by

$$(i, M, L) \in \{(1, B, A); (2, B, C); (3, D, C); (4, D, A)\},$$

see Fig. 7.1. Using (7.20), now the shear part of the stiffness matrix (K_s) can be computed using full integration (2×2 Gauss quadrature); this element will be referred to as MITC.

7.1.4 Curvature strain smoothing

Curvature strain smoothing for plate bending was first proposed in [49] in meshfree methods and in [35] in a finite element framework. Following the derivation in [35], the smoothed bending strains for Mindlin-Reissner plates are given as

$$\tilde{\kappa} = \tilde{B}_b^C q, \tag{7.22}$$

and the corresponding smoothed element bending stiffness matrix is

$$\tilde{\boldsymbol{K}}_b = \int_{\Omega^e} (\tilde{\boldsymbol{B}}_b^C)^T \boldsymbol{D}_b \tilde{\boldsymbol{B}}_b^C \, d\Omega^e = \sum_{C=1}^{nc} \left(\tilde{\boldsymbol{B}}_b^C(\boldsymbol{x}_C)\right)^T \boldsymbol{D}_b \tilde{\boldsymbol{B}}_b^C(\boldsymbol{x}_C) A_C, \tag{7.23}$$

where nc is the number of smoothing cells in the element. As described in [35], the integrands are constant over each smoothing cell domain (Ω_C^e) and the nonlocal curvature displacement matrix is given by

$$\tilde{\boldsymbol{B}}_{bi}^C(\boldsymbol{x}_C) = \frac{1}{A_C} \int_{\Gamma_C} \begin{pmatrix} 0 & N_i n_x & 0 \\ 0 & 0 & N_i n_y \\ 0 & N_i n_y & N_i n_x \end{pmatrix} d\Gamma_C. \tag{7.24}$$

This equation is evaluated using one Gauss point over each boundary cell segment Γ_C^m:

$$\tilde{\boldsymbol{B}}_{bi}^C(\boldsymbol{x}_C) = \frac{1}{A_C} \sum_{m=1}^{mt} \begin{pmatrix} 0 & N_i(\boldsymbol{x}_m^G) n_x(\boldsymbol{x}_m^G) & 0 \\ 0 & 0 & N_i(\boldsymbol{x}_m^G) n_y(\boldsymbol{x}_m^G) \\ 0 & N_i(\boldsymbol{x}_m^G) n_y(\boldsymbol{x}_m^G) & N_i(\boldsymbol{x}_m^G) n_x(\boldsymbol{x}_m^G) \end{pmatrix} l_m^C, \tag{7.25}$$

where $\boldsymbol{x}_m^G$ is the Gauss point (midpoint of segment m), l_m^C is the length of segment m, and mt is the total number of segments. The expression in (7.25) already includes the product of the Jacobian of transformation for a 1D 2-node element ($l_m^C/2$) and the Gauss quadrature weight (2). Combining the curvature strain smoothing and the mixed interpolation of tensorial components gives the following expression for the stiffness matrix of a Mindlin-Reissner plate:

$$\tilde{\boldsymbol{K}} = \tilde{\boldsymbol{K}}_b + \tilde{\boldsymbol{K}}_s, \tag{7.26}$$

$$\tilde{\boldsymbol{K}} = \sum_{C=1}^{nc} \left(\tilde{\boldsymbol{B}}_b^C(\boldsymbol{x}_C)\right)^T \boldsymbol{D}_b \tilde{\boldsymbol{B}}_b^C(\boldsymbol{x}_C) A_C + \int_{\Omega^e} \bar{\boldsymbol{B}}_s^T \boldsymbol{D}_s \bar{\boldsymbol{B}}_s \, d\Omega^e. \tag{7.27}$$

As mentioned in [35], this element will be referred to as MISCk (mixed interpolation and smoothed curvatures) with $k \in \{1, 2, 3, 4\}$ representing the number of smoothing cells (nc). These elements were shown to pass the patch test and to be slightly more accurate than the MITC for regular meshes. The most promising feature was their improved performance for irregular meshes and coarse meshes and their lower computational cost.

7.1.4.1 Smoothing of bending geometric stiffness matrix

Following a similar approach as the one presented by [35,49] and shown above, the bending part of the geometric stiffness matrix given in (7.15) can be written as

$$\tilde{\boldsymbol{K}}_{Gb} = h\int_{\Omega^e} (\tilde{\mathbf{G}}_b^C)^T \hat{\boldsymbol{\sigma}}_0 \tilde{\mathbf{G}}_b^C d\Omega^e = h\sum_{C=1}^{nc} (\tilde{\mathbf{G}}_b^C(\boldsymbol{x}_C))^T \hat{\boldsymbol{\sigma}}_0 \tilde{\mathbf{G}}_b^C(\boldsymbol{x}_C) A_C, \quad (7.28)$$

where

$$\tilde{\mathbf{G}}_{bi}^C(\boldsymbol{x}_C) = \frac{1}{A_C}\int_{\Gamma_C}\begin{pmatrix} N_i n_x & 0 & 0 \\ N_i n_y & 0 & 0 \end{pmatrix} d\Gamma_C. \quad (7.29)$$

Following the same numerical implementation used for (7.24), (7.29) will be evaluated using one Gauss point over each boundary cell segment Γ_C^m:

$$\tilde{\mathbf{G}}_{bi}^C(\boldsymbol{x}_C) = \frac{1}{A_C}\sum_{m=1}^{mt}\begin{pmatrix} N_i(\boldsymbol{x}_m^G)n_x(\boldsymbol{x}_m^G) & 0 & 0 \\ N_i(\boldsymbol{x}_m^G)n_y(\boldsymbol{x}_m^G) & 0 & 0 \end{pmatrix} l_m^C. \quad (7.30)$$

Using this expression in (7.28) corresponds to smoothing the higher-order bending terms given in (7.4). This case will be considered in the present work, in order to study the effect of smoothing higher-order terms within a shear deformable plate formulation. This element will be defined as MISC*k*_b (mixed interpolation and smoothed curvatures of global stiffness and geometric stiffness matrix). Similarly to MISC*k*, the number *k* in MISC*k*_b represents the number of smoothing cells in the element.

7.1.5 Extended finite element method for shear deformable plates

Following a similar enriched approximation for plate bending as presented in [19], deflection and rotations can be approximated as

$$\boldsymbol{w}^h(\boldsymbol{x}) = \sum_{i\in\mathcal{N}^{fem}} N_i(\boldsymbol{x})w_i + \sum_{j\in\mathcal{N}^{crack}} N_j(\boldsymbol{x})\bar{H}_j(\boldsymbol{x})b_j^w + \sum_{k\in\mathcal{N}^{tip}} N_k(\boldsymbol{x})(\sum_{l=1}^{4}\bar{G}_{lk}(r,\theta)c_{kl}^w), \quad (7.31)$$

$$\boldsymbol{\beta}^h(\boldsymbol{x}) = \sum_{i\in\mathcal{N}^{fem}} N_i(\boldsymbol{x})\beta_i + \sum_{j\in\mathcal{N}^{crack}} N_j(\boldsymbol{x})\bar{H}_j(\boldsymbol{x})b_j^\beta + \sum_{k\in\mathcal{N}^{tip}} N_k(\boldsymbol{x})(\sum_{l=1}^{4}\bar{F}_{lk}(r,\theta)c_{kl}^\beta), \quad (7.32)$$

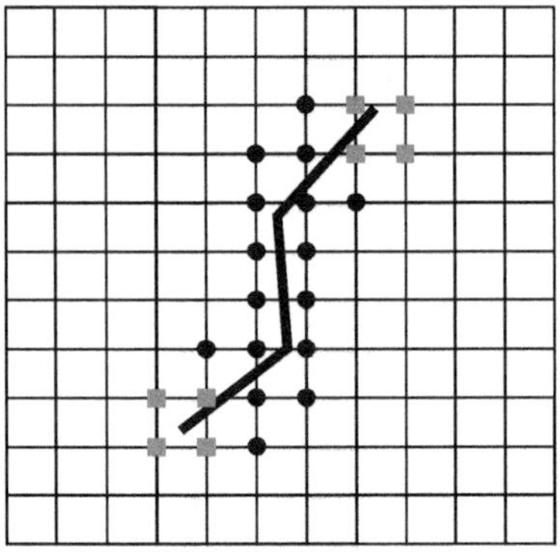

Figure 7.2 Crack on a uniform mesh of bilinear quadrilateral elements. Circle nodes are enriched by jump functions and square nodes by the asymptotic crack tip field.

where $N(\boldsymbol{x})$ denotes the standard bilinear shape functions, w_i and β_i the nodal unknowns associated with the continuous solution, b_j the nodal enriched degrees of freedom associated with the Heaviside function $\bar{H}_j(\boldsymbol{x})$, and c_{kl} the nodal enriched degrees of freedom associated with the elastic asymptotic crack tip functions $\bar{G}_{lk}(r,\theta)$ and $\bar{F}_{lk}(r,\theta)$. In (7.31) and (7.32) N^{fem} is the set of all nodes in the mesh, N^{crack} is the set of nodes whose shape function support is cut by the crack interior (circular nodes in Fig. 7.2), and N^{tip} is the set of nodes whose shape function support is cut by the crack tip (square nodes in Fig. 7.2).

The shifted enrichment functions in (7.31) and (7.32) are given by

$$\bar{H}_i(\boldsymbol{x}) = (H(\boldsymbol{x}) - H(\boldsymbol{x}_i)), \tag{7.33}$$

$$\bar{G}_{li}(\boldsymbol{x}) = (G_l(\boldsymbol{x}) - G_l(\boldsymbol{x}_i)), \bar{F}_{li}(\boldsymbol{x}) = (F_l(\boldsymbol{x}) - F_l(\boldsymbol{x}_i)). \tag{7.34}$$

Shifting the enrichment functions is particularly useful because the influence of the enrichment on the displacement must vanish at the nodes for ease of applying boundary conditions.

The Heaviside enrichment function $H(x)$ in (7.33) is defined by

$$H(x) = \begin{cases} +1 & \text{if the point is above the crack face,} \\ -1 & \text{if the point is below the crack face.} \end{cases} \tag{7.35}$$

Eq. (7.35) is responsible for the description of the interior of the crack (jump in displacements).

The elastic crack tip enrichment functions in (7.34) are defined as:

$$\{G_l(r,\theta)\} \equiv \left\{ r^{3/2}\sin(\frac{\theta}{2}), r^{3/2}\cos(\frac{\theta}{2}), r^{3/2}\sin(\frac{3\theta}{2}), r^{3/2}\cos(\frac{3\theta}{2}) \right\}, \quad (7.36)$$

$$\{F_l(r,\theta)\} \equiv \left\{ \sqrt{r}\sin(\frac{\theta}{2}), \sqrt{r}\cos(\frac{\theta}{2}), \sqrt{r}\sin(\frac{\theta}{2})\sin(\theta), \sqrt{r}\cos(\frac{\theta}{2})\sin(\theta) \right\}. \quad (7.37)$$

Here (r, θ) are polar coordinates with origin at the crack tip. These functions are not only responsible for closing the crack at the tip but they also introduce analytical information in the numerical approximation.

7.1.6 Smoothed extended finite element method

In SmXFEM, smoothing must now be performed on discontinuous and non-polynomial approximations. In the present case, curvature smoothing coupled with partition of unity enrichment can produce a plate element capable of cracking which is significantly more accurate than formerly proposed elements. Enriched smoothed bending strain and enriched shear strain can be written as

$$\tilde{\boldsymbol{\kappa}}^h = \sum_{i\in N^{fem}} \tilde{\boldsymbol{B}}^C_{b_{fem}i}\boldsymbol{q}_i + \sum_{j\in N^{crack}} \tilde{\boldsymbol{B}}^C_{b_{crack}j}\boldsymbol{b}_j + \sum_{k\in N^{tip}} \tilde{\boldsymbol{B}}^C_{b_{tip}k}\boldsymbol{c}_k, \quad (7.38)$$

$$\boldsymbol{\gamma}^h = \sum_{i\in N^{fem}} \bar{\boldsymbol{B}}_{s_{fem}i}\boldsymbol{q}_i + \sum_{j\in N^{crack}} \bar{\boldsymbol{B}}_{s_{crack}j}\boldsymbol{b}_j + \sum_{k\in N^{tip}} \bar{\boldsymbol{B}}_{s_{tip}k}\boldsymbol{c}_k. \quad (7.39)$$

The enriched gradient operators in (7.38) and (7.39) can be explicitly written as

$$\tilde{\boldsymbol{B}}^C_{b_{crack}j}(\boldsymbol{x}_C) = \frac{1}{A_C}\sum_{m=1}^{mt}\begin{pmatrix} 0 & N_j(\boldsymbol{x}^G_m)\bar{H}_j(\boldsymbol{x}^G_m)n_x(\boldsymbol{x}^G_m) & 0 \\ 0 & 0 & N_j(\boldsymbol{x}^G_m)\bar{H}_j(\boldsymbol{x}^G_m)n_y(\boldsymbol{x}^G_m) \\ 0 & N_j(\boldsymbol{x}^G_m)\bar{H}_j(\boldsymbol{x}^G_m)n_y(\boldsymbol{x}^G_m) & N_j(\boldsymbol{x}^G_m)\bar{H}_j(\boldsymbol{x}^G_m)n_x(\boldsymbol{x}^G_m) \end{pmatrix} l^C_m, \quad (7.40)$$

$$\tilde{\boldsymbol{B}}^C_{b_{tip}k}(\boldsymbol{x}_C)\,\big|_{l=1,2,3,4} = \frac{1}{A_C}\sum_{m=1}^{mt}\begin{pmatrix} 0 & N_k\left(\boldsymbol{x}^G_m\right)\bar{F}_{lk}\left(\boldsymbol{x}^G_m\right)n_x\left(\boldsymbol{x}^G_m\right) & 0 \\ 0 & 0 & N_k\left(\boldsymbol{x}^G_m\right)\bar{F}_{lk}\left(\boldsymbol{x}^G_m\right)n_y\left(\boldsymbol{x}^G_m\right) \\ 0 & N_k\left(\boldsymbol{x}^G_m\right)\bar{F}_{lk}\left(\boldsymbol{x}^G_m\right)n_y\left(\boldsymbol{x}^G_m\right) & N_k\left(\boldsymbol{x}^G_m\right)\bar{F}_{lk}\left(\boldsymbol{x}^G_m\right)n_x\left(\boldsymbol{x}^G_m\right) \end{pmatrix} l^C_m \quad (7.41)$$

and

$$\bar{\boldsymbol{B}}_{s_{crack}j} = \boldsymbol{J}^{-1}\begin{bmatrix} [N_j(\boldsymbol{x})\bar{H}_j(\boldsymbol{x})]_{,\xi} & b_j^{11}[N_j(\boldsymbol{x})\bar{H}_j(\boldsymbol{x})]_{,\xi} & b_j^{12}[N_j(\boldsymbol{x})\bar{H}_j(\boldsymbol{x})]_{,\xi} \\ [N_j(\boldsymbol{x})\bar{H}_j(\boldsymbol{x})]_{,\eta} & b_j^{21}[N_j(\boldsymbol{x})\bar{H}_j(\boldsymbol{x})]_{,\eta} & b_j^{22}[N_j(\boldsymbol{x})\bar{H}_j(\boldsymbol{x})]_{,\eta} \end{bmatrix}, \tag{7.42}$$

$$\bar{\boldsymbol{B}}_{s_{tip}k}\Big|_{l=1,2,3,4} = \boldsymbol{J}^{-1}\begin{bmatrix} [N_k(\boldsymbol{x})\bar{G}_{lk}(\boldsymbol{x})]_{,\xi} & b_i^{11}[N_k(\boldsymbol{x})\bar{F}_{lk}(\boldsymbol{x})]_{,\xi} & b_i^{12}[N_k(\boldsymbol{x})\bar{F}_{lk}(\boldsymbol{x})]_{,\xi} \\ [N_k(\boldsymbol{x})\bar{G}_{lk}(\boldsymbol{x})]_{,\eta} & b_i^{21}[N_k(\boldsymbol{x})\bar{F}_{lk}(\boldsymbol{x})]_{,\eta} & b_i^{22}[N_k(\boldsymbol{x})\bar{F}_{lk}(\boldsymbol{x})]_{,\eta} \end{bmatrix}. \tag{7.43}$$

Similarly, for the bending and shear part of the geometric stiffness, the enriched gradient operators can be expressed as

$$\mathbf{G}_{b_{crack}j} = \begin{bmatrix} [N_j(\boldsymbol{x})\bar{H}_j(\boldsymbol{x})]_{,x} & 0 & 0 \\ [N_j(\boldsymbol{x})\bar{H}_j(\boldsymbol{x})]_{,y} & 0 & 0 \end{bmatrix},$$

$$\mathbf{G}_{b_{tip}k}\Big|_{l=1,2,3,4} = \begin{bmatrix} [N_k(\boldsymbol{x})\bar{G}_{lk}(\boldsymbol{x})]_{,x} & 0 & 0 \\ [N_k(\boldsymbol{x})\bar{G}_{lk}(\boldsymbol{x})]_{,y} & 0 & 0 \end{bmatrix}, \tag{7.44}$$

$$\mathbf{G}_{s1_{crack}j} = \begin{bmatrix} 0 & [N_j(\boldsymbol{x})\bar{H}_j(\boldsymbol{x})]_{,x} & 0 \\ 0 & [N_j(\boldsymbol{x})\bar{H}_j(\boldsymbol{x})]_{,y} & 0 \end{bmatrix},$$

$$\mathbf{G}_{s1_{tip}k}\Big|_{l=1,2,3,4} = \begin{bmatrix} 0 & [N_k(\boldsymbol{x})\bar{F}_{lk}(\boldsymbol{x})]_{,x} & 0 \\ 0 & [N_k(\boldsymbol{x})\bar{F}_{lk}(\boldsymbol{x})]_{,y} & 0 \end{bmatrix}, \tag{7.45}$$

$$\mathbf{G}_{s2_{crack}j} = \begin{bmatrix} 0 & 0 & [N_j(\boldsymbol{x})\bar{H}_j(\boldsymbol{x})]_{,x} \\ 0 & 0 & [N_j(\boldsymbol{x})\bar{H}_j(\boldsymbol{x})]_{,y} \end{bmatrix},$$

$$\mathbf{G}_{s2_{tip}k}\Big|_{l=1,2,3,4} = \begin{bmatrix} 0 & 0 & [N_k(\boldsymbol{x})\bar{F}_{lk}(\boldsymbol{x})]_{,x} \\ 0 & 0 & [N_k(\boldsymbol{x})\bar{F}_{lk}(\boldsymbol{x})]_{,y} \end{bmatrix}. \tag{7.46}$$

7.1.7 Integration

The standard (nonenriched) domain integration uses 2×2 Gauss quadrature as it evaluates the bilinear shape functions sufficiently. The only exception to the standard (nonenriched) domain integration is with the shear part of the stiffness matrix, $\boldsymbol{K}_s$ in (7.14), which uses 1×1 Gauss quadrature. For standard surface integration, terms in (7.25) and (7.30), two subcells with one Gauss point per boundary are used; in this way computational efficiency and numerical stability are guaranteed.

For elements that are enriched the integration has to be adapted, so that the weak form on both sides of the crack contributes the correct enriched terms. The most common approach, which is also implemented here, is to subtriangulate the element (for example, Delaunay triangulation) in a way that the triangle edges conform with the discontinuity.

In the case of enriched elements using domain integration the following integration rules are used:

- Tip-blending elements: 16 Gauss points for the total element.
- Split-blending elements: 2 Gauss points for the total element.
- Split-tip-blending elements: 4 Gauss points for each triangular subelement.
- Split elements: 3 Gauss points for each triangular subelement.
- Tip elements: 13 Gauss points for each triangular subelement.

In the case of enriched elements using surface integration (smoothed enriched elements) the following integration rules are used:

- Tip-blending elements: 8 subcells with 1 Gauss points per edge.
- Split-blending elements: 2 subcells with 1 Gauss points per edge.
- Split-tip-blending elements: 4 subelements, each with 4 subcells with 1 Gauss points per edge.
- Split elements: 4 subelements, each with 2 subcells with 1 Gauss points per edge.
- Tip elements: 4 subelements, each with 8 subcells with 1 Gauss points per edge.

As in standard XFEM, there are 6 different types of elements:

- Tip elements are elements that contain the crack tip. All nodes belonging to a tip element are enriched with the near-tip fields ($\bar{G}_l(\boldsymbol{x})$ and $\bar{F}_l(\boldsymbol{x})$).
- Split elements are elements completely cut by the crack. Their nodes are enriched with the discontinuous function $\bar{H}(\boldsymbol{x})$.
- Tip-blending elements are elements neighboring tip elements. They are such that some of their nodes are enriched with the near-tip fields ($\bar{G}_l(\boldsymbol{x})$ and $\bar{F}_l(\boldsymbol{x})$) and others are not enriched at all.
- Split-blending elements are elements neighboring split elements. They are such that some of their nodes are enriched with the discontinuous function, $\bar{H}(\boldsymbol{x})$, and others are not enriched at all.
- Split-tip-blending elements are elements completely cut by the crack and neighboring tip elements. They are such that all of their nodes are enriched with the discontinuous function, $\bar{H}(\boldsymbol{x})$, and some of their nodes are enriched with the near-tip fields ($\bar{G}_l(\boldsymbol{x})$ and $\bar{F}_l(\boldsymbol{x})$).

- Standard elements are elements that are in neither of the above categories. None of their nodes are enriched.

7.2. Fractures in shell and plates using the phantom node method

7.2.1 Phantom node method for the Belytschko-Tsay shell element

Simulation of the fracture of shell structures is engendering considerable interest in the industrial and defense communities. Many components where fracture is of concern, such as windshields, ship hulls, fuel tanks and car bodies are not amenable to three-dimensional solid modeling, for the expense would be enormous. Furthermore, fracture is often an important criterion in determining their performance envelopes.

Here, we describe a finite element method based on the extended finite element method (XFEM) [11,31] for modeling shell structures in explicit finite element programs and illustrate their performance in nonlinear problems involving dynamic fracture. The methodology is based on the Hansbo and Hansbo [25] approach, which has previously been applied by Song et al. [46] and by Areias et al. [4,5]. The equivalence of the Hansbo and Hansbo [25] basis functions to XFEM [11,31] is shown in Areias and Belytschko [2]. The method employs an elementwise progression of the crack, i.e. the crack tip is always on an element edge. Réthoré et al. [45] have reported that this is usually adequate for dynamic crack propagation. We do not use any near-tip enrichment, although Elguedj et al. [23] have achieved good success with near-tip enrichments for static problems.

The literature on dynamic crack propagation in shells is quite limited. Cirak et al. [18] have developed on interelement crack method, where the crack is limited to propagation along the element edges. The method is based on the Kirchhoff shell theory. Penalty functions were used to enforce continuity on all interelement edges. Areias and Belytschko [3] and Areias et al. [4,5] have developed a method for shell fracture based on the extended finite element method for static and implicit time integration.

7.2.1.1 Shell formulation with fracture

The discontinuous shell formulation is based on the degenerated shell concept (Ahmad et al. [1], and Hughes and Liu [26]), which is almost equivalent to the Mindlin-Reissner formulation when the edges connect-

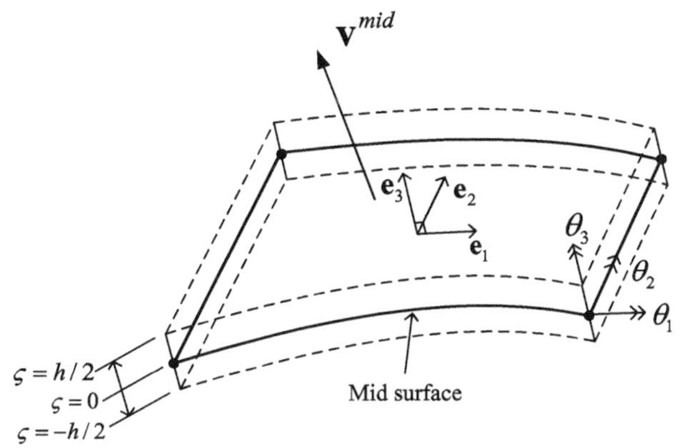

Figure 7.3 The nomenclature for a continuum shell.

ing the top and bottom surfaces are normal to the midsurface. We will use a kinematic theory based on the corotational rate-of-deformation and the corotational Cauchy stress rate. These features are briefly summarized in Section 7.2.1.2, but are well-known, so we will focus on the modifications needed for the XFEM treatment of fracture.

The velocity field is given by

$$\mathbf{v}(\boldsymbol{\xi}, t) = \mathbf{v}^{mid}(\boldsymbol{\xi}, t) - \zeta \boldsymbol{e}_3 \times \boldsymbol{\theta}^{mid}(\boldsymbol{\xi}, t) \tag{7.47}$$

where $\boldsymbol{v}^{mid} \in \mathcal{R}^3$ are the velocities of the shell midsurface, $\boldsymbol{\theta}^{mid} \in \mathcal{R}^3$ are angular velocities of the normals to the midsurface, ζ varies linearly from $-h/2$ to $h/2$ along the thickness, and $\boldsymbol{\xi} = (\xi_1, \xi_2)$ are material coordinates of the manifold that describes the midsurface of the shell; at any point of the shell, we construct tangent unit vectors $\boldsymbol{e}_1$ and $\boldsymbol{e}_2$ so that

$$\boldsymbol{e}_3 = \boldsymbol{e}_1 \times \boldsymbol{e}_2 \tag{7.48}$$

The nomenclature is illustrated in Fig. 7.3.

For the further development of the discontinuous shell formulation, we will limit ourselves to cracks with surfaces normal to the shell midsurfaces as shown in Fig. 7.4. Although this is not an intrinsic limitation of the method, it simplifies several aspects of the formulation.

The discontinuous velocity fields due to a crack in any Mindlin-Reissner theory can be described by

$$\boldsymbol{v}^{mid}(\boldsymbol{\xi}, t) = \boldsymbol{v}^{cont}(\boldsymbol{\xi}, t) + H(f(\boldsymbol{\xi}))\boldsymbol{v}^{disc}(\boldsymbol{\xi}, t) \tag{7.49}$$

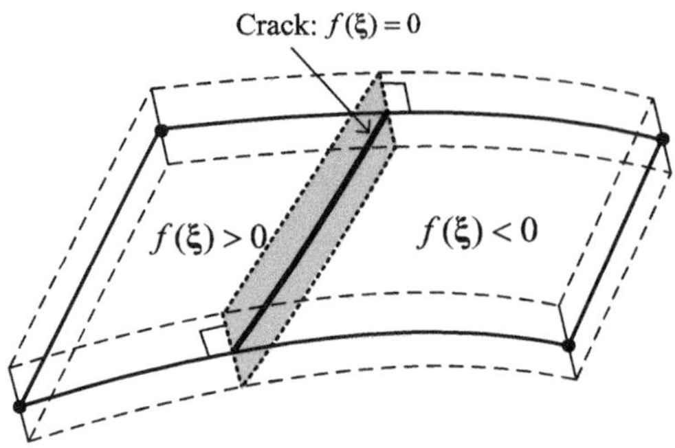

Figure 7.4 Representation of a discontinuity in the reference configuration by a level set function $f(\boldsymbol{\xi})$ in the shell midsurface.

$$\boldsymbol{\theta}^{mid}(\boldsymbol{\xi}, t) = \boldsymbol{\theta}^{cont}(\boldsymbol{\xi}, t) + H(f(\boldsymbol{\xi}))\boldsymbol{\theta}^{disc}(\boldsymbol{\xi}, t) \tag{7.50}$$

where $f(\boldsymbol{\xi}) = 0$ gives the intersection of the crack surface with the midsurface of the shell and $H(\cdot)$ is the Heaviside function given by

$$H(x) = \begin{cases} 1 & x > 0 \\ 0 & x \leq 0 \end{cases} \tag{7.51}$$

In the above, $\boldsymbol{v}^{cont}$ and $\boldsymbol{v}^{disc}$ are continuous functions that are used to model the continuous and the discontinuous parts of the velocity fields, respectively. Similarly, $\boldsymbol{\theta}^{cont}$ and $\boldsymbol{\theta}^{disc}$ are the continuous functions that are used to model continuous and discontinuous parts of the angular velocity fields, respectively. The discontinuities that model the cracks arise from the step function that precedes $\boldsymbol{v}^{disc}$ and $\boldsymbol{\theta}^{disc}$. It can be seen from Eqs. (7.47) and (7.49)-(7.50) that these velocity fields can result in a loss of compatibility and in particular material overlap, as indicated in Fig. 7.5, when there is a significant discontinuity in the angular motion but the crack opening is small. We will deal with this incompatibility by introducing a penalty in the cohesive law.

7.2.1.2 Element formulation

The shell element used here is a 4-node shell originally described in [14] with improvements in [15,46]. The shell element employs an one-point quadrature rule with stabilization [10,12] for computational efficiency.

When the velocity fields given in Eqs. (7.49)-(7.50) are specialized to shell finite elements, the continuous part of the corotational velocity com-

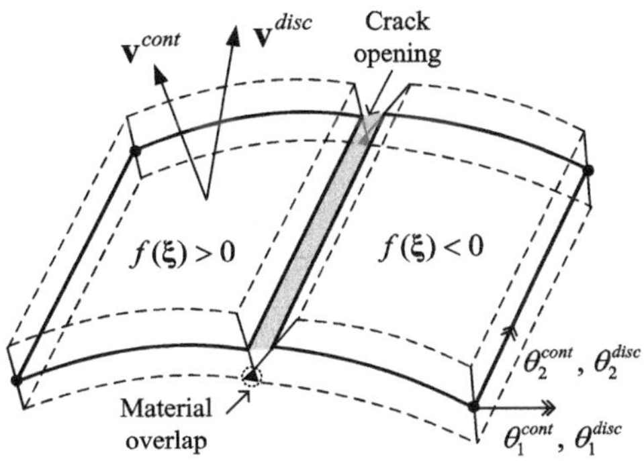

Figure 7.5 Nomenclature of a fractured shell: incompatible material overlapping may occur at the bottom surface due to the opening of the crack.

ponents are given by

$$\hat{v}_x(\boldsymbol{\xi}, t) = N_I(\boldsymbol{\xi})\hat{\boldsymbol{v}}_{xI}(t) + \zeta N_I(\boldsymbol{\xi})\hat{\boldsymbol{\theta}}_{yI}(t) \tag{7.52}$$

$$\hat{v}_y(\boldsymbol{\xi}, t) = N_I(\boldsymbol{\xi})\hat{\boldsymbol{v}}_{xI}(t) - \zeta N_I(\boldsymbol{\xi})\hat{\boldsymbol{\theta}}_{xI}(t) \tag{7.53}$$

where N_I are the conventional 4-node finite element bilinear shape functions and the repeated subscripts I denote summation over all nodes. The corotational components of the rate-of-deformation tensor are given by

$$\hat{D}_{ij} = \frac{1}{2}\left(\frac{\partial \hat{v}_i}{\partial \hat{x}_j} + \frac{\partial \hat{v}_j}{\partial \hat{x}_i}\right) \tag{7.54}$$

Substituting Eqs. (7.52)-(7.53) into (7.54) yields an expression for the rate-of-deformation components

$$\hat{D}_x = \boldsymbol{b}_{\hat{x}I}\hat{\boldsymbol{v}}_{\hat{x}I} + \zeta(\boldsymbol{b}^c_{\hat{x}I}\boldsymbol{v}_{\hat{x}I} + \boldsymbol{b}_{\hat{x}I}\boldsymbol{\theta}_{yI}) \tag{7.55}$$

$$\hat{D}_y = \boldsymbol{b}_{\hat{y}I}\hat{\boldsymbol{v}}_{\hat{y}I} + \zeta(\boldsymbol{b}^c_{\hat{y}I}\boldsymbol{v}_{\hat{y}I} - \boldsymbol{b}_{\hat{y}I}\boldsymbol{\theta}_{xI}) \tag{7.56}$$

$$2\hat{D}_{xy} = \boldsymbol{b}_{\hat{x}I}\hat{\boldsymbol{v}}_{\hat{x}I} + \boldsymbol{b}_{\hat{y}I}\hat{\boldsymbol{v}}_{\hat{y}I} + \zeta(\boldsymbol{b}^c_{\hat{x}I}\boldsymbol{v}_{\hat{x}I} + \boldsymbol{b}^c_{\hat{y}I}\boldsymbol{v}_{\hat{y}I} + \boldsymbol{b}_{\hat{x}I}\boldsymbol{\theta}_{yI} - \boldsymbol{b}_{\hat{y}I}\boldsymbol{\theta}_{xI}) \tag{7.57}$$

where

$$\begin{Bmatrix} \boldsymbol{b}_{\hat{x}I} \\ \boldsymbol{b}_{\hat{y}I} \end{Bmatrix} = \frac{1}{2A}\begin{bmatrix} \hat{y}_{24} & \hat{y}_{31} & \hat{y}_{42} & \hat{y}_{13} \\ \hat{x}_{42} & \hat{x}_{13} & \hat{x}_{24} & \hat{x}_{31} \end{bmatrix} \tag{7.58}$$

$$\begin{Bmatrix} \boldsymbol{b}^c_{\hat{x}I} \\ \boldsymbol{b}^c_{\hat{y}I} \end{Bmatrix} = \frac{2\hat{\gamma}_K\hat{z}_K}{A^2}\begin{bmatrix} \hat{x}_{13} & \hat{x}_{42} & \hat{x}_{31} & \hat{x}_{24} \\ \hat{y}_{13} & \hat{y}_{42} & \hat{y}_{31} & \hat{y}_{24} \end{bmatrix} \tag{7.59}$$

and where $\hat{x}_{IJ} = \hat{x}_I - \hat{x}_J$, A is the area of the element and $\hat{\gamma}_K$ is a projection operator, see Belytschko and Bachrach [10]. A state of plane stress is assumed. In Belytschko et al. [15], two methods are proposed for the evaluation of $\boldsymbol{b}^c$. Here in Eq. (7.59), we adopted the $\hat{z}$ method. In this case, curvature is only coupled with the translations for *a warped element.*

We also have used the shear projection scheme introduced in Belytschko et al. [15]. This shear projection scheme gives the components of the transverse shear strain as

$$\hat{D}_{xz} = \boldsymbol{b}^s_{x1I}\hat{\boldsymbol{v}}_{zI} + \boldsymbol{b}^s_{x2I}\hat{\boldsymbol{\theta}}_{xI} + \boldsymbol{b}^s_{x3I}\hat{\boldsymbol{\theta}}_{yI} \tag{7.60}$$

$$\hat{D}_{yz} = \boldsymbol{b}^s_{y1I}\hat{\boldsymbol{v}}_{zI} + \boldsymbol{b}^s_{y2I}\hat{\boldsymbol{\theta}}_{xI} + \boldsymbol{b}^s_{y3I}\hat{\boldsymbol{\theta}}_{yI} \tag{7.61}$$

where

$$\begin{Bmatrix} \boldsymbol{b}^s_{x1I} & \boldsymbol{b}^s_{x2I} & \boldsymbol{b}^s_{x3I} \\ \boldsymbol{b}^s_{y1I} & \boldsymbol{b}^s_{y2I} & \boldsymbol{b}^s_{y3I} \end{Bmatrix} = \tag{7.62}$$

$$\frac{1}{4}\begin{bmatrix} 2(\bar{x}_{JI} - \bar{x}_{IK}) & (\hat{x}_{JI}\bar{y}_{JI} + \hat{x}_{IK}\bar{y}_{IK}) & -(\hat{x}_{JI}\bar{y}_{JI} + \hat{x}_{IK}\bar{y}_{IK}) \\ 2(\bar{y}_{JI} - \bar{y}_{IK}) & (\hat{y}_{JI}\bar{y}_{JI} + \hat{y}_{IK}\bar{y}_{IK}) & -(\hat{y}_{JI}\bar{y}_{JI} + \hat{y}_{IK}\bar{y}_{IK}) \end{bmatrix}$$

and where $(I, J, K) = \{(I, J, K) \mid (1,2,4),\ (2,3,1),\ (3,4,1),\ (4,1,3)\}$ and $\bar{x}_{IJ} = \hat{x}_{IJ}/||\hat{x}_{IJ}||$.

7.2.1.3 Representation of the discontinuity

The velocity field of a fractured shell element, which is given by Eqs. (7.49)-(7.50), can be approximated in the XFEM by

$$\hat{\boldsymbol{v}}^{mid}(\boldsymbol{\xi}, t) = N_I(\boldsymbol{\xi})\hat{\boldsymbol{v}}_I^{cont}(t) + H(f(\boldsymbol{\xi}))N_I(\boldsymbol{\xi})\hat{\boldsymbol{v}}_I^{disc}(t) \tag{7.63}$$

$$\hat{\boldsymbol{\theta}}^{mid}(\boldsymbol{\xi}, t) = N_I(\boldsymbol{\xi})\hat{\boldsymbol{\theta}}^{cont}(t) + H(f(\boldsymbol{\xi}))N_I(\boldsymbol{\xi})\hat{\boldsymbol{\theta}}_I^{disc}(t) \tag{7.64}$$

However, when element-wise crack propagation is employed, we have found that it is simpler to program the implementation in the Hansbo and Hansbo [25] form, as developed by Song et al. [46]. An element completely cut by a crack is represented by a set of overlapping elements with added phantom nodes as shown in Fig. 7.6, see also Fig. 7.7. The discontinuous velocity field is then constructed by two superimposed velocity fields:

$$\begin{aligned} \hat{\boldsymbol{v}}(\boldsymbol{\xi}, t) &= \hat{\boldsymbol{v}}^{e1}(\boldsymbol{\xi}, t) + \hat{\boldsymbol{v}}^{e2}(\boldsymbol{\xi}, t) \\ &= \sum_{I \in S_1} N_I(\boldsymbol{\xi})H(-f(\boldsymbol{\xi}))\hat{\boldsymbol{v}}_I^{e1}(t) + \sum_{I \in S_2} N_I(\boldsymbol{\xi})H(f(\boldsymbol{\xi}))\hat{\boldsymbol{v}}_I^{e2}(t) \end{aligned} \tag{7.65}$$

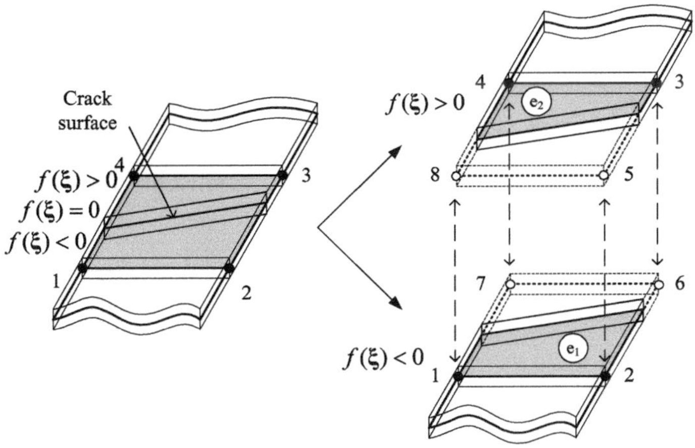

Figure 7.6 The decomposition of a cracked shell element with generic nodes 1–4 into two elements e_1 and e_2; solid and hollow circles denote the original nodes and the added phantom nodes, respectively.

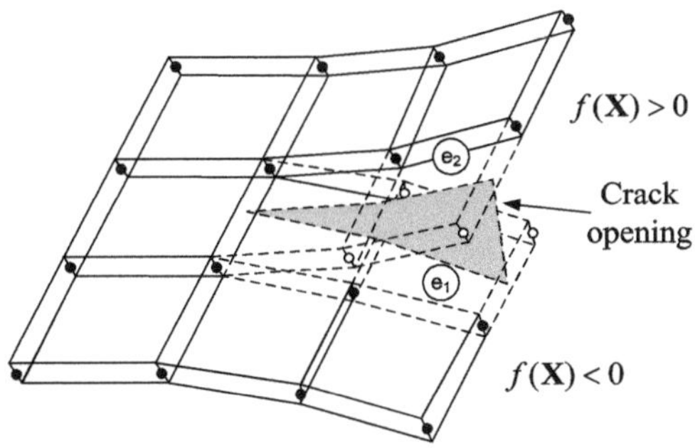

Figure 7.7 Schematic of a crack opening in shell elements with the phantom node method; solid and hollow circles denote the original nodes and the added phantom nodes, respectively.

$$
\begin{aligned}
\hat{\boldsymbol{\theta}}(\boldsymbol{\xi},t) &= \hat{\boldsymbol{\theta}}^{e1}(\boldsymbol{\xi},t)+\hat{\boldsymbol{\theta}}^{e2}(\boldsymbol{\xi},t) \\
&= \sum_{I\in S_1} N_I(\boldsymbol{\xi})H(-f(\boldsymbol{\xi}))\hat{\theta}_I^{e1}(t)+\sum_{I\in S_2} N_I(\boldsymbol{\xi})H(f(\boldsymbol{\xi}))\hat{\theta}_I^{e2}(t)
\end{aligned} \tag{7.66}
$$

where S_1 and S_2 are the sets of the nodes of the overlapping element e_1 and e_2, respectively. Note that velocity fields $\hat{\boldsymbol{v}}^{e1}(\boldsymbol{\xi},t)$ and $\hat{\boldsymbol{v}}^{e2}(\boldsymbol{\xi},t)$ (or $\hat{\boldsymbol{\theta}}^{e1}(\boldsymbol{\xi},t)$ and $\hat{\boldsymbol{\theta}}^{e2}(\boldsymbol{\xi},t)$) are non-zero only for $f(\boldsymbol{\xi})<0$ and $f(\boldsymbol{\xi})>0$, respectively, due to the Heaviside step function $H(x)$ that appears in the above equations.

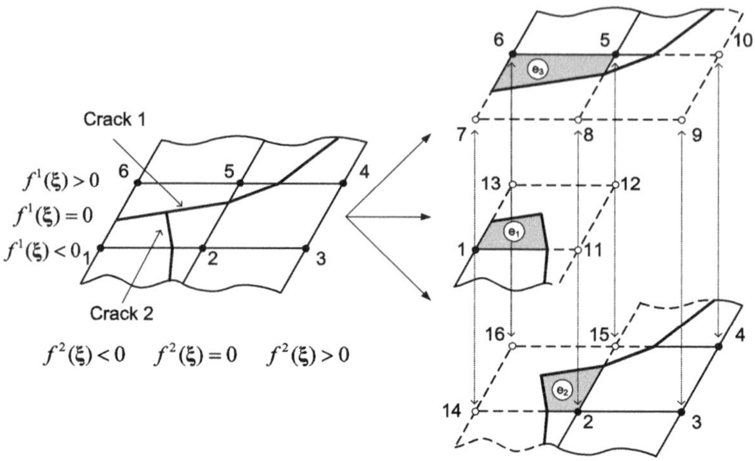

Figure 7.8 The decomposition of an element into three elements e_1, e_2 and e_3 to model crack branching; solid and hollow circles denote the original nodes and the added phantom nodes, respectively.

The phantom nodes are integrated in time by the same central difference explicit method as those of the other nodes.

7.2.1.4 Representation of multiple discontinuities: crack branching

The concept of the overlapping element method can be easily extended to crack branch modeling. When the original crack, *crack 1*, branches into *crack 1* and *crack 2*, as shown in Fig. 7.8, the element in which the crack branches is replaced by three overlapping elements.

Let $f^1(\boldsymbol{\xi}) = 0$ describe the original crack and one branch, and let $f^2(\boldsymbol{\xi}) = 0$ describe the second branch. The discontinuous velocity field is then given by

$$
\begin{aligned}
\hat{\boldsymbol{v}}(\boldsymbol{\xi},t) &= \hat{\boldsymbol{v}}^{e1}(\boldsymbol{\xi},t)+\hat{\boldsymbol{v}}^{e2}(\boldsymbol{\xi},t)+\hat{\boldsymbol{v}}^{e3}(\boldsymbol{\xi},t) \\
&= \sum_{I\in S_1} N_I(\boldsymbol{\xi})H(-f^1(\boldsymbol{\xi}))H(-f^2(\boldsymbol{\xi}))\hat{\boldsymbol{v}}_I^{e1}(t) \\
&+ \sum_{I\in S_2} N_I(\boldsymbol{\xi})H(-f^1(\boldsymbol{\xi}))H(f^2(\boldsymbol{\xi}))\hat{\boldsymbol{v}}_I^{e2}(t) \\
&+ \sum_{I\in S_3} N_I(\boldsymbol{\xi})H(f^1(\boldsymbol{\xi}))H(-f^2(\boldsymbol{\xi}))\hat{\boldsymbol{v}}_I^{e3}(t)
\end{aligned}
\tag{7.67}
$$

7.2.1.5 Discretization

The element nodal forces are developed as in Belytschko et al. [14]. In addition, curvature-translation coupling terms are added and a shear projection operator replaces the previous transverse shear terms. The principle of virtual power is used to derive the relationship for the internal nodal forces. The principle states that:

$$\delta \mathcal{P}^{\text{int}} = \underbrace{A(\boldsymbol{B}_{\text{m}}\delta \boldsymbol{v})^T \begin{Bmatrix} \hat{f}_x^r \\ \hat{f}_y^r \\ \hat{f}_{xy}^r \end{Bmatrix}}_{\text{virtual membrane power}} + \underbrace{A(\boldsymbol{B}_{\text{b}}\delta \boldsymbol{v})^T \begin{Bmatrix} \hat{m}_x^r \\ \hat{m}_y^r \\ \hat{m}_{xy}^r \end{Bmatrix}}_{\text{virtual bending power}} + \underbrace{\bar{\kappa} A(\boldsymbol{B}_{\text{s}}\delta \boldsymbol{v})^T \begin{Bmatrix} \hat{f}_{xz}^r \\ \hat{f}_{yz}^r \end{Bmatrix}}_{\text{virtual transverse shear power}} \tag{7.68}$$

where $\bar{\kappa}$ is the shear reduction factor from the Mindlin shell theory, and $\hat{f}_{ij}^r$ and $\hat{m}_{ij}^r$ are the resultant forces and moments which are integrated through the element thickness.

$$\hat{f}_{ij}^r = \int \hat{\sigma}_{ij} \, d\hat{z} \tag{7.69}$$

$$\hat{m}_{ij}^r = \int \hat{z}\hat{\sigma}_{ij} \, d\hat{z} \tag{7.70}$$

where $\hat{z} = \zeta \frac{h}{2}$.

We substitute Eqs. (7.55)-(7.62) into (7.68) and invoking the arbitrariness of $\delta \boldsymbol{v}$ yields the discretized element nodal forces:

$$\hat{f}_{xI}^{int} = A_e \, (b_{xI}\hat{f}_x^r + b_{yI}\hat{f}_{xy}^r + b_{xI}^c \hat{m}_x^r + b_{yI}^c \hat{m}_{xy}^r \tag{7.71}$$

$$\hat{f}_{yI}^{int} = A_e \, (b_{yI}\hat{f}_y^r + b_{xI}\hat{f}_{xy}^r + b_{yI}^c \hat{m}_y^r + b_{xI}^c \hat{m}_{xy}^r) \tag{7.72}$$

$$\hat{f}_{zI}^{int} = A_e \bar{\kappa} \, (b_{x1I}^s \hat{f}_{xz}^r + b_{y1I}^s \hat{f}_{yz}^r) \tag{7.73}$$

$$\hat{m}_{xI}^{int} = A_e \, [\bar{\kappa}(b_{x2I}^s \hat{f}_{xz}^r + b_{y2I}^s \hat{f}_{yz}^r) - (b_{yI}\hat{m}_y^r + b_{xI}\hat{m}_{xy}^r)] \tag{7.74}$$

$$\hat{m}_{yI}^{int} = A_e \, [\bar{\kappa}(b_{x3I}^s \hat{f}_{xz}^r + b_{y3I}^s \hat{f}_{yz}^r) + (b_{xI}\hat{m}_x^r + b_{yI}\hat{m}_{xy}^r)] \tag{7.75}$$

$$\hat{m}_{zI}^{int} = 0 \tag{7.76}$$

The final form of the element internal forces in the global coordinates can be determined by performing the transformation between the corotational and global coordinates as below:

$$\boldsymbol{f}_e^{int} = \boldsymbol{T}_e^T (\hat{\boldsymbol{f}}_e^{int} + \hat{\boldsymbol{f}}_e^{stab}) \tag{7.77}$$

where $\boldsymbol{T}$ is the transformation matrix between global and corotational components and $\hat{\boldsymbol{f}}_e^{int}$ is the nodal internal force vector in the corotational coordinate systems. In Eq. (7.77), to circumvent the rank deficiency due to one point integration, an hourglass control force, $\hat{\boldsymbol{f}}_e^{stab}$, is added to the internal force vector. For a description of the hourglass control scheme, see [10,15].

For each of the overlapped elements modeling a crack, the nodal forces are given by

$$\boldsymbol{f}_e^{int} = \left(\sum_{k=1}^{N_{ele}^{ovr}} \frac{A_{e_k}}{A_e} \boldsymbol{T}_{e_k}^T \hat{\boldsymbol{f}}_{e_k}^{int}\right) + \boldsymbol{T}_e^T \hat{\boldsymbol{f}}_e^{stab} \tag{7.78}$$

where N_{ele}^{ovr} is the total number of overlapped elements, A_{e_k} is the activated area of the corresponding overlapping elements in the corotational coordinates, $\boldsymbol{f}_e$ is the nodal force vector of a cracked element and $\hat{\boldsymbol{f}}_{e_k}$ is the corotational nodal force vector of the overlapped element e_k. Note that the internal nodal forces of elements e_k can be calculated by multiplying Eqs. (7.71)-(7.76) by the area fraction, A_{e_k}/A_e. A more detailed discussion of the concept of the modification of cracked element nodal forces by area fractions can be found in Song et al. [46].

7.2.2 Phantom node method based on the three-node isotropic triangular MITC shell element

Let us consider a shell model with an arbitrary through-the-thickness crack. Due to the crack, three types of elements need to be distinguished, namely uncracked, cracked, and tip elements; note that the tip elements are partially cut by the crack and we use the tip element technique presented in Section 4.2 in the context of shell analysis.

7.2.2.1 The uncracked element

An uncracked element, on which the displacement field is continuous, can be treated as the standard finite element method. In computational analysis of shells, approaches based on continuum mechanics based shell elements originated from the work of Ahmad et al. [1] are popular due to their simplicity of formulation and implementation in the finite element procedures [50]. Unfortunately, the displacement based elements are too stiff in bending-dominated shell problems when the thickness is small, leading to locking phenomena [29]. To alleviate the locking, there are many methods such as uniform and selected reduced integration, assumed natural strain

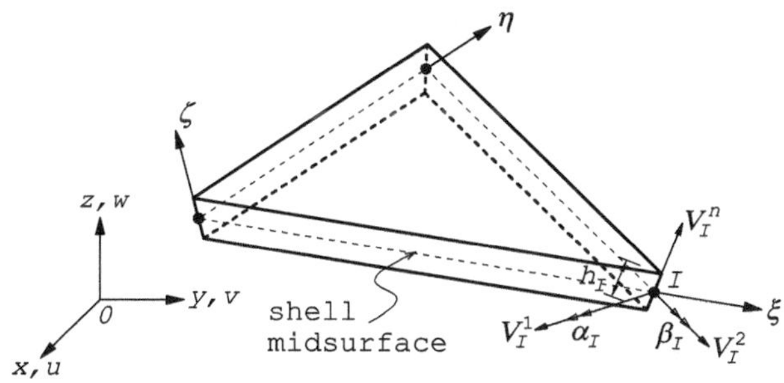

Figure 7.9 Three-node shell finite element [17].

(ANS), enhanced assumed strain (EAS) or so-called mixed interpolation of tensorial components (MITC) [8,22], to name a few.

Here, the uncracked part is discretized by the linear MITC3 element of Lee and Bathe [29]. Particularly, since the three-node element has flat geometry, it only suffers from shear locking. To ameliorate the shear locking and meet the requirement of spatial isotropy, the MITC technique here was designed so that the transverse shear strain variations corresponding to the three edge directions of the element are identical.

In the continuum mechanics based shell elements proposed by [1], the displacement approximation can be obtained by considering geometric approximation mapping of any point $\boldsymbol{x}$ of the shell in the global Cartesian coordinate system (x, y, z) into the natural coordinates (ξ, η, ζ) as defined in

$$\boldsymbol{x}(\xi, \eta, \zeta) = \sum_{I=1}^{3} N_I(\xi, \eta)\boldsymbol{x}_I + \frac{\zeta}{2}\sum_{I=1}^{3} h_I N_I(\xi, \eta)\boldsymbol{V}_I^n \tag{7.79}$$

where $N_I(\xi, \eta)$ is the standard C^0 shape function corresponding to the surface $\zeta = \boldsymbol{constant}$; $\boldsymbol{x}_I$ is the nodal coordinates in the global Cartesian system; and h_I and $\boldsymbol{V}_I^n$ denote the shell thickness and the director vector, respectively. The subscript I indicates the values at node I.

The displacement approximation $\boldsymbol{u}$ of the element is then given by

$$\boldsymbol{u}(\xi, \eta, \zeta) = \sum_{I=1}^{3} N_I(\xi, \eta)\boldsymbol{u}_I + \frac{\zeta}{2}\sum_{I=1}^{3} h_I N_I(\xi, \eta)(-\boldsymbol{V}_I^2\alpha_I + \boldsymbol{V}_I^1\beta_I) \tag{7.80}$$

wherein, $\boldsymbol{V}_I^1$ and $\boldsymbol{V}_I^2$ are unit vectors orthogonal to $\boldsymbol{V}_I^n$ and each other to create a nodal coordinate system at node I, see Fig. 7.9; $\boldsymbol{u}_I = \{u_I, v_I, w_I\}^T$ is the translational displacements in the global Cartesian coordinate system,

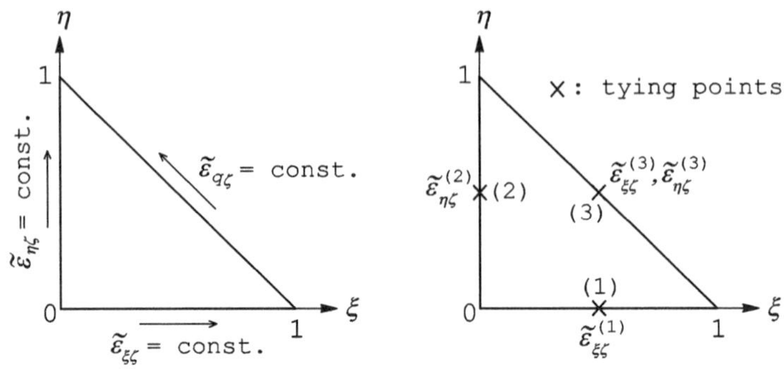

Figure 7.10 Constant transverse shear strain along edges and positions of tying points.

and (α_I, β_I) are the rotational displacements of the director vector $\boldsymbol{V}_I^n$ about $\boldsymbol{V}_I^1$ and $\boldsymbol{V}_I^2$, respectively, in the nodal coordinate system.

The purely displacement-based three-node shell elements suffer from a severe "shear locking" phenomenon as the shell thickness decreases. To circumvent the shear locking, the covariant transverse shear strains in the MITC3 element are separately interpolated from values of the covariant transverse shear strains evaluated at "tying points" which are the center of the isotropic element edges as shown in Fig. 7.10. To satisfy the isotropic property of the transverse shear strain fields, their interpolations were constructed by [29] as follows

$$\tilde{\varepsilon}_{\xi\zeta}^{MITC3} = \tilde{\varepsilon}_{\xi\zeta}^{(1)} + c\eta \tag{7.81a}$$

$$\tilde{\varepsilon}_{\eta\zeta}^{MITC3} = \tilde{\varepsilon}_{\eta\zeta}^{(2)} - c\xi \tag{7.81b}$$

in which,

$$c = \tilde{\varepsilon}_{\eta\zeta}^{(2)} - \tilde{\varepsilon}_{\xi\zeta}^{(1)} - \tilde{\varepsilon}_{\eta\zeta}^{(3)} + \tilde{\varepsilon}_{\xi\zeta}^{(3)} \tag{7.82}$$

and the covariant strain components are derived as

$$\tilde{\varepsilon}_{ij} = \frac{1}{2}\left(\frac{\partial \boldsymbol{x}}{\partial \xi_i}\cdot\frac{\partial \boldsymbol{u}}{\partial \xi_j} + \frac{\partial \boldsymbol{x}}{\partial \xi_j}\cdot\frac{\partial \boldsymbol{u}}{\partial \xi_i}\right) \tag{7.83}$$

with $\xi_1 = \xi$, $\xi_2 = \eta$, $\xi_3 = \zeta$.

Next, we use the relationships between the components of the covariant strain tensor $\tilde{\varepsilon}_{ij}$ and those of the global Cartesian strain tensor ε_{mn}

$$\tilde{\varepsilon}_{ij}\boldsymbol{g}^i \otimes \boldsymbol{g}^j = \varepsilon_{mn}\boldsymbol{e}_m \otimes \boldsymbol{e}_n \tag{7.84}$$

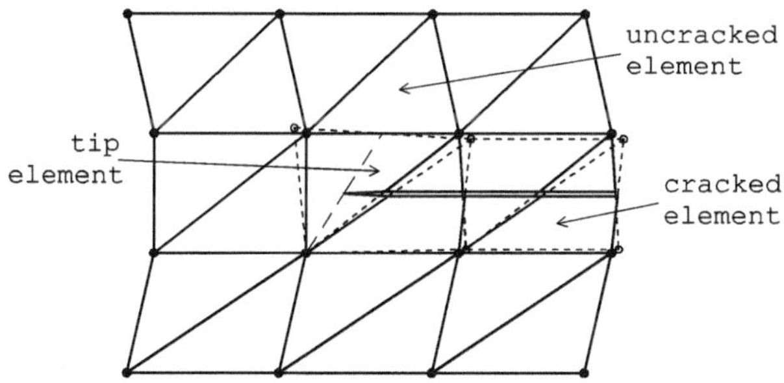

Figure 7.11 Displacement jump described by the phantom-node method. Solid circles are physical nodes; empty circles are phantom nodes.

in which $\boldsymbol{g}^i$ are the contravariant base vectors and satisfy $\boldsymbol{g}_i \cdot \boldsymbol{g}^j = \delta_i^j$, the (mixed) Kronecker delta, and $\boldsymbol{e}_m$ are the unit base vectors defined in the global Cartesian coordinate system, to transform the in-plane and transverse strain components respectively given in Eqs. (7.83) and (7.81) into those measured in the global Cartesian coordinate system. The strain-displacement matrix $\boldsymbol{B}$ can then be formulated.

7.2.2.2 Overlapping paired elements for cracked elements

In a cracked element, which is completely cut by a crack, the displacement field is discontinuous across the crack but independently continuous on each part of the element. Hence, the displacement field can be superimposed by two separate displacement fields, each of which is continuous on its own part of the element as illustrated in Fig. 7.11. This technique is given in the paper of Hansbo and Hansbo [25]. While it has been proved to be equivalent to the XFEM [2], this approximation of the displacement field for the cracked element exhibits more advantages since there are no discontinuous enrichments required. The gradient of the displacement field is continuous on each part of the cracked element. The shell formulations can then be simply implemented on each part of the cracked element as a pair of overlapping standard MITC3 elements.

Let a cracked element Ω_e be divided into two complementary parts, Ω_e^+ and Ω_e^-, by a crack. The displacement field in the cracked element can be described as

$$\boldsymbol{u}_{cr} = \begin{cases} \boldsymbol{u}^+ & \text{in } \Omega_e^+ \\ \boldsymbol{u}^- & \text{in } \Omega_e^- \end{cases} \tag{7.85}$$

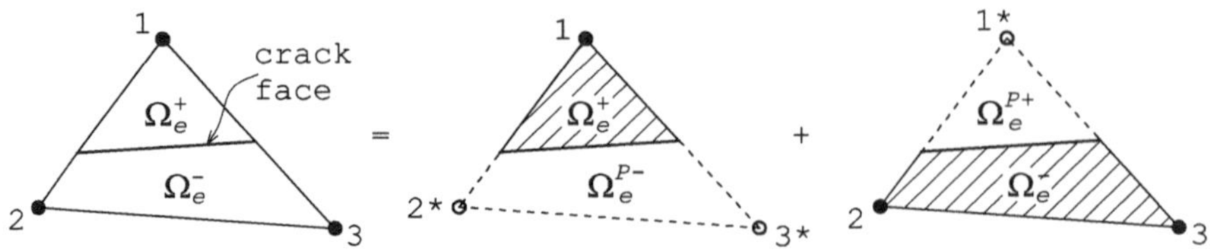

Figure 7.12 Additional phantom nodes and phantom domains for cracked elements.

To use the standard approximation of the displacement field on each part of the cracked element, the real parts Ω_e^+ and Ω_e^- are extended to their opposite sides as Ω_e^{P-} and Ω_e^{P+}, respectively, by additionally introducing the local duplication of homologous nodes called phantom-nodes, as shown in Fig. 7.12. These nodes have the same director vectors as the real nodes. As a result, the continuous displacement field on each part of the cracked element can be approximated similarly to that of the continuum mechanics based shell element

$$\boldsymbol{u}_{cr}(\xi,\eta,\zeta) = \begin{cases} \sum_{K=1}^{3} N_K(\xi,\eta)\boldsymbol{u}_K \\ \quad + \frac{\zeta}{2}\sum_{K=1}^{3} h_K N_K(\xi,\eta)(-\boldsymbol{V}_K^2\alpha_K + \boldsymbol{V}_K^1\beta_K) & \textit{in } \Omega_e^+ \\ \sum_{L=1}^{3} N_L(\xi,\eta)\boldsymbol{u}_L \\ \quad + \frac{\zeta}{2}\sum_{L=1}^{3} h_L N_L(\xi,\eta)(-\boldsymbol{V}_L^2\alpha_L + \boldsymbol{V}_L^1\beta_L) & \textit{in } \Omega_e^- \end{cases} \tag{7.86}$$

where $K \in \{1, 2^*, 3^*\}$ and $L \in \{1^*, 2, 3\}$ are nodes belonging to $\Omega_e^+ \cup \Omega_e^{P-}$ and $\Omega_e^- \cup \Omega_e^{P+}$, respectively.

The MITC technique, given in Eqs. (7.81) and (7.82), is now applied straightforwardly to the continuous displacement fields in Eq. (7.86) of the overlapping paired elements of which the domains are $(\Omega_e^+ \cup \Omega_e^{P-})$ and $(\Omega_e^- \cup \Omega_e^{P+})$. Similarly to the formulations for the MITC3 element in Section 7.2.2, the corresponding strain-displacement matrices can be obtained for the displacement fields in Eq. (7.86) realized only on the real parts, i.e. Ω_e^+ and Ω_e^-.

7.2.2.3 Constrained overlapping paired elements for tip elements

The displacement field in tip elements jumps across the partially intersected crack does not jump at the tip as shown in Fig. 7.13. To fulfill the required displacement field, the classical XFEM uses branch enrichments which are basic functions of the asymptotic displacement fields near the crack tip [31].

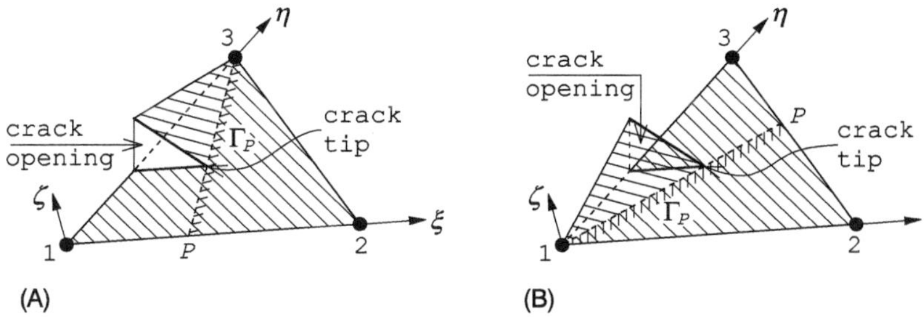

Figure 7.13 Two possibilities of discontinuous displacement field (hatched area) with crack opening in a tip element.

However, the analytical solution is known only in a few cases and therefore the branch enrichments, note needed in the phantom node method, add complexity but do not necessarily improve the results.

In another approach, Zi and Belytschko [52] managed to satisfy the condition by using the step enrichment only and a different shape function for the enriched field of the displacement approximation, which results in the crack-opening displacement vanishing at edge Γ_P, see Fig. 7.13. By omitting the branch enrichments, numerical computation is less costly since it avoids the problems related to blending and integrating the singularity and the non-polynomial terms. This approach was extended to derive a simple tip element for the phantom-node method used in two-dimensional problems [43].

The three-node triangular shell element is an extension of the work in [43] to shell elements. Assume that a partial crack cuts edge $\overline{13}$ (in natural coordinates) of the tip element. There are two possibilities of the discontinuous displacement fields in the tip element as illustrated in Fig. 7.13. Consider the displacement field required in Fig. 7.13A. To provide a set of full interpolation bases for the displacement field of the small triangle, phantom nodes 1^* and 2^* are added at the position of nodes 1 and 2, respectively, to create a completely new three-node triangular shell element. The tip element is now decomposed into two elements $\Delta 123$ and $\Delta 1^*2^*3$, which share node 3 (see Fig. 7.14A). As a result, for each overlapping paired element the formulations of the MITC3 element can be implemented straightforwardly similarly to that for cracked elements.

Additionally, since the crack-opening displacement at the tip vanishes, the displacement approximations of the overlapping paired elements must be identical along the Γ_P edge, as demonstrated in Fig. 7.14. To satisfy this kinematical condition, the nodal displacements of the overlapping paired

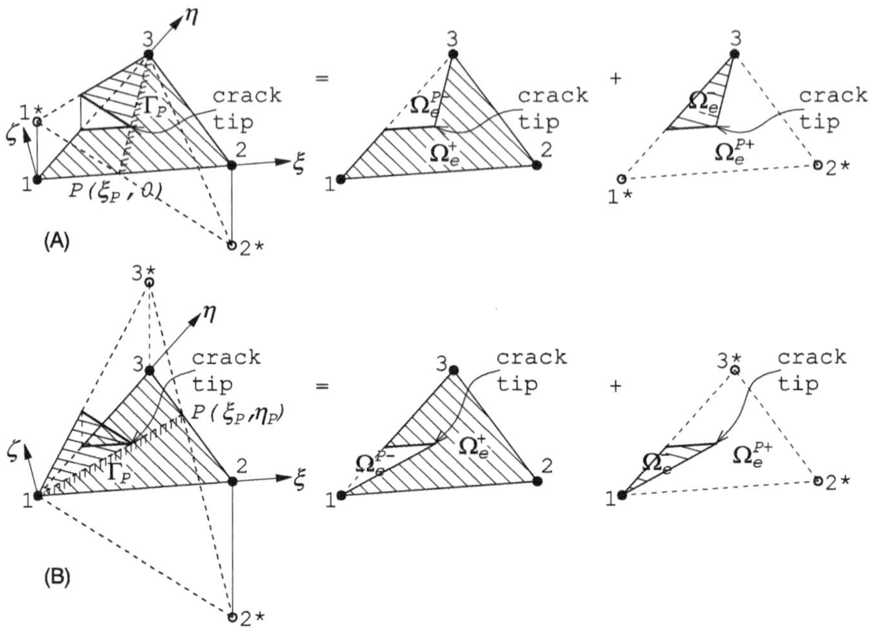

Figure 7.14 Additional phantom nodes (empty circle) determined by the kinematical constraints for the two possibilities of displacement field given in Fig. 7.13.

elements must be constrained. For the case given in Fig. 7.14A, using the similarity of triangles $\Delta 1P1^*$ and $\Delta 2P2^*$, the displacements of phantom node 2^* are constrained as

$$\begin{aligned} \xi_P(\boldsymbol{q}_2 - \boldsymbol{q}_{2^*}) &= (1-\xi_P)(\boldsymbol{q}_{1^*} - \boldsymbol{q}_1) \\ \Rightarrow \boldsymbol{q}_{2^*} &= \frac{1-\xi_P}{\xi_P}\boldsymbol{q}_1 + \boldsymbol{q}_2 - \frac{1-\xi_P}{\xi_P}\boldsymbol{q}_{1^*} \end{aligned} \tag{7.87}$$

here, $\boldsymbol{q}_i = \{\boldsymbol{u}_i^T, \alpha_i, \beta_i\}^T$ is the nodal translational and rotational displacements.

Or, the nodal displacements of the overlapping paired element are constrained in the matrix form as

$$\begin{Bmatrix} \boldsymbol{q}_{1^*} \\ \boldsymbol{q}_{2^*} \\ \boldsymbol{q}_3 \end{Bmatrix} = \boldsymbol{T}^* \boldsymbol{q}_e^{tip-} \tag{7.88}$$

with

$$\boldsymbol{T}^{*}=\begin{bmatrix} \mathbf{0} & \mathbf{0} & \mathbf{0} & \boldsymbol{I} \\ \frac{1-\xi_P}{\xi_P}\boldsymbol{I} & \boldsymbol{I} & \mathbf{0} & -\frac{1-\xi_P}{\xi_P}\boldsymbol{I} \\ \mathbf{0} & \mathbf{0} & \boldsymbol{I} & \mathbf{0} \end{bmatrix} \quad \text{and} \quad \boldsymbol{q}_e^{tip-}=\begin{Bmatrix} \boldsymbol{q}_1 \\ \boldsymbol{q}_2 \\ \boldsymbol{q}_3 \\ \boldsymbol{q}_{1^*} \end{Bmatrix} \tag{7.89}$$

For the other case in Fig. 7.14B in which node 1 is shared, the kinematical constraint of phantom node 2* gives

$$\begin{aligned} &\xi_P(\boldsymbol{q}_2-\boldsymbol{q}_{2^*})=(1-\xi_P)(\boldsymbol{q}_{3^*}-\boldsymbol{q}_3) \\ &\Rightarrow \boldsymbol{q}_{2^*}=\boldsymbol{q}_2+\frac{1-\xi_P}{\xi_P}\boldsymbol{q}_3-\frac{1-\xi_P}{\xi_P}\boldsymbol{q}_{3^*} \end{aligned} \tag{7.90}$$

In the matrix form, the relationship between the nodal displacements of the overlapping paired elements in Fig. 7.14B is

$$\begin{Bmatrix} \boldsymbol{q}_1 \\ \boldsymbol{q}_{2^*} \\ \boldsymbol{q}_{3^*} \end{Bmatrix}=\boldsymbol{T}^{*}\boldsymbol{q}_e^{tip-} \tag{7.91}$$

with

$$\boldsymbol{T}^{*}=\begin{bmatrix} \boldsymbol{I} & \mathbf{0} & \mathbf{0} & \mathbf{0} \\ \mathbf{0} & \boldsymbol{I} & \frac{1-\xi_P}{\xi_P}\boldsymbol{I} & -\frac{1-\xi_P}{\xi_P}\boldsymbol{I} \\ \mathbf{0} & \mathbf{0} & \mathbf{0} & \boldsymbol{I} \end{bmatrix} \quad \text{and} \quad \boldsymbol{q}_e^{tip-}=\begin{Bmatrix} \boldsymbol{q}_1 \\ \boldsymbol{q}_2 \\ \boldsymbol{q}_3 \\ \boldsymbol{q}_{3^*} \end{Bmatrix} \tag{7.92}$$

Similarly, we can obtain the constrained matrix $\boldsymbol{T}^{*}$ for the other cases where the partial crack cuts edges $\overline{12}$ or $\overline{23}$ of the tip shell element.

7.2.2.4 Equilibrium equations and numerical integration

Discretized equilibrium equations

A shell model Ω containing a crack Γ_{cr} is given. The model is constrained on the necessary boundary Γ_u and sustains an external load of traction \τ0 on the natural boundary, Γ_t. Note that $\Gamma_u \cup \Gamma_t = \Omega$ and $\Gamma_u \cap \Gamma_t = \emptyset$. It is assumed that the crack surface is free of traction and the material is linear elastic.

As the conventional finite element procedure, the discretized equations of equilibrium can be obtained as

$$\boldsymbol{f}^{int}=\boldsymbol{K}\boldsymbol{q}=\boldsymbol{f}^{ext} \tag{7.93}$$

where $\boldsymbol{f}^{int}$ and $\boldsymbol{f}^{ext}$ are the discrete internal and external forces, respectively, $\boldsymbol{K}$ is the stiffness matrix, and $\boldsymbol{q}$ are the generalized nodal displacements, including all physical and phantom nodes. The expressions for $\boldsymbol{f}^{int}$ and $\boldsymbol{f}^{ext}$ are given by

$$\begin{aligned} \boldsymbol{f}^{int} = & \sum_{\substack{uncracked\\elements}} \int_{\Omega_e} \boldsymbol{B}^T \boldsymbol{C}\boldsymbol{B} \mathrm{d}\Omega \boldsymbol{q}_e + \sum_{\substack{cracked\\elements}} \left(\int_{\Omega_e^+} \boldsymbol{B}^T \boldsymbol{C}\boldsymbol{B} \mathrm{d}\Omega \boldsymbol{q}_e^{cr+} + \int_{\Omega_e^-} \boldsymbol{B}^T \boldsymbol{C}\boldsymbol{B} \mathrm{d}\Omega \boldsymbol{q}_e^{cr-} \right) \\ & + \sum_{\substack{tip\\elements}} \left(\int_{\Omega_e^+} \boldsymbol{B}^T \boldsymbol{C}\boldsymbol{B} \mathrm{d}\Omega \boldsymbol{q}_e^{tip+} + \int_{\Omega_e^-} \boldsymbol{T}^{*T} \boldsymbol{B}^T \boldsymbol{C}\boldsymbol{B}\boldsymbol{T}^* \mathrm{d}\Omega \boldsymbol{q}_e^{tip-} \right) \end{aligned} \tag{7.94}$$

$$\boldsymbol{f}^{ext} = \sum_{\substack{all\\elements}} \int_{\Gamma_e} \boldsymbol{N}^T \backslash \tau_0 \mathrm{d}\Gamma \tag{7.95}$$

where $\boldsymbol{B}$ is the strain-displacement matrix of the MITC3 shell element established on physical nodes for the uncracked elements and nodes belonging to the overlapped paired elements $\Omega_e^+ \cup \Omega_e^{P-}$ or $\Omega_e^- \cup \Omega_e^{P+}$ for the cracked or tip elements; $\boldsymbol{q}_e$ is the nodal displacements of an uncracked element; $\boldsymbol{q}_e^{cr+}$ and $\boldsymbol{q}_e^{cr-}$ are the nodal displacements of the overlapped paired elements $\Omega_e^+ \cup \Omega_e^{P-}$ and $\Omega_e^- \cup \Omega_e^{P+}$, respectively, for the cracked elements; $\boldsymbol{q}_e^{tip+}$ and $\boldsymbol{q}_e^{tip-}$ are the displacements of all physical nodes and physical nodes plus a phantom node, respectively, of the overlapping paired tip elements; $\boldsymbol{T}$ is the constrained matrix similar to Eqs. (7.89) or (7.92); $\mathbf{C}$ is the constitutive matrix representing the linear elastic stress-strain law in the Cartesian coordinates and given by

$$\mathbf{C} = \mathbf{Q}^T \left(\frac{E}{1-\nu^2} \begin{bmatrix} 1 & \nu & 0 & 0 & 0 & 0 \\ & 1 & 0 & 0 & 0 & 0 \\ & & 0 & 0 & 0 & 0 \\ & & & \frac{1-\nu}{2} & 0 & 0 \\ & \boldsymbol{sym.} & & & k\frac{1-\nu}{2} & 0 \\ & & & & & k\frac{1-\nu}{2} \end{bmatrix} \right) \mathbf{Q} \tag{7.96}$$

in which, E, ν, and k are the Young modulus, the Poisson's ratio, and the shear correction factor, respectively; $\mathbf{Q}$ is a matrix that transforms the stress-strain law from a Cartesian shell-aligned coordinate system to the global Cartesian coordinate system.

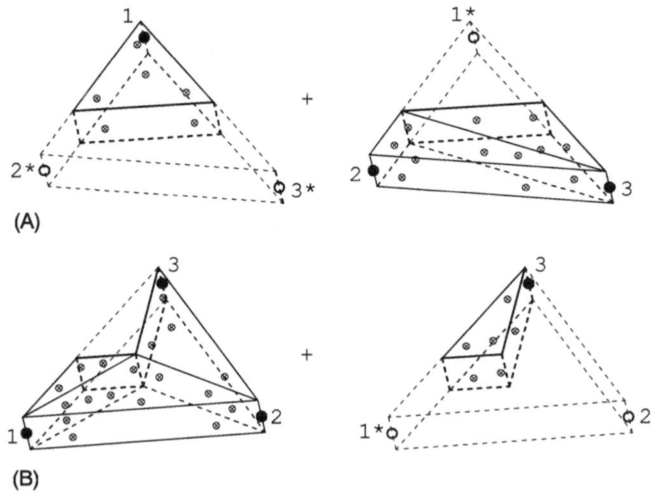

Figure 7.15 Sub-triangles for real parts of elements containing discontinuity: (A) crack element; (B) tip element. Each sub-triangle has (3 × 2) Gaussian quadrature points (cross circles).

Numerical integration

For elements not cut by a crack, the standard Gaussian method with (3 × 2) quadrature points, i.e. 3 in the in-plane and 2 in the thickness direction, is applied to numerically evaluate definite integrals.

To obtain a description of discontinuity by overlapping paired elements, the definite integrals need to be integrated on their own real parts only. To this end, the real parts are subdivided into triangular domains. The standard Gaussian quadratures and weights are modified by mapping into the subtriangular domains [31] (see Fig. 7.15).

7.2.2.5 Calculation of fracture parameters

In this section, we describe the domain form of the path-independent J-integral for continuum mechanics based shell models. Additionally, the formulations for extraction of mixed-mode stress intensity factors in shell models with an arbitrary crack are derived based on the domain form of the interaction integral.

Domain form for J-integral calculation

Let us consider a through-the-thickness crack with the crack front normal to the mid-surface of the shell model as shown in Fig. 7.16. The J-integral

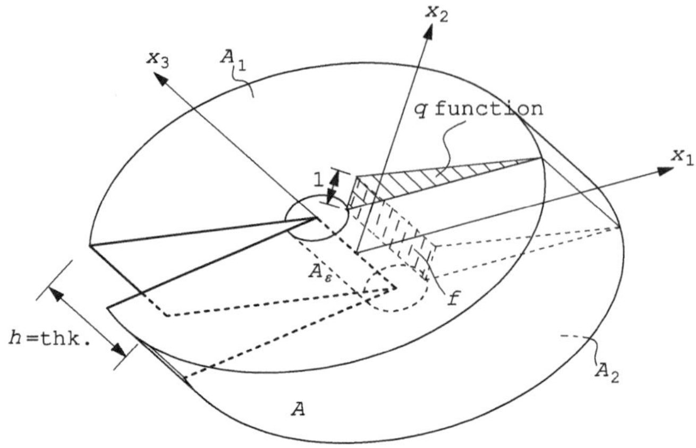

Figure 7.16 Definition of J-domain, function q, and local coordinates at the tip of a crack in a shell.

is redefined as

$$fJ=\int_S\left(\sigma_{ij}\frac{\partial u_i}{\partial x_1}-W\delta_{1j}\right)n_j q\,\mathrm{d}S \tag{7.97}$$

in which, $W=\frac{1}{2}\sigma_{ij}\varepsilon_{ij}$ is the strain energy density, σ_{ij} is the stress, ε_{ij} is the strain, u_i is the displacement, n_j is a component of the unit normal vector to the surface S of a volume V surrounding the crack front, δ_{1j} is the Kronecker delta, q is an arbitrary but continuous function which is equal to zero on A and non-zero on A_ε, and f is the area under the q function curve along the crack front, see Fig. 7.16.

Here, the value of J, which is equivalent to the energy release rate G in the frame of linear elastic fracture mechanics (LEFM), is computed in the direction x_1 of a local crack front coordinate system (x_1, x_2, x_3) at the tip.

Applying the divergence theorem to Eq. (7.97), we obtain the following domain form of J-integral in the case of LEFM

$$fJ=\int_V\left(\sigma_{ij}\frac{\partial u_i}{\partial x_1}-W\delta_{1j}\right)\frac{\partial q}{\partial x_j}\mathrm{d}V-\int_{A_1+A_2}\left(\sigma_{ij}\frac{\partial u_i}{\partial x_1}-W\delta_{1j}\right)n_j q\,\mathrm{d}S \tag{7.98}$$

For shell models discretized by the continuum mechanics based MITC3 elements, the domain form can be considered as a cylinder containing a crack front through the tip and normal to the midsurface. The cylinder is limited by the upper and lower faces, A_1 and A_2 of the shells. The local coordinates (x_1, x_2, x_3) are constructed by x_1 normal to the crack front or tangent to the midsurface, x_3 normal to the midsurface and x_2 orthogonal to x_1 and x_3 following the right-handed rule (Fig. 7.16). We simply assume that A_1 and A_2 are equal and orthogonal to the crack front. Then,

$$\begin{aligned} n_1 = n_2 = 0, \quad n_3 = 1 \quad \text{on } A_1 \\ n_1 = n_2 = 0, \quad n_3 = -1 \quad \text{on } A_2 \end{aligned}$$

and the q function is taken to be constant through the shell thickness, i.e. $f = h$, and equal to 1 on A_ε. Eq. (7.98) can be rewritten as

$$hJ = \int_V \left(\sigma_{ij} \frac{\partial u_i}{\partial x_1} - \frac{1}{2} \sigma_{ij} \varepsilon_{ij} \delta_{1j} \right) \frac{\partial q}{\partial x_j} \mathrm{d}V - \int_{A_1 + A_2} \sigma_{i3} \frac{\partial u_i}{\partial x_1} n_3 q \mathrm{d}S \tag{7.99}$$

In Eq. (7.99), the stress, strain, and derivatives of displacements with respect to the local coordinate system (x_1, x_2, x_3) at the tip can be obtained by transforming those that are post-processing values of a finite element solution to the global Cartesian coordinate system (x, y, z).

The last term of Eq. (7.99) is calculated using the integration points $\zeta = \pm 1/\sqrt{3}$ through the shell thickness:

$$\begin{aligned} hJ = & \int_V \left(\sigma_{ij} \frac{\partial u_i}{\partial x_1} - \frac{1}{2} \sigma_{ij} \varepsilon_{ij} \delta_{1j} \right) \frac{\partial q}{\partial x_j} \mathrm{d}V \\ & - \sqrt{3} \int_{A_1} \left[F(\zeta = \frac{1}{\sqrt{3}}) - F(\zeta = \frac{-1}{\sqrt{3}}) \right] q \mathrm{d}S \end{aligned} \tag{7.100}$$

with $F = \sigma_{i3}(\partial u_i / \partial x_1)$.

For finite element implementation, the domain V is a set of elements which has at least one node in and one node out of a circular cylinder with the central axis being the crack front and radius r_d. In this chapter, the value of $r_d = 2.5 h_{avg}$ will be used, wherein h_{avg} is the mean value of the square-roots of the cracked elements' areas [31].

The q function evaluated inside any element can be interpolated by means of finite element shape functions as

$$q = \sum_{I=1}^{3} N_I(\xi, \eta) q_I \tag{7.101}$$

in which $N_I(\xi, \eta)$ is the standard C^0 shape function, and q_I is the nodal value of function q, which is equal to 1 if node I is inside or equal to 0 if I is outside the circular cylinder.

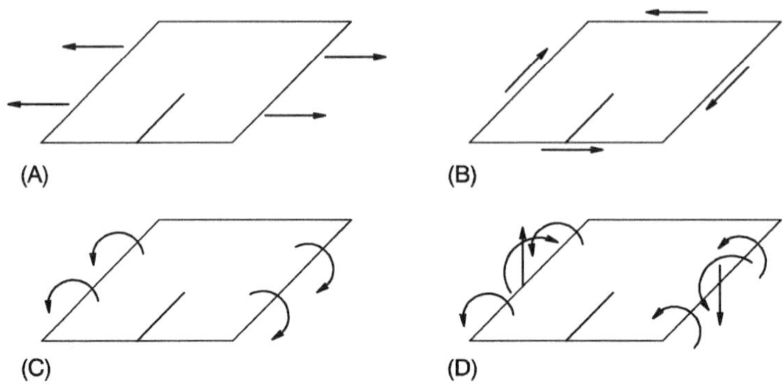

Figure 7.17 Stress intensity factors related to each mode of loading; (A) Symmetric membrane loading, K_I; (B) Antisymmetric membrane loading, K_{II}; (c) Symmetric bending: Kirchhoff theory, k_1, Reissner theory, K_1; (D) Antisymmetric bending and shear: Kirchhoff theory, k_2, Reissner theory, K_2, K_3.

Extraction of mixed-mode stress intensity factors

In LEFM, the stress intensity factors (SIFs) in shell models under combined in-plane and out-of-plane loadings are defined as a superposition of the plane stress and plate theory fields. The SIFs associated with each of the loadings in Fig. 7.17 can be summarized as follows:

- Membrane loadings
  - Symmetric membrane loading: $K_I = \lim_{r\to 0} \sqrt{2\pi r}\sigma_{\theta\theta}(r, 0)$.
  - Antisymmetric membrane loading: $K_{II} = \lim_{r\to 0} \sqrt{2\pi r}\sigma_{r\theta}(r, 0)$.
- Kirchhoff theory's bending loadings
  - Symmetric loading (bending): $k_1 = \lim_{r\to 0} \sqrt{2r}\sigma_{\theta\theta}(r, 0, \frac{h}{2})$.
  - Antisymmetric loading (twisting): $k_2 = \lim_{r\to 0} \frac{3+\nu}{1+\nu}\sqrt{2r}\sigma_{r\theta}(r, 0, \frac{h}{2})$.
- Reissner theory's bending loadings
  - Symmetric loading (bending): $K_1 = \lim_{r\to 0} \sqrt{2r}\sigma_{\theta\theta}(r, 0, \frac{h}{2})$.
  - Antisymmetric loading (twisting): $K_2 = \lim_{r\to 0} \sqrt{2r}\sigma_{r\theta}(r, 0, \frac{h}{2})$.
  - Antisymmetric loading (bending): $K_3 = \lim_{r\to 0} \sqrt{2r}\sigma_{\theta 3}(r, 0, 0)$.

where (r, θ) are the polar coordinates at a crack tip.

The relationship between the J-integral value and SIFs in the cases of mixed-mode loadings is given as

$$J = G = \frac{1}{E}\left(K_I^2 + K_{II}^2\right) + \frac{\pi}{3E}\left(\frac{1+\nu}{3+\nu}\right)\left(k_1^2 + k_2^2\right) \qquad \text{for Kirchhoff theory} \qquad (7.102a)$$

$$J = G = \frac{1}{E}\left(K_I^2 + K_{II}^2\right) + \frac{\pi}{3E}\left[K_1^2 + K_2^2 + K_3^2\frac{8(1+\nu)}{5}\right] \quad \text{for Reissner theory} \tag{7.102b}$$

To calculate SIF for a particular mode of loading in cases of mixed-mode loadings, we follow the scheme which was derived in [31,32] for two- or three-dimensional problems. Consider two states of a cracked model corresponding to the present state $(\sigma_{ij}^{(1)}, \varepsilon_{ij}^{(1)}, u_i^{(1)})$ and an auxiliary stage $(\sigma_{ij}^{(2)}, \varepsilon_{ij}^{(2)}, u_i^{(2)})$ which will be chosen as the asymptotic fields given in the Appendix. From Eq. (7.99), the domain form of the J-integral for the superposition of the two states is

$$\begin{aligned} J^{(1+2)} = & \frac{1}{h}\int_V \left((\sigma_{ij}^{(1)} + \sigma_{ij}^{(2)}) \frac{\partial (u_i^{(1)} + u_i^{(2)})}{\partial x_1} \right. \\ & \left. - \frac{1}{2}(\sigma_{ij}^{(1)} + \sigma_{ij}^{(2)})(\varepsilon_{ij}^{(1)} + \varepsilon_{ij}^{(2)})\delta_{1j} \right) \frac{\partial q}{\partial x_j} \mathrm{d}V \\ & - \frac{1}{h}\int_{A_1+A_2} (\sigma_{i3}^{(1)} + \sigma_{i3}^{(2)}) \frac{\partial (u_i^{(1)} + u_i^{(2)})}{\partial x_1} n_3 q \mathrm{d}S \end{aligned} \tag{7.103}$$

Expanding and arranging terms in Eq. (7.103) gives

$$J^{(1+2)} = J^{(1)} + J^{(2)} + I^{(1,2)} \tag{7.104}$$

in which $I^{(1,2)}$, the so-called interaction integral, is

$$\begin{aligned} I^{(1,2)} = & \frac{1}{h}\int_V \left(\sigma_{ij}^{(1)} \frac{\partial u_i^{(2)}}{\partial x_1} + \sigma_{ij}^{(2)} \frac{\partial u_i^{(1)}}{\partial x_1} - \sigma_{ij}^{(1)}\varepsilon_{ij}^{(2)}\delta_{1j} \right) \frac{\partial q}{\partial x_j} \mathrm{d}V \\ & - \frac{1}{h}\int_{A_1+A_2} \left(\sigma_{ij}^{(1)} \frac{\partial u_i^{(2)}}{\partial x_1} + \sigma_{ij}^{(2)} \frac{\partial u_i^{(1)}}{\partial x_1} \right) n_3 q \mathrm{d}S \end{aligned} \tag{7.105}$$

Using Eq. (7.102), the J-integral for the combination of the present and auxiliary states is

For Kirchhoff theory,

$$\begin{aligned} J^{(1+2)} = & J^{(1)} + J^{(2)} + \frac{2}{E}\left(K_I^{(1)}K_I^{(2)} + K_{II}^{(1)}K_{II}^{(2)}\right) \\ & + \frac{2\pi}{3E}\left(\frac{1+\nu}{3+\nu}\right)\left(k_1^{(1)}k_1^{(2)} + k_2^{(1)}k_2^{(2)}\right) \end{aligned} \tag{7.106a}$$

For Reissner theory,

$$J^{(1+2)} = J^{(1)} + J^{(2)} + \frac{2}{E}(K_I^{(1)}K_I^{(2)} + K_{II}^{(1)}K_{II}^{(2)}) + \frac{2\pi}{3E}\left[K_1^{(1)}K_1^{(2)} + K_2^{(1)}K_2^{(2)} + K_3^{(1)}K_3^{(2)}\frac{8(1+\nu)}{5}\right] \tag{7.106b}$$

From Eqs. (7.104) and (7.106), we have the following relationship between the interaction integral $I^{(1,2)}$ in Eq. (7.105) and the SIFs

For Kirchhoff theory,

$$I^{(1,2)} = \frac{2}{E}(K_I^{(1)}K_I^{(2)} + K_{II}^{(1)}K_{II}^{(2)}) + \frac{2\pi}{3E}\left(\frac{1+\nu}{3+\nu}\right)(k_1^{(1)}k_1^{(2)} + k_2^{(1)}k_2^{(2)}) \tag{7.107a}$$

For Reissner theory,

$$I^{(1,2)} = \frac{2}{E}(K_I^{(1)}K_I^{(2)} + K_{II}^{(1)}K_{II}^{(2)}) + \frac{2\pi}{3E}\left[K_1^{(1)}K_1^{(2)} + K_2^{(1)}K_2^{(2)} + K_3^{(1)}K_3^{(2)}\frac{8(1+\nu)}{5}\right] \tag{7.107b}$$

By choosing the auxiliary state as the pure symmetric membrane loading, meaning that the asymptotic fields have only $K_I^{(2)} = 1$ and the other SIFs are equal to zero, the stress intensity factor $K_I^{(1)}$ for the present state in terms of the interactive integral $I^{(1,K_I)}$ is given as

$$K_I^{(1)} = \frac{E}{2}I^{(1,K_I)} \tag{7.108}$$

Similarly, we can determine the other stress intensity factors.

7.3. Extended meshfree methods for fracture in shells

Meshfree methods are well suited to model thin shells based on the Kirchhoff-Love theory due to their higher order continuity. In this section, we will propose a thin shell method that does not require rotational degrees of freedom or the discretization of the director field. The crack is modeled in a very easy way through partition-of-unity enrichment. The meshfree thin shell combines classical shell theory with a continuum based shell.

The kinematic assumptions of classical KL shell theory is adopted. From the continuum description, we make use of the generality provided by the strain energy, so that constitutive models developed for continua are easily applicable to shells. The formulation is valid for finite strains.

7.3.1 Shell model

7.3.1.1 Kinematics

Consider a body Ω with material points $\mathbf{X} \in \Omega_0$ of the shell in the reference configuration with discontinuities, e.g. cracks, on lines Γ_0^c. The boundary is denoted by Γ_0, where Γ_0^u and Γ_0^t are the complementary boundaries on which, respectively, displacements and tractions are prescribed. We consider a surface parametrized by two independent variables θ^α, $\alpha = 1$ to 2; the surface in the reference (initial) configuration is described by $\mathbf{R}(\theta^\alpha)$, $\mathbf{R} \in \Re^3$. The material points in the reference configuration are given by

$$\mathbf{X}(\theta^i) = \mathbf{R}(\theta^\alpha) + \theta^3 \frac{d}{2} \mathbf{N}(\theta^\alpha) \tag{7.109}$$

where θ^i are curvilinear coordinates, $-1 \leq \theta^3 \leq 1$, d is the thickness of the shell, $\mathbf{N}$ is the shell normal and $\mathbf{R}$ is a point on the mid-surface of the shell in the reference configuration $\mathcal{S}_0$. Upper Latin and Greek indices range from 1 to 3 and from 1 to 2, respectively and refer to quantities in the cartesian or curvilinear coordinate system. The current configuration is given by

$$\mathbf{x}(\theta^i, t) = \mathbf{r}(\theta^\alpha, t) + \theta^3 \frac{d}{2} \mathbf{n}\ (\theta^\alpha, t) \tag{7.110}$$

where t is the time, $\mathbf{n}$ is the director field and $\mathbf{r}$ is a point on the mid-surface position in the current configuration. The first and second fundamental forms are given by

$$A_{\alpha\beta} = \mathbf{R}_{,\alpha} \cdot \mathbf{R}_{,\beta} \tag{7.111}$$

$$B_{\alpha\beta} = \mathbf{R}_{,\alpha,\beta} \cdot \mathbf{N} = -\mathbf{R}_{,\alpha} \cdot \mathbf{N}_{,\beta} \tag{7.112}$$

The curvilinear coordinates θ^α are such that $\mathbf{R}_{,\alpha}$ form a basis for the tangent space in $\mathbf{X} \in \mathcal{S}_0$. For arbitrary θ^3, we define a family of surfaces $\mathcal{S}(\theta^3)$ with $\mathcal{S}_0 = \mathcal{S}(0)$ for which the tangent basis is established from (7.109) as $\boldsymbol{X}_{,\alpha} = \boldsymbol{R}_{,\alpha} + \frac{\theta^3}{2}\boldsymbol{N}_\alpha$. We extend this basis by including $\boldsymbol{N} = \boldsymbol{X}_{,3}\frac{2}{d}$. The resulting basis spans Ω_0, and we can then define the metric of Ω_0 as $G_{ij} = \boldsymbol{X}_{,i} \cdot \boldsymbol{X}_{,j}$. The dual basis is given by $\mathbf{G}^i = G^{ij}\boldsymbol{X}_{,j}$ with $\left[G^{ij}\right] = \left[G_{ij}\right]^{-1}$.

The Cauchy-Green tensor is

$$\boldsymbol{C} = \boldsymbol{F}^T\boldsymbol{F} = \underbrace{\left(\boldsymbol{x}_{,i} \cdot \boldsymbol{x}_{,j}\right)}_{C_{ij}} \mathbf{G}^i \otimes \mathbf{G}^j \tag{7.113}$$

where $\mathbf{x}$ are the spatial position coordinates and $\boldsymbol{F}$ is the deformation gradient.

The Kirchhoff-Love hypothesis is imposed by requiring that $\mathbf{n}$ is perpendicular to $\mathbf{r}_{,\alpha}$, $\alpha = 1, 2$:

$$\mathbf{n} = \frac{\mathbf{r}_{,1} \times \mathbf{r}_{,2}}{\|\mathbf{r}_{,1} \times \mathbf{r}_{,2}\|} \tag{7.114}$$

The mid-surface position and director field in the reference configuration are denoted by $\mathbf{R}$ and $\mathbf{N}$, respectively. The relation (7.114) is also valid for the reference configuration.

7.3.1.2 Virtual work

The weak form of the momentum equation is written with the principle of virtual work (see e.g. [16]): find $\mathbf{r} \in \mathcal{V}$ such that

$$\delta W = \delta W_{int} - \delta W_{ext} + \delta W_{kin} - \delta W_E = 0 \quad \forall \delta\mathbf{r} \in \mathcal{V}_0 \tag{7.115}$$

where

$$\begin{aligned} \mathcal{V} &= \left\{\mathbf{r}(\cdot, t) | \mathbf{r}(\cdot, t) \in \mathcal{H}^2(\Omega_0 / \Gamma_0^c),\ \mathbf{r}(\cdot, t) = \bar{\mathbf{r}}(t) \text{ on } \Gamma_0^u, \right. \\ &\qquad \left. \mathbf{r} \text{ discontinuous on } \Gamma_0^c\right\} \\ \mathcal{V}_0 &= \left\{\delta\mathbf{r} | \delta\mathbf{r} \in \mathcal{V},\ \delta\mathbf{r} = 0 \text{ on } \Gamma_0^u,\ \delta\mathbf{r} \text{ discontinuous on } \Gamma_0^c\right\} \end{aligned} \tag{7.116}$$

$$\delta W_{int} = \int_{\Omega_0 \backslash \Gamma_0^c} \left\{s^{\alpha\beta} \mathbf{x}_{,\alpha} \cdot \delta\mathbf{x}_{,\beta} \left[\mathbf{G}_1 \cdot (\mathbf{G}_2 \times \mathbf{G}_3)\right]\right\} d\Omega \tag{7.117}$$

$$\delta W_{ext} = \int_{\Omega_0 \backslash \Gamma_0^c} \varrho_0\, \mathbf{b} \cdot \delta\mathbf{u}\ d\Omega_0 + \int_{\Gamma_0^t} \bar{\mathbf{t}}_0 \cdot \delta\mathbf{u}\ d\Gamma_0 \tag{7.118}$$

$$\delta W_E = \int_{\Gamma_c} \bar{\mathbf{t}}_c \cdot \delta[\![\mathbf{u}]\!]\ d\Gamma \tag{7.119}$$

$$\delta W_{kin} = \int_{\Omega_0 \backslash \Gamma_0^c} \varrho_0\ \delta\mathbf{u} \cdot \ddot{\mathbf{u}}\ d\Omega_0 \tag{7.120}$$

where the prefix δ identifies the test function and W_{ext} is the external energy, W_{int} designates the internal energy, W_E is the crack cohesive energy and W_{kin} the kinetic energy, ϱ_0 is the density, $\mathbf{s}$ is the Kirchhoff stress, $\mathbf{b}$ is the body force and $\bar{\mathbf{t}}_0$ the prescribed traction; superposed dots denote material time derivatives. For details of the discrete equations see [37–39,42].

7.3.1.3 Discretization

The approximation of the shell surface is given by

$$\mathbf{r}(\theta^\alpha, t) = \sum_{I \in \mathcal{W}} \Phi_I(\theta^\alpha)\ \mathbf{r}_I(t) \tag{7.121}$$

where $\Phi_I(\theta^\alpha)$ are the shape functions and $\mathcal{W}$ is the domain of influence of the corresponding particle.

We require C^1 displacement continuity in the meshfree method. So there is no need to discretize the director field $\mathbf{n}$ and $\mathbf{n}$ is readily obtained from Eq. (7.114). The variation of the motion $\mathbf{x}$ (and their spatial derivatives) is given by

$$\delta\mathbf{x} = \delta\mathbf{r} + \theta^3 \frac{d}{2} \delta\mathbf{n} \tag{7.122}$$

The variation of the normal can be expressed in terms of $\mathbf{r}$:

$$\delta\mathbf{n} = a_1 \boldsymbol{a}_4 \cdot \boldsymbol{a}_2 \tag{7.123}$$

where $a_1 = \frac{1}{\|\boldsymbol{r}_{,1} \times \boldsymbol{r}_{,2}\|}$, $\boldsymbol{a}_2 = \boldsymbol{r}_{,1} \times \delta\boldsymbol{r}_{,2} - \boldsymbol{r}_{,2} \times \delta\boldsymbol{r}_{,1}$ and $\boldsymbol{a}_4 = (\boldsymbol{I} - \boldsymbol{n} \otimes \boldsymbol{n})$.

The derivatives of the motion $\mathbf{x}$ are given by

$$\mathbf{x}_{,\alpha} = \underbrace{\sum_{I \in \mathcal{W}} \Phi_{I,\alpha} \mathbf{r}_I}_{\mathbf{r}_{,\alpha}} + \theta^3 \frac{d}{2} \mathbf{n}_{,\alpha} \tag{7.124}$$

with

$$\mathbf{n}_{,\gamma} = a_1 \boldsymbol{a}_4 \cdot \boldsymbol{a}_5 \tag{7.125}$$

with $\boldsymbol{a}_5 = \boldsymbol{r}_{,1} \times \boldsymbol{r}_{,2\gamma} - \boldsymbol{r}_{,2} \times \boldsymbol{r}_{,1\gamma}$.

To obtain the variation $\delta\mathbf{x}_{,\alpha}$, the derivatives of the variation of the normal have to be computed. Considering Eq. (7.123), the derivatives are:

$$\begin{aligned} \delta\mathbf{n}_{,\gamma} = & - a_1^3[(\boldsymbol{r}_{,1} \times \boldsymbol{r}_{,2}) \cdot \boldsymbol{a}_5](\boldsymbol{a}_4 \cdot \boldsymbol{a}_2) - 2a_1(\boldsymbol{n} \otimes \boldsymbol{a}_5)^{\mathrm{S}} \cdot \boldsymbol{a}_2 + \\ & a_1 \boldsymbol{a}_4 \cdot \delta\boldsymbol{a}_5 \end{aligned} \tag{7.126}$$

Table 7.1 The return mapping in the material setting; encapsulation of the small strain case.

Make $\boldsymbol{C}_0^p = \boldsymbol{I}$ (and therefore $\boldsymbol{C}_0^{p^{-1}} = \boldsymbol{I}$) and $\boldsymbol{\xi}_0 = \mathbf{0}$ for all quadrature points
For each quadrature point at time-step n, perform the following calculations

1. Using the current position field $\boldsymbol{x}$ calculate $\boldsymbol{C}_{n+1} = (\boldsymbol{x}_{,\alpha} \cdot \boldsymbol{x}_{,\beta})\mathbf{G}^\alpha \otimes \mathbf{G}^\beta$ with $\mathbf{G}^\alpha = G^{\alpha\beta}\boldsymbol{X}_{,\beta}$ and $[G^{\alpha\beta}] = [\boldsymbol{X}_{,\alpha} \cdot \boldsymbol{X}_{,\beta}]^{-1}$
2. Calculate the trial of the elastic measure $\boldsymbol{C}_{n\star}^E = \boldsymbol{C}_{n+1}\boldsymbol{C}_n^{p^{-1}}$
3. Use a second order Padé approximation to calculate $\boldsymbol{\varepsilon}_{n\star} = \frac{1}{2}\ln \boldsymbol{C}_{n\star}^E$
4. Using a *modified* (unsymmetric) small strain return-mapping algorithm, update $\boldsymbol{\xi}_n$ and calculate $\boldsymbol{\varepsilon}_{n+1}$ $\Delta\alpha$, $\boldsymbol{\Sigma}$ and the small strain consistent modulus $\mathbb{C}$
5. Calculate the new plastic metric inverse as $\boldsymbol{C}_{n+1}^{p-1} = \boldsymbol{C}_{n+1}^{-1}\exp[2\boldsymbol{\varepsilon}_{n+1}]$ using a first order Padé approximation for the exponential function
6. Calculate the contravariant components of the stress as

$$s^{\alpha\beta} = \mathbf{G}^\alpha \cdot \left(\mathbf{C}^{-1}\boldsymbol{\Sigma}\mathbf{G}^\beta\right)$$

and the contravariant components of the tangent modulus as

$$C^{\alpha\beta\gamma\delta} = G^{\alpha i}G^{\beta j}G^{\gamma k}G^{\delta l}\mathfrak{C}_{ijkl}$$

where $\mathfrak{C}_{ijkl}$ is the elasticity tensor and $G^{\alpha i} = \mathbf{G}^\alpha \cdot \boldsymbol{e}_i$ for any $i = 1, 2, 3$ and $\alpha = 1, 2$. $\boldsymbol{\xi}_0$ are a set of internal variables, α is the plastic parameter of the yield surface and $\boldsymbol{\Sigma}$ are the elastic stresses.

We use the EFG shape functions for the meshfree discretization with a quartic polynomial basis in order to suppress membrane locking:

$$\begin{aligned}\mathbf{p}(\theta^\alpha) = \big(&1, \theta^1, \theta^2, (\theta^1)^2, \theta^1\theta^2, (\theta^2)^2, (\theta^1)^3, (\theta^1)^2\theta^2, (\theta^2)^2\theta^1, \ldots \\ &\ldots (\theta^2)^3, (\theta^1)^4, (\theta^1)^3\theta^2, (\theta^1)^2(\theta^2)^2, \theta^1(\theta^2)^3, (\theta^2)^4\big)\end{aligned} \qquad (7.127)$$

Note, that we choose to express all quantities in curvilinear coordinates. For rectangular plates and for cylinders, the Jacobian is constant and hence θ^i are linear combinations of $\mathbf{x}$ and linear completeness is guaranteed. It is noted that the same shape functions are employed in shape and displacement approximations to guarantee strain-free states in rigid body motion, see Krysl and Belytschko [27].

7.3.2 Continuum constitutive models

For the constitutive model, we adopt the algorithm of Table 7.1. We use a two-dimensional non-symmetric radial return and rotate so that the 3-3

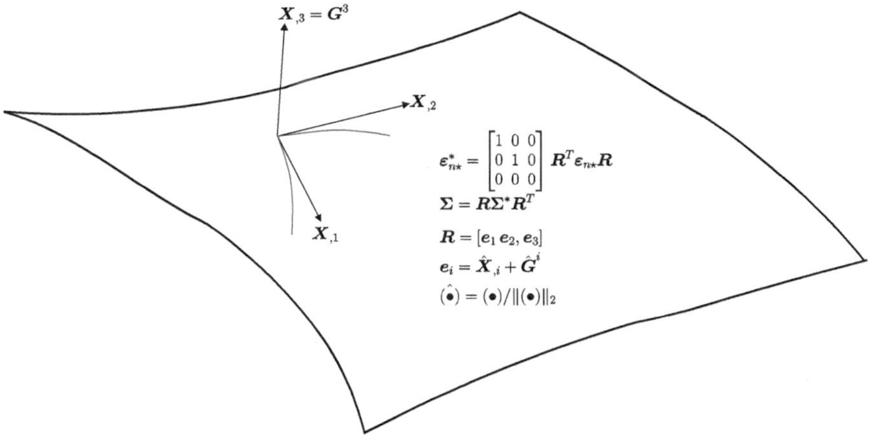

Figure 7.18 Return mapping: filtering of the normal strain and stress.

component corresponds to the normal. The normal strain (and consequently the normal stress) is filtered, according to Fig. 7.18. This way, we can easily adopt continuum constitutive models.

7.3.3 Crack model

The approach is extended to shells with cracks by enriching the mid-surface motion $\mathbf{r}(\theta^\alpha, t)$ with a discontinuous function. The jump in the director field is then obtained directly via the discontinuous part of $\mathbf{r}(\theta^\alpha, t)$. The fact that there is no need to discretize the director field, facilitates the incorporation of discontinuities in shells.

The force introduced here corresponds to the resistance to opening, which is a function of the opening displacement. The opening displacement can be written as a function of the mid-surface position *and* the director.

We use two methods; both were already used to model cracks in continua, see Rabczuk and Belytschko [37,39], Ventura et al. [48]. The last one was used in linear fracture mechanics and is here modified to deal with cohesive cracks.

7.3.3.1 Method 1: cracked particles

To model cracks, the approximation for $\mathbf{r}$ is enriched with a discontinuous function, so

$$\mathbf{r}(\theta^\alpha, t) = \mathbf{r}^{\text{cont}}(\theta^\alpha, t) + \mathbf{r}^{\text{enr}}(\theta^\alpha, t) \tag{7.128}$$

where $\mathbf{r}^{cont}(\theta^\alpha, t)$ is given by Eq. (7.136). Let $\mathcal{N}$ be the total set of particles in the discretization and $\mathcal{N}_c$ the set of cracked particles. The set of cracked particles consists of the particles where a fracture nucleation or propagation criterion has been met. To model the discontinuous part of the displacement, the test and trial functions are enriched with sign functions which are parametrized by $\delta\mathbf{q}_I$ and $\mathbf{q}_I$, respectively. The crack surface is assumed to be normal to the reference surface $\mathcal{S}_0$, so that the trial and test functions are:

$$\mathbf{r}(\theta^\alpha, t) = \sum_{I \in \mathcal{N}} \Phi_I(\theta^\alpha)\, \mathbf{r}_I(t) + \sum_{I \in \mathcal{N}_c} \Phi_I(\theta^\alpha)\, H(f_I(\theta^\alpha))\, \mathbf{q}_I(t) \tag{7.129}$$

$$\delta\mathbf{r}(\theta^\alpha) = \sum_{I \in \mathcal{N}} \Phi_I(\theta^\alpha)\, \delta\mathbf{r}_I + \sum_{I \in \mathcal{N}_c} \Phi_I(\theta^\alpha)\, H(f_I(\theta^\alpha))\, \delta\mathbf{q}_I \tag{7.130}$$

where $\mathcal{N}$ is the total set of particles, $\mathcal{N}_c$ is the set of cracked particles and $f_I(\theta^\alpha)$ is given by

$$f_I(\theta^\alpha) = \mathbf{m} \cdot (\theta^\alpha - \theta_I^\alpha) \tag{7.131}$$

where $\mathbf{m}$ is the normal to the crack. The sign function $H(f(\theta^\alpha))$ is defined as:

$$H(f_I(\theta^\alpha)) = \begin{cases} 1 & \forall f_I(\theta^\alpha) > 1 \\ -1 & \forall f_I(\theta^\alpha) < 1 \end{cases} \tag{7.132}$$

Note that, in general, different shape functions can be used for the continuous and discontinuous parts. Since at least second order complete basis polynomials have to be used, the domains of influence are large and the cracked particles will influence more particles than in the continuum version of this method [37]. Note also, that in [37,39], the method was developed for a stress point integration where stresses are evaluated at nodes and stress points. In this approach, the nodal stresses are obtained by MLS fits.

7.3.3.2 Method 2: local partition of unity approach

Alternative 1

In this approach, we enrich the test and trial functions with additional unknowns so that the crack is continuous and includes branch functions at the tip as in Moes et al. [31], Ventura et al. [48]. Therefore, the test and

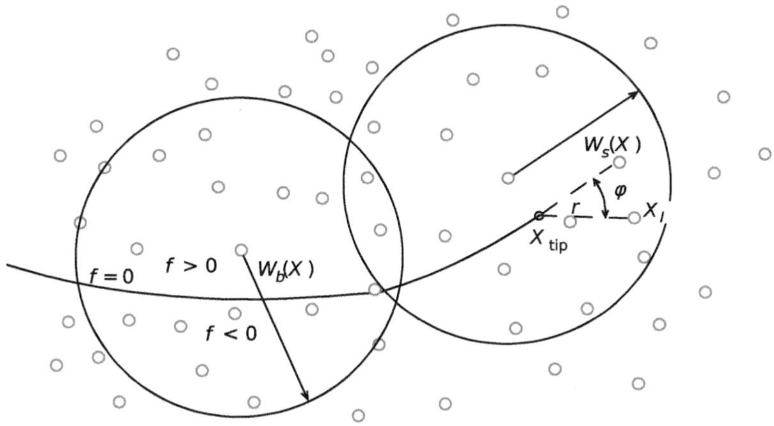

Figure 7.19 Crack with partially cut and complete cut domain of influence particles.

trial functions in terms of a signed distance function f, see Fig. 7.19, are

$$\mathbf{r}(\theta^\alpha,t)=\sum_{I\in W(\theta^\alpha)}\Phi_I(\theta^\alpha)\,\mathbf{r}_I(t)+\sum_{I\in W_b(\theta^\alpha)}\Phi_I(\theta^\alpha)\,H(f_I(\theta^\alpha))\,\mathbf{q}_I(t)$$
$$+\sum_{I\in W_s(\theta^\alpha)}\Phi_I(\theta^\alpha)\sum_K\mathbf{B}_K(\theta^\alpha)\,\mathbf{b}_{KI}(t) \tag{7.133}$$

$$\delta\mathbf{r}(\theta^\alpha)=\sum_{I\in W(\theta^\alpha)}\Phi_I(\theta^\alpha)\,\delta\mathbf{r}_I+\sum_{I\in W_b(\theta^\alpha)}\Phi_I(\theta^\alpha)\,H(f_I(\theta^\alpha))\,\delta\mathbf{q}_I$$
$$+\sum_{I\in W_s(\theta^\alpha)}\Phi_I(\theta^\alpha)\sum_K\mathbf{B}_K(\theta^\alpha)\,\delta\mathbf{b}_{KI} \tag{7.134}$$

The first term on the RHS of Eq. (7.133) is the usual approximation where Φ_I are the shape functions, the second and third term is the enrichment. The coefficients $\mathbf{q}_I$ and $\mathbf{b}_I$ are additional degrees of freedom. Only nodes which are located in the domain $W_b(\theta^\alpha)$ are enriched with the additional unknowns. The third term of Eqs. (7.134) and (7.133) is applied around the crack tip $W_s(\theta^\alpha)$.

For cohesive cracks, the crack tip enrichment is usually omitted and the cohesive forces depend only on the additional unknowns $\mathbf{q}_I$. In XFEM, the omission of the third term in Eq. (7.134) is straightforward since it is easily possible to impose the appropriate boundary conditions, see Fig. 7.20A. However, in meshfree methods, this technique cannot be applied analogously, see Fig. 7.20B, since there will always be particles with a partially cut domain of influence. Therefore, we introduce the branch function en-

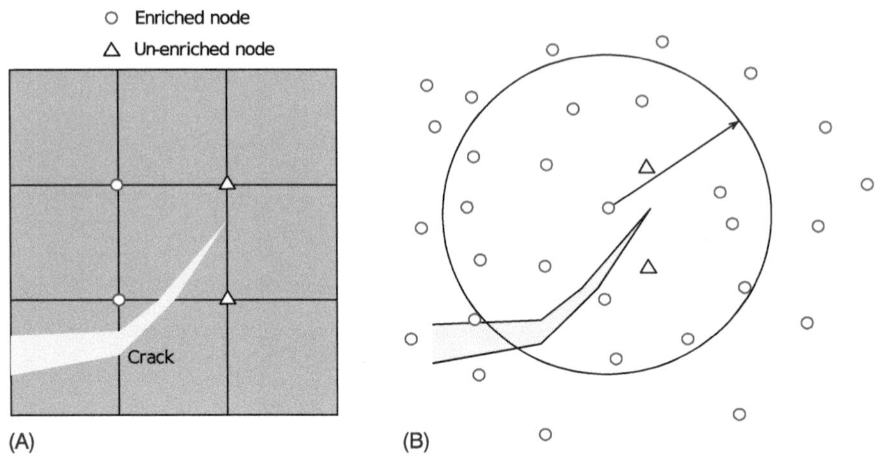

Figure 7.20 Crack with enriched nodes in XFEM and meshfree methods.

richment and use [13]:

$$\mathbf{B} = \left(r \, sin \, \frac{\phi}{2} \right) \tag{7.135}$$

where r is the distance from the crack tip and ϕ the angle as shown in Fig. 7.19.

We would like to mention that for particles in the blending region, i.e. the particles whose domain of influence is not cut but that are influenced by the "enriched" particles, only the usual approximation (first term on the RHS of Eqs. (7.134) and (7.133)), is considered in the approximation of the test and trial functions. For crack propagation, we control the crack length and propagate the crack through an entire background cell.

Alternative 2

In meshfree methods, there are different ways to increase the order of completeness. In the element-free Galerkin method, the order of completeness is determined intrinsically by the polynomial basis $\mathbf{p}$ while in the $h-p$-cloud method [20,21] or in the partition of unity finite element method (PUFEM) [30], the order of completeness is determined by an extrinsic basis:

$$u^h(\mathbf{X}, t) = \sum_{J \in S} \Phi_J(\mathbf{X}) \left(u_J(t) + \sum_{I=1}^{N} \mathbf{p}_I(\mathbf{X}) \, \mathbf{a}_{JI} \right) \tag{7.136}$$

where $\Phi_J(\mathbf{X})$ are the shape functions, $\mathbf{p}$ is the third order polynomial basis, $\mathbf{u}$ and $\mathbf{a}$ are nodal parameters, S is the set of nodes where $\Phi_J(\mathbf{X}) \neq 0$ and N is the number of the polynomial basis. There are two advantages of using an extrinsic basis in the context of thin shell analysis. First of all, we found that adding additional degrees of freedom in order to increase the order of polynomial completeness is computationally less expensive in comparison to using a quartic polynomial basis. And secondly, a quartic polynomial basis smoothes the discontinuity over a wide region. Therefore, we have developed an alternative approximation to Eq. (7.133):

$$\begin{aligned}\mathbf{r}(\theta^\alpha, t) = &\sum_{I \in \mathcal{N}} \Phi_I(\theta^\alpha) \left(\mathbf{r}_I(t) + \sum_{J=1}^{N} \mathbf{p}_J(\theta^\alpha) \mathbf{a}_{IJ} \right) \\ &+ \sum_{I \in \mathcal{N}_c} \Phi_I(\theta^\alpha)\ S[f_I(\theta^\alpha)] \\ &\left(\mathbf{q}_I(t) + \sum_{J=1}^{N} \tilde{\mathbf{p}}_J(\theta^\alpha) \mathbf{b}_{IJ} \right)\end{aligned} \tag{7.137}$$

The first line of the RHS of Eq. (7.137) represents the continuous part. Hereby, $\mathbf{a}$ are additional unknowns that are introduced to increase the order of completeness. The second and third line of Eq. (7.137) is the enrichment. We found that the extrinsic basis for the enrichment $\tilde{\mathbf{p}}$ can be of lower order-second order completeness – than the continuous extrinsic basis $\mathbf{a}$. Alternatively, one could also use an intrinsic basis for the enrichment:

$$\begin{aligned}\mathbf{r}(\theta^\alpha, t) = &\sum_{I \in \mathcal{N}} \Phi_I(\theta^\alpha) \left(\mathbf{r}_I(t) + \sum_{J=1}^{N} \mathbf{p}_J(\theta^\alpha) \mathbf{a}_{IJ} \right) \\ &+ \sum_{I \in \mathcal{N}_c} \tilde{\Phi}_I(\theta^\alpha)\ S[f_I(\theta^\alpha)]\ \mathbf{q}_I(t)\end{aligned} \tag{7.138}$$

where $\tilde{\Phi}$ are second order complete MLS shape functions. In [47], it was shown that it is admissible to use different order of shape functions for the continuous and discontinuous part. Accordingly, the test functions can be expressed as:

$$\delta\mathbf{r}(\theta^\alpha, t) = \sum_{I \in \mathcal{N}} \Phi_I(\theta^\alpha) \left(\delta\mathbf{r}_I(t) + \sum_{J=1}^{N} \mathbf{p}_J(\theta^\alpha) \delta\mathbf{a}_{IJ} \right)$$

$$+\sum_{I\in\mathcal{N}_c}\Phi_I(\theta^\alpha)\ S[f_I(\theta^\alpha)]\left(\delta\mathbf{q}_I(t)+\sum_{J=1}^{N}\tilde{\mathbf{p}}_J(\theta^\alpha)\delta\mathbf{b}_{IJ}\right) \tag{7.139}$$

or

$$\delta\mathbf{r}(\theta^\alpha,t)=\sum_{I\in\mathcal{N}}\Phi_I(\theta^\alpha)\left(\delta\mathbf{r}_I(t)+\sum_{J=1}^{N}\mathbf{p}_J(\theta^\alpha)\delta\mathbf{a}_{IJ}\right)+\sum_{I\in\mathcal{N}_c}\tilde{\Phi}_I(\theta^\alpha)\ S[f_I(\theta^\alpha)]\ \delta\mathbf{q}_I(t) \tag{7.140}$$

7.4. An immersed particle method for fluid-structure interaction

In this section, we extend the meshfree thin shell formulation to coupled problems involving fluid-structure interaction. Initially, the fluid may be on one or both sides of the structure. The structural domain is a thin shell and is denoted by Ω^S; its reference configuration is denoted by Ω_0^S. The subscript "0" denotes a reference state, which are here considered to be initial states for any variable. The fluid domain is denoted by Ω^F. The superscript "S" and "F" will be used to denote quantities related to the structure and the fluid, respectively. The boundaries of the structure and fluid are denoted by Γ^S and Γ^F, respectively. They are decomposed into two mutually exclusive sets: Γ_u where displacement boundary conditions $\bar{\mathbf{u}}$ are prescribed and Γ_t where tractions $\bar{\mathbf{t}}$ are applied. The structure contains cracks Γ_c with cohesive tractions $\mathbf{t}^{coh}$. Both the fluid and the structure are subject to a body load $\mathbf{b}$. We will denote by $\mathbf{P}^F$ and $\mathbf{P}^S$ the nominal stresses of the fluid and structure, respectively. The densities of the fluid and structure are denoted by ρ^F and ρ^S, respectively.

Both the fluid and the structure are discretized by the sets $\mathcal{S}^S$ and $\mathcal{S}^F$ of meshfree particles. The FSI constraint to be enforced is a point-wise version of the constraint

$$\mathbf{u}^S(\mathbf{X})=\mathbf{u}^F(\mathbf{X})\,,\forall\,\mathbf{X}\in\Omega_0^S, \tag{7.141}$$

where $\mathbf{u}^S$ and $\mathbf{u}^F$ are the displacement fields of the structure and the fluid, respectively. Constraint (7.141) enforces the no slip condition. We can allow

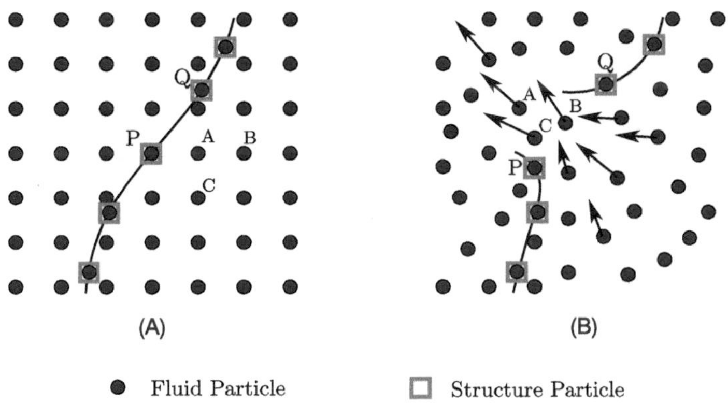

Figure 7.21 Illustration of the meshfree FSI model of a thin shell structure immersed within a fluid. (A) shows the initial undeformed configuration where fluid particles have been place at the location of all structure particles. (B) shows the configuration of the fluid and structure particle after the structure has fractured. Fluid particles are shown to flow through the crack [44].

slip of the fluid in the direction tangent to the surface of the structure by enforcing the alternative constraint

$$\mathbf{n}^S \cdot \left(\mathbf{u}^S(\mathbf{X}) - \mathbf{u}^F(\mathbf{X})\right) = 0, \forall \mathbf{X} \in \Omega_0^S, \tag{7.142}$$

where $\mathbf{n}^S$ is the normal to the structure.

The no slip constraint (7.141) is illustrated in Fig. 7.21. In the model shown, fluid particles have been placed at the locations of the structural particles – this is done for illustration purposes only and is not a requirement of the model. In the model described here, the fluid is allowed to flow between the surfaces of a crack. Initially the structure is unfractured, as shown in Fig. 7.21A. Suppose that the structure deforms and fractures (e.g. between structure particles P and Q). Once the structure has failed fluid particles (e.g. fluid particles $A - C$) can flow between the crack surfaces, as shown in Fig. 7.21B.

Most FSI methods are design for interactions with unfractured structures and require modification to deal effectively with the case of fluid flow through a crack. The FSI method described here treats FSI within a single framework for fractured and unfractured structures alike.

In the following, we assume a compressible fluid model. The fluid is assumed to be inviscid, since we are concerned with high pressure, impulsive loadings where viscous effects are insignificant; it is considered to be compressible and it is described via an equation of state (EOS). We assume that the shell structure is completely immersed in the fluid and has zero thickness. So when the no slip constraint (7.141) is imposed, the fluid displacement field is continuous through the thickness of the shell.

The discrete equation governing the FSI model are obtained from the Principle of Virtual Work: find $\mathbf{u}^F \in \mathcal{U}^F$ and $\mathbf{x}^S \in \mathcal{U}^S$ such that

$$\begin{aligned}&\int_{\Omega_0^F} \delta u_{i,j}^F P_{ji}^F d\Omega_0^F + \int_{\Omega_0^F} \delta u_i^F \rho_0^F \ddot{u}_i^F d\Omega_0^F - \int_{\Gamma_{0t}^F} \delta u_i^F \bar{\mathbf{t}}^F d\Gamma_0^F \\ &+ \int_{\Omega_0^S} \delta F_{ij}^S P_{ji}^S d\Omega_0^S + \int_{\Omega_0^S} \delta x_i^S \rho_0^S \ddot{u}_i^S d\Omega_0^S - \int_{\Gamma_{0t}^S} \delta x_i^S \bar{t}_i^S d\Gamma_0^S \\ &- \int_{\Gamma_{0c}} \delta x_i^S t_i^{coh} d\Gamma_{0c} = 0, \ \forall \, \delta u_i^S \in \mathcal{U}_0^S, \ \forall \, \delta u_i^F \in \mathcal{U}_0^F \end{aligned} \tag{7.143}$$

subject to the condition (7.141), where

$$\mathcal{U}^F = \{\mathbf{u} | \mathbf{u} \in \mathcal{H}^1(\Omega^F), \mathbf{u} = \bar{\mathbf{u}}^F \text{ on } \Gamma_{0u}^F\} \tag{7.144}$$

$$\mathcal{U}_0^F = \{\mathbf{u} | \mathbf{u} \in \mathcal{H}^1(\Omega^F), \mathbf{u} = 0 \text{ on } \Gamma_{0u}^F\} \tag{7.145}$$

$$\mathcal{U}^S = \{\mathbf{u} | \mathbf{u} \in \mathcal{H}^2(\Omega^S), \mathbf{u} \text{ is discontinuous on } \Gamma_{0c}, \ \mathbf{u} = \bar{\mathbf{u}}^S \text{ on } \Gamma_{0u}^S\} \tag{7.146}$$

$$\mathcal{U}_0^S = \{\mathbf{u} | \mathbf{u} \in \mathcal{H}^2(\Omega^S), \mathbf{u} \text{ is discontinuous on } \Gamma_{0c}, \ \mathbf{u} = 0 \text{ on } \Gamma_{0u}^S\} \tag{7.147}$$

Note that these definitions are consistent, i.e. the fluid flow can be continuous in the presence of the discontinuity in the structural motion, as can be seen approximately in the flow shown in Fig. 7.21.

We desired to satisfy (7.143) under the constraint (7.141) using a master-slave coupling technique. In our model the structure particles are slaves to the fluid particles. This means that the displacement of the structure will be driven by the motion of the fluid particles. The structure will then in turn contribute to the internal, external and kinematic forces of the fluid. It is therefore necessary to express the degrees of freedom of the structure in terms to those of the fluid.

The imposition of the constraint (7.141) is quite awkward because the MLS shape functions are not interpolatory, i.e. they do not satisfy the Kronecker delta property. A good approximation to the constraint (7.141) can be obtained by minimizing the following norm Π with respect to $\mathbf{u}_J^S$ and $\mathbf{q}_K^S$:

$$\Pi = \frac{1}{2} \sum_{I \in \mathcal{S}^S} \left(\mathbf{u}^S(\boldsymbol{\theta}_I^S) - \mathbf{u}^F(\mathbf{X}^S(\boldsymbol{\theta}_I^S))\right)^2$$

$$= \frac{1}{2} \sum_{I \in \mathcal{S}^S} \left(\sum_{J \in \mathcal{S}^S} N_{JI}^S \mathbf{u}_J^S + \sum_{K \in \mathcal{S}_{pum}^S} N_{KI}^S H_I \mathbf{q}_K^S - \sum_{L \in \mathcal{S}^{FSI}} N_{LI}^F \mathbf{u}_L^F \right)^2 \tag{7.148}$$

where $\boldsymbol{\theta}_I^S$ is the curvelinear coordinates of structural particle I, $\mathcal{S}^{FSI}$ is the set of all fluid particles with shape function supports containing at least one structural particle in Ω_0^S, $N_{JI}^S = N_J^S(\boldsymbol{\theta}_I^S)$, $N_{JI}^F = N_J^F(\mathbf{X}^S(\boldsymbol{\theta}_I^S))$ and $H_I = H(f(\boldsymbol{\theta}_I^S))$.

Minimization of Π with respect to $\mathbf{u}_J^S$ and $\mathbf{q}_K^S$ leads to the following system of equations

$$\begin{bmatrix} \overline{\mathbf{M}}^{uu} & \overline{\mathbf{M}}^{uq} \\ \overline{\mathbf{M}}^{uq\top} & \overline{\mathbf{M}}^{qq} \end{bmatrix} \begin{Bmatrix} \mathbf{d}_u^S \\ \mathbf{d}_q^S \end{Bmatrix} = \begin{bmatrix} \mathbf{D}^u \\ \mathbf{D}^q \end{bmatrix} \left\{ \mathbf{d}^{FSI} \right\} \tag{7.149}$$

where $\mathbf{d}_u^{S\top} = \{\mathbf{u}_1^S, \mathbf{u}_2^S, \mathbf{u}_{n^S}^S\}$, $\mathbf{d}_q^{S\top} = \{\mathbf{q}_1^S, \mathbf{q}_2^S, \mathbf{q}_{m^S}^S\}$ and n^S and m^S are the number of particles in sets $\mathcal{S}^S$ and $\mathcal{S}_{pum}^S$, respectively. The vector $\mathbf{d}^{FSI}$ is the vector of the degrees of freedom of the fluid particles in $\mathcal{S}^{FSI}$. The matrices are

$$\overline{M}_{JjKk}^{uu} = \sum_{I \in \mathcal{S}^S} N_{JI}^S N_{KI}^S \delta_{jk},\ J, K \in \mathcal{S}^S, \tag{7.150}$$

$$\overline{M}_{JjKk}^{uq} = \sum_{I \in \mathcal{S}^S} N_{JI}^S N_{KI}^S H_I \delta_{jk},\ J \in \mathcal{S}^S,\ K \in \mathcal{S}_{pum}^S, \tag{7.151}$$

$$\overline{M}_{JjKk}^{qq} = \sum_{I \in \mathcal{S}^S} N_{JI}^S N_{KI}^S H_I \delta_{jk},\ J, K \in \mathcal{S}_{pum}^S, \tag{7.152}$$

$$D_{JjKk}^u = \sum_{I \in \mathcal{S}^S} N_{JI}^S N_{KI}^F \delta_{jk},\ J \in \mathcal{S}^S,\ K \in \mathcal{S}^{FSI}. \tag{7.153}$$

$$D_{JjKk}^q = \sum_{I \in \mathcal{S}^S} N_{JI}^S H_I N_{KI}^F \delta_{jk},\ J \in \mathcal{S}_{pum}^S,\ K \in \mathcal{S}^{FSI}. \tag{7.154}$$

By solving Eq. (7.149), we can drive the displacements of the structure in terms of those of the fluid, i.e.

$$\mathbf{d}^S = \overline{\mathbf{M}}^{-1} \mathbf{D} \mathbf{d}^{FSI} = \mathbf{T} \mathbf{d}^{FSI} \tag{7.155}$$

where $\mathbf{d}^{S\top} = \{\mathbf{d}_u^S, \mathbf{d}_q^S\}$, $\mathbf{T} = \overline{\mathbf{M}}^{-1} \mathbf{D}$ is the coupling matrix,

$$\overline{\mathbf{M}} = \begin{bmatrix} \overline{\mathbf{M}}^{uu} & \overline{\mathbf{M}}^{uq} \\ \overline{\mathbf{M}}^{uq\top} & \overline{\mathbf{M}}^{qq} \end{bmatrix} \text{ and } \mathbf{D} = \begin{bmatrix} \mathbf{D}^u \\ \mathbf{D}^q \end{bmatrix}. \tag{7.156}$$

For convenience we decompose $\mathbf{T}$ as

$$\left\{\begin{array}{c}\mathbf{d}_u^S\\ \mathbf{d}_q^S\end{array}\right\}=\left[\begin{array}{c}\mathbf{T}^u\\ \mathbf{T}^q\end{array}\right]\left\{\mathbf{d}^{FSI}\right\}. \tag{7.157}$$

Each of the submatrices can be diagonalized by the row-sum technique without much loss of accuracy, so this is done here. The time derivatives of $\mathbf{d}^S$ and $\mathbf{d}^F$ are related by the same matrix $\mathbf{T}$, i.e.

$$\dot{\mathbf{d}}^S=\mathbf{T}\dot{\mathbf{d}}^{FSI} \tag{7.158}$$

$$\ddot{\mathbf{d}}^S=\mathbf{T}\ddot{\mathbf{d}}^{FSI} \tag{7.159}$$

The discrete equations are obtained by substituting the displacement approximations into (7.143) along with the corresponding test functions

$$\delta\mathbf{u}^F(\mathbf{X})=\sum_{I\in\mathcal{S}^F}N_I^F(\mathbf{X})\,\delta\mathbf{u}^F \tag{7.160}$$

$$\delta\mathbf{x}^S(\boldsymbol{\theta})=\sum_{I\in\mathcal{S}^S}N_I^S(\boldsymbol{\theta})\;\delta\mathbf{u}_I^S+\sum_{I\in\mathcal{S}^S_{pum}}N_I^S(\boldsymbol{\theta})\;H(f(\boldsymbol{\theta}))\;\delta\mathbf{q}_I^S$$
$$+\theta^3\frac{d}{2}\delta\mathbf{n}\left(\delta\mathbf{u}_I^S,\;\delta\mathbf{q}_I^S\right) \tag{7.161}$$

which gives

$$\sum_{I\in\mathcal{S}^F}\delta u_{Ii}^F\left(\sum_{J\in\mathcal{S}^F}M_{IJ}^F\ddot{u}_{Ji}^F-f_{Ii}^{F,int}+f_{Ii}^{F,ext}\right)$$
$$+\sum_{I\in\mathcal{S}^S}\delta u_{Ii}^S\left(\sum_{J\in\mathcal{S}^S}M_{IJ}^{S,uu}\ddot{u}_{Ji}^S+\sum_{K\in\mathcal{S}^S_{pum}}M_{IK}^{S,uq}\ddot{q}_{Ki}^S-f_{Ii}^{S,int}f_{Ii}^{S,ext}\right)$$
$$+\sum_{I\in\mathcal{S}^S_{pum}}\delta q_{Ii}^S\left(\sum_{J\in\mathcal{S}^S}M_{JI}^{S,uq}\ddot{u}_{Ji}^S+\sum_{K\in\mathcal{S}^S_{pum}}M_{IK}^{S,qq}\ddot{q}_{Ki}^S-\bar{f}_{Ii}^{S,int}+\bar{f}_{Ii}^{S,ext}\right)$$
$$=0\,,\forall\,\delta u_{Ii}^F,\;\forall\,\delta u_{Ii}^S,\;\forall\,\delta q_{Ii}^S \tag{7.162}$$

where the mass matrix and force vectors of the fluid are

$$M_{IJ}^F=\rho_0^F\int_{\Omega_0^F}N_I^FN_J^F\;d\Omega_0^F\,, \tag{7.163}$$

$$f_{Ii}^{F,int}=\int_{\Omega_0^F}\frac{\partial N_I^F}{\partial X_j}P_{ji}\left(\mathbf{u}^F\right)\;d\Omega_0^F\,,I\in\mathcal{S}^F \tag{7.164}$$

and

$$f_{Ii}^{F,ext} = \int_{\Omega_0^F} N_I^F b_i \, d\Omega_0^F + \int_{\Gamma_{0t}^F} N_I^F \bar{t}_i \, d\Gamma_0^F \, , I \in \mathcal{S}^F \, , \tag{7.165}$$

respectively.

The mass matrices $\mathbf{M}^{S,uu}$, $\mathbf{M}^{S,uq}$ and $\mathbf{M}^{S,qq}$ along with the force vectors $\mathbf{f}^{S,int}$, $\mathbf{f}^{S,ext}$, $\bar{\mathbf{f}}^{S,int}$, and $\bar{\mathbf{f}}^{S,ext}$ of the structure are defined in [41].

Eqs. (7.162) can be rewritten in matrix form as

$$\delta \mathbf{d}^{F^{\top}} \left(\mathbf{M}^F \ddot{\mathbf{d}}^F - \mathbf{f}^F \right) + \delta \mathbf{d}^{S^{\top}} \left(\mathbf{M}^S \ddot{\mathbf{d}}^S - \mathbf{f}^S \right) = 0 \tag{7.166}$$

where

$$\mathbf{f}^F = \mathbf{f}^{F,int} - \mathbf{f}^{F,ext} \, , \tag{7.167}$$

$$\mathbf{f}^S = \left\{ \begin{array}{c} \mathbf{f}^{S,int} \\ \bar{\mathbf{f}}^{S,int} \end{array} \right\} - \left\{ \begin{array}{c} \mathbf{f}^{S,ext} \\ \bar{\mathbf{f}}^{S,ext} \end{array} \right\} \tag{7.168}$$

and

$$\mathbf{M}^S = \left[\begin{array}{cc} \mathbf{M}^{S,uu} & \mathbf{M}^{S,uq} \\ \mathbf{M}^{S,uq^{\top}} & \mathbf{M}^{S,qq} \end{array} \right] . \tag{7.169}$$

To introduce the master-slave coupling between the structure and the fluid we substitute (7.157), (7.159) and

$$\delta \mathbf{u}_I^S = \sum_{J \in \mathcal{S}^{FSI}} T_{IJ}^u \delta \mathbf{u}_J^F \tag{7.170}$$

$$\delta \mathbf{q}_I^S = \sum_{J \in \mathcal{S}^{FSI}} T_{IJ}^q \delta \mathbf{u}_J^F \tag{7.171}$$

into (7.166) which gives

$$\delta \mathbf{d}^{F^{\top}} \left(\mathbf{M}^F \ddot{\mathbf{d}}^F - \mathbf{f}^F \right) + \delta \mathbf{d}^{FSI^{\top}} \mathbf{T}^{\top} \left(\mathbf{M}^S \mathbf{T} \ddot{\mathbf{d}}^{FSI} - \mathbf{f}^S \right) = 0 \, , \, \forall \, \delta \mathbf{d}^F \tag{7.172}$$

where we have made use of the fact that $\delta \mathbf{d}^{FSI} \subseteq \delta \mathbf{d}^F$. Invoking the arbitrariness of $\delta \mathbf{u}_I^F$ in (7.172) gives the coupled semi-discrete equations:

$$\left(\mathbf{M}^F + \mathbf{T}^{\top} \mathbf{M}^S \mathbf{T} \right) \ddot{\mathbf{d}}^F - \mathbf{f}^F - \mathbf{T}^{\top} \mathbf{f}^S = 0 \, . \tag{7.173}$$

Eqs. (7.173) are integrated in time using the central differencing scheme.

Remark. The coupling forces from the structure only effect a small subset of fluid particles, $\mathcal{S}^{FSI} \subset \mathcal{S}^F$.

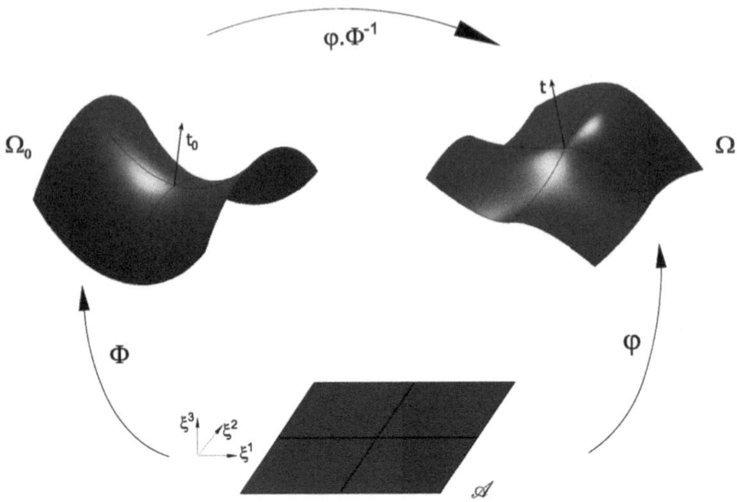

Figure 7.22 Shell geometry in the reference and the deformed configurations.

7.5. XIGA models for plates and shells

CAD basis functions such as NURBS are also higher order continuous and therefore ideally suited for thin shell analysis based on Kirchhoff-Love theory. The XIGA shell [36] presented subsequently is based on the thin-shell formulation from [34].

7.5.1 Kinematics of the shell

The deformation of a thin shell can be fully described by the deformation of its mid-surface, which is a two-dimensional manifold embedded in three-dimensional space. The mapping of the shell mid-surface is parametrized by curvilinear coordinates $\xi^1, \xi^2 \in \mathcal{A} \subset \mathbb{R}^2$ (see Fig. 7.22).

The displacement of the points of the shell mid-surface is defined by

$$\mathbf{u}\left(\xi^1, \xi^2\right) = \mathbf{x}\left(\xi^1, \xi^2\right) - \mathbf{X}\left(\xi^1, \xi^2\right) \tag{7.174}$$

where $\mathbf{X}$ is the position vector of a material point of the shell mid-surface in the reference configuration, and $\mathbf{x}$ is the position vector of the same point in the deformed geometry. The position vector of a material point in the reference geometry is given by

$$\mathbf{\Phi}\left(\xi^1, \xi^2, \xi^3\right) = \mathbf{X}\left(\xi^1, \xi^2\right) + \xi^3 \mathbf{G}_3\left(\xi^1, \xi^2\right) \ , \tag{7.175}$$

and similar for the deformed geometry

$$\boldsymbol{\varphi}\left(\xi^1,\xi^2,\xi^3\right)=\mathbf{x}\left(\xi^1,\xi^2\right)+\xi^3\mathbf{g}_3\left(\xi^1,\xi^2\right) , \tag{7.176}$$

where ξ^3 denotes the coordinate in thickness direction, $-0.5t \leqslant \xi^3 \leqslant 0.5t$, with t as the shell thickness.

The deformation gradient is given by

$$\mathbf{F}=\nabla\boldsymbol{\varphi}\cdot(\nabla\boldsymbol{\Phi})^{-1} \tag{7.177}$$

The covariant base vectors in the reference and current configurations are obtained from the derivatives of the respective position vector fields with respect to the curvilinear coordinates, as follows

$$\mathbf{G}_\alpha=\mathbf{X}_{,\alpha} \quad ; \qquad \mathbf{g}_\alpha=\mathbf{x}_{,\alpha} \tag{7.178}$$

where Greek indices take the values 1, 2 and refer to the parametric directions on the shell mid-surface, and $(\cdot)_{,\alpha}:=\partial(\cdot)/\partial\xi^\alpha$. The covariant metric coefficients of the surface are defined as

$$G_{\alpha\beta}=\mathbf{G}_\alpha\cdot\mathbf{G}_\beta \quad ; \qquad g_{\alpha\beta}=\mathbf{g}_\alpha\cdot\mathbf{g}_\beta , \tag{7.179}$$

and the contravariant base vectors are computed by

$$\mathbf{G}^\alpha=G^{\alpha\beta}\mathbf{G}_\beta \ ; \qquad G^{\alpha\beta}=\left[G_{\alpha\beta}\right]^{-1} . \tag{7.180}$$

The unit normal vectors of the middle surface are calculated by

$$\mathbf{G}_3=\frac{\mathbf{G}_1\times\mathbf{G}_2}{|\mathbf{G}_1\times\mathbf{G}_2|}=\mathbf{t}^0 \ ; \qquad \mathbf{g}_3=\frac{\mathbf{g}_1\times\mathbf{g}_2}{|\mathbf{g}_1\times\mathbf{g}_2|}=\mathbf{t} . \tag{7.181}$$

The Green-Lagrange strain tensor is given by

$$\mathbf{e}=\frac{1}{2}\left(\mathbf{F}^T\mathbf{F}-\mathbf{I}\right) , \tag{7.182}$$

where $\mathbf{I}$ is the identity tensor. The covariant strain components are given by

$$E_{\alpha\beta}=\varepsilon_{\alpha\beta}+\xi^3\kappa_{\alpha\beta} \tag{7.183}$$

The membrane strains $\varepsilon_{\alpha\beta}$ are given by

$$\varepsilon_{\alpha\beta}=\frac{1}{2}\left(\mathbf{g}_\alpha\cdot\mathbf{g}_\beta-\mathbf{G}_\alpha\cdot\mathbf{G}_\beta\right) \tag{7.184}$$

and the bending strains $\kappa_{\alpha\beta}$, measuring the changes in curvature due to bending, are

$$\kappa_{\alpha\beta} = \mathbf{g}_{\alpha} \cdot \mathbf{g}_{3,\beta} - \mathbf{G}_{\alpha} \cdot \mathbf{G}_{3,\beta} \tag{7.185}$$

Expanding Eq. (7.184) and Eq. (7.185) using Eq. (7.174) and omitting nonlinear terms, we obtain

$$\varepsilon_{\alpha\beta} = \frac{1}{2}\left(\mathbf{G}_{\alpha} \cdot \mathbf{u}_{,\beta} - \mathbf{G}_{\alpha} \cdot \mathbf{u}_{,\alpha}\right) \tag{7.186}$$

and

$$\begin{aligned}\kappa_{\alpha\beta} = {} & -\mathbf{u}_{,\alpha\beta} + \frac{1}{\sqrt{\bar{j}}}\left[\mathbf{u}_{,1} \cdot \left(\mathbf{G}_{\alpha,\beta} \times \mathbf{G}_2\right) + \mathbf{u}_{,2} \cdot \left(\mathbf{G}_1 \times \mathbf{G}_{\alpha,\beta}\right)\right] \\ & + \frac{\mathbf{G}_3 \cdot \mathbf{G}_{\alpha,\beta}}{\sqrt{\bar{j}}}\left[\mathbf{u}_{,1} \cdot (\mathbf{G}_2 \times \mathbf{G}_3) + \mathbf{u}_{,2} \cdot (\mathbf{G}_3 \times \mathbf{G}_1)\right]\end{aligned} \tag{7.187}$$

where $\bar{j} = |\mathbf{G}_1 \times \mathbf{G}_2|$.

7.5.2 Weak form

Applying the principle of virtual work, we seek $\mathbf{u} \in \vartheta$ such that

$$\delta\Pi = \delta\Pi_{\text{int}} + \delta\Pi_{ext} = 0 \qquad \forall \delta\mathbf{u} \in \vartheta_0 \tag{7.188}$$

in which

$$\begin{aligned}\vartheta &= \left\{\mathbf{u} | \mathbf{u} \in H^2\left(\Omega_0/\boldsymbol{\Gamma}_0^c\right), \mathbf{u} = \bar{\mathbf{u}} \text{ on } \boldsymbol{\Gamma}_0^u, \ \mathbf{u} \text{ discontinuous on } \boldsymbol{\Gamma}_0^c\right\} \\ \vartheta_0 &= \left\{\delta\mathbf{u} | \delta\mathbf{u} \in H^2\left(\Omega_0/\boldsymbol{\Gamma}_0^c\right), \ \delta\mathbf{u} = 0 \text{ on } \boldsymbol{\Gamma}_0^u, \ \delta\mathbf{u} \text{ discontinuous on } \boldsymbol{\Gamma}_0^c\right\}\end{aligned}$$

where Γ_0^c is the crack surface (or crack boundary) in the initial configuration; Γ_0^u and Γ_0^t (introduced later) are the partitions of the boundary where the Dirichlet and Neumann boundary conditions are enforced, respectively.

The internal virtual work can be expressed as

$$\delta\Pi_{\text{int}} = \int_{\Omega_0}\int_{-t/2}^{t/2} \boldsymbol{\sigma} : \left[\delta\mathbf{x}_{,\alpha} \otimes \mathbf{g}^{\alpha} + \delta\mathbf{t} \otimes \mathbf{g}^{3} + \xi^3 \delta\mathbf{t}_{,\alpha} \otimes \mathbf{g}^{\alpha}\right] \det(\nabla\Phi)\, d\xi^3 d\Omega_0 \tag{7.189}$$

where the prefix δ identifies the test function, and $\boldsymbol{\sigma}$ is the Cauchy stress tensor.

The external virtual work is given by

$$\delta\Pi_{\text{ext}} = \int_{\Omega_0/\Gamma_0^c} \rho_0 \mathbf{b}\cdot\delta\mathbf{u}\, d\Omega_0 + \int_{\Gamma_0^t} \bar{\mathbf{t}}_0\cdot\delta\mathbf{u}\, d\Gamma_0 \tag{7.190}$$

where ρ_0 is the initial density, $\mathbf{b}$ is the body force and $\bar{\mathbf{t}}_0$ is the prescribed traction.

Integrating the Cauchy stress tensor $\boldsymbol{\sigma}$ over the thickness yields

$$\begin{aligned} {}^{\alpha} &= \frac{1}{\bar{j}} \int_{-t/2}^{t/2} \boldsymbol{\sigma}\mathbf{g}^{\alpha} \det(\nabla\Phi)\, d\xi^3 \\ \mathbf{m}^{\alpha} &= \frac{1}{\bar{j}} \int_{-t/2}^{t/2} \xi^3 \boldsymbol{\sigma}\mathbf{g}^{\alpha} \det(\nabla\Phi)\, d\xi^3 \end{aligned} \tag{7.191}$$

where $^{\alpha}$ denotes the resultant membrane stress vector and $\mathbf{m}^{\alpha}$ the result bending stress vector.

Substituting Eqs. (7.191)-(7.191) into Eq. (7.189), we obtain

$$\begin{aligned} \delta\Pi_{\text{int}} &= \int_{\Omega_0} \left({}^{\alpha}\cdot\delta\mathbf{x}_{,\alpha} + \mathbf{m}^{\alpha}\cdot\delta\mathbf{t}_{,\alpha}\right) d\Omega_0 \\ &= \int_{\Omega_0} \left(\overbrace{{}^{\alpha\beta}\mathbf{x}_{,\beta}\cdot\delta\mathbf{u}_{,\alpha} + \mathbf{m}^{\alpha\beta}\mathbf{t}_{,\beta}\cdot\delta\mathbf{u}_{,\alpha}}^{{}^{\alpha}\cdot\delta\mathbf{x}_{,\alpha}} + \overbrace{\mathbf{m}^{\alpha\beta}\mathbf{x}_{,\beta}\cdot\delta\mathbf{t}_{\alpha}}^{\mathbf{m}^{\alpha}\cdot\delta\mathbf{t}_{,\alpha}} \right) d\Omega_0 \\ &= \int_{\Omega_0} \left(\tilde{n}^{\alpha\beta}\cdot\delta\varepsilon_{\alpha\beta} + \tilde{m}^{\alpha\beta}\cdot\delta\kappa_{\alpha\beta}\right) d\Omega_0 \end{aligned}$$

For a linear elastic material, the resultant effective membrane stress and bending stress can be expressed by

$$\tilde{n}^{\alpha\beta} = tD^{\alpha\beta\gamma\delta}\varepsilon_{\gamma\delta}\ ; \qquad \tilde{m}^{\alpha\beta} = \frac{t^3}{12} D^{\alpha\beta\gamma\delta}\kappa_{\gamma\delta} \tag{7.192}$$

$$\begin{aligned} D^{\alpha\beta\gamma\delta} = \frac{E}{1-\nu}\Big[&\nu(\mathbf{X}_{,\alpha}\mathbf{X}_{,\beta})(\mathbf{X}_{,\gamma}\mathbf{X}_{,\delta}) + \frac{1}{2}(1-\nu)(\mathbf{X}_{,\alpha}\mathbf{X}_{,\gamma})(\mathbf{X}_{,\delta}\mathbf{X}_{,\beta}) \\ &+ \frac{1}{2}(1-\nu)(\mathbf{X}_{,\alpha}\mathbf{X}_{,\delta})(\mathbf{X}_{,\gamma}\mathbf{X}_{,\beta})\Big] \end{aligned} \tag{7.193}$$

Using Voigt's notation one can write

$$\tilde{\mathbf{n}} = \begin{pmatrix} \tilde{n}_{11} \\ \tilde{n}_{22} \\ \tilde{n}_{12} \end{pmatrix} = t\tilde{\mathbf{D}}\boldsymbol{\varepsilon}\ ; \quad \tilde{\mathbf{m}} = \begin{pmatrix} \tilde{m}_{11} \\ \tilde{m}_{22} \\ \tilde{m}_{12} \end{pmatrix} = \frac{t^3}{12}\tilde{\mathbf{D}}\boldsymbol{\kappa}\ ;$$

$$\boldsymbol{\varepsilon} = \begin{pmatrix} \varepsilon_{11} \\ \varepsilon_{22} \\ 2\varepsilon_{12} \end{pmatrix} \quad ; \quad \boldsymbol{\kappa} = \begin{pmatrix} \kappa_{11} \\ \kappa_{22} \\ 2\kappa_{12} \end{pmatrix} \tag{7.194}$$

where $\tilde{\mathbf{D}}$ is the constitutive tensor of the isotropic material, E is the Young's modulus and ν is the Poisson's ratio.

$$\tilde{\mathbf{D}} = \frac{E}{1-\nu^2} \begin{bmatrix} \left(a_0^{11}\right)^2 & \nu a_0^{11} a_0^{22} + (1-\nu)\left(a_0^{12}\right)^2 & \cdots \\ \vdots & \left(a_0^{22}\right)^2 & \\ symm & \cdots & \end{bmatrix}$$

$$\cdots \begin{bmatrix} a_0^{11} a_0^{12} \\ a_0^{22} a_0^{12} \\ \frac{1}{2}\left[(1-\nu)\, a_0^{11} a_0^{22} - (1-\nu)\left(a_0^{12}\right)^2\right] \end{bmatrix} \tag{7.195}$$

where $a_0^{\alpha\beta} = \mathbf{X}_{,\alpha} \cdot \mathbf{X}_{,\beta}$. Correspondingly, Eq. (7.192) can be rewritten in a compact form as

$$\delta\Pi_{\text{int}} = \int_{\Omega_0} \left(\tilde{\boldsymbol{n}} \cdot \delta\boldsymbol{\varepsilon} + \tilde{\mathbf{m}} \cdot \delta\boldsymbol{\kappa}\right) d\Omega_0 \tag{7.196}$$

7.5.3 Discretization of the displacement field and enrichment

7.5.3.1 In-plane enrichment

The displacement and membrane stress components at a crack tip can be expressed in polar coordinates (r, θ) as follows:

$$\begin{Bmatrix} u_1 \\ u_2 \end{Bmatrix} = \frac{K_I}{2\mu}\sqrt{\frac{r}{2\pi}} \begin{Bmatrix} \cos\frac{\theta}{2}\left(2\frac{1-\nu}{1+\nu} + 2\sin^2\frac{\theta}{2}\right) \\ \sin\frac{\theta}{2}\left(\frac{4}{1+\nu} - 2\cos^2\frac{\theta}{2}\right) \end{Bmatrix} + \frac{K_{II}}{2\mu}\sqrt{\frac{r}{2\pi}} \begin{Bmatrix} \sin\frac{\theta}{2}\left(\frac{4}{1+\nu} + 2\cos^2\frac{\theta}{2}\right) \\ -\cos\frac{\theta}{2}\left(2\frac{1-\nu}{1+\nu} - 2\sin^2\frac{\theta}{2}\right) \end{Bmatrix} \tag{7.197}$$

$$\begin{Bmatrix} \sigma_{11} \\ \sigma_{12} \\ \sigma_{22} \end{Bmatrix} = \frac{K_I}{\sqrt{2\pi r}}\cos\frac{\theta}{2} \begin{Bmatrix} 1 - \sin\frac{\theta}{2}\sin\frac{3\theta}{2} \\ \sin\frac{\theta}{2}\cos\frac{3\theta}{2} \\ 1 + \sin\frac{\theta}{2}\sin\frac{3\theta}{2} \end{Bmatrix} + \frac{K_{II}}{\sqrt{2\pi r}}\cos\frac{\theta}{2} \begin{Bmatrix} \sin\frac{\theta}{2}\left(2 + \cos\frac{\theta}{2}\cos\frac{3\theta}{2}\right) \\ cos\frac{\theta}{2}\left(1 - \sin\frac{\theta}{2}\sin\frac{3\theta}{2}\right) \\ \sin\frac{\theta}{2}\cos\frac{\theta}{2}\cos\frac{3\theta}{2} \end{Bmatrix} \tag{7.198}$$

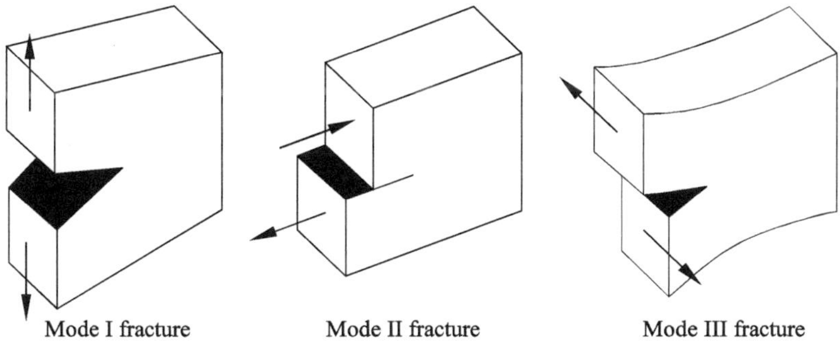

Figure 7.23 Fracture modes.

where $\mu = E/2(1+\nu)$ is the shear modulus, and K_I and K_{II} are membrane stress intensity factors (SIFs) of mode I and mode II fracture, respectively. The different fracture modes are illustrated in Fig. 7.23.

The membrane stress intensity factors are given by

$$K_I = \lim_{r \to 0} \sqrt{2\pi r}\,(\sigma_{\theta\theta}(r,0)) \quad ; \qquad K_{II} = \lim_{r \to \infty} \sqrt{2\pi r}\,(\sigma_{r\theta}(r,0)) \tag{7.199}$$

where $\sigma_{\theta\theta}$ and $\sigma_{r\theta}$ are stress components in polar coordinates.

It can be shown that the in-plane tip enrichment functions

$$\boldsymbol{\psi}(r,\theta) = \left\{ \sqrt{r}\sin\left(\frac{\theta}{2}\right), \sqrt{r}\cos\left(\frac{\theta}{2}\right), \sqrt{r}\sin\left(\frac{\theta}{2}\right)\sin(\theta), \sqrt{r}\cos\left(\frac{\theta}{2}\right)\sin(\theta) \right\} \tag{7.200}$$

represent the asymptotic near crack tip solution.

7.5.3.2 Out-of-plane enrichments

The internal stresses due to bending can be computed by

$$\begin{Bmatrix} \sigma_{rr} \\ \sigma_{r\theta} \\ \sigma_{\theta\theta} \end{Bmatrix} = \frac{k_1}{\sqrt{2r}} \frac{x_3}{2h} \frac{1}{3+\nu} \begin{Bmatrix} (3+5\nu)\cos\frac{\theta}{2} - (7+\nu)\cos\frac{3\theta}{2} \\ -(1-\nu)\sin\frac{\theta}{2} + (7+\nu)\sin\frac{3\theta}{2} \\ (5+3\nu)\cos\frac{\theta}{2} + (7+\nu)\cos\frac{3\theta}{2} \end{Bmatrix}$$
$$+ \frac{k_2}{\sqrt{2r}} \frac{x_3}{2h} \frac{1}{3+\nu} \begin{Bmatrix} -(3+5\nu)\sin\frac{\theta}{2} + (5+3\nu)\sin\frac{3\theta}{2} \\ (-1+\nu)\cos\frac{\theta}{2} + (5+3\nu)\cos\frac{3\theta}{2} \\ -(5+3\nu)\left(\sin\frac{\theta}{2} + \sin\frac{3\theta}{2}\right) \end{Bmatrix} \tag{7.201}$$

where x_3 is the out of plane component of the current coordinate vector. The out-of-plane displacement is given

$$u_3 = \frac{(2r)^{\frac{3}{2}}\left(1-\nu^2\right)}{2Eh(3+\nu)}\left\{k_1\left[\frac{1}{3}\left(\frac{7+\nu}{1-\nu}\right)\cos\frac{3\theta}{2} - \cos\frac{\theta}{2}\right]\right.$$
$$\left. + k_2\left[-\frac{1}{3}\left(\frac{5+3\nu}{1-\nu}\right)\sin\frac{3\theta}{2} + \sin\frac{\theta}{2}\right]\right\} \tag{7.202}$$

where k_1 and k_2 are the bending and twisting stress intensity factors:

$$k_1 = \lim_{r\to 0}\sqrt{2r}\sigma_{\theta\theta}\left(r, 0, \frac{h}{2}\right) \quad ; \quad k_2 = \lim_{r\to 0}\frac{3+\nu}{1+\nu}\sqrt{2r}\sigma_{r\theta}\left(r, 0, \frac{h}{2}\right) \tag{7.203}$$

The corresponding out of plane tip enrichment functions are given by

$$\mathbf{F}(r,\theta) = \left\{\sqrt{r}\sin\left(\frac{\theta}{2}\right), r^{\frac{3}{2}}\sin\left(\frac{\theta}{2}\right), r^{\frac{3}{2}}\cos\left(\frac{\theta}{2}\right), r^{\frac{3}{2}}\sin\left(\frac{3\theta}{2}\right), r^{\frac{3}{2}}\cos\left(\frac{3\theta}{2}\right)\right\} \tag{7.204}$$

7.5.3.3 Discretization of the displacement field

Similar to the enrichment strategy used in XFEM, the XIGA displacement field of the cracked shell surface can be expressed as

$$\left\{\begin{array}{c} u_1 \\ u_2 \end{array}\right\} = \sum_{I=1}^{NI} R^I \left\{\begin{array}{c} \hat{u}_1^I \\ \hat{u}_2^I \end{array}\right\} + \sum_{J=1}^{NJ} R^J\left(H(\mathbf{x}) - H\left(\mathbf{x}_J\right)\right)\left\{\begin{array}{c} \hat{a}_1^J \\ \hat{a}_2^J \end{array}\right\}$$
$$+ \sum_{K=1}^{NK} R^K \sum_{L=1}^{LK}\left(\psi_K^L(\mathbf{x}) - \psi_K^L(\mathbf{x}_K)\right)\left\{\begin{array}{c} \hat{b}_{1K}^L \\ \hat{b}_{2K}^L \end{array}\right\} \tag{7.205}$$

$$u_3 = \sum_{I=1}^{NI} R^I \hat{u}_3^I + \sum_{J=1}^{NJ} R^J\left(H(\mathbf{x}) - H\left(\mathbf{x}_J\right)\right)\hat{a}_3^J$$
$$+ \sum_{K=1}^{NK} R^K \sum_{L=1}^{LK}\left(F_K^L(\mathbf{x}) - F_K^L(\mathbf{x}_K)\right)\hat{b}_{3K}^L \tag{7.206}$$

where R are NURBS basis functions, $\hat{\mathbf{u}}_1$, $\hat{\mathbf{u}}_2$, $\hat{\mathbf{u}}_3$ are the standard DOFs, $\hat{\mathbf{a}}_1$, $\hat{\mathbf{a}}_2$, $\hat{\mathbf{a}}_3$ and $\hat{\mathbf{b}}_1$, $\hat{\mathbf{b}}_2$, $\hat{\mathbf{b}}_3$ are additional DOFs, $\boldsymbol{\psi}$ and $\mathbf{F}$ are the tip-enrichment functions, and H is Heaviside function. There are two different procedures for choosing the tip enriched area [9]: topological enrichment and geometrical enrichment. Topological enrichment refers to the strategy in which the size of the tip enriched area is proportional to the mesh size. By

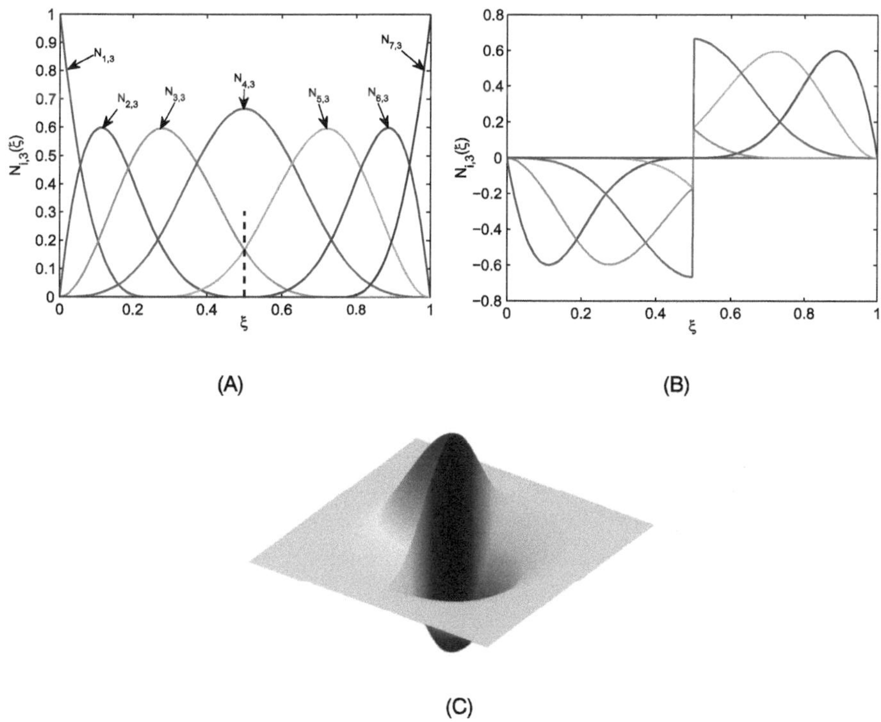

Figure 7.24 (A) Cubic B-spline basis functions with knot vector $\xi = [0, 0, 0, 0, 1/4, 1/2, 3/4, 1, 1, 1, 1]$. (B) Step-enriched basis functions for a crack at $\xi = 1/2$. (C) Step-enriched basis functions with $N_{4,3}$ in 2D.

contrast, in geometrical enrichment a constant tip enriched area independent of the mesh size is considered. In order to obtain optimal convergence rates in crack applications, it is crucial to use the geometrical enrichment. The quartic B-spline basis functions and the step-enriched shape functions affected by the crack are shown for a one-dimensional discretization in Fig. 7.24. Fig. 7.25 presents the enriched control points and the sub-triangulation needed for integration for a two-dimensional discretization.

7.5.3.4 Essential boundary conditions, numerical integration and compatibility enforcement

Imposition of essential boundary conditions: As NURBS do not satisfy the Kronecker Delta property, direct imposition of essential boundary condition on control points can result in significant errors with deteriorated rates of convergence. Consequently, specific techniques for imposing essen-

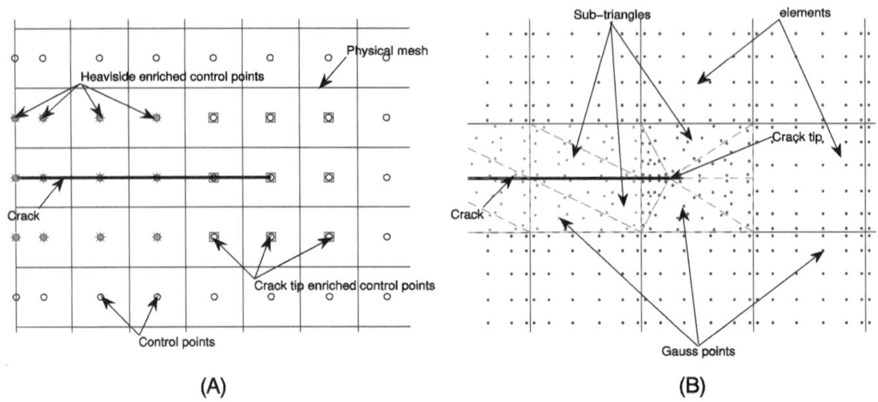

Figure 7.25 (A) Illustration of enriched control points for a quadratic NURBS mesh. (B) Sub-triangulation and quadrature points for step-enriched and tip-enriched nodes.

tial boundary conditions are needed. The penalty method, Lagrange multiplier method, Nitsche method (see [33,40] for a review and comparison of these techniques in the context of meshfree methods) and transformation method are among the available techniques. A least-squares minimization method for the weak imposition of essential boundary conditions in XIGA will be exploited. The main idea is to determine the boundary control point displacements $\boldsymbol{q}_i$ by minimizing a least-squares error function $\mathbb{F}$ on a set of interpolating points $\boldsymbol{x}_k$ defined on the essential boundary:

$$\mathbb{F} = \frac{1}{2}\sum_k \|\boldsymbol{u}(\boldsymbol{x}_k) - \bar{\boldsymbol{u}}(\boldsymbol{x}_k)\|^2 = \frac{1}{2}\sum_k \left\| \sum_i R_i(\xi_k)\boldsymbol{q}_i - \bar{\boldsymbol{u}}(\boldsymbol{x}_k) \right\|^2 \tag{7.207}$$

where $\boldsymbol{u}$ and $\bar{\boldsymbol{u}}$ are approximated and prescribed displacement fields, respectively. Minimizing $\mathbb{F}$ with respect to $\boldsymbol{q}$ gives,

$$\frac{\partial \mathbb{F}}{\partial \boldsymbol{q}_j} = 0 \implies \sum_k \left(\sum_i R_i(\xi_k)\boldsymbol{q}_i - \bar{\boldsymbol{u}}(\boldsymbol{x}_k) \right) R_j(\xi_k) = \boldsymbol{0} \tag{7.208}$$

which results in following linear system of equations:

$$\mathbf{A}\mathbf{q} = \boldsymbol{C}$$

$$\mathbf{A} = \sum_k \begin{pmatrix} R_1(\boldsymbol{\xi}_k)R_1(\boldsymbol{\xi}_k) & \cdots & R_1(\boldsymbol{\xi}_k)R_{n_c}(\boldsymbol{\xi}_k) \\ \vdots & \ddots & \vdots \\ R_{n_c}(\boldsymbol{\xi}_k)R_1(\boldsymbol{\xi}_k) & \cdots & R_{n_c}(\boldsymbol{\xi}_k)R_{n_c}(\boldsymbol{\xi}_k) \end{pmatrix};$$

$$\mathbf{C}=\sum_{k}\begin{pmatrix}\bar{u}_x(\boldsymbol{x}_k)R_1(\boldsymbol{\xi}_k) & \bar{u}_y(\boldsymbol{x}_k)R_1(\boldsymbol{\xi}_k)\\ \vdots & \vdots\\ \bar{u}_x(\boldsymbol{x}_k)R_{n_c}(\boldsymbol{\xi}_k) & \bar{u}_y(\boldsymbol{x}_k)R_{n_c}(\boldsymbol{\xi}_k)\end{pmatrix};$$

$$\mathbf{q}=\begin{pmatrix}q_1^x & q_1^y\\ \vdots & \vdots\\ q_{n_c}^x & q_{n_c}^y\end{pmatrix} \tag{7.209}$$

where n_c is the number of control points located on the essential boundary.

Numerical integration: In crack modeling, there are two types of elements where the standard Gauss integration could lead to inaccurate results and therefore special integration strategies are needed. For elements completely cut by the crack, partitioning the cut element into triangular, rectangular or polygonal sub-cells and employing standard Gauss quadrature in each sub-cell will significantly increase the accuracy. For elements containing the crack tip, an "almost polar integration" technique proposed in [28] can cancel the $r^{-1/2}$ singularity at the crack tip and give very good results for high-order NURBS. An alternative to the almost polar integration is the strain smoothing technique that does not require the integration of the singular terms since no derivatives of the shape functions need to be computed.

Compatibility enforcement: In order to get optimal convergence rates in XIGA for problems in linear fracture mechanics, a special treatment of blending elements is essential. Blending elements (partially enriched elements) in XFEM are posing two major problems: (1) Due to lack of partition of unity property in these elements, enrichment functions cannot be reproduced exactly. (2) Some undesirable terms are introduced in the approximations which cannot be compensated by the standard FE part of approximation. These problems can severely decrease the accuracy and rate of convergence in XFEM [28]. Two common techniques to ensure the compatibility between two different enriched sub-domains and increase the convergence rate are shifting and blending. For the shifting technique, two different strategies called *basic shifting* and *improved shifting* were discussed. Super optimal convergence rates for high-order NURBS have been obtained by employing an improved shifting technique for enrichment functions and adding C^0 lines along the boundaries of fully enriched and un-enriched sub-domains. These C^0 lines are introduced by knot insertion. However, C^1 continuity required for thin shell analysis no longer holds. See Fig. 7.26.

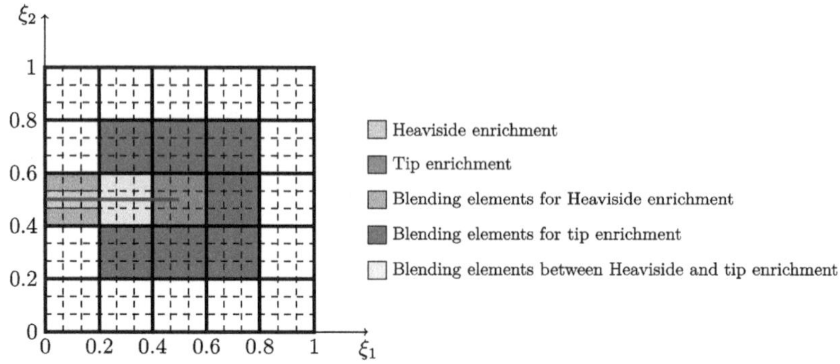

Figure 7.26 Enrichment strategy. From N. Nguyen-Thanh et al. An extended isogeometric thin shell analysis based on Kirchhoff-Love theory, Computer Methods in Applied Mechanics and Engineering, Volume 284, 1 February 2015, Pages 265-291, Originally adapted from E. De Luycker, D.J. Benson, T. Belytschko, Y. Bazilevs, M.C. Hsu, X-FEM in isogeometric analysis for linear fracture mechanics, Internat. J. Numer. Methods Engrg., 87 (2011), pp. 541-565.

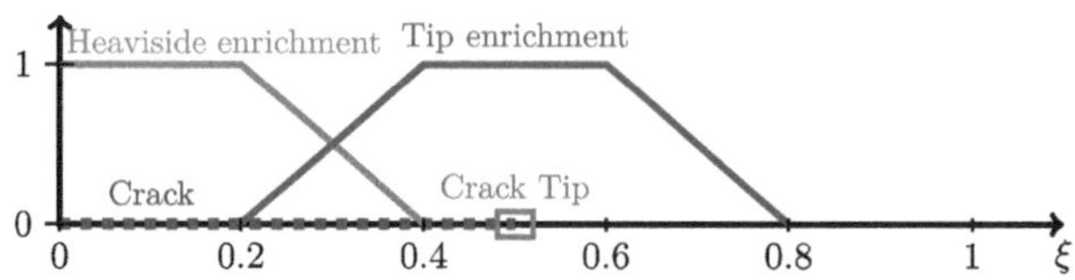

Figure 7.27 Heaviside blending function $B^{\mathcal{H}}$ and tip enrichment blending function $B^{\mathcal{T}}$.

For a two-dimensional continuum, the convergence rates can be further improved by the use of blending functions $B_{\mathcal{H}}$ and $B_{\mathcal{T}}$ for the Heaviside and tip enrichment functions, respectively. The XIGA approximation in this 2D continuum formulation is given as:

$$\left\{ \begin{array}{c} u_1 \\ u_2 \end{array} \right\} = \sum_{I=1}^{NI} R^I \left\{ \begin{array}{c} \hat{u}_1^I \\ \hat{u}_2^I \end{array} \right\} + B_{\mathcal{H}} \sum_{J=1}^{NJ} R^J \left(H(\mathbf{x}) - \chi^J \right) \left\{ \begin{array}{c} \hat{a}_1^J \\ \hat{a}_2^J \end{array} \right\}$$
$$+ B_{\mathcal{T}} \left\{ \begin{array}{c} \hat{b}_1 \\ \hat{b}_2 \end{array} \right\} \sum_{K=1}^{NK} R^K \left(\Psi(\mathbf{x}) - \chi^K \right) \tag{7.210}$$

where $\chi^J = H(\mathbf{x}_J)$ and χ^K is obtained by a least-squares minimization of $\left\| \Psi(\mathbf{x}) - \sum_K R^K \chi^K \right\|$. Fig. 7.27 represents a 1D illustration of blending functions.

Remark. The tip enrichment function used here is the first term of the displacement expansion near the crack tip for a pure mode I crack with unit stress-intensity factor.

For the blending technique, a projected blending function which is compatible with high-order NURBS functions can give the best results.

The displacement approximation used in the projected blending strategy is given as

$$\left\{\begin{array}{c} u_1 \\ u_2 \end{array}\right\} = \sum_{I=1}^{NI} R^I \left\{\begin{array}{c} \hat{u}_1^I \\ \hat{u}_2^I \end{array}\right\} + \sum_{J=1}^{NJ} R^J B_{\mathcal{H}}^J H(\boldsymbol{x}) \left\{\begin{array}{c} \hat{a}_1^J \\ \hat{a}_2^J \end{array}\right\} + \sum_{K=1}^{NK} R^K B_{\mathcal{T}}^K \Psi(\boldsymbol{x}) \left\{\begin{array}{c} \hat{b}_1^K \\ \hat{b}_2^K \end{array}\right\} \tag{7.211}$$

where $B_{\mathcal{H}}^J$ and $B_{\mathcal{T}}^K$ being projected blending coefficients associated with the Heaviside and tip enrichment, respectively. These coefficients are obtained by projecting a bilinear blending function on the NURBS basis functions in a way that $\sum_J R^J(\boldsymbol{\xi}) B_{\mathcal{H}}^J$ and $\sum_K R^K(\boldsymbol{\xi}) B_{\mathcal{T}}^K$ closely approximates the blending functions $B_{\mathcal{H}}$ and $B_{\mathcal{T}}$.

7.5.4 Discrete system of equations

Substituting the discretization of the trial functions, Eqs. (7.205), (7.206), their derivatives, as well as the test functions which have a similar structure to the trial functions into the weak form, Eq. (7.188), the following linear system of equations is obtained:

$$\mathbf{K}\mathbf{u} = \mathbf{f} \tag{7.212}$$

with the global stiffness matrix

$$\mathbf{K}_{ij} = \int_A \left(t \left(\mathbf{B}_i^m\right)^T \mathbf{D} \mathbf{B}_j^m + \frac{t^3}{12} \left(\mathbf{B}_i^b\right)^T \mathbf{D} \mathbf{B}_j^b \right) \bar{j} \mathrm{d}\xi^1 \mathrm{d}\xi^2 \tag{7.213}$$

in which $\mathbf{B}^m$ and $\mathbf{B}^b$ are the membrane and bending-strain matrices, respectively.

The standard membrane and bending strain matrices associated with node I are given by

$$\mathbf{B}_I^m = \begin{bmatrix} R_{,1}^I \mathbf{G}_1 \cdot \mathbf{e}_1 & R_{,1}^I \mathbf{G}_1 \cdot \mathbf{e}_2 & R_{,1}^I \mathbf{G}_1 \cdot \mathbf{e}_3 \\ R_{,2}^I \mathbf{G}_1 \cdot \mathbf{e}_1 & R_{,2}^I \mathbf{G}_1 \cdot \mathbf{e}_2 & R_{,2}^I \mathbf{G}_1 \cdot \mathbf{e}_3 \\ \left(R_{,1}^I \mathbf{G}_2 + R_{,2}^I \mathbf{G}_1\right) \cdot \mathbf{e}_1 & \left(R_{,1}^I \mathbf{G}_2 + R_{,2}^I \mathbf{G}_1\right) \cdot \mathbf{e}_2 & \left(R_{,1}^I \mathbf{G}_2 + R_{,2}^I \mathbf{G}_1\right) \cdot \mathbf{e}_3 \end{bmatrix} \tag{7.214}$$

and

$$\mathbf{B}_I^b = \begin{bmatrix} \mathbf{B}_1^{bI} \cdot \mathbf{e}_1 & \mathbf{B}_1^{bI} \cdot \mathbf{e}_2 & \mathbf{B}_1^{bI} \cdot \mathbf{e}_3 \\ \mathbf{B}_2^{bI} \cdot \mathbf{e}_1 & \mathbf{B}_2^{bI} \cdot \mathbf{e}_2 & \mathbf{B}_2^{bI} \cdot \mathbf{e}_3 \\ \mathbf{B}_3^{bI} \cdot \mathbf{e}_1 & \mathbf{B}_3^{bI} \cdot \mathbf{e}_2 & \mathbf{B}_2^{bI} \cdot \mathbf{e}_3 \end{bmatrix}. \tag{7.215}$$

The terms $\mathbf{e}_1$, $\mathbf{e}_2$, $\mathbf{e}_3$ are the global basis vectors of an orthonormal coordinate reference frame, and

$$\mathbf{B}_1^{bI} = -R^I_{,11}\mathbf{G}_3 + \frac{1}{\sqrt{\bar{j}}}\left[R^I_{,1}\mathbf{G}_{1,1} \times \mathbf{G}_2 + R^I_{,2}\mathbf{G}_1 \times \mathbf{G}_{1,1}\right] + \frac{\mathbf{G}_3 \cdot \mathbf{G}_{1,1}}{\sqrt{\bar{j}}}\left[R^I_{,1}\mathbf{G}_2 \times \mathbf{G}_3 + R^I_{,2}\mathbf{G}_3 \times \mathbf{G}_1\right] \tag{7.216}$$

$$\mathbf{B}_2^{bI} = -R^I_{,22}\mathbf{G}_3 + \frac{1}{\sqrt{\bar{j}}}\left[R^I_{,1}\mathbf{G}_{2,2} \times \mathbf{G}_2 + R^I_{,2}\mathbf{G}_1 \times \mathbf{G}_{2,2}\right] + \frac{\mathbf{G}_3 \cdot \mathbf{G}_{2,2}}{\sqrt{\bar{j}}}\left[R^I_{,1}\mathbf{G}_2 \times \mathbf{G}_3 + R^I_{,2}\mathbf{G}_3 \times \mathbf{G}_1\right] \tag{7.217}$$

$$\mathbf{B}_3^{bI} = -R^I_{,12}\mathbf{G}_3 + \frac{1}{\sqrt{\bar{j}}}\left[R^I_{,1}\mathbf{G}_{1,2} \times \mathbf{G}_2 + R^I_{,2}\mathbf{G}_1 \times \mathbf{G}_{1,2}\right] + \frac{\mathbf{G}_3 \cdot \mathbf{G}_{1,2}}{\sqrt{\bar{j}}}\left[R^I_{,1}\mathbf{G}_2 \times \mathbf{G}_3 + R^I_{,2}\mathbf{G}_3 \times \mathbf{G}_1\right] \tag{7.218}$$

The force contribution of the ith node is

$$\mathbf{f}_i = \int_{\mathcal{A}} \mathbf{b}R_i\bar{j}\mathrm{d}\xi^1\mathrm{d}\xi^2 + \int_{\partial\mathcal{A}} \mathbf{p}R_i\|X_{,t}\|\mathrm{d}l_\xi \tag{7.219}$$

where $\mathbf{p}$ are the forces per unit length on the boundary of the middle surface and dl_ξ is the line element of the boundary of the middle surface.

7.5.4.1 Membrane B-matrix for XIGA

The $\mathbf{B}^m_{\mathcal{H}}$ associated to the Heaviside enrichment is given by

$$\mathbf{B}_{\mathcal{H}}^{mJ} = \begin{bmatrix} \mathbf{B}_{\mathcal{H}_1}^{mJ} \cdot \mathbf{e}_1 & \mathbf{B}_{\mathcal{H}_1}^{mJ} \cdot \mathbf{e}_2 & \mathbf{B}_{\mathcal{H}_1}^{mJ} \cdot \mathbf{e}_3 \\ \mathbf{B}_{\mathcal{H}_2}^{mJ} \cdot \mathbf{e}_1 & \mathbf{B}_{\mathcal{H}_2}^{mJ} \cdot \mathbf{e}_2 & \mathbf{B}_{\mathcal{H}2}^{mJ} \cdot \mathbf{e}_3 \\ \mathbf{B}_{\mathcal{H}_3}^{mJ} \cdot \mathbf{e}_1 & \mathbf{B}_{\mathcal{H}_2}^{mJ} \cdot \mathbf{e}_2 & \mathbf{B}_{\mathcal{H}_3}^{mJ} \cdot \mathbf{e}_3 \end{bmatrix} \tag{7.220}$$

where

$$\begin{aligned}
\mathbf{B}^{mJ}_{\mathcal{H}_1} &= R^J_{,1}\left(H(\boldsymbol{x}) - H(\boldsymbol{x}_J)\right)\mathbf{G}_1 \\
\mathbf{B}^{mJ}_{\mathcal{H}_2} &= R^J_{,2}\left(H(\boldsymbol{x}) - H(\boldsymbol{x}_J)\right)\mathbf{G}_2 \\
\mathbf{B}^{mJ}_{\mathcal{H}_3} &= \left(R^J_{,1}\mathbf{G}_2 + R^J_{,2}\mathbf{G}_1\right)\left(H(\boldsymbol{x}) - H(\boldsymbol{x}_J)\right)
\end{aligned} \tag{7.221}$$

The $\mathbf{B}^m_{\mathcal{T}}$ associated to the tip enrichment is

$$\mathbf{B}^{mKL}_{\mathcal{T}} = \begin{bmatrix} \mathbf{B}^{mKL}_{\mathcal{T}_1}\cdot\mathbf{e}_1 & \mathbf{B}^{mKL}_{\mathcal{T}_1}\cdot\mathbf{e}_2 & \mathbf{B}^{mKL}_{\mathcal{T}_1}\cdot\mathbf{e}_3 \\ \mathbf{B}^{mKL}_{\mathcal{T}_2}\cdot\mathbf{e}_1 & \mathbf{B}^{mKL}_{\mathcal{T}_2}\cdot\mathbf{e}_2 & \mathbf{B}^{mKL}_{\mathcal{T}_2}\cdot\mathbf{e}_3 \\ \mathbf{B}^{mKL}_{\mathcal{T}_3}\cdot\mathbf{e}_1 & \mathbf{B}^{mKL}_{\mathcal{T}_3}\cdot\mathbf{e}_2 & \mathbf{B}^{mKL}_{\mathcal{T}_3}\cdot\mathbf{e}_3 \end{bmatrix} \tag{7.222}$$

where

$$\begin{aligned}
\mathbf{B}^{mKL}_{\mathcal{T}_1} &= \left[R^K_{,1}\left(\psi^L_K(\xi) - \psi^L_K(\xi_K)\right) + R^K\psi^L_{K,1}(\xi)\right]\mathbf{G}_1 \\
\mathbf{B}^{mKL}_{\mathcal{T}_2} &= \left[R^K_{,2}\left(\psi^L_K(\xi) - \psi^L_K(\xi_K)\right) + R^K\psi^L_{K,2}(\xi)\right]\mathbf{G}_2 \\
\mathbf{B}^{mKL}_{\mathcal{T}_3} &= \left(R^K_{,1}\mathbf{G}_2 + R^K_{,2}\mathbf{G}_1\right)\left(\psi^L_K(\xi) - \psi^L_K(\xi_K)\right) \\
&\quad + R^K\left(\psi^L_{K,1}(\xi)\mathbf{G}_2 + \psi^L_{K,2}(\xi)\mathbf{G}_1\right)
\end{aligned} \tag{7.223}$$

7.5.4.2 Bending B-matrix for XIGA

The $\mathbf{B}^{bJ}_{\mathcal{H}}$-matrix associated with the Heaviside-enriched DOFs is given by

$$\mathbf{B}^{bJ}_{\mathcal{H}} = \left(H(\xi) - H(\xi^J)\right)\begin{bmatrix} \mathbf{B}^{bJ}_{\mathcal{H}_1}\cdot\mathbf{e}_1 & \mathbf{B}^{bJ}_{\mathcal{H}_1}\cdot\mathbf{e}_2 & \mathbf{B}^{bJ}_{\mathcal{H}_1}\cdot\mathbf{e}_3 \\ \mathbf{B}^{bJ}_{\mathcal{H}_2}\cdot\mathbf{e}_1 & \mathbf{B}^{bJ}_{\mathcal{H}_2}\cdot\mathbf{e}_2 & \mathbf{B}^{bJ}_{\mathcal{H}_2}\cdot\mathbf{e}_3 \\ \mathbf{B}^{bJ}_{\mathcal{H}_3}\cdot\mathbf{e}_1 & \mathbf{B}^{bJ}_{\mathcal{H}_3}\cdot\mathbf{e}_2 & \mathbf{B}^{bJ}_{\mathcal{H}_3}\cdot\mathbf{e}_3 \end{bmatrix} \tag{7.224}$$

where $\mathbf{B}^{bJ}_{\mathcal{H}_1}$, $\mathbf{B}^{bJ}_{\mathcal{H}_2}$, $\mathbf{B}^{bJ}_{\mathcal{H}_3}$ are the same as in the standard part.

The contribution of the crack tip enrichment is calculated by

$$\mathbf{B}^{bKL}_{\mathcal{T}} = \begin{bmatrix} \mathbf{B}^{bKL}_{1}\cdot\mathbf{e}_1 & \mathbf{B}^{bKL}_{\mathcal{T}_1}\cdot\mathbf{e}_2 & \mathbf{B}^{bKL}_{\mathcal{T}_1}\cdot\mathbf{e}_3 \\ \mathbf{B}^{bKL}_{\mathcal{T}_2}\cdot\mathbf{e}_1 & \mathbf{B}^{bKL}_{\mathcal{T}_2}\cdot\mathbf{e}_2 & \mathbf{B}^{bKL}_{2}\cdot\mathbf{e}_3 \\ \mathbf{B}^{bKL}_{\mathcal{T}_3}\cdot\mathbf{e}_1 & \mathbf{B}^{bKL}_{\mathcal{T}_3}\cdot\mathbf{e}_2 & \mathbf{B}^{bKL}_{\mathcal{T}_3}\cdot\mathbf{e}_3 \end{bmatrix} \tag{7.225}$$

where $\mathbf{B}^{bKL}_{\mathcal{T}_1}$, $\mathbf{B}^{bKL}_{\mathcal{T}_2}$, $\mathbf{B}^{bKL}_{\mathcal{T}_3}$ are similar to the standard bending terms, but replacing $R^I_{,11}$, $R^I_{,22}$, $R^I_{,12}$, $R^I_{,1}$, $R^I_{,2}$ by the following terms

$$\begin{aligned}
R^I_{,11} &\rightarrow R^K_{,11}\left(\psi^L_K(\xi) - \psi^L_K(\xi_K)\right) + 2R^I_{,1}\psi^L_{K,1}(\xi) + R^I\psi^L_{K,11}(\xi) \\
R^I_{,22} &\rightarrow R^K_{,22}\left(\psi^L_K(\xi) - \psi^L_K(\xi_K)\right) + 2R^I_{,2}\psi^L_{K,2}(\xi) + R^I\psi^L_{K,22}(\xi)
\end{aligned}$$

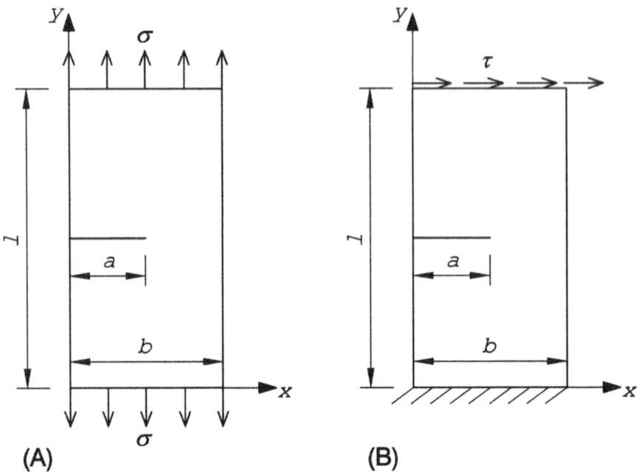

Figure 7.28 Geometry of edge cracked plates under: (A) the tension, (B) the shear. The dimensions are $b=7$, $a=3.5$, and $l=16$.

$$
\begin{aligned}
&R^I_{,12} \rightarrow R^K_{,12}\left(\psi^L_K(\xi)-\psi^L_K(\xi_K)\right)+R^I_{,1}\psi^L_{K,2}(\xi)+R^I_{,2}\psi^L_{K,1}(\xi)+R^I\psi^L_{K,12}(\xi)\\
&R^I_{,1} \rightarrow R^K_{,1}\left(\psi^L_K(\xi)-\psi^L_K(\xi_K)\right)+R^I\psi^L_{K,1}(\xi)\\
&R^I_{,2} \rightarrow R^K_{,2}\left(\psi^L_K(\xi)-\psi^L_K(\xi_K)\right)+R^I\psi^L_{K,2}(\xi).
\end{aligned}
\tag{7.226}
$$

7.5.5 Edge cracked plates under tension or shear

Consider a plane stress plate of width $b=7$, height $l=16$, and thickness $h=1$ with an edge crack length of $a=b/2$. The material properties are $E=3\times 10^7$, and $\nu=0.25$. The plate is subjected to a tension $\sigma=1$ at the top and bottom edges as shown in Fig. 7.28A, or is clamped onto the bottom edge and sustains a shear $\tau=1$ on the top edge (see Fig. 7.28B).

The analytical solution for the plate under tension is given in [31]

$$
K_I = C\sigma\sqrt{\pi a} \tag{7.227}
$$

with $C=1.12-0.231(a/b)+10.55(a/b)^2-21.72(a/b)^3-30.39(a/b)^4$.

The values of K_I and K_{II} for the shear case in [51] are used as the reference solutions:

$$
K_I = 34.0, \qquad K_{II} = 4.55 \tag{7.228}
$$

7.5.5.1 Results of the phantom node MITC3 elements

The plate is modeled by various structured meshes ($n_x \times n_y$) defined as number of elements along x and y axis (see Fig. 7.28), respectively. These are

Table 7.2 Normalized K_I of the tension case for various meshes and integral domain sizes.

Method	**Phantom-node**			**XFEM**		
r_d/h_{avg}	mesh1	mesh2	mesh3	mesh1	mesh2	mesh3
1.5	0.872	0.953	0.959	1.045	0.972	0.958
2.0	0.882	0.953	0.960	1.104	1.006	0.991
2.5	0.878	0.953	0.960	1.114	1.002	0.988
3.0	0.879	0.953	0.961	1.113	1.003	0.989

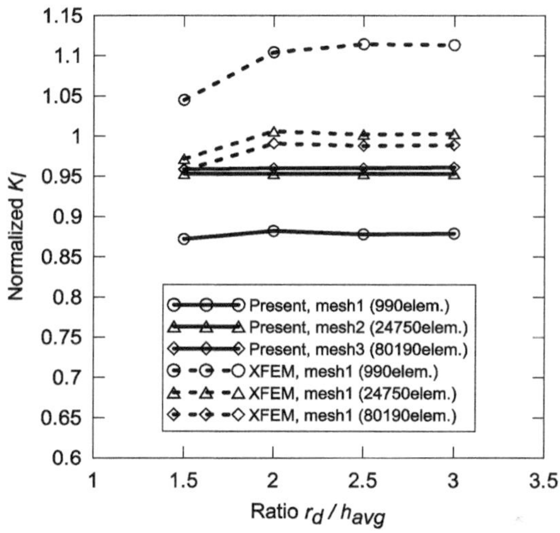

Figure 7.29 Normalized K_I for the edge cracked plate under tension with different meshes and J-integral domain sizes: phantom-node method (continuous lines) and XFEM (dash lines).

three structured meshes of (15 × 32), (75 × 165), and (135 × 297) MITC3 shell elements. In the following tables and figures, these meshes are denoted as mesh1, mesh2, and mesh3. The numerical results normalized by the reference solutions are given in Table 7.2 and Fig. 7.29 for the tension, and Tables 7.3 and 7.4 and Figs. 7.30 and 7.31 for the shear. The structure mesh and displacement fields are illustrated in Fig. 7.32.

The results show that as mesh density increases the stress intensity factors approach the reference solutions and are mostly independent of the domain sizes of the interaction integral. Reasonable values of r_d/h_{avg} range from 2.0 to 3.0, where r_d and h_{avg} were defined in Section 7.2.2.5. We also present the SIFs computed by the XFEM using three-node triangular elements with the asymptotic enrichment functions at the tip elements in

Table 7.3 Normalized K_I of the shear case for various meshes and integral domain sizes.

Method	Phantom-node			XFEM		
r_d/h_{avg}	mesh1	mesh2	mesh3	mesh1	mesh2	mesh3
1.5	0.799	0.962	0.986	0.989	0.986	0.984
2.0	0.812	0.967	0.985	1.039	1.021	1.021
2.5	0.808	0.966	0.984	1.047	1.019	1.017
3.0	0.809	0.966	0.985	1.046	1.019	1.018

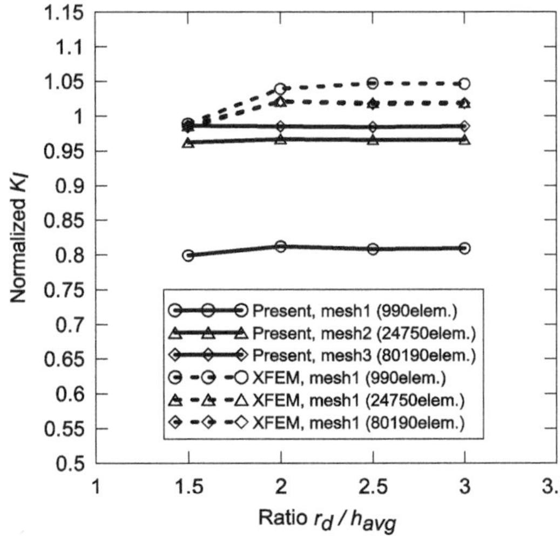

Figure 7.30 Normalized K_I for the edge cracked plate under shear with different meshes and J-integral domain sizes: phantom-node method (continuous lines) and XFEM (dash lines).

Table 7.4 Normalized K_{II} of the shear case for various meshes and integral domain sizes.

Method	Phantom-node			XFEM		
r_d/h_{avg}	mesh1	mesh2	mesh3	mesh1	mesh2	mesh3
1.5	0.803	1.004	1.051	0.653	0.990	1.104
2.0	0.827	0.975	1.031	0.901	1.000	1.034
2.5	0.846	0.977	1.014	0.886	0.985	1.012
3.0	0.847	0.979	1.013	0.898	0.988	1.013

the plane stress condition [53]. Compared to the XFEM, the phantom-node method requires fine mesh to obtain acceptable results. However, the asymptotic enrichment functions are known for this kind of problem. For other more complex problems where the solution is not known, asymp-

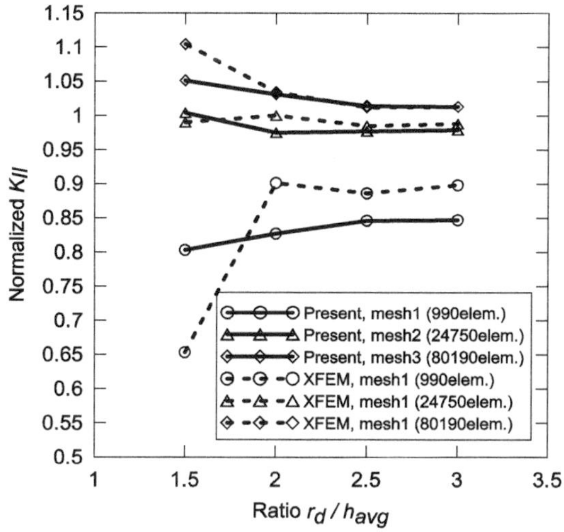

Figure 7.31 Normalized K_{II} for the edge cracked plate under shear with different meshes and J-integral domain sizes: phantom-node method (continuous lines) and XFEM (dash lines).

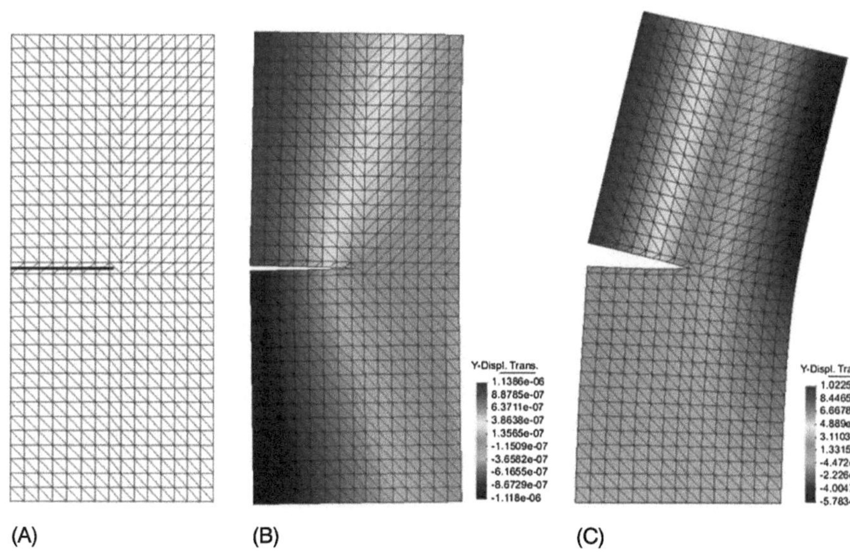

Figure 7.32 (A) Structured mesh with (15 × 32) MITC3 shell elements for the edge cracked plate; Vertical displacement field: (B) under the tension, (C) under the shear. The deformation factor is 10^5.

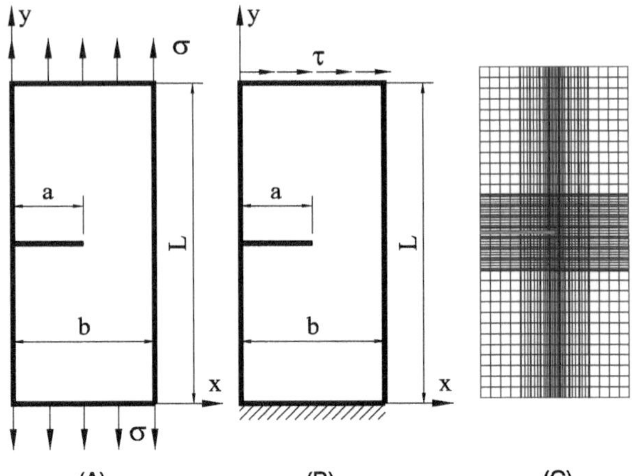

Figure 7.33 Geometry of edge cracked plates under (A) tension and (B) shear loading; (C) Refined non-uniform mesh.

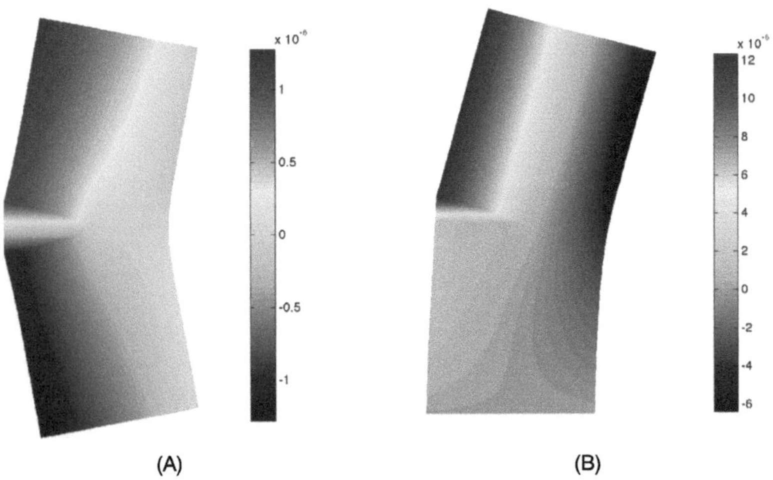

Figure 7.34 Contour plots displacement of the plate under (A) tension and (B) shear loading.

totic enrichment does not necessarily provide more accurate results but introduces additional complexity and parameters to calibrate.

7.5.5.2 Results obtained by XIGA

Fig. 7.33, Fig. 7.34 and Fig. 7.35 show contour plots of one displacement and stress component of the edge cracked plates under tension and shear loading, respectively. Fig. 7.36 and Fig. 7.37 present the normalized SIFs

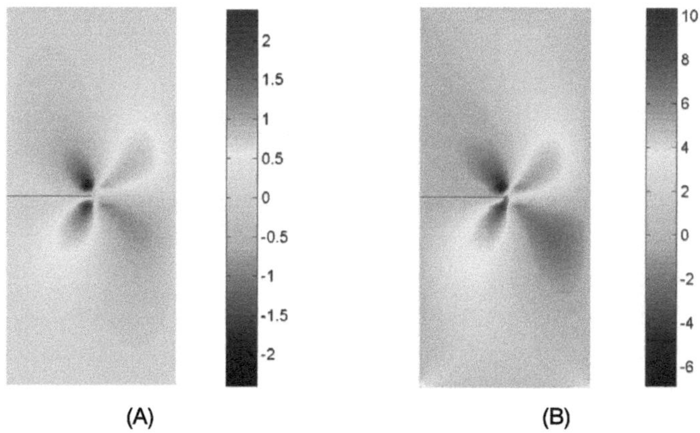

Figure 7.35 Contour plots stress of the plate under (A) tension and (B) shear loading.

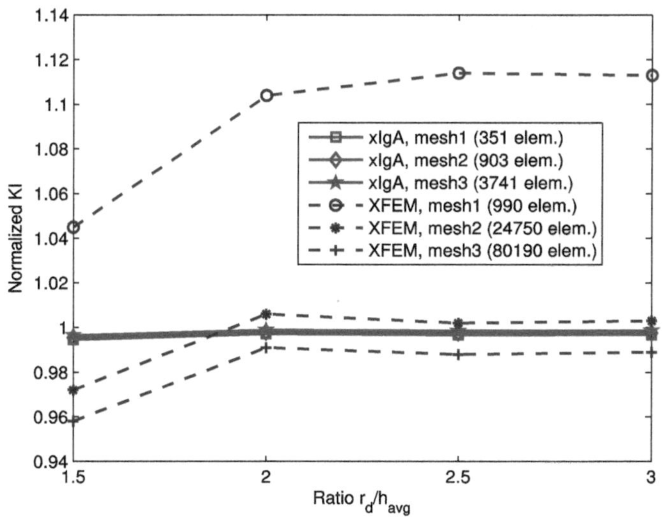

Figure 7.36 Normalized K_I for the edge cracked plate with a tension loading.

K_I for the edge cracked plate under tension and shear loading, respectively. The normalized SIF K_{II} under shear loading is displayed in Fig. 7.38. On the x-axis, the radius value $r_d = 2.5h_{avg}$, and h_{avg} is the average value of the square-roots of the cracked element's areas. Note that the XIGA formulation is based on cubic polynomials while a (bi-)linear XFEM formulation is used. The results show that, as the mesh density increases, the stress intensity factors approach the reference solution and are independent of the domain

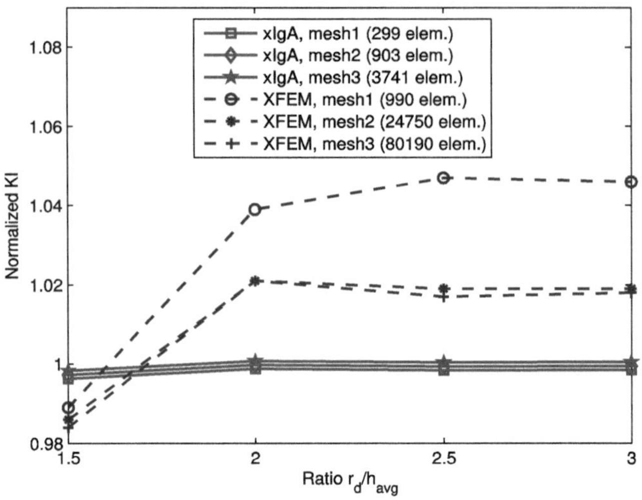

Figure 7.37 Normalized K_I for the edge cracked plate with the shear loading.

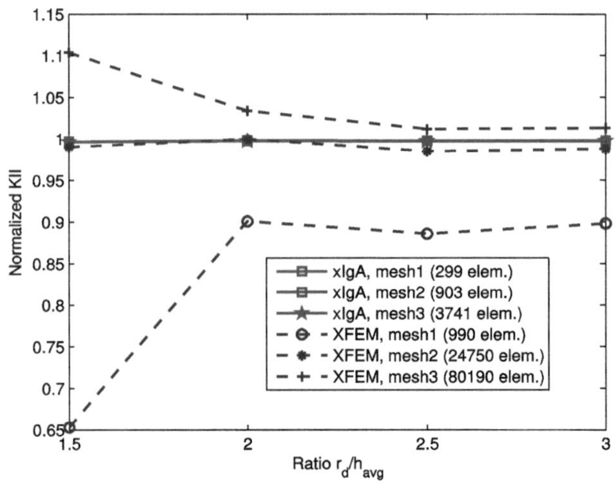

Figure 7.38 Normalized K_{II} for the edge cracked plate with the shear loading.

sizes of the interaction integral. The XIGA results are more accurate with less elements than XFEM.

7.5.6 Pressurized cylinder with an axial crack

A thin walled cylinder with the mean radius $R = 20$, thickness $h = 0.25$, and length $l = 100$ containing an axial through-the-thickness crack of length $2a$

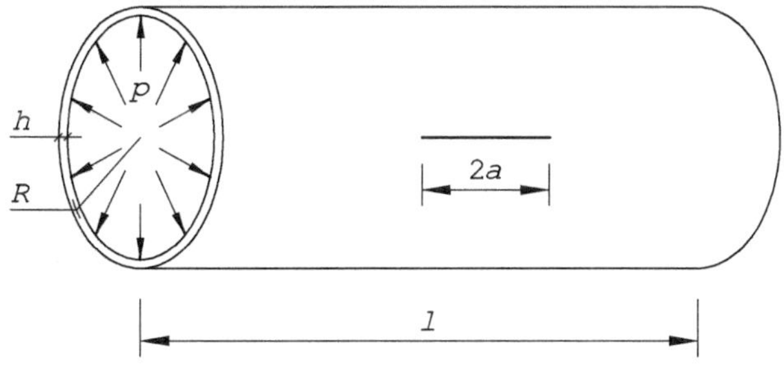

Figure 7.39 Geometry of a cylinder with an axial crack under internal pressure.

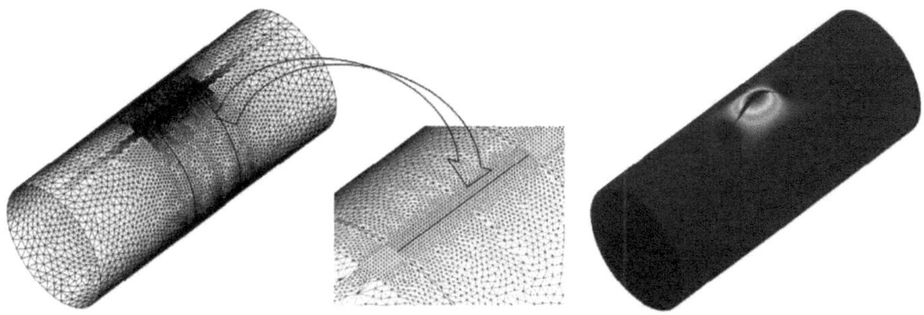

(A) (B) (C)

Figure 7.40 (A) Typical mesh of a cylinder with an axial crack; (B) Regular and fine mesh near the crack; (C) Displacement field with crack opening (deformation factor 0.05).

is subjected to an internal pressure $p = 1.0$ (see Fig. 7.39). E is 1000, and $\nu = 0.3$. We support open-end conditions.

7.5.6.1 Results of the phantom node MITC3 element

The mesh used for calculations is regular and fine around the crack area but irregular and gradually coarse in the other area remote from the crack. The typical mesh and displacement field with crack opening are depicted in Fig. 7.40 for the cylinder with a crack length of $a = 9.841$.

The interaction integral I for the Kirchhoff theory is employed to extract the stress intensity factor K_I of symmetric membrane loading. Because the stress intensity factor is mostly influenced by the stress and strain fields near the crack tip, various regular meshes around the crack, which differ in the length of triangular elements, are done to investigate the dependence of accuracy on the meshes. Fig. 7.41 presents the relationship between the length of elements in the regular meshes around the crack and the rela-

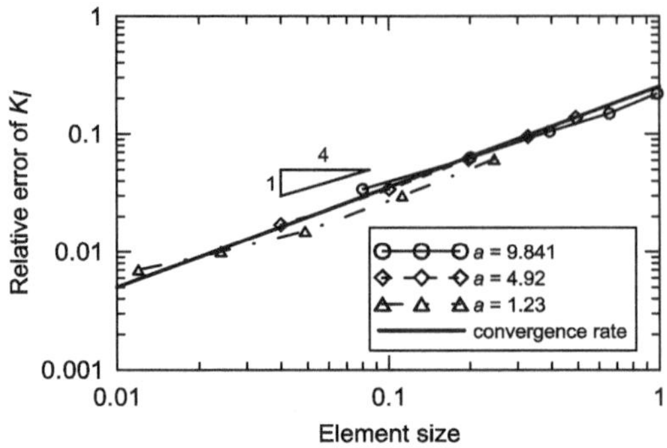

Figure 7.41 Relative errors in the K_I vs. element size obtained by the phantom-node method.

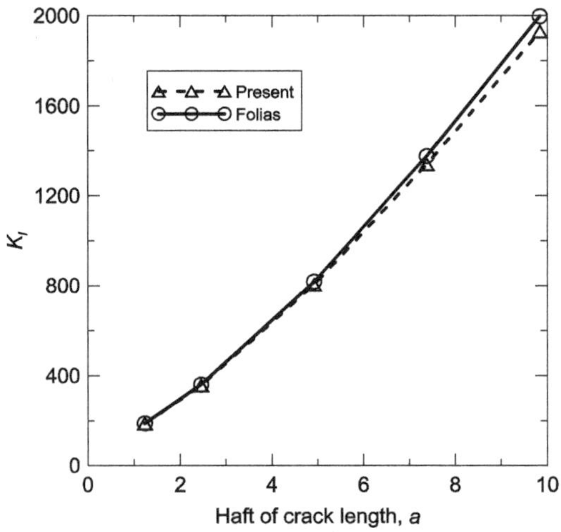

Figure 7.42 Stress intensity factor K_I corresponding to membrane symmetric loading.

tive error of K_I which is obtained based on the analytical solution given by Folias. In this figure, the convergence rate of K_I is about 1.0 and K_I can be calculated within 4% accuracy using triangular elements of length $a/123$.

Fig. 7.42 shows values of K_I for various lengths of the axial crack. In all the cases, elements of length $a/123$ are used for structured mesh around the crack. These results are in good agreement with the analytical solution of Folias.

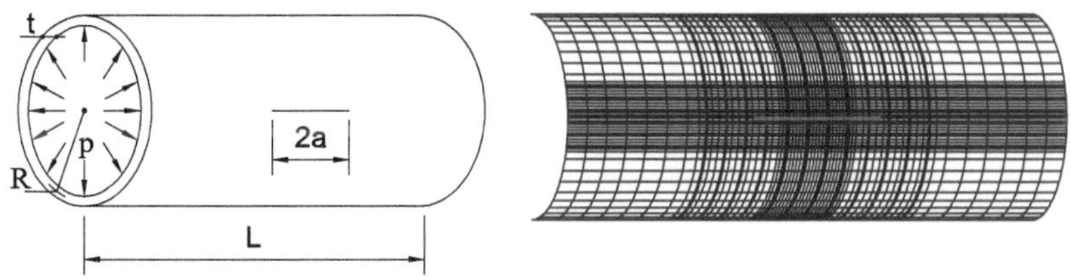

Figure 7.43 Geometry of a cylinder under internal pressure with a longitudinal crack and refined non-uniform NURBS mesh.

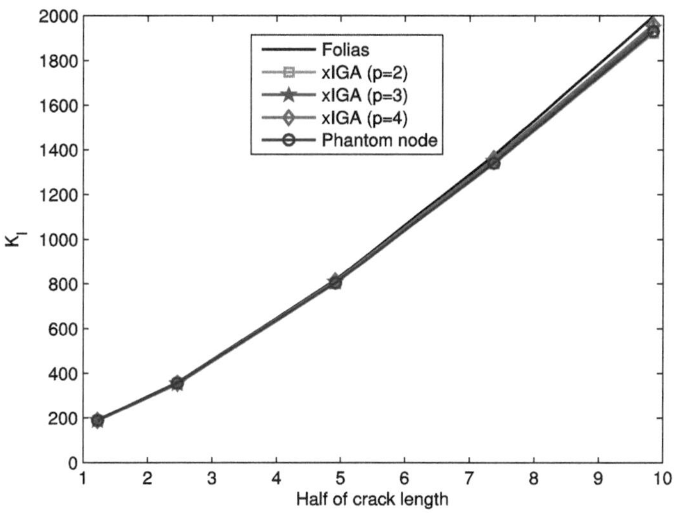

Figure 7.44 Stress intensity factor K_I corresponding to membrane symmetric loading.

7.5.6.2 Results obtained by XIGA

The geometry and the mesh are illustrated in Fig. 7.43. Fig. 7.44 shows the stress intensity factor K_I for various axial cracks of lengths $2a = 2.46, 4.92, 9.84, 14.76, 19.682$. The relative error of the results for different number of elements is compared (see Fig. 7.45). It is observed that the expected convergence rate for XIGA is achieved. The convergence rate of the XIGA is higher than the convergence rate obtained with the phantom node method. However, linear triangular elements are used for the phantom node method while the XIGA is based on quadratic, cubic and quartic elements.

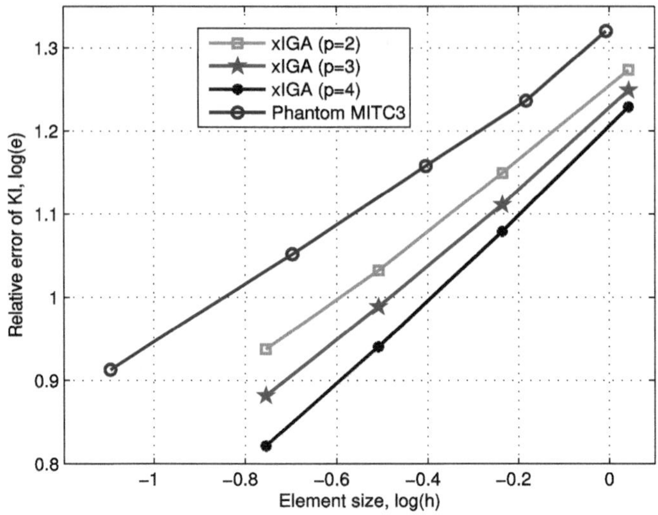

Figure 7.45 Relative error of the K_I versus element size of thin cylinder with crack of length $2a = 9.84$.

References

[1] S. Ahmad, B.B. Irons, O.C. Zienkiewicz, Analysis of thick and thin shell structures by curved finite elements, International Journal for Numerical Methods in Engineering 2 (1970) 419–451.

[2] P.M. Areias, T. Belytschko, A comment on the article "A finite element method for simulation of strong and weak discontinuities in solid mechanics" by A. Hansbo and P. Hansbo [Comput. Methods Appl. Mech. Engrg. 193 (2004) 3523–3540]. Letter to the Editor, 2006, Computer Methods in Applied Mechanics and Engineering 195 (2006) 1275–1276.

[3] P.M.A. Areias, T. Belytschko, Non-linear analysis of shells with arbitrary evolving cracks using XFEM, International Journal for Numerical Methods in Engineering 62 (2005) 384–415.

[4] P.M.A. Areias, J.-H. Song, T. Belytschko, A finite-strain quadrilateral shell element based on discrete Kirchhoff–Love constraints, International Journal for Numerical Methods in Engineering 64 (1) (2005) 1166–1206.

[5] P.M.A. Areias, J.H. Song, T. Belytschko, Analysis of fracture in thin shells by overlapping paired elements, Computer Methods in Applied Mechanics and Engineering 195 (2006) 5343–5360.

[6] P. Baiz, S. Natarajan, S. Bordas, P. Kerfriden, T. Rabczuk, Linear buckling analysis of cracked plates by SFEM and XFEM (SmXFEM), Journal of Mechanics of Materials and Structures 6 (9–10) (2011) 1213–1238.

[7] K.-J. Bathe, E.N. Dvorkin, A four-node plate bending element based on Mindlin/Reissner plate theory and a mixed interpolation, International Journal for Numerical Methods in Engineering 21 (2) (1985) 367–383.

[8] K.-J. Bathe, E.N. Dvorkin, A formulation of general shell elements – the use of mixed interpolation of tensorial components, International Journal for Numerical Methods in Engineering 22 (3) (1986) 697–722.
[9] E. Bechet, H. Minnebo, N. Moës, B. Burgardt, Improved implementation and robustness study of the X-FEM for stress analysis around cracks, International Journal for Numerical Methods in Engineering 64 (2005) 1033–1056.
[10] T. Belytschko, W.E. Bachrach, Efficient implementation of quadrilaterals with high coarse-mesh accuracy, Computer Methods in Applied Mechanics and Engineering 54 (1986) 279–301.
[11] T. Belytschko, T. Black, Elastic crack growth in finite elements with minimal remeshing, International Journal for Numerical Methods in Engineering 45 (1999) 601–620.
[12] T. Belytschko, I. Leviathan, Physical stabilization of the 4-node shell element with one point quadrature, Computer Methods in Applied Mechanics and Engineering 113 (1994) 321–350.
[13] T. Belytschko, Y. Lu, Element-free Galerkin methods for static and dynamic fracture, International Journal of Solids and Structures 32 (1995) 2547–2570.
[14] T. Belytschko, J.I. Lin, C.S. Tsay, Explicit algorithms for the nonlinear dynamics of shells, Computer Methods in Applied Mechanics and Engineering 42 (1984) 225–251.
[15] T. Belytschko, B.L. Wong, H.Y. Chiang, Advances in one-point quadrature shell elements, Computer Methods in Applied Mechanics and Engineering 96 (1992) 93–107.
[16] T. Belytschko, W.K. Liu, B. Moran, Nonlinear Finite Elements for Continua and Structures, John Wiley and Sons, Chichester, 2000.
[17] T. Chau-Dinh, G. Zi, P.-S. Lee, T. Rabczuk, J.-H. Song, Phantom-node method for shell models with arbitrary cracks, Computers & Structures 92–93 (2012) 242–256.
[18] F. Cirak, M. Ortiz, A. Pandolfi, A cohesive approach to thin-shell fracture and fragmentation, Computer Methods in Applied Mechanics and Engineering 194 (2005) 2604–2618.
[19] J. Dolbow, N. Moes, T. Belytschko, Modeling fracture in Mindlin-Reissner plates with the extended finite element method, International Journal of Solids and Structures 37 (6) (2000) 7161–7183.
[20] C. Duarte, J. Oden, An h-p adaptive method using clouds, Computer Methods in Applied Mechanics and Engineering 139 (1–4) (1996) 237–262.
[21] C. Duarte, J. Oden, H-p clouds—an h-p meshless method, Numerical Methods for Partial Differential Equations 12 (6) (1996) 673–705.
[22] E.N. Dvorkin, K.-J. Bathe, A continuum mechanics based four-node shell element for general non-linear analysis, Engineering Computations 1 (1) (1984) 77–88.
[23] T. Elguedj, A. Gravouil, A. Combescure, Appropriate extended functions for X-FEM simulation of plastic fracture mechanics, Computer Methods in Applied Mechanics and Engineering 195 (2006) 501–515.
[24] F. Gruttmann, W. Wagner, A stabilized one-point integrated quadrilateral Reissner-Mindlin plate element, International Journal for Numerical Methods in Engineering 61 (2004) 2273–2295.
[25] A. Hansbo, P. Hansbo, A finite element method for the simulation of strong and weak discontinuities in solid mechanics, Computer Methods in Applied Mechanics and Engineering 193 (2004) 3523–3540.
[26] T.J.R. Hughes, W.K. Liu, Nonlinear finite element analysis of shells: Part II, two dimensional shells, Computer Methods in Applied Mechanics and Engineering 26 (1981) 167–181.

[27] P. Krysl, T. Belytschko, Analysis of thin shells by the element-free Galerkin method, International Journal of Solids and Structures 33 (20–22) (1996) 3057–3078.

[28] P. Laborde, J. Pommier, Y. Renard, M. Salaun, High-order extended finite element method for cracked domains, International Journal for Numerical Methods in Engineering 64 (2005) 354–381.

[29] P.-S. Lee, K.-J. Bathe, Development of MITC isotropic triangular shell finite elements, Computers & Structures 82 (11) (2004) 945–962.

[30] J.M. Melenk, I. Babuska, The partition of unity finite element method: basic theory and applications, Computer Methods in Applied Mechanics and Engineering 139 (1996) 289–314.

[31] N. Moes, J. Dolbow, T. Belytschko, A finite element method for crack growth without remeshing, International Journal for Numerical Methods in Engineering 46 (1) (1999) 133–150.

[32] N. Moes, A. Gravouil, T. Belytschko, Non-planar 3d crack growth by the extended finite element and level sets—part I: mechanical model, International Journal for Numerical Methods in Engineering 53 (2002) 2549–2568.

[33] V. Nguyen, T. Rabczuk, S. Bordas, M. Duflot, Meshless methods: a review and computer implementation aspects, Mathematics and Computers in Simulation 79 (2008) 763–813.

[34] N. Nguyen-Thanh, J. Kiendl, H. Nguyen-Xuan, R. Wüchner, K. Bletzinger, Y. Bazilevs, T. Rabczuk, Rotation free isogeometric thin shell analysis using PHT-splines, Computer Methods in Applied Mechanics and Engineering 200 (47–48) (2011) 3410–3424.

[35] H. Nguyen-Xuan, T. Rabczuk, S. Bordas, J.F. Debongnie, A smoothed finite element method for plate analysis, Computer Methods in Applied Mechanics and Engineering 197 (13) (2008) 1184–1203.

[36] N. Nguyen-Thanh, N. Valizadeh, M. Nguyen, H. Nguyen-Xuan, X. Zhuang, P. Areias, G. Zi, Y. Bazilevs, L. De Lorenzis, T. Rabczuk, An extended isogeometric thin shell analysis based on Kirchhoff-Love theory, Computer Methods in Applied Mechanics and Engineering 284 (2015) 265–291.

[37] T. Rabczuk, T. Belytschko, Cracking particles: a simplified meshfree method for arbitrary evolving cracks, International Journal for Numerical Methods in Engineering 61 (13) (2004) 2316–2343.

[38] T. Rabczuk, T. Belytschko, Application of meshfree particle methods to static fracture of reinforced concrete structures, International Journal of Fracture 137 (1–4) (2006) 19–49.

[39] T. Rabczuk, T. Belytschko, A three dimensional large deformation meshfree method for arbitrary evolving cracks, Computer Methods in Applied Mechanics and Engineering 196 (29–30) (2007) 2777–2799.

[40] T. Rabczuk, S. Xiao, M. Sauer, Coupling of meshfree methods with finite elements: basic concepts and test results, Communications in Numerical Methods in Engineering 22 (10) (2006) 1031–1065.

[41] T. Rabczuk, P. Areias, T. Belytschko, A meshfree thin shell method for non-linear dynamic fracture, International Journal for Numerical Methods in Engineering 72 (5) (2007) 524–548.

[42] T. Rabczuk, P. Areias, T. Belytschko, A simplified meshfree method for shear bands with cohesive surfaces, International Journal for Numerical Methods in Engineering 69 (5) (2007) 993–1021.

[43] T. Rabczuk, G. Zi, A. Gerstenberger, W. Wall, A new crack tip element for the phantom node method with arbitrary cohesive cracks, International Journal for Numerical Methods in Engineering 75 (2008) 577–599.

[44] T. Rabczuk, R. Gracie, J.-H. Song, T. Belytschko, Immersed particle method for fluid-structure interaction, International Journal for Numerical Methods in Engineering 81 (1) (2010) 48–71.

[45] J. Rethore, A. Gravouil, A. Combescure, An energy-conserving scheme for dynamic crack growth using the extended finite element method, International Journal for Numerical Methods in Engineering 63 (2005) 631–659.

[46] J.H. Song, P.M.A. Areias, T. Belytschko, A method for dynamic crack and shear band propagation with phantom nodes, International Journal for Numerical Methods in Engineering 67 (2006) 868–893.

[47] T. Strouboulis, K. Copps, I. Babuška, The generalized finite element method: an example of its implementation and illustration of its performance, International Journal for Numerical Methods in Engineering 47 (8) (2000) 1401–1417.

[48] G. Ventura, J. Xu, T. Belytschko, A vector level set method and new discontinuity approximations for crack growth by EFG, International Journal for Numerical Methods in Engineering 54 (6) (2002) 923–944.

[49] D. Wang, J.-S. Chen, Locking-free stabilized conforming nodal integration for meshfree Mindlin-Reissner plate formulation, Computer Methods in Applied Mechanics and Engineering 193 (12) (2004) 1065–1083.

[50] H.T.Y. Yang, S. Saigal, A. Masud, R.K. Kapania, A survey of recent shell finite elements, International Journal for Numerical Methods in Engineering 47 (2000) 101–127.

[51] J. Yau, S. Wang, H. Corten, A mixed-mode crack analysis of isotropic solids using conservation laws of elasticity, Journal of Applied Mechanics 47 (1980) 335–341.

[52] G. Zi, T. Belytschko, New crack-tip elements for XFEM and applications to cohesive cracks, International Journal for Numerical Methods in Engineering 57 (2003) 2221–2240.

[53] G. Zi, J.H. Song, S.H.L.E. Budyn, T. Belytschko, A method for growing multiple cracks without remeshing and its application to fatigue crack growth, Modelling and Simulation in Materials Science and Engineering 12 (2004) 901–915.

CHAPTER EIGHT

Fracture criteria and crack tracking procedures

8.1. Fracture criteria

This chapter is devoted to cracking criteria and cohesive models. We will first describe criteria when to introduce and propagate a crack and also discuss the orientation of the crack. Though the criteria for crack initiation and crack propagation are basically the same, one should keep in mind that the initiation of a crack is from an implementational point of view different than crack propagation, at least if crack path continuity is desired.

8.2. Cracking criteria

8.2.1 Criteria in LEFM

There are basically four major cracking or crack propagation criteria in LEFM:

- Maximum hoop stress criterion or maximum principal stress criterion.
- Minimum strain energy density criterion, Shih [63].
- Maximum energy release rate criterion, Wu [77].
- The zero K_{II} criterion (vanishing in-plane SIF (K_{II}) in shear mode for infinitesimally small crack extension), Goldstein and Salganik [24].

The first two criteria predict the direction of the crack trajectory from the stress state prior to the crack extension. The last two criteria require stress analysis for virtually extended cracks in various directions to find the appropriate crack-growth direction. For medium mixed-mode problems, all criteria lead almost to the same results, Duflot and Hung [16]. However, according to Shen and Stephansson [62], only the maximum energy release rate criterion allows to predict secondary cracks in compressed specimens. According to the knowledge of the author, the first of the above mentioned criteria is mostly used in numerical simulations. The crack is propagated in an angle of θ_c from the crack tip. Note, that these criteria will give only the orientation of the crack but not its length. To determine the crack length, these criteria have to be checked in different distances around the crack tip. Often a constant crack propagation speed is assumed.

Extended Finite Element and Meshfree Methods
https://doi.org/10.1016/B978-0-12-814106-9.00014-0
Copyright © 2020 Elsevier Inc.
All rights reserved.

In the maximum hoop stress or maximum principal stress criterion, a crack is oriented perpendicular to the direction of the maximum principal stress. The maximum circumferential stress $\sigma_{\theta\theta}$, often called hoop stress, in the polar coordinate system around the crack tip corresponds to the maximum principal stress and is given for a crack propagating with constant velocity v_c by:

$$\sigma_{\theta\theta} = \frac{K_I}{\sqrt{2\pi r}} f_h^I(\theta, v_c) + \frac{K_{II}}{\sqrt{2\pi r}} f_h^{II}(\theta, v_c) \tag{8.1}$$

where the functions f_h^I and f_h^{II} represent the angular variation of stress for different values of crack-tip speed v_c. When the maximum hoop stress is larger equal a critical hoop stress $\sigma_{\theta\theta}^c$ then the crack is propagated in the direction perpendicular to the maximum hoop stress. For pure mode I fracture, $\sigma_{\theta\theta}^c$ is given by

$$\sigma_{\theta\theta}^c = \frac{K_I^c}{\sqrt{2\pi r}} \tag{8.2}$$

with the fracture toughness K_I^c that is obtained from experiments.

In LEFM, the local direction of the crack growth is determined by the condition that the local shear stress is zero that leads to the condition:

$$K_I \sin\theta_c + K_{II}(3\cos\theta_c - 1) = 0 \tag{8.3}$$

that results in the crack propagation angle

$$\theta_c = 2\arctan\left(\frac{K_I - \sqrt{K_I^2 + 8K_{II}^2}}{4K_{II}}\right) \tag{8.4}$$

The minimum strain energy density criterion is based on a critical strain-energy-density factor S_r^c. The basic assumption is that crack initiation occurs when the interior minimum of the strain-energy-density factor S reached S_r^c. The strain energy density factor S represents the strength of the elastic energy field in the vicinity of the crack tip which is singular of the order r^{-1}. The factor S_r^c is assumed to be a material parameter and can be used as a measure of the fracture toughness under mixed mode conditions. For pure mode I fracture, S_r^c can be directly expressed in terms of K_I^c:

$$S_r^c = (\kappa - 1)\frac{\pi^2 (K_I^c)^2}{8G} \tag{8.5}$$

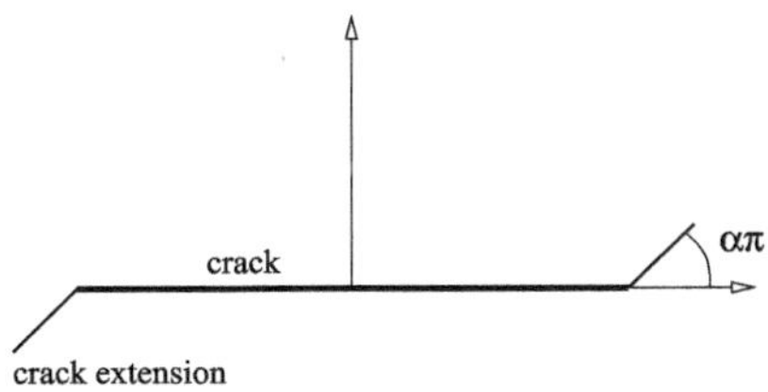

Figure 8.1 Crack propagation with the energy release rate criterion.

where G denotes the shear modulus and κ is the Kolosov constant. For general mixed mode fracture, S_r^c is:

$$S_r^c = a_{11}K_I^2 + 2a_{12}K_IK_{II} + a_{22}K_{II}^2 \quad \forall\theta = \theta_0 \tag{8.6}$$

with $-\pi < \theta_0 < \pi$ and

$$\begin{aligned} a_{11} &= \frac{(1+\cos\theta)(\kappa - \cos\theta)}{16G} \\ a_{12} &= \frac{(2\cos\theta - \kappa + 1)\sin\theta}{16G} \\ a_{22} &= \frac{(\kappa+1)(1-\cos\theta) + (1+\cos\theta)(3\cos\theta - 1)}{16G} \end{aligned} \tag{8.7}$$

The parameters a_{ij} are obtained from the strain-energy density function. For further details, see Shih [63]. Note, that the minimum strain energy density criterion is a local criterion since fracture occurs when the energy density in a volume element near the crack tip reaches a critical value while the classical fracture mechanics theory is based on global energy balance.

In the maximum energy release rate, the crack propagates in the direction defined by the angle $\alpha = \alpha_c(\sigma)$ where α_c satisfies the condition:

$$\begin{aligned} \frac{\partial \mathcal{G}}{\partial \alpha}\Big|_{\alpha=\alpha_c} &= 0 \\ \frac{\partial^2 \mathcal{G}}{\partial \alpha^2}\Big|_{\alpha=\alpha_c} &\leq 0 \end{aligned} \tag{8.8}$$

where α is defined in Fig. 8.1 and $\mathcal{G}$ denotes the energy release rate. Crack propagation occurs at a stress state $\sigma = \sigma_c$ when $\mathcal{G}(\sigma_c, \alpha_c(\sigma_c)) = \mathcal{G}_g$ where $\mathcal{G}_g$ is a material parameter. According to Wu [77], the energy release rate can be decomposed into an anti-plane (anti-plane shear load) part $\mathcal{G}_A$ and a plane (plane load) part $\mathcal{G}_P$:

$$\mathcal{G}(\sigma, \alpha) = \mathcal{G}_P(\sigma, \alpha) + \mathcal{G}_A(\sigma, \alpha) \tag{8.9}$$

For the anti-plane shear load, $\mathcal{G}_A$ is given by

$$\mathcal{G}_A(\sigma, \alpha) = \frac{\pi}{2}\sigma_{23}^2 \left(\frac{1-\alpha^{\alpha}}{1+\alpha}\right) \tag{8.10}$$

For plane strain conditions, Wu [77] gave an expression for $\mathcal{G}_P$:

$$\mathcal{G}_P(\sigma, \alpha) = \frac{1-\nu}{2}\sigma_{23}^2 \left(K_I^2 + K_{II}^2\right) \tag{8.11}$$

8.2.2 Global energy criteria

Meschke and Dummerstorf [39], Dummerstorf and Meschke [17] have presented a method in the context of the two-dimensional XFEM. Miehe and Guerses [40] presented a similar method in the context of r-adaptive interelement-separation method in two-dimension. The basic idea is that fracture is considered as a global energy minimization problem and crack orientation and crack length are introduced as additional unknowns in the variational formulation if applied in an XFEM-context. The nice feature of this method is that the orientation and length of the crack is the direct outcome of the analysis. Meschke and Dummerstorf [39] presented their method in a two-dimensional static setting for cohesive and cohesionless crack. The application of these methods to 3D seem to be cumbersome though.

8.2.3 Rankine criterion

For a Rankine material, a crack is introduced when the principal tensile stress reaches the uniaxial tensile strength at a particle. The crack is initiated perpendicular to the direction of the principal tensile stress. Usually, some kind of smoothing technique is applied that generally either averages the crack normal or the stress tensor, Gasser and Holzapfel [21], Mariani and Perego [34], Mergheim et al. [37], Rabczuk et al. [56], Wells and Sluys [76]. This is done in order to improve the reliability of the computed stresses in front of the crack tip. Otherwise, awkward crack pathes can be obtained. For example, Wells and Sluys [76] and Mergheim et al. [37] use an averaged stress tensor of the form:

$$\boldsymbol{\sigma}_m = \int_{\Omega_c} w\, \boldsymbol{\sigma}\, d\Omega \tag{8.12}$$

where Ω_c is a certain domain around the crack tip and w is a weighting function:

$$w = \frac{1}{(2\pi)^{3/2}\, l^3} \exp\left(-\frac{r^2}{2l^2}\right) \tag{8.13}$$

where l determines how fast the weight function decays from the crack tip. Gasser and Holzapfel [21] and Rabczuk et al. [56] smooth the crack normal instead of the stress tensor that leads to very similar results.

The Rankine criterion is applicable to brittle materials and works well for mode I-fracture.

8.2.4 Loss of material stability condition

A crack or a shear band results from a material instability. A classical definition of material stability is based on the so-called Legendre-Hadamard condition, which establishes that for any non zero vectors **n** and **h** the following point-wise inequality must hold:

$$e = min_{\mathbf{n},\mathbf{h}}(\mathbf{n} \otimes \mathbf{h}) : \mathbf{A} : (\mathbf{n} \otimes \mathbf{h}) > 0 \tag{8.14}$$

where $\mathbf{A} = \mathbf{C}^t + \boldsymbol{\sigma} \otimes \boldsymbol{\delta}$, $\mathbf{C}^t$ is the constitutive tangent operator and e is the so-called ellipticity indicator. For a linear comparison solid, if non-propagating singular surfaces do not occur, the Legendre-Hadamard condition is identical to the strong-ellipticity condition, used e.g. in Simo et al. [66]. In case, Eq. (8.14) is no longer fullfilled, the material looses stability and **n** defines the direction of propagation, and **h** is the *polarization* of the wave. This condition ensures that the speed of propagating waves in a solid remains real. Equality in expression (8.14) is the necessary condition for stationary waves. Belytschko et al. [3] give a textbook description for obtaining condition (8.14) by means of a stability analysis of the momentum equation when a perturbation of the form $\boldsymbol{u} = \boldsymbol{h}\exp(i\omega t + k\boldsymbol{n} \cdot \boldsymbol{x})$ is applied. The Legendre-Hadamard condition is occasionally called strong ellipticity of the constitutive relation (see Marsden and Hughes [35]). In the dynamical case, the Legendre-Hadamard condition implies the hyperbolicity of the IBVP. For a rate-dependent material, we talk about loss-of-material stability. The reader is referred to Silhavy [64] or Ogden [44] for details about the concepts mentioned above.

Loss of hyperbolicity (or loss of ellipticity or loss of material stability, respectively) is determined by minimizing e with respect to **n** and **h**; if e is negative for any combination of **n** and **h**, the PDE has lost hyperbolicity at that material point. In 2D, e can be expressed as a function of the two

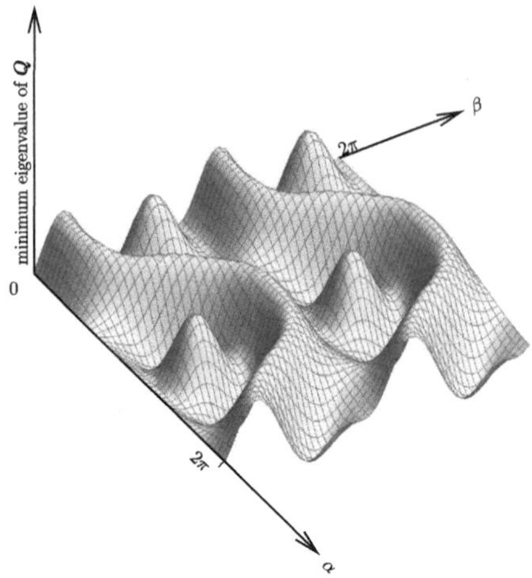

Figure 8.2 Minimum eigenvalue of the acoustic tensor as a function of the two angles for one material point.

angles θ and ϕ where $\mathbf{n} = [\cos(\theta)\ \sin(\theta)]$ and $\mathbf{h} = [\cos(\phi)\ \sin(\phi)]$. It can be shown, that this condition is equivalent to condition (8.15). Therefore, based on expression in (8.14), let us define for a given material point of a solid at a given time,

$$\mathbf{Q} = \mathbf{n} \cdot \mathbf{A} \cdot \mathbf{n} \tag{8.15}$$

where $\mathbf{Q}$ is the acoustic tensor with components $Q_{ik} = n_j n_l A_{ijkl}$. We say that a material point is stable whenever the minimum eigenvalue of $\mathbf{Q}$ is strictly positive, and unstable otherwise.

One difficulty is that the analysis of the acoustic tensor generally will give us two directions $\mathbf{n}$ from which one direction has to be chosen. At times the loss-of-material-stability criterion becomes ambiguous; an example of the minimum eigenvector of $\mathbf{Q}$ as a function of the angle θ is shown in Fig. 8.3; simplified for a two-dimensional example. For uniaxial tension, two angles are obtained, but they correspond to the same plane. In other stress states, four angles corresponding to two planes, are sometimes obtained. A typical eigenvalue landscape as function of the crack angles in three dimension is shown in Fig. 8.2 for one material point and isotropic J_2-plasticity with strain softening. We choose the direction of the maximum displacement gradient by maximizing

$$g = \underbrace{\max}_{l} \left(\mathbf{n}_l^T \cdot (\nabla \mathbf{u} \cdot \mathbf{h}_l) \right), \quad l = 1, 2 \tag{8.16}$$

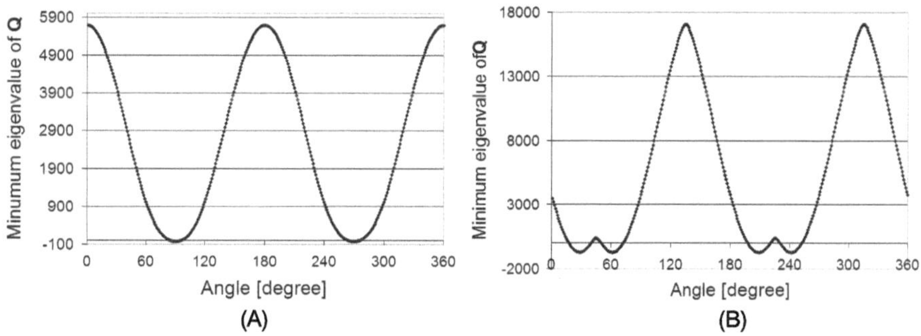

Figure 8.3 Minimum eigenvalue of the acoustic tensor **Q** versus different angles of the orientation of **n** for (A) uniaxial tension, (B) an arbitrarily directed load.

where the normals $\mathbf{n}_l$ correspond to minima of $\mathbf{Q}$, Eq. (8.15); $\mathbf{h}$ is the corresponding eigenvector of $\mathbf{Q}$. Oliver et al. [47] has shown that an anisotropic continuum model will lead to a unique angle θ in the discontinuous bifurcation analysis.

Since the normal $\mathbf{n} = [\cos\alpha\ \cos\varphi, \cos\alpha\ \sin\varphi, \sin\alpha]$ depends on two angles in three dimensions, the procedure of finding the minimum eigenvalue of $\mathbf{Q}$ can become computationally expensive. A bisection method can reduce computational cost. Another efficient way to compute the eigenvalues of the acoustic tensor is given by Ortiz et al. [48]. We have used parallelization to compute keep computational cost low. We would like to mention that for several problems in static crack propagation, loss-of-material-stability is only checked within a certain domain around the crack tip.

We would also like to mention that the loss-of-material stability criterion as given above is not applicable to problems with constraints. For thin shells for example, the crack state has to be compatible with the Kirchhoff-Love (KL) constraint. For multi-field problems, the criterion has to be adjusted as well.

8.2.5 Rank-one-stability criterion

The rank-one-stability criterion

$$\dot{\mathbf{F}} : \mathbf{A} : \dot{\mathbf{F}} \leq 0 \ \forall \ \dot{\mathbf{F}} \tag{8.17}$$

is a stricter but computational cheaper criterion than the loss-of material stability criterion. However, the rank-one-stability criterion does not provide the orientation of the crack.

8.2.6 Determining the crack orientation

There are numerous criteria for crack initiation or propagation, often based on variables such as damage or effective plastic strain, etc. The main difficulty is to obtain the orientation of the crack. In case of a crack propagation, the crack can be propagated from its crack front in the direction of the material point where the cracking criteria is met.

8.2.6.1 *Energy minimization*

The orientation of the crack can be obtained by global energy minimization. For different orientations of the crack, the global energy is computed and the crack is propagated in the direction where the global energy has its minimum. This global fracture criterion can be expressed as:

$$\mathcal{J} = \int_{\Omega} W(\mathbf{u})\ d\Omega + \int_{\Gamma} \mathcal{G}_t\ d\Gamma \tag{8.18}$$

where $\mathcal{G}_t$ is the surface toughness that characterizes the surface energy necessary to create the new crack surface.

8.2.6.2 *Micro plane models*

The *microplane model* closes the gap between smeared crack models and continuum damage models. They can be used to obtain the localization direction as well. In microplane models, several crack planes are assumed at a given material point. A one-dimensional stress-strain relation is applied at the different crack-plane levels. With an integral equilibrium consideration at a sphere using the principle of the virtual displacements, a relation is established between global stresses and strains. Softening occurs automatically in the appropriate crack plane. Hence, the orientation of the crack is obtained from the softening of the micro-plane.

8.2.7 Computation of the crack length

The computation of the crack length is closely related to the cracking criterion. The easiest way is to control the crack length as described e.g. in Section 8.3.3. The crack length is often related to the size of a background cell, Voronoi cell or the size of the domain of influence due to implementational reasons.

Another approach is based on the hyperbolicity indicator $e = \mathbf{h} \cdot \mathbf{Q} \cdot \mathbf{h}$ which must vanish at the crack tip:

$$\frac{\partial e}{\partial t} + \mathbf{v}^c \cdot \nabla e = 0 \text{ with } \mathbf{v}^c = v^c \, \mathbf{s} \tag{8.19}$$

where v^c is the crack speed, $\mathbf{s}$ gives its direction which must fulfill the condition $\mathbf{n} \cdot \mathbf{s} = 0$. To obtain the crack length, Eq. (8.19) has to be solved for v^c.

8.3. Crack surface representation and tracking the crack path

We will focus now on algorithms to track the crack path. This is necessary for all methods that model the crack continuously. Despite the numerous contribution in the different methods (XFEM, Embedded Finite Element Method, meshfree methods), there are only very few publications that are concerned about the three-dimensional implementation of these methods. In the following, we will address methods to describe and track the crack in a very general view. In Section 8.3.2, we will focus on crack tracking procedures in three-dimensions.

There are numerous ways to track the crack path and to represent the crack surface. The crack surface can be represented explicitly, meaning by introducing another mesh that represents the crack surface. Most commonly, the crack surface is described by piecewise linear crack segments requiring C^0 continuous crack surfaces though it is also possible to describe the crack surface smoothly, e.g. with B-splines or NURBS (Non-uniform rational B-Spline). Here it might be obvious that meshfree methods are well suited. The crack surface can also be described implicitly with the help of level sets or signed distance functions.

Crack path tracking algorithms can be classified into three classes, Gasser and Holzapfel [22]: global methods, local methods [18,73], and the level set method [25].

The basic idea of the global crack tracking procedure is to define a linear thermal problem to be solved each time step of the original mechanical problem. Therefore, two vector fields $\mathbf{a}$ and $\mathbf{b}$ are introduced that have to fulfill the following condition:

$$\mathbf{a} \cdot \mathbf{n}_0 = \mathbf{b} \cdot \mathbf{n}_0 = 0 \tag{8.20}$$

where $\mathbf{n}_0$ is the crack normal in the initial configuration. The family of surfaces, enveloping both vector fields $\mathbf{a}$ and $\mathbf{b}$ can be described by a temperature-like function $\mathcal{T}(\mathbf{X})$ if

$$\begin{aligned} \mathbf{a} \cdot \nabla_0 \mathcal{T} = \nabla_0 \mathcal{T} \cdot \mathbf{a} = 0 \text{ in } \Omega_0 \\ \mathbf{b} \cdot \nabla_0 \mathcal{T} = \nabla_0 \mathcal{T} \cdot \mathbf{b} = 0 \text{ in } \Omega_0 \end{aligned} \tag{8.21}$$

holds. Eq. (8.21) can be rephrased as an anisotropic heat conduction problem [45,59]. The drawback of this method is that the heat conduction problem has to be solved at every time step, that makes the method computationally very expensive.

The partial domain crack tracking algorithm [18] is a modification of the global crack tracking procedure. Instead of globally constructing the scalar field $\mathcal{T}$, it is only constructed for a subset of the domain that is affected by the crack.

With local crack tracking algorithms, the alignment of the crack surfaces is enforced with respect to its neighborhood. Local crack tracking algorithms are usually characterized by recursively "cutting" elements and are especially effective in three dimensions. These techniques were pursued by Gasser and Holzapfel [21,22] and Areias and Belytschko [2]. We will propose a three dimensional method in Section 8.3.2.1 that falls into that category as well. In that method, only the crack normal to the corresponding crack segment is needed as well as the coordinates of the piecewise linear crack front. The signed distance function is then used to completely describe the crack surfaces. In contrast to the method of Gasser and Holzapfel [22] who considered only the propagation of a single crack front, we are also able to handle crack initiation, crack branching, multiple cracking and crack junctions. Jäger et al. [26,27] report difficulties of local tracking algorithms in the context of XFEM and state that they are not capable of capturing curved cracks in 3D. Those problems do not occur in meshfree methods due to the absence of a mesh, even when the crack segments are plane as shown by Bordas et al. [6] Rabczuk et al. [56,58]. Also, when a background mesh for integration is employed, a different (refined) mesh for representing the crack by plane crack segments can always be constructed and refined such that curved cracks can be modeled with ease.

A very powerful tool that is used for tracking the crack path is the level set method [49]. It was initially applied to track interfaces and later on extended to tracking cracks. Since the level set method is a very hot topic of current research (especially with respect to crack surface representation and crack tracking), we have devoted the next section to this subject. Afterwards, we fill explain three dimensional crack tracking procedures. Crack propagation in 3D is one of the most challenging "computational" tasks. We will first give a brief overview of existing methods without claim of completeness. Then, we will describe two different crack tracking algorithms that we used in our computations. The first 3D technique can be assigned to the category of local crack tracking algorithms and was our most applied method for three dimensional crack problems with crack path continuity. The other technique is based on level sets for smooth crack surfaces. We will then propose an adaptive method that is based on the explicit representation of the crack using the visibility criterion. This technique was only implemented in two dimensions. We will not explain crack tracking procedures for the two dimensional case in more detail and refer the interested reader to the literature, Belytschko et al. [4], Gravouil et al. [25], Moes et al. [41], Ventura et al. [74] describe crack tracking techniques with level sets. Other two-dimensional crack tracking procedures (e.g. using the visibility method) can be found in Li and Simonson [29], Li et al. [30, 31], Simkins and Li [65]. Finally, we will propose different criteria to determine the crack length before we complete this section with some final comments.

8.3.1 The level set method to trace the crack path

The level set method uses signed distance functions to describe the crack. The initial crack is represented by discretizing these signed distance functions, i.e. the crack surface. Note that when the discontinuity moves, the level set deviates from the signed distance function. Usually, a reinitialization of the level set is performed that is stays as close to the signed distance as possible, Cho [1]. As noted before, the level set does not necessarily have to be the signed distance.

The accuracy of the crack surface representation depends on the discretization (shape functions) of the level set functions. If initially curved cracks are assumed, linear finite element shape functions will fail to represent the crack surface accurately. Usually, the same discretization is employed for the mechanical properties and the level set function that makes the method very attractive and elegant from an implementational point of view.

Initially, the level set method was applied for tracking interfaces as they occur e.g. in two-phase flow problems. As noted e.g. by Ventura et al. [75],

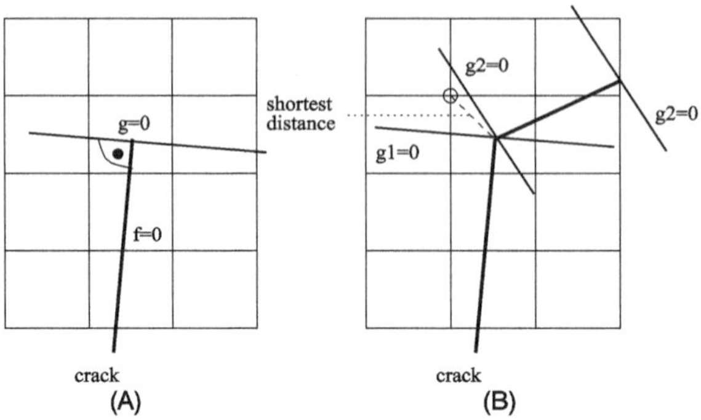

Figure 8.4 Determining the shortest distance to the crack line for crack tips and kinked cracks in two dimensions.

the original level set method [49] is not ideally suited to track the crack path due to three major reasons. First, the zero level set must be updated behind the tips to take into account the fact that once a material point is cracked, it remains cracked. Second, in contrast to interface tracking algorithms, the level set functions are not updated with the speed of an interface in the direction normal to itself but with the speed at the crack fronts. Hence, the equations for updating the interface

$$\frac{\partial f}{\partial t} + \mathbf{v} \cdot \nabla f = 0 \tag{8.22}$$

cannot be adopted to model crack propagation. In Eq. (8.22), f denotes the level set and $\mathbf{v}$ the velocity of the interface. And third, the crack is an open surface that grows from its crack front. Therefore, an additional level set has to be introduced to be able to completely describe the crack surface. This additional level set function (at the crack front) is perpendicular to the original level set function (and perpendicular to the current crack front) and hence has to be updated as well.

For open surfaces and kinks, there are certain difficulties involved in finding the shortest distance of a point to the crack surface. Therefore, let us consider Fig. 8.4A that shows a crack in two dimensions where f is the signed distance function to the crack surface and g is a function perpendicular to f. In Fig. 8.4A, every point knows to which domain it belongs ($f > 0 \wedge g > 0$, $f > 0 \wedge g < 0$, $f < 0 \wedge g < 0$ or $f < 0 \wedge g > 0$). In this figure, it becomes obvious why two level sets are needed to uniquely defined the position of a particle with respect to the crack.[1] While for the crack segment 2, node 2 will be on one side of the crack, Ω^+, for crack segment 1, node 2 will be on the opposite side of the crack, Ω^-. Moreover, the closest distance to the crack is not necessarily $\mathbf{n} \cdot (\mathbf{x} - \mathbf{x}_c)$ anymore where $\mathbf{x}_c$ is a point on the crack surface[2] and $\mathbf{n}$ is the normal to the corresponding crack surface. It can also be the distance from the kink, $\mathbf{x}_{kink}$, to a particle. If the crack is described smoothly, this problem vanishes somehow. If no crack tip enrichment is used, it is "only" necessary to find out on which side of the crack surface a node is located. Errors that occur due to the wrong computation of the shortest distance won't affect the results unless the node is detected on the correct side of the crack. However, if a crack tip enrichment is employed, the shortest distance to the crack directly influences the solution. We would like to mention again, that while it is relatively simple to determine the correct signed distance of a particle to the crack, the angle θ in the tip enrichment is not uniquely defined at kinks. In the literature, this problematic is barely addressed.

Crack propagation with level sets can be modeled by different techniques that can be classified into four groups [15]. In the first group, the level set is updated by the solution of differential equations, similar to Eq. (8.22) [25] where the level set functions are the unknowns. These methods require the discretization of the level set functions. In [25], the same discretization is used for the mechanical properties and the level set function. This is not mandatory but will complicate the implementation otherwise. Compared to interface tracking algorithms, additional complexity is added since (as noted above) two level set functions have to be introduced for every crack front that have to be perpendicular to each other. Usually, a reinitialization is done after updating the level set functions. This guarantees that the zero level set function is indeed the signed distance function.

The second group is defined on algebraic relations between the coordinates of a given point, the coordinates of the crack front and the crack advance vector, see e.g. Stolarska et al. [67]. It is not easy extendable into three dimensions.

The Vector level set method [74,75] is defined in terms of geometric transformations. In the vector level set method, the distance to the crack surface is stored in addition to the signed distance function. This facilitates

[1] I.e. to determine on which side of the crack the particle is located.

[2] In Fig. 8.4, it will be a point on crack segment 1 and 2.

implementation since there is no need to solve a PDE to update the level set.

The last class of methods are based on algebraic and trigonometric equations involving the initial value of the level set functions and the crack advance vector. Some of them also require the description of the crack front. They are well suited in three dimensional applications. Duflot [15] suggested a mixed method defined with differential and non-differential equations. He also discussed the level set method with respect to its application in computing the J integral that is used to compute SIFs. He found inaccuracies for most existing level set techniques. Since we are mainly interested in cohesive cracks where there is no need to compute SIFs and hence we do not exploit the level set method to compute the J integral, we did not encounter such difficulties. The only requirement needed is that f has to be perpendicular to g. This is not difficult to meet even in three dimensions. However, the paper of Duflot [15] gives an excellent overview on state-of-the-art level set techniques.

An advantage of the level set method is its potential of representing a *smooth* crack surface. This has certain advantages over methods that describe the crack front with piecewise linear lines since the local crack coordinate system at the crack front is not defined at kinks as mentioned above. The only requirement for a smooth representation of the crack surface is higher order continuity of the shape functions for the approximation of the level set function. Here, meshfree methods are ideally suited. We would like to emphasize again, that the big advantage is to use the same discretization for the level set functions and the mechanical properties. If another discretization is introduced for the level set, the method will become more cumbersome not only due to the mapping procedures. Nevertheless, other authors suggest the use of different meshes for the level set function and the mechanical properties, see e.g. Prabel et al. [51] or Duddu et al. [13] who applied XFEM to biofilm growth.

The level set technique couples also well for methods that use a crack tip enrichment since they can be exploited to compute the distance to the crack front $r = \sqrt{f^2 + g^2}$ and the angle $\theta = \arctan(f/g)$, see Fig. 5.33 for a two dimensional illustration. The level sets are especially useful for the approximation of θ since they guarantee that $\theta = \pm\pi$ on the crack surface since $f = 0$. In other methods some kind of mapping is needed to compute θ for kinked cracks, see e.g. Fleming et al. [19].

8.3.2 Tracking the crack path in 3D

While there are numerous papers on above mentioned methods, there exist only a few paper that are concerned about the three-dimensional implementation of such methods. The most difficult issue in a three-dimensional setting is how to describe and track the crack surface, at least if crack path continuity is required. Most methods require a C^0 continuous crack surface while there are only a few papers that are concerned with higher order continuous crack surfaces, Gravouil et al. [25], Moes et al. [42]. Feist and Hofstetter [18], Mosler and Meschke [43] and Oliver et al. [45,46] proposed three-dimensional methods within the context of the EFEM. Duarte et al. [11,12] describe concepts for through-the-thickness cracks within the generalized finite element method [68,69]. Though their method is based on hierarchical interpolation, the crack surface is still C^0 continuous. In [10], they extend their 3D cracking within GFEM based on global-local enrichment strategies. Within the context of the XFEM, Sukumar et al. [71], Moes et al. [42] and Gravouil et al. [25] were the first who presented a three-dimensional implementation with application to linear elastic fracture mechanics (LEFM) that was extended to finite deformation theory [33]. Chopp and Sukumar [8], Lo et al. [32], Sukumar et al. [72] presented interesting methods for multiple cracks in three dimension. The few articles that describe the application of the element-free Galerkin method to three dimensional crack problems are the papers by Krysl and Belytschko [28], Sukumar et al. [70], Duflot [14] and Bordas et al. [6] Rabczuk et al. [56].

With respect to their implementation, there are countless ways to track the crack path in three dimensions. In two dimensions, the crack is simply a line and propagation is just the extension of this line. In three dimensions, we are dealing with a crack surface that can be concave or convex. The crack surface can be represented by segments (different mesh or equations of the crack segments) or by other methods such as level sets and signed distance functions. It is still a tricky task to propagate the crack front in three dimensions and many assumptions have to be made. Another difficulty is to model crack junctions and crack branching if simultaneously crack path continuity should be guaranteed. We will give first a non-excessive overview on 3D-cracking and then focus on two algorithms that we used.

Three dimensional modeling of crack initiation and propagation still poses essential difficulties, especially when crack path continuity should be enforced. Efficient methods that don't rely on representing the crack as continuous surface are given in the meshfree context by Rabczuk and Belytschko [52], Rabczuk et al. [55] and in the context of embedded elements by Sancho et al. [60]. We note that Sancho et al. [60] used a certain amount of crack adaptability, i.e. the crack orientation is allowed to change once the crack propagates, similar to rotating crack models. They reported that this simple trick is sufficient in order to avoid stress locking and secondary cracking without introducing global crack tracking algorithms and exclusion zones, i.e. zones where no crack is allowed to initiate. In the following we will address methods that do enforce crack path continuity.

There are very few approaches so far in the literature that can handle arbitrary crack propagation in three dimensions and preserve crack path continuity at the same time. One of the first methods that can handle arbitrary crack propagation in three dimensions was developed by Martha et al. [36] in the finite element method. Many methods are restricted to planar crack growths and linear elastic fracture mechanics or were applied only for such cases. The methods of Xu and Ortiz [78], Xu et al. [79] and Galdos [20] are models that can handle planar three dimensional cracks in elastostatics.

Cervenka [7] proposed a method to model three dimensional cracks based on a fracture mechanics concept. Within the embedded finite element method, the work of Feist and Hofstetter [18] and Mosler and Meschke [43] is explicitly mentioned. Through-the-thickness cracks within the generalized finite element method were proposed by Duarte et al. [12].

For linear elastic fracture mechanics within the extended finite element concept, Moes et al. [42] and Gravouil et al. [25] published two subsequent papers how to treat arbitrary non-planar three dimensional cracks. The crack surface was represented smoothly and is tracked with level sets. Chopp and Sukumar [8], Lo et al. [32] presented interesting methods for multiple cracks in three dimension for LEFM. Areias and Belytschko [2] employed XFEM to three dimensional crack problems for nonlinear materials. The crack was represented by triangles. A similar approach was pursued by Gasser and Holzapfel [21] within the Partition of Unity finite element method (PUFEM). Though reasonable results were presented, they did not enforce crack path continuity completely in their original paper (only across certain lines). In a subsequent paper, they [22] improved their method in the sense that crack path continuity is completely preserved. Their method is based on a two-step algorithm. First the crack normal is computed according to the Rankine criterion. Then a standard smoothing operation of the crack normal is done:

$$\mathbf{n} = \frac{\sum_{J \in \mathcal{S}} \Phi_J(\mathbf{X})\ \mathbf{n}_J}{|\sum_{J \in \mathcal{S}} \Phi_J(\mathbf{X})\ \mathbf{n}_J|} \tag{8.23}$$

In [23], they applied their model to 3D cracking in bones and combined their local crack tracking algorithm with a global crack tracking scheme. Other interesting work dealing with three-dimensional crack growth within the XFEM can be found e.g. in Bordas [5], Mergheim et al. [38] and Pereira et al. [50].

More recently, Oliver et al. [46] published a paper where the embedded element method is compared to XFEM for three dimensional crack problems. For several examples, they showed that embedded elements can give almost the same results as XFEM.

The few articles that describe the application of the element-free Galerkin method to three dimensional crack problems are the papers by Krysl and Belytschko [28], Sukumar et al. [70], Duflot [14], Zhuang and Augarde [80] and Bordas et al. [6], Rabczuk et al. [56]. In the first three papers, the visibility or the diffraction method was used to describe the crack surface. While the two latter papers [14,70] deal with LEFM cracks in statics, the first paper by Krysl and Belytschko [28] proposes a general method in elastodynamics. Bordas et al. [6], Rabczuk et al. [56] proposed a local PU-based method for three-dimensional crack problems for nonlinear materials in dynamics. The approach was also applied in statics in large deformation settings [57].

8.3.2.1 Tracking the crack path in 3D with plane segments

Crack initiation

We propose now a simple method to model three dimensional crack initiation and crack growth in meshfree method when crack path continuity is desired. We used this method mainly in combination with the loss-of-material-stability criterion or the Rankine criterion as transition criterion to the discontinuity in all our computations. The direction of the crack is then given by the discontinuous bifurcation analysis and completely arbitrary. The movement of the crack front at each time step is governed by the size of the background cell of meshfree methods, i.e. the crack front is always aligned with the edges of background cells that is constructed using many tetrahedrons, to ensure the crack path continuity. Therefore the crack is represented by piecewise continuous three dimensional plates, each of which is a cross section of a tetrahedron.

Due to the higher order continuous meshfree approximations, better stress distributions are obtained compared to linear finite elements. Hence, no special techniques are generally needed to smooth the crack path.[3] However, even in meshfree methods, a smoothing technique is sometimes needed to guarantee crack path continuity of a propagating crack front, especially for cracks with high curvature and when coarse discretizations are used. To guarantee crack path continuity of a propagating crack front, we smooth the crack normal obtained from the localization analysis. The smoothed, nonlocal crack normal used to determine the crack direction at a given Gauss point is given by the normalized weighted average of the neighboring normals, where the weight functions are the meshfree shape functions:

$$\mathbf{n}_{\text{nonlocal}} = \frac{\sum_{J\in\mathcal{S}} \Phi_J(\mathbf{X})\, \mathbf{n}_J}{\left\| \sum_{J\in\mathcal{S}} \Phi_J(\mathbf{X})\, \mathbf{n}_J \right\|}, \tag{8.24}$$

where $\mathcal{S}$ is the set of cells neighboring the cell of interest.

If loss of material stability is detected at a Gauss point, an "almost" – this will be explained in the following – penny shaped crack is introduced. We control the crack length, so that the radius of the newly initiated penny-shaped crack is the interpolation radius of the closest particle. If a background cell is crossed by the circle defined by the intersection of the ball of influence of the closest particle and the cracking plane, the cell is assumed cracked. This is illustrated in three dimensions in Fig. 8.5 and, for clarity, in two dimensions, in Fig. 8.6. Possible intersections of a tetrahedron by the crack plane are given in Fig. 8.7. Due to the fact that the entire background cell is cracked, the crack will not be exactly penny shaped but will depend on the arrangement of the background cells; as is evident in Fig. 8.5. At this point, it becomes obvious that crack initiation depends also on the size of the tetrahedra background cell (and the meshfree dilation parameter). Accuracy is improved significantly by adaptive refinement.

It may happen that several Gauss points lose material stability in a given cell. In this case, the cracking plane is assumed to go through the iso-bar center of the cracked Gauss points, and the normal is taken as the average of the normals associated with each cracked Gauss point. An example of this is shown in Fig. 8.8. If the normals computed at the failing Gauss points

[3] In linear FEM, the crack path tends to oscillate if no smoothing techniques are applied.

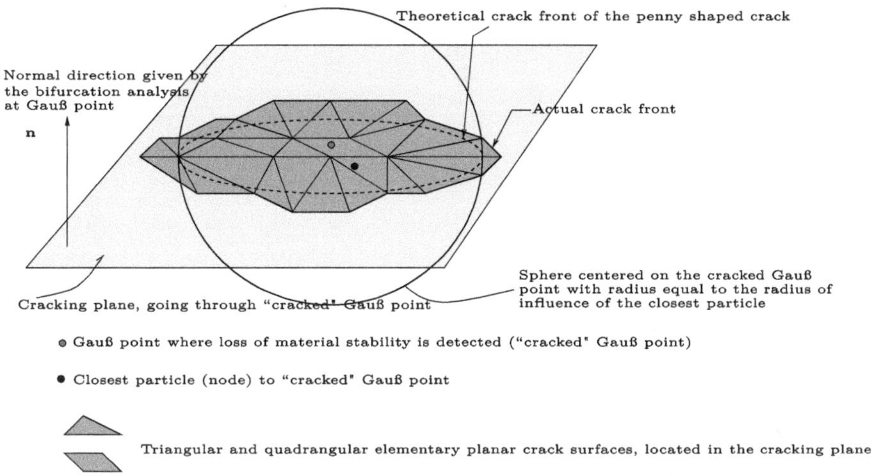

Figure 8.5 Representation of a penny-shaped crack as defined upon crack initiation. Note that the crack is not exactly circular. *h* is the dilation parameter, i.e. the diameter of the domain of influence of the particle closest to the Gauss point where material instability is detected. Note that the normal, **n** is in fact an MLS average of neighboring normals, themselves computed through the discontinuous bifurcation analysis.

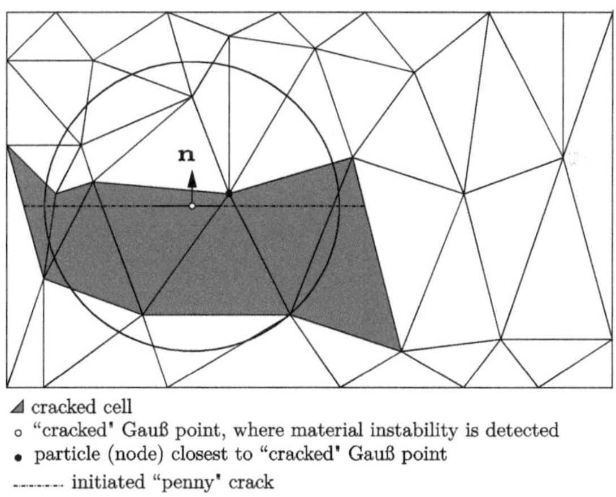

Figure 8.6 Two-dimensional representation of a penny-shaped crack (here a line segment) as defined upon crack initiation. Note that the crack length is not exactly equal to the radius of influence of the closest node, since all background cells that are intersected by the crack segment are assumed *completely* cracked. *h* is the dilation parameter, i.e. the diameter of the domain of influence of the particle closest to the Gauss point where material instability is detected. Note that the normal, **n** is in fact an MLS average of neighboring normals, themselves computed through the discontinuous bifurcation analysis.

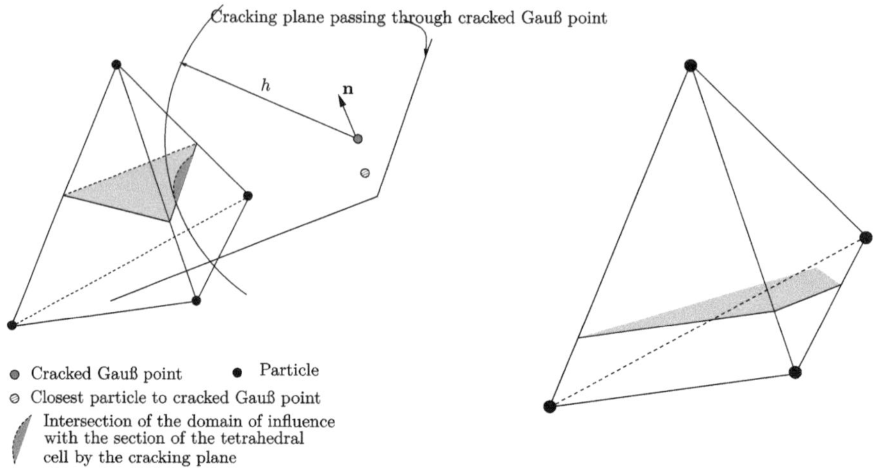

Figure 8.7 Elementary triangular and quadrangular sections of a tetrahedral cell by the cracking plane, defined by its normal obtained by the bifurcation analysis.

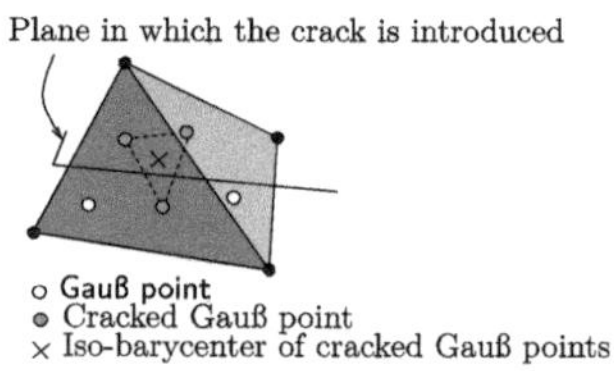

Figure 8.8 Crack initiation in the case of several cracked Gauss points in a given cell.

are too different, then, this is a sign that a crack is trying to branch in the cell.

Note that the piecewise linear line used to define the front (tip) enrichment functions is aligned with the edges of the background cells where the crack front closes. A possible front is shown for illustration in Fig. 8.9. The union of the spherical enriched domains centered on points on the crack front forms an enriched tube around the front. We found it advantageous to place enriched nodes at the end of each crack front line. Difficulties arise at sharp corners or kinks at the crack front, since a local coordinate system cannot be defined uniquely at each of the vertices on the front. We use the enrichment to the closest distance to the crack front and did not observe any major difficulties. A smoothing of the crack front shape can alleviate these problems. Here, a smooth representation of the crack surface

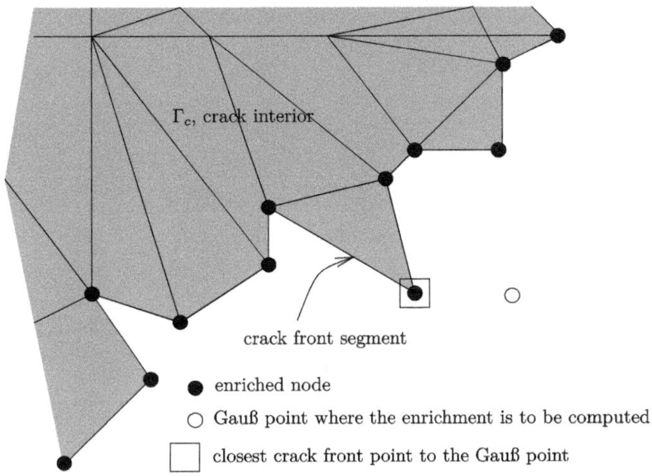

Figure 8.9 A possible piecewise linear crack front.

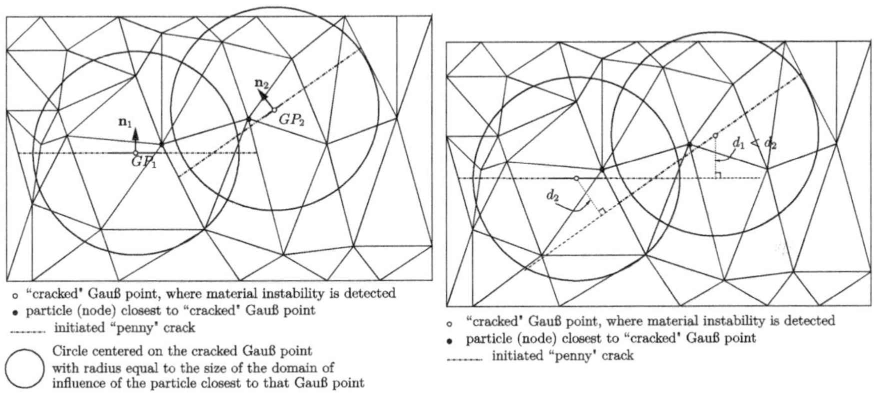

Figure 8.10 Two cracks initiating and joining simultaneously. All cells crossed by the circle, and by the line with normal **n** and passing through the failed Gauss point are considered cracked.

by level sets or non-uniform rational B-splines (NURBS) can definitely be advantageous.

Another possible case is that of cracks initiating and joining simultaneously, as shown in Figs. 8.10 and 8.11. Due to computational efficiency, this algorithm was modified in [58] such that simultaneously initiating cracks are allowed to intersect and cross. It was applied to multiple cracking with large deformations.

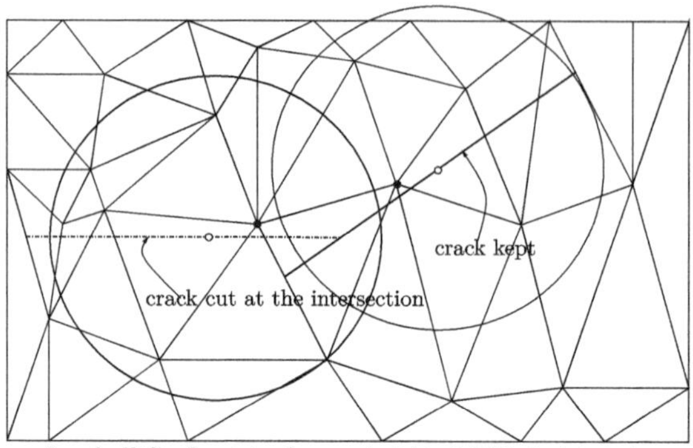

Figure 8.11 Two cracks initiating and joining simultaneously, the one further from the failed Gauss point is cut at the intersection with the other initiating crack.

Crack front propagation and crack junctions

A difficult task in three-dimensions is to track the crack path, at least if crack path continuity is desired. We propose a simple scheme to propagate and join cracks in three dimensions; crack branching is not explicitly considered as e.g. in Daux et al. [9] and Belytschko et al. [4].

The crack propagation procedure is the same as the crack nucleation technique defined in the previous section. We would like to mention again, that it is important to maintain crack path continuity if the crack is described as single surface since otherwise, the solution will depend on the mesh. If the crack normals[4] of adjacent integration cells differ severely, the crack front can be erratic that can lead to erroneous crack paths (see Fig. 8.12 and Fig. 8.13). This occurs mainly in dynamic simulations when too coarse meshes are used and for high crack curvatures. Though arbitrary cracks can be modeled with enriched meshfree methods, there is still a coupling between the crack and the discretization. As discussed before, a simple smoothing technique can alleviate such problems.

There are two possible elemental planar crack surface shapes: triangular and quadrangular, see Fig. 8.7. When two cracks are joining, we assume the propagating crack to stop at the intersection line, as outlined in Section 8.3.2.1 and depicted in Fig. 8.14. In case two propagating or newly created cracks are joining, we always stop the newly created crack and cut

[4] Obtained from the physical cracking criterion.

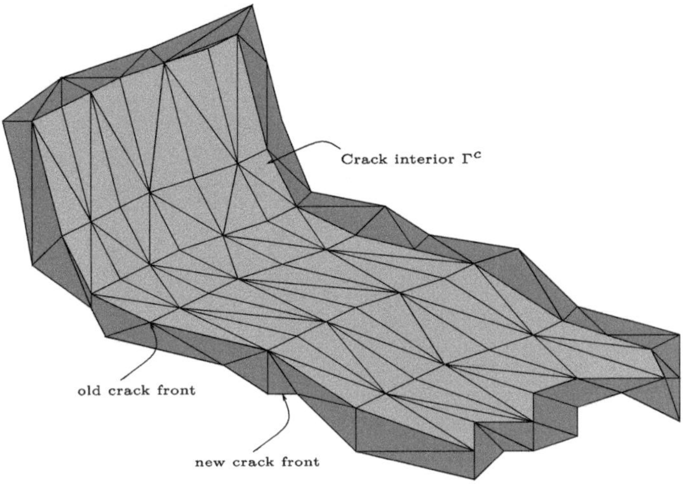

Figure 8.12 Propagation of the crack front.

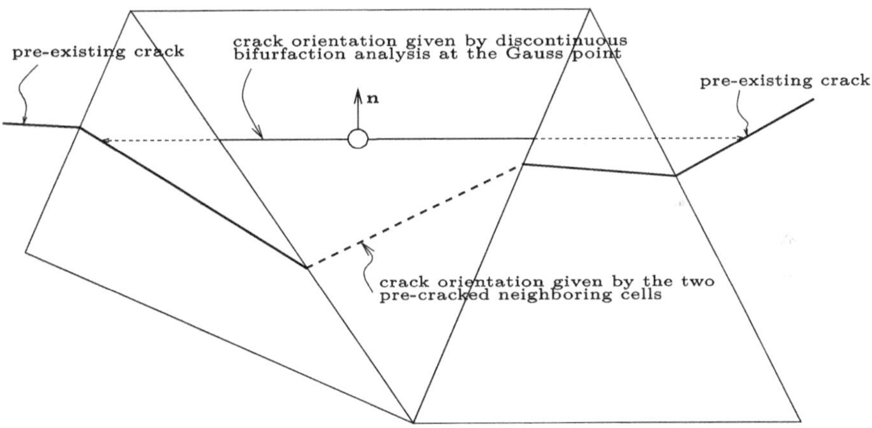

Figure 8.13 Possible incompatibilities between neighboring crack normals, and normals given by the discontinuous bifurcation analysis.

the crack at its intersecting line. In rare cases, where two newly initiated cracks join in a tetrahedral cell, as illustrated in Fig. 8.16 simultaneously we compute the distance of the midpoint of the crack plane to the material points where hyperbolicity is lost. The crack surface with smaller distance is kept while the other one will be cut at the intersection, as is shown in Figs. 8.10 and 8.11. The general algorithm is illustrated in Fig. 8.15.

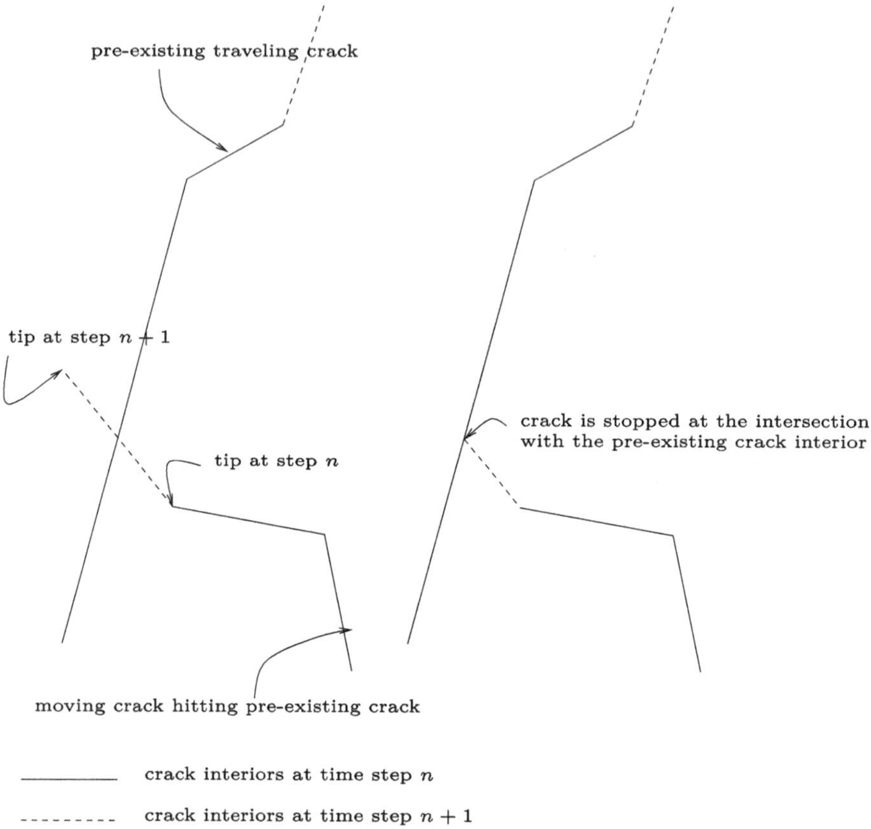

Figure 8.14 When two cracks intersect, the moving crack is stopped at the intersection with the existing crack. Configurations such as that of the left of the figure are not allowed.

Rabczuk et al. [58] extended the algorithm to include crack branching. Crack branching is realized as outline in Section 5.6.3.

8.3.2.2 Smooth crack path representation with level sets

We also would like to remark that a smooth crack surface causes difficulties due to integration. We have not done any arrangements and just increased the number of Gauss points though this is not the most elegant way.

Moreover, the determination of the normal to the crack surface is complicated since the normal varies at any point at the crack surface. While it is relatively simple to represent a pre-given crack surface smoothly with level sets, it is quite complicated to advance the crack surface smoothly. At this

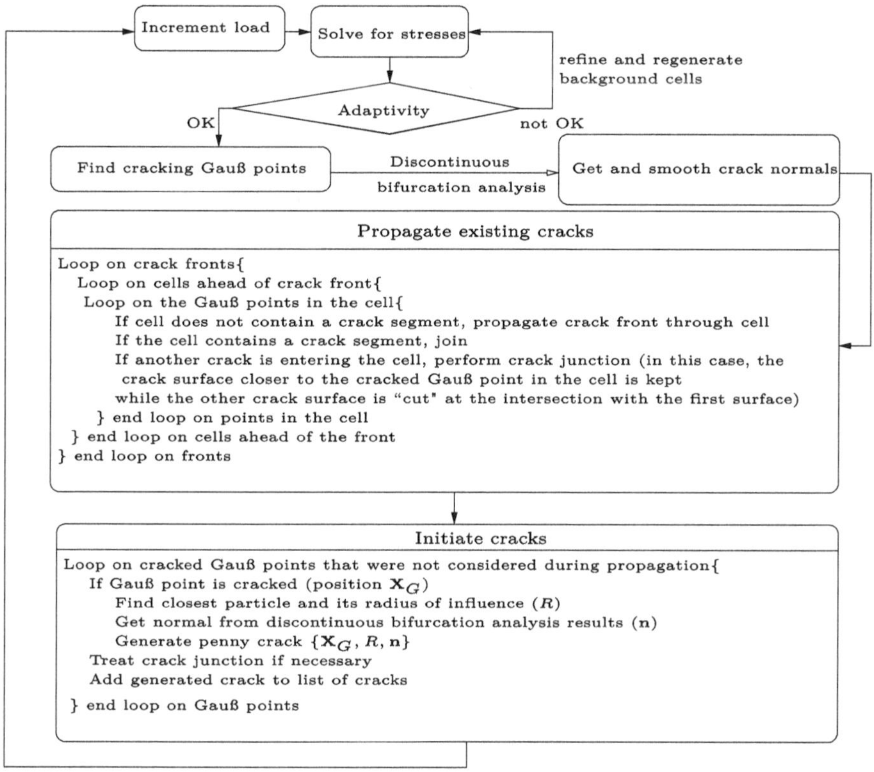

Figure 8.15 Algorithm for crack propagation, initiation and junction.

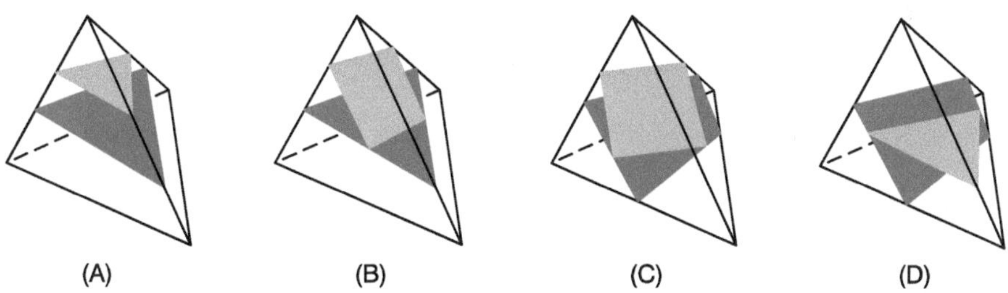

Figure 8.16 Crack junction.

point, we would like to mention that the crack initiation and propagation criterion influences the representation of the crack surfaces. The loss-of-material-stability criterion for examples directly determines the normal to the crack. Hence, if the crack surface is smooth, it might be not consistent with the cracking criterion. Moreover, it will complicate the implemen-

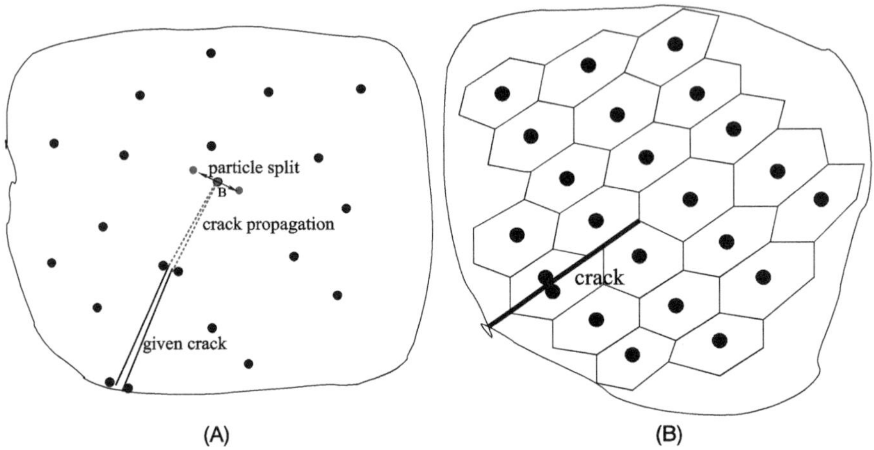

Figure 8.17 (A) Scheme of crack propagation and particle split; (B) Voronoi cells for a particle arrangement with a crack.

tation since the loss of material stability criterion is checked at the Gauss points, nodes or stress points. To obtain a smooth representation of the crack using the loss-of-material-stability criterion, MLS fitting is needed to obtain a spatial distribution of the crack normal. In a next step, the zero iso-level of the level set needs to be discretized.

8.3.3 Adaptive crack propagation technique

We will propose now a two dimensional adaptive cracking algorithm that we used in combination with the visibility method though it is also applicable to other techniques such as the diffraction and transparency method. Therefore, imagine a given crack as shown in Fig. 8.17. Suppose that the cracking criterion is met for particle B close to the crack tip. The crack will propagate in the direction of this particle. We treat the crack by two adjacent surfaces as illustrated in Fig. 8.17. Hence, particle B is split into two new particles. The particle split requires the recomputation of the new particle masses. They might be computed according to a Voronoi diagram where the new crack boundary has to be taken into account, see Fig. 8.17B. More simply, the masses can be halved when a particle is split. Since an adaptive refinement is used to obtain good resolution near the crack, the masses of all affected particles have to be recomputed. Therefore, we compute the consistent mass matrix after every adaptation step. The diagonal

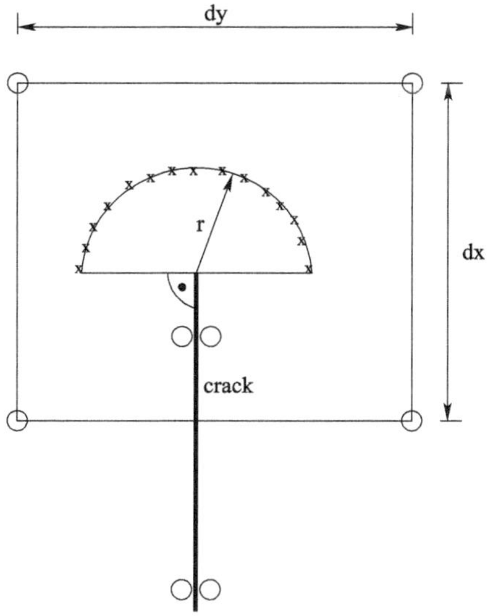

Figure 8.18 Scheme for the circular refinement.

mass matrix is obtained by a row sum technique as described in [3]. All other data are kept from the original particle.

To obtain good resolution near the crack and to insure that the crack is propagated in the correct direction, high particle resolution near the crack, particularly the crack tip, is necessary. Therefore, an adaptive refinement is used at locations with high strain gradients, that is along the crack.

Moreover, we used another technique to better determine the orientation of the crack. In addition to the 'usual' adaptive refinement, particles are added adaptively in a half circle around the crack tip as illustrated in Fig. 8.18. They are distinguished from the other particles by a superimposed x. All data is interpolated from the neighbor particles which are denoted by a superimposed o. The stresses and strains for such particles are:

$$\mathbf{F}^x = \sum_J \nabla\Phi(\mathbf{X}^x - \mathbf{X}_J^o, h)\ \mathbf{u}_J^o\ ,\ \mathbf{P}^{x,t+dt} = \mathbf{P}^{x,t} + \mathbf{E}_t^x : \mathbf{F}^x \tag{8.25}$$

The stresses $\mathbf{P}^{x,t}$ are interpolated from the original particles. The stresses $\mathbf{P}^{x,t+dt}$ can be obtained directly from the total deformation tensor $\mathbf{F}$ or by interpolation.

A crucial point is the choice of the radius r of the half circle. It is chosen as the minimum particle distance $\Delta x = \sqrt{dx^2 + dy^2}$ to $r = \alpha \min \Delta x$, with

$0.25 < \alpha < 1$. The particle at x, the previous crack tip, is kept and split. All other particles associated with this point are removed in the next step. This is necessary since with such excessive refinement, very small particle masses and volumes would be obtained. A small value r also destroys the stable time step in an explicit time integration scheme. The distance between the new (adaptively added) particles and the old particles is checked, too. If the distance undershoots a given value, the corresponding old particle is deleted. This ensures a larger stable time step.

For quasistatic behavior, r plays a secondary role. For dynamic behavior, r has to be chosen carefully, since the crack speed might be influenced.

8.3.4 Comments

Modeling the crack and tracking the crack path in 3D and for problems involving branching and joining cracks is extremely difficult and cumbersome. Moreover, reliable criteria for branching cracks are still missing. Crack branching typically occurs when cracking is determined at a material point within a certain radius around the crack front while crack initiation occurs if cracking is detected outside that given radius. While this is unproblematic in quasi static applications with few cracks, it can cause severe difficulties in dynamic applications with many cracks. It causes even more trouble with finite elements since crack branching has to be considered explicitly as in Daux et al. [9] and Belytschko et al. [4]. Therefore, consider Fig. 8.19A. If the crack should bifurcate, the element in front of the crack tip has to be modified such that it can handle two bifurcating cracks.

One difficulty is to distinguish between crack propagation and crack initiation. Another difficulty in dynamic applications is to decide when and whether a crack branches or not. Crack *propagation* typically occurs when cracking is detected at a material point with a certain radius around the crack front while crack *initiation* occurs if cracking is detected outside that given radius. While this is unproblematic in quasi static applications with few cracks, it can cause severe difficulties in dynamic applications with many cracks.

To consider branching cracks, let us consider exemplary Fig. 8.19A. First of all, crack branching within a single element, Fig. 8.19B, requires the design of special elements as described e.g. in Daux et al. [9] and Belytschko et al. [4]. Secondly, to the best of our knowledge, only empirical models exist in order to decide whether a crack branches or not. These models have to be used with care. The authors for example have used in [53,54] the deviation in the crack orientation in front of the crack tip as criteria for

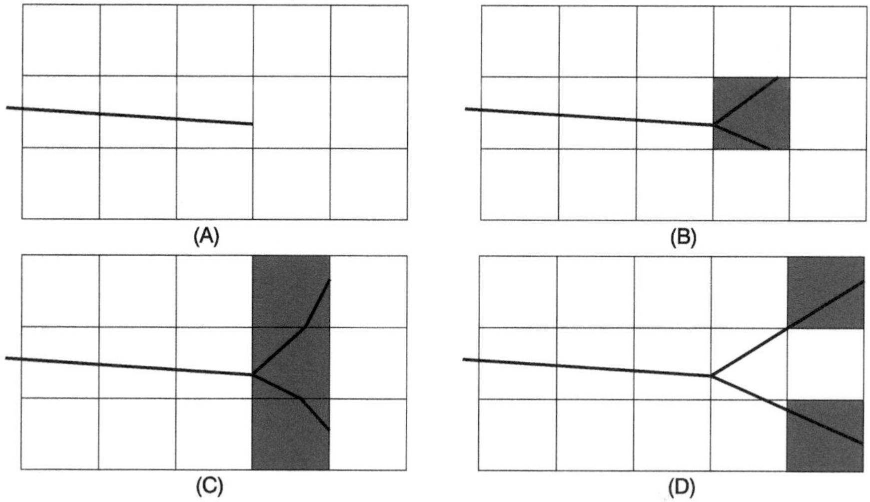

Figure 8.19 Difficulties that occur when crack branching occurs and crack path continuity has to be enforced.

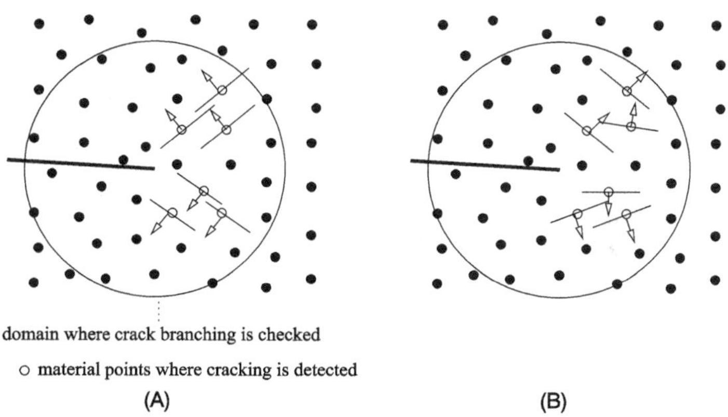

Figure 8.20 Deciding if branching occurs in methods with crack path continuity, (A) compatible crack normal, (B) incompatible crack normal.

crack branching. Fig. 8.20 shows two types of crack branching in methods with crack path continuity with and without compatible crack normals. The main difficulty in dynamic simulations with high loading rate is that the cracking criteria is often met at several sampling points in front of the crack tip simultaneously. For finite elements, the jumps at the 'oscillating' stress field around the crack tip require smoothing techniques. Meshfree methods have a clear advantage due to their higher order continuity and nonlocal interpolation character but we observed 'oscillating' stress fields around the crack tip even in meshfree methods.

Sharon and Fineberg [61] found in brittle fracture, that the crack surface is non-smooth and that there are many micro-cracks in front of the major macro-crack. He showed that the crack speed is drastically diminished due to the daughter micro-crack branches. This way, he was able to explain why methods such as XFEM that model only the major crack overpredicts the crack speed. Though it is principally possible to capture excessive micro-crack branches also with methods that model the crack path continuously, it is extremely cumbersome.

References

[1] D. Chopp, Esam 495 Notes: Numerical Methods for Moving Interfaces, 2007.

[2] P. Areias, T. Belytschko, Analysis of three-dimensional crack initiation and propagation using the extended finite element method, International Journal for Numerical Methods in Engineering 63 (2005) 760–788.

[3] T. Belytschko, W.K. Liu, B. Moran, Nonlinear Finite Elements for Continua and Structures, John Wiley and Sons, Chichester, 2000.

[4] T. Belytschko, N. Moes, S. Usui, C. Parimi, Arbitrary discontinuities in finite elements, International Journal for Numerical Methods in Engineering 50 (4) (2001) 993–1013.

[5] S. Bordas, Extended Finite Element and Level Set Methods With Applications to Growth of Cracks and Biofilms, Phd thesis, Northwestern University, December 2003.

[6] S. Bordas, T. Rabczuk, G. Zi, Three-dimensional crack initiation, propagation, branching and junction in non-linear materials by an extended meshfree method without asymptotic enrichment, Engineering Fracture Mechanics 75 (5) (2008) 943–960.

[7] J. Cervenka, Discrete Crack Modeling in Concrete Structures, Phd thesis, University of Colorado, 1994.

[8] D. Chopp, N. Sukumar, Fatigue crack propagation of multiple coplanar cracks with the coupled extended finite element/fast marching method, International Journal of Engineering Science 41 (8) (2003) 845–869.

[9] C. Daux, N. Moes, J. Dolbow, N. Sukumar, T. Belytschko, Arbitrary branched and intersection cracks with the extended finite element method, International Journal for Numerical Methods in Engineering 48 (2000) 1731–1760.

[10] C. Duarte, D. Kim, Analysis and applications of a generalized finite element method with global-local enrichment functions, Computer Methods in Applied Mechanics and Engineering 197 (6–8) (2008) 487–504.

[11] C. Duarte, L. Reno, A. Simone, A generalized finite element method for the simulation of three-dimensional dynamic crack propagation, Computer Methods in Applied Mechanics and Engineering 119 (15–17) (2001) 2227–2262.

[12] C. Duarte, L. Reno, A. Simone, A high-order generalized FEM for through-the-thickness branched cracks, International Journal for Numerical Methods in Engineering 72 (3) (2007) 325–351.

[13] R. Duddu, S. Bordas, D. Chopp, B. Moran, A combined extended finite element and level set method for biofilm growth, International Journal for Numerical Methods in Engineering 74 (5) (2008) 848–870.

[14] M. Duflot, A meshless method with enriched weight functions for three-dimensional crack propagation, International Journal for Numerical Methods in Engineering 65 (12) (2006) 1970–2006.
[15] M. Duflot, A study of the representation of cracks with level set, International Journal for Numerical Methods in Engineering 70 (11) (2007) 1261–1302.
[16] M. Duflot, N. Hung, A meshless method with enriched weight functions for fatigue crack growth, International Journal for Numerical Methods in Engineering 59 (2004) 1945–1961.
[17] P. Dummerstorf, G. Meschke, Crack propagation criteria in the framework of X-FEM-based structural analyses, International Journal for Numerical and Analytical Methods in Geomechanics 31 (2007) 239–259.
[18] C. Feist, G. Hofstetter, Three-dimensional fracture simulations based on the SDA, International Journal for Numerical and Analytical Methods in Geomechanics 31 (2007) 189–212.
[19] M. Fleming, Y. Chu, B. Moran, T. Belytschko, Enriched element-free Galerkin methods for crack tip fields, International Journal for Numerical Methods in Engineering 40 (8) (1997) 1483–1504.
[20] R. Galdos, A finite element technique to simulate the stable shape evolution of planar cracks with an application to a semi-elliptical surface crack in a bimaterial finite solid, International Journal for Numerical Methods in Engineering 40 (1997) 905–917.
[21] T. Gasser, G. Holzapfel, Modeling 3D crack propagation in unreinforced concrete using PUFEM, Computer Methods in Applied Mechanics and Engineering 194 (25–26) (2005) 2859–2896.
[22] T. Gasser, G. Holzapfel, 3D crack propagation in unreinforced concrete: a two-step algorithm for tracking 3D crack paths, Computer Methods in Applied Mechanics and Engineering 195 (37–40) (2006) 5198–5219.
[23] T. Gasser, G. Holzapfel, A numerical framework to model 3-D fracture in bone tissue with application to failure of the proximal femur, in: Discretization Methods for Evolving Discontinuities, Springer Verlag, Heidelberg, 2007.
[24] R. Goldstein, R. Salganik, Brittle fracture of solids with arbitrary cracks, International Journal of Fracture 10 (1974) 507–527.
[25] A. Gravouil, N. Moes, T. Belytschko, Non-planar 3D crack growth by the extended finite element and level sets – part II: level set update, International Journal for Numerical Methods in Engineering 53 (2002) 2569–2586.
[26] P. Jäger, P. Steinmann, E. Kuhl, On local tracking algorithms for the simulation of three-dimensional discontinuities, Computational Mechanics 42 (2008) 395–406.
[27] P. Jäger, P. Steinmann, E. Kuhl, Towards the treatment of boundary conditions for global crack path tracking in three-dimensional brittle fracture, Computational Mechanics 45 (2009) 91–107.
[28] P. Krysl, T. Belytschko, The element free Galerkin method for dynamic propagation of arbitrary 3-D cracks, International Journal for Numerical Methods in Engineering 44 (6) (1999) 767–800.
[29] S. Li, B.C. Simonson, Meshfree simulation of ductile crack propagation, International Journal for Computational Methods in Engineering Science and Mechanics 6 (2004) 1–19.
[30] S. Li, W. Hao, W. Liu, Mesh-free simulations of shear banding in large deformation, International Journal of Solids and Structures 37 (2000) 7185–7206.

[31] S. Li, W. Hao, W. Liu, Numerical simulations of large deformation of thin shell structures using meshfree methods, Computational Mechanics 25 (2–3) (2000) 102–116.
[32] S. Lo, C. Dong, Y. Cheung, Integral equation approach for 3d multiple crack problems, Engineering Fracture Mechanics 72 (12) (2005) 1830–1840.
[33] S. Loehnert, D. Mueller-Hoppe, P. Wriggers, 3D corrected XFEM approach and extension to finite deformation theory, International Journal for Numerical Methods in Engineering 86 (4–5) (2011) 431–452, https://doi.org/10.1002/nme.3045.
[34] S. Mariani, U. Perego, Extended finite element method for quasi-brittle fracture, International Journal for Numerical Methods in Engineering 58 (2003) 103–126.
[35] J.E. Marsden, T. Hughes, Mathematical Foundations of Elasticity, Dover Publications Inc., New York, 1983.
[36] L. Martha, P. Wawrzynek, A. Ingraffea, Arbitrary crack representation using solid modeling, Engineering With Computers 9 (1993) 63–82.
[37] J. Mergheim, E. Kuhl, P. Steinmann, A finite element method for the computational modelling of cohesive cracks, International Journal for Numerical Methods in Engineering 63 (2005) 276–289.
[38] J. Mergheim, E. Kuhl, P. Steinmann, Towards the algorithmic treatment of 3d strong discontinuities, Communications in Numerical Methods in Engineering 23 (2007) 97–108.
[39] G. Meschke, P. Dummerstorf, Energy-based modeling of cohesive and cohesionless cracks via X-FEM, Computer Methods in Applied Mechanics and Engineering 196 (2007) 2338–2357.
[40] C. Miehe, E. Guerses, A robust algorithm for configurational-force-driven brittle crack propagation with r-adaptive mesh alignment, International Journal for Numerical Methods in Engineering 72 (2007) 127–155.
[41] N. Moes, J. Dolbow, T. Belytschko, A finite element method for crack growth without remeshing, International Journal for Numerical Methods in Engineering 46 (1) (1999) 133–150.
[42] N. Moes, A. Gravouil, T. Belytschko, Non-planar 3-D crack growth by the extended finite element method and level sets, part I: mechanical model, International Journal for Numerical Methods in Engineering 53 (11) (2002) 2549–2568.
[43] J. Mosler, G. Meschke, 3d modeling of strong discontinuities in elastoplastic solids: fixed and rotating localization formulations, International Journal for Numerical Methods in Engineering 57 (11) (2003) 1553–1576.
[44] R.W. Ogden, Non-linear Elastic Deformations, Halsted Press, New York, 1984.
[45] J. Oliver, A. Huespe, E. Samaniego, E. Chaves, Continuum approach to the numerical simulation of material failure in concrete, International Journal for Numerical and Analytical Methods in Geomechanics 28 (2004) 609–632.
[46] J. Oliver, A. Huespe, P. Sanchez, A comparative study on finite elements for capturing strong discontinuities: E-FEM vs. X-FEM, Computer Methods in Applied Mechanics and Engineering 195 (2006) 4732–4752.
[47] J. Oliver, D. Linero, A. Huespe, O. Manzoli, Two-dimensional modeling of material failure in reinforced concrete by means of a continuum strong discontinuity approach, Computer Methods in Applied Mechanics and Engineering 197 (2008) 332–348.
[48] M. Ortiz, Y. Leroy, A. Needleman, Finite element method for localized failure analysis, Computer Methods in Applied Mechanics and Engineering 61 (2) (1987) 189–214.
[49] S. Osher, J.A. Sethian, Fronts propagating with curvature-dependent speed: algorithms based on Hamilton-Jacobi transformations, Journal of Computational Physics 79 (1) (1988) 12–49.

[50] J. Pereira, C. Duarte, X. Jiao, D. Guoy, Generalized finite element method enrichment functions for curved singularities in 3d fracture mechanics problems, Computational Mechanics 44 (2009) 73–92.
[51] B. Prabel, A. Combescure, A. Gravouil, S. Marie, Level set XFEM non-matching meshes: application to dynamic crack propagation in elasto-plastic media, International Journal for Numerical Methods in Engineering 69 (8) (2007) 1553–1569.
[52] T. Rabczuk, T. Belytschko, A three dimensional large deformation meshfree method for arbitrary evolving cracks, Computer Methods in Applied Mechanics and Engineering 196 (29–30) (2007) 2777–2799.
[53] T. Rabczuk, G. Zi, A meshfree method based on the local partition of unity for cohesive cracks, Computational Mechanics 39 (6) (2007) 743–760.
[54] T. Rabczuk, P. Areias, T. Belytschko, A meshfree thin shell method for non-linear dynamic fracture, International Journal for Numerical Methods in Engineering 72 (5) (2007) 524–548.
[55] T. Rabczuk, P. Areias, T. Belytschko, A simplified meshfree method for shear bands with cohesive surfaces, International Journal for Numerical Methods in Engineering 69 (5) (2007) 993–1021.
[56] T. Rabczuk, S. Bordas, G. Zi, A three dimensional meshfree method for static and dynamic multiple crack nucleation/propagation with crack path continuity, Computational Mechanics 40 (3) (2007) 473–495.
[57] T. Rabczuk, G. Zi, S. Bordas, H. Nguyen-Xuan, A geometrically nonlinear three-dimensional cohesive crack method for reinforced concrete structures, Engineering Fracture Mechanics 75 (2008) 4740–4758.
[58] T. Rabczuk, G. Zi, S. Bordas, On three-dimensional modelling of crack growth using partition of unity methods, Computers & Structures 88 (2010) 1391–1411.
[59] E. Samaniego, X. Oliver, A. Huespe, Contributions to the Continuum Modelling of Strong Discontinuities in Two-Dimensional Solids, PhD thesis, Monograph CIMNE No. 72, International Center for Numerical Methods in Engineering, Barcelona, Spain, 2003.
[60] J. Sancho, J. Planas, A. Fathy, J. Galvez, D. Cendon, Three-dimensional simulation of concrete fracture using embedded crack elements without enforcing crack path continuity, International Journal for Numerical and Analytical Methods in Geomechanics 31 (2007) 173–187.
[61] E. Sharon, J. Fineberg, Microbranching instability and the dynamic fracture of brittle materials, Physical Review. B 54 (10) (1996) 7128–7139.
[62] B. Shen, O. Stephansson, Modification of the g-criterion for crack propagation subjected to compression, Engineering Fracture Mechanics 47 (1994) 177–189.
[63] G. Shih, Strain energy density factor applied to mixed-mode crack problems, International Journal of Fracture 10 (1974) 305–321.
[64] M. Silhavy, The Mechanics and Thermodinamics of Continuous Media, Springer Verlag, Berlin, 1997.
[65] D.J. Simkins, S. Li, Meshfree simulations of ductile failure under thermal-mechanical loads, Computational Mechanics 38 (3) (2006) 235–249.
[66] J.C. Simo, J. Oliver, F. Armero, An analysis of strong discontinuities induced by strain-softening in rate-independent inelastic solids, Computational Mechanics 12 (1993) 277–296.
[67] M. Stolarska, D.L. Chopp, N. Moës, T. Belytschko, Modelling crack growth by level sets in the extended finite element method, International Journal for Numerical Methods in Engineering 51 (2001) 943–960.

[68] T. Strouboulis, I. Babuška, K. Copps, The design and analysis of the generalized finite element method, International Journal for Numerical Methods in Engineering 181 (2000) 43–69.
[69] T. Strouboulis, K. Copps, I. Babuška, The generalized finite element method: an example of its implementation and illustration of its performance, International Journal for Numerical Methods in Engineering 47 (8) (2000) 1401–1417.
[70] N. Sukumar, B. Moran, T. Black, T. Belytschko, An element-free Galerkin method for three-dimensional fracture mechanics, Computational Mechanics 20 (1997) 170–175.
[71] N. Sukumar, N. Moes, B. Moran, T. Belytschko, Extended finite element method for three-dimensional crack modelling, International Journal for Numerical Methods in Engineering 48 (2000) 1549–1570.
[72] N. Sukumar, D. Chopp, E. Bechet, N. Moes, Three-dimensional non-planar crack growth by a coupled extended finite element and fast marching method, International Journal for Numerical Methods in Engineering 76 (5) (2008) 727–748.
[73] T. Most, C. Bucher, Energy-based simulation of concrete cracking using an improved mixed-mode cohesive crack model within a meshless discretization, International Journal for Numerical Methods in Engineering 31 (2007) 285–305.
[74] G. Ventura, J. Xu, T. Belytschko, A vector level set method and new discontinuity approximations for crack growth by EFG, International Journal for Numerical Methods in Engineering 54 (6) (2002) 923–944.
[75] G. Ventura, E. Budyn, T. Belytschko, Vector level sets for description of propagating cracks in finite elements, International Journal for Numerical Methods in Engineering 58 (2003) 1571–1592.
[76] G.N. Wells, L.J. Sluys, A new method for modelling cohesive cracks using finite elements, International Journal for Numerical Methods in Engineering 50 (2001) 2667–2682.
[77] C.-H. Wu, Fracture under combined loads by maximum energy release rate criterion, Journal of Applied Mechanics 45 (1978) 553–558.
[78] G. Xu, M. Ortiz, A variational boundary integral method for the analysis of 3d cracks of arbitrary geometry modellled as continuous distributions of dislocation loops, International Journal for Numerical Methods in Engineering 36 (1993) 3675–3701.
[79] G. Xu, F. Bower, M. Ortiz, An analysis of non-planar crack growth under mized mode loading, International Journal of Solids and Structures 31 (1994) 2167–2193.
[80] X. Zhuang, C. Augarde, Aspects of the use of orthogonal basis functions in the element-free Galerkin method, International Journal for Numerical Methods in Engineering 81 (3) (2010) 366–380.

CHAPTER NINE

Multiscale methods for fracture

In computational materials design, multiscale methods are powerful for extracting material parameters based on the fine-scale details. While numerous multiscale methods (see e.g. [26,46,47]) were developed for intact materials, far fewer methods are applicable for fracture simulations. Multiscale methods can be categorized into hierarchical, semi-concurrent and concurrent methods [8], Fig. 9.1. In hierarchical multiscale methods, information is passed from the fine-scale to the coarse-scale but not vice versa. Computational homogenization [22] is a classical up-scaling technique. Hierarchical multiscale approaches are very efficient. However, their extension to model fracture is complex, in particular for materials involving strain softening. One basic assumption for the application of homogenization theories is the existence of disparate length scales [29]: $\mathcal{L}_{Cr} \ll \mathcal{L}_{RVE} \ll \mathcal{L}_{Spec}$ where $\mathcal{L}_{Cr}$, $\mathcal{L}_{RVE}$ and $\mathcal{L}_{Spec}$ are the crack length, the representative volume element (RVE)- and specimen-size, respectively. For problems involving fracture, the first condition is violated as $\mathcal{L}_{Cr}$ is of the order of $\mathcal{L}_{RVE}$. Moreover, periodic boundary conditions (PBC) often used at the fine-scale, cannot be used when a crack touches a boundary as the displacement jump in that boundary violates the PBC.

The basic idea of semi-concurrent multiscale methods is illustrated in Fig. 9.1B. In semi-concurrent multiscale methods, information is passed from the fine-scale to the coarse-scale and vice versa. Semi-concurrent multiscale methods are of the same computational efficiency as concurrent multiscale methods. The key advantage of semi-concurrent multiscale methods over concurrent multiscale methods is their flexibility, i.e. their ability to couple two different software packages, e.g. MD software to FE software. Parallelization is generally simple as well. A classical semi-concurrent multiscale method is the FE^2 [15] originally developed for intact materials. Kouznetsova [22] was the first who extended this method to problems involving material failure, see also Kouznetsova et al. [23] or later contributions by Verhoosel et al. [44] and Belytschko et al. [7].

Numerous concurrent multiscale methods [1,26,46,47] have been developed that can be classified into 'Interface' coupling methods and 'Handshake' coupling methods. Interface coupling methods seem to be less effective for dynamic applications as avoiding spurious wave reflections at the 'artificial' interface seem to be more problematic. Some of the concurrent

Extended Finite Element and Meshfree Methods
https://doi.org/10.1016/B978-0-12-814106-9.00015-2
Copyright © 2020 Elsevier Inc.
All rights reserved.

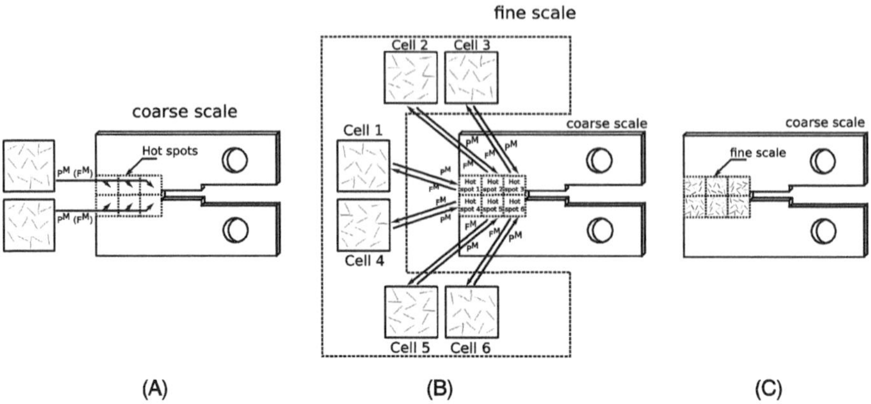

Figure 9.1 Schematic of a (A) hierarchical, (B) semi-concurrent and (C) concurrent multiscale methods.

multiscale methods have been extended to modeling fracture [9,10,19,21, 37,41–43,48].

Another multiscale solution approach for solving problems with highly dependency on fine scale phenomena is the 'Global-Local' method [12,13]. In this method, a local problem is constructed for each node of the global mesh whose support intersects an area of interest (e.g. discontinuity). Since communication between local problems is not required, this method can be efficiently parallelized. Another advantage of this method is that it can be easily implemented in the partition of unity concept such as generalized finite element method (GFEM).

In this book, we present two concurrent multiscale methods for fracture which exploit extended finite element and meshfree methods to model fracture at least on one scale: (1) The Extended Bridging Domain Method/Extended Arlequin Method, which is based on a handshake coupling and therefore developed for dynamics and (2) The Extended Bridging Scale Method which is an efficient interface coupling method. This approach is developed in statics.

9.1. Extended Bridging Domain Method

The Extended Bridging Domain Method or Extended Arlequin Method is an extension of the Bridging Domain Method or Arlequin Method, respectively, for problems involving fracture. Therefore, consider the domain $\Omega = \Omega^{fs} \cup \Omega^{cs}$, the superscripts *fs* and *cs* denoting the fine-scale and coarse-scale, respectively. The outer boundary of Ω^{cs} is denoted by $\partial\Omega^{cs}$ with $\partial\Omega^{cs} = \partial\Omega_t^{cs} \cup \partial\Omega_u^{cs} \cup \partial\Omega_{cz}^{cs}$ and $\partial\Omega_t^{cs} \cap \partial\Omega_u^{cs} = \emptyset$, $\partial\Omega_{cz}^{cs} \cap \partial\Omega_u^{cs} = \emptyset$, $\partial\Omega_t^{cs} \cap \partial\Omega_{cz}^{cs} = \emptyset$; the subscripts u, t and cz indicate 'displacement-', 'traction-' and 'cohesive zone-', respectively. Note that $\partial\Omega_{cz} = \partial\Omega_{int} \cup \partial\Omega_c$ consists of the crack boundary $\partial\Omega_c$ and the interface between two materials M1 and M2 $\partial\Omega_{int}$. The area in front of the crack tip (crack front in 3D) is of particular interest and is therefore modeled by a fine-scale containing features of the micro-structure of the material. The fine scale can be a 'micro-structure' which is commonly based on continuum mechanics and hence modeled by finite elements, meshfree methods. However, it can also be based on other approaches such as molecular dynamics. The coarse scale is based on a homogenized continuum and discretized with finite elements, meshfree methods, etc. accordingly. The Extended Bridging Domain Method will subsequently be described for coupling two continuum models though the coupling between continuum and atomistic models is similar, see e.g. [42] which presents a coupling the FEM code PERMIX with LAMMPS.

In the Bridging Domain/Arlequin method, the coarse-scale is continuously blended into a fine-scale in a so-called handshake domain $\Omega^h = \Omega^{cs} \cap \Omega^{fs}$. The Arlequin method is based on a linear weighting of the energy of the coarse-scale and fine-scale in Ω^h:

$$W(\mathbf{u}) = w(\mathbf{X})\, W^{cs} + (1 - w(\mathbf{X}))\, W^{fs} + W^h \tag{9.1}$$

where W^{cs} is the energy in the coarse-scale, W^{fs} the energy in the fine-scale and W^h denotes the energy due to the coupling of the two scales that is described later. The weighting function $w(\mathbf{X})$ is required to fulfill the following requirements:

$$w(\boldsymbol{X}) = \begin{cases} 1 & \forall \mathbf{X} \in \Omega^{cs} \setminus \Omega^{fs} \\ [0, 1] & \forall \mathbf{X} \in \Omega^h \\ 0 & \forall \mathbf{X} \in \Omega^{fs} \setminus \Omega^{cs} \end{cases} \tag{9.2}$$

In order to define $w(\boldsymbol{X})$, we use a normalized distance function

$$w(\mathbf{X}) = \frac{l(\mathbf{X})}{l_0} \tag{9.3}$$

where $l(\boldsymbol{X})$ is the orthogonal projection of $\boldsymbol{X}$ on the interior boundary of the coarse-scale subdomain and l_0 is the length of this orthogonal projection to the fine-scale boundary as shown in Fig. 9.2 [47].

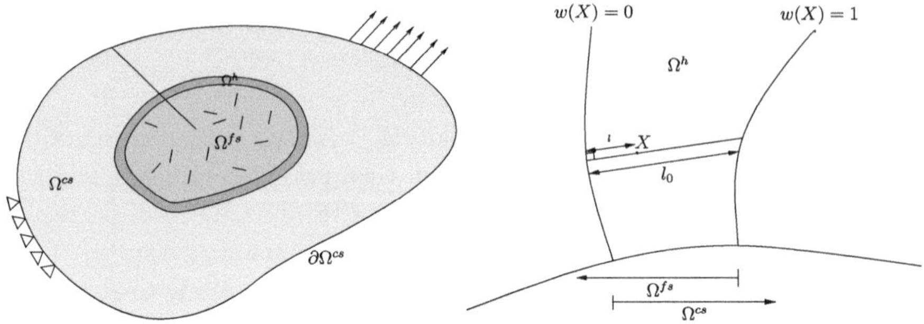

Figure 9.2 (A) The relation between fine- and coarse-scale areas and (B) the geometry of handshake domain, from [47].

9.1.1 Concurrent coupling of two models at different length scales

9.1.1.1 *Variational formulation*

The governing equation is the equilibrium equation that can be expressed in variational formulation: Find $\mathbf{u}^i \in \mathcal{U}\ \forall \delta\mathbf{u}^i \in \mathcal{U}_0$ $(i = fs, cs)$ such that the variation in the energy equals zero:

$$\delta W = w(\mathbf{X})\delta W^{cs} + (1 - w(\mathbf{X}))\delta W^{fs} + \delta W^h = 0 \tag{9.4}$$

with the approximation spaces of admissible trial functions $\mathbf{u}^i$ and test functions $\delta\mathbf{u}^i$

$$\begin{aligned} \mathcal{U} &= \left\{\mathbf{u}^i | \mathbf{u}^i \in \mathcal{H}^1, \mathbf{u}^i = \bar{\mathbf{u}}^i \text{ on } \partial\Omega_u, \mathbf{u}^i \text{ discontinuous on } \partial\Omega_c\right\} \\ \mathcal{U}_0 &= \left\{\delta\mathbf{u}^i | \delta\mathbf{u}^i \in \mathcal{H}^1, \delta\mathbf{u}^i = 0 \text{ on } \partial\Omega_u, \delta\mathbf{u}^i \text{ discontinuous on } \partial\Omega_c\right\} \end{aligned} \tag{9.5}$$

in which δW^{cs} and δW^{fs} can be defined as:

$$\delta W^{cs} = \delta W^{cs}_{int}(\boldsymbol{u}^{cs}, \boldsymbol{u}^{fs}) - \delta W^{cs}_{ext}(\boldsymbol{u}^{cs}) - \delta W^{cs}_{coh}(\boldsymbol{u}^{cs}) \tag{9.6}$$

$$\delta W^{fs} = \delta W^{fs}_{int}(\boldsymbol{u}^{cs}, \boldsymbol{u}^{fs}) - \delta W^{fs}_{ext}(\boldsymbol{u}^{fs}) - \delta W^{fs}_{coh}(\boldsymbol{u}^{fs}) \tag{9.7}$$

with

$$\begin{aligned} \delta W^{cs}_{int}(\boldsymbol{u}^{cs}, \boldsymbol{u}^{fs}) &= \int_{\Omega^{cs}} \boldsymbol{\sigma}(\mathbf{u}^{cs}, \mathbf{u}^{fs}) : \text{grad}^{\text{sym}}(\delta\mathbf{u}^{cs}) \text{d}\Omega \\ \delta W^{cs}_{ext}(\boldsymbol{u}^{cs}) &= \int_{\Omega^{cs}} \mathbf{f} \cdot \delta\mathbf{u}^{cs} \text{d}\Omega + \int_{\partial\Omega^{cs}_t} \mathbf{t} \cdot \delta\mathbf{u}^{cs} \text{d}\partial\Omega \\ \delta W^{cs}_{coh}(\boldsymbol{u}^{cs}) &= \int_{\partial\Omega^{cs}_{disc}} \mathbf{t}^c \cdot [\![\delta\mathbf{u}^{cs}]\!] \text{d}\Gamma \end{aligned} \tag{9.8}$$

$$\delta W_{int}^{fs}(\boldsymbol{u}^{cs},\boldsymbol{u}^{fs}) = \int_{\Omega^{fs}} \boldsymbol{\sigma}(\mathbf{u}^{cs},\mathbf{u}^{fs}) : \text{grad}^{\text{sym}}(\delta\mathbf{u}^{fs})\text{d}\Omega$$
$$\delta W_{ext}^{fs}(\boldsymbol{u}^{fs}) = \int_{\Omega^{fs}} \mathbf{f}\cdot\delta\mathbf{u}^{fs}\text{d}\Omega$$
$$\delta W_{coh}^{fs}(\boldsymbol{u}^{fs}) = \int_{\partial\Omega_{disc}^{fs}} \mathbf{t}^{c}\cdot[\![\delta\mathbf{u}^{fs}]\!]\text{d}\Gamma + \int_{\partial\Omega_{int}^{fs}} \mathbf{t}^{int}\cdot\delta\boldsymbol{\Delta}\text{d}\Gamma \tag{9.9}$$

$$\delta\boldsymbol{\Delta} = \mathbf{u}^{M1}\mid_{\delta\Omega_{int}^{fs}} - \mathbf{u}^{M2}\mid_{\delta\Omega_{int}^{fs}} \tag{9.10}$$

where $\mathbf{u}^{M1}\mid_{\delta\Omega_{int}^{fs}}$ and $\mathbf{u}^{M2}\mid_{\delta\Omega_{int}^{fs}}$ denote the relative displacement between two materials (M1 and M2) at their interface at the fine scale[1]; $\mathbf{t}^{c}$ and $\mathbf{t}^{int}$ are the cohesive tractions for cracks and for the interfaces between different materials, respectively; $\mathbf{f}$ are the body forces and $\mathbf{t}$ are the tractions (von Neumann boundary conditions on $\partial\Omega_t^{cs}$); no von Neumann boundary conditions are imposed on the fine-scale boundary. Note that the stress tensor $\boldsymbol{\sigma}$ in the handshake domain depends on both the fine-scale and the coarse-scale displacement field. Note also the presence of cohesive forces on both scales. However, the up-scaling procedure for cohesive cracks in the context of a concurrent multiscale method is far from simple; in particular when cohesive forces are considered on both scales. Choosing the size of the fine-scale domain such that no cohesive forces are required on the coarse-scale simplifies Eq. (9.6):

$$\delta W^{cs} = \delta W_{int}^{cs}(\boldsymbol{u}^{cs},\boldsymbol{u}^{fs}) - \delta W_{ext}^{cs}(\boldsymbol{u}^{cs}) \tag{9.11}$$

The size of the fine-scale domain needs to be at least of the order of the process zone of the homogenized material which might be computationally challenging depending on the material. The term δW^h refers to the variation in the energy due to the coupling of the coarse-scale to the fine-scale and will be addressed next.

9.1.1.2 Coupling method

The coarse and the fine-scale domains in the Arlequin method is commonly coupled by Lagrange multipliers $\boldsymbol{\lambda}$. Hence, the term δW^h is expressed as

$$\delta W^h = \int_{\Omega^h} \delta\boldsymbol{\lambda}_I^T\boldsymbol{g} + \int_{\Omega^h} \boldsymbol{\lambda}^T\delta\mathbf{g} \tag{9.12}$$

[1] A classical application for multiscale methods for fracture are composites.

with $\mathbf{g} = \mathbf{u}^{cs} - \mathbf{u}^{fs}$. The Lagrange multiplier field $\boldsymbol{\lambda}$ (and $\delta\boldsymbol{\lambda}$) can be discretized according to the discretization of the displacement field. If a purely Heaviside enriched discretization of the displacement field is employed, it will yield the following expression of the Lagrange multiplier discretization:

$$\boldsymbol{\lambda}^h(\mathbf{X}) = \sum_{I \in \mathcal{N}} N_I(\mathbf{X})\,\boldsymbol{\Lambda}_I + \sum_{I \in \mathcal{N}_b} N_I(\mathbf{X})\,\bar{H}\left(f_I(\mathbf{X})\right)\boldsymbol{\Lambda}_I \tag{9.13}$$

$$\delta\boldsymbol{\lambda}^h(\mathbf{X}) = \sum_{I \in \mathcal{N}} N_I(\mathbf{X})\,\delta\boldsymbol{\Lambda}_I + \sum_{I \in \mathcal{N}_b} N_I(\mathbf{X})\,\bar{H}\left(f_I(\mathbf{X})\right)\delta\boldsymbol{\Lambda}_I \tag{9.14}$$

Gracie et al. [19] noted in the context of coupling an atomistic model to a continuum model through the Arlequin method that by omitting the enriched part for the discretization of the Lagrange multiplier field, i.e. the second term on the RHS of Eqs. (9.13) and (9.14), more accurate results are obtained (i.e. when comparing the results of the coupled model to results of atomistic simulations). They argue that the missing surface energy in the continuum coupling justifies such a coupling. Indeed, one of the key advantages of an adaptive concurrent multiscale method is that the coarse-scale contains less fine-scale features for computational savings. Or in other words: The kinematics of the coarse-scale cannot and should not contain complex crack features of the fine-scale, see Fig. 9.3. In the opinion of the authors, enforcing the same crack kinematics on the coarse-scale and the fine-scale defeats the purpose of the coupling method. A computationally cheaper option might be to transform the domain integral in Eqs.[2] (9.8) and (9.9) to a boundary integral as done in the smoothed finite element method and then enforce displacement compatibility in a weak sense (or least square sense) over the element edges of the coarse-scale elements in Ω^h.

Instead of the L_2 coupling described above, Guidault et al. [21] proposed an H^1 coupling[3]:

$$\delta W^h = \int_{\Omega^h} \delta\boldsymbol{\lambda}_I^T \mathbf{g} + \int_{\Omega^h} \boldsymbol{\lambda}^T \delta\mathbf{g} + \int_{\Omega_h} l^2\left(\delta\boldsymbol{\Lambda}^T\mathbf{G} + \boldsymbol{\Lambda}\delta\mathbf{G}\right)\, d\Omega \tag{9.15}$$

with $\boldsymbol{\Lambda} = \nabla \otimes \boldsymbol{\lambda}$ and $\mathbf{G} = \nabla \otimes \mathbf{u}^{cs} - \nabla \otimes \mathbf{u}^{fs}$. Their results show only marginal differences to the L_2 coupling though it allows the development of error estimators which can drive adaptive refinement and coarsening schemes.

[2] First line.

[3] Requiring additionally compatibility in the strain field.

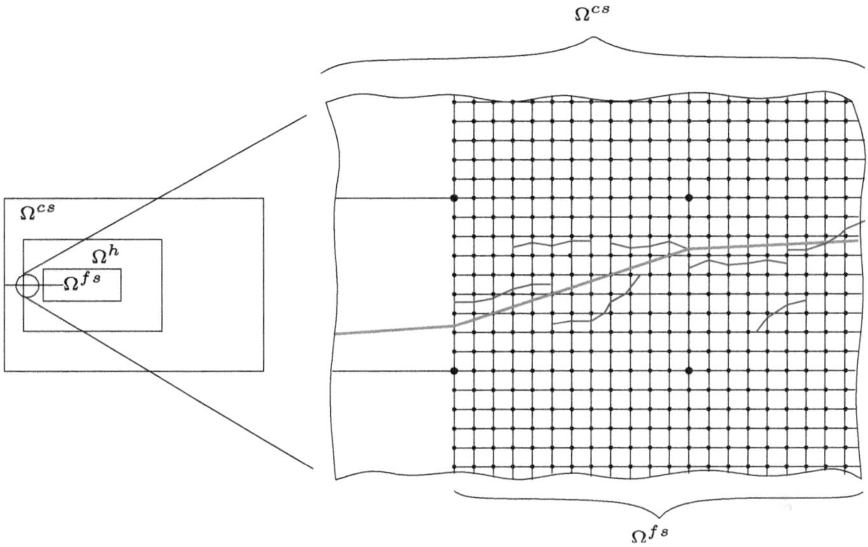

Figure 9.3 Crack kinematics on the coarse-scale and the fine-scale.

9.1.1.3 Adaptive coarsening and refinement

When the crack grows, the size of the fine-scale domain should be adjusted automatically in the course of the simulation for sake of computational efficiency. Consequently, an appropriate coarsening and refinement strategy is needed. Baumann et al. [2], Oden et al. [30] developed adaptive coarsening and refinement algorithms for coupling atomistic models to continuum models. However, they provided such a scheme only for propagation of a single dislocation in XFEM. An efficient adaptive coarsening and refinement scheme for coupled continuum-atomistic models have been presented by the authors of this book in [10] which is presented in more detail in Section 9.2.

One important aspect is to find the 'coarse-scale' elements in the handshake domain that should contain the macroscopic cracks. The fine-scale will contain initial defects and it is expected that some of those defects will grow even in areas where no macroscopic crack is needed. To determine the elements to be coarsened or refined, the consistent material tangent stiffness in the coarse-scale elements can be estimated based on fine-scale features:

$$\mathbf{C}^t = \frac{1}{V_{el}^{cs}} \int_{V_{el}^{cs}} \frac{\partial \boldsymbol{\sigma}^{fs}}{\partial \boldsymbol{\epsilon}^{fs}} \, dV \tag{9.16}$$

with the volume of the coarse-scale element V_{el}^{cs}. When the 'coarse-scale' consistent material tangent stiffness looses positive definiteness, a macroscopic crack is introduced. In order to obtain the initial nodal parameters in the coarse-scale, [7] proposed a scheme that enforces compatibility in the coarse-scale strains and fine-scale strains in an average sense:

$$\langle \boldsymbol{\epsilon}_{cs} \rangle = \langle \boldsymbol{\epsilon}_{fs} \rangle \tag{9.17}$$

in which $\boldsymbol{\epsilon}_{fs}$ and $\boldsymbol{\epsilon}_{cs}$ can be defined as:

$$\langle \boldsymbol{\epsilon}_{fs} \rangle \stackrel{def}{=} \frac{1}{\Omega_{fs}} \int_{\Omega_{fs}} \nabla \mathbf{u}_{fs} \, d\Omega \tag{9.18}$$

$$\langle \boldsymbol{\epsilon}_{cs} \rangle \stackrel{def}{=} \frac{1}{\Omega_{cs}} \int_{\Omega_{cs}} \nabla \mathbf{u}_{cs} \, d\Omega \tag{9.19}$$

In order to obtain the 'initial' nodal values of the additional degrees of freedom on the coarse-scale, the energy in the fine-scale and the coarse-scale can be equated yielding

$$\langle \sigma_{ij} : \epsilon_{ij} \rangle_{\Omega_{fs}} = \langle \sigma_{ij} : \epsilon_{ij} \rangle_{\Omega_{cs}} \tag{9.20}$$

As mentioned above, the adaptively coarse-grained domain should not contain all fine-scale features for computational reasons. Developing efficient procedures particularly for complex fracture patterns still remains a challenge for multiscale methods for fracture.

9.1.1.4 Size of the fine-scale domain

Choosing (adaptively) the size of the fine-scale domain is not a trivial task. The size of the fine-scale domain can be obtained by error estimators. The size of the fine-scale also directly influences the equations of the Arlequin method. If the fine-scale includes the entire process zone, then the cohesive zone term in δW_{ext}^{cs} on the coarse-scale, see Eq. (9.6), can be omitted. In other words, when the minimum crack opening on the coarse-scale exceeds the crack opening displacement w_{max} (see Fig. 9.4) with $\mathbf{t}_c = 0$, then no cohesive force term is needed on the coarse-scale. This facilitates the up-scaling procedure. The characteristic length of the process zone can be estimated according to e.g. [14,27] by

$$l_{ch} = \frac{EG_f}{f_t^2} \quad \text{or} \quad l_{ch} = \frac{EG_f}{(1-\vartheta^2) f_t^2} \tag{9.21}$$

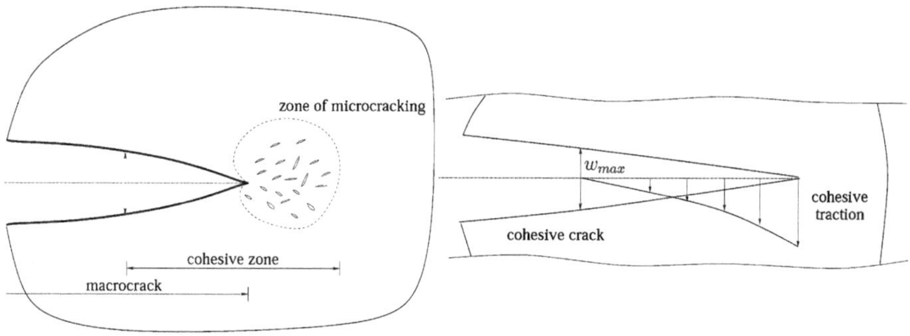

Figure 9.4 Cohesive crack model.

Note that some handshake coupling methods, e.g. the multiscale projection method [7], do not offer this flexibility as the fine-scale domain overlaps the entire coarse-scale domain. In the context of a semi-concurrent multiscale method, a procedure to extract a coarse-scale cohesive zone model from the fine-scale was proposed by [44].

9.1.2 Consistency of material properties

It still needs to be assured that the homogenized coarse-scale material behaves as the fine-scale material when the material is still intact. This can be achieved for instance by computational homogenization in the context of hierarchical upscaling if the material response is linear. For nonlinear materials, a semi-concurrent FE^2 approach seem suitable. If the fine-scale domain is atomistic, the Cauchy-Born rule is typically adopted in BDM and XBDM.

9.2. Extended bridging scale method

The Bridging Scale Method belongs to the category of concurrent interface coupling methods and has been extended to fracture in [9,33]; see also the formulation in a meshless context in [48]. Compared to the Extended Bridging Domain Method, the XBSM has the following distinctive features/differences:

- The coarse and the fine scale is coupled by enforcing the displacement boundary conditions on so-called ghost atoms (no handshake domain).
- No derivatives of the shape functions are needed as will be shown subsequently.

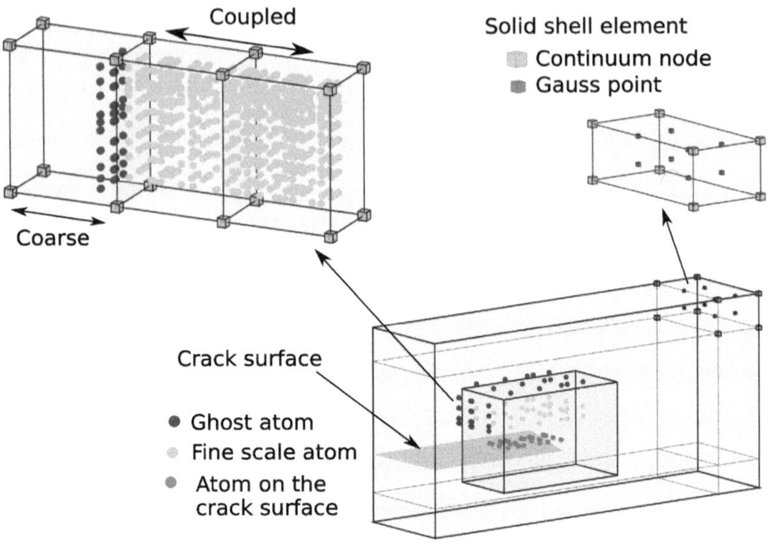

Figure 9.5 Schematic of a three-dimensional coupled continuum-atomistic model showing the mechanics of coarse scale domain modeled with solid shell element.

- No Cauchy Born rule is used. The coarse-scale material model is obtained through a so-called virtual atom cluster based on a quasi-continuum approach (energy equivalence between the fine-scale and the coarse-scale).
- The coarse scale problem and the fine scale problem can be solved independently.
- The XBSM has been developed only for coupling a continuum (coarse-scale) domain with an atomistic (fine-scale) domain.

Consider a three-dimensional multiscale model shown in Fig. 9.5 for the adaptive simulation of crack growth. To model fracture in the coarse scale region, the methods presented in the earlier chapters such as XFEM or the phantom node method can be employed. Fracture in the fine-scale region happens 'naturally' due to breaking bonds between atoms.

For efficiency reasons, it is important to model only the area around the crack tip/front base on the fine-scale approach as suggested in Fig. 9.5. An initial crack in the fine scale region is commonly modeled by deleting the bonds between the atoms on the crack surface and updating the neighbor list accordingly. Ghost atoms located on the boundary of the coarse region but within the cutoff radius of the atoms in the fine region, are used to enforce the boundary conditions for the fine scale solution. In the two scale model, the total displacement field $\mathbf{u}_\alpha$ of an atom α is decomposed into coarse and fine scale components:

$$\mathbf{u}_\alpha = \mathbf{u}_\alpha^{\mathrm{C}} + \mathbf{u}_\alpha^{\mathrm{A}} \tag{9.22}$$

where $\mathbf{u}_\alpha^{\mathrm{C}}$ is the coarse scale component and $\mathbf{u}_\alpha^{\mathrm{A}}$ is the fine scale component. The fine scale component $\mathbf{u}_\alpha^{\mathrm{A}}$ is the difference between the actual position of an atom α and the interpolated position of the coarse scale. Therefore, $\mathbf{u}_\alpha^{\mathrm{A}}$ is insignificant in the regions far away from the crack tip, and hence, $\mathbf{u}_\alpha^{\mathrm{C}}$ is sufficient to model the deformation in the coarse scale region. On the other hand, in the fine scale region, both coarse and fine scale components are required. Let the coarse scale displacement $\mathbf{u}_\alpha^{\mathrm{C}}$ of an atom α be represented by a set of FEM basis functions defined over a set of n^{C} nodal points,

$$\mathbf{u}_\alpha^{\mathrm{C}} = \sum_{I=1}^{n^{\mathrm{C}}} N_I(\mathbf{X}_\alpha)\mathbf{u}_I^{\mathrm{C}} \tag{9.23}$$

where $N_I(\mathbf{X}_\alpha)$ is the shape functions defined at node I, estimated at the αth atom with the material coordinate $\mathbf{X}_\alpha$, and $\mathbf{u}_I^{\mathrm{C}}$ is the continuum displacement vector at node I.

In the bridging scale method, the coupling conditions are realized by enforcing the displacement boundary conditions on the ghost atoms, see Fig. 9.5. The positions of the ghost atoms are interpolated from the coarse scale solution. Let β be the index of the ghost atoms; the corresponding ghost atom displacements are estimated as:

$$\mathbf{u}_\beta^{\mathrm{C}} = \sum_{I=1}^{n^{\mathrm{C}}} N_I(\mathbf{X}_\beta)\mathbf{u}_I^{\mathrm{C}} \tag{9.24}$$

where $N_I(\mathbf{X}_\beta)$ are the shape functions defined at node I, estimated at the βth atom with material coordinates $\mathbf{X}_\beta$.

9.2.1 Consistency of material properties

Consistency of material properties between the fine-scale and coarse-scale in the coarse-scale area (i.e. the area far away from the crack tip) is realized through a quasi continuum approach taking advantage of a so-called virtual atom cluster (VAC). The VAC assumes symmetry of a crystal structure, where a cluster of atoms is taken as the representative model of the whole lattice structure [34,35]. Consequently, this approach is limited to certain

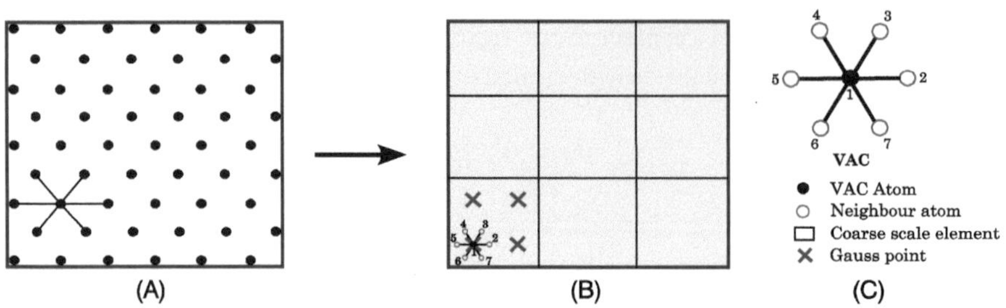

Figure 9.6 A demonstration of VAC coarse scale model in two dimensions. (A) Atomistic model with triangular lattice as on the (111) plane of an fcc material. (B) Equivalent continuum model with the VAC being placed at a particular Gauss point. (C) Details of the VAC [11].

materials. However, all the calculations can be performed with reference to the representative cluster instead of the whole lattice, which leads to improved computational efficiency. Since the locations of atoms in the cluster do not represent the exact locations of the atoms, the representative cluster is called a virtual atom cluster (VAC). The same inter atomic potential as in the full scale atomistic model is used for the VAC [9,33,35]. A full scale atomistic model will be realized when the VAC assumes the structure of the underlying lattice. In other words, the VAC can be regarded as an efficient coarse graining technique.

A typical VAC based coarse scale model in two dimensions is shown in Fig. 9.6. The total potential energy of a fine scale system as shown in Fig. 9.6A is given by the sum of all bond potentials ϕ_α. Consider an equivalent coarse scale model based on the VAC, illustrated in Fig. 9.6B. Since the fine scale and coarse scale models are equivalent, their potential energy must be equal. This is achieved by defining a distributed energy density function ϕ_ρ [35].

Considering the periodic nature of the lattice, ϕ_ρ is defined as the potential energy of a VAC divided by the volume of the VAC. For a homogeneous lattice, each VAC consists of a single atom and its volume is that of the unit cell of the lattice. Therefore, the distributed energy density function ϕ_ρ for a triangular lattice (see Fig. 9.6) can be defined as [9]:

$$\phi_\rho = \frac{\phi_{\mathrm{VAC}}}{\mathrm{V}_0} = \frac{1}{2}\sum_{\beta=2}^{7} \frac{V(r_{1\beta})}{\sqrt{3}a^2/2} = \frac{1}{2}\sum_{\beta=2}^{7} \phi_{1\beta} \tag{9.25}$$

Using the definition of ϕ_ρ from Eq. (9.25), the internal nodal forces can be expressed as [9]:

$$\mathbf{F}_I^{\text{int}} \approx -\sum_{\text{G}} w_{\text{G}} \frac{\partial \phi_\rho^{\text{G}}}{\partial \mathbf{u}} \frac{\partial \mathbf{u}}{\partial \mathbf{u}_I^{\text{C}}} \approx -\sum_{\text{G}} w_{\text{G}} \sum_{\alpha=1}^{7} \frac{\partial \phi_\rho^{\text{G}}}{\partial \mathbf{u}_\alpha^{\text{C}}} \frac{\partial \mathbf{u}_\alpha^{\text{C}}}{\partial \mathbf{u}_I^{\text{C}}}$$
$$= -\sum_{\text{G}=1}^{n^{\text{G}}} w_{\text{G}} \sum_{\alpha=1}^{7} \frac{\partial \phi_\rho^{\text{G}}}{\partial \mathbf{u}_\alpha^{\text{C}}} N_I(\mathbf{X}_\alpha). \tag{9.26}$$

The term $\frac{\partial \phi_\rho}{\partial \mathbf{u}_\alpha^{\text{C}}}$ in Eq. (9.26) is evaluated for each atom α in the VAC as given below:

$\alpha = 1$

$$\frac{\partial \phi_\rho}{\partial u_{1i}^{\text{C}}} = \frac{\partial \phi_{12}}{\partial r_{12}} \frac{r_{12i}}{r_{12}} + \frac{\partial \phi_{13}}{\partial r_{13}} \frac{r_{13i}}{r_{13}} + \frac{\partial \phi_{14}}{\partial r_{14}} \frac{r_{14i}}{r_{14}} + \frac{\partial \phi_{15}}{\partial r_{15}} \frac{r_{15i}}{r_{15}} + \frac{\partial \phi_{16}}{\partial r_{16}} \frac{r_{16i}}{r_{16}} + \frac{\partial \phi_{17}}{\partial r_{17}} \frac{r_{17i}}{r_{17}} \tag{9.27}$$

$\alpha = 2$ to 7

$$\frac{\partial \phi_\rho}{\partial u_{\alpha i}^{\text{C}}} = -\frac{\partial \phi_{1\alpha}}{\partial r_{1\alpha}} \frac{r_{1\alpha i}}{r_{1\alpha}} \tag{9.28}$$

where i is the index of the coordinate axes. Knowing the internal nodal forces in Eq. (9.26), the minimization problem can be solved for the coarse scale solution by minimizing the potential energy for the given boundary conditions.

9.2.2 Upscaling and downscaling

For computational efficiency, it is of utmost importance to propagate the fine-scale region when cracks and dislocations propagate. This requires efficient techniques for upscaling and downscaling in the case of propagating fractures. Transforming coarse-scale domains into fine-scale domains (downscaling) is usually done in regions in front of the crack tip/front when the material is intact at both scales. Hence, obtaining the positions of the fine-scale atoms from the coarse-scale is quite simple. Particularly challenging is upscaling in the case of fracture, i.e. transforming part of the fine-scale domain into an 'equivalent' coarse-scale domain. Upscaling for fracture requires basically two key ingredients: (1) Detection of 'fine-scale' fractures and (2) an efficient and accurate upscaling approach.

9.2.2.1 Detection of 'fine-scale' fractures

Energy criteria as well as the centro symmetry parameter (CSP) are commonly employed to detect vacancies, dislocations and 'cracks' in an atomistic model.

Energy criteria

The total potential energy of an atomistic system is estimated as the sum of all bond potentials ϕ_α. The bond potential of a particular atom depends on the distance between the atom (α) and its neighbors (β). In the initial configuration, all the atoms are assumed to possess the same potential energy. The initial crack is created by deleting the bonds between the atoms and updating the neighbor list accordingly. Continuous increase in the external load leads to the stretching of the bonds of the atoms around the crack tip. Increase in bond length/distance between the atoms leads to increase in the system potential energy. A bond breaks when the bond length reaches a certain threshold, transferring the load to the immediate neighbors.

Therefore, the atoms around the crack tip possess the highest energy in the entire lattice. This is in agreement with continuum theory, where stress concentrations are observed around the crack tip. Hence, the potential energy provides an indication of the location of the crack tip. The energy criterion has been successfully applied to detect the locations of the crack tip [9] and the core of the dislocation [20]. Let $\mathcal{E}_n^{\mathrm{HE}}$ be the set of elements containing at least one atom with high potential energy, i.e.

$$\mathcal{E}_n^{\mathrm{HE}} = \{\mathrm{e} \in \mathcal{E}_n^{\mathrm{A}} \mid \text{energy of an atom in e} > \mathrm{tol}^{\mathrm{E}}\} \tag{9.29}$$

where $\mathcal{E}_n^{\mathrm{A}}$ is the set of total atoms and $\mathrm{tol}^{\mathrm{E}}$ is the specified energy tolerance. As a guideline, $\mathrm{tol}^{\mathrm{E}}$ can be specified in the range of 15 and 30% higher than the energy of an atom in equilibrium in a perfect lattice.

Centro symmetry parameter (CSP)

Another common criterion to detect 'fractures' in an atomistic model is based on the centro symmetry parameter of an atom α which is defined as [31]:

$$\mathrm{CSP}_\alpha = \sum_{\beta=1}^{n^{\mathrm{nb}}/2} |\mathbf{r}_{\alpha\beta} + \mathbf{r}_{\alpha(\beta+n^{\mathrm{nb}}/2)}|^2 \tag{9.30}$$

where $\mathbf{r}_{\alpha\beta}$ and $\mathbf{r}_{\alpha(\beta+n^{\mathrm{nb}}/2)}$ are the distance between the atoms α and β and α and $(\beta + n^{\mathrm{nb}}/2)$, respectively and n^{nb} are the total number of nearest neighbors of atom α. Consider an atom α in the fine scale region containing face centered cubic (fcc) lattice structure. Let β denote the neighbors of α. In an fcc lattice structure every atom α is surrounded by 6 nearest neighbors

Table 9.1 Range of centro symmetry parameter for various defects, normalized by square of the lattice parameter a_0^2.

Defect	$\mathrm{csp}_\alpha/a_0^2$	**Range** $\Delta\mathrm{csp}_\alpha/a_0^2$
Perfect lattice	0.0000	$\mathrm{csp}_\alpha < 0.1$
Partial dislocation	0.1423	$0.01 \leq \mathrm{csp}_\alpha < 2$
Stacking fault	0.4966	$0.2 \leq \mathrm{csp}_\alpha < 1$
Surface atom	1.6881	$\mathrm{csp}_\alpha > 1$

(n^{nb}). Therefore, the CSP of the atom α in the fcc lattice is given by:

$$\mathrm{CSP}_\alpha = \sum_{\beta=1}^{3} |\mathbf{r}_{\alpha\beta} + \mathbf{r}_{\alpha(\beta+3)}|^2 \tag{9.31}$$

From Eq. (9.31), the CSP of an atom α in the fcc lattice, is the summation of square of the total distance between the opposing neighbors. In other words, the CSP of an atom in a periodic perfect lattice structure with symmetric atomic arrangement is zero and the CSP values of the atoms on the defect surface/stacking fault is not equal to zero. This criterion is used to separate the atoms on the crack surface. Normalized CSP values for various defects are listed in Table 9.1. From Table 9.1, atoms on the crack surface can be distinguished as the atoms possessing normalized CSP values greater than or equal to 1.6881.

9.2.2.2 Upscaling, downscaling and adaptivity

To improve the computational efficiency, the fine scale region is adaptively enlarged (downscaling) with the defect propagation and the region behind the core of the defect (e.g. crack tip) is coarse-grained. Adaptive multiscale methods have been proposed by several researchers in the context of different methods, see e.g. the contributions in [33,45]. Consider a fine scale domain embedded within the 'boundaries' of the nodes/particles around the crack tip. The refinement algorithm should be activated sufficiently often such that a buffer layer of elements/'regions' is always maintained between the crack tip and the coupling boundary. The 'regions' refer to the area/volume generated by connecting the immediate neighboring particles in meshless methods, such that they resemble the elements in the mesh based techniques. Secondly, to ensure that the refinement operation is not activated in the first load step itself, at least one layer of elements/regions is commonly considered between the crack tip and the buffer element layer.

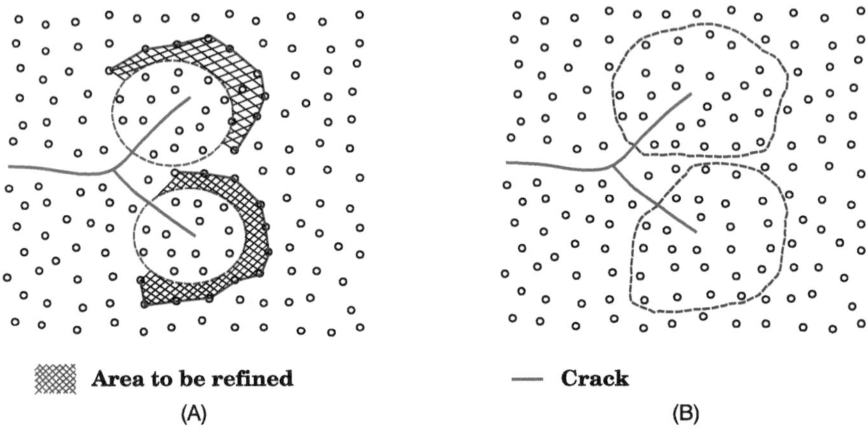

Figure 9.7 Sketch of the adaptive refinement operation. (A) Flagged particles to be refined are hashed. (B) Increased atomistic region after the refinement operation. Picture reproduced with permission from [48].

Finally, the crack tip element layer is sandwiched by at least one layer of elements/regions in the transverse direction. The adaptivity scheme consists of an adaptive refinement (downscaling) and coarse graining operations (upscaling) which can be summarized as follows:

1. Estimate the region in the coarse scale domain Ω^{C} to be refined. A refinement operation involves the expansion of the fine scale region by converting the estimated coarse region into a fine region, Fig. 9.7.
2. Estimate the region in the fine scale domain Ω^{A} to be coarsened. In a coarse graining operation the coarse region is expanded by converting the estimated fine region into a coarse region, see Fig. 9.8.

In the above steps, when the sizes of the regions refined and coarse grained are similar, the net change in the size of the fine scale domain is almost zero. As a result, the fine scale region is adaptively moved with the propagation of the defect. As mentioned before, (adaptively) choosing the size of the fine-scale domain remains a challenge.

Downscaling-adaptive refinement

The major steps of the refinement (Fig. 9.7) procedure for a multiscale method based on an atomistic fine scale model can be described as follows:

1. Identify the region to be refined (Ω_{ref}).
2. Create and initialize the atoms in Ω_{ref}.

3. Identify and update the newly cracked atoms.
4. Update the fine and coarse scale regions.

Fig. 9.7A shows the region identified for a refinement operation. The fine scale region after the refinement is depicted in Fig. 9.7B. Let the nodes/particles (before a refinement operation) in the fine, coarse and completely cracked regions are indicated by, $\mathcal{P}_n^{\mathrm{A}}$, $\mathcal{P}_n^{\mathrm{C}}$ and $\mathcal{P}_n^{\mathrm{split}}$, respectively. The region containing split elements indicates the completely cracked region. The steps of a refinement operation are:

- Calculate the atoms on the crack surface based on the CSP and store the regions containing the atoms on the crack surface into the set $\mathcal{P}_n^{\mathrm{csp}}$.
- Estimate the neighbors of the regions containing the atoms on the crack surface in $\mathcal{P}_n^{\mathrm{csp}}$ and store them in $\mathcal{P}_{n+1}^{\mathrm{minA}}$.
- Calculate the regions to be refined, $\mathcal{P}_{n+1}^{\mathrm{refine}}$ by removing the fine scale region $\mathcal{P}_n^{\mathrm{A}}$ from the set $\mathcal{P}_{n+1}^{\mathrm{minA}}$.
- Flag the regions to be refined and increase the atomistic domain by creating the atoms in the flagged elements.
- Initialize the positions of the newly created atoms through interpolation based on the coarse scale solution.
- Update the fine and coarse regions after a refinement operation. Update the neighbor list ($nlist_{n+1}$) of the fine scale atoms in the current load step ($n+1$).
- Identify the newly cracked particles in the fine scale region and update the split and tip nodes and the nodal connectivity table.

A detailed algorithm of selecting the particles to be refined, initializing the newly created atoms in the region identified for refinement and propagating the crack in the coarse scale region in a multiscale framework is explained in [9,48].

Upscaling-adaptive coarse graining

The major steps for the coarse graining operation (Fig. 9.8) are:

1. Identify the fine scale region to be coarse grained (Ω_{coa}).
2. Estimate the equivalent coarse scale region of Ω_{coa}.
3. Delete the atoms in the region to be coarsened.
4. Update the fine and coarse scale particles/nodes.

The process of an adaptive coarse graining operation is explained in Fig. 9.8. Let $\mathcal{P}_{\mathrm{n}}^{\mathrm{CS}}$ be the regions containing atoms on the crack surface at load step n. Let $\mathcal{P}_{\mathrm{n}}^{\mathrm{BA}}$ be the regions lying in the fine scale domain and attached to the coupling 'boundary'. The particles/nodes to be coarsened are the particles/nodes which are in both set $\mathcal{P}_{\mathrm{n}}^{\mathrm{CS}}$ and the set $\mathcal{P}_{\mathrm{n}}^{\mathrm{BA}}$ in front

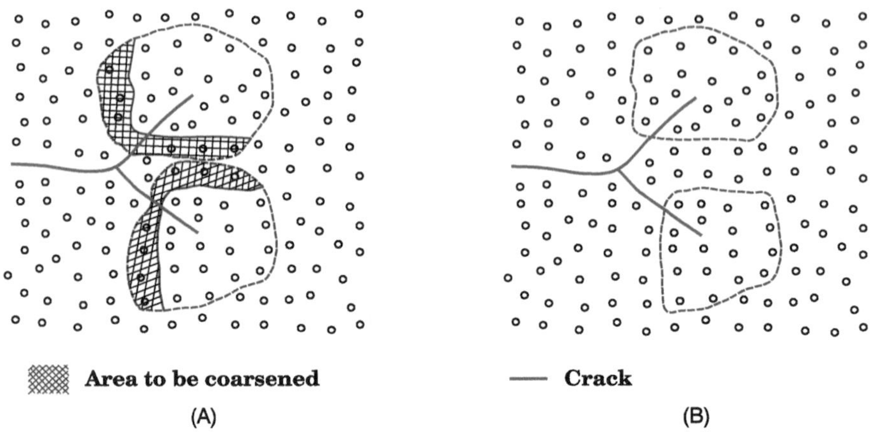

Figure 9.8 Schematic of the adaptive coarsening operation. (A) Flagged particles to be coarsened are hashed. (B) Reduced atomistic region after the coarsening operation. Picture reproduced with permission from [48].

of the crack tip, $\mathcal{P}_{\mathrm{n}}^{\mathrm{coarsen}} = \mathcal{P}_{\mathrm{n}}^{\mathrm{CS}} \cap \mathcal{P}_{\mathrm{n}}^{\mathrm{BA}}$. The steps of an adaptive coarsening operation are:

- Estimate and store the regions containing the elements on the crack surface (far away from the crack tip) into $\mathcal{P}_n^{\mathrm{LE}}$.
- Find the fine scale regions attached to the coupling boundary, $\mathcal{P}_n^{\mathrm{BA}}$.
- The regions to be coarse grained ($\mathcal{P}_{n+1}^{\mathrm{coarsen}}$) are given by $\mathcal{P}_n^{\mathrm{LE}} \cap \mathcal{P}_n^{\mathrm{BA}}$.
- Flag the regions to be coarsen grained and delete the atoms in the flagged regions.
- Update the particle/nodal set in the fine and coarse scale regions and the neighbor list of the fine scale atoms, after a coarsening operation.

Upscaling the fracture related material information from the fine scale to the coarse scale is a major difficulty in multiscale methods for fracture, particularly for complex crack patterns. A robust and simple coarse graining technique in the context of multiscale modeling for fracture is developed by Budarapu et al., [10]. The major steps in [10] to develop an equivalent model of the $\Omega_{\mathrm{def}}^{\mathrm{A}}$, the coarse graining (CG) method (Fig. 9.9) are:

1. Determine the atoms on the crack surface, e.g., using the CSP.
2. Identify the regions containing atoms on the crack surface, based on the positions of the atoms on the crack surface and the positions of the particles/nodes of the background discretization, see Fig. 9.9B.

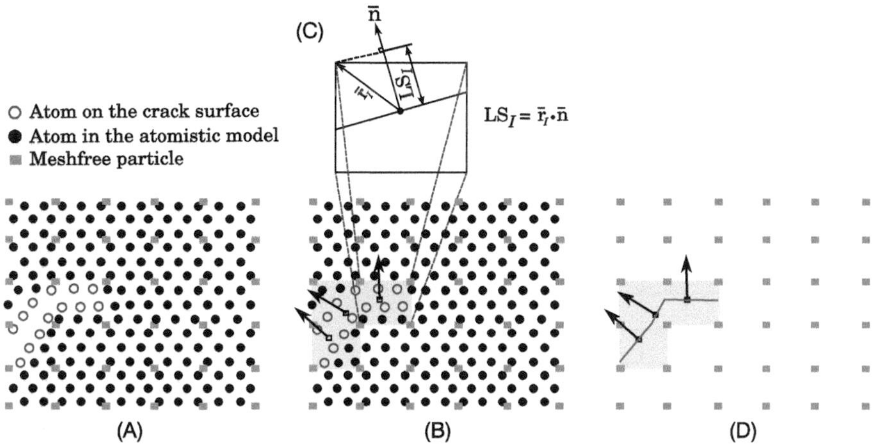

Figure 9.9 Schematic of meshless equivalent coarse scale model. (A) Meshfree particles superimposed on the atomistic model, (B) regions containing the atoms on the crack surface are highlighted along with the normals of the crack surface in each region, (C) calculation of level sets and (D) approximation of crack surface by joining the crack path in the regions containing the crack. Picture reproduced from [10] with permission.

3. Estimate the normal and center of gravity (CoG) of the atoms on the crack surface. Calculate the effective CoG of a crack region by averaging the CoGs of the atoms on the crack surface in the considered crack region.
4. Approximate the crack path in each crack region by joining the effective normal and CoG of the atoms on the crack surface, refer Fig. 9.9D and Section 9.2.2.
5. Estimate the nodes or particles on either side of the crack surface or around the tip, see Fig. 9.9C.

Crack surface orientation

Consider a deformed configuration of the fine scale model, superimposed with a discretized coarse scale model as shown in Fig. 9.9A. The atoms in the fine region can be separated into small rectangular cells surrounded by four nodes/particles in the coarse region. The center of gravity of a cell containing the atoms on the crack surface can be calculated by averaging the positions of center of gravities of the atoms on the crack surface ($\mathbf{r}_\alpha^{\text{cog}}$) in that cell [10]:

$$\mathbf{r}_{\text{cell}}^{\text{cog}} = \frac{\sum_{\alpha=1}^{n^{\text{cacr}}} \mathbf{r}_\alpha^{\text{cog}}}{n^{\text{cs}}} \tag{9.32}$$

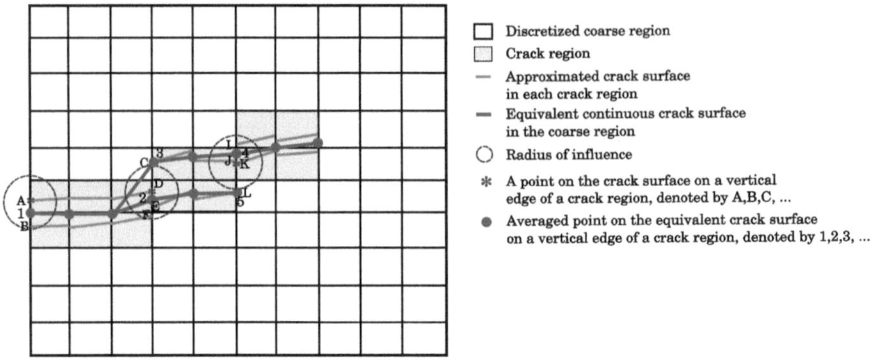

Figure 9.10 Schematic of averaging the approximated individual crack surface orientation in each crack region, to generate a smooth continuous equivalent crack surface. Picture reproduced from [10] with permission.

where $\mathbf{r}_{\text{cell}}^{\text{cog}}$ is the approximated position of the center of gravity of the atoms on the crack surface and n^{cs} are the total number of atoms on the crack surface, in a crack region. The normal of the approximated crack surface in the crack region is computed as the average of the normals of the atoms on the crack surface:

$$\mathbf{n}_{\text{cell}}^{\text{cog}} = \frac{\sum_{\alpha=1}^{n^{\text{cs}}} \mathbf{n}_{\alpha}^{\text{cog}}}{n^{\text{cs}}} \tag{9.33}$$

where $\mathbf{n}_{\text{cell}}^{\text{cog}}$ is the normal vector of the approximated crack surface in a crack region. Therefore, the crack surfaces in the crack regions is obtained based on the planes passing through $\mathbf{r}_{\text{cell}}^{\text{cog}}$, whose normals are estimated from Eq. (9.33). Finally, the approximated crack surface in the CG model is obtained by joining the crack surfaces in each crack region.

In order to generate a smooth and continuous crack surface in the CG domain, the start/end positions of the crack surfaces on the vertical edges of the crack regions are averaged, as illustrated in the schematic Fig. 9.10. As a rule of thumb, a cell containing at least 12 atoms on the crack surface is observed to be considered as crack region [10]. Therefore, the minimum size of the cell can be adopted as 13 times the lattice parameter. The cell size or the size of fine-scale domain in general could be determined by a-posteriori error estimators. An example of generation of a continuous crack surface in the coarse region is demonstrated in Fig. 9.10. Consider the vertical edge containing points C, D, E, and F. The points D and E correspond to end points of two crack surfaces and the points C and F are the starting

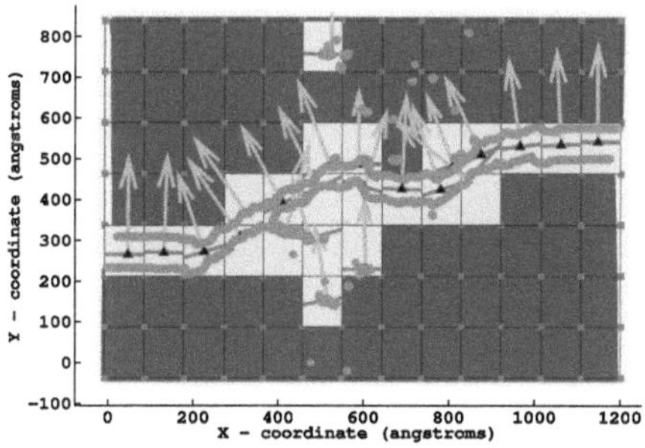

(A)

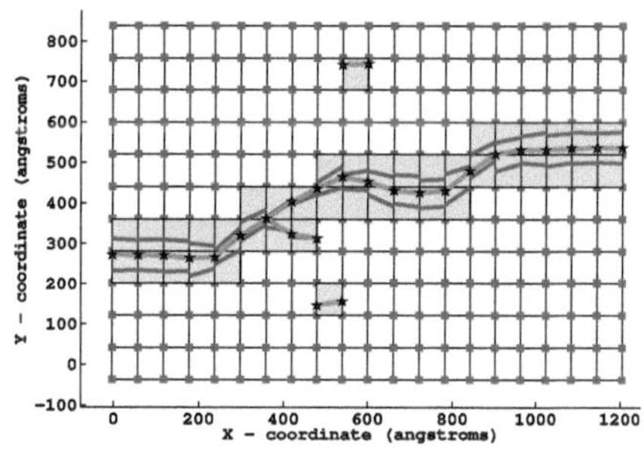

(B)

Figure 9.11 Development of an equivalent coarse scale model of a given fine scale model, for a dynamic crack propagation of double edge crack model. (A) Deformed configuration after 108 pico-seconds along with the highlighted atoms on the crack surface, crack regions and their normals. (B) Approximated crack surface showing the corresponding approximated equivalent crack surfaces.

points of new crack surfaces. The largest distance between these points is the distance between the points C and F which is larger than the domain of influence. Thus there exists more than one point on the equivalent crack surface on this particular edge. The total number of points on the equivalent crack surface on this vertical edge can be estimated by recursively checking if the distance between the neighbors of points C, D, E and F falls within the domain of influence. Fig. 9.11 shows the equivalent coarse grained model of an atomistic model [10]. The deformed configuration of the atomistic model for a dynamic double edge crack propagation after 108 pico-seconds is shown in Fig. 9.11A. The corresponding equivalent coarse grained model is shown in Fig. 9.11B.

9.3. Multiscale aggregating discontinuity (MAD) method

9.3.1 Overview of the method

Consider a crack which nucleates at the center of unit cell as shown in Fig. 9.12A, then grows to the left, and then penetrates the side AB as shown in Fig. 9.12B. The standard procedure for prescribing the displacements of the boundary of the unit cell is

$$\boldsymbol{u}^m(\boldsymbol{X}^m) = \left(\boldsymbol{F}^M(\boldsymbol{X}^M) - \boldsymbol{I}\right) \cdot \boldsymbol{X}^m + \boldsymbol{\omega}(\boldsymbol{X}^m) \qquad \boldsymbol{X}^m \in \Gamma_0^m \quad \text{and} \quad \boldsymbol{X}^M \in \Omega_0^M \tag{9.34}$$

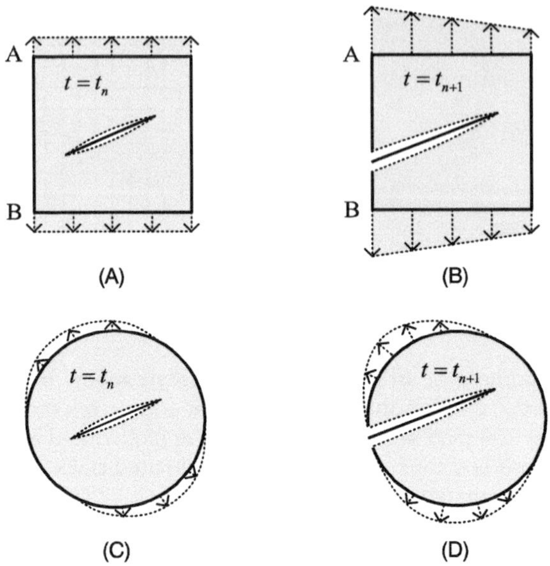

Figure 9.12 Crack growth with the crack penetrating edge AB for time $t_{n+1} > t_n$ for: (A) and (B) square unit cell, and (C) and (D) circular unit cells.

where $\boldsymbol{u}^m$ is the microscale displacement, $\boldsymbol{F}^M$ is the macroscale deformation gradient, $\boldsymbol{I}$ is the second order unit tensor, $\boldsymbol{\omega}$ is the fluctuation in the microdisplacement field, and $\boldsymbol{X}$ is the material coordinate; superscripts m and M refer to the microscale and macroscale, respectively. The microscale displacement $\boldsymbol{u}^m$ in Eq. (9.34) does not account for the jump in displacement at the intersection of the crack with the edge, so it will be inconsistent with the microscale displacement field given in Eq. (9.34). Therefore, methods that account for situations where cracks penetrate the surface of the unit cell need to be introduced.

We have shown two types of unit cells in Fig. 9.12: square unit cells and circular unit cells. The latter were also used in Belytschko et al. [7]. Circular unit cells are of particular advantage in coarse graining of micromodels with crack growth because they avoid the difficulties arising from corners and provide a way to characterize in more detail cracks that do not span the unit cell.

The second difficulty is illustrated in Fig. 9.13. In Belytschko et al. [7], cracks in the micromodel were coarse-grained by a single discontinuity with constant crack opening and smoothing schemes were used to construct a smoothly opening crack at the coarse scale. Here, we propose a method where the crack opening that is passed to the coarse-grained

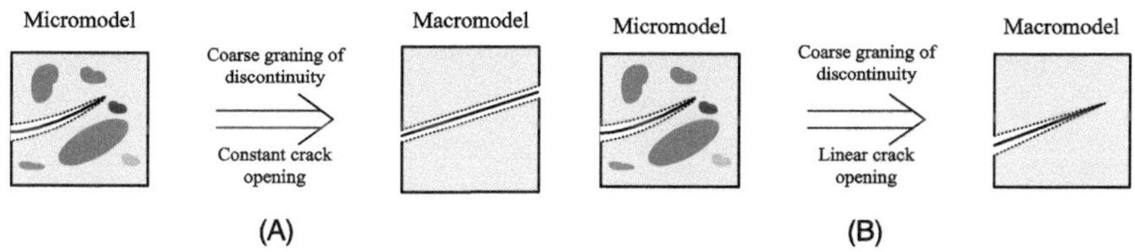

Figure 9.13 Schematic of coarse-graining of a crack at the macroscale: (A) according to the original MAD method [7], and (B) the proposed MAD method.

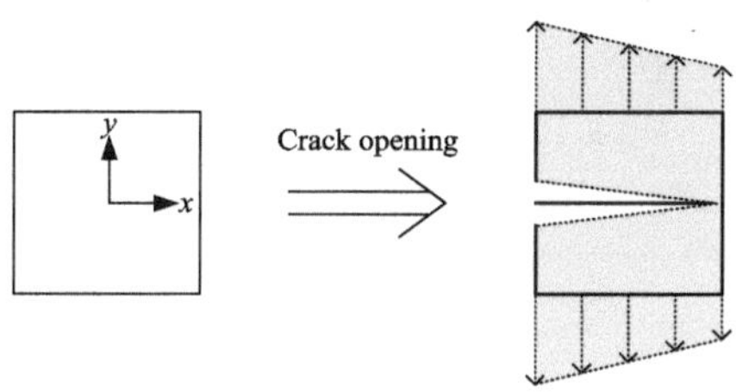

Figure 9.14 Schematic of crack opening; showing the importance of the hourglass mode.

model can vary linearly in the macroelement. Moreover, we have added techniques to model nucleating cracks. These techniques provide a substantially better representation of the discontinuities at the macroscale.

The motivation for the third objective is illustrated in Fig. 9.14. When a crack opens and grows, the deformation of a unit cell is approximately that shown in Fig. 9.14. This mode of deformation cannot be effectively represented with Eq. (9.34) for the deformation associated with an opening crack is a bilinear displacement field, often called an hourglass mode or bilinear mode in the finite element literature [4,18]. The inadequacy of constant deformation modes should be clear from the schematic of the constant deformation gradient modes shown in Fig. 9.15. Although the constant mode F_{yx} has similarities to the hourglass mode shown in Fig. 9.14, it is characterized by shear, whereas the crack-opening mode shown in Fig. 9.14 should be a tensile mode with linearly varying F_{yy}.

The bilinear displacement mode shown in Fig. 9.14 has a linearly varying extensional deformation gradient, F_{yy} and this is what has been added in this work. To obtain this mode, the boundary displacements must vary like XY, which is the hourglass mode. In many situations, particularly those involving a single crack or a dominant crack near percolation, effective modeling of crack growth requires that the hourglass mode be included in the deformation of the unit cell. Incidentally, we have only shown the

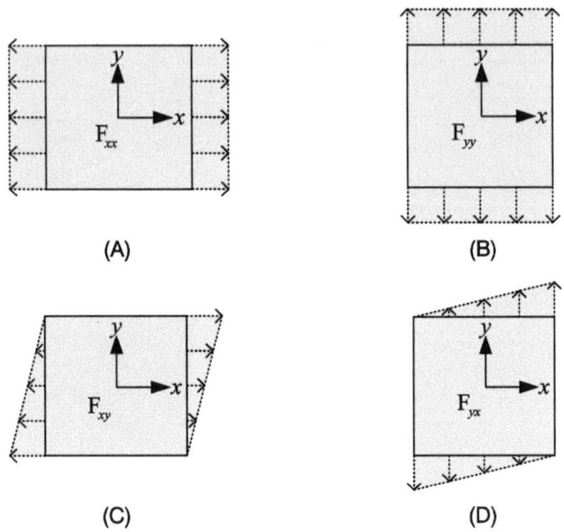

Figure 9.15 Schematic of constant deformation gradient modes.

y-hourglass mode in Fig. 9.14; a similar x-hourglass mode, i.e. for the x-component of the displacement field, is also considered.

As in Belytschko et al. [7], we employ two key concepts for coarse-graining failure phenomena:

1. all averaging operations are performed over a "perforated" unit cell that excludes all subdomains that lose convexity (to be defined later); these can be loosely considered subdomains where material stability is lost (areas of material instability, of course, include cracks),
2. a formula is developed whereby the discontinuous and localized deformation in a unit cell is replaced by a single equivalent discontinuity.

It is assumed that the size scale the microscale models, l^c, i.e. the unit cells, are of order h, where h is the length of a typical finite element in the macromodel. In contrast to representative volume elements in homogenization theories, the method is not independent of l^c.

9.3.2 Coarse graining method

The developments in this section will be described for two-dimensional problems, although they can be extended to three dimensions. In many cases, we give equations applicable to both two or three dimensions, but we limit the detailed formulation to two dimensions.

As before, the superscripts M and m refer to the macroscale and microscale, respectively. The method can easily be translated for an analysis at several scales by letting $M = K$, $m = K + 1$ for whatever pair of scales is being considered, which is the notation used in [7]. The reference domain of the macromodel is denoted by Ω_0^M and its boundary by Γ_0^M, and for convenience, the origin of the coordinate system is assumed to be at the center.

The domain of the perforated unit cell is denoted by $\tilde{\Omega}_0^m$ so

$$\tilde{\Omega}_0^m = \Omega_0^m \setminus \Omega_0^{uns} \tag{9.35}$$

where Ω_0^{uns} is the subdomain of the unit cell where the material is not convex; this could be an area of localization of strain. Any crack is excluded from Ω_0, even though it is a set of measure zero, since the material must lose convexity as the stress goes to zero on the crack plane. The implications of this are rather delicate for cracks, and are not discussed here. We just note that as a consequence, any cohesive stress is not included in the averaging operation.

All averaging operations are performed over the perforated unit cell, so denoting the averaging operation by $\langle \cdot \rangle$, we have for any function $f(\boldsymbol{X}^m)$:

$$\langle f(\boldsymbol{X}^m) \rangle = \frac{1}{|\tilde{\boldsymbol{\Omega}}_0^m|} \int_{\tilde{\Omega}_0^m} f(\boldsymbol{X}^m) \, \mathrm{d}\Omega_0 \tag{9.36}$$

where $| \cdot |$ denotes the measure of the domain, such as the area in two dimensions or the volume in three dimensions. The macroscale deformation gradient $\boldsymbol{F}^M$ and the macroscale first Piola-Kirchhoff stress $\boldsymbol{P}^M$ are defined as the averages of the microscale deformation gradient $\boldsymbol{F}^m$ and the microscale first Piola-Kirchhoff stress $\boldsymbol{P}^m$ over the perforated unit cell, respectively, so

$$\langle \boldsymbol{F}^m \rangle = \frac{1}{|\tilde{\boldsymbol{\Omega}}_0^m|} \int_{\tilde{\Omega}_0^m} \boldsymbol{F}^m \, \mathrm{d}\Omega_0 \tag{9.37}$$

$$\langle \boldsymbol{P}^m \rangle = \frac{1}{|\tilde{\boldsymbol{\Omega}}_0^m|} \int_{\tilde{\Omega}_0^m} \boldsymbol{P}^m \, \mathrm{d}\Omega_0 \tag{9.38}$$

To treat the hourglass modes, two generalized hourglass strains $\boldsymbol{q} = [q_1, q_2]$ and two corresponding generalized stresses $\mathbf{Q} = [Q_1, Q_2]$ are added to the kinematic and kinetic descriptions at the macroscale. These extra generalized stresses and strains are assumed to be energetically consistent

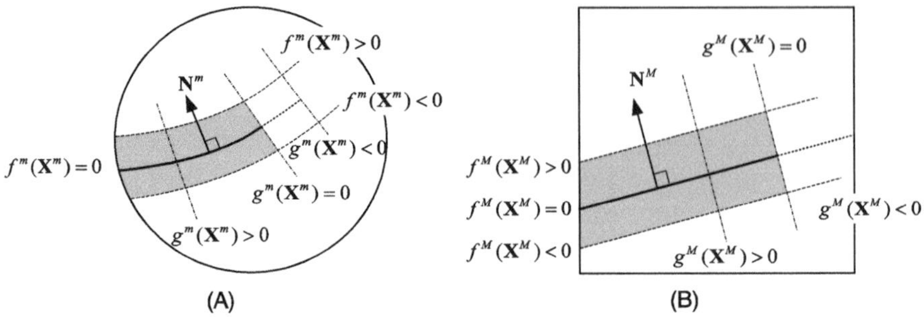

Figure 9.16 Relation of cracks at: (A) the microscale, and (B) the macroscale.

with the work in the perforated unit cell so that

$$\boldsymbol{P}^M : \delta\boldsymbol{F}^M + \mathbf{Q}\cdot\delta\boldsymbol{q} = \int_{\tilde{\Omega}_0^m} \boldsymbol{P}^m : \delta\boldsymbol{F}^m \, d\Omega_0 \tag{9.39}$$

The introduction of additional modes is in the same spirit as in Kouznetsova et al. [23], but only the first higher order generalized stresses and strains are considered.

The macrocrack is an approximation to either a single crack or a group of cracks at the microscale. The cracks at the microscale are described by

$$f_\beta^m(\mathbf{X}^m) = 0 \quad \text{and} \quad g_\beta^m(\mathbf{X}^m) > 0 \tag{9.40}$$

where $f_\beta^m(\mathbf{X}^m)$ describes the surface of the crack β and $g_\beta^m(\mathbf{X}^m) > 0$ describes its extent. The crack path at the microscale may be jagged, but it is assumed that the crack path penetrates the walls of the unit cells at no more than two points. If the front of crack β is within the unit cell, it is given by

$$f_\beta^m(\mathbf{X}^m) = g_\beta^m(\mathbf{X}^m) = 0 \tag{9.41}$$

A typical crack at the microscale and its macroscale equivalent is shown in Fig. 9.16.

The geometry of the equivalent macro crack in a neighborhood corresponding to the unit cell is described by an affine level set function

$$f^M(\mathbf{X}^M) = \alpha_0 + \alpha_\beta X_\beta^M = 0 \qquad \beta = 1 \text{ to } 2 \tag{9.42}$$

where α_0 and α_β are parameters obtained from the coarse graining. In addition, to describe the ends of the crack, we introduce a second level set function $g^M(\mathbf{X}^M)$ and define it so that on the crack $g^M(\mathbf{X}^M) > 0$.

The motion $\boldsymbol{\phi}^m(\boldsymbol{X}^m)$ on the outside boundary of the unit cell is given by

$$\boldsymbol{\phi}^m(\boldsymbol{X}^m) = \mathfrak{F}^M \cdot \boldsymbol{X}^m + \boldsymbol{q}XY + \boldsymbol{\omega}(\boldsymbol{X}^m) \qquad \boldsymbol{X}^m \in \Gamma_0^m \tag{9.43}$$

where in two dimensions $\boldsymbol{q}^T = [q_x, q_y]$. The last term in Eq. (9.43) accounts for the hourglass modes; $\boldsymbol{q}$ is obtained from the macroscale deformation as described later. The second term is one of the key differences from the previously presented MAD method [7]. Henceforth we drop the fluctuations $\boldsymbol{\omega}(\boldsymbol{X})$ since the MAD method is not used with periodic boundary conditions.

Eq. (9.43) is not consistent with crack penetrating the surface, i.e. in the neighborhood of $f^m(\boldsymbol{X}^m) = 0$, unless the crack displacement on the surface is considered part of the fluctuations. Therefore, if the motion of the boundary is completely described by (9.43) the motion will not be consistent with a cracking unit cell. In fact, the motion prescribed by Eq. (9.43) will tend to arrest the crack as it approaches the boundary.

To avoid this effect, we switch from a prescribed displacement boundary condition to a prescribed traction boundary condition in a neighborhood of the crack, i.e. intersection of the shaded region with the periphery of the circle in Fig. 9.16A. If we denote this portion of the boundary by Γ_0^{mP} and the prescribed displacement portion by Γ_0^{mF}, then the boundary condition becomes

$$\bar{\boldsymbol{\phi}}^m(\boldsymbol{X}^m) = \mathfrak{F}^M \cdot \boldsymbol{X}^m + \boldsymbol{q}XY \qquad \boldsymbol{X}^m \in \Gamma_0^{mF} \tag{9.44}$$

$$\bar{\boldsymbol{t}}^m(\boldsymbol{X}^m) = \langle \boldsymbol{P}^m \rangle \cdot \boldsymbol{n} \qquad \boldsymbol{X}^m \in \Gamma_0^{mP} \tag{9.45}$$

where $\bar{\boldsymbol{\phi}}^m$ and $\bar{\boldsymbol{t}}^m$ are the prescribed displacement and traction along Γ_0^{mF} and Γ_0^{mP}, respectively, and $\boldsymbol{n}$ is the traction boundary normal. Note that the above requires an iterative procedure, since the stress $\langle \boldsymbol{P}^m \rangle$ is not known until the solution for the unit cell has been obtained. This procedure consists of obtaining $\langle \boldsymbol{P}^m \rangle^{new}$, and then solving the unit cell. The procedure converged in our examples, but requires further study.

We next describe how the magnitude of the discontinuity and the normal to the discontinuity at the macroscale are extracted from the unit cell deformation. We will use the following equation for this purpose

$$|\tilde{\boldsymbol{\Omega}}_0^m| \left(\langle \boldsymbol{F}^m \rangle - \mathfrak{F}^M \right) + \int_{\Gamma_0^c} [\![\boldsymbol{\phi}^m \otimes \boldsymbol{N}^m]\!] \, d\Gamma_0 = 0 \tag{9.46}$$

This equation holds only if a crack does not penetrate through the walls of the unit cell, but we will use it for the mixed conditions described above.

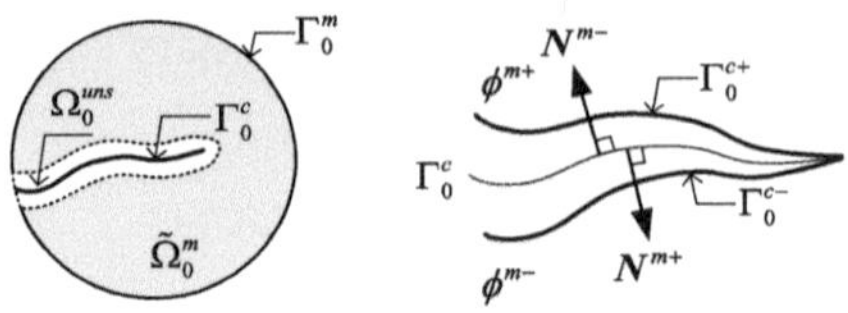

Figure 9.17 Nomenclature for crack surfaces and normals.

To obtain this equation, we note that

$$\langle \boldsymbol{F}^m \rangle = \frac{1}{|\tilde{\boldsymbol{\Omega}}_0^m|}\int_{\tilde{\Omega}_0^m} \boldsymbol{F}^m \, \mathrm{d}\Omega_0 = \frac{1}{|\tilde{\boldsymbol{\Omega}}_0^m|}\int_{\tilde{\Omega}_0^m} \nabla_0 \otimes \boldsymbol{\phi}^m \, \mathrm{d}\Omega_0 \tag{9.47}$$

By the divergence theorem

$$\int_{\tilde{\Omega}_0^m} \nabla_0 \otimes \boldsymbol{\phi}^m \, \mathrm{d}\Omega_0 = \int_{\Gamma_0^m \backslash \Gamma_0^c} \boldsymbol{\phi}^m \otimes \boldsymbol{N}^m \, \mathrm{d}\Gamma_0 - \int_{\Gamma_0^c} [\![\boldsymbol{\phi}^m \otimes \boldsymbol{N}^m]\!] \, \mathrm{d}\Gamma_0 \tag{9.48}$$

where

$$[\![\boldsymbol{\phi}^m \otimes \boldsymbol{N}^m]\!] = \boldsymbol{\phi}^{m+}\boldsymbol{N}^{m-} + \boldsymbol{\phi}^{m-}\boldsymbol{N}^{m+} \tag{9.49}$$

and

$$\boldsymbol{\phi}^{m+} = \phi^m(\boldsymbol{X}^m_{\Gamma_0^c} + \epsilon \boldsymbol{N}^{m-}), \qquad \boldsymbol{\phi}^{m-} = \phi(\boldsymbol{X}^m_{\Gamma_0^c} + \epsilon \boldsymbol{N}^{m+}), \qquad \epsilon \to 0 \tag{9.50}$$

where $\boldsymbol{N}^{m+}$ and $\boldsymbol{N}^{m-}$ are the normals to the top and the bottom surfaces of the microcrack; see Fig. 9.17.

Substituting Eq. (9.48) into (9.47) and using (9.43) gives

$$\begin{aligned} |\tilde{\boldsymbol{\Omega}}_0^m| \langle \boldsymbol{F}^m \rangle = & \int_{\Gamma_0^m \backslash \Gamma_0^c} \mathfrak{F}^M \cdot \boldsymbol{X}^m \otimes \boldsymbol{N}^m \, \mathrm{d}\Gamma_0 \\ & + \underbrace{\int_{\Gamma_0^m \backslash \Gamma_0^c} \mathfrak{q}XY \otimes \boldsymbol{N}^m \, \mathrm{d}\Gamma_0}_{=0} - \int_{\Gamma_0^c} [\![\boldsymbol{\phi}^m \otimes \boldsymbol{N}^m]\!] \, \mathrm{d}\Gamma_0 \end{aligned} \tag{9.51}$$

where we have indicated in the above that since the origin of the coordinate system of the unit cell is taken to the center, the second term of the right hand side vanishes. The first term can be simplified by the following steps:

$$\int_{\Gamma_0^m \backslash \Gamma_0^c} \mathfrak{F}^M \cdot \boldsymbol{X}^m \otimes \boldsymbol{N}^m \, \mathrm{d}\Gamma_0 = \mathfrak{F}^M \int_{\Gamma_0^m \backslash \Gamma_0^c} \boldsymbol{X}^m \otimes \boldsymbol{N}^m \, \mathrm{d}\Gamma_0 \tag{9.52}$$

$$= \mathfrak{F}^M \int_{\tilde{\Omega}_0^m} \nabla_0 \boldsymbol{X}^m \, \mathrm{d}\Omega_0 \tag{9.53}$$

$$= |\tilde{\boldsymbol{\Omega}}_0^m| \mathfrak{F}^M \tag{9.54}$$

Hence

$$-\int_{\Gamma_0^c} [\![\boldsymbol{\phi}^m \otimes \boldsymbol{N}^m]\!] \, \mathrm{d}\Gamma_0 = |\tilde{\boldsymbol{\Omega}}_0^m| \left(\langle \boldsymbol{F}^m \rangle - \mathfrak{F}^M \right) \equiv |\tilde{\boldsymbol{\Omega}}_0^m| \boldsymbol{F}_E^M \tag{9.55}$$

which concludes the derivation of Eq. (9.46). This relation is identical to the relation for the standard unit cell; it is independent of $\boldsymbol{q}$. The above is a generalization of Hill's averaging theorem which is given in Loehnert [24] and Belytschko et al. [7].

In coarse-graining of the crack, we wish to find a $\boldsymbol{U}^M(\xi^M) = [\![\boldsymbol{\phi}^M(\xi^M)]\!]$ such that

$$\int_{\Gamma_0^c} \boldsymbol{U}^M \otimes \boldsymbol{N}^M \, \mathrm{d}\Gamma_0 = |\tilde{\boldsymbol{\Omega}}_0^m| \boldsymbol{F}_E^M \tag{9.56}$$

where

$$\boldsymbol{F}_E^M = \langle \boldsymbol{F}^m \rangle - \mathfrak{F}^M \tag{9.57}$$

Since the left hand side of Eq. (9.56) is of rank 1, whereas the right-hand side is of rank 3, in most cases, a vector $\boldsymbol{U}^M$ that satisfies Eq. (9.56) exactly can not be found. Therefore, we obtain an estimate of $\boldsymbol{U}^M$ by minimizing

$$J^D = \left\| \int_{\Gamma_0^M} \boldsymbol{U}^M \otimes \boldsymbol{N}^M \, \mathrm{d}\Gamma_0 - |\tilde{\boldsymbol{\Omega}}_0^m| \boldsymbol{F}_E^M \right\|^2 \tag{9.58}$$

The integrand in the above depends on the topology of the crack at the macroscale. In all cases, we assume that the crack is planar within a macroelement, so $\boldsymbol{N}^M$ is constant.

Some of the topologies are considered:

1. if the crack has penetrated the left edge L, but not the right edge, then we assume a linear variation of the crack opening at the macroscale.

$$\boldsymbol{U}^M(\xi^M) = \frac{\boldsymbol{U}_L^M}{2}(1 - \xi^M) \tag{9.59}$$

where ξ^M is defined in Fig. 9.18. A linear variation in the magnitude of the discontinuity is assumed because that is all the information that can be incorporated at the macroscale model. The quadratic form J^D is

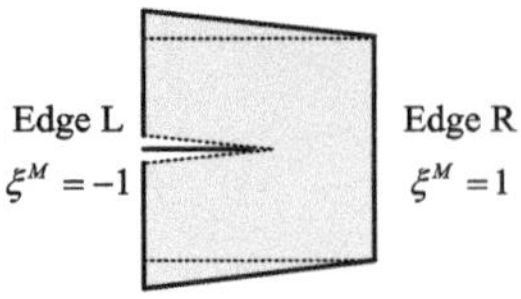

Figure 9.18 Definition of ξ^M at the macros element.

then given by substituting Eq. (9.59) into Eq. (9.58). Similar formulas can easily be developed for cracks that penetrate the right edge, the bottom or the top.

2. If the crack penetrates both sides then

$$\boldsymbol{U}^M(\xi^M) = \frac{1}{2}\left(\boldsymbol{U}_L^M(1-\xi^M) + \boldsymbol{U}_R^M(1+\xi^M)\right) \tag{9.60}$$

In that case, we use Eq. (9.58) to obtain $\boldsymbol{U}^M = \frac{1}{2}\left(\boldsymbol{U}_L^M + \boldsymbol{U}_R^M\right)$ and obtain $\boldsymbol{U}_L^M$ and $\boldsymbol{U}_R^M$ as described in Section 9.3.3.

9.3.3 Micro-macro linkage

Both the micro scale and macro crack models are solved by the XFEM [5, 28,38] approach, but the methodology applies to other methods for modeling cracks. An overview of the linkage is shown in Fig. 9.19. As shown, the micro model passes the stresses and the magnitude and direction of the discontinuity to the corresponding macro element, $\boldsymbol{U}^M$ and $\boldsymbol{N}^M$, respectively. The discontinuity is directly injected into the macro model as long as the micro model is not completely cut in two, i.e. prior to percolation. The stress is passed to the quadrature point in the macro element. The equations of equilibrium (or the momentum equation for dynamic processes) are then solved. In the solution process, the motion of the nodes adjust for the injected discontinuity.

Once percolation has occurred in the micro model, only the direction of the crack is passed to the macro model. The magnitude of the discontinuity then becomes an unknown which is determined.

In cracked macroscale elements, the displacement field is given by

$$\boldsymbol{u}^h(\boldsymbol{X}^M, t) = \sum_I \Psi_I(\boldsymbol{X}^M)\boldsymbol{u}_I(t) + \sum_J \Psi_J(\boldsymbol{X}^M) H(f^M(\boldsymbol{X}^M) - f^M(\boldsymbol{X}_J^M))\boldsymbol{a}_J(t) \tag{9.61}$$

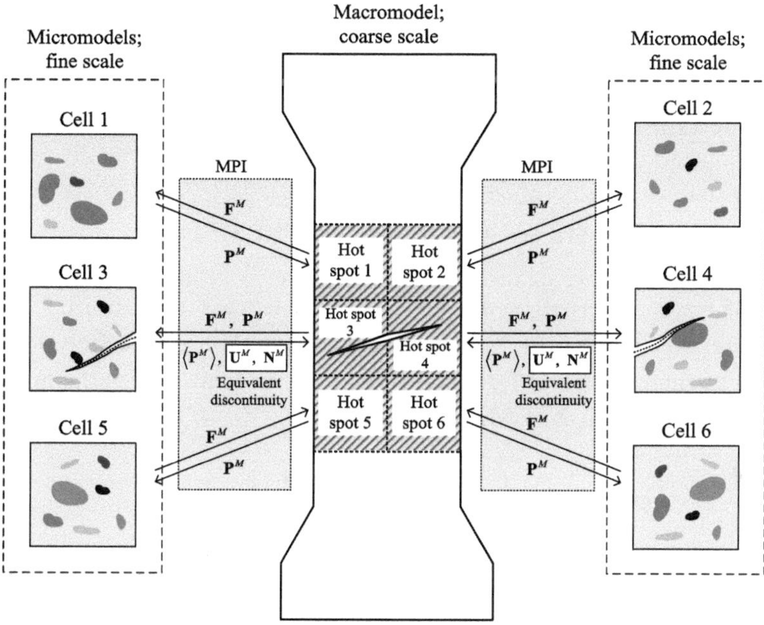

Figure 9.19 Schematic of macro-micro linkages of the MAD method.

where $\Psi_I(\boldsymbol{X})$ are the element shape functions and $H(\cdot)$ is the Heaviside function. The level set function $f^M(\boldsymbol{X}^M)$ must be chosen so that

$$\nabla_0 f^M(\boldsymbol{X}^M) = \bar{\boldsymbol{N}}^M \tag{9.62}$$

The condition in Eq. (9.62) does not suffice to determine the level set function $f^M(\boldsymbol{X}^M)$. The procedure for any element e is as follows:

1. if a crack exists in an element adjacent to element e, the crack path, i.e. the level set function, is set so it is continuous between the two elements;
2. if a crack nucleates in element e, it is centered in the element.

The amplitude of crack opening, which is specified by $a_J(t)$ in Eq. (9.61), also requires matching between adjacent elements. The matching constraints for continuity of crack opening at interfaces between elements are written as a linear equation

$$\boldsymbol{Da} = \boldsymbol{b} \tag{9.63}$$

where $\boldsymbol{D}$ depends on the position of the crack and $\boldsymbol{b}$ is a linear function of $\boldsymbol{U}_L$ and $\boldsymbol{U}_R$ in the cracked elements. Note that the crack opening depends only on $\boldsymbol{a}$, and the above conditions are linear in $\boldsymbol{a}$.

The equations of motion then incorporate the above as constraints by Lagrange multipliers. The modified equation of motion is

$$\boldsymbol{M}\ddot{\boldsymbol{d}} = \boldsymbol{f} + \boldsymbol{D}^T\boldsymbol{\lambda} = \boldsymbol{f}^{ext} - \boldsymbol{f}^{int} + \boldsymbol{D}^T\boldsymbol{\lambda} \tag{9.64}$$

where $\boldsymbol{M}$ is the mass matrix, $\boldsymbol{d}$ is the matrix of degrees of freedom, $\boldsymbol{f}^{int}$ is the nodal internal forces and $\boldsymbol{f}^{ext}$ is the nodal external forces, see Belytschko et al. [6]. The matrices $\boldsymbol{d}$ and $\boldsymbol{f}$ consist of the parts associated with the enriched parts emanating from XFEM, i.e.

$$\boldsymbol{d}_I^T = \begin{Bmatrix} \boldsymbol{u}_I \\ \boldsymbol{a}_I \end{Bmatrix}, \quad \boldsymbol{f}_I^T = \begin{Bmatrix} \boldsymbol{f}_I \\ \boldsymbol{A}_I \end{Bmatrix} \tag{9.65}$$

$$\boldsymbol{d}^T = \{\boldsymbol{d}_I\}_{I=1}^{n_N}, \quad \boldsymbol{f}^T = \{\boldsymbol{f}_I\}_{I=1}^{n_N} \tag{9.66}$$

where n_N is the number of nodes.

Crack nucleation can be treated by a new method similar to an s-method [16,17], and the cracking particle method by Rabczuk and Belytschko [36]. The XFEM methods for crack nucleation were previously developed by Bellec and Dolbow [3] for elastic fracture mechanics. They use blending methods to combine two sets of the branch functions given in [3]. In this work, we model nucleating cracks by bubble functions.

The bubble function is developed as follows. Suppose that a nucleated crack with normal $\bar{\boldsymbol{N}}$ and length l^c appears in element e. We then let the displacement field in element e be

$$\boldsymbol{u}^h(\boldsymbol{X}^M, t) = \sum_I \Psi_I(\boldsymbol{X}^M)\boldsymbol{u}_I(t) + \sum \Psi_J(\boldsymbol{X})\Phi(\xi^M)H(\bar{\boldsymbol{N}} \cdot (\boldsymbol{X}^M - \bar{\boldsymbol{X}}_e^M))\boldsymbol{a}_J(t) \tag{9.67}$$

where $\bar{\boldsymbol{X}}_e^M$ is the center of the element e, $\Phi(\xi)$ is a cubic spline function given by

$$\Phi(\xi) = \begin{cases} 4(\frac{\xi}{l_c} - 1)(\frac{\xi}{l_c})^2 + \frac{2}{3} & 0 < \xi < 0.5\, l_c \\ \frac{4}{3}(1 - \frac{\xi}{l_c})^3 & 0.5\, l_c \le \xi \le l_c \\ 0 & \text{otherwise} \end{cases} \tag{9.68}$$

9.4. Crack opening in unit cells with the hourglass mode

As we described in the previous section, the motion of unit cells during failure processes cannot be solely driven by the constant deformation gradients that emanate from coarse scale models, because the unit cell when crack opening takes place deforms primarily in a bilinear mode. also called the hourglass modes. A constant deformation gradient on the boundaries cannot represent this high order motion.

Here, we will briefly describe a scheme for linking the bilinear mode from the coarse scale model that is discretized with 4-node quadrilateral elements with the micromodel. The scheme for extracting the bilinear modes are based on Flanagan and Belytschko [18], and Belytschko and Bachrach [4]. In the works [4,18], the extracted hourglass modes are used to control hourglass modes due to one-point quadrature 4-node quadrilateral elements. Here we use them to link the higher order modes of the coarse-scale model with the fine scale model.

Following Flanagan and Belytschko [18], the bilinear mode is computed by

$$q_i = \sum_I u_{iI}\gamma_I \tag{9.69}$$

where $\boldsymbol{u}_I$ is the nodal displacement of the finite element, and $\boldsymbol{\gamma}_I$ is the hourglass mode projection operator defined by

$$\boldsymbol{\gamma}_I = \frac{1}{4}\left\{\boldsymbol{h}_I - \left(\sum_J \boldsymbol{h}_J \boldsymbol{X}_J\right)\boldsymbol{b}_I^X - \left(\sum_J \boldsymbol{h}_J \boldsymbol{Y}_J\right)\boldsymbol{b}_I^Y\right\} \tag{9.70}$$

where $\boldsymbol{X}_I$ and $\boldsymbol{Y}_J$ is the X and Y components of the current nodal coordinates of the finite element, respectively, and $\boldsymbol{h}$ and $\boldsymbol{b}$ are defined as

$$\boldsymbol{h}^T = [1 \quad -1 \quad 1 \quad -1] \tag{9.71}$$

$$\begin{bmatrix} \boldsymbol{b}_I^X \\ \boldsymbol{b}_I^Y \end{bmatrix} = \begin{bmatrix} \partial\Psi_I(\mathbf{0})/\partial X \\ \partial\Psi_I(\mathbf{0})/\partial Y \end{bmatrix} \tag{9.72}$$

Note that strictly speaking $\boldsymbol{q}$ is the strength of the bilinear mode in the referential coordinates, but we ignore this difference.

The macro stresses are linked to the unit cell as follows. We use the generalized Hill-Mandel energetic relations in Eq. (9.39). Substituting the displacement field in Eq. (9.43) into Eq. (9.39), we obtain the following expressions for the macro stresses

$$\boldsymbol{P}^M = \frac{1}{|\tilde{\Omega}_0^m|}\int_{\Gamma_0^m} \boldsymbol{P}^m \cdot \boldsymbol{N} \otimes \boldsymbol{X}\, \mathrm{d}\Gamma_0 \tag{9.73}$$

$$\mathbf{Q} = \frac{1}{|\tilde{\Omega}_0^m|}\int_{\Gamma_0^m} \boldsymbol{P}^m XY \cdot \boldsymbol{N}\, \mathrm{d}\Gamma_0 \tag{9.74}$$

The expressions for the nodal forces of a 4-node quadrilateral element with consistent stabilization [18] are

$$f_{iI}^{int} = \int_{\Omega_0^M} \frac{\partial \Psi_I}{\partial X_j} P_{ij}^M \, \mathrm{d}\Omega_0 + f_{iI}^{HG} \tag{9.75}$$

For a one-point quadrature element, the above can be written as

$$f_{iI}^{int} = A_e B_{jI}^T(\mathbf{0}) P_{ji}^M + f_{iI}^{HG} \tag{9.76}$$

where A_e is the area of element e, and $\boldsymbol{B}(\mathbf{0})$ and $\boldsymbol{f}^{HG}$ are defined, respectively, as:

$$B_{jI}(\mathbf{0}) = \frac{\partial \Psi_I(\mathbf{0})}{\partial X_j} \tag{9.77}$$

$$f_{iI}^{HG} = A_e Q_i \gamma_I \tag{9.78}$$

For element that are cut by a crack, the integration method proposed in Song et al. [38] are used.

9.5. Stability of the macromaterial

The issue of the ellipticity of the macromodel is more intricate than for the linkage involving only the standard terms, $\boldsymbol{F}^M$ and $\boldsymbol{P}^m$. In fact, it becomes impossible to show the ellipticity of the macromaterial in the standard sense, although it is possible to show that the macroelement stiffness matrix is positive definite.

In order to examine this in more detail, we will first examine the consequences of strict ellipticity within the perforated unit cell on the ellipticity if the macromaterial. We will then show that strict ellipticity of the material in the perforated unit cell implies the positive definiteness of the tangent stiffness for the stabilized one-point quadrature element used for the macromodel. This insures that the discrete problem for the bulk material is stable.

We first note that convexity of the micromaterial and macromaterial requires, respectively, that

$$\dot{\boldsymbol{F}} : \boldsymbol{C}^m : \dot{\boldsymbol{F}} > 0 \quad \forall \ \dot{\boldsymbol{F}} \tag{9.79}$$

$$\dot{\boldsymbol{F}} : \boldsymbol{C}^M : \dot{\boldsymbol{F}} > 0 \quad \forall \ \dot{\boldsymbol{F}} \tag{9.80}$$

where $\boldsymbol{C}^m$ and $\boldsymbol{C}^M$ are tangent material matrices of that respectively relate $\dot{\boldsymbol{P}}$ to $\dot{\boldsymbol{F}}$ by

$$\dot{\boldsymbol{P}}^m = \boldsymbol{C}^m : \dot{\boldsymbol{F}}^m \tag{9.81}$$

$$\dot{\boldsymbol{P}}^M = \boldsymbol{C}^M : \dot{\boldsymbol{F}}^M \tag{9.82}$$

The rank-one stability condition, often called the ellipticity condition is

$$\boldsymbol{g} \otimes \boldsymbol{h} : \boldsymbol{C}^M : \boldsymbol{g} \otimes \boldsymbol{h} > 0 \quad \forall \ \boldsymbol{g} \text{ and } \boldsymbol{h} \tag{9.83}$$

The above condition is equivalent to the condition for the ellipticity of the governing partial differential equation.

Then invoking the tangent material relation for the micromaterial (9.79), we have

$$\int_{\tilde{\Omega}_0^m} \dot{\boldsymbol{F}}^m : \dot{\boldsymbol{P}}^m \, \mathrm{d}\Omega = \int_{\tilde{\Omega}_0^m} \dot{\boldsymbol{F}}^m : \boldsymbol{C}^m : \dot{\boldsymbol{F}}^m \, \mathrm{d}\Omega > 0 \tag{9.84}$$

where the inequality follows from the assumption that the material in the perforated unit cell is convex, i.e. it satisfies Eq. (9.79). From the general form of the Hill-Mandel inequality and Eq. (9.81), it follows then that

$$\dot{\boldsymbol{F}}^M : \dot{\boldsymbol{P}}^M + \alpha \dot{\boldsymbol{q}} \cdot \dot{\boldsymbol{Q}} > 0 \quad \forall \ \dot{\boldsymbol{F}}^M \text{ and } \dot{\boldsymbol{q}} \tag{9.85}$$

where

$$\alpha = \begin{cases} 0 & \text{for MAD method without hourglass multiscale coupling} \\ 1 & \text{for MAD method with hourglass multiscale coupling} \end{cases} \tag{9.86}$$

We note first that the above implies convexity at the macroscale if we do not consider the hourglass modes. In that case $\dot{\boldsymbol{q}} = 0$, and we have

$$\dot{\boldsymbol{F}}^M : \dot{\boldsymbol{P}}^M = \dot{\boldsymbol{F}}^M : \boldsymbol{C}^M : \dot{\boldsymbol{F}}^M > 0 \quad \forall \ \dot{\boldsymbol{F}}^M \tag{9.87}$$

i.e. the strict ellipticity condition of the macro material.

This also implies the rank-one stability condition. To demonstrate this, we simply note that for any $\boldsymbol{g}$ and $\boldsymbol{h}$, we can let $\dot{\boldsymbol{F}} = \boldsymbol{g} \otimes \boldsymbol{h}$, so Eq. (9.87) implies the rank-one stability condition.

However, neither convexity nor rank-one stability can be deduced if $\alpha = 1$, i.e. for the method proposed here, since Eq. (9.85) then does not imply Eq. (9.87).

Nevertheless, it can be shown that the tangent stiffness matrix of the bulk material is positive definite. We assume that the generalized strain rates and stress rates are related by

$$\left\{ \begin{array}{c} \dot{\boldsymbol{P}} \\ \dot{\boldsymbol{Q}} \end{array} \right\} = \underbrace{\left[\begin{array}{cc} \boldsymbol{C}^M & \boldsymbol{C}_{\sigma\gamma}{}^M \\ (\boldsymbol{C}_{\sigma\gamma}{}^M)^T & \boldsymbol{C}_\gamma^M \end{array} \right]}_{\bar{\boldsymbol{C}}^M} \left\{ \begin{array}{c} \dot{\boldsymbol{F}} \\ \dot{\boldsymbol{q}} \end{array} \right\} \tag{9.88}$$

The macro tangent stiffness is given by

$$\boldsymbol{K}_e^M = A_e [\boldsymbol{B}^T \quad \boldsymbol{\gamma}^T] \bar{\boldsymbol{C}}^M \left\{ \begin{array}{c} \boldsymbol{B} \\ \boldsymbol{\gamma} \end{array} \right\} \tag{9.89}$$

where A_e is the area of element e and $\boldsymbol{B}$ is the matrix defined in Eq. (9.77). Note that these relations give both the material and geometric tangent stiffness, since $\bar{\boldsymbol{C}}^M$ includes the initial stress, so they apply to arbitrary nonlinear problems.

We now wish to show that

$$\dot{\boldsymbol{d}}_e^T \boldsymbol{K}_e^M \dot{\boldsymbol{d}}_e^T > 0 \quad \forall \ \dot{\boldsymbol{d}}_e \neq 0 \tag{9.90}$$

Substituting Eq. (9.89) into Eq. (9.90) and using Eqs. (9.69) and (9.77) gives

$$\dot{\boldsymbol{d}}_e^T \boldsymbol{K}_e^M \dot{\boldsymbol{d}}_e^T = A_e(\dot{\boldsymbol{F}} : \dot{\boldsymbol{P}} + \dot{\boldsymbol{q}} \cdot \dot{\boldsymbol{Q}}) > 0 \tag{9.91}$$

where the inequality follows from Eq. (9.91). Thus the bulk stiffness, which does not include the behavior of the macrocrack, is positive definite. We stress that only the bulk material tangent stiffness is positive definite. The combination of the stiffnesses of the cohesive law acting on the macrocrack and the bulk material stiffness is not positive definite.

As summarized by Marsden and Hughes [25], convexity in $\boldsymbol{F}$ is often considered an unacceptable condition, because

1. it precludes buckling,
2. it is not frame invariant,
3. the behavior as $J \to 0$ is not reasonable.

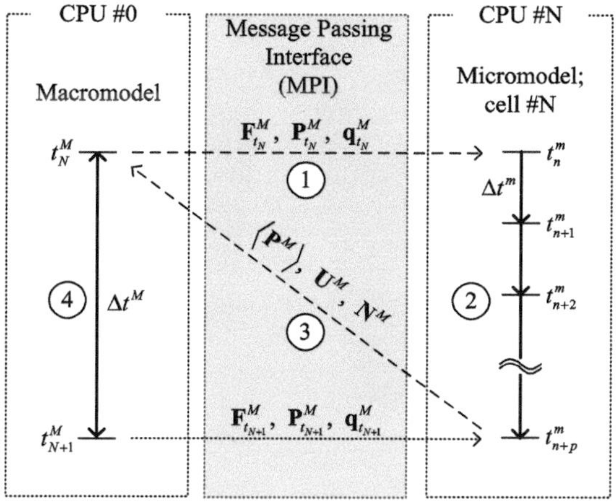

Figure 9.20 The macro-micro coupling scheme of the MAD method; the circled numbers indicate the sequence of the steps.

We note that as already pointed out in Belytschko et al. [7], we are not concerned with problems involving buckling or those where $J \to 0$. The absence of frame invariance is more troubling. However, we emphasize that convexity is used only as a criterion to remove material from the RVE. Therefore, its undesirable properties may not invalidate its use.

9.6. Implementation

The governing equations were integrated by the central difference method. The loads were applied very slowly, so that dynamic effects were small during most of the simulation; the kinetic energy did not exceed 1% of the total energy. This approach bypasses some of difficulties associated with the snapback behavior on the equilibrium path.

A schematic of the approach is shown in Fig. 9.19. As shown, only a few of hot spots are linked with micro models. Note that cells 3 and 4 contain strong discontinuities, so the coarse grained failure information within those unit cells is provided to the associated elements in the macro model. As shown in Fig. 9.20, smaller time step was used for the micromodels.

The method was run on a parallel computer. A message passing interface (MPI) module is used for the communication between the macromodel and the micromodels.

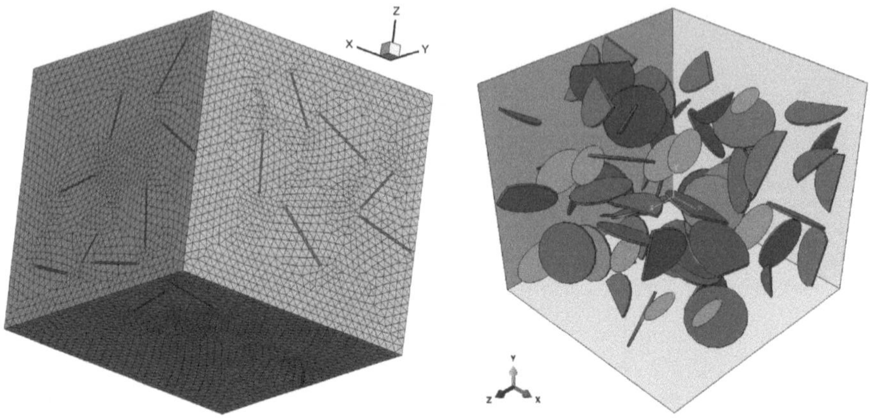

Figure 9.21 Three-dimensional representative volume element (RVE) including randomly oriented and distributed clay particles (red coin shaped objects) and cracks (gray ellipsoids).

9.7. Numerical examples

9.7.1 3D modeling of cracks in a nanocomposite

In this subsection, a nanocomposite with two different material types and several cracks is considered. The model is a continuum representative volume element (RVE) of silicate/epoxy nanocomposites with 2% clay weight ratio. Fig. 9.21 depicts the initial configuration of the RVEs, which includes 10 penny shaped cracks. Fig. 9.21 illustrates the cracks in gray while the disc shaped clay reinforcement is plotted in red.

The model in Fig. 9.21 has 303,470 tetrahedral elements and 54,049 nodes. Both materials are assumed to be linear elastic. The epoxy matrix has a Young's modulus of 1.96 GPa and a Poisson's ratio of 0.3. The clay has a much higher stiffness with Young's modulus of 200.0 GPa and a Poisson's ratio of 0.2. The top face of the RVE is loaded with a pressure of −0.1 and the bottom face is fixed. Fig. 9.22 shows the stress in the X direction in several cross-sections. The crack openings are also visible.

9.7.2 Hierarchical multiscale example

Let us consider the clay/epoxy nanocomposite similar to the previous subsection in order to predict it's homogenized Young's modulus by hierarchical upscaling. Fig. 9.23 shows the initial geometry of epoxy/clay nanocomposite.

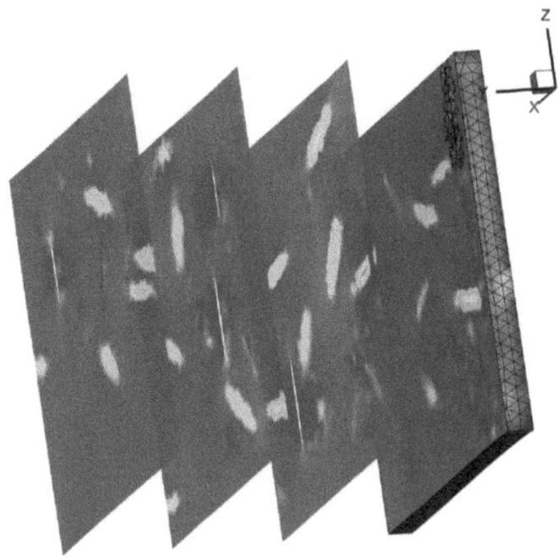

Figure 9.22 The σ_{xx} contour of the loaded nanocomposite RVE.

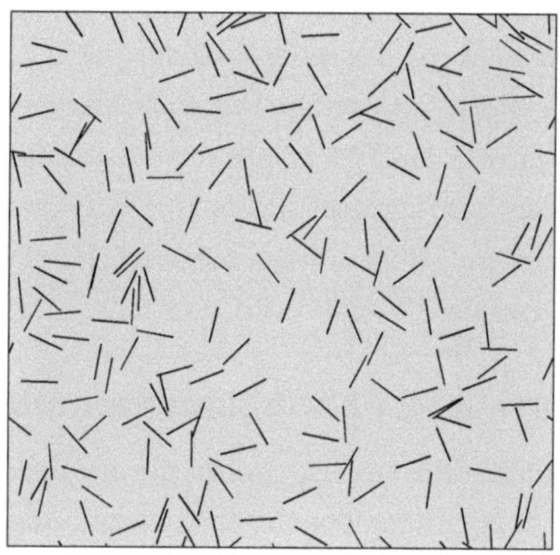

Figure 9.23 Initial configuration of clay/epoxy nanocomposite RVE.

The RVE contains a 2 %wt clay/epoxy ratio and the clay particles are approximated with a rectangular geometry of $l = 100$ nm length and a $t = 1$ nm thickness. The RVE size is 1500×1500 nm^2. Since the bulk composite is isotropic, the RVE size with respect to the clay size should be selected large enough that the RVE will also be isotropic. The material is assumed to be linear elastic material for both clays and epoxy matrix. The Young's modulus, E, and Poisson's ratio, ν, for the clays are 196 GPa and 0.25 respectively. The material properties for the matrix are: Young's modulus $E = 1.96$ GPa, and Poisson's ratio, $\nu = 0.35$. The RVE is

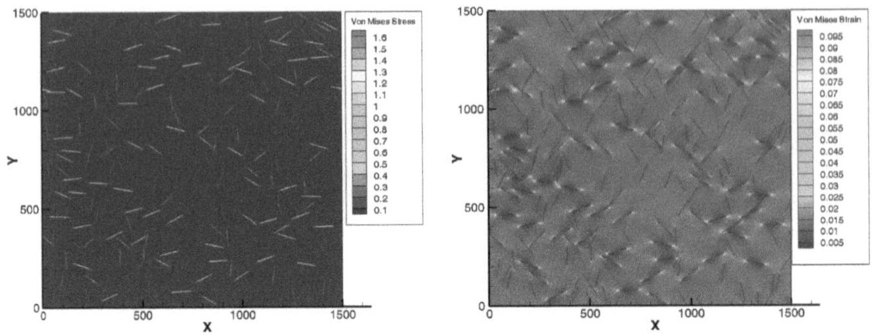

Figure 9.24 The Von Mises stress and Von Mises equivalent strain contours of the RVE.

discretized with four node quadrilateral plain stress elements. Linear displacement (LD) boundary condition are applied resulting in an average strain of 0.016 to the RVE in horizontal direction. This boundary condition satisfies the Hill's energy criterion. Fig. 9.24 shows the Von Mises stress and equivalent strain contours. The predicted elastic modulus and Poisson's ratio for the clay/epoxy nanocomposite is $E = 2.1725$ GPa and $\nu = 0.343$ yielding in an approximate 10% improvement in the elastic modulus of the nanocomposite with only a 2% clay ratio.

9.7.3 Semi-concurrent FE-FE coupling example

Les us apply now a semi-concurrent multiscale method to the clay/epoxy nanocomposite. Fig. 9.25 illustrates the schematic view of the coarse and fine scale models. The geometry and material parameters from the previous section are adopted though we consider a dog-bone sample for the coarse-scale model as depicted in Fig. 9.25. The experiment is done displacement controlled and symmetry is exploited so we only model a quarter of the specimen. The material parameters of the coarse scale model are computed from the fine scale model where the stress tensor is homogenized at each integration point.

A linear displacement is applied to the right edge of the dog-bone sample resulting in a constant strain of 0.11 in the gage section. Fig. 9.26 shows the strain contour in the x direction at the coarse scale. The same figure shows the equivalent Von Mises strain at the fine scale for an arbitrary selected integration point.

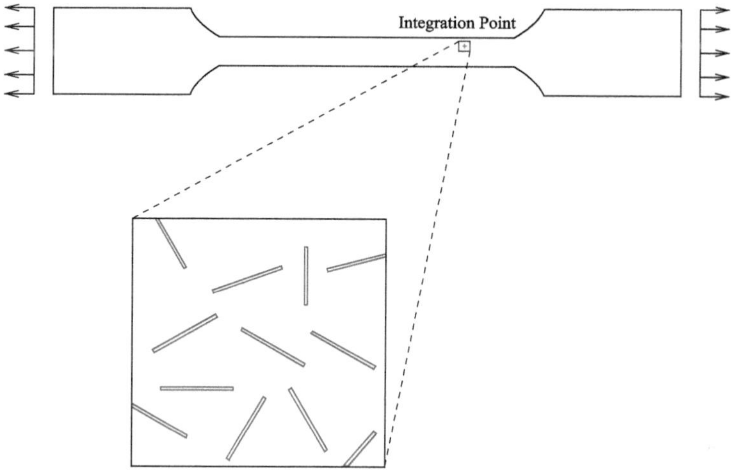

Figure 9.25 FE^2 multiscale analysis of simple tension test for clay nanocomposites.

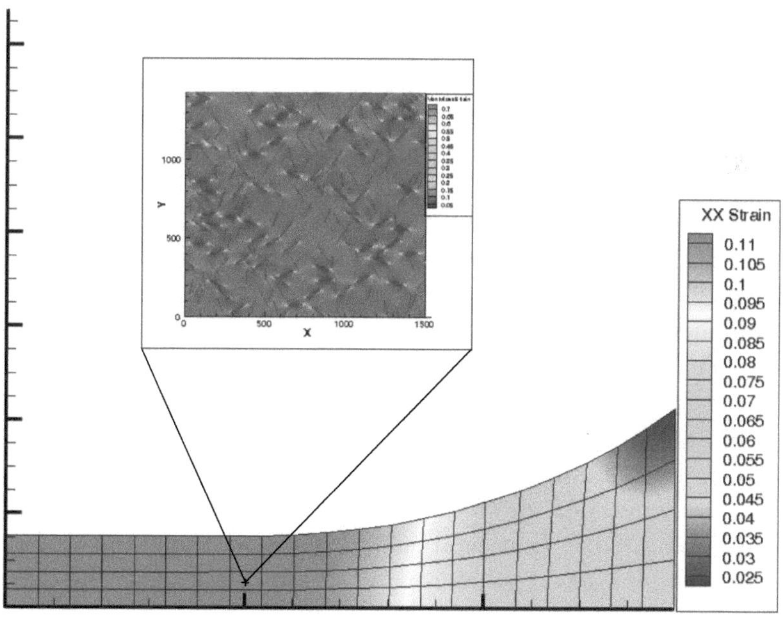

Figure 9.26 ϵ_{xx} contour at the coarse scale and the equivalent Von Mises strain contour at the fine scale for the FE^2 example.

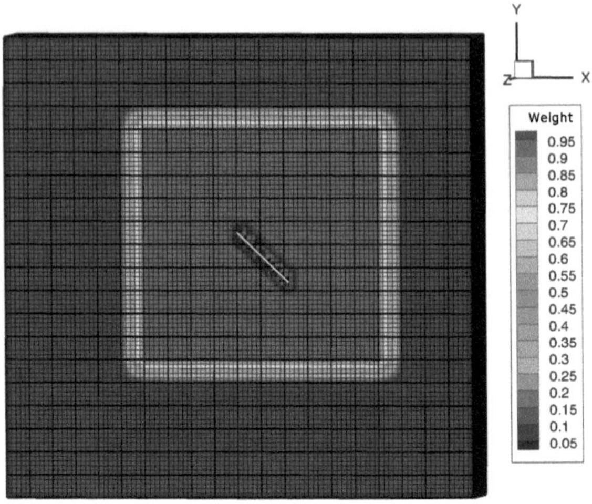

Figure 9.27 Initial configuration of the coupled FE-XFEM example with the weighting function values.

9.7.4 Concurrent FE-XFEM coupling example

Next, we focus on a concurrently coupled model, where an FE mesh and an XFEM model is coupled using the Arlequin method – in explicit dynamics. Therefore, a plate with an inclined crack at the center is loaded with a pressure load at the top and the value of the pressure is -0.25. We model the vicinity of the crack with a very 'fine' mesh and the area far from the crack with a 'coarse' mesh. The dimensions of the plate are $100 \times 100 \times 20$ and the length of the crack is 15. Two finite element parts are created with the same dimensions and different mesh sizes. We then use the fine mesh to compute a reference solution and compare it to the coupled one. Both parts are discretized with eight node brick elements. Fig. 9.27 shows the initial mesh and the weighting parameters (Eq. (9.2)).

Both the coarse and fine-scale models have the same material properties with $E = 26.0$, $\nu = 0.3$ and density, $\rho = 1.0$. The time step, Δt, is 0.05. Stress waves travel from the coarse-scale to the coupling region and to the fine-scale and travel into the coarse-scale again. Fig. 9.28 shows the result of the simulation. On the right hand side, the displacement in the Y direction is shown for both scales. The left hand side of Fig. 9.28 shows the reference solution and the Y displacement contours. We do not observe any artificial wave reflection.

Fig. 9.29 shows the Y displacement at a point located at $(60, 60, 10)$ for the two domains i.e. the reference solution and the coupled one. As is

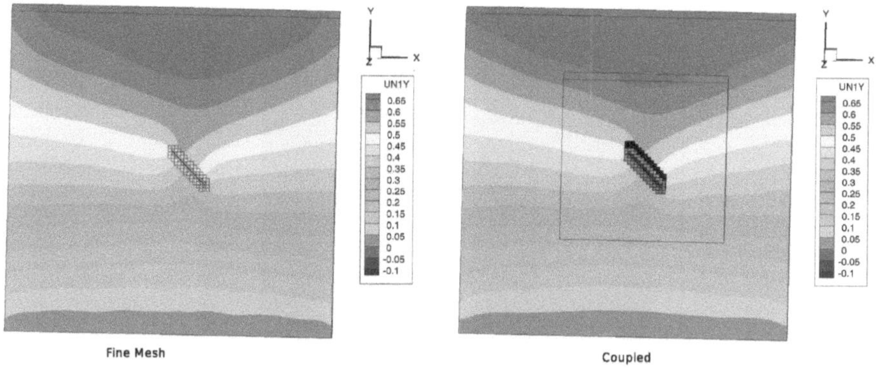

Figure 9.28 The Y displacement at the time 564.0 seconds. Left: reference solution (fine mesh), right: the coupled model.

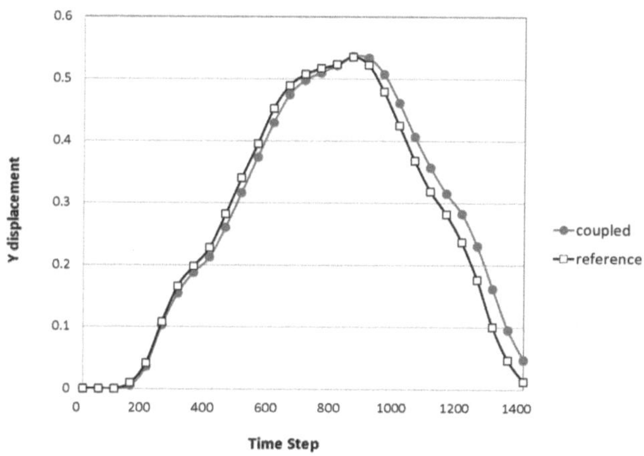

Figure 9.29 The Y displacement versus time for a point (60, 60, 10) at the coupled and the reference solution.

shown in the figure, the discrepancy in the Y displacements versus time is small and due to the displacement approximation in the fine-scale region of significantly higher resolution.

9.7.5 MD-XFEM coupling example

Finally, we consider a two-dimensional single crystal with dimensions of 200×100 units. In this example a straight crack of length 55 units is present in the domain. The continuum model consists of 253 quadrilateral elements and 576 degrees of freedom. The element size is constant over the domain,

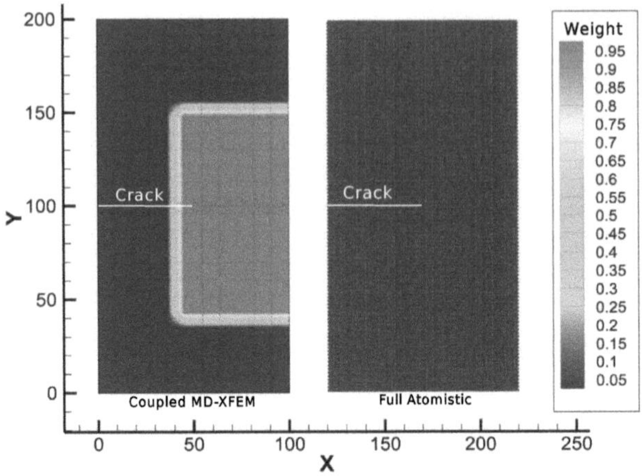

Figure 9.30 Initial configuration of the coupled model and the full atomistic counter part.

about 11 units. An atomistic domain of almost the same size is placed on top of the finite element part. Fig. 9.30 shows a schematic configuration of the system. The open-access code LAMMPS is used for the atomistic system. Since part of the crack is located in the atomistic domain, the crack must be modeled in the atomistic region as well. We do not remove rows of atoms along the crack though as this is somewhat arbitrary and introduces extra parameters in the formulation. Instead, we modify the neighbor list of the atoms to prevent force transmission across the crack faces. This method will produce a crack which is consistent with a sharp crack within XFEM. A second atomistic part is also defined in the same model which has the same configuration without coupling to any finite element mesh. This allows us to directly compare the full atomistic simulation to the coupled one.

The atomistic domain is a two-dimensional lattice from a hexagonal (HEX) crystal lattice with lattice constant 0.91 LJ units extended in the [1 0 0] crystal direction. For style LJ, all quantities are without units and LAMMPS sets the fundamental quantities mass, sigma, epsilon, and the Boltzmann constant = 1. Atomic interactions are modeled by the Lennard-Jones potential with parameters $\sigma = 1.0$ LJ units, $\epsilon = 1.0$, and a cut-off radius of 2.5; the mass of all atoms is taken as 1.0. Before the actual coupled simulation, we minimize the potential energy – by the conjugate gradient (CG) method [32] – in the pure atomistic part to equilibrate the system.

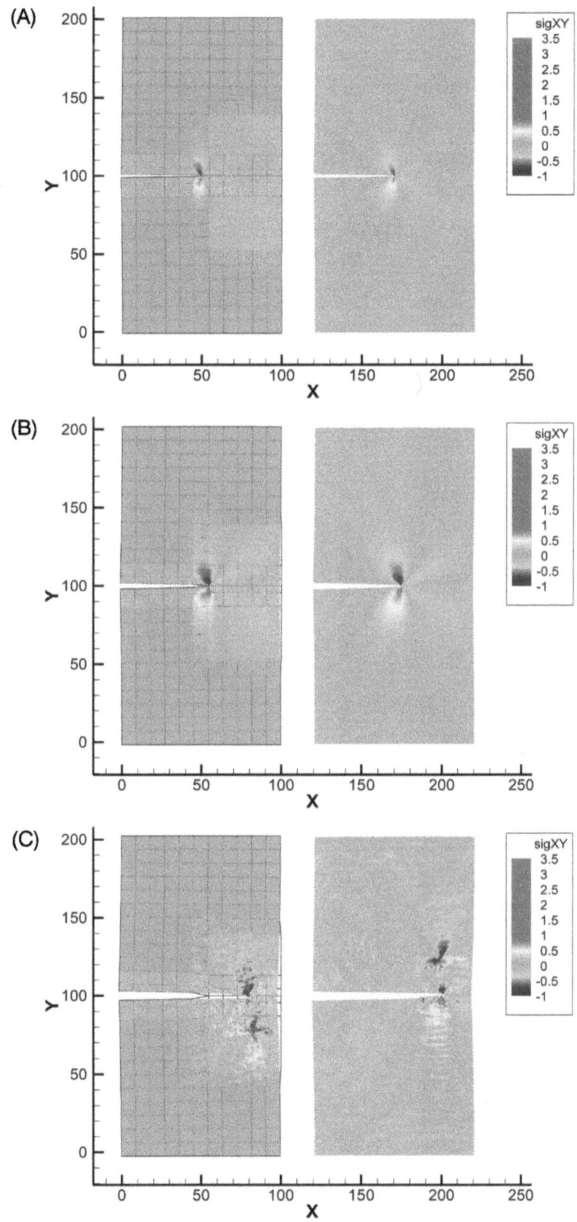

Figure 9.31 Atomistic σ_{xy} contour around the crack for coupled and full atomistic parts at 48 ps (A), 90 ps (B) and 127.5 ps (C).

The coupling of the continuum and atomistic parts is performed within a cubic box of dimensions $65 \times 110 \times 0$ LJ units^2. The elements which are cut by this box are the bridging elements and the atoms which are located inside the bridging elements are the bridging atoms. Consequently, the coupling region is one element wide. The driving force for the system is introduced through a velocity boundary condition on the top and bottom faces of the continuum region. A velocity of 0.02 and −0.02 is imposed on all the nodes belonging to the top and bottom boundary of the continuum domain at each time step, respectively. The time step is 0.003 and the magnitude of the time step is chosen according to the stability criterion in the pure atomistic domain. Fig. 9.30 shows the initial configuration of the body and the weighting of the atoms and nodes.

Fig. 9.31 shows the Virial stresses at four different time steps. The symmetric Virial stress tensor is computed for each atom and for pair potentials [39,40] by

$$\sigma_{ij}^{V} = \frac{1}{V}\sum_{\alpha}\left[\frac{1}{2}\sum_{\beta=1}^{N}\left(R_i^{\beta} - R_i^{\alpha}\right)F_j^{\alpha\beta} - m^{\alpha}v_i^{\alpha}v_j^{\alpha}\right] \tag{9.92}$$

where (i,j) range over the spatial directions, x, y, z; $\beta \in [1,\ldots,N]$ ranges over the N neighbors of atom α, R_i^{α} is the coordinate of atom α in the i direction, $F_j^{\alpha\beta}$ is the force on atom α from atom β along the j direction, V is the total volume, m^{α} is the mass of atom α and v^{α} is the velocity of atom α. In Fig. 9.31A and B a stress concentration is visible, that is initially confined at the crack front; subsequently when propagation occurs, stress waves are emitted from the crack tip. The coupled model can accurately predict the crack propagation behavior with much fewer degrees of freedom.

References

[1] F. Abraham, J. Broughton, N. Bernstein, E. Kaxiras, Spanning the length scales in dynamic simulation, Computational Physics 12 (1998) 538–546.

[2] P. Baumann, J. Oden, S. Prudhomme, Adaptive multiscale modeling of polymeric materials with Arlequin coupling and goals algorithms, Computer Methods in Applied Mechanics and Engineering 198 (2009) 799–818.

[3] J. Bellec, J.E. Dolbow, A note on enrichment functions for modelling crack nucleation, Communications in Numerical Methods in Engineering 19 (2003) 921–932.

[4] T. Belytschko, W.E. Bachrach, Efficient implementation of quadrilaterals with high coarse-mesh accuracy, Computer Methods in Applied Mechanics and Engineering 54 (1986) 279–301.

[5] T. Belytschko, T. Black, Elastic crack growth in finite elements with minimal remeshing, International Journal for Numerical Methods in Engineering 45 (1999) 601–620.

[6] T. Belytschko, W.K. Liu, B. Moran, Nonlinear Finite Elements for Continua and Structures, John Wiley & Sons Ltd., 2000.
[7] T. Belytschko, S. Loehnert, J.-H. Song, Multiscale aggregating discontinuities: a method for circumventing loss of material stability, International Journal for Numerical Methods in Engineering 73 (6) (2008) 869–894.
[8] T. Belytschko, R. Gracie, M. Xu, Concurrent coupling of atomistic and continuum models, in: J. Fish (Ed.), Multiscale Methods Bridging the Scales in Science and Engineering, Oxford Press, 2009, pp. 93–133.
[9] P. Budarapu, R. Gracie, S. Bordas, T. Rabczuk, An adaptive multiscale method for quasi-static crack growth, Computational Mechanics 53 (6) (2014) 1129–1148.
[10] P. Budarapu, R. Gracie, S.-W. Yang, X. Zhuang, T. Rabczuk, Efficient coarse graining in multiscale modeling of fracture, Theoretical and Applied Fracture Mechanics 69 (2014) 126–143.
[11] P.R. Budarapu, T. Rabczuk, Multiscale methods for fracture: a review, Journal of the Indian Institute of Science 97 (3) (2017) 339–376.
[12] C. Duarte, D. Kim, Analysis and applications of a generalized finite element method with global-local enrichment functions, Computer Methods in Applied Mechanics and Engineering 197 (6–8) (2008) 487–504.
[13] C.A. Duarte, I. Babuška, A global-local approach for the construction of enrichment functions for the generalized FEM and its application to propagating three-dimensional cracks, in: V.M.A. Leitão, C.J.S. Alves, C. Armando Duarte (Eds.), ECCOMAS Thematic Conference on Meshless Methods, Lisbon, Portugal, 2005.
[14] M. Elices, G. Guinea, J. Gomez, J. Planas, The cohesive zone model: advantages, limitations and challenges, Engineering Fracture Mechanics 69 (2) (2002) 137–163.
[15] F. Feyel, J.-L. Chaboche, fe^2 multiscale approach for modeling the elastoviscoplastic behavior of long fiber SiC/Ti composite materials, Computer Methods in Applied Mechanics and Engineering 183 (2000) 309–330.
[16] J. Fish, The s-version of the finite element method, Computers & Structures 43 (1992) 539–547.
[17] J. Fish, R. Guttal, The s-version of finite element method for laminated composites, International Journal for Numerical Methods in Engineering 39 (1996) 3641–3662.
[18] D.P. Flanagan, T. Belytschko, A uniform strain hexahedron and quadrilateral with orthogonal hourglass control, International Journal for Numerical Methods in Engineering 17 (1981) 679–706.
[19] R. Gracie, T. Belytschko, Concurrently coupled atomistic and XFEM models for dislocations and cracks, International Journal for Numerical Methods in Engineering 78 (2009) 354–378.
[20] R. Gracie, T. Belytschko, Adaptive continuum-atomistic simulations of dislocation dynamics, International Journal for Numerical Methods in Engineering 86 (4–5) (2011) 575–597.
[21] P.A. Guidault, O. Allix, L. Champaney, C. Cornuault, A multiscale extended finite element method for crack propagation, Computer Methods in Applied Mechanics and Engineering 197 (2008) 381–399.
[22] V. Kouznetsova, Computational Homogenization for the Multi-Scale Analysis of Multi-Phase Materials, PhD Thesis, Netherlands Institute for Metals Research, The Netherlands, 2002.
[23] V. Kouznetsova, M. Geers, W. Brekelsmans, Multi-scale constitutive modeling of heterogeneous materials with a gradient-enhanced computational homogenization

scheme, International Journal for Numerical Methods in Engineering 54 (2002) 1235–1260.
[24] S. Loehnert, Computational Homogenization of Microheterogeneous Materials at Finite Strains Including Damage, PhD thesis, University Hannover, Germany, 2004.
[25] J.E. Marsden, T.J. Hughes, Mathematical Foundations of Elasticity, Dover Publications, New York, 1994.
[26] R.E. Miller, E.B. Tadmor, The quasicontinuum method: overview, applications and current directions, Journal of Computer-Aided Materials Design 9 (2002) 203–239.
[27] N. Moes, T. Belytschko, Extended finite element method for cohesive crack growth, Engineering Fracture Mechanics 69 (2002) 813–834.
[28] N. Moes, J. Dolbow, T. Belytschko, A finite element method for crack growth without remeshing, International Journal for Numerical Methods in Engineering 46 (1999) 131–150.
[29] S. Nemat-Nasser, M. Hori, Micromechanics: Overall Properties of Heterogeneous Materials, Elsevier, Amsterdam, 1993.
[30] J. Oden, S. Prudhomme, A. Romkes, P. Baumann, Multiscale modeling of physical phenomena: adaptive control of models, SIAM Journal on Scientific Computing 28 (6) (2006) 2359–2389.
[31] S. Plimpton, Fast parallel algorithms for short-range molecular dynamics, Journal of Computational Physics 117 (1995) 1–19.
[32] W.H. Press, B.P. Flannery, S.A. Teukolsky, W.T. Vetterling, Numerical Recipes in FORTRAN 77: The Art of Scientific Computing, 2nd edition, Cambridge University Press, Sept. 1992.
[33] D. Qian, S. Chirputkar, Bridging scale simulation of lattice fracture using enriched space-time finite element method, International Journal for Numerical Methods in Engineering 97 (2014) 819–850.
[34] D. Qian, R. Gondhalekar, A virtual atom cluster approach to the mechanics of nanostructures, International Journal of Multiscale Computational Engineering 2 (2) (2004) 277–289.
[35] D. Qian, G. Wagner, W. Liu, A multiscale projection method for the analysis of carbon nanotubes, Computer Methods in Applied Mechanics and Engineering 193 (17–20) (2003) 1603–1632.
[36] T. Rabczuk, T. Belytschko, Cracking particles: a simplified meshfree method for arbitrary evolving cracks, International Journal for Numerical Methods in Engineering 61 (2004) 2316–2343.
[37] M. Silani, H. Talebi, S. Ziaei-Rad, A. Hamouda, G. Zi, T. Rabczuk, A three dimensional extended Arlequin method for dynamic fracture, Computational Materials Science 96 (PB) (2015) 425–431.
[38] J.H. Song, P.M.A. Areias, T. Belytschko, A method for dynamic crack and shear band propagation with phantom nodes, International Journal for Numerical Methods in Engineering 67 (2006) 868–893.
[39] A.K. Subramaniyan, C. Sun, Continuum interpretation of virial stress in molecular simulations, International Journal of Solids and Structures 45 (14) (2008) 4340–4346.
[40] R.J. Swenson, Comments on virial theorems for bounded systems, American Journal of Physics 51 (10) (1983) 940–942.
[41] H. Talebi, M. Silani, S. Bordas, P. Kerfriden, T. Rabczuk, Molecular dynamics/XFEM coupling by a three-dimensional extended bridging domain with applications to dynamic brittle fracture, International Journal of Multiscale Computational Engineering 11 (6) (2013) 527–541.

[42] H. Talebi, M. Silani, S. Bordas, P. Kerfriden, T. Rabczuk, A computational library for multiscale modeling of material failure, Computational Mechanics 53 (5) (2014) 1047–1071.

[43] H. Talebi, M. Silani, T. Rabczuk, Concurrent multiscale modeling of three dimensional crack and dislocation propagation, Advances in Engineering Software 80 (2015), (C) 82–92.

[44] C.V. Verhoosel, J.J.C. Remmers, M.A. Gutierrez, R. de Borst, Computational homogenization for adhesive and cohesive failure in quasi-brittle solids, International Journal for Numerical Methods in Engineering 83 (8–9) (2010) 1155–1179.

[45] F. Vernerey, M. Kabiri, Adaptive concurrent multiscale model for fracture and crack propagation in heterogeneous media, Computer Methods in Applied Mechanics and Engineering 276 (2014) 566–588.

[46] G. Wagner, W. Liu, Coupling of atomistic and continuum simulations using a bridging scale decomposition, Journal of Computational Physics 190 (2003) 249–274.

[47] S. Xiao, T. Belytschko, A bridging domain method for coupling continua with molecular dynamics, Computer Methods in Applied Mechanics and Engineering 193 (2004) 1645–1669.

[48] S.-W. Yang, P. Budarapu, D. Mahapatra, S. Bordas, G. Zi, T. Rabczuk, A meshless adaptive multiscale method for fracture, Computational Materials Science 96 (PB) (2015) 382–395.

CHAPTER TEN

A short overview of alternatives for fracture

10.1. Numerical manifold method (finite cover method)

The numerical manifold method (NMM) is a numerical method developed during the early 1990s based on the concept of "manifold" or "finite cover" [61]. NMM and XFEM share some similarities in terms of introducing additional nodal degrees of freedom by the concept of partition of unity (PU). NMM has been called also other name such as the finite cover method (FCM) [71], however, it is essentially NMM. The NMM has been successfully applied in solving some interesting problems in engineering such as free boundary value problems. The NMM has recently attracted much interest due to its distinctive capability for analyzing problems in finite deformation, discontinuities and cracking where material interfaces or geometries are changing. In this way, NMM embraces the advantages of classical finite element method however it becomes more flexible to offer the possibility of transition to discrete discontinuum based modeling approach such as discontinuous deformation analysis.

NMM is based on a so-called topological manifold and a differentiable manifold with two independent meshes, namely, a mathematical mesh and a physical mesh (see Fig. 10.1). The cover functions constructed by the NMM are free to take different orders of expansion for specific parts of the problem, and different types of analytical expansions in subdomains are possible. The displacements along the interfaces of subdomains are naturally consistent. In this paper, the FCM is extended to elastoplastic problems based on a least squares approximation. High order polynomials are used to construct the physical cover functions based on a least squares approach so that the boundary conditions can be applied as conveniently as with the conventional FEM. Polynomial expansions are commonly used to construct cover functions in the FCM because they are most convenient and straightforward. However, for high order elements, the number of unknowns at each node becomes much larger than the number of physical degrees of freedom. This makes the system of global governing equations much greater than in the FEM, which is computationally undesirable. In

Extended Finite Element and Meshfree Methods
https://doi.org/10.1016/B978-0-12-814106-9.00016-4
Copyright © 2020 Elsevier Inc.
All rights reserved.

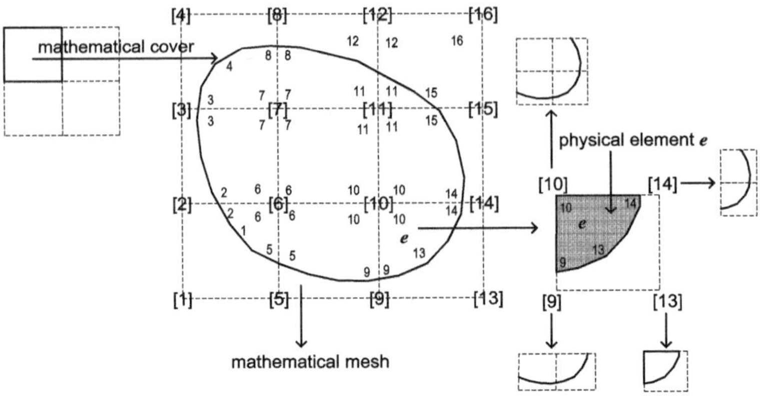

Figure 10.1 The mathematical cover, physical cover and physical element in the FCM.

addition, the imposition of essential boundary conditions is more difficult than for the FEM.

10.1.1 The cover approximation

The *physical cover* is defined by the intersection of *mathematical cover* and physical domain. The mathematical cover is the same as the physical cover if there are no material discontinuities or interfaces. Displacements are introduced at physical cover and are represented as physical cover functions $u^{Li}(\boldsymbol{x})$, where i is the index of physical cover. For example, the physical cover associated with node 10 is shown in Fig. 10.1. The intersection of the physical cover is physical element. Within each element, displacement is approximated from the finite cover functions of its composing nodes. Finally, the finite cover approximation are obtained as

$$u^e(\boldsymbol{x}) = \sum_i^{n_e} w_i u^{Li}(\boldsymbol{x}) \quad i = 1, \cdots, n_e, \tag{10.1}$$

and w_i is the weight function related with the ith physical cover and n_e is the number of associated physical covers. The weight function in Eq. (10.1) should be non-negative and conforms to partition of unity, i.e., $\sum_i^{n_e} w_i = 1$. Here, linear weight functions in 2D, which are exactly the same the standard FEM, are used. For an arbitrary physical cover e shown in Fig. 10.1, the displacement is approximated by nodes 9, 10, 13, 14. The conventional way to construct the physical cover function is to use expansions $\mathbf{p}^T(\boldsymbol{x}) = \{p_1, \cdots, p_m\}$, where m is the number of expansions. Lagrange polynomials are used in this study where for 2D first order $\mathbf{p}(\boldsymbol{x}) = \{1, x, y\}$ and second order $\mathbf{p}(\boldsymbol{x}) = \{1, x, y, x^2, xy, y^2\}$. The cover function over the physical cover is then constructed as

$$u^{Li}(\boldsymbol{x}) = \mathbf{p}^T(\boldsymbol{x})\mathbf{D}_i = \sum_j^m p_j(\boldsymbol{x})d_j \tag{10.2}$$

where $\mathbf{D}_i$ is a vector of unknown coefficients and i is the index for physical cover as in Eq. (10.1). Substituting (10.2) into (10.1), the finite cover approximation is obtained as

$$u^e(\boldsymbol{x}) = \sum_i^{n_e}\sum_j^m p_j(\boldsymbol{x})d_j^i \tag{10.3}$$

10.1.2 The least square-based physical cover functions

The physical cover function developed here is also approximated by polynomials

$$u^{Lc}(\boldsymbol{x}) = \sum_j^m p_j(\boldsymbol{x})a_j = \mathbf{p}^T\mathbf{a} \tag{10.4}$$

A least squares method [56] is used where the total difference between the approximated nodal values and real nodal values is minimized, which gives us

$$u^{Lc}(\boldsymbol{x}) = \sum_I^n \phi_I(\boldsymbol{x})u_I \tag{10.5}$$

where

$$\phi_I(\boldsymbol{x}) = \sum_j^m p_j(\boldsymbol{x})(\mathbf{A}^{-1}\mathbf{B})_{jI}, \quad \mathbf{A} = \sum_I^n p^T(\boldsymbol{x}_I)p(\boldsymbol{x}_I), \quad \mathbf{B} = \sum_I^n p^T(\boldsymbol{x}_I) \tag{10.6}$$

It can be seen that when m increases for higher order expansions, the number of unknowns increases significantly for the conventional FCM while for the LS based FCM, the number of degrees of freedom remains the same. Although there is extra computational cost in inverting $\mathbf{A}$ in (10.6) in order to get the shape functions, the size of the general stiffness matrix is much smaller and also the solution time is largely reduced which is desirable for nonlinear iterative solution procedures.

10.1.3 The imposition of boundary conditions

There are different ways to impose boundary conditions. A dual basis method was proposed in [15]. Substituting Eq. (10.5) into (10.2), we get the approximation on the physical element as

$$u^e(\boldsymbol{x}) = \sum_i^{n_e} w_i \left(\sum_I^{n} \phi_I(\boldsymbol{x}) u_I \right) = \sum_i^{n_e} w_i \boldsymbol{\phi} \mathbf{u} \tag{10.7}$$

Since $\phi_I(\boldsymbol{x})$ from the LS approximation is not interpolatory, imposing prescribed displacements becomes a problem. A solution is provided by modifying the shape functions into the following form

$$\boldsymbol{\phi}^p(\boldsymbol{x}) = [\phi_1(\boldsymbol{x}) - \phi_1(\boldsymbol{x}_c), 1 + \phi_c(\boldsymbol{x}) - \phi_c(\boldsymbol{x}_c), \cdots, \phi_n(\boldsymbol{x}) - \phi_n(\boldsymbol{x}_c)] \tag{10.8}$$

so that $u^e(\boldsymbol{x}_c) = u_c$. Using $\boldsymbol{\phi}^p(\boldsymbol{x})$ for $\boldsymbol{\phi}(\boldsymbol{x})$, then the essential boundary conditions can be treated as easily as in the FEM.

10.1.4 Fracture modeling

NMM was proposed few years after the discontinuous deformation analysis (DDA) from Shi [61]. Both NMM and DDA were motivated by solving problems in rock mechanics and rock engineering. Therefore, many works and test examples were often shown with the application background for slopes, dams, tunnel etc where rocks are handled as brittle linear elastic medium. NMM is claimed to be able for solving continuum and discontinuum in a unified framework without additional treatment, such as the Heaviside functions in XFEM. An example was from Ning et al. [49] (see Fig. 10.2), where NMM was applied in footwall slope stability analysis which predicted the entire failure process of the slope from crack propagation to sliding and showed that NMM can well capture continuum-discontinuum problems in a unified framework. Zheng et al. [78] applied NMM in the analysis of unconfined seepage flow in porous media. We should notify that though NMM works for continuum-discontinuum, the switch from continuum to discontinuum is not automatically handled by NMM. It means one needs to change from an equilibrium quasi-static system to a kinematic system by using parameters to control the computation. NMM is claimed to be used in simulating complex crack patterns. In this context, numerous efforts have been made. Wang and Ge [62] applied NMM in modeling crack growth based on the linear fracture mechanics. Then Wang et al. [26] considered modeling contact-friction of

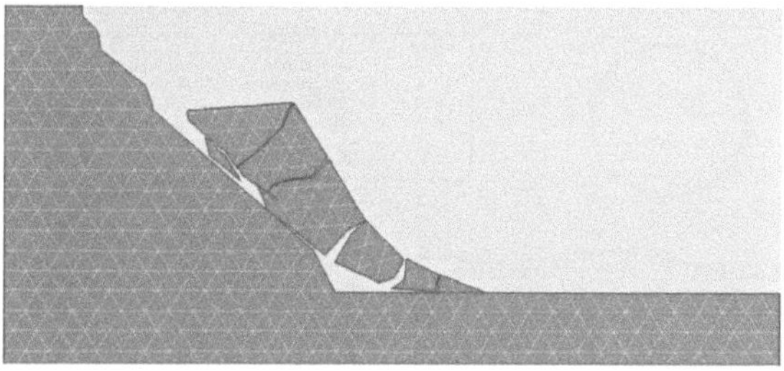

Figure 10.2 Buckling failure process of a footwall slope [49].

the interfaces of materials with improved Lagrangian method calculated stress intensity factors with a path independent contour integral method by NMM, which yielded numerical results similar to experimental results. Tsay et al. [72] developed NMM with a local mesh refinement and auto-remeshing schemes and applied it in the prediction of crack growth with satisfactory accuracy. Zhang et al. [76] studied the initiation and propagation of cracks in a gravity dam and two arch dam abutments with faults and joints by manifold method. Chiou et al. [18] integrated NMM with the virtual crack extension method, to the modeling of mixed mode fracture propagation, which deployed a fictitious crack model to investigate the fracture process zone and set the cohesive normal stress to be linear. Zheng and Xu [77] solved several issues in NMM for solving linear elastic fracture problems, including a new variational principle that suppress the gradient-dependent DOFs to reduce the rank deficiency of stiffness matrix, a new integration scheme for crack analysis, a sign convention of argument in the polar system at the crack tip for kinked crack analysis, and a mesh dependency for strong singularity such as crack problems by NMM. Zheng et al. [78] reduced the static growth of multiple cracks to a nonlinear complementarity problem and a MLS-based NMM with a projection-contraction algorithm was designed to solve the problem, and observed through numerical experiment in Fig. 10.3 that there might be more than one critical point on the full curve of nominal stress vs. strain, which overturned the definition of brittle cracking.

For a physical cover which has no crack tips or inclusion that results in stress concentration, polynomial basis can be used as local basis function as shown in Eq. (10.2). If a crack tip is included in a physical cover, namely ω_i^p, as shown in Fig. 10.4, the first order term of Williams' displacement

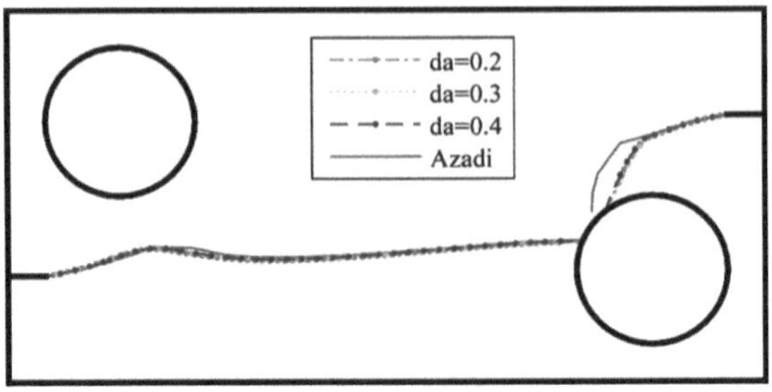

Figure 10.3 Crack growth of a plate with two hole [78].

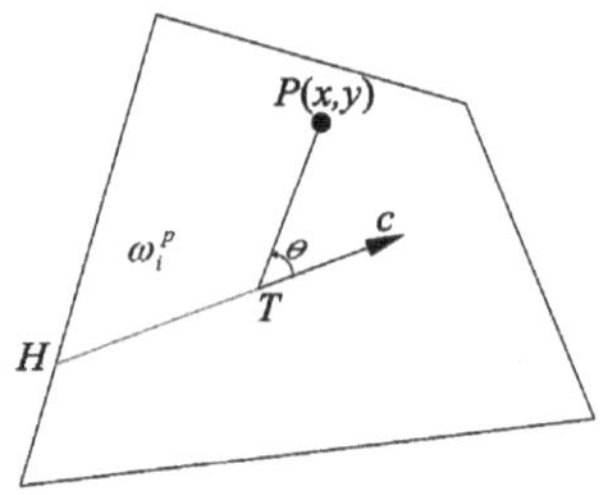

Figure 10.4 A physical cover containing a crack tip.

series [74] near the crack tip are included in

$$\mathbf{B}_{\mathbf{i}}^{\mathrm{T}}(r)=\left[1,\ \sqrt{r}\cos\frac{\theta}{2},\ \sqrt{r}\sin\frac{\theta}{2},\ \sqrt{r}\cos\frac{3\theta}{2},\ \sqrt{r}\sin\frac{3\theta}{2}\right] \tag{10.9}$$

where (r,θ) is the polar coordinates defined in ω_i^p with respect to the crack tip as shown in Fig. 10.4 and $-\pi \leq \theta \leq \pi$.

To show how displacement jump can be modeled by the NMM, we will explain the displacement approximations of a pair of two points lying on either side the crack as shown in Fig. 10.5. Consider points located on the crack denoted by A_1 and A_2, it is conspicuous that both points share the same coordinates. However, A_1 belongs to the lower portion of the crack surface while A_2 belongs to the upper portion of the crack surface A_1 is covered by physical patches ω_{2-1}^p and ω_{2-3}^p so the global approximation at point A_1 can be defined as

$$u^e(\boldsymbol{x}_{A_1})=\omega_{2-1}u^{L_{2-1}}+\omega_{2-3}u^{L_{2-3}} \tag{10.10}$$

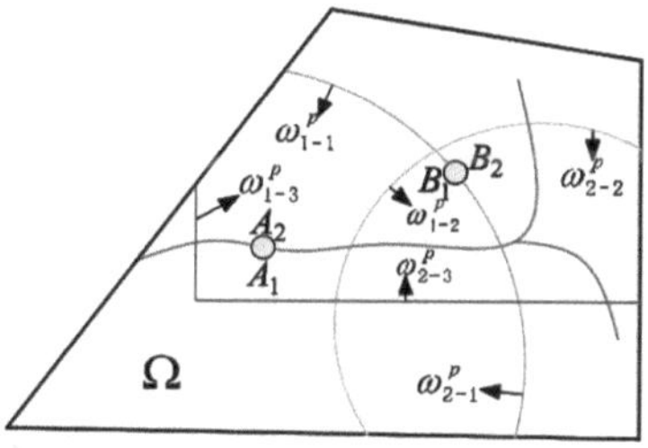

Figure 10.5 A domain covered by six physical patches ($\omega_{1-2}^p \leftarrow$ refers to that boundary segment of mathematical patch ω_1^m pointed by the arrow and the physical domain boundary segment constitute the boundary of physical patch ω_{1-2}^p).

For point A_2, it is covered by physical cover ω_{1-1}^P and ω_{1-3}^P and the global approximation is therefore defined by

$$u^e(\boldsymbol{x}_{A_2}) = \omega_{1-1} u^{L_{1-1}} + \omega_{1-3} u^{L_{1-3}} \ . \tag{10.11}$$

Here, $\omega_{2-1}(A_1) = \omega_{1-1}(A_2) = \omega_1(A_1) = \omega_1(A_2)$ and $\omega_{2-3}(A_1) = \omega_{1-3}(A_2) = \omega_3(A_1) = \omega_3(A_2)$. And $\omega_1(A_1)$ and $\omega_3(A_1)$ correspond to the value of weight functions of mathematical cover at point A_1 and same meaning for $\omega_1(A_2)$ and $\omega_3(A_2)$. The local approximation $u^{L_{1-1}}$ and $u^{L_{2-1}}$ are defined on separate physical cover, namely $u^{L_{1-1}} \neq u^{L_{2-1}}$. Similarly, that $u^{L_{1-3}} \neq u^{L_{2-3}}$. It can be easily seen that $u^e(\boldsymbol{x}_{A_1}) \neq u^e(\boldsymbol{x}_{A_2})$, which demonstrates that the NMM approximation is discontinuous across the crack. It can be seen that the cut of the mathematical cover due to the existence of crack into two physical covers will result in two independent local approximations defined over two physical cover. Hence, the global approximation is discontinuous across the crack. Different from XFEM, which introduces a Heaviside function explicitly to describe the discontinuities, the cutting process used in NMM is naturally introduced through physical covers and mathematical covers. Thus, NMM is more suitable for modeling intersecting cracks.

10.1.5 Geometric and material nonlinear analysis

The NMM has been applied for solving nonlinear problems by using different types of descriptions, namely Lagrangian, Eulerian and Arbitrary Lagrangian Eulerian (ALE) descriptions. With Lagrangian description, Cai et al. [16] proposed the geometric nonlinear analysis in the framework of NMM with updated Lagrangian formulation, which is useful for analyzing discontinuous large deformation problems. Terada et al. [70,71] and Su

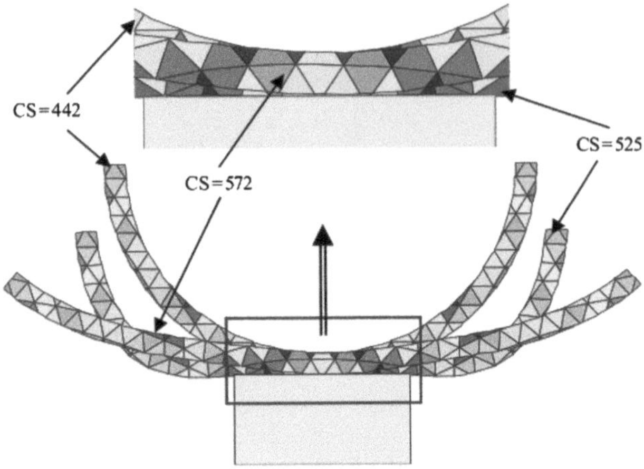

Figure 10.6 Large deformation analysis of a flexible semi-ring using the NMM [20].

et al. [28,69] extended the finite cover method (FCM), which is essentially the numerical manifold method, to the geometric nonlinear analysis involving heterogeneities with nonlinear mechanical behavior due to interfacial debonding. With Eulerian description, [43] studied the variational principles and application of nonlinear numerical method with Eulerian description and applied into the plastic flow. For material nonlinear application, most of applications of NMM in solid mechanics are limited in elastic model. Fan et al. [20] extended the NMM to large deformation and large rotation problems with a strain-rotation decomposition theorem of dynamic formulation as shown in Fig. 10.6. The following work contributed to the development of material nonlinear analysis. Luo [43] also simulated the metal forming process by NMM with rigid plastic constitution relationship. Wei et al. [73] applied NMM in the elastoplastic analysis for rock mass by using the parametric variation principle, which included the solving of non-associate flow shuffling off iterations at every load step.

10.2. Peridynamics and dual-horizon peridynamics

Peridynamics (PD) has recently attracted broad interests for researchers in computational solid mechanics since it shows great flexibility in modeling dynamic fractures. The crack happens naturally, and no attention is paid to describe crack topology in the PD simulation. Hence, complicated fracture patterns i.e. crack branching and coalescence of multiple cracks, can be captured with ease by breaking the bonds between material points. No techniques such as smoothing the normals of the crack surfaces in order to avoid erratic crack paths is necessary in PD, while such techniques are commonly used in the extended finite element method (XFEM) [9], meshless methods [55] or other partition of unity methods (PUM) [7]. Meanwhile, PD can be easily extended from 2D to 3D, which cannot always be done in other methods [17,25]. In addition, an important feature in PD is that it can now be easily coupled with the more classical local continuum mechanics [42,59]. The traditional PD method was proposed by Silling [63] and has been applied onwards for a wide range of problems including impact loading, fragmentation [23,38], composites delamination [51], beam and plate/shell structures [19,47,50], thermal diffusion [11] and flow in porous media [30,34].

In PD, the equation of motion is formulated in an integral form rather than the partial differential form. The material point interacts with other material points in its horizon, a domain associated to the point. The horizon is often represented by circular area (2D) or spherical region (3D) with a radius. There are mainly 3 types of PD formulations, i.e. bond based peridynamics (BB-PD), ordinary state based peridynamics (OSB-PD) and non-ordinary state based peridynamics (NOSB-PD). PD formulation starts with discretizing the solid domain into material points, where any two material points within horizon's radius forms a bond. The first version of PD is called the bond based formulation (BB-PD) [63] as the bonds act like independent springs. One limitation of BB-PD is that only the material with Poisson's ratio of 1/3 in 2D or 1/4 in 3D can be modeled. In order to eliminate the Poisson's ratio restriction, the state based peridynamics (SB-PD) [67] is proposed, where "state" means the bond deformation depends not only on the deformation of the bond itself, but also on collective deformation of other bonds. The SB-PD was extended into two types, namely the OSB-PD and the NOSB-PD. After that, BB-PD is regarded as a special case of OSB-PD. NOSB-PD is capable of simulating the material with advanced constitute models, e.g. viscoplasticity model [21], the Drucker-Prager plasticity model [38] and Johnson-Cook (JC) constitutive model [2].

Aside from the pure PD formulation above mentioned, much work was devoted to the coupling of peridynamics with the continuum methods [6,29,42,59,60]. Such strategies enable to model the fracture domain with PD while other domain modeled with continuum methods. Saleson et al. developed a force-based coupling scheme for peridynamics and classical elasticity in one dimension [59] and higher dimensions [60]. This method satisfies the Newton's third law and Patch test, and no ghost force exists in the blending domain. Lubineau et al. developed a morphing strategy to couple nonlocal to local continuum mechanics. The strategy is based on the domain transition where the strain energy is divided into nonlocal part and local continuum part. In this method, the PD is discretized with finite element and the local integration matrix is assembled.

One basic constraint of traditional PD formulations is that the radii of horizons are required to be constant, otherwise spurious wave reflections shall emerge and ghost forces between material points will deteriorate the results. However, in many applications, for sake of computational efficiency, it is necessary to vary the horizon sizes with respect to the spatial distribution of the material points, e.g. for adaptive refinement, and multiscale modeling. In the implementation of traditional PD discretized with particles, in order to achieve acceptable accuracy, the horizon radius has to be determined with respect to the lowest material point resolution locally required. It should be noted that horizon is guided by the characteristic range of long-range forces [42,63], not the discretization of the computational domain.

The dual-horizon peridynamics (DH-PD) is firstly proposed to solve the issue of spurious wave reflections when variable horizons are adopted. The key idea lies in the definition of dual-horizon, which is defined as the dual term of the horizons centering at each material points. Within the framework of DH-PD, the traditional peridynamics with constant horizon can be derived. The derivation of DH-PD is based on the Newton's third law and other additional techniques such as variational principal or Taylor expansion [34] are not required. The dual-horizon PD not only considers the variable horizons, but also reduce the computational cost, allowing the domain discretized like finite element method (FEM) or finite volume method (FVM), where the domain of interest utilizes dense node distribution and the other domain coarse node distribution.

In this chapter, we review the DH-PD and propose a universal principle in dual-horizon, the dual property of dual-horizon. We also discuss the influence of random material points distribution on crack pattern and multiple materials issue in peridynamics. The content is outlined as follows: we begin with the equation of motion of dual-horizon peridynamics and then discuss three ways to calculate the bond force in DH-PD. Next, we prove the general dual property of dual-horizon, based on which the balance of linear momentum and the balance of angular momentum of the present PD

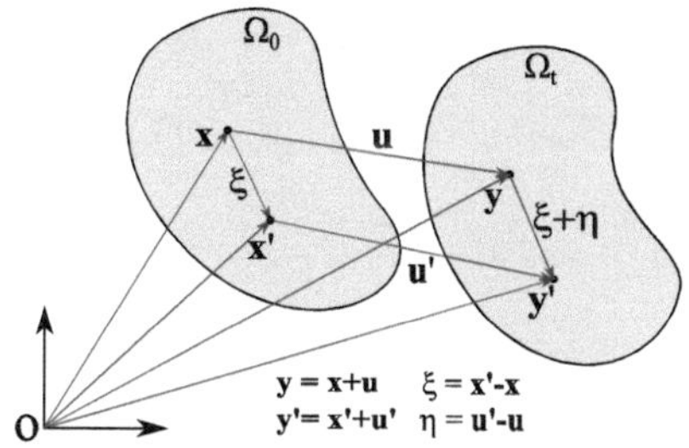

Figure 10.7 The configuration for the deformed body [58].

formulation are re-proved. Then, the issue of multiple materials in peridynamics is discussed with 2 numerical examples. Finally, the simulation of Kalthoff-Winkler experiment is carried out to test the influence of random material points distribution on the crack pattern in peridynamics.

10.2.1 Dual-horizon peridynamics

Consider a solid domain in the initial and current configuration as shown in Fig. 10.7. $\mathbf{x}$ is the material coordinate in initial configuration $\mathbf{\Omega}_0$; and $\mathbf{y} := \mathbf{y}(\mathbf{x}, t)$ and $\mathbf{y}' := \mathbf{y}(\mathbf{x}', t)$ are the spatial coordinates in the current configuration $\mathbf{\Omega}_t$; $\mathbf{u} := \mathbf{u}(\mathbf{x}, t)$ and $\mathbf{u}' := \mathbf{u}(\mathbf{x}', t)$ are the displacement vectors for associated points; $\mathbf{v} := \mathbf{v}(\mathbf{x}, t)$ and $\mathbf{v}' := \mathbf{v}(\mathbf{x}', t)$ are the velocity for associated points; $\xi := \mathbf{x}' - \mathbf{x}$ denotes the initial bond vector; the relative displacement vector of bond ξ is $\boldsymbol{\eta} := \mathbf{u}' - \mathbf{u}$, and the current bond vector is $\mathbf{y}\langle\xi\rangle := \mathbf{y}(\mathbf{x}', t) - \mathbf{y}(\mathbf{x}, t) = \boldsymbol{\xi} + \boldsymbol{\eta}$.

Since the bond ξ starts from $\mathbf{x}$ to $\mathbf{x}'$, an alias $\mathbf{x}\mathbf{x}'$ is used to denote ξ. Obviously, $\mathbf{x}'\mathbf{x}$ is in opposite direction of $\mathbf{x}\mathbf{x}'$. Let $\mathbf{f}_{\mathbf{x}\mathbf{x}'} := \mathbf{f}_{\mathbf{x}\mathbf{x}'}(\boldsymbol{\eta}, \xi)$ denote the force vector density acting on point $\mathbf{x}$ due to bond $\mathbf{x}\mathbf{x}'$; based on Newton's third law, $\mathbf{x}'$ undertakes a reaction force $-\mathbf{f}_{\mathbf{x}\mathbf{x}'}$.

As the conventional peridynamics formulation with constant horizon can be viewed as a special case of the DH-PD [57], we here mainly focus on the DH-PD. There are general three types of peridynamics, as shown in Fig. 10.8 and DH-PD will be applied to these three types.

10.2.1.1 The shortcomings of constant horizons

In conventional peridynamics, the horizon is required to be constant for a homogeneous body, where being homogeneous means that the strain energy density for any uniform deformation must be invariant with respect to changes in δ [66]. When the computational domain is discretized with

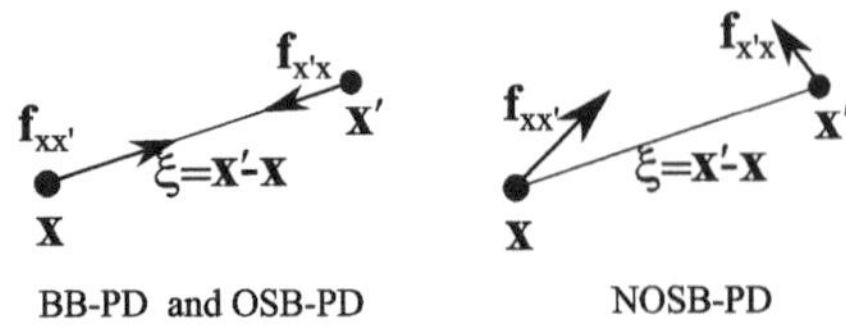

Figure 10.8 Force vectors of BB-PD, OSB-PD and NOSB-PD.

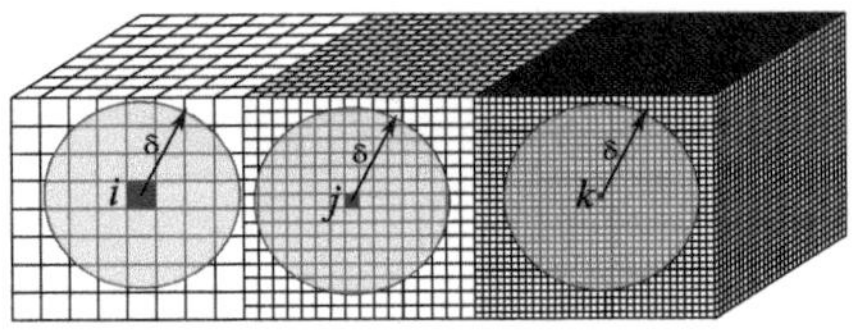

Figure 10.9 The neighbors for points i, j and k with constant horizon radius $\delta = 3r_i$. $r_i = 2r_j = 4r_k$, neighbor number $N_i \approx 200$, $N_j \approx 2^3 N_i \approx 1600$, $N_k \approx 4^3 N_i \approx 12800$.

different material point sizes, as shown in Fig. 10.9, in order to avoid the spurious wave reflection, the number of neighbors in the right side is dramatically increased. The fact of too many neighbors not only increases the computational cost, but also contributes little to the accuracy improvement, as the value in the point is smoothed out by its neighbors.

When different horizon radii are used, spurious wave reflections occur in conventional peridynamics. The reason is that the traditional peridynamics formulation with single horizon is not easy to account for the interaction between two points with different horizon radii, which has been discussed in detail in [57].

10.2.1.2 Horizon *and* dual-horizon

In conventional PD, the *horizon* is a domain centered at each point with a radius of δ. In DH-PD formulation, the *horizon* $H_{\mathbf{x}}$ is the domain where any material point $\mathbf{x}'$ falling inside forms bond $\mathbf{x}\mathbf{x}'$, where the bond $\mathbf{x}\mathbf{x}'$ will exert direct force $\mathbf{f}_{\mathbf{x}\mathbf{x}'}$ on $\mathbf{x}$ and whereas $\mathbf{x}'$ will undertake reaction force $-\mathbf{f}_{\mathbf{x}\mathbf{x}'}$. In this sense, the horizon in DH-PD can be viewed as direct force horizon. The *dual-horizon* in DH-PD is defined as a union of points whose horizons include $\mathbf{x}$, denoted by

$$H'_{\mathbf{x}} = \{\mathbf{x}' | \mathbf{x} \in H_{\mathbf{x}'}\} \tag{10.12}$$

The bond from the dual-horizon is named as dual-bond. For any point $\mathbf{x}' \in H'_{\mathbf{x}}$, the dual-bond $\mathbf{x}'\mathbf{x}$ is actually the bond $\mathbf{x}'\mathbf{x}$ from $H_{\mathbf{x}'}$. Dual-bond

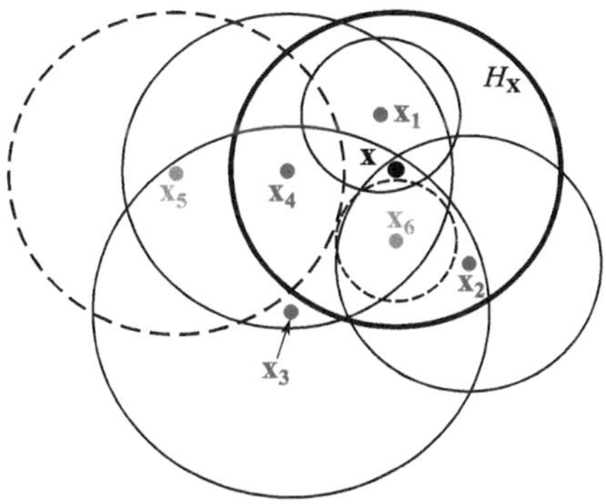

Figure 10.10 The schematic diagram for horizon and dual-horizon, all circles above are horizons. The green points $\{\mathbf{x}_1, \mathbf{x}_2, \mathbf{x}_3, \mathbf{x}_4\} \in H'_{\mathbf{x}}$, whose horizons are denoted by the thin solid line; the red points $\{\mathbf{x}_5, \mathbf{x}_6\} \notin H'_{\mathbf{x}}$, whose horizons are denoted by dashed line.

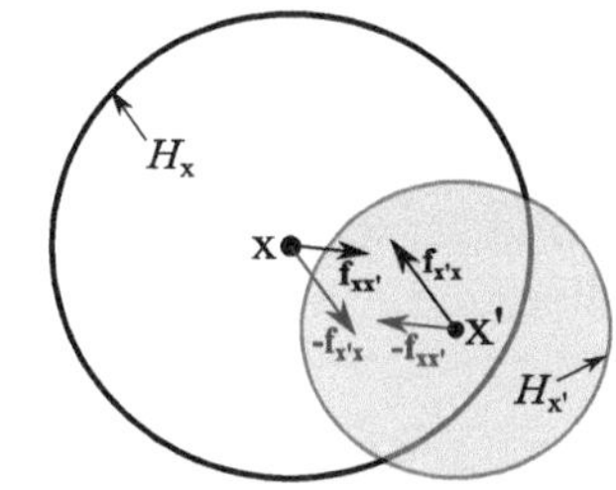

Figure 10.11 The force vectors in dual-horizon peridynamics. Bond $\mathbf{xx}'$ exerts direct force $\mathbf{f}_{\mathbf{xx}'}$ on $\mathbf{x}$ and reaction force $-\mathbf{f}_{\mathbf{xx}'}$ on $\mathbf{x}'$. In this case, $\mathbf{f}_{\mathbf{xx}'} \neq 0$, $\mathbf{f}_{\mathbf{x}'\mathbf{x}} = 0$.

$\mathbf{x}'\mathbf{x}$ exerts reaction force on $\mathbf{x}$. In this sense, the dual-horizon can be viewed as reaction force horizon. One simple example to illustrate the dual-horizon is shown in Fig. 10.10, where $\{\mathbf{x}_5, \mathbf{x}_6\} \notin H'_{\mathbf{x}}$ because $H_{\mathbf{x}_5}$ and $H_{\mathbf{x}_6}$ do not contain $\mathbf{x}$. The direct force and reaction force from horizon and dual-horizon are shown in Fig. 10.11.

Bond and dual-bond are broken independently in the fracture models. There are reasons accounting for it. For example, in BB-PD, the equivalence of energy release rate of homogeneous body in fracture models indicates that different horizon sizes correspond to different critical stretches. This means bond $\mathbf{xx}'$ and dual-bond $\mathbf{x}'\mathbf{x}$ may have different critical stretches. If they are broken at the same bond stretch, the one with larger critical bond stretch is broken in advance, which indicates a lower material strength for that point and thus violates the equivalence of energy release rate.

10.2.1.3 Equation of motion for dual-horizon peridynamics

The internal forces that are exerted at each material point include two parts, the direct forces from the horizon and the reaction forces from the dual-horizon. The other forces applied to a material point include the body force and the inertia force. Let $\Delta V_{\mathbf{x}}$ denote the volume associated with $\mathbf{x}$. The body force for $\mathbf{x}$ can be expressed as $\mathbf{b}(\mathbf{x}, t)\Delta V_{\mathbf{x}}$, where $\mathbf{b}(\mathbf{x}, t)$ is the body force density. The inertia force is denoted by $\rho\ddot{\mathbf{u}}(\mathbf{x}, t)\Delta V_{\mathbf{x}}$, where ρ is the density.

The direct force due to bond $\mathbf{xx}'$ in $H_{\mathbf{x}}$ is

$$\mathbf{f}_{\mathbf{xx}'}(\boldsymbol{\eta}, \boldsymbol{\xi}) \cdot \Delta V_{\mathbf{x}} \cdot \Delta V_{\mathbf{x}'}, \forall \mathbf{x}' \in H_{\mathbf{x}}. \tag{10.13}$$

The reaction force due to dual-bond $\mathbf{x}'\mathbf{x}$ in $H'_{\mathbf{x}}$ is

$$-\mathbf{f}_{\mathbf{x}'\mathbf{x}}(-\boldsymbol{\eta}, -\boldsymbol{\xi}) \cdot \Delta V_{\mathbf{x}'} \cdot \Delta V_{\mathbf{x}}, \forall \mathbf{x}' \in H'_{\mathbf{x}}. \tag{10.14}$$

By summing over all forces on point $\mathbf{x}$, including inertia force, body force, direct force in Eq. (10.13) and reaction force in Eq. (10.14), the equation of motion for DH-PD is obtained.

$$\rho\ddot{\mathbf{u}}(\mathbf{x}, t)\Delta V_{\mathbf{x}} = \sum_{H_{\mathbf{x}}} \mathbf{f}_{\mathbf{xx}'}(\boldsymbol{\eta}, \boldsymbol{\xi})\Delta V_{\mathbf{x}'}\Delta V_{\mathbf{x}} + \sum_{H'_{\mathbf{x}}}(-\mathbf{f}_{\mathbf{x}'\mathbf{x}}(-\boldsymbol{\eta}, -\boldsymbol{\xi})\Delta V_{\mathbf{x}'}\Delta V_{\mathbf{x}}) + \mathbf{b}(\mathbf{x}, t)\Delta V_{\mathbf{x}}. \tag{10.15}$$

As the volume $\Delta V_{\mathbf{x}}$ associated to $\mathbf{x}$ is independent of the summation, we can eliminate $\Delta V_{\mathbf{x}}$ in Eq. (10.15), yielding the governing equation based on $\mathbf{x}$:

$$\rho\ddot{\mathbf{u}}(\mathbf{x}, t) = \sum_{H_{\mathbf{x}}} \mathbf{f}_{\mathbf{xx}'}(\boldsymbol{\eta}, \boldsymbol{\xi})\Delta V_{\mathbf{x}'} - \sum_{H'_{\mathbf{x}}} \mathbf{f}_{\mathbf{x}'\mathbf{x}}(-\boldsymbol{\eta}, -\boldsymbol{\xi})\Delta V_{\mathbf{x}'} + \mathbf{b}(\mathbf{x}, t) . \tag{10.16}$$

When the discretization is sufficiently fine, the discrete form converges to the integral form:

$$\lim_{\Delta V_{\mathbf{x}'} \to 0} \sum_{H_{\mathbf{x}}} \mathbf{f}_{\mathbf{xx}'}(\boldsymbol{\eta}, \boldsymbol{\xi})\Delta V_{\mathbf{x}'} = \int_{H_{\mathbf{x}}} \mathbf{f}_{\mathbf{xx}'}(\boldsymbol{\eta}, \boldsymbol{\xi})\, \mathbf{d}V_{\mathbf{x}'} \tag{10.17}$$

and

$$\lim_{\Delta V_{\mathbf{x}'} \to 0} \sum_{H'_{\mathbf{x}}} \mathbf{f}_{\mathbf{x}'\mathbf{x}}(-\boldsymbol{\eta}, -\boldsymbol{\xi})\Delta V_{\mathbf{x}'} = \int_{H'_{\mathbf{x}}} \mathbf{f}_{\mathbf{x}'\mathbf{x}}(-\boldsymbol{\eta}, -\boldsymbol{\xi})\, \mathbf{d}V_{\mathbf{x}'} . \tag{10.18}$$

Thus, the integral form of the equation of motion in dual-horizon peridynamics (DH-PD) is given as

$$\rho \ddot{\mathbf{u}}(\mathbf{x}, t) = \int_{H_{\mathbf{x}}} \mathbf{f}_{\mathbf{x}\mathbf{x}'}(\boldsymbol{\eta}, \boldsymbol{\xi}) \, \mathbf{d} V_{\mathbf{x}'} - \int_{H'_{\mathbf{x}}} \mathbf{f}_{\mathbf{x}'\mathbf{x}}(-\boldsymbol{\eta}, -\boldsymbol{\xi}) \, \mathbf{d} V_{\mathbf{x}'} + \mathbf{b}(\mathbf{x}, t) \, . \quad (10.19)$$

Note that Eq. (10.19) is also valid for points near the boundary area.

When the horizons are set constant, as $\mathbf{x}' \in H'_{\mathbf{x}} \Leftrightarrow \mathbf{x}' \in H_{\mathbf{x}}$, we have $H'_{\mathbf{x}} = H_{\mathbf{x}}$. Then Eq. (10.19) degenerates to traditional constant horizon peridynamics.

$$\rho \ddot{\mathbf{u}}(\mathbf{x}, t) = \int_{H_{\mathbf{x}}} \mathbf{f}_{\mathbf{x}\mathbf{x}'}(\boldsymbol{\eta}, \boldsymbol{\xi}) - \mathbf{f}_{\mathbf{x}'\mathbf{x}}(-\boldsymbol{\eta}, -\boldsymbol{\xi}) \, \mathbf{d} V_{\mathbf{x}'} + \mathbf{b}(\mathbf{x}, t) \, . \quad (10.20)$$

10.2.1.4 Numerical implementation of DH-PD

We aim at solving Eq. (10.16), the discrete form of DH-PD with the Velocity-Verlet algorithm [53]. After discretization of the domain, each material point's neighbors based on its horizon radius are stored explicitly, while the neighbors in dual-horizon are inferred from other material points' horizons. Every time the bond force from one point's horizon is calculated, the dual-bond force from the other point's dual-horizon is determined according to Newton's third law. For example, assuming the material points with horizons are shown in Fig. 10.10, the horizons of $\mathbf{x}$ is $H_{\mathbf{x}} = \{\mathbf{x}_1, \mathbf{x}_2, \mathbf{x}_4, \mathbf{x}_6\}$, and $H'_{\mathbf{x}} = \{\mathbf{x}_1, \mathbf{x}_2, \mathbf{x}_3, \mathbf{x}_4\}$. Direct forces on $\mathbf{x}$ from its horizon $H_{\mathbf{x}}$ are summed as follows:

$\mathbf{x}_1 \in H_{\mathbf{x}}$, calculate $\mathbf{f}_{\mathbf{x}\mathbf{x}_1}$, add $\mathbf{f}_{\mathbf{x}\mathbf{x}_1} \Delta V_{\mathbf{x}_1} \Delta V_{\mathbf{x}}$ to $\mathbf{x}$, add $-\mathbf{f}_{\mathbf{x}\mathbf{x}_1} \Delta V_{\mathbf{x}_1} \Delta V_{\mathbf{x}}$ to $\mathbf{x}_1$;

$\mathbf{x}_2 \in H_{\mathbf{x}}$, calculate $\mathbf{f}_{\mathbf{x}\mathbf{x}_2}$, add $\mathbf{f}_{\mathbf{x}\mathbf{x}_2} \Delta V_{\mathbf{x}_2} \Delta V_{\mathbf{x}}$ to $\mathbf{x}$, add $-\mathbf{f}_{\mathbf{x}\mathbf{x}_2} \Delta V_{\mathbf{x}_2} \Delta V_{\mathbf{x}}$ to $\mathbf{x}_2$;

$\mathbf{x}_4 \in H_{\mathbf{x}}$, calculate $\mathbf{f}_{\mathbf{x}\mathbf{x}_4}$, add $\mathbf{f}_{\mathbf{x}\mathbf{x}_4} \Delta V_{\mathbf{x}_4} \Delta V_{\mathbf{x}}$ to $\mathbf{x}$, add $-\mathbf{f}_{\mathbf{x}\mathbf{x}_4} \Delta V_{\mathbf{x}_4} \Delta V_{\mathbf{x}}$ to $\mathbf{x}_4$;

$\mathbf{x}_6 \in H_{\mathbf{x}}$, calculate $\mathbf{f}_{\mathbf{x}\mathbf{x}_6}$, add $\mathbf{f}_{\mathbf{x}\mathbf{x}_6} \Delta V_{\mathbf{x}_6} \Delta V_{\mathbf{x}}$ to $\mathbf{x}$, add $-\mathbf{f}_{\mathbf{x}\mathbf{x}_6} \Delta V_{\mathbf{x}_6} \Delta V_{\mathbf{x}}$ to $\mathbf{x}_6$.

We don't calculate $\mathbf{f}_{\mathbf{x}_1\mathbf{x}}$, $\mathbf{f}_{\mathbf{x}_2\mathbf{x}}$, $\mathbf{f}_{\mathbf{x}_3\mathbf{x}}$, $\mathbf{f}_{\mathbf{x}_4\mathbf{x}}$ from dual-horizon $H'_{\mathbf{x}}$. The force from $H'_{\mathbf{x}}$ are determined when calculating forces in $H_{\mathbf{x}_1}$, $H_{\mathbf{x}_2}$, $H_{\mathbf{x}_3}$, $H_{\mathbf{x}_4}$.

$\mathbf{x} \in H_{\mathbf{x}_1}$, calculate $\mathbf{f}_{\mathbf{x}_1\mathbf{x}}$, add $\mathbf{f}_{\mathbf{x}_1\mathbf{x}} \Delta V_{\mathbf{x}} \Delta V_{\mathbf{x}_1}$ to $\mathbf{x}_1$, add $-\mathbf{f}_{\mathbf{x}_1\mathbf{x}} \Delta V_{\mathbf{x}} \Delta V_{\mathbf{x}_1}$ to $\mathbf{x}$;

$\mathbf{x} \in H_{\mathbf{x}_2}$, calculate $\mathbf{f}_{\mathbf{x}_2\mathbf{x}}$, add $\mathbf{f}_{\mathbf{x}_2\mathbf{x}} \Delta V_{\mathbf{x}} \Delta V_{\mathbf{x}_2}$ to $\mathbf{x}_2$, add $-\mathbf{f}_{\mathbf{x}_2\mathbf{x}} \Delta V_{\mathbf{x}} \Delta V_{\mathbf{x}_2}$ to $\mathbf{x}$;

$\mathbf{x} \in H_{\mathbf{x}_3}$, calculate $\mathbf{f}_{\mathbf{x}_3\mathbf{x}}$, add $\mathbf{f}_{\mathbf{x}_3\mathbf{x}}\Delta V_{\mathbf{x}}\Delta V_{\mathbf{x}_3}$ to $\mathbf{x}_3$, add $-\mathbf{f}_{\mathbf{x}_3\mathbf{x}}\Delta V_{\mathbf{x}}\Delta V_{\mathbf{x}_3}$ to $\mathbf{x}$;

$\mathbf{x} \in H_{\mathbf{x}_4}$, calculate $\mathbf{f}_{\mathbf{x}_4\mathbf{x}}$, add $\mathbf{f}_{\mathbf{x}_4\mathbf{x}}\Delta V_{\mathbf{x}}\Delta V_{\mathbf{x}_4}$ to $\mathbf{x}_4$, add $-\mathbf{f}_{\mathbf{x}_4\mathbf{x}}\Delta V_{\mathbf{x}}\Delta V_{\mathbf{x}_4}$ to $\mathbf{x}$.

Hence, in the process of calculating forces in horizons, the forces from dual-horizons are automatically done. Compared with the algorithm for OSB-PD given in [53], force expression: $\mathbf{f}_{\mathbf{x}\mathbf{x}'} - \mathbf{f}_{\mathbf{x}'\mathbf{x}}$ for point $\mathbf{x}$ with bond $\mathbf{x}\mathbf{x}'$ and force expression: $\mathbf{f}_{\mathbf{x}'\mathbf{x}} - \mathbf{f}_{\mathbf{x}\mathbf{x}'}$ for point $\mathbf{x}'$ with bond $\mathbf{x}'\mathbf{x}$ are calculated independently. Compared with the force summation in traditional PD, the bond force for every bond in DH-PD is calculated only once.

10.2.1.5 The bond force density $\mathbf{f}_{\mathbf{x}\mathbf{x}'}$

There mainly exist three types of peridynamics formulations [63,67], namely BB-PD, OSB-PD and NOSB-PD. Their key difference is the way they calculate the bond force density $\mathbf{f}_{\mathbf{x}'\mathbf{x}}$. The applications of DH-PD to the three types of peridynamics are direct [57]. The calibration of constitutive parameters with respect to the continuum model and some issues concerning the implementations will be discussed, for example, in BB-PD.

In BB-PD, the energy per unit volume in the body at a given time t is given by [65]

$$W = \frac{1}{2}\int_{H_{\mathbf{x}}} w(\boldsymbol{\eta}, \boldsymbol{\xi})\mathbf{d}V_{\xi} \ , \tag{10.21}$$

where $w(\boldsymbol{\eta}, \boldsymbol{\xi})$ is the strain energy for bond ξ. For the BB-PD theory, by enforcing the strain energy density being equal to the strain energy density in the classical theory of elasticity [65], we obtain the expression of the microelastic modulus $C(\delta)$ as

$$C(\delta) = \frac{3E}{\pi\delta^4(1-2\nu)} \ , \tag{10.22}$$

where δ is the horizon radius associated to that material point, and ν is the Poisson's ratio. Note that the value of $C(\delta)$ takes half of the microelastic modulus c used in the constant-horizon (CH) BB-PD since the bond energy for varying horizon is determined by both horizon and dual-horizon.

Let $w_0(\xi) = C(\delta)s_0^2(\delta)\xi/2$ denote the work required to break a single bond, where $s_0(\delta)$ is the critical bond stretch. By breaking half of all the bonds connected to a given material point along the fracture surface and equalizing the breaking bonds energy with the critical energy release rate

G_0 [23], we can get the expression between the critical energy release rate G_0 and the critical bond stretch $s_0(\delta)$ in three dimensions

$$s_0(\delta) = \sqrt{\frac{5G_0}{6E\delta}} . \tag{10.23}$$

Both the microelastic modulus $C(\delta)$ and the critical stretch $s_0(\delta)$ depend on the horizon radii for variable horizons.

In the implementation of the bond based peridynamics, fracture is introduced by removing points from the neighbor list once the bond stretch exceeds the critical bond stretch s_0. In order to specify whether a bond is broken or not, a history-dependent scalar valued function μ is introduced [65],

$$\mu(t,\xi) = \begin{cases} 1 & \text{if } s(t',\xi) < s_0 \text{ for all } 0 \le t' \le t, \\ 0 & \text{otherwise.} \end{cases} \tag{10.24}$$

The local damage at $\mathbf{x}$ is defined as

$$\phi(\mathbf{x},t) = 1 - \frac{\int_{H_\mathbf{x}} \mu(\mathbf{x},t,\xi)\mathbf{d}V_\xi}{\int_{H_\mathbf{x}} \mathbf{d}V_\xi} . \tag{10.25}$$

For any material point with dual-horizon ($H'_\mathbf{x}$) and horizon ($H_\mathbf{x}$), the direct force $\mathbf{f}_{\mathbf{xx}'}$ and the reaction force $\mathbf{f}_{\mathbf{x}'\mathbf{x}}$ in Eq. (10.16) or (10.19) are computed by the following expressions, respectively

$$\mathbf{f}_{\mathbf{xx}'} = C(\delta_{\mathbf{x}'}) \cdot s_{\mathbf{xx}'} \cdot \frac{\eta+\xi}{\|\eta+\xi\|}, \qquad \forall \mathbf{x}' \in H_\mathbf{x} \tag{10.26}$$

$$\mathbf{f}_{\mathbf{x}'\mathbf{x}} = C(\delta_{\mathbf{x}}) \cdot s_{\mathbf{xx}'} \cdot \frac{-(\eta+\xi)}{\|\eta+\xi\|}, \qquad \forall \mathbf{x}' \in H'_\mathbf{x} , \tag{10.27}$$

where $C(\delta_{\mathbf{x}'})$ and $C(\delta_\mathbf{x})$ are the microelastic modulus based on $\delta_{\mathbf{x}'}$ and $\delta_\mathbf{x}$ computed from Eq. (10.22) respectively, and $s_{\mathbf{xx}'}$ is the stretch between points $\mathbf{x}$ and $\mathbf{x}'$.

For the other two types of peridynamics [67], comparing the equation of motion in traditional state-based PD with Eq. (10.20), the expression for bond force $\mathbf{f}_{\mathbf{xx}'}$ and $\mathbf{f}_{\mathbf{xx}'}$ can be obtained with ease. For more details please refer to [57].

10.2.1.6 Volume correction

In order to account for the effect of material points partly falling inside the horizon, some volume correction techniques are utilized to improve

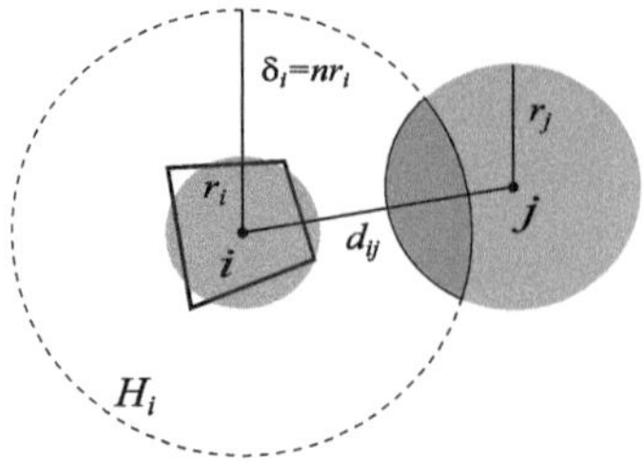

Figure 10.12 Volume correction.

the accuracy of force summation in horizon. The commonly used volume correction method in constant-horizon peridynamics is [53]

$$\nu_{\mathbf{x}_p-\mathbf{x}_i} = \begin{cases} -\frac{||\mathbf{x}_p-\mathbf{x}_i||}{2r_i} + (\frac{\delta}{2r_i} + \frac{1}{2}) & \text{if } \delta - r_i \le ||\mathbf{x}_p - \mathbf{x}_i|| \le \delta, \\ 1 & \text{if } ||\mathbf{x}_p - \mathbf{x}_i|| \le \delta - r_i \ , \\ 0 & \text{otherwise,} \end{cases} \tag{10.28}$$

where $r_i = \Delta/2$, Δ is the spacing between material points, or the material point size. In the dual-horizon peridynamics, as any material point has its unique horizon, the volume ratio of big material point with small material point can be big. For example, one case as shown in Fig. 10.12, the center of material point j is falling outside of i's horizon but with certain nontrivial part (green [dark gray in print version] domain in Fig. 10.12) of j belonging to H_i. In order to improve the accuracy of integration in H_i, the green (dark gray in print version) domain in Fig. 10.12 should be added to calculate the reaction force from j. As the material point is generated from finite element, the shape of each material point is irregular, the material point size can be estimated by its volume. Let r_i denote material point i's radius in the shape of circle or sphere. Let two circles (or spheres) of radii R and r be located along the x-axis centered at $(0, 0, 0)$ and $(d, 0, 0)$, respectively, as shown in Fig. 10.13. The intersection area A (or volume V) between two circles (or spheres) is

$$\begin{aligned} A = {} & r^2 \cos^{-1}(\frac{d^2 + r^2 - R^2}{2dr}) + R^2 \cos^{-1}(\frac{d^2 + R^2 - r^2}{2dR}) \\ & - \frac{1}{2}\left[(-d + r + R)(d + r - R)(d - r + R)(d + r + R)\right]^{1/2}, \end{aligned} \tag{10.29}$$

$$V = \frac{\pi}{12d}(R + r - d)^2(d^2 + 2dr - 3r^2 + 2dR + 6rR - 3R^2). \tag{10.30}$$

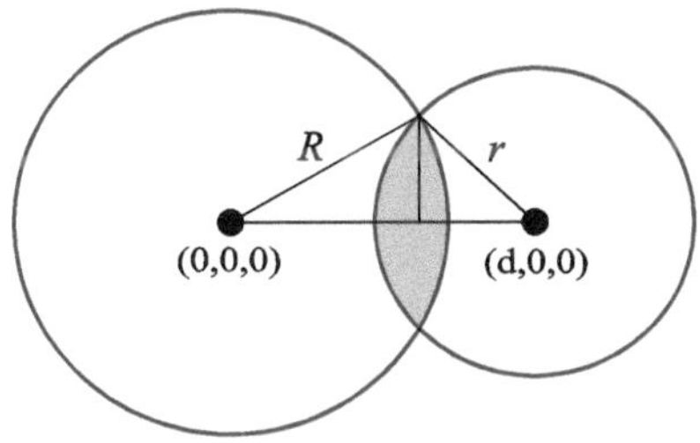

Figure 10.13 Circle to circle intersection.

10.2.2 The dual property of dual-horizon

Let $\mathcal{F}(\mathbf{x},\mathbf{x}')$ be any expression depend on bond $\mathbf{x}x'$. The dual property of dual-horizon is that the double integration of the term $\mathcal{F}(\mathbf{x},\mathbf{x}')$ in dual-horizon can be converted to the double integration of the term $\mathcal{F}(\mathbf{x}',\mathbf{x})$ in horizon, as shown in Eq. (10.31) and Eq. (10.32). The key idea lies in that the term $\mathcal{F}(\mathbf{x},\mathbf{x}')$ can be both interpreted in $H_{\mathbf{x}}$ and $H'_{\mathbf{x}'}$.

$$\sum_{\mathbf{\Omega}}\left(\sum_{H'_{\mathbf{x}}}\mathcal{F}(\mathbf{x},\mathbf{x}')\,\Delta V_{\mathbf{x}'}\right)\Delta V_{\mathbf{x}} = \sum_{\mathbf{\Omega}}\left(\sum_{H_{\mathbf{x}}}\mathcal{F}(\mathbf{x}',\mathbf{x})\,\Delta V_{\mathbf{x}'}\right)\Delta V_{\mathbf{x}} \qquad \text{Discrete form} \tag{10.31}$$

$$\int_{\mathbf{\Omega}}\int_{H'_{\mathbf{x}}}\mathcal{F}(\mathbf{x},\mathbf{x}')\,\mathbf{d}V_{\mathbf{x}'}\mathbf{d}V_{\mathbf{x}} = \int_{\mathbf{\Omega}}\int_{H_{\mathbf{x}}}\mathcal{F}(\mathbf{x}',\mathbf{x})\,\mathbf{d}V_{\mathbf{x}'}\mathbf{d}V_{\mathbf{x}} \qquad \text{Continuum form} \tag{10.32}$$

Proof. Let Ω be discretized with N Voronoi tessellations (or other shape), as shown in Fig. 10.14. Each polygon is denoted with an index $i \in \{1,\cdots,N\}$, $\mathbf{x}_i$ is the coordinate for i's center of gravity, ΔV_i is the volume associated to polygon i, H_i and H'_i are i's horizon and dual-horizon, respectively. So

$$\Omega = \sum_{1\le i\le N}\Delta V_i$$

Let $\mathcal{F}(i,j) := \mathcal{F}(\mathbf{x}_i,\mathbf{x}_j)$ be any expression depend on bond $\mathbf{x}_i\mathbf{x}_j$. Consider the double summation of $\mathcal{F}(i,j)$ on $\mathbf{\Omega}$.

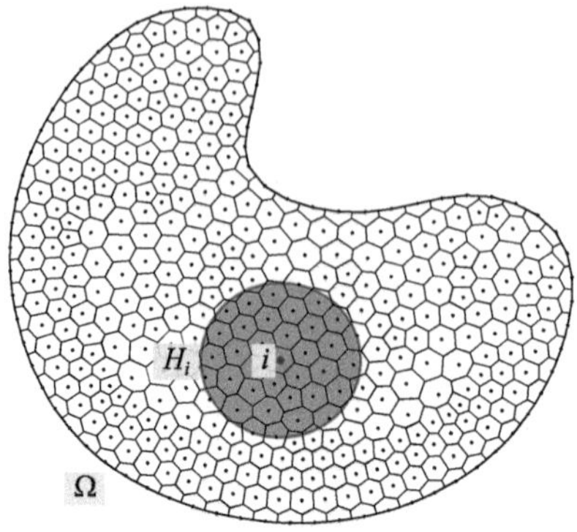

Figure 10.14 The discretization of domain **Ω**.

$$
\begin{aligned}
&\sum_{\Delta V_i \in \Omega} \left(\sum_{\Delta V_j \in H_i'} \mathcal{F}(i,j)\Delta V_j \right) \Delta V_i \\
&= \sum_{1 \le i \le N} \left(\sum_{j \in H_i'} \mathcal{F}(i,j)\Delta V_j \right) \Delta V_i \\
&= \sum_{j \in H_1'} \mathcal{F}(1,j)\Delta V_j \Delta V_1 + \sum_{j \in H_2'} \mathcal{F}(2,j)\Delta V_j \Delta V_2 + \cdots + \sum_{j \in H_N'} \mathcal{F}(N,j)\Delta V_j \Delta V_N
\end{aligned} \tag{10.33}
$$

In the third step, $j \in \{1, \cdots, N\}$ means j belongs to that point's dual-horizon. Each term $\mathcal{F}(i,j)\Delta V_j \Delta V_i$ in H_i' can be interpreted as term $\mathcal{F}(i,j)\Delta V_i \Delta V_j$ in H_j. Let us sum all terms in a way based on point j's horizon H_j, where $j \in \{1, \cdots, N\}$

$$
\begin{aligned}
&\sum_{j \in H_1'} \mathcal{F}(1,j)\Delta V_j \Delta V_1 + \sum_{j \in H_2'} \mathcal{F}(2,j)\Delta V_j \Delta V_2 + \cdots + \sum_{j \in H_N'} \mathcal{F}(N,j)\Delta V_j \Delta V_N \\
&= \sum_{i \in H_1} \mathcal{F}(i,1)\Delta V_1 \Delta V_i + \sum_{i \in H_2} \mathcal{F}(i,2)\Delta V_2 \Delta V_i + \cdots + \sum_{i \in H_N} \mathcal{F}(i,N)\Delta V_N \Delta V_i
\end{aligned} \tag{10.34}
$$

In the second step of Eq. (10.34), for example, $\sum_{i \in H_1} \mathcal{F}(i,1)\Delta V_1 \Delta V_i$ means gathering all terms $j = 1$ in

$$
\sum_{j \in H_2'} \mathcal{F}(2,j)\Delta V_j \Delta V_2 + \sum_{j \in H_3'} \mathcal{F}(3,j)\Delta V_j \Delta V_3 + \cdots + \sum_{j \in H_N'} \mathcal{F}(N,j)\Delta V_j \Delta V_N
$$

$$\sum_{i\in H_1}\mathcal{F}(i,1)\Delta V_1\Delta V_i+\sum_{i\in H_2}\mathcal{F}(i,2)\Delta V_2\Delta V_i+\cdots+\sum_{i\in H_N}\mathcal{F}(i,N)\Delta V_N\Delta V_i$$
$$=\sum_{j\in H_1}\mathcal{F}(j,1)\Delta V_j\Delta V_1+\sum_{j\in H_2}\mathcal{F}(j,2)\Delta V_j\Delta V_2+\cdots+\sum_{j\in H_N}\mathcal{F}(j,N)\Delta V_j\Delta V_N$$
$$=\sum_{1\leq i\leq N}\left(\sum_{j\in H_i}\mathcal{F}(j,i)\Delta V_j\right)\Delta V_i \tag{10.35}$$

In the second step of Eq. (10.35), i and j is swapped, yielding

$$\sum_{1\leq i\leq N}\left(\sum_{j\in H_i'}\mathcal{F}(i,j)\Delta V_j\right)\Delta V_i=\sum_{1\leq i\leq N}\left(\sum_{j\in H_i}\mathcal{F}(j,i)\Delta V_j\right)\Delta V_i$$

When i and j are replaced with $\mathbf{x}$ and $\mathbf{x}'$, respectively, we have

$$\sum_{\mathbf{\Omega}}\left(\sum_{H_{\mathbf{x}}'}\mathcal{F}(\mathbf{x},\mathbf{x}')\,\Delta V_{\mathbf{x}'}\right)\Delta V_{\mathbf{x}}=\sum_{\mathbf{\Omega}}\left(\sum_{H_{\mathbf{x}}}\mathcal{F}(\mathbf{x}',\mathbf{x})\,\Delta V_{\mathbf{x}'}\right)\Delta V_{\mathbf{x}} \tag{10.36}$$

When $N\to\infty$ so that $\Delta V_i\to 0$, we have

$$\lim_{N\to\infty}\sum_{1\leq i\leq N}\left(\sum_{j\in H_i'}\mathcal{F}(i,j)\Delta V_j\right)\Delta V_i=\int_{i\in\Omega}\int_{j\in H_i'}\mathcal{F}(i,j)\mathbf{d}V_j\mathbf{d}V_i$$

Hence, the dual property of dual-horizon in the integral form is

$$\int_{\mathbf{\Omega}}\int_{H_{\mathbf{x}}'}\mathcal{F}(\mathbf{x},\mathbf{x}')\,\mathbf{d}V_{\mathbf{x}'}\mathbf{d}V_{\mathbf{x}}=\int_{\mathbf{\Omega}}\int_{H_{\mathbf{x}}}\mathcal{F}(\mathbf{x}',\mathbf{x})\,\mathbf{d}V_{\mathbf{x}'}\mathbf{d}V_{\mathbf{x}} \tag{10.37}$$

Eq. (10.37) means the double integration of the term in dual-horizon can be converted to the double integration of the term with $\mathbf{x}$ and $\mathbf{x}'$ swapped in horizon.

When $\mathcal{F}(\mathbf{x},\mathbf{x}')=\mathbf{f}_{\mathbf{x}\mathbf{x}'}$, we have

$$\int_{\mathbf{\Omega}}\int_{H_{\mathbf{x}}}\mathbf{f}_{\mathbf{x}\mathbf{x}'}\,\mathbf{d}V_{\mathbf{x}'}\mathbf{d}V_{\mathbf{x}}=\int_{\mathbf{\Omega}}\int_{H_{\mathbf{x}}'}\mathbf{f}_{\mathbf{x}'\mathbf{x}}\,\mathbf{d}V_{\mathbf{x}'}\mathbf{d}V_{\mathbf{x}} \tag{10.38}$$

When $\mathcal{F}(\mathbf{x},\mathbf{x}')=(\mathbf{x}+\mathbf{u})\times\mathbf{f}_{\mathbf{x}\mathbf{x}'}$, we have

$$\int_{\mathbf{\Omega}}\int_{H_{\mathbf{x}}}(\mathbf{x}+\mathbf{u})\times\mathbf{f}_{\mathbf{x}\mathbf{x}'}\,\mathbf{d}V_{\mathbf{x}'}\mathbf{d}V_{\mathbf{x}}=\int_{\mathbf{\Omega}}\int_{H_{\mathbf{x}}'}(\mathbf{x}'+\mathbf{u}')\times\mathbf{f}_{\mathbf{x}'\mathbf{x}}\,\mathbf{d}V_{\mathbf{x}'}\mathbf{d}V_{\mathbf{x}} \tag{10.39}$$

□

10.2.2.1 Proof of basic physical principles

Two basic physical principles include the balance of the linear momentum and the balance of the angular momentum. We showed in [57] that DH-PD fulfills these basic principles. In this manuscript, we provide a shorter proof based on the dual property of the dual-horizon.

Balance of linear momentum

The internal forces shall satisfy the balance of linear momentum for any bounded body $\mathbf{\Omega}$ given in form by

$$\begin{aligned}&\int_{\mathbf{\Omega}}(\rho\ddot{\mathbf{u}}(\mathbf{x},t)-\mathbf{b}(\mathbf{x},t))\mathbf{d}V_{\mathbf{x}}\\&=\int_{\mathbf{\Omega}}\int_{H'_{\mathbf{x}}}\mathbf{f}_{\mathbf{x}\mathbf{x}'}(\boldsymbol{\eta},\boldsymbol{\xi})\,\mathbf{d}V_{\mathbf{x}'}\mathbf{d}V_{\mathbf{x}}-\int_{\mathbf{\Omega}}\int_{H_{\mathbf{x}}}\mathbf{f}_{\mathbf{x}'\mathbf{x}}(-\boldsymbol{\eta},-\boldsymbol{\xi})\,\mathbf{d}V_{\mathbf{x}'}\mathbf{d}V_{\mathbf{x}}\\&=\mathbf{0}\,.\end{aligned}\tag{10.40}$$

Proof. Based on Eq. (10.38), Eq. (10.40) is apparently satisfied, so the linear momentum is conserved. □

Balance of angular momentum

To satisfy the balance of angular momentum for any bounded body $\mathbf{\Omega}$, it is required that

$$\begin{aligned}&\int_{\mathbf{\Omega}}\mathbf{y}\times(\rho\ddot{\mathbf{u}}(\mathbf{x},t)-\mathbf{b}(\mathbf{x},t))\mathbf{d}V_{\mathbf{x}}\\&=\int_{\mathbf{\Omega}}\mathbf{y}\times\left(\int_{H'_{\mathbf{x}}}\mathbf{f}_{\mathbf{x}\mathbf{x}'}(\boldsymbol{\eta},\boldsymbol{\xi})\,\mathbf{d}V_{\mathbf{x}'}-\int_{H_{\mathbf{x}}}\mathbf{f}_{\mathbf{x}'\mathbf{x}}(-\boldsymbol{\eta},-\boldsymbol{\xi})\,\mathbf{d}V_{\mathbf{x}'}\right)\mathbf{d}V_{\mathbf{x}}\\&=\int_{\mathbf{\Omega}}\int_{H'_{\mathbf{x}}}\mathbf{y}\times\mathbf{f}_{\mathbf{x}\mathbf{x}'}(\boldsymbol{\eta},\boldsymbol{\xi})\,\mathbf{d}V_{\mathbf{x}'}\mathbf{d}V_{\mathbf{x}}-\int_{\mathbf{\Omega}}\int_{H_{\mathbf{x}}}\mathbf{y}\times\mathbf{f}_{\mathbf{x}'\mathbf{x}}(-\boldsymbol{\eta},-\boldsymbol{\xi})\,\mathbf{d}V_{\mathbf{x}'}\mathbf{d}V_{\mathbf{x}}\\&=\mathbf{0}\end{aligned}\tag{10.41}$$

For simplicities, let $\mathbf{f}_{\mathbf{x}\mathbf{x}'}$ represent $\mathbf{f}_{\mathbf{x}\mathbf{x}'}(\boldsymbol{\eta},\boldsymbol{\xi})$, and $\mathbf{f}_{\mathbf{x}'\mathbf{x}}$ represent $\mathbf{f}_{\mathbf{x}'\mathbf{x}}(-\boldsymbol{\eta},-\boldsymbol{\xi})$.

Proposition. *In the dual-horizon peridynamics, suppose a constitutive model of the form*

$$\mathbf{f}=\hat{\mathbf{f}}(\mathbf{y},\mathbf{\Lambda})\tag{10.42}$$

where $\hat{\mathbf{f}}:\mathcal{V}\to\mathcal{V}$ is bounded and Riemann-integrable on Hilbert function space and $\mathcal{V}$ is the vector state; $\mathbf{\Lambda}$ denotes all variables other than the current deformation vector state.

If

$$\int_{H_{\mathbf{x}}} \mathbf{y}\langle \mathbf{x}' - \mathbf{x} \rangle \times \mathbf{f}_{\mathbf{x}'\mathbf{x}} \, \mathbf{d} V_{\mathbf{x}'} = 0 \qquad \forall \mathbf{y} \in \mathcal{V}\,, \tag{10.43}$$

then the balance of angular momentum, Eq. (10.41) holds for any deformation of $\boldsymbol{\Omega}$ *for any given constitutive model.*

Proof. Based on Eq. (10.39), Eq. (10.41) can be written as

$$\begin{aligned}
&\int_{\boldsymbol{\Omega}} \mathbf{y} \times (\rho \ddot{\mathbf{u}}(\mathbf{x}, t) - \mathbf{b}(\mathbf{x}, t)) \mathbf{d} V_{\mathbf{x}} \\
&= \int_{\boldsymbol{\Omega}} \mathbf{y} \times \left(\int_{H'_{\mathbf{x}}} \mathbf{f}_{\mathbf{x}\mathbf{x}'}(\boldsymbol{\eta}, \boldsymbol{\xi}) \, \mathbf{d} V_{\mathbf{x}'} - \int_{H_{\mathbf{x}}} \mathbf{f}_{\mathbf{x}'\mathbf{x}}(-\boldsymbol{\eta}, -\boldsymbol{\xi}) \, \mathbf{d} V_{\mathbf{x}'} \right) \mathbf{d} V_{\mathbf{x}} \\
&= \int_{\boldsymbol{\Omega}} \int_{H'_{\mathbf{x}}} \mathbf{y} \times \mathbf{f}_{\mathbf{x}\mathbf{x}'}(\boldsymbol{\eta}, \boldsymbol{\xi}) \, \mathbf{d} V_{\mathbf{x}'} \mathbf{d} V_{\mathbf{x}} - \int_{\boldsymbol{\Omega}} \int_{H_{\mathbf{x}}} \mathbf{y} \times \mathbf{f}_{\mathbf{x}'\mathbf{x}}(-\boldsymbol{\eta}, -\boldsymbol{\xi}) \, \mathbf{d} V_{\mathbf{x}'} \mathbf{d} V_{\mathbf{x}} \\
&= \int_{\boldsymbol{\Omega}} \int_{H_{\mathbf{x}}} \mathbf{y}' \times \mathbf{f}_{\mathbf{x}'\mathbf{x}}(-\boldsymbol{\eta}, -\boldsymbol{\xi}) \, \mathbf{d} V_{\mathbf{x}'} \mathbf{d} V_{\mathbf{x}} - \int_{\boldsymbol{\Omega}} \int_{H_{\mathbf{x}}} \mathbf{y} \times \mathbf{f}_{\mathbf{x}'\mathbf{x}}(-\boldsymbol{\eta}, -\boldsymbol{\xi}) \, \mathbf{d} V_{\mathbf{x}'} \mathbf{d} V_{\mathbf{x}} \\
&= \int_{\boldsymbol{\Omega}} \int_{H_{\mathbf{x}}} (\mathbf{y}' - \mathbf{y}) \times \mathbf{f}_{\mathbf{x}'\mathbf{x}}(-\boldsymbol{\eta}, -\boldsymbol{\xi}) \, \mathbf{d} V_{\mathbf{x}'} \mathbf{d} V_{\mathbf{x}} \\
&= \mathbf{0}
\end{aligned} \tag{10.44}$$

The expression in step 4 is equal to zero in all three types of peridynamics (see proof in [67]), so the angular momentum is conserved. □

It can be seen that the conservation of angular momentum somehow depends **only** on the horizon; the dual-horizon is not involved. The latter is only needed in Eq. (10.16) and Eq. (10.19). This conclusion can be also illustrated for the BB-PD and OS-PD since the internal forces $\mathbf{f}_{\mathbf{x}\mathbf{x}'}$ and $\mathbf{f}_{\mathbf{x}'\mathbf{x}}$ are parallel to the bond vector in the current configuration, see Fig. 10.8.

10.2.3 Wave propagation in 1D homogeneous bar

Consider a 1D bar with length of $L = 1$ m, one third of the bar in the middle is discretized with $\Delta x_1 = 4.19 \times 10^{-3}$ m and the other parts discretized with $\Delta x_2 = 10\Delta x_1$. The horizon size is selected as $\delta_i = 3.015\Delta x_i$, as shown in Fig. 10.15. Both ends of the bar are free. The initial displacement field is given by

$$u(x, 0) = 0.002 \exp(-\frac{(x - 0.5L)^2}{1000}) \tag{10.45}$$

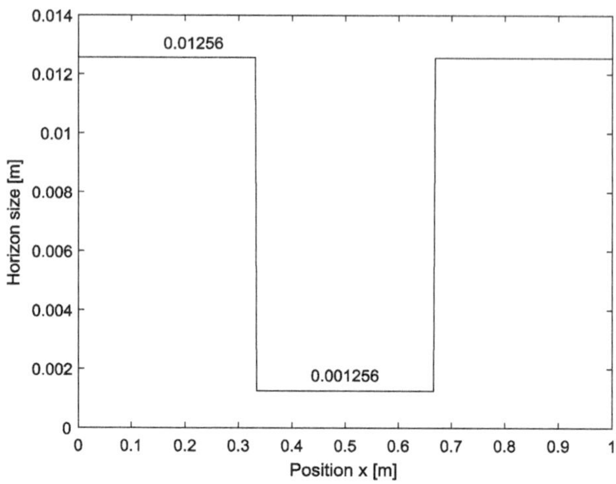

Figure 10.15 The horizon sizes distribution.

The bar is solved with DH-PD (Case I) and traditional PD formulation (Case II). The wave profiles at different times are shown in Fig. 10.16 for DH-PD and Fig. 10.17 for traditional PD. It can be seen the horizon size interfaces have limited influence on the wave profiles in Case I, while the spurious wave emerged for traditional PD formulation. Two points $x = 0.32$ m and $x = 0.34$ m are selected to analyze the wave profile, and the displacements at different time are shown in Fig. 10.18. It shows that the incident wave passed the interface and the magnitude of reflected wave is smaller than 3.17% of the magnitude of incident wave.

On the other hand, for the traditional PD without any treatment on variable horizons, when the incident wave approached the interface, the deformation of the fine material points caused the ghost forces among the coarse material points. The ghost forces give rise to the spurious wave in the domain with coarse material points, as shown in Fig. 10.17 ($T = 0.2$ s). In addition, the wave in the fine material points is hard to pass the interface but to be reflected due to the fact that its horizon size is much smaller than that of coarse material points.

10.2.4 Numerical examples

10.2.4.1 Two-dimensional wave reflection in a rectangular plate

Consider a rectangular plate with dimensions of 0.1×0.04 m^2 (see Fig. 10.19). The Young's modulus, density and the Poisson's ratio used for the plate are $E = 1$, $\rho = 1$ and $\nu = 0$, respectively. Note that this is actu-

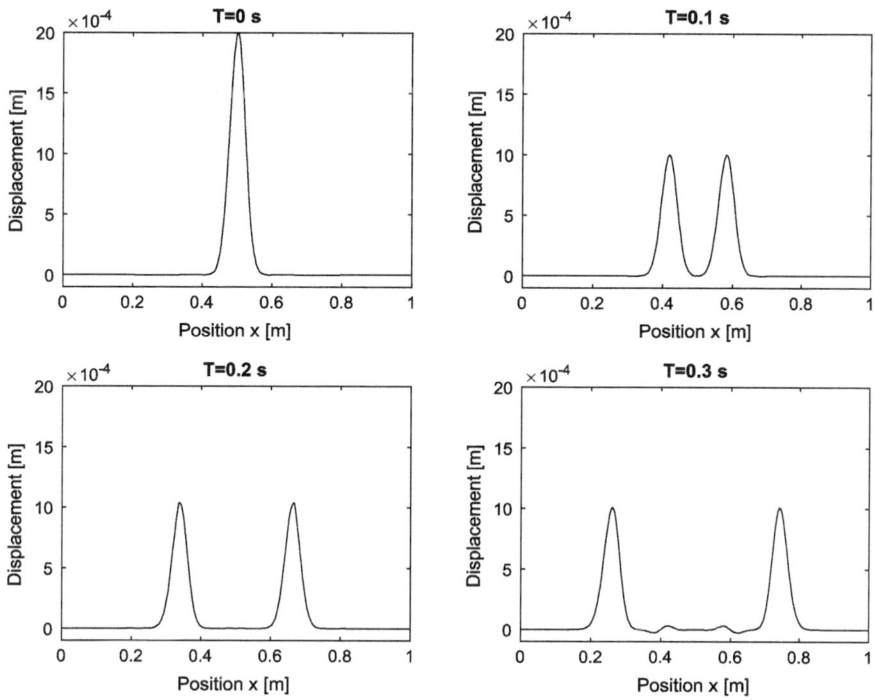

Figure 10.16 The wave profiles for Case I at different times.

ally a 1-D problem solved in 2D, hence, an analytical solution is available. The initial displacement applied to the plate is described by the following equation

$$u_0(x, y) = 0.0002\exp[-(\frac{x}{0.01})^2], \quad v_0(x, y) = 0,$$
$$x \in [0, 0.1], y \in [-0.02, 0.02] , \tag{10.46}$$

where u_0 and v_0 denote the displacement in the x and y directions respectively. The wave speed is $v = \sqrt{E/\rho} = 1$ m/s. At any time step, the L_2 error in the displacement is given by

$$\|\text{err}\|_{L_2} = \frac{\|\mathbf{u}^h - \mathbf{u}_{\text{analytic}}\|}{\|\mathbf{u}_{\text{analytic}}\|} , \tag{10.47}$$

with

$$\|\mathbf{u}\| = \left(\int_{\Omega_0} \mathbf{u} \cdot \mathbf{u} \, d\Omega_0\right)^{\frac{1}{2}} .$$

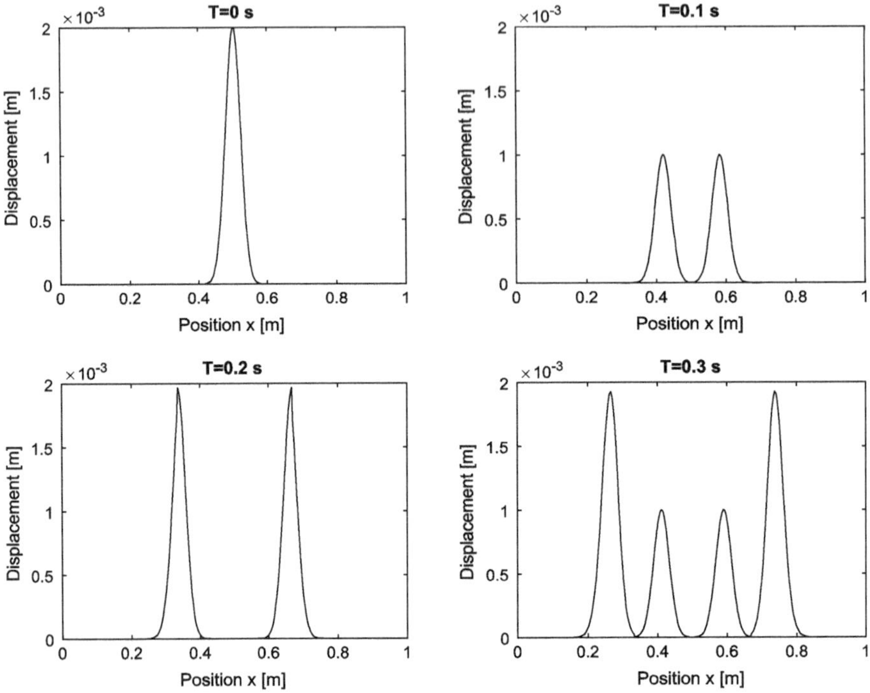

Figure 10.17 The wave profiles for Case II at different times.

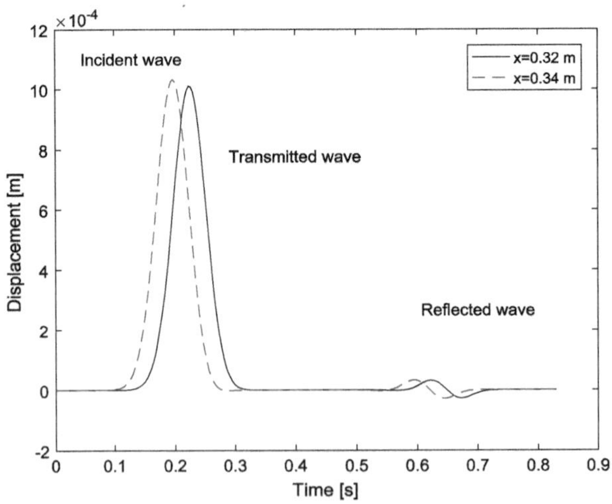

Figure 10.18 The wave for points $x = 0.32$ m and $x = 0.34$ m.

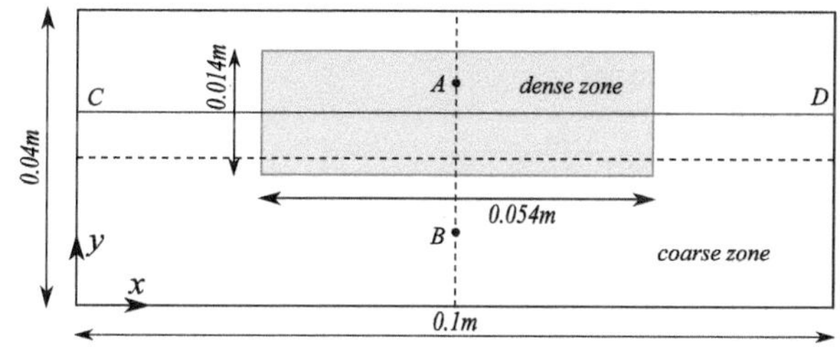

Figure 10.19 Setup of the plate.

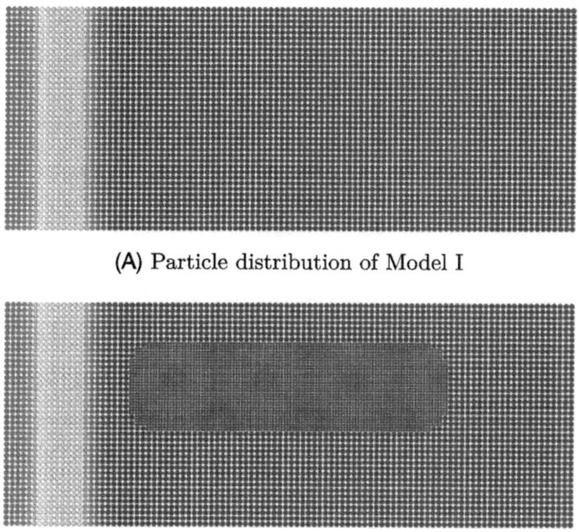

(A) Particle distribution of Model I

(B) Particle distribution of Models II and III

Figure 10.20 The discretizations of Models I, II and III.

Three models for solving this problem, namely model I, II and III, are devised, see Figs. 10.20A and 10.20B. In model I, the plate is discretized with 4000 particles and all particles have the same horizon sizes of 0.001 m, see Fig. 10.20A. For models II and III, we adopted a discretization of varying horizons as shown in Fig. 10.20B. Verlet integration method is adopted for the time integration. The horizon radius associated to each particle is set as 3 times the particle sizes for all models. Therefore, along the interface between the coarse and fine discretization, the horizon sizes vary. The minimal particle size of the 3 models is $\Delta x = 5 \times 10^{-4}$ m, yielding a critical time increment $\Delta t_{max} = \Delta x / v = 5 \times 10^{-4}$ s. The physical computational time step is set 0.5 second with the time increment $\Delta t_{max} = 5 \times 10^{-4}$ s.

Table 10.1 Values of the peridynamic parameters for the mixed model.

Model	$\Delta x = \Delta y$	$10^3 \cdot \delta$	Particle numbers
I	0.001	3	4000
II	0.001, 0.0005	3/1.5	8628
III	0.001, 0.0005	3/1.5	8628

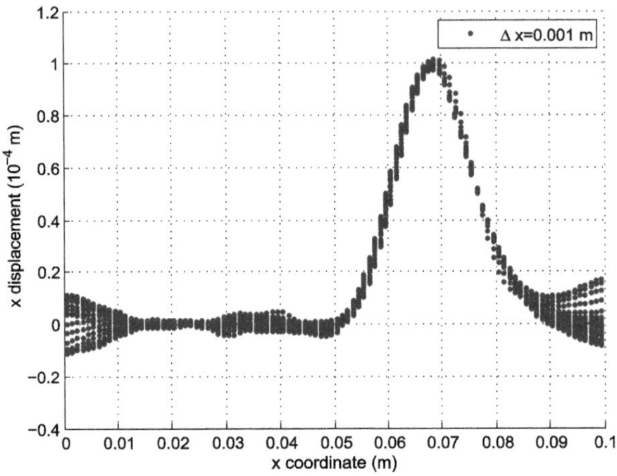

Figure 10.21 Displacement wave u_x versus x coordinate of model I at step 650.

As BB-PD can not simulate the model with Poisson ratio $\nu = 0$, the OSB-PD model is adopted. Model I was solved with the OSB-PD with constant horizons. Model II is the original OSB-PD with variable horizon (without additional treatment for ghost force), and model III is our dual-horizon formulation for OSB-PD with variable horizon. The parameters used for the three models are listed in Table 10.1.

Figs. 10.21, 10.22, 10.23 show the displacement in the x-direction along the x-coordinate of all particles. The red star points are particles with $\Delta x = 0.0005$ m, the blue dot points refer to $\Delta x = 0.001$ m.

The displacement in model III (present dual-horizon PD) is almost identical to that of model I (PD with constant horizon), as shown in Fig. 10.21 and Fig. 10.23. Spurious wave reflections are observed for Model II, as shown in Fig. 10.22. At step 650, several wave peaks are observed in Model II and the maximum displacement is lower than that of Models I and III. Therefore, the displacement wave in Model II were affected by spurious wave reflection and the results deviate greatly from results of I and II.

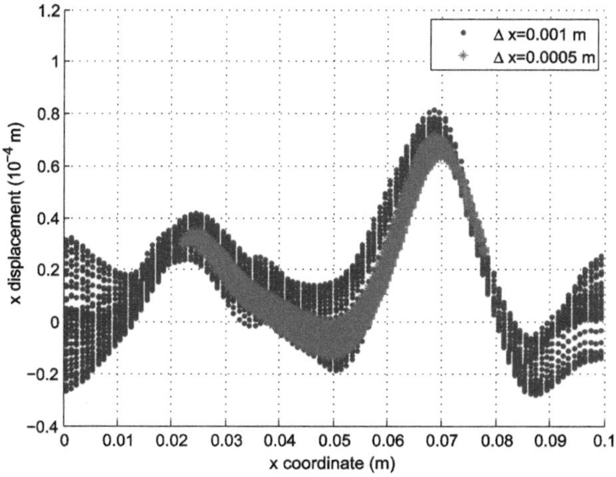

Figure 10.22 Displacement wave u_x versus x coordinate of model II at step 650.

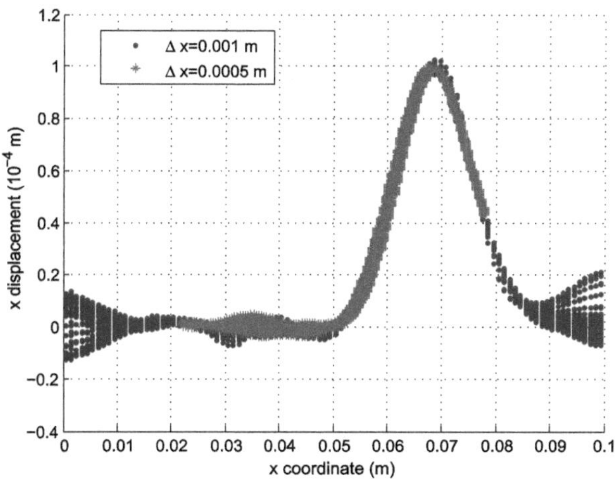

Figure 10.23 Displacement wave u_x versus x coordinate of model III at step 650.

We compute the L_2 errors of the wave profile along the line-CD (see Fig. 10.19). The L_2 error (see Table 10.2) of the present peridynamic formulation, can achieve the almost same accuracy as the conventional peridynamics with constant horizon.

Displacement and velocity of monitor points

Two monitor points (A, B) are selected to evaluate the spurious wave reflection of Models II and III as shown in Fig. 10.19. The displacement

Table 10.2 L_2 error of displacement for the initial condition (Gauss distribution of displacement) before and after the wave reflection.

Model	Step 100	Step 200	Step 650	Step 990
I	0.0410	0.0452	0.1277	0.2498
II	0.1149	0.1738	0.7638	1.3233
III	0.0447	0.0500	0.1253	0.2871

and velocity curves show the spurious wave reflection exists in model II while not exists in model III, see Fig. 10.24.

10.2.4.2 Multiple materials

The material interface problem in peridynamics was formally discussed in [1], where the authors analyzed the convergence of peridynamics for heterogeneous media, and found that the operator of linear peridynamics diverges in the presence of material interfaces. Some approaches have been proposed to deal with this issue. In [67], the interfaces and mixtures between multiple materials are accounted for by using separate influence functions for each material. For example, if $\mathbf{x}$ is near an interface between two materials denoted by (1) and (2), two separate influence functions $\omega^{(1)}$ and $\omega^{(2)}$ can be defined such that each vanishes at points $\mathbf{x}+\xi$ outside its associated material. The force state is given by

$$\mathbf{T}_{\mathbf{x}\mathbf{x}'} = \omega^{(1)}\langle\xi\rangle P_{\mathbf{x}}^{(1)}\mathbf{K}^{-1}\xi + \omega^{(2)}\langle\xi\rangle P_{\mathbf{x}}^{(2)}\mathbf{K}^{-1}\xi \tag{10.48}$$

where $P_{\mathbf{x}}^{(1)}$ and $P_{\mathbf{x}}^{(1)}$ are the first Piola-Kirchhoff stress from the respective constitution models and respective deformation gradient tensors $\mathbf{F}^{(1)}$ and $\mathbf{F}^{(2)}$, $\mathbf{K}$ is the shape tensor. In reference [1], a peridynamic interface model was proposed. However, such issue can be handled in a simpler way in DH-PD. In the DH-PD formulation, any point $\mathbf{x}$ has a force state $\mathbf{T}_{\mathbf{x}\square}$, which acts like a force potential field applied to its nearby neighbors in $H_{\mathbf{x}}$. We assume $\mathbf{x}$ imposes reaction bond forces to its neighbors no matter what the neighbor's material type is. The bond force can be determined by one side or both sides. For examples, in the BB-PD, microelastic modulus $C(\delta)$ and critical material stretch $s(\delta)$ are the primary parameters. For two points i and j with initial coefficient $[C(\delta_i), s(\delta_i)]$ (abbreviated as $[C_i, s_i]$) and $[C(\delta_j), s(\delta_j)]$, respectively, when taking into consideration of the combination of the initial coefficients, the alternate forms can be formulated as the arithmetic average or geometric average of microelastic moduli or critical stretches.

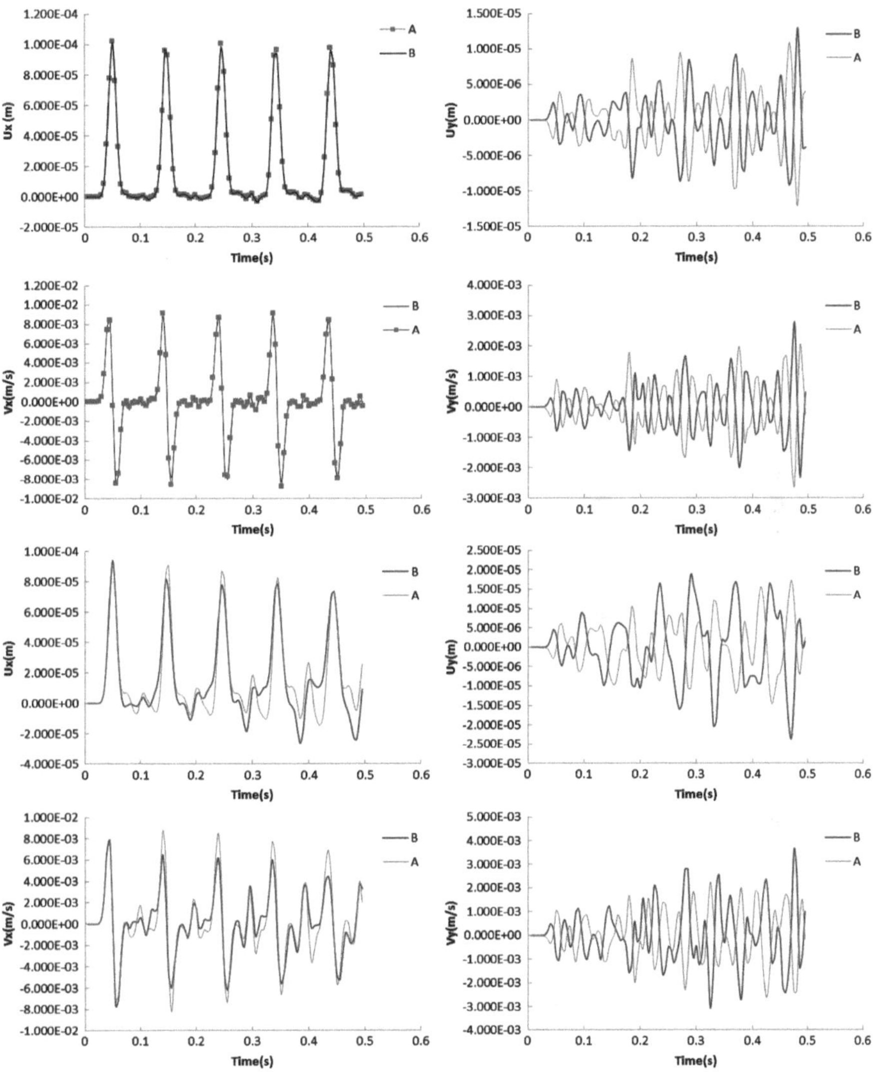

Figure 10.24 The displacement and velocity curve of monitor points *A* and *B* for model II and III.

For the OSB-PD and NOSB-PD in multiple materials, the force state $t_{\mathbf{xx}'}$ and $T_{\mathbf{xx}'}$ can be calculated similarly to the single homogeneous material and hence, it is not necessary to distinguish between homogeneous and heterogeneous materials.

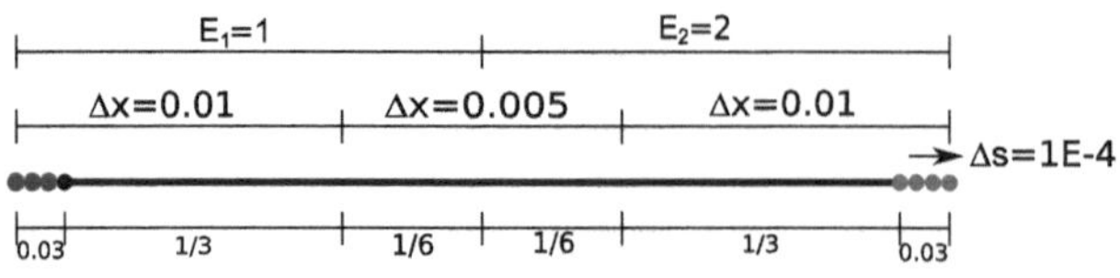

Figure 10.25 The setup of 1D bar.

1D bar tensile test

The length of the bar is 1 m, discretized with grid spacing $\Delta x = 0.01, 0.005$ m. The elastic modulus and density are $E_1 = 1$, $E_2 = 2$ and $\rho = 1$, respectively. Boundary conditions are applied at 3 material points at each end, i.e. fixed boundary condition at left end (blue [dark gray in print version] points in Fig. 10.25) and displacement boundary conditions of Δs = 1e-4 m (red [gray in print version] points in Fig. 10.25). The damping coefficient is α = 5e-3 m. Two cases were tested. Case I is based on the traditional constant-horizon BB-PD but without any treatment for multiple materials, while Case II on dual-horizon BB-PD with variable horizons. The equilibrium state is obtained at $t = 0.8$ s. The bar internal sectional force is approximated by summing the bond force at one side given by

$$f_i = \delta_i \sum_{j \in H_i, j>i} C_i s_{ij} \Delta V_j \tag{10.49}$$

where s_{ij} is the bond stretch for bond ij, δ_i is the radius of horizon H_i. Due to the error of discretization, Eq. (10.49) cannot sum the sectional force from material point i accurately; when right half of H_i contains just 3 neighbors and volume correction is considered, there are 2.5 material points in Eq. (10.49), which corresponds to 5/6 of the theoretical sectional force. For Case I, the force discontinuity in the interface of different point spacing is observed in Fig. 10.26. The force solution diverts from the analytical solution. The reason for the result is that the microelastic modulus is determined by one point. Consider a point i close to the interface of two materials with microelastic modulus C_i, the equilibrium of i requires the total force from left side being equal to that from right side. As a result, the displacement field in left half and right half horizon are anti-symmetrical with respect to i. Hence, there is no discontinuity in the strain field at the interface. At last, the spurious equilibrium displacement and force field is obtained.

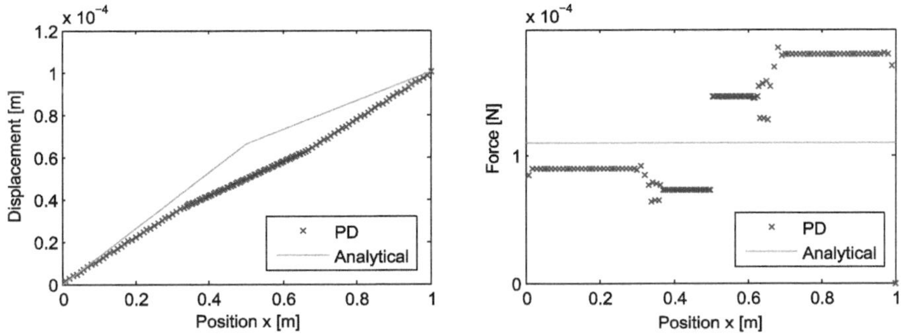

Figure 10.26 The displacement and sectional force for Case 1.

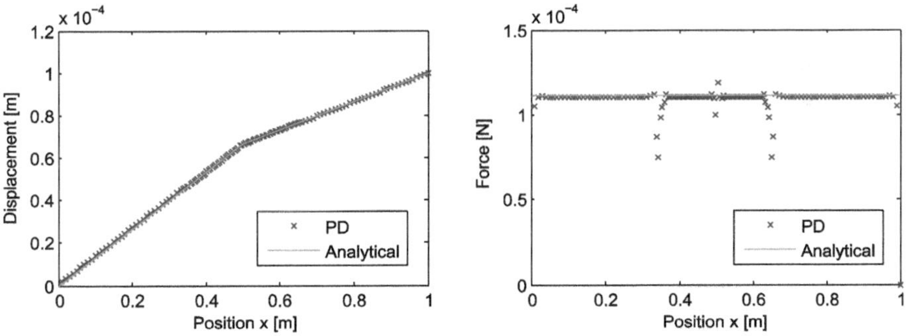

Figure 10.27 The displacement and sectional force for Case 2.

However, for Case II modeled with DH-PD, the force solution and displacement solution agree well with the analytical solutions, respectively, see Fig. 10.27. Both the material interface and point spacing interface have minimal influence on the final solution, though small variation is observed at the interfaces of material or grid spacing. The variation of the sectional force arises from the nonlocality in peridynamics. In PD, the force can transmit over finite distance, the equilibrium of a point means the forces from horizon and dual-horizon are canceled. In this sense, the equilibrium is determined by the domains of horizon and dual-horizon, two nearby material points may have different domains, therefore, the jumping of internal sectional force is reasonable. This is essentially different from the ghost force which is unbalanced.

In conclusion, we show for a simple 1D example that DH-PD has the potential to effectively deal with multiple materials.

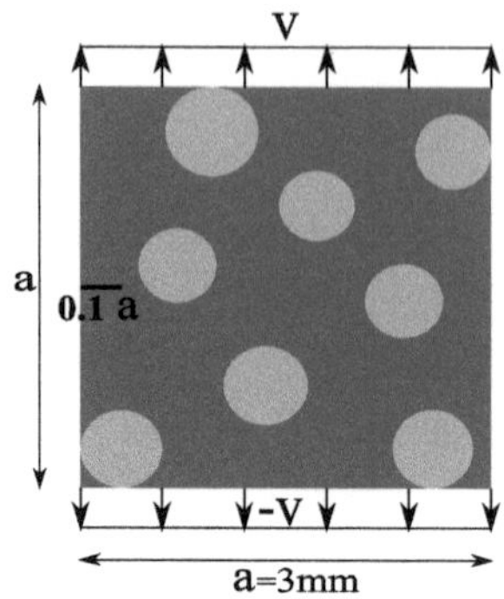

Figure 10.28 The setup of the plate with inclusions.

Crack propagation in heterogeneous materials

Consider a plate subjected to constant velocity traction at both the top and down edges, as shown in Fig. 10.28. The matrix is filled with many randomly distributed inclusions; all inclusions have the same material parameters. The material parameters for matrix include $E_1 = 72$ GPa, $\nu_1 = 1/3$, $G_0 = 40\ \mathrm{J/m^2}$. The parameters for the inclusions are $E_2 = 144$ GPa, $\nu_2 = 1/3$, $G_0 = 80\ \mathrm{J/m^2}$. The plate is discretized into 40,000 material points with point spacing of Δx = 1.5e-5 m. The horizon radius is $\delta = 3.015\Delta x$. The material's critical bond stretch is calculated based on Eq. (10.23), i.e. $s_1 = s_2 \approx$ 4.64e-3. Three cases are conducted to show the capability of DH-PD in dealing with composite materials. The velocity of traction on boundary is $v_{bc} = 0.5$ m/s for Case I, II and $v_{bc} = 2$ m/s for Case III. Two different heterogeneous structures are tested; the heterogeneous structure for Case II and III coincide.

We use a simple damage rule to take into account the crack propagation in heterogeneous materials. In the framework of DH-PD, the bond and dual-bond are broken separately. For example, as shown in Fig. 10.29, bond *ji* is interpreted as *i* exerting force on *j*, and the strength of bond *ji* is assumed to depend only on *i* parameters, e.g. *i*'s critical stretch; the same applies to bond *ij*. Fig. 10.30B and Fig. 10.30C show when initial crack tip is in the inclusion (Fig. 10.30C), the crack will firstly propagate in the inclusions; when initial crack tip is in the matrix, the crack firstly propagates in the matrix. We also note two phenomena when the load rate is increased:

1. The number of crack branches is increased. Such a phenomenon is well known in dynamic fracture.

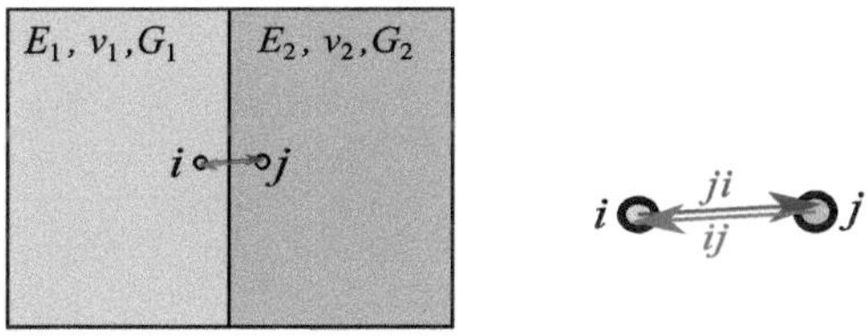

(A) (B)

Figure 10.29 The bonds in composite materials. Bond *ji* depends on *i*. When calculating *i*, all neighbors in H_i are pretended to have the same parameters as *i*. Bond *ij* depends on *j*, when calculating *j*, all neighbors in H_j are assumed to have the same parameters as *j*. Bond *ij* and *ji* are broken separately.

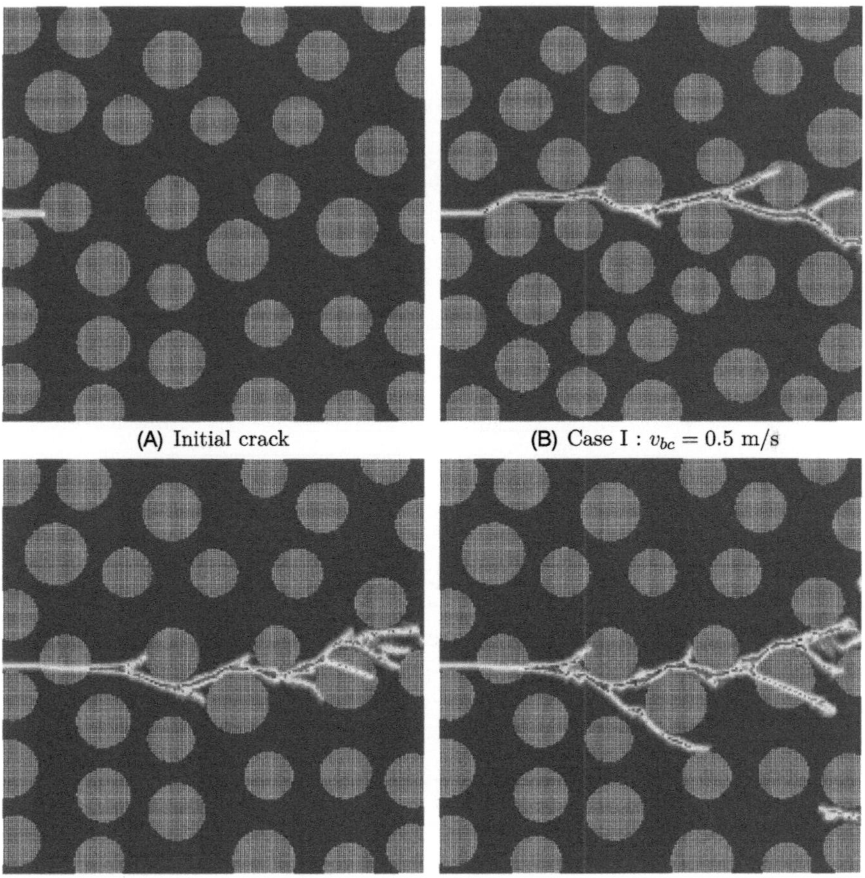

(A) Initial crack
(B) Case I : $v_{bc} = 0.5$ m/s
(C) Case II : $v_{bc} = 0.5$ m/s
(D) Case III : $v_{bc} = 2$ m/s

Figure 10.30 The crack patterns in multiple-material plate.

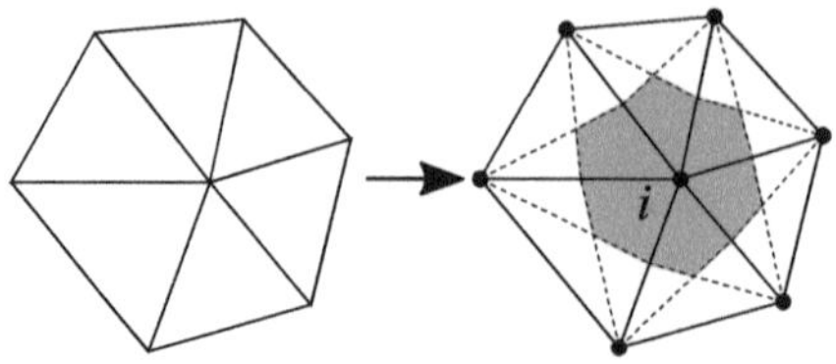

Figure 10.31 Triangle elements to material points.

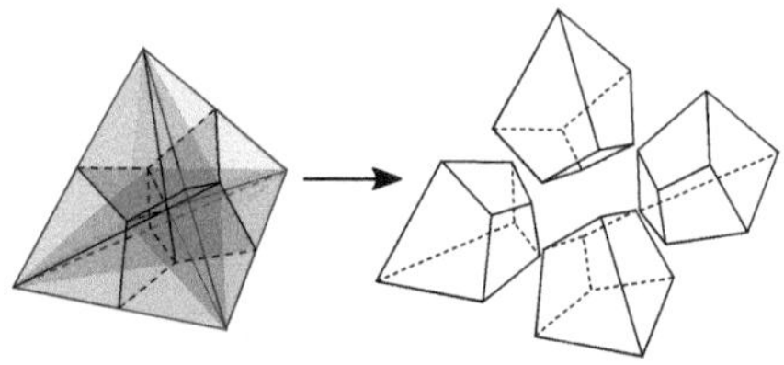

Figure 10.32 Tetrahedron element to material points.

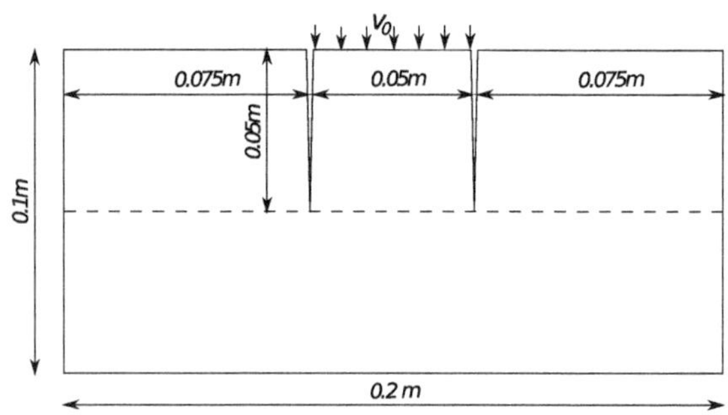

Figure 10.33 Kalthoff-Winkler's experimental setup.

2. The crack propagates through the inclusions. This tendency has been also observed experimentally, i.e. the number of aggregates/particle cracks in a weaker matrix material is increased with increasing strain rate/load rate.

10.2.4.3 Simulation of the Kalthoff-Winkler experiment

The Kalthoff-Winkler experiment is a classical benchmark problems studied by [8,10,40,54,55,68,75]. The setup is depicted in Fig. 10.33; the thickness of the specimen is 0.01 m. In the experiment, the evolution of crack pattern was observed to be dependent on the impact loading velocity. For a plate made of steel 18Ni1900 subjected to an impact loading at the

Table 10.3 The parameters in 4 simulations of Kalthoff-Winkler experiment.

Case	$\delta_{max}/\delta_{min}$	Type	Volume correction
I	1	CH-PD	Yes
II	3.1	DH-PD	Yes
III	1	CH-PD	Yes
IV	4.4	DH-PD	Yes
V	4.4	CH-PD	Yes
VI	4.4	DH-PD	No

speed of $v_0 = 32$ m/s, brittle fracture was observed [64]. The crack propagates from the end of the initial crack at an angle around 70° with respect to the initial crack direction. In [64], the BB-PD with constant horizon was used to test this example. For convenience of comparison, the present dual-horizon formulation is also applied to the bond based peridynamics to test the example. The material parameters used are the same as in [64], i.e. the elastic modulus $E = 190$ GPa, $\rho = 7800$ kg/m^3, $\nu = 0.25$ and the energy release rate $G_0 = 6.9 \times 10^4$ J/m^2. The impact loading was imposed by applying an initial velocity at $v_0 = 22$ m/s to the first three layers of material points in the domain as shown in Fig. 10.33; all other boundaries are free. For brevity, the surface correction method [44] or fictitious boundary [52] are not employed.

In order to create the irregular material points distribution in the computational domain, we utilize the Abaqus software to mesh the geometrical model with different mesh seeds setting, and then convert the finite element mesh into material points. For example, in the case of triangle element or tetrahedron element, each element's area (or volume) is evenly allocated to its nodes, as shown in Fig. 10.31 and Fig. 10.32. Every node (acted as material point) is associated with a unique volume or mass. The material point size Δx is the equivalent diameter, which is calculated based on its volume by assuming the material point is circle or sphere. Six cases as shown in Table 10.3, are tested. Case I, II are in two dimensions with the same material points distribution where the ratio $\Delta x_{max}/\Delta x_{min} = 3.1$, while Case III-VI in three dimensions with particle size ratio $\Delta x_{max}/\Delta x_{min} = 4.4$. Case I and III are based on the traditional constant-horizon peridynamics (CH-PD), while Case II and Case IV are carried out with the DH-PD. Case V is conducted to test the influence of spurious wave reflections when variable horizons is used in traditional PD. Case VI is used to test the effect of volume correction given in Appendix A. Two different discretizations

Table 10.4 The angle of crack propagation with respect to the original crack.

	Case I	Case II	Case III	Case IV	Case V	Case VI
Left	56.1°	67.6°	56.4°	61.4°	62.5°	58.2°
Right	57.6°	64.7°	56.3°	62.7°	76.1°	65.3°

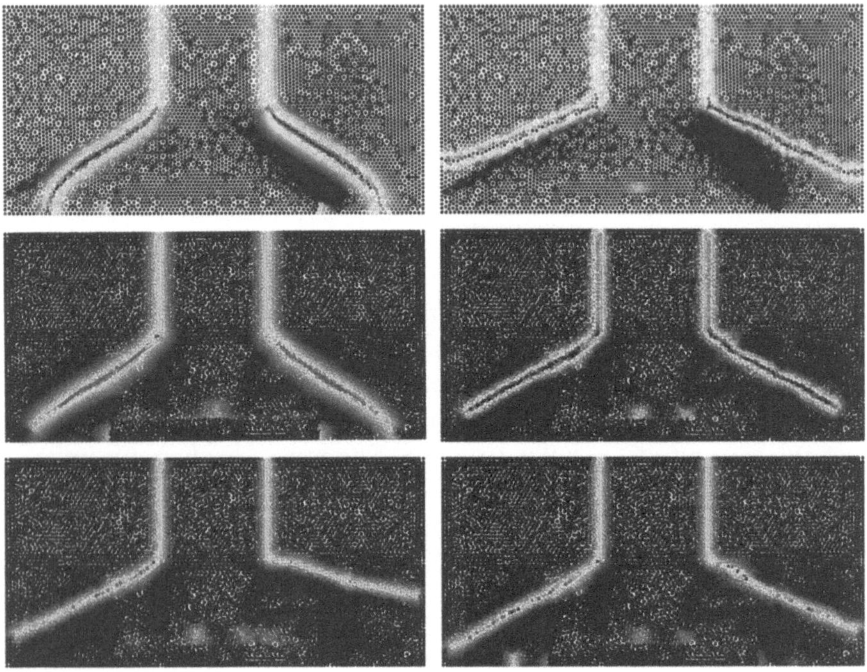

Figure 10.34 The crack patterns in the simulations of Kalthoff Winkler experiment.

(for I, II and III-VI, respectively) with a bias towards a 'wrong' crack path. In Case II and IV-VI, the radius of each material point's horizon is selected as three times of the material point size, e.g. $\delta_i = 3.015\,\Delta x_i$, while in Case I and III $\delta = 3.015\,\Delta x_{max}$. The bond microelastic modulus $C(\delta_\mathbf{x})$ is calculated based on Eq. (10.22).

The angle of the crack propagation with respect to the original crack direction is shown in Table 10.4. For relatively low speed impact (e.g. $\leq$30 m/s), the crack angle in [31,32] is approximately 70°. Table 10.4 shows that result by DH-PD is better than that by CH-PD. The results in Fig. 10.34 and Table 10.4 show that the crack pattern both in 2D and 3D cases by the traditional constant-horizon bond based peridynamics differs from the experiment, while the crack pattern obtained by the DH-PD

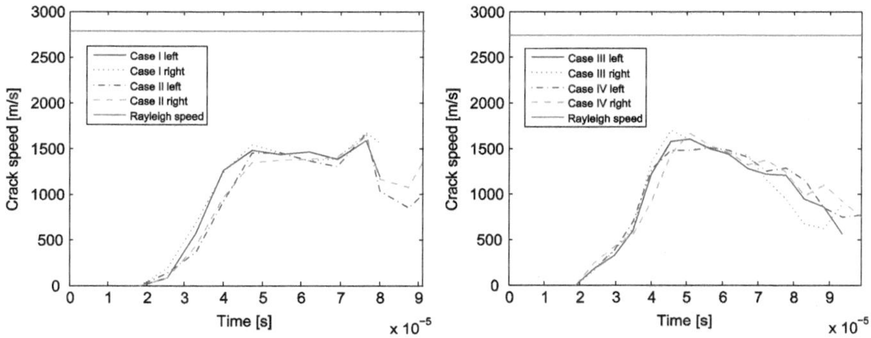

Figure 10.35 The crack propagation speed in the simulations of Kalthoff Winkler experiment.

agrees well with the experiment besides the introduction of a mesh bias towards an 'incorrect' crack path. The comparison of Case IV and Case V shows the crack pattern is affected by the spurious wave reflections. The comparison of Case IV and Case VI shows the volume correction has some positive effect on the crack propagation, therefore, the volume correction is recommended.

The crack propagation speed is computed by

$$V_{l-0.5} = \frac{\|\mathbf{x}_l - \mathbf{x}_{l-1}\|}{t_l - t_{l-1}} \,, \tag{10.50}$$

where $\mathbf{x}_l$ and $\mathbf{x}_{l-1}$ are the positions of the crack tip at the times t_l and t_{l-1}, respectively. An expression for the Rayleigh wave speed c_R [24] is given as

$$\frac{c_R}{c_s} = \frac{0.87 + 1.12\nu}{1 + \nu} \,, \tag{10.51}$$

where ν is the Poisson's ratio, $c_s = \sqrt{\mu/\rho}$ is the shear wave speed and μ is the shear modulus. The crack starts to propagate at 20 μs. The crack speed in all four cases simulation is shown in Fig. 10.35. The maximum crack propagation speeds are 1669 m/s, 1658 m/s, 1699 m/s, 1664 m/s for Case I, II, III, IV, respectively. All of them are within the limit of 75% of Rayleigh speed. The crack angle of Kalthoff Winkler experiment by other methods [10,55] are 65.1° and 63.8° as shown in Fig. 10.36A and Fig. 10.36B, respectively. Therefore, it can be concluded that with the present formulation, the random material points distribution has limited influence on the crack propagation when simulated by DH-PD.

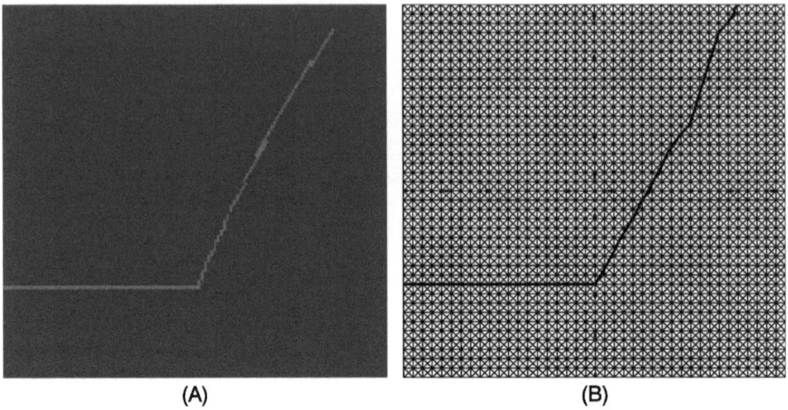

Figure 10.36 The crack pattern of Kalthoff Winkler experiment by other methods. (A) From T. Rabczuk, G. Zi, S. Bordas, H. Nguyen-Xuan, A simple and robust three-dimensional cracking-particle method without enrichment, Comput. Methods Appl. Mech. Engrg., 199 (37) (2010), pp. 2437-2455. (B) From T. Belytschko, H. Chen, J. Xu, G. Zi, Dynamic crack propagation based on loss of hyperbolicity and a new discontinuous enrichment, Internat. J. Numer. Methods Engrg., 58 (12) (2003), pp. 1873-1905.

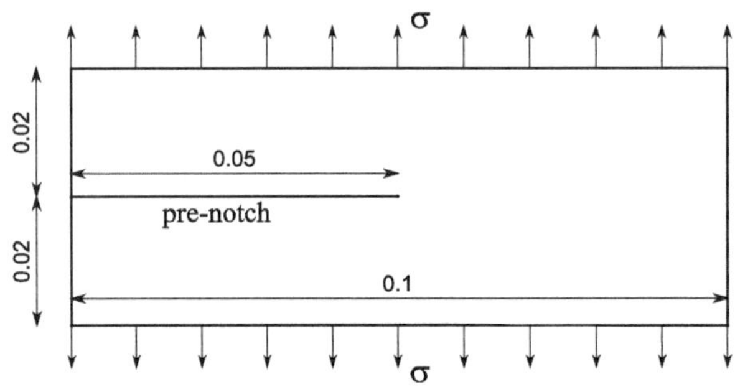

Figure 10.37 Setup of the pre-cracked plate under traction load.

10.2.4.4 Plate with pre-crack subjected to traction

In the last example, we will test the capability of the new formulation in modeling crack branching. Hence, a plate with pre-crack subjected to traction is considered as shown in Fig. 10.37. The traction remains constant during the entire simulation. The same example has been well studied in [27] with the BB-PD, in [10] using XFEM.

The material parameters of the plate (soda-lime glass) are $E = 72$ GPa, $\rho = 2440\ \text{kg/m}^3$ and critical energy release rate being $G_0 = 135\ \text{J/m}^2$ [27]. Two models, namely Case 1 and Case 2, are set up for comparison. Case 1 is solved by using the traditional BB-PD with constant horizons, while Case 2

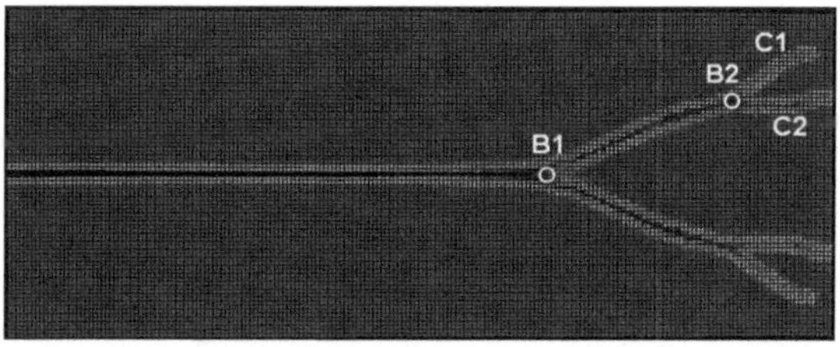

Figure 10.38 The crack pattern of pre-cracked plate with uniform horizons.

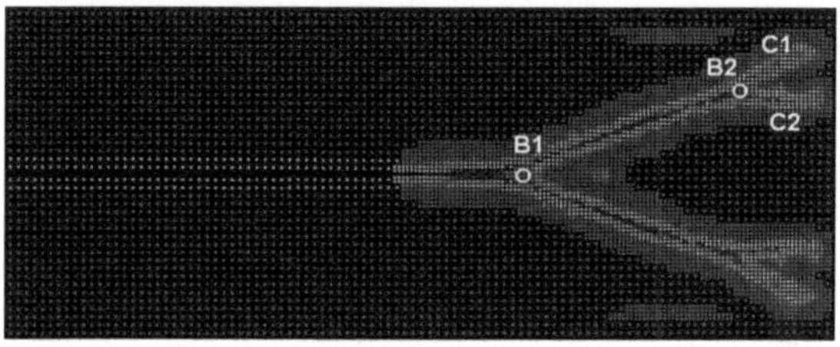

Figure 10.39 The crack pattern of adaptive refined pre-cracked plate.

is modeled by an adaptively refined BB-PD using the present dual-horizon formulation. The plate is assumed under plane stress condition and all simulations are carried out in 2D. The particle sizes chosen are 5×10^{-4} m for Case 1 and 1×10^{-3} m for Case 2, respectively. Case 2 uses an adaptively refined model, and once a particle is refined, the minimal particle size will become the same as in Case 1. 16,000 particles are for Case 1 and 4000-6424 for Case 2, respectively. The Rayleigh speed for the material is 3102 m/s according to Eq. (10.51). The maximum crack speed is 1881.4 m/s (60.6% of c_R) for Case 1 and 2184.1 m/s (70.4% of c_R) for Case 2, respectively (see Fig. 10.40). One possible reason for the different maximum crack speeds is the deviation of the mapping method.

For Case 1, the crack starts to propagate at 11.9 μs, and the first crack branching point $B1$ occurs at 24.5 μs at the speed of 1043.6 m/s and the second crack branch point $B2$ takes place at 40.9 μs at the speed of 1147.1 m/s. It is observed once branching initiates, the crack propagation speed decreases, since the energy release to create more fracture surfaces decelerates the propagation speed. Afterward, the crack speed increases. For Case 2, the initial crack starts to propagate at 10.3 μs, first branching point $B1$ at 19.4 μs with 1247.6 m/s, and second branching point $B2$ at 34.2 μs with 1330.6 m/s. The crack patterns for Case 1 and Case 2 are similar, see Figs. 10.38 and 10.39.

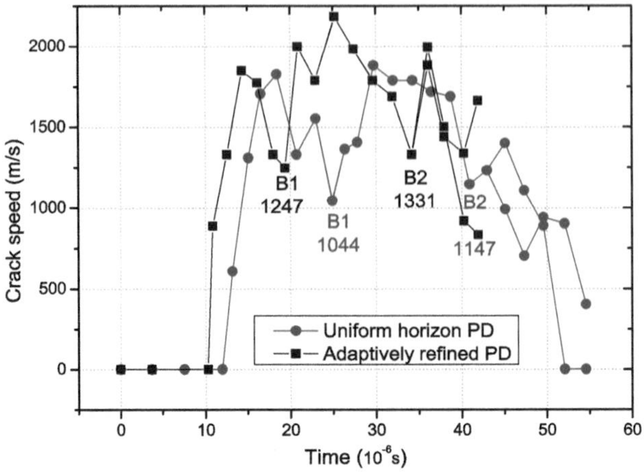

Figure 10.40 The crack speed of pre-cracked plate under traction.

10.3. Phase field models

Phase-field models for modeling material failure have recently gained significant attention. They fall under the broader class of continuous approaches to fracture, but they differ from most earlier models in their reliance on a variational theory of fracture. This is itself a relatively recent concept, being developed in the late 90s by Francfort and Marigo [22] to overcome certain limitations associated with the classical Griffith theory such as its inability to predict crack initiation and branching, as well as to handle more complex crack paths. In the variational theory, the total energy potential is assumed to include a surface term associated with the energy required for the formation of a crack, i.e.

$$E(\boldsymbol{u}, \Gamma) = \int_{\Omega} \Psi^{b}(\boldsymbol{\varepsilon})\mathrm{d}\Omega + G_c \int_{\Gamma_s} \mathrm{d}\Gamma \tag{10.52}$$

where $\Psi^b(\boldsymbol{\varepsilon})$ is an energy density, G_c is the critical energy release rate and $\Gamma_s \in \Omega$ denotes an admissible crack set. This potential is minimized with respect to the displacement field $\mathbf{u}$ and the crack geometry, with the idea being that the fracture topology evolution should occur in such a way that results in minimal potential energy. Further developments were introduced by Bourdin et al. [13,14], the most notable being the regularization of the sharp crack topology into a diffuse entity defined via a scalar variable (later termed the phase-field) which serves to interpolate between cracked and intact states of the material. Bourdin et al. [13] utilized an evolution equation for the phase-field derived from the Mumford-Shah functional for image segmentation. Kuhn and Müller [35] alternatively proposed an evolution equation of the Ginzburg-Landau type and mentioned that the application of a phase field approach to crack propagation is possible after accounting for the irreversible character of crack propagation. Noting that the steady state problem associated with the phase-field evolution equation yields a solution of exponential type, they [36] subsequently developed a finite element formulation based on exponential shape functions and demonstrated their superior performance compared to standard Lagrange elements in cases where the crack path aligns with the element edges. A modification of the original functional proposed in [13] was made by Lancioni and Royer-Carfagni [39] to handle problems involving deviatoric-type fracture. However, while their model correctly does not allow for material inter-penetration it is also unable to model crack-opening, hence limiting its applicability to cases where the dominant stress behavior is compressive. Amor et al. [5] improved the prior model to be able to handle crack opening by allowing stiffness degradation corresponding to the spherical component of the stress tensor when the value of said component is positive (signifying volumetric expansion). Significant contributions to the theory were also made by Miehe et al. [45,46] in the form of a thermodynamically consistent framework for brittle mode-I fracture for the case of rate-independent as well as viscous material response. In particular, their formulation ensures local irreversibility of the phase-field which permits the correct simulation of material behavior for cyclic loading. In [46] the coupled system consisting of stress equilibrium equations and the phase-field evolution is solved simultaneously (with local growth of the phase field imposed via a penalty term). A staggered scheme is employed in [45], in which a local energy history field is introduced as a state variable to ensure irreversible crack growth. In the past few years, the phase field approach to fracture has been extended and applied to numerous engineering applications and problems. The beauty of the formulation lies in the thermodynamic consistent framework, which allows a straightforward implementation into a finite element framework. A good overview on those applications and the progress on phase field models can be found in the review paper of [3].

10.3.1 Concepts

Let us consider the 1D problem consisting of an infinitely long cylindrical bar (constant cross-sectional area A) that is aligned along the x-axis and

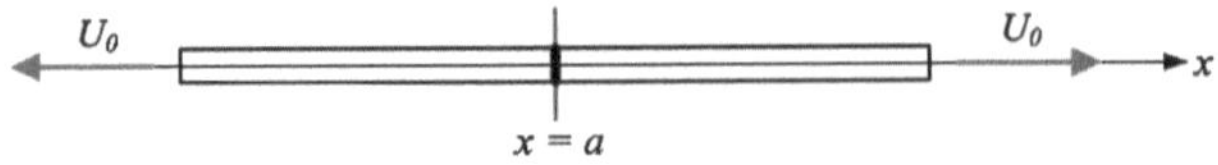

Figure 10.41 1D bar with a crack at $x = a$.

fully cut by a crack at $x = a$ as shown in Fig. 10.41. The surface energy dissipated by the formation of the crack may be calculated simply as the product of the critical energy release rate G_c and the crack surface equal to the cross sectional area of the bar, i.e.

$$\Psi^s = \int_{\Gamma_s} G_c \, d\Gamma = G_c \, A \tag{10.53}$$

Unfortunately, such a calculation implies knowledge of the crack path, and while this is trivial in 1D, the actual crack topology is needed in higher dimensions to be able to compute the value of A. An alternative is to smooth out the crack such that the surface energy can be calculated as a domain integral. As demonstrated in [46], this can be accomplished by introducing a scalar field variable to represent the diffuse crack of the form

$$\phi(x) = \exp\left(\frac{-|x-a|}{l_0}\right) \tag{10.54}$$

where $\phi(x)$ is denoted as the crack phase field, and l_0 is a parameter that controls the amount of crack diffusion. It can be seen that $\phi(x) \in [0, 1]$; with 0 representing a completely intact state of the material while 1 indicates a fully broken state. The crack phase field for the 1D problem described above is shown in Fig. 10.42 for different values of l_0. It should be emphasized that l_0 does not represent the actual width of the diffuse crack. On the contrary, the form of Eq. (10.54) means that the crack is smoothed out over the entire domain, since $\phi(x)$ is never actually zero. Furthermore, while (10.54) may not be the only possible expression for describing the diffuse crack, the choice of functions is not entirely arbitrary. In particular, $\phi(x)$ should be symmetric and monotonically decaying to zero as we move away from the crack location, i.e.

$$\phi(x) \to 0 \text{ as } x \to \pm\infty \tag{10.55}$$

$$\frac{\partial \phi(x)}{\partial x} \to 0 \text{ as } x \to \pm\infty \tag{10.56}$$

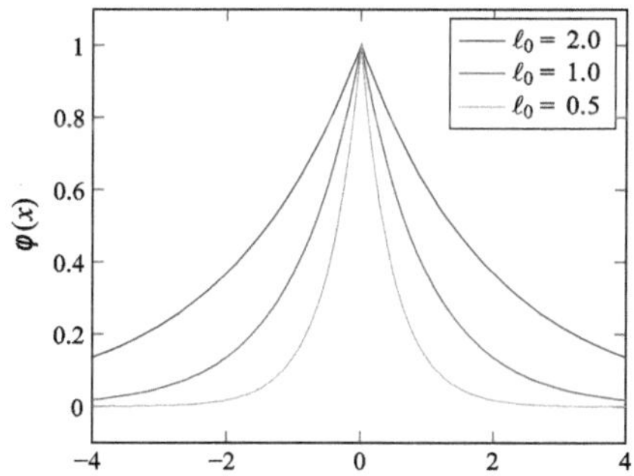

Figure 10.42 Crack phase-field ϕ for different values of l_0.

Taking derivatives of Eq. (10.54) yields

$$\frac{\partial \phi(x)}{\partial x} = -\frac{\operatorname{sgn}(x-a)}{l_0} \exp\left(\frac{-|x-a|}{l_0}\right) \tag{10.57}$$

$$\frac{\partial^2 \phi(x)}{\partial x^2} = \left[\frac{\operatorname{sgn}(x-a)}{l_0}\right]^2 \exp\left(\frac{-|x-a|}{l_0}\right) \tag{10.58}$$

These derivatives are undefined at $x = a$ but for $x \neq a$, the expression in (10.54) satisfies the ODE

$$-\frac{\partial^2 \phi(x)}{\partial x^2} + \frac{1}{l_0^2}\phi(x) = 0 \tag{10.59}$$

The weak formulation corresponding to the above equation along with the derivative boundary conditions given in (10.56) is

$$0 = \int_{-\infty}^{\infty} \frac{\partial v}{\partial x}\frac{\partial \phi}{\partial x} + \frac{1}{l_0^2} v\, \phi \; dx \tag{10.60}$$

v denoting some test functions. The associated quadratic functional is then given by

$$I(\phi(x)) = \int_{-\infty}^{\infty} 0.5\left[\left(\frac{\partial \phi}{\partial x}\right)^2 + \frac{1}{l_0^2}\phi^2\right] dx \tag{10.61}$$

The construction of $I(\phi)$ implies that (10.57) along with (10.56) constitute the Euler-Lagrange equations associated with the variational problem

$$\phi^* = \arg \underbrace{\inf}_{\phi \in \mathcal{W}} \{I(\phi)\} \tag{10.62}$$

in which $\mathcal{W}$ is the set of functions satisfying (10.56) along with the internal condition $\phi(a) = 1$. Furthermore, ϕ^* is none other than the function given in Eq. (10.54). Plugging this into the functional in (10.61) yields

$$I(\phi^*) = \int_{-\infty}^{\infty} \frac{1}{l_0^2} \exp\left(\frac{-2|x-a|}{l_0}\right) dx = \frac{1}{l_0^2} \tag{10.63}$$

Defining a function

$$\gamma(\phi) = 0.5\left[l_0\left(\frac{\partial \phi}{\partial x}\right)^2 + \frac{1}{l_0}\phi^2\right] \tag{10.64}$$

which is simply the integrand of (10.61) scaled by l_0 so that the regularized surface energy may now be written as a domain integral:

$$\Psi^s = G_c \int_\Omega \gamma(x) d\Omega = G_c \int_A \int_{-\infty}^{\infty} \gamma(x)\ dx\ dA = G_c A \tag{10.65}$$

Correspondingly, the bulk energy can be defined as

$$\Psi^b = \int_\Omega g(\phi)\psi^e d\Omega \tag{10.66}$$

in which $g(\phi)$ is a stress degradation function given for instance by $g(\phi) = (1-\phi)^2 + k$, k being a parameter chosen to be as small as possible while at the same time keeping the system of equations well-conditioned; for a linear elastic solid ψ^e is the strain energy density. It is worthwhile to note that the parameter a representing the crack location does not appear in (10.65), which means that no preliminary estimation of the crack path is required. This endows the phase-field model with enormous flexibility, such as the automatic nucleation, branching and merging of cracks. The expression given in (10.64) may be readily extended to 2D and 3D, with the relevant form being

$$\gamma(\phi) = 0.5\left[l_0\left(\frac{\partial \phi}{\partial x_i}\frac{\partial \phi}{\partial x_i}\right)^2 + \frac{1}{l_0}\phi^2\right] \tag{10.67}$$

where $i = 1, \ldots, d$ with d being the number of dimensions.

In order to ensure that cracks occur only under tension but not under compression, the elastic energy can be decomposed into a positive and negative part: $\Psi_0 = \Psi_0^+ + \Psi_0^-$ [45,46] which in turn is based on the spectral decomposition of the linear strain tensor $\boldsymbol{\varepsilon}$: $\boldsymbol{\varepsilon} = \sum_{I=1}^{3} \varepsilon_I \boldsymbol{n}_I \otimes \boldsymbol{n}_I$, $\{\varepsilon\}_{I=1}^{3}$ and

$\{\boldsymbol{n}\}_{I=1}^{3}$ being the principal strains and their directions. The decomposed elastic energy reads

$$\Psi_0^{\pm}(\boldsymbol{\varepsilon}) = \frac{1}{2}\lambda\langle \mathrm{tr}(\boldsymbol{\varepsilon})\rangle_{\pm}^{2} + \mu \mathrm{tr}(\boldsymbol{\varepsilon}_{\pm}^{2}) \tag{10.68}$$

where $\boldsymbol{\varepsilon}_{\pm} = \sum_{I=1}^{3} \langle \varepsilon_I \rangle_{\pm} \boldsymbol{n}_I \otimes \boldsymbol{n}_I$. The modified elastic energy Ψ_e used in the total energy functional can now be expressed as

$$\Psi_e = \left[(1-\phi)^2 + k)\right] \Psi_0^{+} + \Psi_0^{-} \tag{10.69}$$

The governing equations $\nabla \cdot \boldsymbol{\sigma} = \boldsymbol{0}$ and the stress tensor is expressed as

$$\boldsymbol{\sigma}(\boldsymbol{u}, \phi) = \left[(1-\phi)^2 + k\right] \frac{\partial \Psi_0^{+}}{\partial \boldsymbol{\varepsilon}} + \frac{\partial \Psi_0^{-}}{\partial \boldsymbol{\varepsilon}} \tag{10.70}$$

while the evolution equation of the phase field in this model is given by

$$\xi \dot{\phi} = 2(1-\phi)\Psi_0^{+}(\boldsymbol{\varepsilon}) + \frac{G_c}{l_0}\left(\phi - l_0^2 \Delta \phi\right) \tag{10.71}$$

where ξ is a viscosity parameter. Note that other energy splits have been performed, see for instance the interesting contribution in [5]. Note that if $l_0 = 2\epsilon$ and $\xi = \frac{1}{M}$, the structure of Eqs. (10.70) and (10.71) resembles those equations in above mentioned contributions. However, the energy decomposition of Miehe et al. [45,46] is different from that of Amor et al. [5] as they only distinguish compressive and tensile parts and imposed a degradation of the compressive energy component while the deviatoric stress tensor is considered in Amor et al. [5]. Another feature of Miehe's model is that Eq. (10.70) results in a highly nonlinear stress-strain relationship leading to a higher computational cost compared to former phase field approaches.

In order to ensure a monotonically increasing phase-field, the irreversibility condition must be imposed as a constraint during compression or unloading. One of the simplest approaches is based on a history-field H defining the loading process:

$$H(\boldsymbol{x}, t) = \max_{\tau \in [0,t]} \psi_0^{+}\left(\boldsymbol{\varepsilon}(\boldsymbol{x}, \tau)\right) \tag{10.72}$$

Replacing Ψ_0^{+} in Eq. (10.71) by H, the finally used evolution equation reads

$$\xi \dot{\phi} = 2(1-\phi)H + \frac{G_c}{l_0}\left(\phi - l_0^2 \Delta \phi\right) \tag{10.73}$$

As mentioned previously, $g(\phi)$ is a degradation function that models the degradation of the elastic energy at fracture initiation. It also models the decrease in the stiffness in a fractured material when the phase field $\phi \rightarrow 0$. Furthermore, $g(\phi)$ must monotonically increase with $g(1)=1$ and $g(0)=0$. Because the derivative $g'(\phi)$ enters the evolution equation of the phase field, $g'(0)=0$ must be imposed to eliminate the elastic driving term $\frac{\partial \Psi}{\partial \phi}=\frac{1}{2}g'(\phi)\boldsymbol{\varepsilon}\mathbb{C}:\boldsymbol{\varepsilon}$ when $\phi=0$. Besides above mentioned degradation functions, a series of other approaches have been proposed in the literature such as the quartic function $g(\phi)=4\phi^3-3\phi^4$ by [33] or the cubic function $g(\phi)=3\phi^2-2\phi^3$ by Kuhn et al. [37].

10.3.2 Governing equations

The governing equations can be derived from the total potential energy functional which can be written as

$$\Psi(\phi,\mathbf{u}) = \int_\Omega \left[(1-\phi)^2+k\right]\psi(\boldsymbol{\varepsilon})\, d\Omega + \int_\Omega 0.5G_c\left[l_0\nabla\phi\cdot\nabla\phi+\frac{1}{l_0}\phi^2\right] d\Omega \tag{10.74}$$

with $\psi(\boldsymbol{\varepsilon})=0.5\lambda\varepsilon_{kk}\varepsilon_{ll}+\mu\varepsilon_{ij}\varepsilon_{ij}$ for a linear-elastic solid, λ and μ denoting the Lamé constants. Assuming small strain theory, the governing equations are obtained as follows: we first define the first variation of the external work increment (dropping terms that evaluate to zero) as

$$\delta W_{ext} = \int_\Omega \delta\mathbf{u}\cdot\mathbf{b}\, d\Omega + \int_{\Gamma_t}\delta\mathbf{u}\cdot\bar{\mathbf{t}}\, d\Gamma \tag{10.75}$$

where $\mathbf{b}$ and $\bar{\mathbf{t}}$ denote the body force and applied traction. The first variation of the internal energy increment is given by

$$\delta W_{int} = \delta\Psi = \frac{\partial \Psi}{\partial \boldsymbol{\varepsilon}}:\boldsymbol{\varepsilon} = \frac{\partial \Psi}{\partial \phi}\delta\phi \tag{10.76}$$

which is given for the energy functional in Eq. (10.74) as

$$\delta\Psi = \int_\Omega \left[(1-\phi)^2+k\right]\boldsymbol{\sigma}:\delta\boldsymbol{\varepsilon}\, d\Omega - 2\int_\Omega (1-\phi)\,\delta\phi\,\psi(\boldsymbol{\varepsilon})\, d\Omega + \int_\Omega G_c\left[l_0\nabla\phi\cdot\nabla\delta\phi+\frac{1}{l_0}\delta\phi\right] d\Omega \tag{10.77}$$

By appropriate transfer of differentiation between variables, it can be shown that the above expression is equivalent to

$$\begin{aligned}
\delta\Psi = & \int_{\Gamma_t} \left[(1-\phi)^2 + k\right] \mathbf{n} \cdot \boldsymbol{\sigma} \cdot \delta\mathbf{u}\, d\Gamma - \int_{\Omega} \left[(1-\phi)^2 + k\right] \nabla \cdot \boldsymbol{\sigma} \cdot \delta\mathbf{u}\, d\Omega \\
& - 2\int_{\Omega} (1-\phi)\, \delta\phi\, \psi(\boldsymbol{\varepsilon})\, d\Omega + \int_{\Gamma_s} \nabla\phi \cdot \mathbf{n}\delta\phi\, d\Gamma \\
& + \int_{\Omega} G_c \left[l_0 \nabla\phi \cdot \nabla\delta\phi + \frac{1}{l_0}\delta\phi \right] d\Omega
\end{aligned} \tag{10.78}$$

where $\mathbf{n}$ denotes the normal to the surface Γ. Since the first variation $\delta W_{ext} - \delta W_{int}$ needs to vanish and hold for arbitrary values of the test functions, the strong form of the governing equations finally reads:

$$\begin{aligned}
\left[(1-\phi)^2 + k\right] \nabla \cdot \boldsymbol{\sigma} + \mathbf{b} = 0 \ \forall \mathbf{x} \in \Omega \\
\left[(1-\phi)^2 + k\right] \mathbf{n} \cdot \boldsymbol{\sigma} = \bar{\mathbf{t}} \text{ on } \Gamma_t \\
\mathbf{u} = \bar{\mathbf{u}} \text{ on } \Gamma_u \\
G_c\, l_0 \nabla\phi \cdot \nabla\phi + \left[\frac{G_c}{l_0} + 2\psi(\boldsymbol{\varepsilon}) \right] \phi = 2\psi(\boldsymbol{\varepsilon}) \\
\nabla\phi \cdot \mathbf{n} = 0 \text{ on } \Gamma_s
\end{aligned} \tag{10.79}$$

The above shows a coupled system consisting of the modified stress equilibrium and the phase-field evolution equation. The second and third equations are the natural and essential boundary conditions associated with mechanical equilibrium, while the last equation is the natural boundary condition associated with the phase field evolution equation. Note that while satisfaction of the last equation seems trivial, it implies an upper limit to the choice of the regularization parameter l_0 with respect to the domain such that the phase-field sufficiently decays to zero at the boundary.

10.3.3 Discretization

Using standard vector-matrix notation, the discretization of the displacement and phase field can be expressed as

$$\boldsymbol{u} = \boldsymbol{N}_u \boldsymbol{d}, \qquad \phi = \boldsymbol{N}_\phi \hat{\boldsymbol{\phi}} \tag{10.80}$$

where $\boldsymbol{u}_i$ and ϕ_i are the nodal values of the displacement and phase field; $\boldsymbol{d}$ and $\hat{\boldsymbol{\phi}}$ are the vectors consisting of nodal values $\boldsymbol{u}_i$ and ϕ_i while $\boldsymbol{N}_u$ and

$\boldsymbol{N}_\phi$ are shape function matrices, respectively, given by

$$\boldsymbol{N}_u = \begin{bmatrix} N_1 & 0 & 0 & \dots & N_n & 0 & 0 \\ 0 & N_1 & 0 & \dots & 0 & N_n & 0 \\ 0 & 0 & N_1 & \dots & 0 & 0 & N_n \end{bmatrix},$$
$$\boldsymbol{N}_\phi = \begin{bmatrix} N_1 & N_2 & \dots & N_n \end{bmatrix} \tag{10.81}$$

where n is the node number in one element and N_i is the corresponding shape function. The same discretization is assumed for the test functions within a Bubnov Galerkin method:

$$\delta \boldsymbol{u} = \boldsymbol{N}_u \delta \boldsymbol{d}, \qquad \delta \phi = \boldsymbol{N}_\phi \delta \hat{\boldsymbol{\phi}} \tag{10.82}$$

where $\delta \boldsymbol{d}$ and $\delta \hat{\boldsymbol{\phi}}$ are the vectors containing the nodal values of the test functions. These associated gradients are expressed as

$$\boldsymbol{\varepsilon} = \boldsymbol{B}_u \boldsymbol{d}, \qquad \nabla \phi = \boldsymbol{B}_\phi \hat{\boldsymbol{\phi}}, \qquad \delta \varepsilon = \boldsymbol{B}_u \delta \boldsymbol{d}, \qquad \nabla \phi = \boldsymbol{B}_\phi \delta \hat{\boldsymbol{\phi}} \tag{10.83}$$

where $\boldsymbol{B}_u$ and $\boldsymbol{B}_\phi$ are the derivatives of the shape functions defined by

$$\boldsymbol{B}_u = \begin{bmatrix} N_{1,x} & 0 & 0 & \dots & N_{n,x} & 0 & 0 \\ 0 & N_{1,y} & 0 & \dots & 0 & N_{n,y} & 0 \\ 0 & 0 & N_{1,z} & \dots & 0 & 0 & N_{n,z} \\ N_{1,y} & N_{1,x} & 0 & \dots & N_{n,y} & N_{n,x} & 0 \\ 0 & N_{1,z} & N_{1,y} & \dots & 0 & N_{n,z} & N_{n,y} \\ N_{1,z} & 0 & N_{1,x} & \dots & N_{n,z} & 0 & N_{n,x} \end{bmatrix},$$
$$\boldsymbol{B}_\phi = \begin{bmatrix} N_{1,x} & N_{2,x} & \dots & N_{n,x} \\ N_{1,y} & N_{2,y} & \dots & N_{n,y} \\ N_{1,z} & N_{2,z} & \dots & N_{n,z} \end{bmatrix} \tag{10.84}$$

Substituting the test and trial functions as well as their respective derivatives into the weak form yields the final system of equations:

$$-(\delta \boldsymbol{d})^{\mathrm{T}} \left[\int_\Omega \boldsymbol{B}_u^{\mathrm{T}} \boldsymbol{D}_e \boldsymbol{B}_u \mathrm{d}\Omega \boldsymbol{d} \right] + (\delta \boldsymbol{d})^{\mathrm{T}} \left[\int_\Omega \boldsymbol{N}_u^{\mathrm{T}} \boldsymbol{b} \mathrm{d}\Omega + \int_{\Omega_h} \boldsymbol{N}_u^{\mathrm{T}} \boldsymbol{f} \mathrm{d}S \right] = 0 \tag{10.85}$$

$$- (\delta \hat{\boldsymbol{\phi}})^{\mathrm{T}} \int_\Omega \left\{ \boldsymbol{B}_\phi^{\mathrm{T}} G_c l_0 \boldsymbol{B}_\phi + \boldsymbol{N}_\phi^{\mathrm{T}} \left[\frac{G_c}{l_0} + 2(1-k)H \right] \boldsymbol{N}_\phi \right\} \mathrm{d}\Omega \hat{\boldsymbol{\phi}}$$
$$+ (\delta \hat{\boldsymbol{\phi}})^{\mathrm{T}} \int_\Omega 2(1-k) H \boldsymbol{N}_\phi^{\mathrm{T}} \mathrm{d}\Omega = 0 \tag{10.86}$$

Because of admissible arbitrary test functions, Eqs. (10.85) and (10.86), the final system to be solved is given by

$$\underbrace{\int_{\Omega} \boldsymbol{B}_u^{\mathrm{T}} \boldsymbol{D}_e \boldsymbol{B}_u \mathrm{d}\Omega \boldsymbol{d}}_{\boldsymbol{F}_u^{int}=\boldsymbol{K}_u \boldsymbol{d}} + \underbrace{\int_{\Omega} \boldsymbol{N}_u^{\mathrm{T}} \boldsymbol{b} \mathrm{d}\Omega + \int_{\Omega_h} \boldsymbol{N}_u^{\mathrm{T}} \boldsymbol{f} \mathrm{d}S}_{\boldsymbol{F}_u^{ext}} = 0 \tag{10.87}$$

$$\underbrace{-\int_{\Omega} \left\{ \boldsymbol{B}_\phi^{\mathrm{T}} G_c l_0 \boldsymbol{B}_\phi + \boldsymbol{N}_\phi^{\mathrm{T}} \left[\frac{G_c}{l_0} + 2(1-k)H \right] \boldsymbol{N}_\phi \right\} \mathrm{d}\Omega \hat{\boldsymbol{\phi}}}_{\boldsymbol{F}_\phi^{int}=\boldsymbol{K}_\phi \hat{\boldsymbol{\phi}}}$$
$$+ \underbrace{\int_{\Omega} 2(1-k) H \boldsymbol{N}_\phi^{\mathrm{T}} \mathrm{d}\Omega}_{\boldsymbol{F}_\phi^{ext}} = 0 \tag{10.88}$$

where $\boldsymbol{F}_u^{int}$ and $\boldsymbol{F}_u^{ext}$ are the internal and external forces for the displacement field and $\boldsymbol{F}_\phi^{int}$ and $\boldsymbol{F}_\phi^{ext}$ are the internal and external force terms of the phase field, respectively. The stiffness matrices read

$$\boldsymbol{K}_u = \int_{\Omega} \boldsymbol{B}_u^{\mathrm{T}} \boldsymbol{D}_e \boldsymbol{B}_u \mathrm{d}\Omega$$
$$\boldsymbol{K}_\phi = \int_{\Omega} \left\{ \boldsymbol{B}_\phi^{\mathrm{T}} G_c l_0 \boldsymbol{B}_\phi + \boldsymbol{N}_\phi^{\mathrm{T}} \left[\frac{G_c}{l_0} + 2(1-k)H \right] \boldsymbol{N}_\phi \right\} \mathrm{d}\Omega. \tag{10.89}$$

Most of the phase field approaches have been proposed within the framework of the finite elements and hence the shape functions shown above are commonly based on Lagrange polynomials. However, there are also several contributions based on other shape functions including meshfree shape functions or basis functions commonly used in IGA such as NURBS or PHT-splines. While the above mentioned phase field approach is based on two coupled second order partial differential equations, the use of IGA-based or meshfree shape functions does not seem to be advantageous. However, there are higher-order phase field approaches where indeed the higher order continuity of IGA and meshfree shape functions can be exploited. In fact, it has been shown e.g. in [4] that a higher model accuracy is achieved for a fourth-order phase field approach compared to its second-order counterpart. The fourth-order model has been originally proposed by Borden et al. [12]. Based on energy balance, thermodynamically consistent governing equations are derived for the fourth-order phase-field model by using the variational fracture approach. The higher-order model has higher regularity in solving for the exact phase field and improves the convergence rate of the numerical solution. The modified energy functional in the model of Borden et al. [12] is expressed as

$$E(\boldsymbol{u},\phi)=\int_{\Omega}\left((\phi^2+\eta)\Psi_0^+(\boldsymbol{\varepsilon})+\Psi_0^-(\boldsymbol{\varepsilon})\right)\mathrm{d}\Omega \\ +G_c\int_{\Omega}\left[\frac{1}{4\epsilon}(1-\phi)^2+\frac{1}{2}\epsilon|\nabla\phi|^2+\frac{1}{4}\epsilon^3(\Delta\phi)^2\right]\mathrm{d}\Omega \tag{10.90}$$

By replacing Ψ_0^+ by the history field H, the resulting evolution equation of the higher-order phase field model is

$$\frac{4\epsilon\phi H}{G_c}+\phi-2\epsilon^2\Delta\phi+\epsilon^4\Delta(\Delta\phi)=0 \tag{10.91}$$

10.3.4 Solution schemes

There are two approaches to solve the coupling equations: monolithic and staggered schemes. In a monolithic scheme, the displacement and phase field are solved simultaneously while in staggered schemes, the phase field and displacement are solved independently. At a given time step, the displacement is solved first at a fixed phase field. By using the updated displacement, the phase-field equation is solved subsequently.

10.3.4.1 Monolithic scheme

For a monolithic scheme, the implicit Hughes-Hibert-Taylor (HHT) time integration can be adopted in dynamic calculations (see Liu et al. [41]). Other integration methods such as the implicit BDF (backward differentiation formulas) and the generalized-α methods have also been successfully used. As an example, the implicit HHT method is presented herein. Letting i be the Newton iteration, the residual force vectors now are re-written as [41]:

$$(\boldsymbol{R}^u)_i^{t+\Delta t}=(\boldsymbol{F}_u^{ext})_i^{t+\Delta t}-(\boldsymbol{F}_u^{ine})_i^{t+\Delta t}-(1+\alpha)(\boldsymbol{F}_u^{int})_i^{t+\Delta t}+\alpha(\boldsymbol{F}_u^{int})_i^t \tag{10.92}$$

$$(\boldsymbol{R}^\phi)_i^{t+\Delta t}=(\boldsymbol{F}_\phi^{ext})_i^{t+\Delta t}-(1+\alpha)(\boldsymbol{F}_\phi^{int})_i^{t+\Delta t}+\alpha(\boldsymbol{F}_\phi^{int})_i^t \tag{10.93}$$

where $\alpha\in[\frac{1}{3},0]$ is the HHT integration operator and $\Delta t=t_{n+1}-t_n$ is the time step size. If the material damping effect is neglected, the tangent matrix in one element is given by [41]:

$$\boldsymbol{S}_{ij}=\boldsymbol{M}_{ij}\frac{\mathrm{d}\ddot{\boldsymbol{u}}}{\mathrm{d}\boldsymbol{u}}+(1+\alpha)\boldsymbol{K}_{ij} \tag{10.94}$$

$$\frac{\mathrm{d}\ddot{\boldsymbol{u}}}{\mathrm{d}\boldsymbol{u}}=\frac{1}{\beta\Delta t^2},\qquad \beta=\frac{(1-\alpha)^2}{4} \tag{10.95}$$

10.3.4.2 Staggered scheme

In a staggered scheme, both implicit and explicit time schemes can be used. In an implicit approach, the procedure of solving for the displacement field is similar to those in a monolithic scheme. For an explicit method, e.g. the explicit central-difference time-integration method, some modification is needed [41]:

$$(\boldsymbol{u})_i^{t+\Delta t} = (\boldsymbol{u})_i^t + (\dot{\boldsymbol{u}})_i^{t+0.5\Delta t}\Delta t \tag{10.96}$$

$$(\dot{\boldsymbol{u}})_i^{t+0.5\Delta t} = (\dot{\boldsymbol{u}})_i^{t-0.5\Delta t} + (\ddot{\boldsymbol{u}})_i^t\Delta t \tag{10.97}$$

and

$$(\ddot{\boldsymbol{u}})_i = (\boldsymbol{M}_{ij})^{-1}\left((\boldsymbol{F}_u^{ext})_i - (\boldsymbol{F}_u^{int})_i\right) \tag{10.98}$$

10.3.5 Implementations

Phase field models can be implemented into existing finite element codes in a straightforward manner and there are a few contributions of such implementations into commercial software.

10.3.5.1 Abaqus implementation

Liu et al. [41] Msekh et al. [48] presented implementations of isotropic and anisotropic phase field approaches in Abaqus. For the monolithic scheme, a UEL subroutine has been employed. The time integration and element discretization are implemented into "Abaqus standard" and the UEL subroutine is called for each iteration in a given increment. For a staggered scheme, a UMAT or VUMAT subroutine has been used. For an implicit time integration scheme, a UMAT subroutine is employed to exchange the local history field H and the phase field ϕ at integration points. Subsequently, the stress $\boldsymbol{\sigma}$ and its tangent modulus with $\boldsymbol{\varepsilon}$ must be updated in every iteration. For an explicit time integration scheme, a VUMAT subroutine has been developed to exchange the local history field H and the phase field ϕ instead, and only the stress $\boldsymbol{\sigma}$ needs to be updated.

10.3.5.2 COMSOL implementation

In Zhou et al. [79–81], anisotropic phase field models have been implemented in the multi-field simulator-COMSOL Multiphysics. The implementation in COMSOL can be easily extended to coupled (multi-field) problems. The modules established in COMSOL are shown in

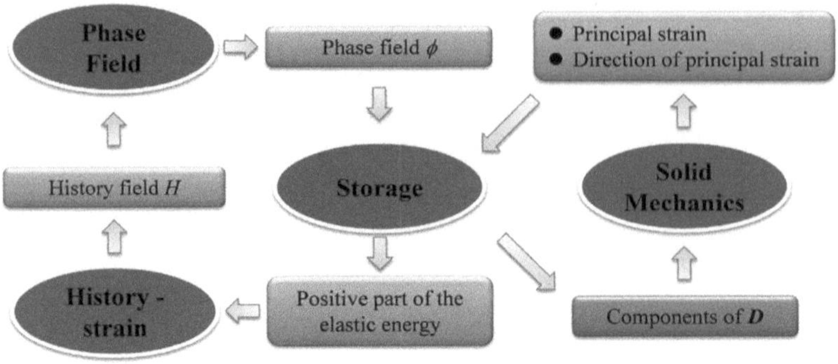

Figure 10.43 Established modules in COMSOL for the phase field modeling.

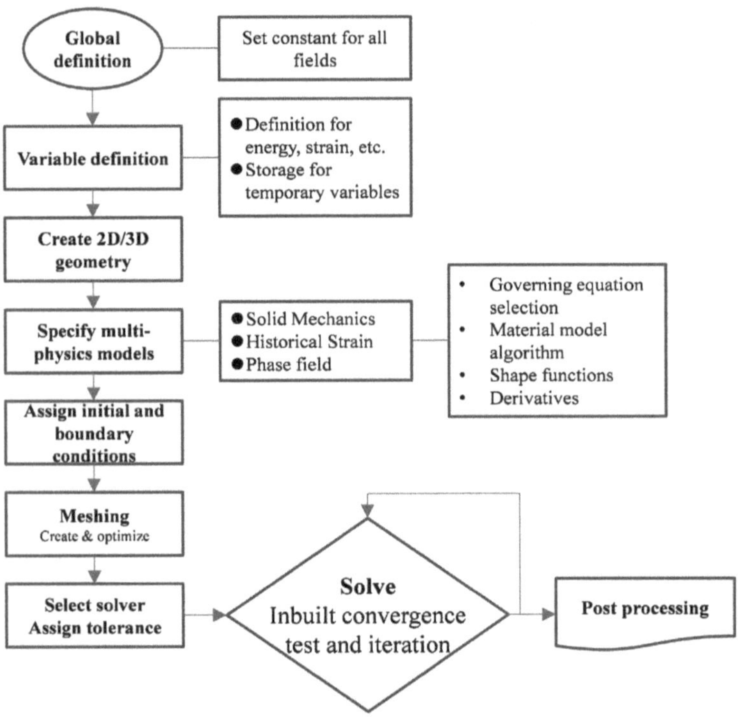

Figure 10.44 Implementation procedure in COMSOL for the phase field modeling [79].

Fig. 10.43 and the implementation procedure in COMSOL is summarized in Fig. 10.44. As shown in Fig. 10.43, the "Storage Module" stores the results obtained from the "Solid Mechanics Module", such as the principal strains and their directions. The positive part of the elastic energy is then calculated to update the local history strain field. Subsequently, the updated history strain is used to solve the phase-field. Due to the high nonlinearity, the elasticity matrix in the "Solid Mechanics Module" must be modified by the updated phase-field solution and varying principal strains and their corresponding directions. Note that only staggered schemes are used in Zhou et al. [79–81]. In addition, the implicit generalized-α method is used for time integration. The Anderson acceleration technology has been exploited increasing the convergence rate. Open source codes of the phase field models in COMSOL can be found at https://sourceforge.net/projects/phasefieldmodelingcomsol/.

References

[1] B. Alali, M. Gunzburger, Peridynamics and material interfaces, Journal of Elasticity 120 (2) (2015) 225–248.

[2] J. Amani, E. Oterkus, P. Areias, G. Zi, T. Nguyen-Thoi, T. Rabczuk, A non-ordinary state-based peridynamics formulation for thermoplastic fracture, International Journal of Impact Engineering 87 (2016) 83–94.

[3] M. Ambati, T. Gerasimov, L. De Lorenzis, A review on phase-field models of brittle fracture and a new fast hybrid formulation, Computational Mechanics 55 (2) (Feb 2015) 383–405.

[4] F. Amiri, D. Millán, M. Arroyo, M. Silani, T. Rabczuk, Fourth order phase-field model for local max-ent approximants applied to crack propagation, in: Phase Field Approaches to Fracture, Computer Methods in Applied Mechanics and Engineering 312 (2016) 254–275.

[5] H. Amor, J.-J. Marigo, C. Maurini, Regularized formulation of the variational brittle fracture with unilateral contact: numerical experiments, Journal of the Mechanics and Physics of Solids 57 (8) (2009) 1209–1229.

[6] Y. Azdoud, F. Han, G. Lubineau, A morphing framework to couple non-local and local anisotropic continua, International Journal of Solids and Structures 50 (9) (2013) 1332–1341.

[7] I. Babuška, J. Melenk, The Partition of the Unity Finite Element Method, Technical Report BN-1185, Institute for Physical Science and Technology, University of Maryland, Maryland, 1995.

[8] R. Batra, M. Ravinsankar, Three-dimensional numerical simulation of the Kalthoff experiment, International Journal of Fracture 105 (2) (2000) 161–186.

[9] T. Belytschko, N. Moës, S. Usui, C. Parimi, Arbitrary discontinuities in finite elements, International Journal for Numerical Methods in Engineering 50 (4) (2001) 993–1013.

[10] T. Belytschko, H. Chen, J. Xu, G. Zi, Dynamic crack propagation based on loss of hyperbolicity and a new discontinuous enrichment, International Journal for Numerical Methods in Engineering 58 (12) (2003) 1873–1905.

[11] F. Bobaru, M. Duangpanya, A peridynamic formulation for transient heat conduction in bodies with evolving discontinuities, Journal of Computational Physics 231 (7) (2012) 2764–2785.

[12] M. Borden, T. Hughes, C. Landis, C. Verhoosel, A higher-order phase-field model for brittle fracture: formulation and analysis within the isogeometric analysis framework, Computer Methods in Applied Mechanics and Engineering 273 (2014) 100–118.

[13] B. Bourdin, G. Francfort, J. Marigo, Numerical experiments in revisited brittle fracture, Journal of the Mechanics and Physics of Solids 48 (4) (2000) 797–826.
[14] B. Bourdin, G. Francfort, J. Marigo, The variational approach to fracture, Journal of Elasticity 91 (2008) 5–148.
[15] Y. Cai, X. Zhuang, C. Augarde, A new partition of unity finite element free from linear dependence problem and processing delta property, Computer Methods in Applied Mechanics and Engineering 199 (2010) 1036–1043.
[16] W. Cao, B. Su, Geometric nonlinear manifold method and its application for analysing discontinuous deformation, China Mining Magazine 5 (2000).
[17] K. Cheng, T. Fries, Higher-order XFEM for curved strong and weak discontinuities, International Journal for Numerical Methods in Engineering 82 (5) (2010) 564–590, https://doi.org/10.1002/nme.2768.
[18] Y.-J. Chiou, Y.-M. Lee, R.-J. Tsay, Mixed mode fracture propagation by manifold method, International Journal of Fracture 114 (4) (2002) 327–347.
[19] S.R. Chowdhury, P. Roy, D. Roy, J. Reddy, A peridynamic theory for linear elastic shells, International Journal of Solids and Structures 84 (2016) 110–132.
[20] H. Fan, H. Zheng, S. He, S–R decomposition based numerical manifold method, Computer Methods in Applied Mechanics and Engineering 304 (2016) 452–478.
[21] J. Foster, S. Silling, W. Chen, Viscoplasticity using peridynamics, International Journal for Numerical Methods in Engineering 81 (10) (2010) 1242–1258.
[22] G. Francfort, J. Marigo, Revisiting brittle fracture as an energy minimization problem, Journal of the Mechanics and Physics of Solids 46 (8) (1998) 1319–1342.
[23] W. Gerstle, N. Sau, S. Silling, Peridynamic modeling of concrete structures, Nuclear Engineering and Design 237 (12) (2007) 1250–1258.
[24] K. Graff, Wave Motion in Elastic Solids, Clarendon, Oxford, 1975.
[25] A. Gravouil, N. Moes, T. Belytschko, Non-planar 3D crack growth by the extended finite element and level sets – part II: level set update, International Journal for Numerical Methods in Engineering 53 (2002) 2569–2586.
[26] S. Guang, G. Xiurun, Z. Guang, Numerical analysis of crack propagation under compression, Chinese Journal of Rock Mechanics and Engineering 6 (1999).
[27] Y. Ha, F. Bobaru, Studies of dynamic crack propagation and crack branching with peridynamics, International Journal of Fracture 162 (2010) 229–244.
[28] S. Haidong, Study on numerical manifold method with fixed meshes, Chinese Journal of Theoretical and Applied Mechanics 43 (1) (2011) 169.
[29] F. Han, G. Lubineau, Y. Azdoud, Adaptive coupling between damage mechanics and peridynamics: a route for objective simulation of material degradation up to complete failure, Journal of the Mechanics and Physics of Solids 94 (2016) 453–472.
[30] R. Jabakhanji, R.H. Mohtar, A peridynamic model of flow in porous media, Advances in Water Resources 78 (2015) 22–35.
[31] J. Kalthoff, S. Winkler, Failure mode transition at high rates of shear loading, in: Impact Loading and Dynamic Behavior of Materials, vol. 1, DGM Informationsgesellschaft mbH, 1988, pp. 185–195.
[32] J.F. Kalthoff, Failure methodology of mode-II loaded cracks, Strength, Fracture and Complexity 1 (3) (2003) 121–138.
[33] A. Karma, D. Kessler, H. Levine, Phase-field model of mode III dynamic fracture, Physical Review Letters 87 (2001) 045501.
[34] A. Katiyar, J.T. Foster, H. Ouchi, M.M. Sharma, A peridynamic formulation of pressure driven convective fluid transport in porous media, Journal of Computational Physics 261 (2014) 209–229.

[35] C. Kuhn, R. Müller, A continuum phase field model for fracture, Engineering Fracture Mechanics 77 (18) (2010) 3625–3634.
[36] C. Kuhn, R. Müller, A new finite element technique for a phase field model of brittle fracture, Journal of Theoretical and Applied Mechanics 49 (4) (2011) 1115–1133.
[37] C. Kuhn, A. Schlüter, R. Müller, On degradation functions in phase field fracture models, Computational Materials Science 108 (2015) 374–384.
[38] X. Lai, B. Ren, H. Fan, S. Li, C. Wu, R.A. Regueiro, L. Liu, Peridynamics simulations of geomaterial fragmentation by impulse loads, International Journal for Numerical and Analytical Methods in Geomechanics 39 (12) (2015) 1304–1330.
[39] G. Lancioni, G. Royer-Carfagni, The variational approach to fracture mechanics. A practical application to the French Panthéon in Paris, Journal of Elasticity 95 (1–2) (2009) 1–30.
[40] S. Li, W.K. Liu, A.J. Rosakis, T. Belytschko, W. Hao, Mesh-free Galerkin simulations of dynamic shear band propagation and failure mode transition, International Journal of Solids and Structures 39 (5) (2002) 1213–1240.
[41] G. Liu, Q. Li, M.A. Msekh, Z. Zuo, Abaqus implementation of monolithic and staggered schemes for quasi-static and dynamic fracture phase-field model, Computational Materials Science 121 (2016) 35–47.
[42] G. Lubineau, Y. Azdoud, F. Han, C. Rey, A. Askari, A morphing strategy to couple non-local to local continuum mechanics, Journal of the Mechanics and Physics of Solids 60 (6) (2012) 1088–1102.
[43] S. Luo, X. Zhang, Y. Cai, The variational principles and application of nonlinear numerical manifold method, Applied Mathematics and Mechanics 21 (12) (2000) 1401–1406.
[44] E. Madenci, E. Oterkus, Peridynamic Theory and Its Applications, Springer, 2014.
[45] C. Miehe, M. Hofacker, F. Welschinger, A phase field model for rate-independent crack propagation: robust algorithmic implementation based on operator splits, Computer Methods in Applied Mechanics and Engineering 199 (2010) 2765–2778.
[46] C. Miehe, F. Welschinger, M. Hofacker, Thermodynamically consistent phase-field models of fracture: variational principles and multi-field FE implementations, International Journal for Numerical Methods in Engineering 83 (10) (2010) 1273–1311.
[47] E. Moyer, M. Miraglia, Peridynamic solutions for Timoshenko beams, Engineering 6 (2014) 46262.
[48] M.A. Msekh, J.M. Sargado, M. Jamshidian, P.M. Areias, T. Rabczuk, Abaqus implementation of phase-field model for brittle fracture, in: Special Issue Polymeric Composites, Computational Materials Science 96 (2015) 472–484.
[49] Y. Ning, X. An, G. Ma, Footwall slope stability analysis with the numerical manifold method, International Journal of Rock Mechanics and Mining Sciences 48 (6) (2011) 964–975.
[50] J. O'Grady, J. Foster, Peridynamic plates and flat shells: a non-ordinary, state-based model, International Journal of Solids and Structures 51 (2014) 4572–4579.
[51] E. Oterkus, E. Madenci, O. Weckner, S. Silling, P. Bogert, A. Tessler, Combined finite element and peridynamic analyses for predicting failure in a stiffened composite curved panel with a central slot, Composite Structures 94 (2012) 839–850.
[52] S. Oterkus, E. Madenci, A. Agwai, Fully coupled peridynamic thermomechanics, Journal of the Mechanics and Physics of Solids 64 (2014) 1–23.
[53] M.L. Parks, P. Seleson, S.J. Plimpton, S. Silling, R.B. Lehoucq, Peridynamics With Lammps: a User Guide v0.3 beta, Sandia report, Sandia National Laboratories, 2011.

[54] T. Rabczuk, T. Belytschko, A three dimensional large deformation meshfree method for arbitrary evolving cracks, Computer Methods in Applied Mechanics and Engineering 196 (29–30) (2007) 2777–2799.
[55] T. Rabczuk, G. Zi, S. Bordas, On three-dimensional modelling of crack growth using partition of unity methods, Computers & Structures 88 (2010) 1391–1411.
[56] S. Rajendran, B. Zhang, A "FE-meshfree" QUAD4 element based on partition of unity, Computer Methods in Applied Mechanics and Engineering 197 (2007) 128–147.
[57] H. Ren, X. Zhuang, Y. Cai, T. Rabczuk, Dual-horizon peridynamics, International Journal for Numerical Methods in Engineering 108 (12) (2016) 1451–1476.
[58] H. Ren, X. Zhuang, T. Rabczuk, Dual-horizon peridynamics: a stable solution to varying horizons, Computer Methods in Applied Mechanics and Engineering 318 (2017) 762–782.
[59] P. Seleson, S. Beneddine, S. Prudhomme, A force-based coupling scheme for peridynamics and classical elasticity, Computational Materials Science 66 (2013) 34–49.
[60] P. Seleson, Y.D. Ha, S. Beneddine, Concurrent coupling of bond-based peridynamics and the Navier equation of classical elasticity by blending, International Journal of Multiscale Computational Engineering 13 (2) (2015).
[61] G.-H. Shi, Manifold Method of Material Analysis, Technical report, Army Research Office Research Triangle Park NC, 1992.
[62] W. Shuilin, G. Xiurun, Application of manifold method in simulating crack propagation, Chinese Journal of Rock Mechanics and Engineering 5 (1997).
[63] S. Silling, Reformulation of elasticity theory for discontinuities and long-range forces, Journal of the Mechanics and Physics of Solids 48 (1) (2000) 175–209.
[64] S. Silling, Dynamic fracture modeling with a meshfree peridynamic code, in: Computational Fluid and Solid Mechanics, 2003, pp. 641–644.
[65] S. Silling, E. Askari, A meshfree method based on the peridynamic model of solid mechanics, Computers & Structures 83 (17) (2005) 1526–1535.
[66] S. Silling, D. Littlewood, P. Seleson, Variable Horizon in a Peridynamic Medium, Technical report, Sandia National Laboratories (SNL-NM), Albuquerque, NM (United States), 2014.
[67] S.A. Silling, M. Epton, O. Weckner, J. Xu, E. Askari, Peridynamic states and constitutive modeling, Journal of Elasticity 88 (2) (2007) 151–184.
[68] J.-H. Song, P.M. Areias, T. Belytschko, A method for dynamic crack and shear band propagation with phantom nodes, International Journal for Numerical Methods in Engineering 67 (6) (2006) 868–893.
[69] H. Su, X. Xie, Preliminary research on solving large displacement problems by numerical manifold method with fixed mathematical meshes, in: M.M. MacLaughlin, N. Sitar (Eds.), Proceedings of ICADD-7 Seventh International Conference on the Analysis of Discontinuous Deformation, Honolulu, Hawaii, US, 12 2005, pp. 175–182.
[70] K. Terada, M. Asai, Finite cover method for physically and geometrically nonlinear problems, in: Meshfree Methods for Partial Differential Equations II, 2005, pp. 169–190.
[71] K. Terada, M. Asai, M. Yamagishi, Finite cover method for linear and non-linear analyses of heterogeneous solids, International Journal for Numerical Methods in Engineering 58 (9) (2003) 1321–1346.
[72] R.-J. Tsay, Y.-J. Chiou, W.-L. Chuang, Crack growth prediction by manifold method, Journal of Engineering Mechanics 125 (8) (1999) 884–890.

[73] W. Wei, Q. Jiang, A modified numerical manifold method for simulation of finite deformation problem, Applied Mathematical Modelling 48 (2017) 673–687.
[74] M. Williams, On the stress distribution at the base of a stationary crack, Journal of Applied Mechanics 24 (1) (1957) 109–114.
[75] X.-P. Xu, A. Needleman, Numerical simulations of fast crack growth in brittle solids, Journal of the Mechanics and Physics of Solids 42 (9) (1994) 1397–1434.
[76] G. Zhang, Y. Sugiura, H. Hasegawa, K. Saito, Propagation of cracks in dam and rock foundation by manifold method, Journal of Japan Society of Dam Engineers 11 (2) (2001) 95–105.
[77] H. Zheng, D. Xu, New strategies for some issues of numerical manifold method in simulation of crack propagation, International Journal for Numerical Methods in Engineering 97 (13) (2014) 986–1010.
[78] H. Zheng, F. Liu, C. Li, Primal mixed solution to unconfined seepage flow in porous media with numerical manifold method, Applied Mathematical Modelling 39 (2) (2015) 794–808.
[79] S. Zhou, T. Rabczuk, X. Zhuang, Phase field modeling of quasi-static and dynamic crack propagation: COMSOL implementation and case studies, Advances in Engineering Software 122 (2018) 31–49.
[80] S. Zhou, X. Zhuang, T. Rabczuk, A phase-field modeling approach of fracture propagation in poroelastic media, Engineering Geology 240 (2018) 189–203.
[81] S. Zhou, X. Zhuang, H. Zhu, T. Rabczuk, Phase field modelling of crack propagation, branching and coalescence in rocks, Theoretical and Applied Fracture Mechanics 96 (2018) 174–192.

CHAPTER ELEVEN

Implementation details

11.1. Computer implementation of enriched methods

It is well known that the standard finite element method based on Lagrange polynomials have been developed for more than 50 years and have a rich history of both theoretical and practical improvements. By contrast, the enriched or partition-of-unity based methods, particularly non-standard basis functions are much newer and although have been intensively studied in recent years they have not reached the same level of acceptance and usage in computational mechanics. However, we mention that XFEM in particular has been natively implemented in some commercial software such as Abaqus and ANSYS. Abaqus also has built-in modules for the discrete element method (DEM) and of smoothed-particle hydrodynamics (SPH). It also allows the implementation of user-defined elements, which has been used to develop plug-ins for various other methods, including isogeometric analysis (IGA) [17]. IGA has also been subject to intensive research and it has been incorporated in some commercially available software such as LS-DYNA [3,4], FEAP [27], and Altair Radioss [20]. Among open-source software, we mention OOFEM [23] for XFEM and IGA, and various research codes such as OpenXFEM++ [6] and ElemFreGalerkin [19] for XFEM and meshfree methods, as well as GeoPDEs [29], IGAFEM [18], PetIGA [10], igatools [21], and IGAPack [2] for isogeometric analysis implementations.

While established commercial and community-based projects have the advantage of better reliability, robustness and user support, they are not always well suited for learning the underlying methods. This is particularly true for large open-source and commercial projects which already contain many thousands of lines of code since one has to understand the already existing structure. Even for modular or object-oriented software that allows the development of customized user elements, extending the existing code to new capabilities is not always straightforward. From this point of view, shorter implementations written in a high-level language such as MATLAB® can be better suited for the understanding the basics of the implemented method and for extending it to different applications.

Regardless of the details of the method and application considered, a finite element program typically consists of three stages: pre-processing,

https://doi.org/10.1016/B978-0-12-814106-9.00017-6
Copyright © 2020 Elsevier Inc.
All rights reserved.

processing, and post-processing. In the pre-processing stage, the problem data such domain geometry, material properties, boundary conditions and other model parameters are initialized. In a mesh-based discretization, the mesh that describes the domain geometry is generated along with the node-element connectivities. For mesh-free methods, only a set of nodes (particles) is generated inside the domain, however some mesh-free methods use a background mesh for integration. In the case of strong or weak discontinuities which are tracked by enriched methods, typically a level set representation of the discontinuity must be generated as detailed in Section 1.4. In the processing stage of the program, the stiffness matrices and load vectors are assembled, the boundary conditions are imposed and a linear system is solved by either direct or iterative methods. In the case of nonlinear finite elements, typically a Newton-Raphson iteration is employed to obtain the solution. Finally, in the post-processing stage the quantities of interest such as displacements, strains, stresses are computed and plotted based on the solution vector. In the case of crack-propagation studies, the stress intensity factors are usually computed. Other post-processing tasks may involve the computation of error estimators or indicators (as detailed in Section 6.5 for hierarchical splines) and marking of elements or domain regions for refinement.

In this section, we will give a general overview of the three stages mentioned above, highlighting some of differences and commonalities between the different methods considered. Later in the chapter, more in-depth details for XFEM, enriched meshfree and XIGA with numerical examples will be given. We will focus on the implementation details for the MATLAB/Octave language, although the same ideas can be implemented in other programming languages as well.

11.1.1 Pre-processing

The most important part of this stage of a finite element program (and often the most time-consuming) is the mesh generation. Fortunately, mesh generators have improved quite significantly over the years and for most standard geometries the meshing tasks can be automated to a great extent. Still, manual intervention can be required which makes the meshing process very time-intensive in particular for problems involving moving boundaries or design optimization and has motivated the development of alternative discretizations as detailed in this book.

Among the meshing methods, we differentiate between simplex meshes (which use triangles in 2D and tetrahedra in 3D) and quadrilateral/hexahedral meshes. The former are easier to generate for arbitrary geometries and fairly robust implementations exist even in open-source programs such as Gmsh [14] or Tetgen [24]. In MATLAB, simplex meshes in 2D and 3D can be generated with the `generateMesh` command, which is part of the PDE toolbox. A drawback of triangular and tetrahedral meshes is that the linear elements tend to be inaccurate (too stiff, particularly in bending-dominated problems). Therefore, it is recommended to use quadratic or higher order elements, although this results in a significant increase in the number of degrees of freedom. Alternatively, hexahedral meshes can be used although they are typically more difficult to generate and the results can be quite sensitive to element distortion.

In the case of meshfree methods, the nodal points can be generated from some structured grid, using some randomized algorithm such as latin-hypercube sampling (LHS), or even by considering the element vertices resulting from a mesh-based approach as in [28]. As stated before, the benefits of meshfree approaches include more flexibility in generating the nodal points and insensitivity to mesh distortion. However, there are some computational costs associated with calculating the neighbor lists and numerical integration which may require a background mesh and an increased number of quadrature points.

When considering the standard (tensor product) isogeometric analysis, the mesh requirements are in some ways similar to those for structured of quadrilateral and hexahedral meshes. However, the knot-vector and control-point structures which are typically used in IGA to define the domain geometry are different from those used in finite elements. A bridge that can allow easier integration of IGA concepts into existing finite element codes has been realized through the concept of Bézier extraction [7] as described in Section 6.1.2. This concept is also particularly useful in the case of hierarchical or non-tensor product discretizations, such as the PHT splines. We mention that there has also been interest the integration of unstructured mesh discretizations based on Bézier triangles [13], tetrahedra [31] and quadrilaterals [5,16].

For isogeometric analysis, a useful tool for generating parametrized geometries is the Octave NURBS toolbox [11,25], which is also compatible with MATLAB. An example for generating the plate with a hole geometry is given in Listing 11.1. In this file, we generate the model for a given radius of the hole R and a plate size L by creating the boundary segments, and parametrizing the interior with the Coon patch method [9] via the `nrbcoons` command. Degree elevation (p-refinement) to quadratic

NURBS and knot-insertion (h-refinement) are handled by the `nrbdegelev` and `nrbkntins` commands respectively. We refer to the documentation of the NURBS toolbox for more details on these commands. The plots showing the control points and the elements for a mesh with 10×10 knot-spans are given in Fig. 11.1.

```
%script plotPlateWithHole.m
%generates the geometry for a quarter plate with a hole
%uses the NURBS toolbox ( https://octave.sourceforge.io/nurbs/index.html )
addpath('./nurbs-1.3.13/inst')

R = 1; %radius of the hole
L = 4; %size of the modeled plate

%create the curved boundary and the other four linear boundary segments
arc = nrbcirc(R, [0, 0], pi/2, pi);
sideBottom = nrbline([-R, 0], [-L, 0]);
sideRight = nrbline([0, R], [0, L]);
pointUpRight = [0; L; 0];
pointUpLeft = [-L; L; 0];
pointDownLeft = [-L; 0; 0];
knotUpLeft = [0, 0, 1/2, 1, 1];
sideUpLeft = nrbmak([pointUpRight, pointUpLeft, pointDownLeft], knotUpLeft);

%degree elevate the linear side and create a 2-element volumetric model
sideBottom = nrbdegelev(sideBottom, 1);
plateModel = nrbcoons(arc, sideUpLeft, sideRight, sideBottom);

%perform knot insertions to refine the model
newKnotsU = [0.1:0.1:0.4, 0.6:0.1:0.9];
newKnotsV = [0.1:0.1:0.9];
plateModel = nrbkntins(plateModel, {newKnotsU, newKnotsV});

%plot the model and the control points
figure
subplot(1,2,1), nrbctrlplot(plateModel)
view(2)
subplot(1,2,2), nrbkntplot(plateModel)
view(2)
```

Listing 11.1: Matlab code for generating the plate with a hole model.

In the framework of enriched methods, additional structures are needed to represent the sets of enriched nodes and the type of enrichment. A unified way to implement this concept is to define an array corresponding to each node that determines its set of enrichment functions. For the functions that are not enriched, this array will consist of just the unity function $\{\xi_1(\mathbf{x}) = 1\}$. For other functions, for example the Heaviside-enriched basis, the array will also include corresponding enrichments, e.g. $\{\xi_1(\mathbf{x}) = 1,\ \xi_2(\mathbf{x}) = H(\mathbf{x})\}$. Then the derivatives of the basis functions of the form $\phi_i \xi_{ji}$ can be computed by the product rule from the derivatives of ϕ_i, $i = 1, \ldots, N$, and ξ_{ji}, $j = 1, \ldots, n_i$ where N is the number of (unenriched)

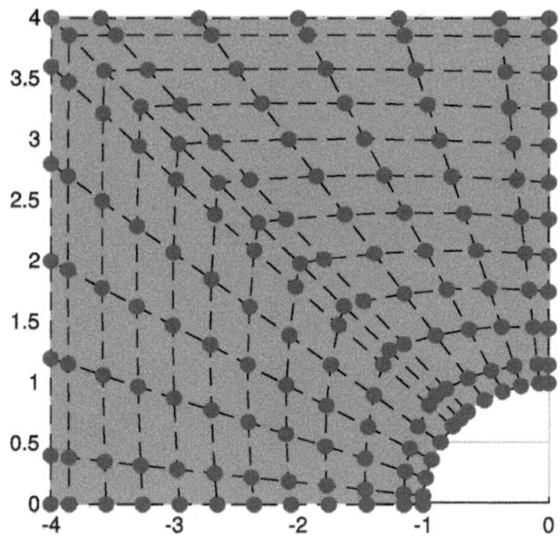

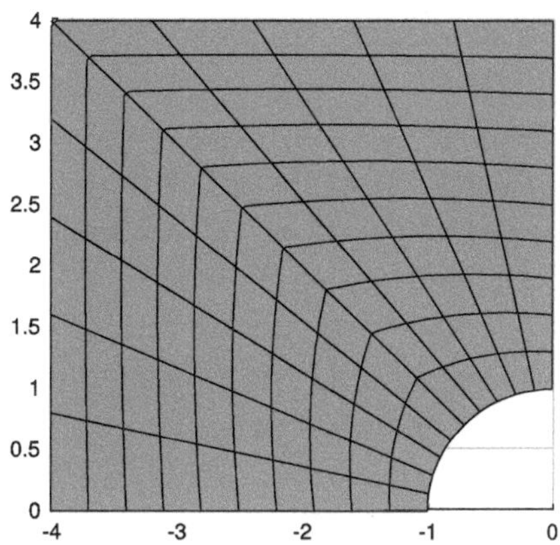

Figure 11.1 Plate with a hole model generated by Listing 11.1, showing the control points on the left and the refined mesh on the right.

global shape functions and n_i is the number of enrichments (including the unity) for the ith basis function.

Finally, in the case level-sets are used to track the strong or weak discontinuities, a discretization for the level-sets is needed, usually in terms of the global basis functions. We refer to the included codes and Sections 1.4 and 8.3 for additional implementation details.

11.1.2 Processing

The processing stage of a finite element program consists of constructing and solving a linear system which is obtained by assembling the stiffness matrices and load vector and by imposing the boundary conditions. The assembly stage is typically conducted as an outer loop over the elements (or integration cells in the case of meshfree methods) and an inner loop over the quadrature Gauss points. In the case of a MATLAB implementation, it is preferable that the values of all the shape functions with support at a particular Gauss point are output as a vector by the basis function subroutine. Then the process of assembling the local stiffness matrices can be vectorized, resulting in a significant performance gain.

Another MATLAB-specific speedup is the use of the so-called "triplet arrays" for the assembly of the sparse-format stiffness matrices. The idea is to replace matrix update statements of the form `K(rowIndex,colIndex)=K(rowIndex,colIndex)+localK` with statements like the ones in Listing 11.2. The reason is that the sparse matrices are stored in the compressed sparse column (CSC) format [15], for which adding additional non-zero entries requires expensive resorting operations. Therefore, it is much more efficient to store the non-zero entries in triplet arrays and build the sparse matrix at the end. A drawback of this procedure is that the triplet arrays consume more memory than the sparse matrix, particularly when there are many repeated row and column indices whose corresponding value needs to be summed. For large matrices (with more than 100,000 degrees of freedom) or when there is a lot of overlap between the basis functions, it may be required to break up the assembly into subdomains and sum up the assembled stiffness matrices. This procedure can also be helpful in the parallelization of the assembly subroutine.

```
iCounter = 0
%Allocate space for the arrays rowSet, colSet, valSet
%...
for indexElem = 1:numElements
   %set colIndex and rowIndex from the connectivity arrays and
   %assemble the local stiffness matrix by looping over Gauss points
   %...
   numEnt = length(rowIndex);
   rowSet(iCounter+1:iCounter+numEnt^2) = repmat(rowIndex,1,numEnt);
   colIndexSet = repmat(colIndex,numEnt,1);
   colSet(iCounter+1:iCounter+numEnt^2) = reshape(colIndexSet,1,numEnt^2);
   valSet(iCounter+1:iCounter+numEnt^2) = reshape(localK,1,nument^2);
   iCounter = iCounter + numEnt^2;
end
K = sparse(rowSet,colSet,valSet,dimK,dimK);
```

Listing 11.2: Snippet of Matlab code for triplet matrix assembly.

For the imposition of the boundary conditions, we distinguish between Dirichlet (e.g. prescribed displacement) and Neumann (e.g. prescribed traction) boundary conditions. The Dirichlet boundary conditions are imposed by changing the stiffness matrix and right-hand side vector, while for Neumann boundary conditions only the right-hand side vector is changed. In the case of finite element method with Lagrange polynomials, a common procedure is to the zero-out the rows of the stiffness matrix corresponding to the degrees of freedom on the Dirichlet boundaries, set the diagonal entries in these rows to one and set the corresponding entries in the right-hand side vector to the prescribed value. This has the drawback that the resulting stiffness matrix is no longer symmetric, which makes the process of solving the linear system less efficient. Symmetry can be restored by condensing out the prescribed degrees of freedom. To give a simple example, suppose we have a matrix with 4 degrees of freedom, of which the second one is given by $x_2 = \bar{u}$, with $\bar{u}$ known. The initial linear symmetric system can be written as:

$$\begin{bmatrix} k_{11} & k_{12} & k_{13} & k_{14} \\ k_{12} & k_{22} & k_{23} & k_{24} \\ k_{13} & k_{23} & k_{33} & k_{34} \\ k_{14} & k_{24} & k_{34} & k_{44} \end{bmatrix} \begin{bmatrix} x_1 \\ x_2 \\ x_3 \\ x_4 \end{bmatrix} = \begin{bmatrix} f_1 \\ f_2 \\ f_3 \\ f_4 \end{bmatrix} \tag{11.1}$$

After setting $x_2 = \bar{u}$, we now have:

$$\begin{bmatrix} k_{11} & k_{12} & k_{13} & k_{14} \\ 0 & 1 & 0 & 0 \\ k_{13} & k_{23} & k_{33} & k_{34} \\ k_{14} & k_{24} & k_{34} & k_{44} \end{bmatrix} \begin{bmatrix} x_1 \\ x_2 \\ x_3 \\ x_4 \end{bmatrix} = \begin{bmatrix} f_1 \\ \bar{u} \\ f_3 \\ f_4 \end{bmatrix} \tag{11.2}$$

To restore the symmetry, we substitute $x_2 = \bar{u}$ in the equations corresponding to the rows 1, 3 and 4 and move the known quantities to the right hand side:

$$\begin{bmatrix} k_{11} & 0 & k_{13} & k_{14} \\ 0 & 1 & 0 & 0 \\ k_{13} & 0 & k_{33} & k_{34} \\ k_{14} & 0 & k_{34} & k_{44} \end{bmatrix} \begin{bmatrix} x_1 \\ x_2 \\ x_3 \\ x_4 \end{bmatrix} = \begin{bmatrix} f_1 - k_{12}\bar{u} \\ \bar{u} \\ f_3 - k_{23}\bar{u} \\ f_4 - k_{24}\bar{u} \end{bmatrix} \tag{11.3}$$

This procedure can naturally be applied for the case of multiple prescribed degrees of freedom. A MATLAB implementation of this algorithm is given in Listing 11.3. In this code, additionally the diagonal entries on the stiffness matrix corresponding to the Dirichlet degrees of freedom and the entries in the load vector are multiplied by the mean value of the diagonal of the original matrix to improve conditioning.

We note that in the case of IGA or mesh methods, the shape functions on the boundary do not, in general, satisfy the Kronecker-Delta property. However, the B-Spline and NURBS functions, as well as some meshfree basis such as the local maximum entropy (LME) basis functions on convex domains or generalized meshfree (GMF) approximations [30] do satisfy a weak Kronecker-delta property. In this case, the shape functions corresponding to the interior nodes (or control points) do not have support on the boundary, however the value of the basis functions on the boundary may not reach 1 at the corresponding nodes as in the strong Kronecker-delta case. This requires an interpolation or projection of the boundary date on the approximation space spanned by the boundary basis functions. In the case where interior shape functions also have support on the bound-

```
function [kk, ff]=applyDirichletSym(kk, ff, bcdof, bcval)
% Apply the Dirichlet constraints to matrix equation [kk]{x}={ff}
% while outputting a symmetric matrix kk
% Input:
%     kk - system matrix before applying constraints
%     ff - system vector before applying constraints
%     bcdof - a vector containging constrained d.o.f
%     bcval - a vector containing constrained value
% Output:
%     kk - the new system matrix after applying the constraints
%     ff - the new load vector after applying the constraints

bcwt=mean(diag(kk)); % a measure of the average size of an element in K
ff=ff-kk(:,bcdof)*bcval'; % modify the force vector
ff(bcdof) = bcwt*bcval;
kk(bcdof,:)=0; % zero out the rows and columns of the K matrix
kk(:,bcdof) = 0;
kk(bcdof,bcdof)=bcwt*speye(length(bcdof)); % put ones*bcwt on the diagonal
```

Listing 11.3: Matlab code for imposing Dirichlet boundary conditions.

ary (such as in the case of MLS or RKP meshfree methods), then other techniques such as Lagrange multipliers are required.

After the linear system is changed to take into account the boundary conditions, the solution vector can be obtained by either a direct or iterative solver. Direct solvers (such as the MATLAB backslash "\" operator) tend to be more robust for ill-conditioned linear systems, but the memory requirements can become very high, particularly for 3D problems with a lot of degrees of freedom. The reason is due to the "fill-in" which occurs when the zero entries within the bandwidth of a matrix become non-zero when the matrix is factorized. Iterative (Krylov subspace) solvers rely on matrix vector multiplications and can be used to solve very large linear systems efficiently.

Examples of iterative solvers are pre-conditioned conjugate gradient (`pcg` command in MATLAB) which is efficient for symmetric positive definite matrices or the generalized minimal residual method (GMRES, `gmres` command in MATLAB) for general matrices. These solvers work better when a pre-conditioner is applied to improve the convergence. Standard preconditioners are incomplete Cholesky factorizations (`ichol` command in MATLAB) for symmetric positive definite matrices and the incomplete LU factorization (`ilu` command in MATLAB) for general sparse matrices. The idea of incomplete factorization is that only the non-zero entries in the original matrix can be non-zero in the factorization, the other entries are set to zero. Other preconditioners, such as Jacobi preconditioners which perform a diagonal scaling of the matrices, or multi-grid preconditioners can be used as well.

The solution of linear systems also appears in nonlinear analysis, where usually Newton-Raphson iterations are performed to solve a linearized model. In this case, on the left-hand side we have the "tangent-stiffness" matrix, as described e.g. in Section 3.6.1.3 for hydro-mechanical coupled problems, and on the left-hand side we have a residual vector. Solving linear systems is also required for time-dependent problems with implicit time-integration (where a system involving the stiffness matrix is solved) or with explicit time-integration where the consistent mass matrix is used. Therefore, for large simulations, the use of efficient solvers or of accurate lumped mass methods is of great importance in keeping the computational times within acceptable limits.

11.1.3 Post-processing

In the post-processing stage, various quantities of interest can be computed and displayed based on the solution vector. A correct and detailed visualization of the output date is very important, particularly for identifying any errors in the model such as incorrect imposition of boundary conditions. In the case of XFEM or basis functions which do not satisfy the (strong) Kronecker-delta property, the displacements cannot be simply inferred from the values of the solution vector. Instead, the basis functions (including the enrichments) need to be evaluated at the chosen points (e.g. element vertices) and the values multiplied by the corresponding coefficients of the solution. The strain and stress values are calculated in the same way, by considering the derivatives of the basis functions and of the enrichments. For meshless methods, typically a *visualization mesh* needs to be constructed for example from a Delaunay triangulation of the domain. Moreover, to precisely show the discontinuities in a fracture simulation, the visualization mesh should be conforming to the crack path.

For many examples, we use output files in the VTK (*Visualization Toolkit*) format. These files can be opened with open-source viewers such as Paraview, and are particularly convenient for 3D visualization as one can easily manipulate the graphs and interactively plot the deformations, stresses and other quantities of interest on structured or unstructured meshes. High-level output subroutines such as `vtkwrite`[1] are available to make the output of scalar or vector quantities relatively straightforward.

[1] https://github.com/joe-of-all-trades/vtkwrite.

For fracture propagation problems, a crack increment is computed according to some failure as well as propagation magnitude and angle criteria. In XFEM (and other enriched methods), stress intensity factors are often used for the computation of these criteria. They can be calculated in 2D by the interaction integral methods [26], which are derived from the J-integral [22], as discussed in Sections 5.9.1.4 and 7.2.2.5. This method considers a contour around the crack tip and is converted to a domain integral for improved integration accuracy. Since the resulting code is quite lengthy, we refer to the included examples and the documentation for more details.

Finally, even though the idea of enriched methods is that the mesh is independent of the crack, often times the results can be improved by considering a finer mesh around the crack tips. In the framework of hierarchical splines, this can be conveniently implemented with a recovery-type error estimator, as detailed in [2]. This type of estimator is quite effective at refining in regions with high stress concentrations.

Another aspect to be considered is the selection of a marking scheme and the amount of elements being marked at each refinement step. When the mesh is coarse, the error can be quite large even relatively far from the regions with non-smooth solution such as the crack tip. As the mesh is refined and the singular regions are better resolved, the areas with large error can shrink considerably. Therefore, it is more economical in terms of degrees of freedom to refine as few elements as possible at each step, however this leads to more refinement steps which increases the computational cost. A strategy that works well and has been implemented in [2] is the Dörfler marking scheme [12], which refines elements are in the 100θ percentile of the largest error. For problems with singularity, we typically choose $\theta = 0.5$, while for problems with smoother solutions (but which can still benefit from adaptivity) $\theta = 0.75$ is more efficient.

11.2. Numerical examples

In this section we provide a few numerical examples of benchmark problems in XFEM, including that of a coupled hydro-mechanical model. Further examples and documentation, including 3D, plate and shell models can be found in the open-source repository IGA(X)FEM at https://sourceforge.net/projects/cmcodes.

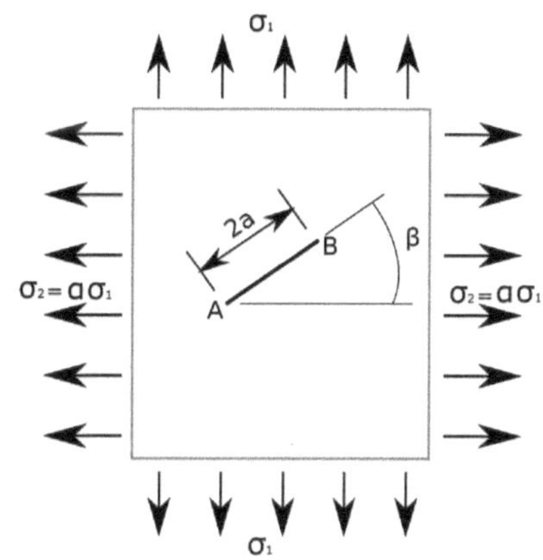

Figure 11.2 Schematic illustration of mechanical loading.

Table 11.1 Material parameters for model under pure mechanical loading.

Young's modulus	$E = 25.85$ GPa
Poisson's ratio	$\nu = 0.18$
uniform stress field	$\sigma_1 = 1.0e7$ Pa
crack length	$2a = 2.15$ m

11.2.1 Crack propagation angle

This section presents a square domain with an inclined center crack under pure mechanical loading (Fig. 11.2). Hereafter, the initial propagation angles of the induced crack with different pre-defined angle β and loading factor α are studied. The material parameters are list in Table 11.1. Fig. 11.3 shows the stress profiles of domains with pre-defined angle β ranging from 10 to 80 degree and loading factor α is set to 0. The comparison between numerical calculation and theoretical model [1] in terms of the propagation angle of the induced crack with different loading ratio α is shown in Fig. 11.4A. Furthermore, the convergence study (case $\beta = 10$) of stiffness intensity factor K_1 of mode I respect to total number of elements is presented in Fig. 11.4B.

11.2.2 Hydro-mechanical model with center cracks

This section presents the domain with inner fractures under fluid flux loading. The geometry and boundary conditions are shown in Fig. 11.5. The material parameters are listed in Table 11.2. Domain with single and five

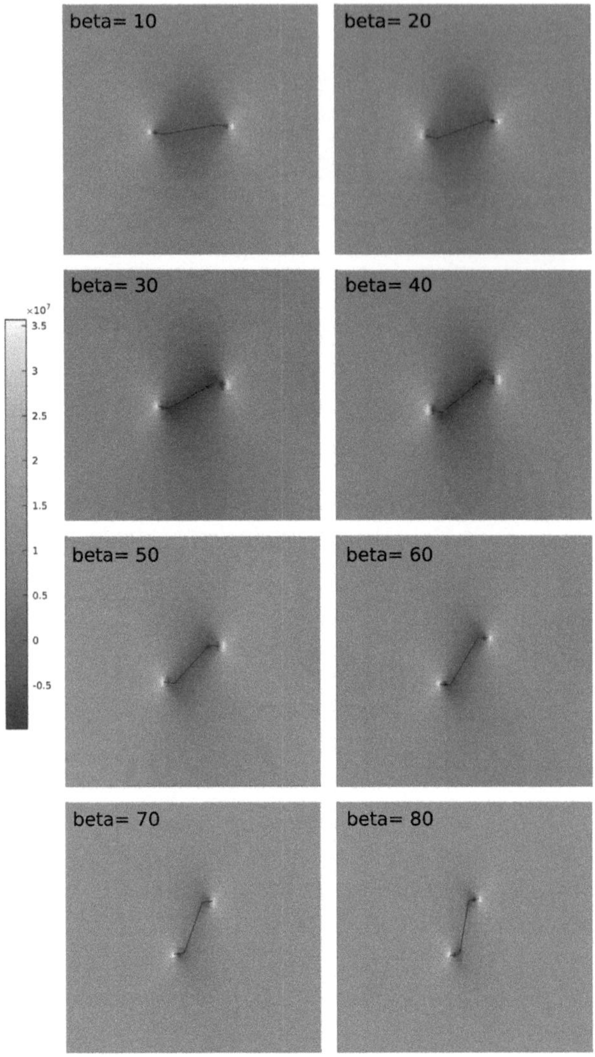

Figure 11.3 Stress profile in *x* direction with different pre-defined angle βs.

randomly distributed inner cracks are studied (Fig. 11.6). Fig. 11.6 clearly indicates that the existing crack has great impact on deformation, pressure profiles of the whole domain.

11.2.3 Hydro-mechanical model with edge crack

This section simulates the rupture of a plate under plane-strain conditions. As shown in Fig. 11.7A, a pre-existing fracture is located at the symmetry

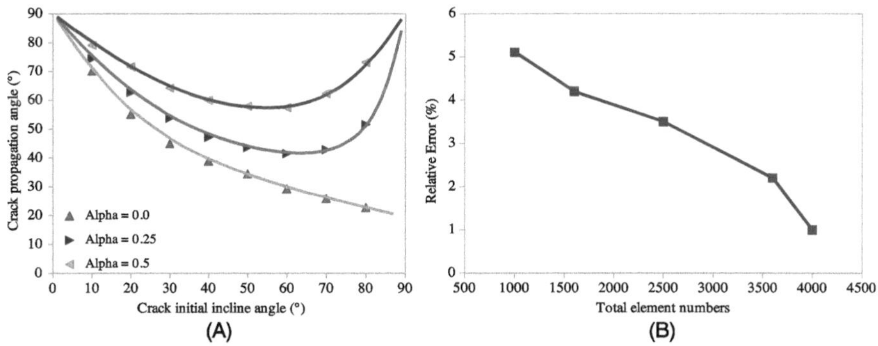

Figure 11.4 (A) Comparison results of the propagation angle of the induced crack between numerical calculation and theoretical model with different loading ratio α, solid lines represent the result from theoretical model. (B) Convergence study of stiffness intensity factor K_1 of mode I respect to total element number.

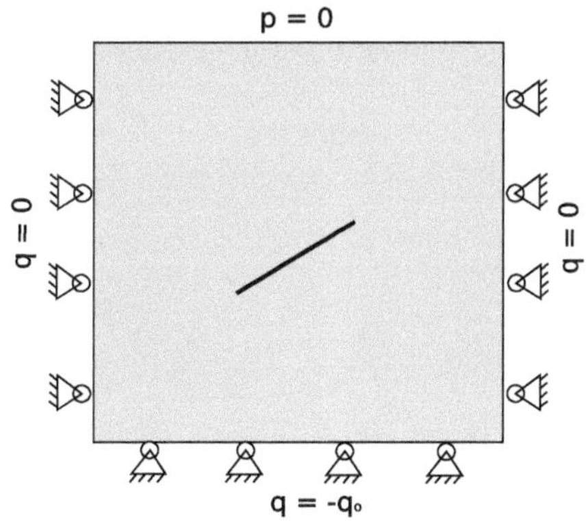

Figure 11.5 Schematic illustration of geometry and boundary conditions. p represents pressure, q represents fluid flux.

Table 11.2 Material parameters for hydro-mechanical model with inner cracks.

Young's modulus	$E = 25.85$ GPa
Poisson's ratio	$\nu = 0.18$
Initial porosity	$n = 0.2$
Intrinsic permeability	$k = 2.78\text{e-}21\ \text{m}^2$
Biot's coefficient	$\alpha = 1$
Solid bulk modulus	$K_s = 13.46$ GPa
Water bulk modulus	$K_f = 0.2$ GPa
Water density	$\rho = 1000\ \text{kg/m}^3$
Water dynamic viscosity	$\mu_w = 5\text{e-}4\ \text{N/m}^2\ \text{s}$
Irreducible saturation	$S_{irr} = 0.0$
Reference pressure	$R_{ref} = 18.6$ MPa
Porous index	$l = 0.4396$

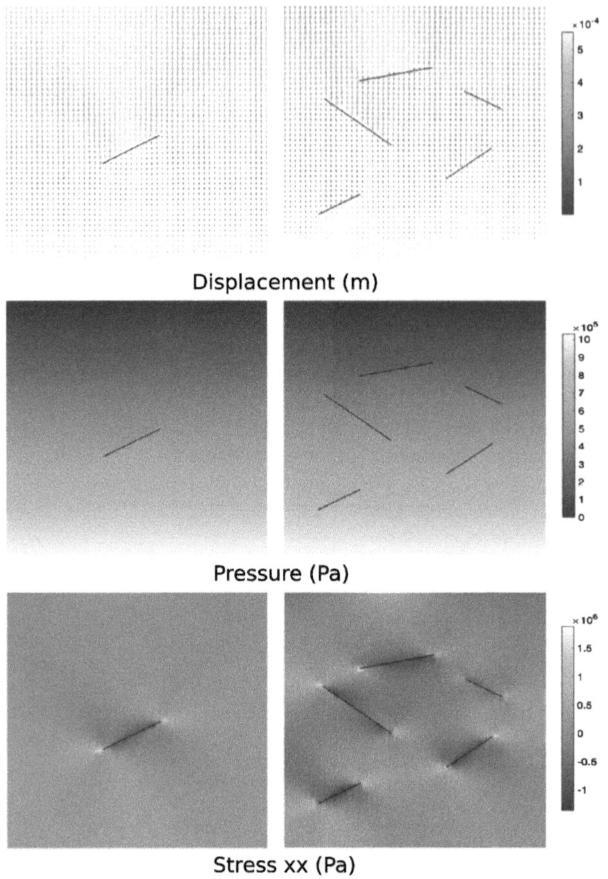

Figure 11.6 Displacement, pressure and stress profile for domain with single and five cracks.

axis of the plate. A fixed vertical velocity $v = 2.35 \times 10^{-2}$ μm/s is prescribed in an opposite direction at the bottom and at the top of the plate (tensile loading). All boundaries of the plate are assumed to be impervious. The induced pressure profile is presented in Fig. 11.7B.

11.2.3.1 Hydro-mechanical model with edge crack under fluid flux loading

This subsection follows an experimental setup [8] (Fig. 11.8). The setup of this experiment can be summarized as follows: a square sample (Fig. 11.8A) with two ports, high pore pressure is introduced to the lower port region by injecting pressurized fluid and maintain the pressure, hereafter a wedge

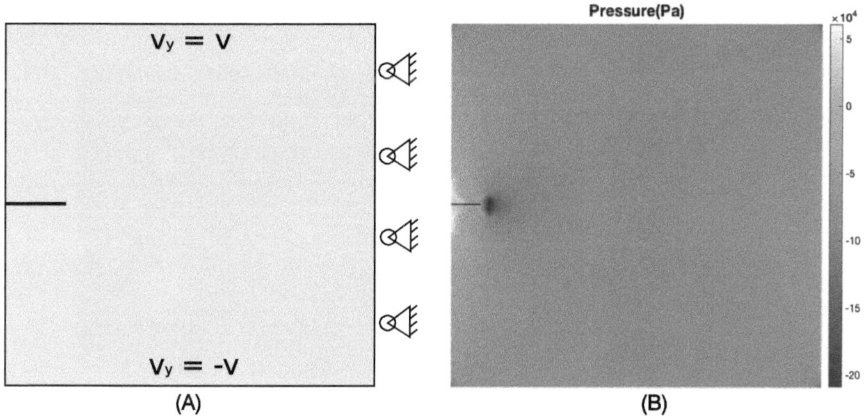

Figure 11.7 (A) Schematic illustration of geometry and boundary conditions, (B) pressure profile.

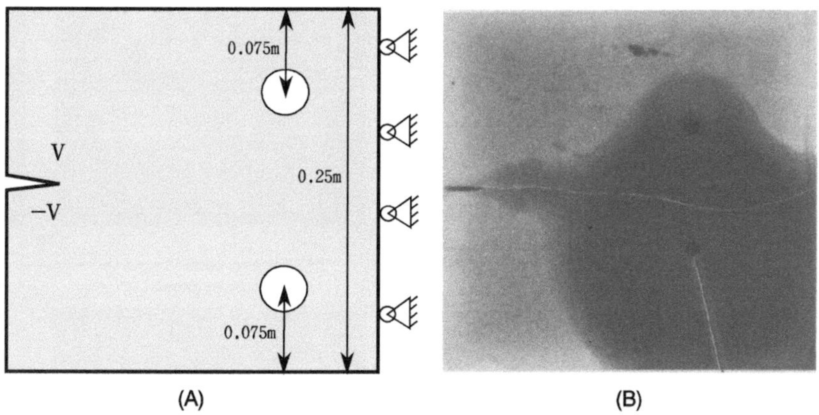

Figure 11.8 (A) Schematic illustration of geometry and boundary conditions, (B) experimental results.

is driven into the pre-shaped concave edge to induce fracture and fracture propagation. The crack propagation path is then identified. Further more, the impact of the permeability of the solid on the crack propagation path is also studied (Fig. 11.9). The fracture development path, displacement of the domain and stress profiles are shown in Fig. 11.9 for porous medias (under identical loading) with different intrinsic permeabilities (k1, k2, k3). Fig. 11.9D-F clearly demonstrates that the fracture paths are attracted to the fluid pressurized port. Furthermore, we found that under identical injection rate, porous media with lower permeability (e.g. k1) will induce

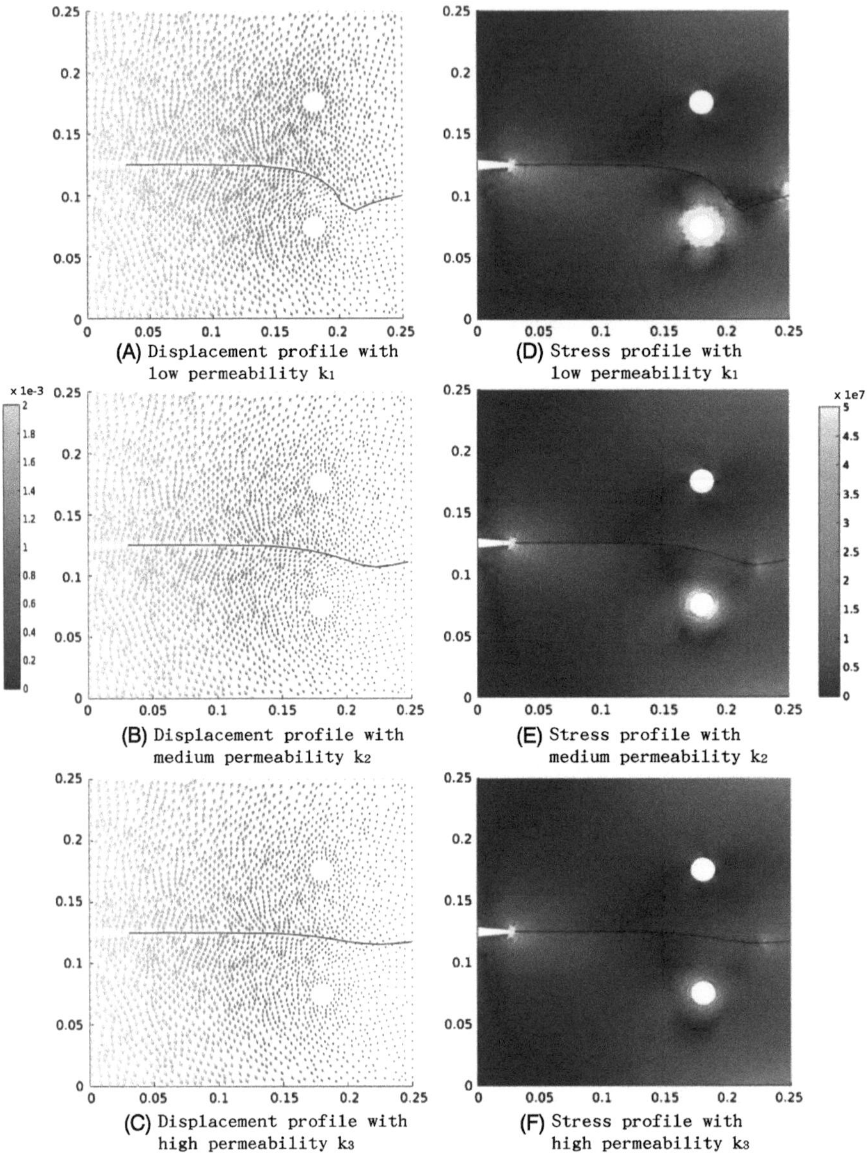

Figure 11.9 Displacement (m) plot of porous media with the intrinsic permeabilities (A) $k_1 = 2.78\text{e-}18\ \text{m}^2$; (B) $k_2 = 1.52\text{e-}17\ \text{m}^2$; (C) $k_3 = 2.78\text{e-}17\ \text{m}^2$ and (D)-(F) are the stress (Pa) plots in cylinder coordinate accordingly.

higher stress/pressure state around the injection port comparing to porous media with higher permeability (e.g. k3), which, in return influences the propagation direction of the fracture.

References

[1] T. Anderson, Fracture Mechanics: Fundamentals and Applications, fourth edition, CRC Press, 2017.

[2] C. Anitescu, M.N. Hossain, T. Rabczuk, Recovery-based error estimation and adaptivity using high-order splines over hierarchical t-meshes, Computer Methods in Applied Mechanics and Engineering 328 (2018) 638–662.

[3] D. Benson, Y. Bazilevs, M. Hsu, T. Hughes, Isogeometric shell analysis: the Reissner–Mindlin shell, Computer Methods in Applied Mechanics and Engineering 199 (5) (2010) 276–289.

[4] D. Benson, Y. Bazilevs, M.-C. Hsu, T. Hughes, A large deformation, rotation-free, isogeometric shell, Computer Methods in Applied Mechanics and Engineering 200 (13) (2011) 1367–1378.

[5] M. Bercovier, T. Matskewich, Smooth Bézier Surfaces Over Unstructured Quadrilateral Meshes, Springer International Publishing, 2017.

[6] S. Bordas, V. Nguyen, C. Dunant, H. Nguyen-Dang, A. Guidoum, An extended finite element library, International Journal for Numerical and Analytical Methods in Engineering 71 (6) (2007) 703–732.

[7] M.J. Borden, M.A. Scott, J.A. Evans, T.J.R. Hughes, Isogeometric finite element data structures based on Bézier extraction of NURBS, International Journal for Numerical Methods in Engineering 87 (1–5) (2011) 15–47.

[8] M. Bruno, F. Nakagawa, Pore pressure influence on tensile fracture propagation in sedimentary rock, International Journal of Rock Mechanics and Mining Sciences & Geomechanics Abstracts 28 (4) (1991) 261–273.

[9] S.A. Coons, Surfaces for Computer-Aided Design of Space Forms, Technical report, Cambridge, MA, USA, 1967.

[10] L. Dalcin, N. Collier, P. Vignal, A. Côrtes, V. Calo, PetIGA: a framework for high-performance isogeometric analysis, Computer Methods in Applied Mechanics and Engineering 308 (2016) 151–181.

[11] C. de Falco, A. Reali, R. Vázquez, GeoPDEs: research tool for isogeometric analysis of PDEs, Advances in Engineering Software 42 (12) (2011) 1020–1034.

[12] W. Dörfler, A convergent adaptive algorithm for Poisson's equation, SIAM Journal on Numerical Analysis 33 (3) (1996) 1106–1124.

[13] L. Engvall, J.A. Evans, Isogeometric triangular Bernstein–Bézier discretizations: automatic mesh generation and geometrically exact finite element analysis, Computer Methods in Applied Mechanics and Engineering 304 (2016) 378–407.

[14] C. Geuzaine, J.-F. Remacle, Gmsh: a 3-D finite element mesh generator with built-in pre- and post-processing facilities, International Journal for Numerical Methods in Engineering 79 (11) (2009) 1309–1331.

[15] J. Gilbert, C. Moler, R. Schreiber, Sparse matrices in MATLAB: design and Implementation, SIAM Journal on Matrix Analysis and Applications 13 (1) (1992) 333–356.

[16] M. Kapl, G. Sangalli, T. Takacs, An isogeometric C1 subspace on unstructured multi-patch planar domains, Computer Aided Geometric Design 69 (2019) 55–75.

[17] Y. Lai, Y.J. Zhang, L. Liu, X. Wei, E. Fang, J. Lua, Integrating CAD with Abaqus: a practical isogeometric analysis software platform for industrial applications, in: High-Order Finite Element and Isogeometric Methods 2016, Computers & Mathematics with Applications 74 (7) (2017) 1648–1660.

[18] V.P. Nguyen, C. Anitescu, S.P. Bordas, T. Rabczuk, Isogeometric analysis: an overview and computer implementation aspects, Mathematics and Computers in Simulation 117 (2015) 89–116.

[19] V.P. Nguyen, T. Rabczuk, S. Bordas, M. Duflot, Meshfree methods: a review and key computer implementation aspects, Mathematics and Computers in Simulation 79 (3) (2008) 763–813.

[20] M. Occelli, T. Elguedj, S. Bouabdallah, L. Morançay, LR B-Splines implementation in the Altair Radioss^{TM} solver for explicit dynamics IsoGeometric Analysis, Advances in Engineering Software 131 (2019) 166–185.

[21] M. Pauletti, M. Martinelli, N. Cavallini, P. Antolin, Igatools: an isogeometric analysis library, SIAM Journal on Scientific Computing 37 (4) (2015) C465–C496.

[22] J.R. Rice, A path independent integral and the approximate analysis of stress concentration by notches and cracks, Journal of Applied Mechanics 35 (1968) 379–386.

[23] D. Rypl, B. Patzák, From the finite element analysis to the isogeometric analysis in an object oriented computing environment, in: CIVIL-COMP, Advances in Engineering Software 44 (1) (2012) 116–125.

[24] H.Si. TetGen, A Delaunay-based quality tetrahedral mesh generator, ACM Transactions on Mathematical Software 41 (2) (2015) 1–36.

[25] M. Spink, D. Claxton, C. de Falco, R. Vázquez, The NURBS toolbox, http://octave.sourceforge.net/nurbs/index.html.

[26] M. Stern, E. Becker, R. Dunham, A contour integral computation of mixed-mode stress intensity factors, International Journal of Fracture 12 (3) (1976) 359–368.

[27] R.L. Taylor, FEAP - Finite Element Analysis Program, 2014.

[28] N. Valizadeh, Y. Bazilevs, J. Chen, T. Rabczuk, A coupled IGA–meshfree discretization of arbitrary order of accuracy and without global geometry parameterization, Computer Methods in Applied Mechanics and Engineering 293 (2015) 20–37.

[29] R. Vázquez, A new design for the implementation of isogeometric analysis in Octave and Matlab: GeoPDEs 3.0, Computer Mathematics and Its Applications 72 (3) (2016) 523–554.

[30] C. Wu, C. Park, J. Chen, A generalized approximation for the meshfree analysis of solids, International Journal for Numerical Methods in Engineering 85 (6) (2011) 693–722.

[31] S. Xia, X. Qian, Isogeometric analysis with Bézier tetrahedra, in: Special Issue on Isogeometric Analysis: Progress and Challenges, Computer Methods in Applied Mechanics and Engineering 316 (2017) 782–816.

PART ONE

Appendices

Derivation of shape derivative for the nanoelasticity problem

Firstly, the total potential energy objective function is considered. The objective function and its constraints are as follows,

Minimize

$$J(\Omega) = \int_{\Omega} \mathbf{u}.\mathbf{b}\, d\Omega + \int_{\Gamma_N} \mathbf{u}.\mathbf{t}\, d\Gamma$$

subject to:

$$a(\mathbf{u}, \delta\mathbf{u}, \Omega) + a_s(\mathbf{u}, \delta\mathbf{u}, \Omega) = -l_s(\mathbf{u}, \Omega) + l(\mathbf{u}, \Omega)$$

(i.e.)

$$\int_{\Omega} \epsilon(\delta\mathbf{u}) : \mathbb{C}^{bulk} : \boldsymbol{\epsilon}(\mathbf{u})\, d\Omega + \int_{\Gamma} (\mathbf{P}\boldsymbol{\epsilon}(\delta\mathbf{u})\mathbf{P}) : \boldsymbol{\tau}_{\mathbf{s}}\, d\Gamma +$$
$$\int_{\Gamma} (\mathbf{P}\boldsymbol{\epsilon}(\delta\mathbf{u})\mathbf{P}) : \mathbb{C}^{s} : (\mathbf{P}\boldsymbol{\epsilon}(\mathbf{u})\mathbf{P})\, d\Gamma = \int_{\Omega} \mathbf{u}.\mathbf{b}\, d\Omega + \int_{\Gamma_N} \mathbf{u}.\mathbf{t}\, d\Gamma.$$

$$\begin{aligned}
&\dot{a}(\mathbf{u}, \mathbf{w}, \Omega) + \dot{a}_s(\mathbf{u}, \mathbf{w}, \Omega) + \dot{l}_s(\mathbf{w}, \Omega) \\
&= \int_{\Omega} \boldsymbol{\epsilon}(\mathbf{u}') : \mathbb{C}^{bulk} : \boldsymbol{\epsilon}(\mathbf{w}) d\Omega + \\
&\int_{\Omega} \boldsymbol{\epsilon}(\mathbf{u}) : \mathbb{C}^{bulk} : \boldsymbol{\epsilon}(\mathbf{w}') d\Omega + \\
&\int_{\Gamma} \boldsymbol{\epsilon}(\mathbf{u}) : \mathbb{C}^{bulk} : \boldsymbol{\epsilon}(\mathbf{w})\, V_n d\Gamma + \\
&\int_{\Gamma} \mathbf{P}\boldsymbol{\epsilon}(\mathbf{w}')\mathbf{P} : \boldsymbol{\tau}_{\mathbf{s}} d\Gamma + \int_{\Gamma} (\nabla_s(\mathbf{P}\boldsymbol{\epsilon}(\mathbf{w})\mathbf{P} : \boldsymbol{\tau}_{\mathbf{s}}) \cdot \mathbf{n} \\
&+ \kappa(\mathbf{P}\boldsymbol{\epsilon}(\mathbf{w})\mathbf{P} : \boldsymbol{\tau}_{\mathbf{s}}))\, V_n d\Gamma + \\
&\int_{\Gamma} (\mathbf{P}\boldsymbol{\epsilon}(\mathbf{u}')\mathbf{P}) : \mathbb{C}^{s} : (\mathbf{P}\boldsymbol{\epsilon}(\mathbf{w})\mathbf{P}) d\Gamma + \\
&\int_{\Gamma} (\mathbf{P}\boldsymbol{\epsilon}(\mathbf{u})\mathbf{P}) : \mathbb{C}^{s} : (\mathbf{P}\boldsymbol{\epsilon}(\mathbf{w}')\mathbf{P}) d\Gamma + \\
&\int_{\Gamma} (\nabla_s(\mathbf{P}\boldsymbol{\epsilon}(\mathbf{u})\mathbf{P}) : \mathbb{C}^{s} : \mathbf{P}\boldsymbol{\epsilon}(\mathbf{w})\mathbf{P}) \cdot \mathbf{n}\, V_n\, d\Gamma +
\end{aligned} \tag{A.1}$$

Extended Finite Element and Meshfree Methods
https://doi.org/10.1016/B978-0-12-814106-9.00019-X
Copyright © 2020 Elsevier Inc.
All rights reserved.

$$\int_{\Gamma} (\mathbf{P}\boldsymbol{\epsilon}(\mathbf{u})\mathbf{P}) : \mathbb{C}^s : \nabla_s(\mathbf{P}\boldsymbol{\epsilon}(\mathbf{w})\mathbf{P})) \cdot \mathbf{n}\, V_n\, d\Gamma +$$

$$\int_{\Gamma} \kappa(\mathbf{P}\boldsymbol{\epsilon}(\mathbf{u})\mathbf{P} : \mathbb{C}^s : \mathbf{P}\boldsymbol{\epsilon}(\mathbf{w})\mathbf{P})\, V_n\, d\Gamma,$$

$$\begin{aligned} \dot{l}(\mathbf{w}, \Omega) = & \int_{\Omega} \mathbf{w}'.\mathbf{b} d\Omega + \int_{\Gamma} \mathbf{w}.\mathbf{b}\, V_n d\Gamma + \int_{\Gamma_N} \mathbf{w}'.\mathbf{t} d\Gamma \\ & + \int_{\Gamma_N} (\nabla(\mathbf{w}.\mathbf{t}).\mathbf{n} + \kappa \mathbf{w}.\mathbf{t})\, V_n d\Gamma \end{aligned} \tag{A.2}$$

$$\dot{J} = \int_{\Omega} \mathbf{u}'.\mathbf{b}\, d\Omega + \int_{\Gamma} \mathbf{u}.\mathbf{b}\, V_n d\Gamma + \int_{\Gamma_N} (\nabla(\mathbf{u}.\mathbf{t}).\mathbf{n} + \kappa \mathbf{u}.\mathbf{t})\, V_n\, d\Gamma. \tag{A.3}$$

The Lagrangian of the objective functional is,

$$L = J + l(\mathbf{w}, \Omega) - a(\mathbf{u}, \mathbf{w}, \Omega) - a_s(\mathbf{u}, \mathbf{w}, \Omega) - l_s(\mathbf{w}, \Omega). \tag{A.4}$$

The material derivative of the Lagrangian is defined as,

$$\dot{L} = \dot{J} + \dot{l}(\mathbf{w}, \Omega) - \dot{a}(\mathbf{u}, \mathbf{w}, \Omega) - \dot{a}_s(\mathbf{u}, \mathbf{w}, \Omega) - \dot{l}_s(\mathbf{w}, \Omega). \tag{A.5}$$

All the terms that contain $\mathbf{w}'$ in the material derivative of Lagrangian are collected and the sum of these terms is set to zero, to get the weak form of the state equation,

$$\begin{aligned} \int_{\Omega} \mathbf{w}'.\mathbf{b}\, d\Gamma + \int_{\Gamma_N} \mathbf{w}'.\mathbf{t}\, d\Gamma = & \int_{\Omega} \boldsymbol{\epsilon}(\mathbf{u}) : \mathbb{C}^{bulk} : \boldsymbol{\epsilon}(\mathbf{w}')\, d\Omega \\ & + \int_{\Gamma} \mathbf{P}\boldsymbol{\epsilon}(\mathbf{w}')\mathbf{P} : \boldsymbol{\tau}_{\mathbf{s}}\, d\Gamma \\ & + \int_{\Gamma} (\mathbf{P}\boldsymbol{\epsilon}(\mathbf{u})\mathbf{P} : \mathbb{C}^s : \mathbf{P}\boldsymbol{\epsilon}(\mathbf{w}')\mathbf{P}) \end{aligned} \tag{A.6}$$

Derivation of the adjoint problem for the nanopiezoelectricity problem

The objective function and its constraints are as follows,

Minimize

$$J(\Psi) = \frac{1}{k^2} = \frac{\Pi_e}{\Pi_d} \tag{B.1}$$

subject to:

$$\int_\Omega d\Omega - \bar{V} = 0 \tag{B.2}$$

$$a(\mathbf{u}, \boldsymbol{\phi}, \delta\mathbf{u}, \boldsymbol{\delta\phi}) + a_s(\mathbf{u}, \boldsymbol{\phi}, \delta\mathbf{u}, \boldsymbol{\delta\phi}) = -l_s(\delta\mathbf{u}, \boldsymbol{\delta\phi}) + l(\delta\mathbf{u}, \boldsymbol{\delta\phi}) \tag{B.3}$$

$$\begin{aligned}
&\dot{a}(\mathbf{u}, \boldsymbol{\phi}, \mathbf{w}, \boldsymbol{\psi}) = \\
&\int_\Omega \boldsymbol{\epsilon}^T(\mathbf{u}') : \mathbb{C}^{bulk} : \boldsymbol{\epsilon}(\mathbf{w}) d\Omega + \int_\Omega \boldsymbol{\epsilon}^T(\mathbf{u}) : \mathbb{C}^{bulk} : \boldsymbol{\epsilon}(\mathbf{w}') d\Omega \\
&- \int_\Omega \boldsymbol{\epsilon}(\boldsymbol{u}')^T : \boldsymbol{e}^T \cdot \boldsymbol{E}(\boldsymbol{\psi})\, d\Omega - \int_\Omega \boldsymbol{\epsilon}(\boldsymbol{u})^T : \boldsymbol{e}^T \cdot \boldsymbol{E}(\boldsymbol{\psi}')\, d\Omega \\
&- \int_\Gamma \boldsymbol{\epsilon}(\boldsymbol{u})^T : \boldsymbol{e}^T \cdot \boldsymbol{E}(\boldsymbol{\psi}) \cdot V_n\, d\Gamma - \int_\Gamma \boldsymbol{\epsilon}^T(\mathbf{u}) : \mathbb{C}^{bulk} : \boldsymbol{\epsilon}(\mathbf{w}) \cdot V_n d\Gamma \\
&- \int_\Omega \boldsymbol{E}(\boldsymbol{\phi}')^T \cdot \boldsymbol{e}^{bulk} : \boldsymbol{\epsilon}(\boldsymbol{w})\, d\Omega - \int_\Omega \boldsymbol{E}(\boldsymbol{\phi})^T \cdot \boldsymbol{e}^{bulk} : \boldsymbol{\epsilon}(\boldsymbol{w}')\, d\Omega \\
&- \int_\Gamma \boldsymbol{E}(\boldsymbol{\phi})^T \cdot \boldsymbol{e}^{bulk} : \boldsymbol{\epsilon}(\boldsymbol{w}) \cdot V_n\, d\Gamma - \int_\Omega \boldsymbol{E}(\boldsymbol{\phi}')^T \cdot \boldsymbol{\kappa} \cdot \boldsymbol{E}(\boldsymbol{\psi}) d\Omega \\
&- \int_\Omega \boldsymbol{E}(\boldsymbol{\phi})^T \cdot \boldsymbol{\kappa} \cdot \boldsymbol{E}(\boldsymbol{\psi}') d\Omega - \int_\Gamma \boldsymbol{E}(\boldsymbol{\phi})^T \cdot \boldsymbol{\kappa} \cdot \boldsymbol{E}(\boldsymbol{\psi}) \cdot V_n d\Gamma
\end{aligned} \tag{B.4}$$

$$\begin{aligned}
&\dot{a}_s(\mathbf{u}, \boldsymbol{\phi}, \mathbf{w}, \boldsymbol{\psi}) = \\
&\int_\Gamma \boldsymbol{\epsilon}_s(\mathbf{u}')^T : \boldsymbol{C}_s : \boldsymbol{\epsilon}_s(\boldsymbol{w}) d\Gamma + \int_\Gamma \boldsymbol{\epsilon}_s(\boldsymbol{u})^T : \boldsymbol{C}_s : \boldsymbol{\epsilon}_s(\boldsymbol{w}') d\Gamma \\
&+ \int_\Gamma [\nabla(\boldsymbol{\epsilon}_s(\boldsymbol{u}) : \boldsymbol{C}_s : \boldsymbol{\epsilon}_s(\boldsymbol{w})) \cdot n + (\boldsymbol{\epsilon}_s(\boldsymbol{u}) : \boldsymbol{C}_s : \boldsymbol{\epsilon}_s(\boldsymbol{w})) \cdot \eta] \cdot V_n\, d\Gamma
\end{aligned}$$

Extended Finite Element and Meshfree Methods
https://doi.org/10.1016/B978-0-12-814106-9.00020-6
Copyright © 2020 Elsevier Inc.
All rights reserved.

$$
\begin{aligned}
&-\int_\Gamma \boldsymbol{\epsilon}_s(\boldsymbol{u}')^T : \boldsymbol{e}_s^T \cdot \boldsymbol{E}_s(\boldsymbol{\psi})\, d\Gamma - \int_\Gamma \boldsymbol{\epsilon}_s(\boldsymbol{u})^T : \boldsymbol{e}_s^T \cdot \boldsymbol{E}_s(\boldsymbol{\psi}')\, d\Gamma \\
&-\int_\Gamma [\nabla(\boldsymbol{\epsilon}_s(\boldsymbol{u})^T : \boldsymbol{e}_s^T \cdot E_s(\boldsymbol{\psi}))) \cdot n + (\boldsymbol{\epsilon}_s(\boldsymbol{u})^T : \boldsymbol{e}_s^T \cdot \boldsymbol{E}_s(\boldsymbol{\psi}))\eta] \cdot V_n\, d\Gamma \\
&-\int_\Gamma \boldsymbol{E}_s(\boldsymbol{\phi}')^T \cdot \boldsymbol{e}_s : \boldsymbol{\epsilon}_s(\boldsymbol{w})\, d\Gamma - \int_\Gamma \boldsymbol{E}_s(\boldsymbol{\phi})^T \cdot \boldsymbol{e}_s : \boldsymbol{\epsilon}_s(\boldsymbol{w}')\, d\Gamma \\
&-\int_\Gamma [\nabla(\boldsymbol{E}_s(\boldsymbol{\phi})^T \cdot \boldsymbol{e}_s : \boldsymbol{\epsilon}(\boldsymbol{w})) \cdot n + (\boldsymbol{E}_s(\boldsymbol{\phi})^T \cdot \boldsymbol{e}_s : \boldsymbol{\epsilon}(\boldsymbol{w}))\eta] \cdot V_n\, d\Gamma \\
&-\int_\Gamma \boldsymbol{E}_s(\boldsymbol{\phi}')^T \cdot \boldsymbol{\kappa}_s \cdot \boldsymbol{E}_s\boldsymbol{\psi}\, d\Gamma - \int_\Gamma \boldsymbol{E}_s(\boldsymbol{\phi})^T \cdot \boldsymbol{\kappa}_s \cdot \boldsymbol{E}_s(\boldsymbol{\psi}')\, d\Gamma \\
&-\int_\Gamma [\nabla(\boldsymbol{E}_s(\boldsymbol{\phi})^T \cdot \boldsymbol{\kappa}_s \cdot \boldsymbol{E}_s(\boldsymbol{\psi}) \cdot n + (\boldsymbol{E}_s(\boldsymbol{\phi})^T \cdot \boldsymbol{\kappa}_s \cdot \boldsymbol{E}_s(\boldsymbol{\psi})) \cdot \eta]\, V_n\, d\Gamma
\end{aligned}
\tag{B.5}
$$

$$
\begin{aligned}
\dot{l}_s(\mathbf{w}) = &\int_\Gamma \boldsymbol{\epsilon}_s(\mathbf{w}')\mathbf{P} : \boldsymbol{\tau}_\mathbf{s} d\Gamma + \int_\Gamma (\nabla_s(\boldsymbol{\epsilon}_s(\mathbf{w}) : \boldsymbol{\tau}_\mathbf{s}) \cdot \mathbf{n} \\
&+ \eta(\boldsymbol{\epsilon}_s(\mathbf{w}) : \boldsymbol{\tau}_\mathbf{s}))\, V_n d\Gamma
\end{aligned}
\tag{B.6}
$$

$$
\begin{aligned}
\dot{l}(\mathbf{w}) = &\int_\Omega \mathbf{w}'.\mathbf{b} d\Omega + \int_\Gamma \mathbf{w}.\mathbf{b}\, V_n d\Gamma + \int_{\Gamma_N} \mathbf{w}'.\mathbf{t} d\Gamma \\
&+ \int_{\Gamma_N} (\nabla(\mathbf{w}.\mathbf{t}).\mathbf{n} + \eta\mathbf{w}.\mathbf{t})\, V_n d\Gamma
\end{aligned}
\tag{B.7}
$$

The Lagrangian of the objective functional is,

$$L = J + l(\mathbf{w}) - a(\mathbf{u}, \boldsymbol{\phi}, \mathbf{w}, \boldsymbol{\psi}) - a_s(\mathbf{u}, \boldsymbol{\phi}, \mathbf{w}, \boldsymbol{\psi}) - l_s(\mathbf{w}). \tag{B.8}$$

The material derivative of the Lagrangian is defined as,

$$\dot{L} = \dot{J} + \dot{l}(\mathbf{w}) - \dot{a}(\mathbf{u}, \boldsymbol{\phi}, \mathbf{w}, \boldsymbol{\psi}) - \dot{a}_s(\mathbf{u}, \boldsymbol{\phi}, \mathbf{w}, \boldsymbol{\psi}) - \dot{l}_s(\mathbf{w}). \tag{B.9}$$

All the terms that contain $\mathbf{u}'$ and $\boldsymbol{\phi}'$ in the material derivative of Lagrangian are collected and the sum of these terms is set to zero, to get the weak form of the adjoint equation,

$$
\begin{aligned}
&\int_\Omega \boldsymbol{\epsilon}^T(\mathbf{u}') : \mathbb{C}^{bulk} : \boldsymbol{\epsilon}(\mathbf{w}) d\Omega - \int_\Omega \boldsymbol{\epsilon}(\boldsymbol{u}')^T : \boldsymbol{e}^T \cdot \boldsymbol{E}(\boldsymbol{\psi})\, d\Omega - \\
&\int_\Gamma \boldsymbol{\epsilon}_s(\mathbf{u}')^T : \boldsymbol{C}_s : \boldsymbol{\epsilon}_s(\boldsymbol{w}) d\Gamma - \int_\Gamma \boldsymbol{\epsilon}_s(\boldsymbol{u}')^T : \boldsymbol{e}_s^T \cdot \boldsymbol{E}_s(\boldsymbol{\psi})\, d\Gamma \\
&= \int_\Omega 2C_1\, \boldsymbol{\epsilon}(\boldsymbol{u}')^T : \mathbb{C}^{bulk} : \boldsymbol{\epsilon}(\boldsymbol{u})\, d\Omega
\end{aligned}
\tag{B.10}
$$

$$\int_{\Omega} \mathbf{E}(\boldsymbol{\phi}')^T \cdot e^{bulk} : \epsilon(\boldsymbol{w})\, d\Omega - \int_{\Omega} \mathbf{E}(\boldsymbol{\phi}')^T \cdot \kappa \cdot \mathbf{E}(\boldsymbol{\psi}) d\Omega -$$
$$\int_{\Gamma} \mathbf{E}_s(\boldsymbol{\phi}')^T \cdot e_s : \epsilon_s(\boldsymbol{w})\, d\Gamma - \int_{\Gamma} \mathbf{E}_s(\boldsymbol{\phi}')^T \cdot \kappa_s \cdot \mathbf{E}_s \boldsymbol{\psi}\, d\Gamma \qquad \text{(B.11)}$$
$$= \int_{\Omega} 2C_2\, \mathbf{E}(\boldsymbol{\phi}')^T \cdot \kappa \cdot \mathbf{E}(\boldsymbol{\phi}) d\Omega$$

Index

D

E

Q

R

S

T

U

V

W

X

FSC
www.fsc.org
MIX
Papier | Fördert
gute Waldnutzung
FSC® C083411

Kolibri 360 GmbH
Ferdinand-Jühlke-Straße 7
99095 Erfurt, Deutschland
produktsicherheit@kolibri360.de